THE ATLAS OF LIFE ON
EARTH

Academic Consultant
Professor Michael J. Benton
University of Bristol, UK

Project Director Ayala Kingsley
Project Editor Lauren Bourque
Art Editors Ayala Kingsley, Martin Anderson
Cartographic Manager Richard Watts
Cartographic Editor Tim Williams
Paleogeography Dougal Dixon
Additional Design Roger Hutchins
Picture Research Alison Floyd
Picture Management Claire Turner
Production Director Clive Sparling
Proofreader Lynne Wycherley
Index Ann Barrett

Illustrators Julian and Janet Baker, Robert and Rhoda Burns, Felicity Cole, Dougal Dixon, Bill Donohoe, Brin Edwards, Samantha Elmhurst, David Hardy, Ron Hayward, Karen Hiscock, Ruth Lindsay, Malting Patnership, Denys Ovenden, Colin Rose, David Russell, John Sibbick

BORDERS®
PRESS

Published in 2004 by
Borders Press, a division of Borders, Inc
100 Pheonix Drive
Ann Arbor
Michigan, 48108

AN ANDROMEDA BOOK

The Brown Reference Group plc.
(incorporating Andromeda Oxford Limited)
8 Chapel Place
Rivington Street
London EC2A 3DQ

ISBN 0-681-15308-3

Printed & bound in China

10 9 8 7 6 5 4 3 2 1

THE ATLAS OF LIFE ON EARTH

DOUGAL DIXON • IAN JENKINS
RICHARD MOODY • ANDREY ZHURAVLEV

BORDERS®
PRESS

CONTENTS

INTRODUCTION

IN OUR MODERN AGE, most people are familiar with the broad outlines of the origin of the Earth, the rise of life in the sea, the age of the dinosaurs, early humans, and the ice ages. But it is astonishing that this story has been put together in only 200 years from scattered rocks exposed in quarries and on beaches, and from fossils picked up by chance.

Scientific observation of the natural world was practiced by the ancient Greeks and Romans, among others. In the year 1200, the Chinese naturalist and poet Zhu Xi wrote, "I have seen shells in the high mountains...the shells must have lived in water. The low places are now elevated high, and the soft material turned into hard stone." In Europe 600 years later, however, natural science had been coming along rather slowly. A few key points had been resolved, but the prevailing view was still of an Earth of very recent vintage, produced by a supreme creator.

Gradually, this view began to be challenged. In 1788 James Hutton, a Scottish landowner and amateur geologist, made a strong case for a hitherto unsuspected antiquity of truly stunning magnitude. In Scotland he oberved the processes of erosion and sediment deposition in the rivers and on the shores. He looked at ancient rock sequences, whose immense thickness suggested to him that they represented huge spans of time: "No vestige of a beginning, no prospect of an end." Hutton invented the principle of uniformitarianism, expressed in the axiom "The present is the key to the past," which means that the laws of nature are constant over time. With this, he laid to rest the medieval approach to geology and established it as a science.

In Hutton's lifetime, fossils moved from curiosities in private collections to the crux of the debate about the origins of life. Until 1750, most naturalists (who included many clergymen) had assumed that the Earth's plants and animals were as they had always been, and always would be: an extinction would mean that the Creator had made a serious mistake. However, with exploration and industrial digging, the remains of unknown plants and animals had begun to pile up.

Early explorers in North America sent their finds back to Europe for study. Shells and fern fronds did not pose a real problem, but around 1750 shipments of huge bones and teeth from surface deposits in the new territory of Ohio arrived in London and Paris. European scholars saw that these parts belonged to some kind of

elephant, but not the modern Indian or African elephant. Perhaps, they reasoned, some other elephant—they dubbed it *Incognitum* ("unknown")—still lived in the remote west of North America? This argument became impossible to sustain as explorers ventured further west without ever spotting a living elephant. By 1795, the great French anatomist and paleontologist Georges Cuvier announced that the American *Incognitum* was an extinct animal, the mastodon. He described other large beasts, known only from their fossilized bones, which were clearly extinct as well; they included Siberian mammoths and the giant South American ground sloth *Megatherium*.

It was in the early 1800s that the intricate relationship of time, rocks, and fossils began to be deciphered.

Cuvier attributed the disappearance of these animals to global catastrophes that wiped out all life. This view was consistent with biblical tales of flood and plague, and was supported by traditionalists opposed to uniformitarianism, which argued for more gradual, steady change. As the science of geology developed, all evidence seemed to favor uniformitarianism, and until as recently as the 1960s most geologists were "ultra-uniformitarians," rejecting any process that could not still be observed in the present world. In fact, the catastrophists were right in many ways; mass extinctions may be attributed to events such as meteorite strikes and ice ages. However, even these are now known to be natural phenomena.

Evidence for the uniformitarian view came from the principles of stratigraphy—the ordering of rocks—which were largely worked out in the 1820s and 1830s. Hutton established a time frame for rocks; his successors noticed that particular distributions of rocks were repeated in many parts of the Earth. In addition, particular rock formations contained predictable sets of fossils. A rock unit from southern England could be correlated with another unit in Scotland or in France that had the same suite of fossils. Having correlated one unit, a geologist could predict what lay below and above. One by one, though not in chronological order, the key divisions of geological time—the Carboniferous, the Jurassic, the Cretaceous, the Silurian, and so on—were defined and named.

But what to make of the fossils? These had clearly changed over time. Did they represent a series of successive creations and extinctions, as Georges Cuvier argued, or were they connected through the ages? Philosophers in England and France discussed this topic in the early part of the nineteenth century, but it was Charles Darwin who finally set out the principle and the mechanism in 1859. He showed that the diversity of life today could only have arisen by the splitting of evolutionary lines (speciations) over time, and that all life could be traced to common ancestors from unimaginably long ago. With this model, supported by more fossil finds, paleontologists in the nineteenth century drew a detailed picture of the history of life that has required little modification. Discoveries in the twentieth century shed further light, especially once the role of genetics was understood, but most contemporary work in paleontology consists of plugging gaps in the broad picture.

The discovery of a more modern animal such as a rabbit in 550-million-year-old rocks would upset the whole theory of evolution, but no such discovery has ever been made.

Geological science was revolutionized by two major advances around 1915. First came radiometric dating: the application of the principles of radioactive decay, discovered by Marie and Pierre Curie in the 1890s, to rocks. For the first time, geologists were able to establish absolute ages for rock sequences, and to put dates on the geological timescale that had been established in the 1830s.

The second revolution was the theory of continental drift. Until 1915, most geologists accepted that the Earth was stable. A few people had noted the jigsaw-puzzle fit between Africa and South America on maps, and others had noted similarities between fossils from widely separated locations. It was the German geologist Alfred Wegener who first insisted that none of this was coincidental. He argued that the continents had been part of the same giant landmass some 250 million years ago in the Permian and Triassic periods—and that the continents were still moving. Most geologists ridiculed his idea, quoting eminent geophysicists who declared that the Earth was solid and that there was no mechanism that could make the continents move.

Wegener's theory was only proven conclusively in the 1950s and early 1960s with discoveries in the deep oceans. The "motor" of continental drift is plate tectonics. Continents and oceans lie on separate plates that are pushed apart in the middle of the oceans by new crust welling up from the Earth's mantle. As the new rock emerges, the oceanic plates are pushed apart equally, but in opposite directions. In other places, to accommodate the new crust, plates dive down under each other, or push into each other. These movements raise mountains like the Andes and the Himalayas.

Geologists cannot study the events of every year throughout the history of the Earth, nor can paleontologists examine a fossil of every organism that has ever lived. But there is abundant evidence to reconstruct four and a half billion years of the Earth's history, and there are almost no unexpected or inconsistent discoveries. For example, thousands of observations show that much of North America and Europe was shrouded in ice one million years ago. No one has found million-year-old Canadian desert rocks or tropical reefs. Among the millions of new fossils uncovered each year, paleontologists have yet to be shocked by finding a rabbit in Cambrian shale, or a human among dinosaurs. Building on prior knowledge, it is possible to predict the gaps, and what might be discovered to fill them. This view might be said to be complacent. But until it is disproved by some extraordinary new find, rocks and fossils may be considered the keepers of the true history of the Earth and all its variety of life.

In this book you will read about the latest discoveries in geology and paleontology. The principles of stratigraphy, dating, and plate tectonics provide the framework, and detailed paleogeographic maps show the astonishing transformations of our planet. Supporting this is all the evidence accumulated by geologists of all nationalities as they piece together the habitats and climates of every corner of the world.

PARTS 1 & 2 begin with the origin of the Earth, moving through the gradual changes that made it fit for life, up to the appearance of early life forms. PARTS 3 & 4 continue the story with mountain-building as the continents shuffled and reshuffled, the development of forests, and a parade of animals from amphibians to dinosaurs and birds. PARTS 5 & 6 present the rapid transformations of more recent time, the rise of mammals, including humans, and our unprecedented effect on the world. It's a story even more awesome than James Hutton could have imagined 200 years ago.

Michael Benton **University of Bristol, UK**

THE GEOLOGICAL TIMESCALE

TRUE SCALE

This geological timescale has been designed as a spiral in order to offer a true linear measurement. It is therefore possible to compare the units of time, from the Archean Eon which lasted 2.05 billion years to the recent Pleistocene Epoch which lasted less than two million.

UP TO DATE?

Nothing in geology is more subject to change than dates, which are the focus of continual review. The data we have used throughout is largely based on the 1998 timetable by Bilal U. Haq and Frans W.B. van Eysinga (Elsevier Science BV) except where we have been otherwise advised.

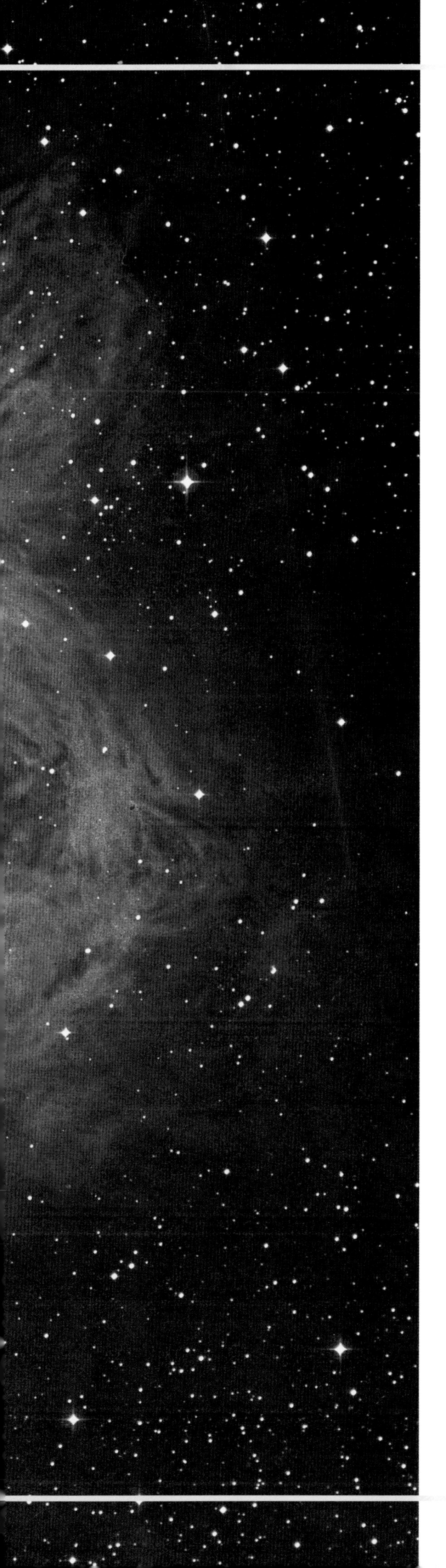

The early history of Earth

PART I

IN THE BEGINNING

4550–545 MILLION YEARS AGO

QUESTIONS ABOUT the age and size of the Earth, and its relation to the Sun and other planets, must have puzzled thousands of generations of our ancestors. To the ancient Greeks, the planets, Sun, and Moon were imagined as separate spheres orbiting the Earth. The Greek mathematician Pythagoras recognized in the sixth century BC that the Earth, also, is a sphere and that its seasons are caused by the tilt of its axis. Aristarchus of Samos was one of the first to argue that the Earth orbits the Sun, but this view was not accepted until after 1543 when the Polish astronomer Nicolas Copernicus published his model of the Solar System, with the Sun at its center. Tycho Brahe and Johannes Kepler refined this model, and Kepler formulated the laws of planetary motion. With the invention of the telescope, Galileo Galilei confirmed the Copernican model, and Isaac Newton gave a theoretical explanation of the physical characteristics of the Solar System.

In Europe, Biblical accounts of Creation shaped beliefs about the Earth's age, which was estimated by a leading seventeenth-century theologian to be less than 6000 years. The question was debated intensely from the late eighteenth through the nineteenth centuries. James Hutton, a Scottish amateur geologist used the evidence of rock strata to argue that the Earth must be unimaginably old. Early in the twentieth century, the discovery of radioactive isotopes allowed the age of rocks to be accurately measured, and it was found that the oldest rocks on Earth were 4.6 billion years old.

OUR SUN and planets are not the oldest entities in the Universe. Long before they existed, many stars had been created from atoms of simple elements such as hydrogen and helium. The Universe itself first came into being in a massive explosion that took place about 15 billion years ago. In the aftermath, clouds of debris drifted through space, forming stars, which burned until they died out—or exploded violently—or collided with other objects. It was from such clouds of debris that the Earth originated approximately 4.55 billion years ago.

Astronomers can now observe many of the processes involved in the formation of planetary systems in deep space, using advanced technology such as the Hubble Telescope to study interstellar clouds and developing stars and planets. They have established both a timescale and a sequence for the events that culminated in the birth of our planet.

Meteorites, moon rocks, and the oldest rocks and minerals on Earth, together with the first fossils, provide vital evidence of the first billion years of Earth history. Data on these early years are limited, however, because the first rocks on our planet were destroyed by an extended bombardment of extraterrestrial debris, causing internal melting, which in turn melted all the surface rocks. Then, approximately 4.51 billion years ago, the Earth was hit by an object the size of Mars, destroying the crusts of both, and forming the Moon. The precise dating of this event is derived from samples of lunar rock brought back to Earth by astronauts since the late 1960s.

As the Earth cooled, it became a less violent place. The early continents began to form; water, carbon dioxide and nitrogen appeared, and the first microbial organisms evolved.

Vast stores of energy can be locked up within bodies of small mass; the impact of a meteorite only 33ft (10m) across, for instance, releases as much energy as a moderately strong earthquake. The greater the original mass of the planet, the greater the amount of energy stored, and the longer it lasts. Untold energy would have been locked up in the Earth by the repeated impacts of planetary fragments in the first billion years. Although the debris was probably frozen when it arrived, heat from these collisions became the driving mechanism in forming the concentric layers (core, mantle, and so on) into which the Earth is divided.

Such intense heat was generated that Earth became a heat engine. Heat and fluid flow started an internal "dynamo," generating convection currents deep in the heart of the planet. In time these would become the driving forces behind plate tectonics, the force that moves continents across the surface of the globe. The internal dynamo and mechanical flow would also lead to the development of a magnetic dynamo and the generation of a magnetic field. Evidence of this magnetic field exists in rocks as old as 3.5 billion years. It is not constant but changes direction periodically, the evidence being "frozen" into the volcanic rocks of the seafloor at the time of their formation.

DURING the first 500 million years of Earth's history, higher temperatures and violent disruption at the surface restricted the development of crust and of viable landmasses. As the mantle cooled and convection slowed, between 4 and 3.5 billion years ago, granitic rocks appeared, forming the cores of the first microcontinents. Another crucial development was the accumulation of water. Water carried to Earth in space debris was gradually released through the chemical and physical processes that were forming the planet, such as volcanism. From salt-rich pools, great oceans developed. They became the medium that eventually supported the evolution of life, the molecules of which may also have arrived initially from deep space. The oceans helped cool the surface further; surface temperatures have probably fluctuated relatively little since.

One of the last features to form was the atmosphere. Whatever gases surrounded the Earth at first must have been blown away by violent winds from the Sun. Gradually they were replaced by a new atmosphere rich in carbon dioxide, nitrogen, and water. This envelope of gases provided a vital shield against ultraviolet light and kept the Earth warm. Convection currents were activated as the gases differentiated into layers and were acted on by different temperatures, and air began to circulate across the globe. There was no oxygen yet, as evident from the lack of oxidation of iron-rich deposits formed at this time. Primitive life forms evolved nonetheless, and increased oxygen in the atmosphere when they began to photosynthesize 3.5 billion years ago.

Richard T.J. Moody

THE ORIGIN AND NATURE OF THE EARTH

*W*ITHIN *our solar system, Earth is unique. Elsewhere in deep space, in galaxies unexplored, there may be thousands of planets that formed at the right distance from a central parent sun, so that they are neither too hot nor too cold; like Earth, they may have protective layers of gas cutting out harmful rays and deflecting or destroying galactic debris. Such planets may also retain vast stretches of water and support a thriving biosphere. However, Earth is still the only known planet that possesses all these attributes.*

The story of Earth began between 10 and 15 billion years ago with the Big Bang. Before this event—the equivalent of an unimaginable number of nuclear explosions—all the matter in the universe had been concentrated into a single, incredibly dense point. The Big Bang produced all the chemical elements, beginning with hydrogen and helium—the two lightest elements, which account for more than 90 percent of all visible matter. These elements are the basic components of the Universe, and thus of the Earth and of life itself.

THE UNIVERSE began 10 to 15 billion years ago in the cataclysmic event in space known as the Big Bang. For several seconds all that existed were vast amounts of energy and tiny subatomic particles such as electrons, protons, and neutrons. The temperature was a staggering 10 billion degrees.

Hydrogen, a light gas, may have been the primordial material of the Universe. It is the lightest and most abundant element, from which all other elements can be made.

Within a few minutes, the temperature fell to less than a billion degrees and the particles began to coalesce, forming the nuclei of light elements. After about a million years, when temperatures had dropped down to a few thousand degrees, atoms began to form. The first element was probably hydrogen (H), the lightest, followed by helium (He).

Early in the Big Bang, hydrogen and helium were blown out into space, away from the point of explosion. They were pulled together by gravity to form dense gas clouds called nebulae, in which stars and galaxies were born. In the thermonuclear processes that take place in stars, hydrogen and its isotopes combine to form helium. In this sense, helium is the product of burning hydrogen.

The burning of other elements, such as helium, produces carbon (C) and oxygen (O). These are the first stages of evolution of a star, and the continued process results in the production of sodium (Na), magnesium (Mg), sulfur (S), phosphorus (P), and silicon (Si). Through continued nuclear fusion, the burning stars generated exceptionally high temperatures and intense activity, producing elements as heavy as iron (Fe). In this manner, most of the elements that comprise the universe gradually appeared.

At some unknown time the early stars exhausted their sources of energy and, effectively, imploded. Additional gravitational forces came into play, and neutron and proton bombardment became intense. New elements heavier than iron were produced, with the collapsed star suddenly terminating as a supernova explosion—blowing more material out into space. Astronomers can observe this ancient cycle of star life and death throughout space today.

Gases and dust generated by supernova explosions are dispersed as more nebulae. In the case of our own solar system, they concentrated into a hot, highly compressed central ball, the proto-Sun. It was surrounded by a rotating disc of various elements that gradually cooled, giving rise to ever larger bodies of material. Planetary fragments called planetesimals collided to form protoplanets or satellite moons. Some fragments continued to orbit the Sun, where there is still a belt of asteroids.

KEYWORDS

ATMOSPHERE
CORE
CRUST
DEGASSING
DIFFERENTIATION
IGNEOUS ROCK
LITHOSPHERE
MANTLE
METAMORPHIC ROCK
PHOTOSYNTHESIS
SEDIMENTARY ROCK
SEQUENCE
TECTONIC PLATE
UNCONFORMITY
UNIFORMITARIANISM

See Also

THE ORIGIN AND NATURE OF LIFE: *Cyanobacteria, Oxygen, Photosynthesis*
THE ARCHEAN: *Early Continents, The Oldest Rocks*
THE PROTEROZOIC: *Development of Complex Life Forms*

12,000 mya (million years ago) | 11,000 | 10,000 | 9000 | 8000

The Big Bang — CREATION OF THE PRESENT UNIVERSE — Nebula cloud contracts

Matter is mainly composed of hydrogen (80%) and helium (15%)

Approximately 2 to 3 billion years after the concentration of the original nebulae, the solar system had taken shape. The Sun began to shine when temperatures in its core reached 10 to 15 million degrees, causing hydrogen to fuse into helium. For almost 10 million years, violent winds created by this fusion blew huge amounts of volatile elements farther out into space. Some of these elements were drawn from the protoplanets, which eventually formed the inner planets of our solar system: Mercury, Venus, Earth, and Mars. Out in space they accreted to form the giant outer planets: Jupiter, Saturn, Uranus, and Neptune. These planets have remained gaseous.

1 The Big Bang, origin of the Universe
2 The solar nebula begins to contract
3 The nebula begins to rotate and flatten

4 Most of the mass sweeps to the center to form the Sun
5 The planets accrete in orbit around the Sun, the solar system
6 Earth and the other terrestrial planets

The accumulation of planetesimals and the loss of volatile elements resulted in an Earth consisting of an unsegregated mass of molten rock with no atmosphere. Gravitational forces quickly prevailed, however, and continued collisions and the release of energy through radioactivity generated further heat and melting. A giant impact between the Earth and a large planetesimal created even more heat and led to the formation of the Moon. This Earth-shattering event took place about 4.5 billion years ago; the oldest rocks on the Moon, retrieved by astronauts, have been dated at 4.4 billion years, older than any known rocks sampled from the Earth's crust.

13 BILLION YEARS

The Big Bang occurred about 13 billion years ago. Eventually the solar nebula formed, and in a few billion years the Sun's accretion disc became the planets of the solar system. About 4.5 billion years ago, a Mars-sized body collided with the Earth to form the Moon.

7 A planetesimal collides with the Earth
8 Fragments in orbit
9 They coalesce to form the Moon

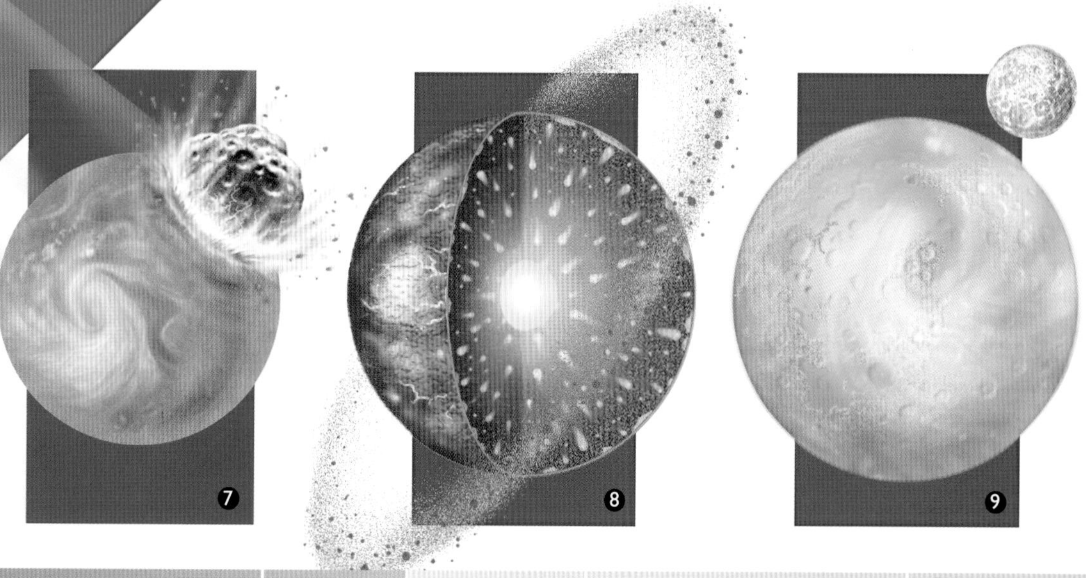

7000 | 6000 | 5000 | **4550** | 4000 | **ARCHEAN** | 3000

Solar system takes shape

Formation of Earth and Moon

Heavier elements synthesize

Differentiation of layers

• Oldest earth rocks

• Oldest moon rocks

IN ITS infancy the Earth consisted of an unsorted mixture of silicates and metals, so that the composition at the surface was the same as in the center. The heat and radioactivity produced as the planet accreted continued to raise the temperature, causing the materials to separate into layers. During this process of differentiation, iron and subsidiary heavy minerals sank towards the center to form a relatively soft core. Calculations show that the sinking of metallic iron within the Earth would have begun when one-eighth of the planet's mass had accrued—well before the entire Earth had accreted. Lighter elements, including gases, moved upwards, with the lightest gases escaping into space. Initially, at about 3.5 billion years ago, the core was heated to a semi-molten state; but as pressure increased, the inner core solidified. The outer core, which was subjected to less pressure, remained liquid.

No one knows exactly how the Earth formed. But scientists believe that core formation had begun before accretion stopped, and that the interior was hot enough to melt the iron core.

A thick ocean of magma or molten rock surrounded the core in the early stages of differentiation. Over a period of 400 million years, as this cooled and lighter materials continued to separate from heavier ones, the mantle and crust formed. While the process of differentiation was taking place, the early Earth was still being bombarded by planetesimals, which drove into the magma ocean, further raising the temperature.

THE compositions of the planets reflect the compositions of the gaseous nebulae from which they originated. Accumulation of materials from space and the continuing loss of lighter gases ensured that the planets nearest to the Sun were made of the heavier elements such as silicon and metals, whereas the outer planets were composed mainly of hydrogen and helium gas. The inner planets probably took approximately 100 million years to form. This period marks the interval between the formation of the solar system and the formation of a crust on the Moon. As the planets formed, they developed internal pressures and rising temperatures. With melting, heavy and light elements separated, the heavier materials moving towards the core while the lighter materials floated to the

Earth and its neighbors formed at about the same time from the same materials, but there are important differences between them.

BIG & SMALL PLANETS

(Right and bottom)The four small planets nearest to the Sun are predominantly rocky. Venus and Earth are similar in size; Mars and Mercury are much smaller, and the Earth's Moon is about the size of Mercury. Jupiter is the biggest of the large gaseous outer planets. The most distant planet, Pluto, is small and rocky like the inner planets, but much colder; it is covered in methane frost.

1 Mercury
2 Venus
3 Earth
4 Mars
5 Jupiter
6 Saturn
7 Uranus
8 Neptune
9 Pluto

DIFFERENTIATION OF THE EARTH

(Right) The early Earth was a homogeneous mixture of silicates and metals. Iron and nickel collected at the center to form the first core. Then oxygen and other metals combined with the silicates to form a mantle around it. Finally, the lightest materials solidified as the outer crust.

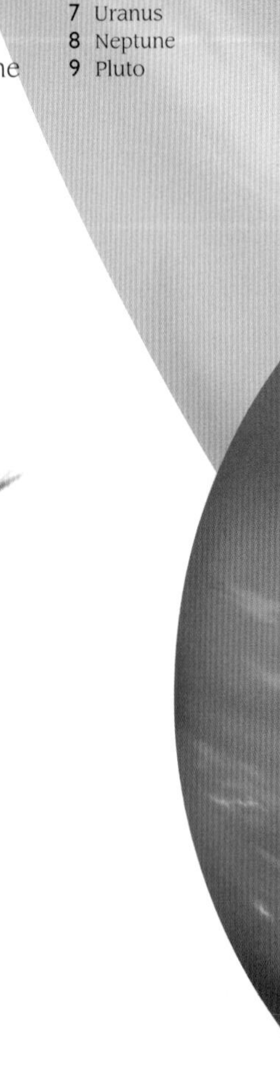

1 Accretion processes heat up the Earth; molten iron sinks to form a core
2 The less dense silicates rise to the surface
3 Differentiation into core, crust, and mantle is complete

THE SUN'S FAMILY

The Sun is a giant nuclear reactor. Fusion reactions in the interior convert hydrogen into helium, releasing vast amounts of energy. At the surface, masses of charged matter erupt as solar flares (right). Other charged particles stream off into space. As the Sun condensed at the center of the solar nebula these solar winds pushed volatile gases—hydrogen, helium, methane, ammonia, and water—out into space, where they formed the low-density outer planets, while in the hot center silicates, oxides, and metals condensed to form the terrestrial planets.

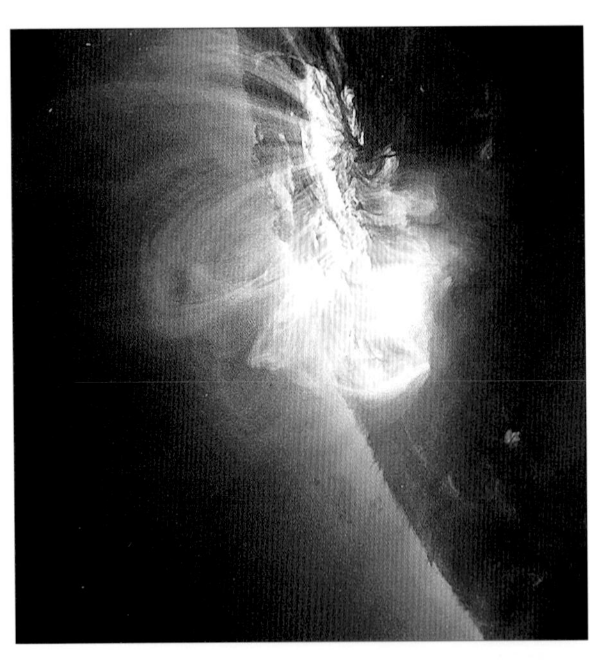

Solar wind

1 2 3 4 5 6 7

Outer core: compounds of carbon, nitrogen, oxygen

Inner core: silicate and iron

Liquid metallic hydrogen and helium

Jupiter/Saturn

Very dense atmosphere

Ocean of liquid hydrogen and helium, plus some methane, ammonia, and water

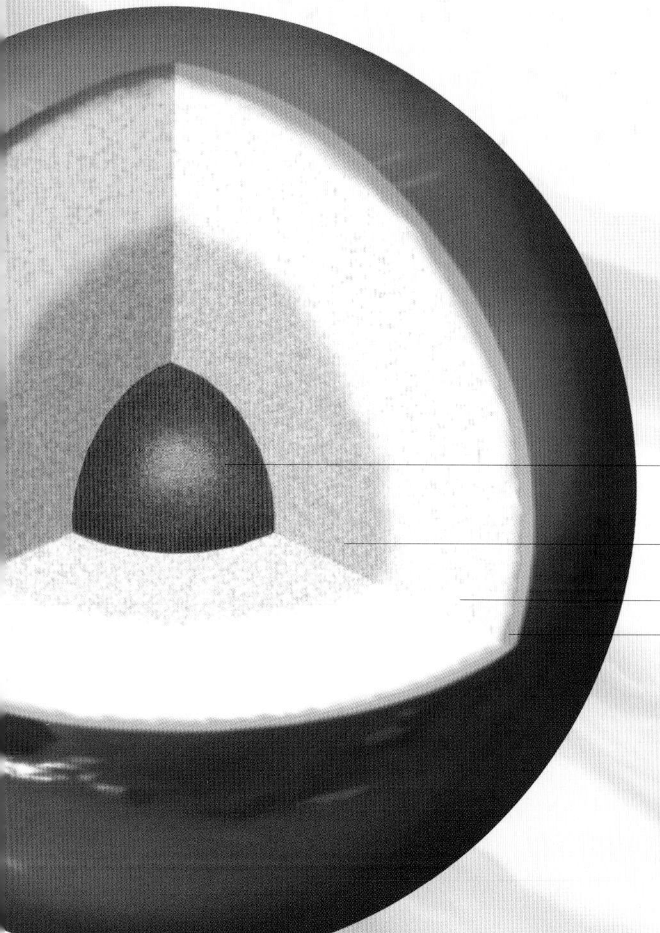

PLANETARY INTERIORS

The internal structure of the planets varies; the outer planets, the gas giants, have silicate and iron cores, whereas the cores of the rocky inner planets are principally iron and nickel. The much greater size and mass of the gas giants mean that they have enough gravity to retain even the lightest gases, helium and hydrogen, as liquid layers below their very thick atmospheres. Uranus and Neptune, with lower masses, contain less hydrogen and helium than Jupiter and Saturn. All the outer planets lack the mantle and crust of silicate minerals that characterize the terrestrial planets.

Uranus/Neptune

Core of silicate compounds and iron

Protoplanetary debris: compounds of carbon, nitrogen, and oxygen

Ocean of liquid hydrogen and helium

Dense atmosphere

surface. But there are important differences between the Earth and its three nearest neighbors.

Mercury is the smallest of the terrestrial, or rocky, planets. It has almost the same mass as Earth, but its large iron core makes up 75 percent of its radius and 80 percent of its mass. The outer layer is comparatively thin, torn away during a collision with a giant planetesimal. Mercury and the Moon are of similar size; both are dead, showing few signs of internal heat or volcanic activity.

Mars is larger in size, but it is of lower mass than the Earth. Its core is either smaller or less dense; unlike that of the Earth, it is cool and not fluid. Venus is the most similar to Earth in its size, composition, and mass. The outer planets, with huge mass, have enough gravity to retain even the lightest elements, hydrogen and helium.

All three of the inner planets are comparatively small, so that they cooled much more rapidly than the Earth. Without internal heat and volcanic activity they were unable to retain viable atmospheres. Their size and low gravity led to the loss of gases and water vapor. Because it is larger, the Earth exerted a greater gravitational effect on the gases and vapors produced by its internal processes. Water brought in as a component of extraterrestrial bodies was also retained. These substances eventually formed the Earth's oceans and atmosphere.

METEORITES—DEBRIS OF THE SOLAR SYSTEM

When meteorites collide with the earth they save us great sums of research money, for they give us important clues about the age and nature of the solar system. Most are small, luckily, but huge craters, such as Meteor Crater in Arizona (0.75mi/1.2km across, 558ft/170m deep), indicate that giants weighing many tons have occasionally hit the surface of our planet. There are three main types of meteorite: iron, stony (chondrites and achondrites), or stony-iron, depending on the proportions of silicates and iron they contain. Most were formed by the melting and recrystallization of primitive planetary material. However, the rare chondritic meteorites have not melted since their formation and have been found to contain organic compounds and simple amino acids, the building blocks of life.

8

9

The core accounts for 31 percent of the Earth's mass. Seismic data indicate that it is composed primarily of iron, probably with sulfur (10 percent) as a dilutant. The outer core, 1400mi (2270km) thick, is molten, with a density more than 10 times that of water. The increased pressures associated with the inner core render it solid. It is about 1500mi (2400km) in diameter, with a density up to 13 times that of water (about the same as the liquid metal mercury). At the center, the temperature may exceed 7200°F (4000°C) with pressures up to 4 million atmospheres. This temperature is maintained by heat from radioactivity in the surrounding rocks, augmented by gravitational compression. Radioactivity is a long-lasting heat source: the Earth's core, approximately 4.5 billion years old, still generates plenty of heat.

THE EARLY Earth was too hot to develop a proto-crust. With differentiation, cooling, and the precipitation of rainwater, its surface must at one time have resembled a crystal mush with the texture of crushed or granulated ice. Continuous remelting of this material would have taken place as long as the Earth remained hot, but when cooling began, gradually a patchy solid crust would have developed. This crust would have been made of compounds of silicon, aluminum, and other light solid elements with low melting temperatures that floated up from the denser core during differentiation.

In order for crust to form, the Earth had to cool down: this stopped the process of remelting and allowed rock to form. However, the core remained partly molten, as it is today.

Today the crust is 18 to 25 miles (30 to 40km) thick under the continents but much thinner under the oceans; this is much thinner than the crusts of the Moon, Venus, or Mars. Below the crust, molten rock forms the Earth's mantle—an intermediate layer between the core and the crust that remained when the heavier elements sank towards the core and the lighter ones rose towards the planet's surface. The mantle extends to a depth of nearly 1850mi (3000km) and still surrounds the core. Occasionally the molten material forces its way up to the surface through volcanic eruptions.

By about 3.8 billion years ago, sufficient rocks had formed and been preserved at the Earth's surface to form the cores of the protocontinents. As the crust formed, it differentiated into continental and oceanic layers, each composed of characteristic types of rock. From these rocks the composition of the crust can be determined. To do this, geologists collect and analyze specific rock types at different levels, estimate their relative abundances, and produce an average for either the oceanic or the continental layers. The quoted figures may include oxygen but, in either case, they indicate light materials that are depleted in iron.

The density of the crust is about three times that of water, and under the continents it consists mainly of sed-

MOLTEN MANTLE

Molten rock in the Earth's mantle might resemble the slow-moving ropy lava known as pahoehoe. Volcanoes provide a direct route for the molten rock in the mantle to reach the Earth's surface. A magma chamber forms beneath a weakness in the crust. As the pressure builds up, the magma forces its way up a fissure, and then bursts forth. The result is a flow of lava, which cools and solidifies into masses of igneous rock.

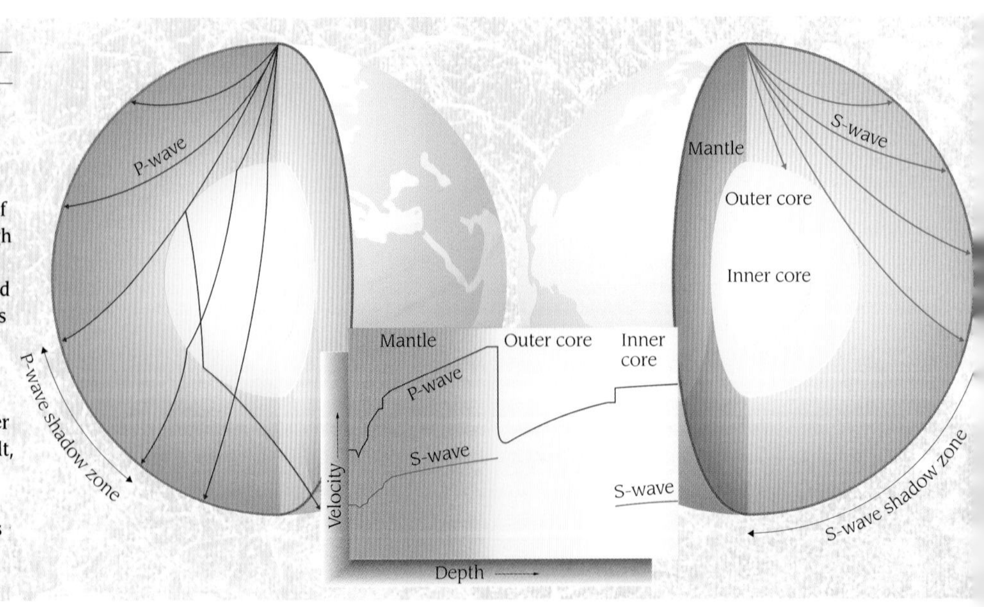

PROBING THE EARTH'S INTERIOR

Scientists use analysis of seismic waves produced by an earthquake to investigate the Earth's inner structure. Shock waves radiate outwards from the earthquake's source near the surface; body waves penetrate the inner layers of the Earth, at a speed depending on the energy of the earthquake and on the density of the material through which the waves travel. The faster P-waves are waves of compression that can travel through both solid and liquid material. The slower S-waves are transverse shear waves that can pass only through solids.

At a depth of approximately 1800mi (2900km), S-waves disappear, and the speed of P-waves is greatly reduced. This depth marks a boundary between the lower mantle and outer core where dense crystalline rocks melt, changing to liquid iron. Seismologists can calculate the time taken for waves to reach the surface and use this information to map the thicknesses of the various layers and to estimate their densities.

Crust
Continental crust: 25mi
Oceanic crust: 4mi
Ocean: 3mi
Upper mantle: 400mi
Lower mantle: 1400mi
Outer core: 1400mi
Inner core: 750mi
Mesosphere: 1650mi
Hydrosphere: 3mi
Lithosphere: 45mi
Asthenosphere: 125mi

INSIDE THE EARTH

The Earth's interior consists of concentric layers. Above the core lies the mantle, an 1800mi (2800km) thick layer of molten rock. The upper mantle and the solid crust make up the lithosphere. The crust is 18–25mi (30–40km) thick under the continents—thicker under mountain ranges—but only about 4mi (6km) thick under the oceans.

EARTH'S ELEMENTS

The crust, mantle, and core have different compositions as well as thicknesses. The upper crust (sial) is largely silica and aluminum; the lower (sima) consists mainly of rocks high in silica and magnesium. Iron- and magnesium-rich rocks dominate the mantle, and the core is nearly all iron.

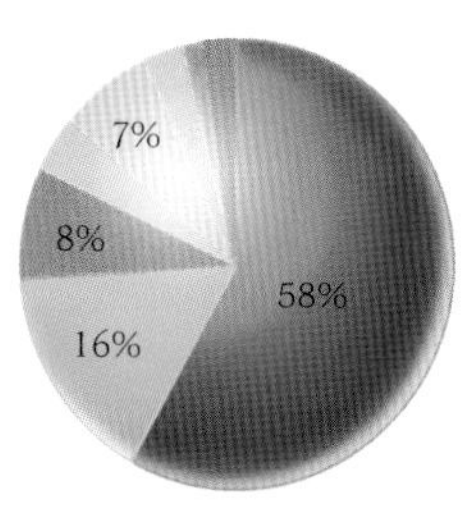

Crust

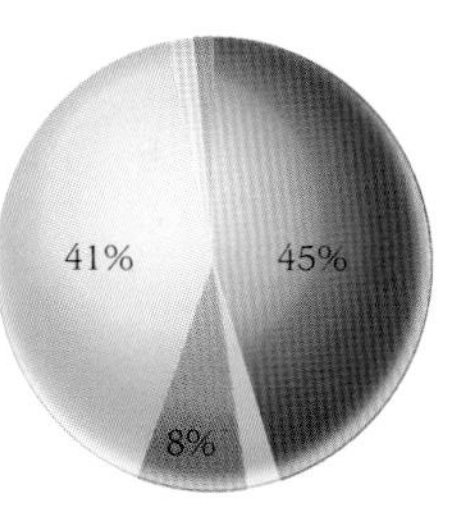

Mantle

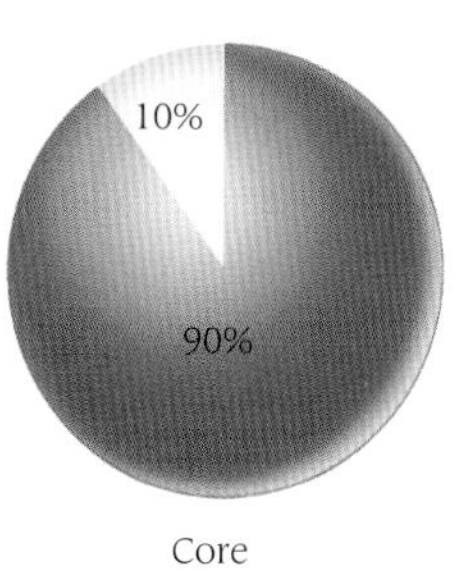

Core

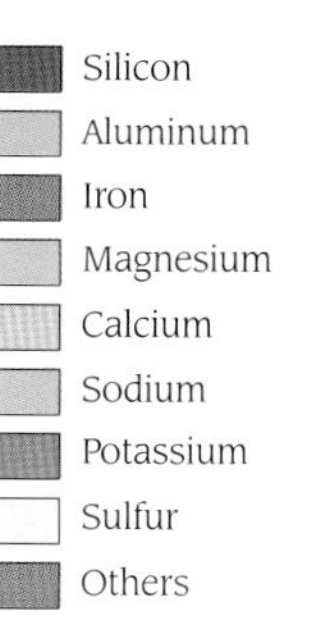

imentary rocks such as sandstone and limestone on a base of granite (an igneous rock). Under the oceans, the igneous rock also includes basalt. The composition of the mantle is more difficult to determine, but the analysis of ocean-floor magmas and kimberlite pipes (kimberlite is a type of rock) provides important clues. The pipes are associated with diamond exploration, and often contain fragments of mantle material, with abundant olivine, pyroxene, and garnet— all characterized by containing silicates of magnesium and iron. Diamond is a high-pressure form of carbon, and the mineral assemblages of the diamond-bearing kimberlites suggest that they originated 90mi (150km) below the surface of the Earth. Silicon and magnesium are the main elements of the upper mantle, whose density increases with depth from about 3.3 to 3.5 times the density of water.

The upper part of the mantle and the crust form the lithosphere, a rocky layer 30 to 45mi (50 to 70km) thick. Huge slabs of this material "float" on the semi-molten asthenosphere. They are segmented into plates, which can move with the motion of the asthenosphere.

AT FIRST the Earth was an inhospitable place. It was hot, even molten, surrounded by steamy clouds of hydrogen and helium residual from the Big Bang. The first 50 million years or so saw little change, but the impact of what became the Moon caused this early layer, and any surface water, to vanish into outer space. The gases that remained—mostly hydrogen, nitrogen, and carbon dioxide—came from various sources. Some were part of the gas and dust from the original solar nebula. Incoming meteorites contained other gases, as did comets, which are composed almost entirely of frozen gases and ice. At first the gases were mixed with all the other components of the protoplanet. But just as gravity caused differentiation into core, mantle, and crust, so it affected the composition of the atmosphere.

Earth's atmosphere has undergone many significant changes since the planet was formed. About 80 percent was carbon dioxide, and there was no free oxygen.

As the Earth continued to cool, water vapor in the atmosphere began to condense and fall as rain. At first it vaporized again on impact with the hot, molten surface of the Earth. Eventually, however, a fragmented blocky crust was formed. Water derived from meteorites, volcanic activity, and precipitation eventually accumulated as lakes, seas, and oceans. Gases were also freed when minerals were melted during differentiation and the reworking of crustal materials.

The size of the Earth was an important factor that enabled the buildup and retention of an atmosphere. The smaller terrestrial planets have low surface gravities, and even the heaviest gases leak into space. Earth held on to these heavier gases and the atmosphere would have been rich in carbon dioxide, nitrogen, and water as the original, lighter hydrogen and helium were removed by solar winds. There was no oxygen, because any that was outgassed reacted immediately with the easily oxidized metals in the primitive crust at the planet's surface.

There were several continuing cycles, with degassing taking place through volcanic emissions and magmatic activity, accompanied by evaporation and precipitation. The lighter gases were drawn into space by the heat of the Sun. It is likely that, during the first few hundred million years, the Earth's atmosphere resembled the mix of gases that emerge from volcanic vents. As the surface cooled and stabilized, water in the atmosphere became affected by ultraviolet light from the Sun, which split it into its component elements of hydrogen and oxygen. This must have been a particularly important process prior to the formation of the ozone layer, with lighter hydrogen passing into space and an ongoing buildup of oxygen modifying the atmosphere.

EVOLUTION OF THE ATMOSPHERE

The Earth's atmosphere was, initially, a steamy mixture of hydrogen and helium (1). The impact of a large asteroid blew these volatile gases away (2). A secondary atmosphere and oceans then formed (3) as volcanoes emitted gases from the interior and photochemical reactions took place.

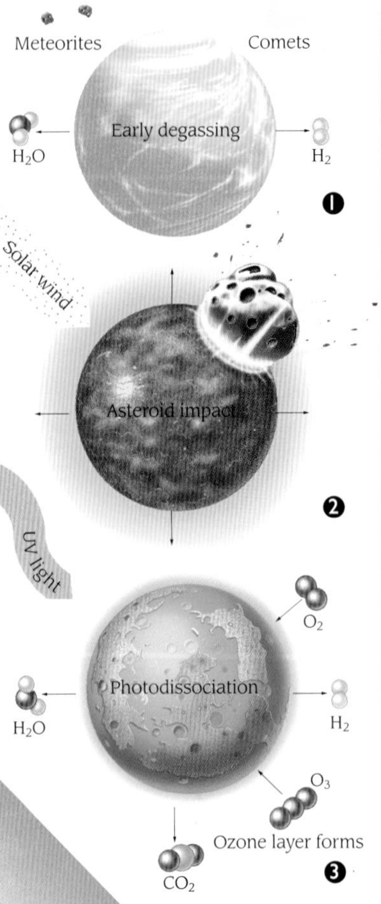

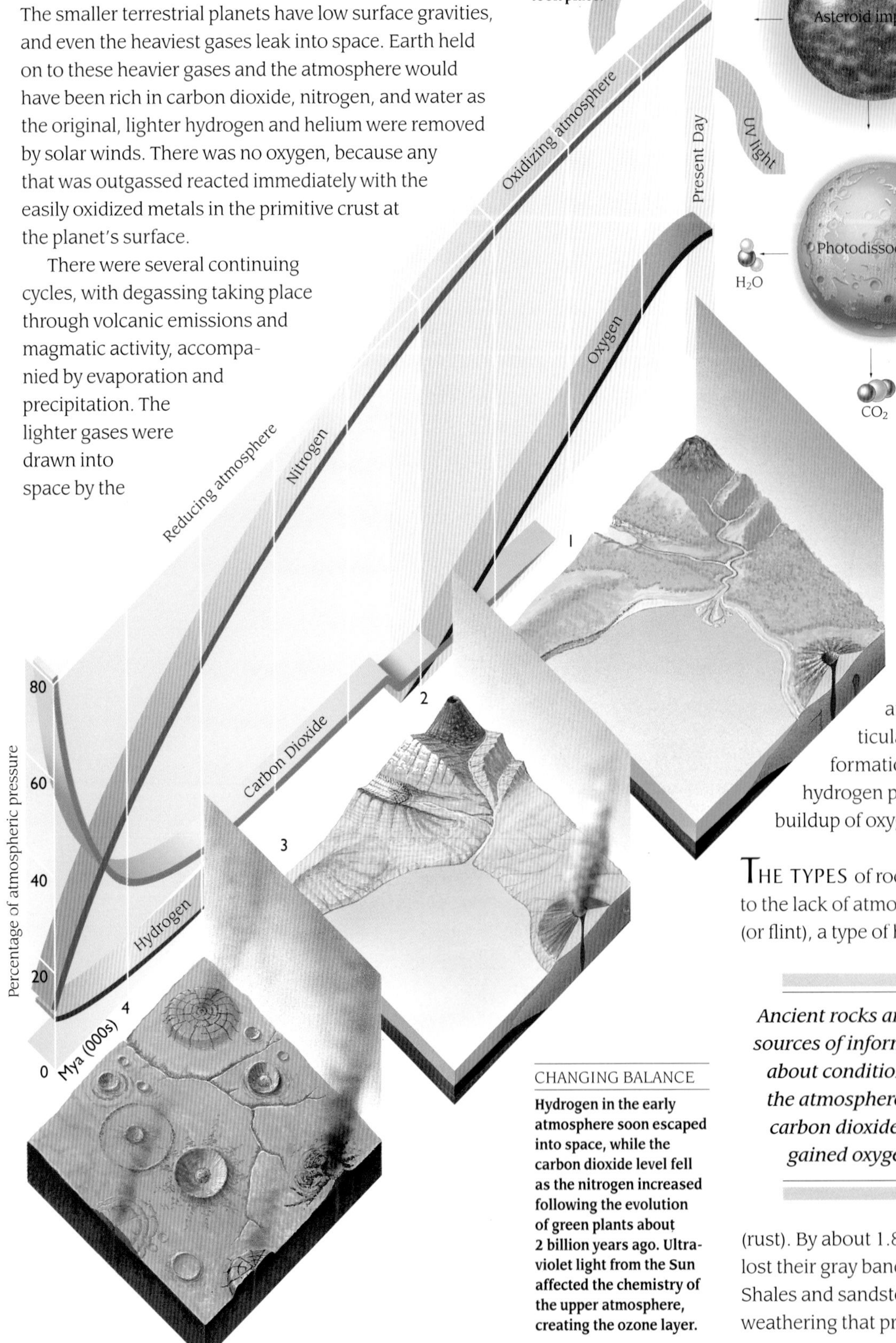

CHANGING BALANCE

Hydrogen in the early atmosphere soon escaped into space, while the carbon dioxide level fell as the nitrogen increased following the evolution of green plants about 2 billion years ago. Ultraviolet light from the Sun affected the chemistry of the upper atmosphere, creating the ozone layer.

THE TYPES of rock deposited at this time bear witness to the lack of atmospheric oxygen. Among them is chert (or flint), a type of hard sedimentary rock that forms in a low-oxygen, acidic environment such as the Earth was once. Gradual increases in the level of oxygen are shown by another kind of rock called banded ironstone, in which alternating bands of gray and rust show the fluctuation of oxygen from low (gray) to higher (rust). By about 1.8 billion years ago, iron deposits had lost their gray bands, indicating the presence of oxygen. Shales and sandstone were also present, and the weathering that produces these rocks could not have

Ancient rocks are rare sources of information about conditions as the atmosphere lost carbon dioxide and gained oxygen.

occurred without an atmosphere. Similarly, the appearance of dolomite and limestone, which are alkaline, shows that the atmosphere was no longer acidic due to a high carbon dioxide content.

Scientists postulate that a combination of electrical storms, glaciation, and magmatic activity gave rise to the chemical compounds essential to life. An oxygen-rich atmosphere would have inhibited the formation of these compounds, but change again played an important role in the Earth's destiny. Once the first organisms had appeared, an irreversible chain of events was set in motion. Different kinds of bacteria and, much later, green plants evolved to modify our world through the oxygen they released as they made their own food. It was only during the Proterozoic, which began 2.5 billion years ago, that the atmosphere came to resemble today's. Nitrogen now constitutes 78 percent of the atmosphere by total volume and oxygen 21 percent. Carbon dioxide comprises only 1 percent of the air we breathe.

While the atmosphere was still rich in carbon dioxide, rain falling and accumulating in the first oceans was much more acidic than at the present time. These acid waters reacted with rocks, dissolving compounds of calcium and magnesium. The composition of the early oceans would also have reflected the fact that some of their waters originated from volcanic emissions and, perhaps, comets. Their chemistry built up gradually, with more salts being added as products of erosion. Scientists believe that the salinity of the oceans has changed little over the last 4 billion years. Today seawater contains an average of 3.5 percent of dissolved salts, almost all of it common salt (sodium chloride), but with significant

A THIN ENVELOPE

The gases of the atmosphere (above) form a thin layer that lets in some—but not too much—warmth and light from the Sun. It extends for about 450mi (700km) above the surface before blending into outer space. Pressure falls off rapidly with altitude, so the first 5mi (8km) above the surface hold three-quarters of the atmosphere's total mass.

ANATOMY OF THE ATMOSPHERE

The layers of the atmosphere (right) differ mainly in the prevailing temperature, which ranges from –130°F (–90°C)in the upper mesosphere to about 2400°F (1300°C)in the upper thermosphere. The lowest layer is the troposphere, site of most clouds and weather systems. Above this is the stratosphere, in which the ozone layer filters out much of the harmful ultraviolet radiation from the Sun that is not reflected back into space at the thermosphere. Without this, the Earth would rapidly become too hot to support life.

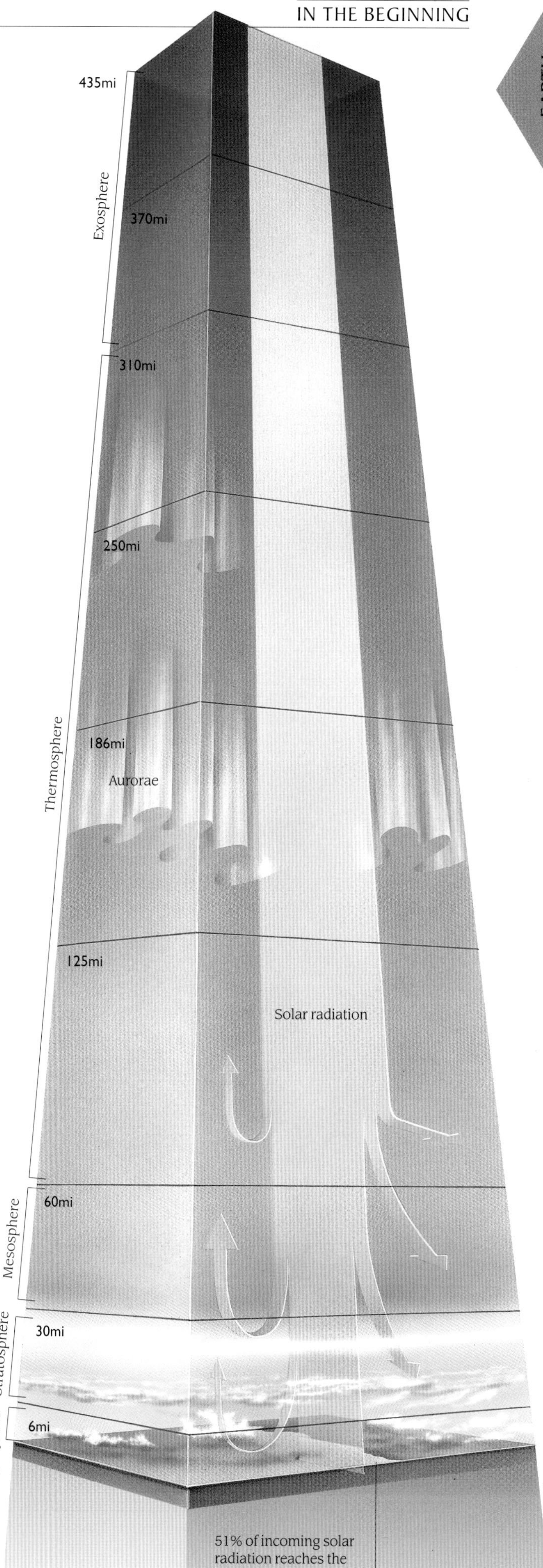

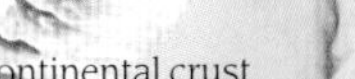

amounts of calcium and magnesium salts.
Heat continues to be generated within the Earth's interior, and heat flow, with its associated volcanic activity, is an important factor in the evolution of the moving slabs of lithosphere that geologists call plates. The heat generated by internal processes, such as radioactivity of the settling of the heaviest material, initially flows upwards and outwards through simple conduction. This is particularly evident within the crust, with much of the heat passing through cool, rigid rocks and escaping to the atmosphere.

Seismic data reveal that below the lithosphere (the crust and brittle upper part of the mantle) the underlying asthenosphere is partially molten. Below about 120mi (200km) the mantle is solid; but at the boundary with the core—1800mi (2900km) down—the internal temperature rises from about 2700°F (1500°C) to over 7000°F (3900°C). Pressure also increases with depth, and closer and closer atomic packing can also be determined from seismic information. It is this pressure and density that stop some of the hottest parts of the inner Earth from melting into a bubbling pool.

Solid matter creeps when subjected to high temperatures and pressures, and the greater the temperature and pressure, the more convection is likely to occur. Convection currents, with the hotter material rising up and then spreading sideways as it reaches the cooler surface levels, create convection cells of circulating rock. These cells—and even layers of convection cells—affect the mantle. The upwelling of hot material spreads out below the mosaic of rigid plates that make up the lithosphere.

The convection of mantle material functions like a giant engine, pushing the plates horizontally but with a distinct rotational movement, too. This action gives rise to the phenomenon of continental drift, by which continents move slowly but continuously across the surface of the planet, borne on the plates which in turn float on top of the molten mantle.

MODERN Earth has seven major tectonic plates and many smaller ones. Each plate rotates around a point called the Euler pole—a point on the Earth's surface from which an imaginary line passes through the center. Plates that slide past each other sideways along faults in the Earth's crust do so in small circles, whereas plates moving away from each other, at mid-ocean ridges, describe larger circles. The junction between sideways-moving plates is a transform boundary, whereas plates moving apart on the ocean floor are the site of seafloor spreading. The San Andreas fault zone on the Pacific coast of North America is an example of a transform boundary between plates, while the Mid-Atlantic Ridge separates the African and South American plates where they are slowly moving apart.

Earthquakes, volcanoes, and mountain-building occur in mobile zones, where plates are moving. If everywhere on Earth was in a mobile zone, there would be no surviving continents and no ancient rocks to study, because they would be repeatedly destroyed.

FORCES ON A PLATE

Stresses on the Earth's crust give rise to folds and faults in surface rocks, and to earthquakes and volcanic activity. Greater forces move whole continents. The crust is divided into huge slabs called tectonic plates. Heat from the mantle sets up convection currents, bringing molten rock up towards the crust, where it cools and descends again. Below the lithosphere, the action of magma moves the plates floating on it. As plates move apart, new crust forms from spreading magma. Where plates come together (destructive margins), crust is pushed underneath them and recycled. These forces build mountains and deep ocean ridges.

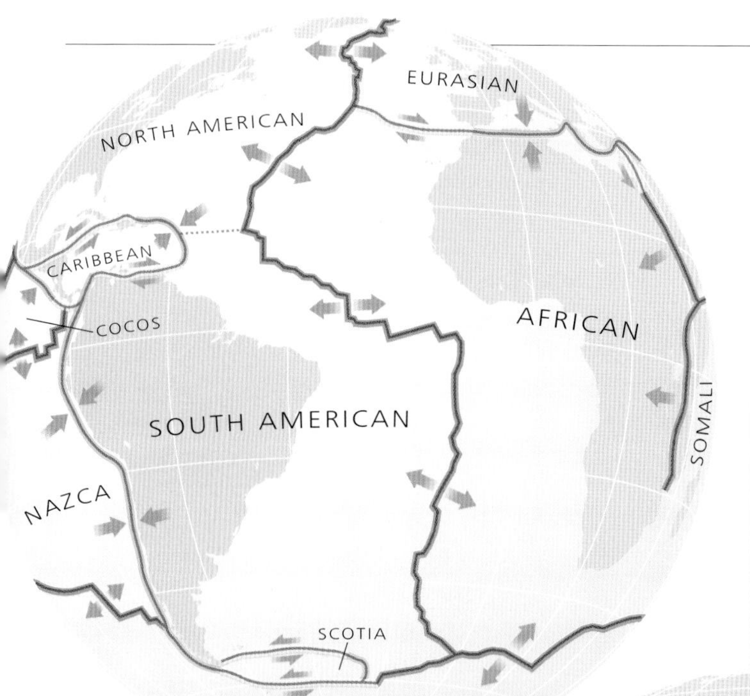

— plate boundary
······ diffuse plate boundary
– – uncertain plate boundary
divergent plate margin
convergent plate margin
transform plate margin
relative movement of plates

At the other major type of plate boundary, one plate descends and passes under the other. During this process, the descending plate angles down and melts into the hot mantle below. Deep ocean trenches often form at such subduction zones, which are also associated with deep-focus earthquakes and volcanic activity.

Geologists can now use satellites to measure the movements of the plates across the surface of the Earth to an accuracy of 0.4in (1cm). Movement of the ancient plates is less well documented. Magnetic reversal data and fossils, however, provide clues to the age of the seafloor and the links between the continents. The Earth's liquid outer core of iron, moved by convection currents that generate electricity, is the source of a magnetic field that changes polarity about every 500,000 years. New rock being laid down at mid-ocean ridges records the prevailing polarity and therefore these changes. The boundaries at constructive plate margins are characterized by a distinct "striping" of magnetic patterns known as seafloor anomalies, which can be used to establish a detailed magneto-stratigraphy.

Not all volcanic activity occurs at mid-oceanic ridges or plate margins. Hot spots and plateau lavas—huge flood basalts—are products of the mantle, indicating gigantic underground plumes associated with convection. The Deccan Traps of India, 0.6mi (1km) thick, are plateau lavas that formed at the end of the Cretaceous.

UNDER THE OCEAN

The Earth's tectonic plates are in a continuous cycle of creation and destruction. Mantle material wells up at mid-ocean ridges (above), producing new lithosphere. At convergent boundaries, marked by mountain ranges and deep ocean trenches, the crust is destroyed as the plate slides back into the mantle; earthquakes and volcanoes are common.

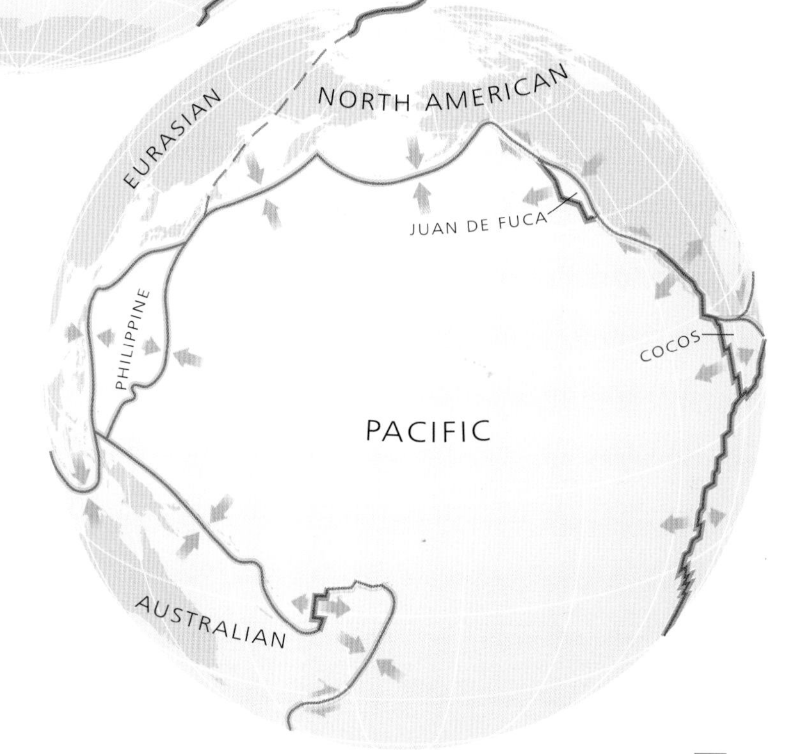

THE EARTH'S crust is made of rock. Rock is a collection of minerals; a mineral is a naturally occurring compound (or sometimes element) that always has the same chemical composition no matter where it is found. A rock is an aggregate of one or more minerals, whose composition may vary depending on the conditions under which it was formed. Thus quartz is a mineral, whereas granite is a rock. There are three major kinds of rock, which differ in their places of origin and in the geological processes by which they form. Molten rock, or magma, rising up fissures in the crust solidifies at the surface; other magma is spewed out of volcanoes before it solidifies. This solidifed magma gives rise to igneous rocks through the process of crystallization. The grain of the rock may be coarse or fine depending on how hot the magma was and how long it took to solidify. Geologists describe the first crystalline igneous rocks of the early Earth as primary rocks. These included basalts and other fine-grained products of volcanic activity. Coarser crystalline rocks were formed in the cooling, thickening crust, with their component minerals reflecting the different temperatures at which they crystallized. The Earth's earliest crust would have consisted exclusively of igneous rock.

The processes that formed the first rocks on Earth can still be studied today as the three kinds of rock—igneous, sedimentary, and metamorphic—are continuously recycled.

At the surface, wind, rain, and other conditions eroded these early rocks, producing detrital minerals. These formed

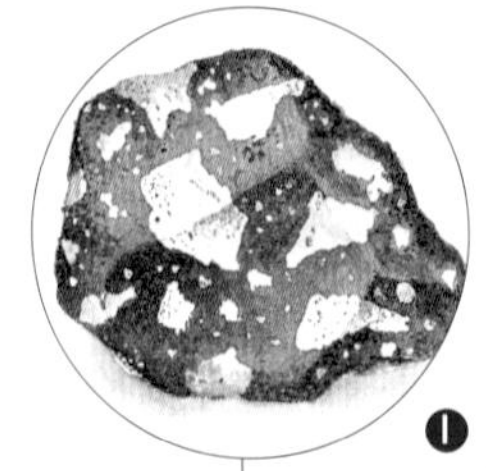

THE GEOLOGICAL PROCESS

Rocks are continuously destroyed and renewed. Weathering breaks down surface rocks. The debris washes away, is buried, compressed, and becomes sedimentary rock. The rock may be uplifted again; it may, through pressure and heat, turn into metamorphic rock, or be carried deep into the Earth to melt and form igneous rock.

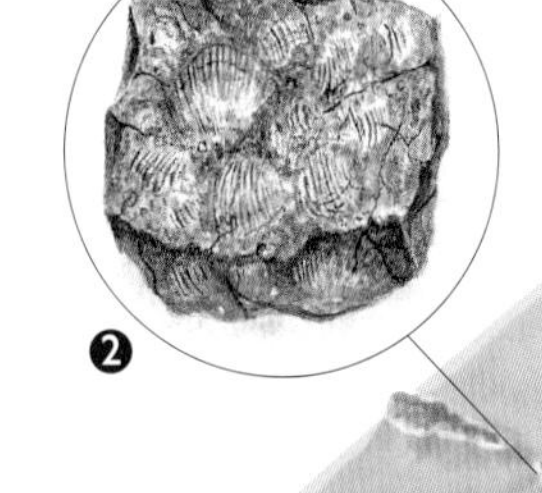

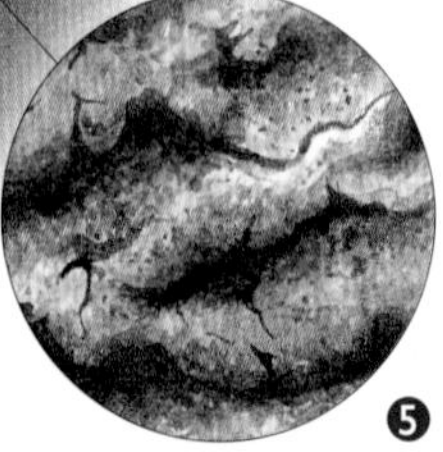

SEDIMENTARY ROCKS

Sediments that settle out of water, ice, or air-blown dust consolidate to form sedimentary rocks such as limestone and sandstone. This can take place under water or on land.

1 Breccia
2 Reef limestone
3 Halite (salt)
4 Sandstone
5 Turbidite

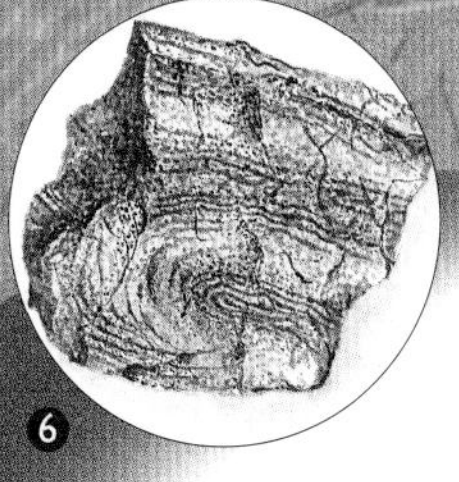

IGNEOUS ROCKS

Igneous rocks form when magma cools and crystallizes. Magma that reaches the surface through extrusions and volcanoes (lava) forms "volcanic" rocks, while rocks that form at depth are "plutonic."

7 Pumice
8 Basalt
9 Granite

METAMORPHIC ROCKS

Extremes of temperature and pressure cause existing rocks to recrystallize, forming metamorphic rocks such as gneiss. This happens deep below the surface of the Earth.

6 Gneiss

THE ROCK CYCLE

(Right) Rocks undergo endless change from one sort into another. Magma from the mantle crystallizes to form igneous rocks, which are subsequently eroded, deposited and consolidated into sedimentary rocks. Heat and pressure below ground convert sedimentary rocks into metamorphic rocks. Rock carried below the mantle melts, forming new magma.

the first sandstones, gravels, and beach conglomerates. Erosion and weathering wore surface rocks into small particles that got washed away by water—rain, river, or seawater. There they settled in layers and compacted together to form sedimentary rocks such as shale, limestone, and sandstone.

If sedimentary rocks are buried deep underground, very high temperatures and pressures transform them into metamorphic rocks. Contact with hot igneous rocks can also bring about such changes; an example of this is intrusion, when molten magma leaks into a layer of sedimentary rock. Baking, melting, folding, and fracturing produced groups of metamorphic rocks ranging from low-grade slate to high-grade schist and gneiss. Grade is a measure of the intensity of deformation or degree of baking: the higher the grade, the greater the heat or pressure that formed the rock. Metamorphic rocks, in turn, weather and erode to form again as sedimentary rocks, thus perpetuating the rock cycle.

This cycle does not always occur in strict order, and interruptions may occur. For example, if igneous rock remains buried in the crust instead of being exposed at the surface, it may be subjected to intense compression and heating associated with plate tectonic activity. In this case, the igneous rock will be transformed into metamorphic rock.

It is ancient metamorphic rock that makes up the continental shields or cratons that are at the heart of continents. The suturing of the ancient continents and the subduction of oceanic plates beneath them thrust up chains of high mountains. Mountain streams developed into rivers, ice fields into glaciers, and lakes into inland seas—all carrying away the sediment shed by the mountains as they wore down under exposure to the elements. Erosion initially produced sedimentary rocks rich in silica: feldspars (which are abundant in the Earth's crust) and, later, clay minerals. The first occurrences of these rocks have been erased from the geological record because of volcanic activity and the constant melting and remelting of rocky materials. Sedimentary rock that formed later accounts for about 70 percent of the rock visible on the Earth's surface. This is the rock that weathered, eroded, and combined with air, water, and organic materials over millions of years to produce soil.

Geologists have a saying: "The present is the key to the past." It means that the fundamental processes involved in the deposition of sediment, such as gravel in a riverbed or till in a glacier, have changed little over time. They can study these processes today to interpret environments that existed through geological time. This principle, called uniformitarianism, helps them understand the Earth's history.

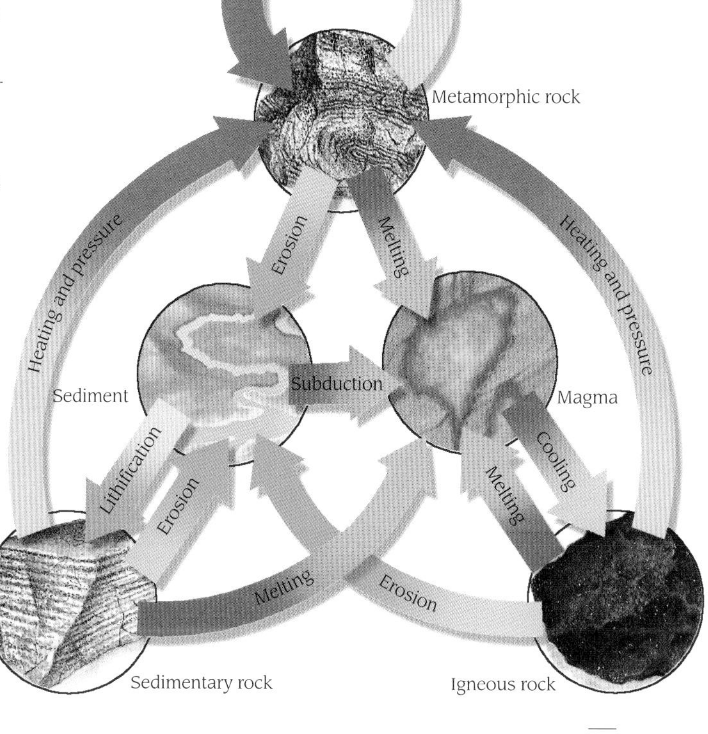

IGNEOUS and metamorphic rock are often formed deep inside the Earth, so the events that made them are difficult to study. Sedimentary rock, however, forms at the surface, and acts as a record of the conditions at the time. It would be comparatively easy to analyze this record if rock formations were preserved in an unbroken sequence but this is rarely the case. Although rock strata are always laid down with the oldest layers at the bottom and the youngest layers at the top there are often instances where a formation is missing, due to erosion for example; such a boundary between two layers in a broken sequence is termed an unconformity. Earth movements may displace strata, tilting them from their original horizontal orientation. Sometimes tectonic action has produced folded strata, which have then been eroded flat and overlain by a layer of new sediment; the resulting arrangement of non-parallel bedding planes is known as an angular unconformity. If the erosional surface developed on an undeformed set of lower beds the term disconformity applies, while an unconformity in which older igneous or metamorphic rocks are overlain by sedimentary strata is a nonconformity.

Interpreting the history of a layer of rock is complicated by the fact that some of it is almost guaranteed to be missing.

RECORD IN THE ROCKS

The 5000ft (1500m) deep Grand Canyon provides a vertical section through layers of rock that date back 2.2 billion years. It was formed when Arizona's Colorado River carved out a gorge in an extensive tableland. The oldest rocks were formed from sediments laid down in shallow water, then metamorphosed by mountain-building processes into what is now known as the Vishnu Schist. This was invaded by intrusions of igneous rock, forming the Zoroaster Granite. The mountains were gradually leveled by erosion and covered by layers of lava flow and sedimentary rock (the Grand Canyon Supergroup). These are separated by the "Great Unconformity" from the layers of Paleozoic sedimentary rocks above.

THE RELATIVE DATING OF ROCKS

The most basic principle of relative dating is that of *superposition*, stated by Nicolaus Steno in 1669: in an undisturbed sequence of sedimentary strata, each bed is younger than the one below it and older than the one above. Steno's second principle is that of *original horizontality*, which pronounces simply that strata are laid down in a horizontal plane (more or less). The understanding that strata are originally flat and unbroken gave Steno his third law, the principle of *original lateral continuity*. This explains how similar rocks can occur on either side of a valley or other interruption.

Strata do not remain undisturbed over the passage of time. Over millions of years the surface of the earth has been constantly subject to erosion, uplift, folding, faulting, and the intrusion or inclusion of other rock bodies. It is assumed that such features are younger than the rocks affected. This is the principle of *cross-cutting relationships*. Undisturbed layers of rock are called *conformable* and breaks in the rock record *unconformities*. An *angular unconformity* occurs when tilted or folded sedimentary rocks are overlaid by flat, younger beds; a *disconformity* is when younger beds overlie an older, eroded, flat-lying surface; and a *nonconformity* separates flat-lying beds from older metamorphic or igneous rocks.

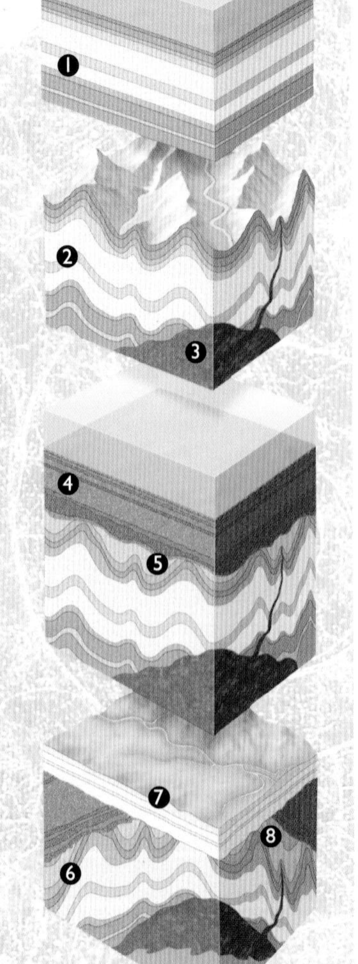

1 Superposition of sediments
2 Metamorphism, deformation and uplift
3 Intrusion of granite pluton
4 Subsidence and marine deposition (shales)
5 Eroded surface preserved as an unconformity
6 Tilting and uplift
7 Deposition of sandstone sediments over eroded surface
8 Angular unconformity

Deposition of sediment tends to be heavier near continental margins, the final destination of material that has been weathered from surface rock, eroded, and carried by water until it reaches the sea. More limited deposition occurs closer to mid-ocean ridges. Land is periodically flooded during periods of marine transgression when the sea level rises, depositing marine sediment in its wake. Large deposits of this sediment form rock sequences representing tens of millions of years of time. They are divided by unconformities, when erosion rather than deposition was taking place and the sequence of deposition was interrupted.

Each layer of sedimentary rock within a sequence has its own characteristic color, grain size, mineralogy, and structure; it may also have fossils unique to the time and place in which the rock formed. Unconformities are frequently marked by the presence of coarse conglomeratic sediments, which may mark a rise in sea level. In this case, they represent the flooding and erosion of an older surface. The type of sediment at the base of the younger, overlying sequence varies laterally because of changes in the environment at the time of deposition. Erosion on land or in a coastal area may occur over the same time frame as deposition in an offshore environment.

The Grand Canyon in Arizona is one of the most spectacular examples of an exposed stratigraphic sequence. Igneous and metamorphic rocks from mountain building activity in the region over two billion years ago form the basement. The rocks were hidden until about 7 million years ago, when floods caused by melting glaciers carved out the canyon along the Colorado River.

Sequences of sedimentary rocks may be turned completely upside-down by the action of folding and faulting. There are several types of folds and many types of faults. Upfolds, or arches of layered rocks, with the oldest strata in the center, are called anticlines; downfolds, or troughs (often associated with anticlines), with the youngest strata in the center, are called synclines. A recumbent fold is when a fold is overturned so that its axis is horizontal. Faults are fractures in the crust along which the flanking

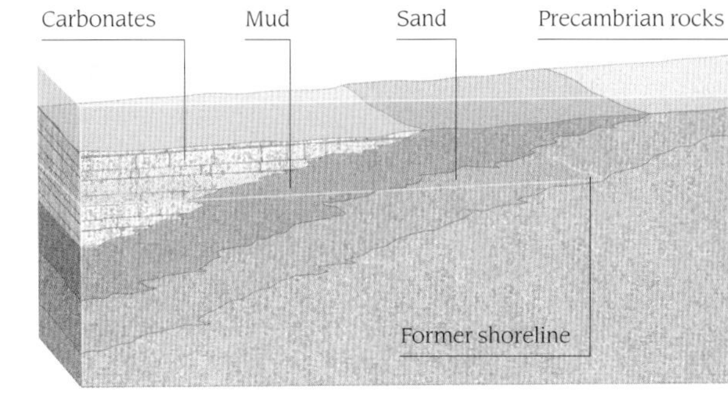

MOVING SHORELINE

The ages of the upper and lower boundaries of rock units can vary. During the Cambrian, the Tapeats Sandstone was deposited along the coast of western North America. As sea levels rose, this shoreline moved eastwards and marine sediments overlaid the nearshore sands, a pattern known as transgression.

FOSSIL EVIDENCE

The fact that some overlying rocks in the Grand Canyon were formed at the same time, in neighboring environments, is supported by fossil finds which cross the formation boundaries. Trilobites from the early Cambrian occur in both the Tapeats Sandstone and the lower part of the Bright Angel Shale.

Kaibab Limestone
Marine fossils

Toroweap Formation
Marine fossils

Coconino Sandstone
Vertebrate animal tracks

Hermit Shale

Supai Group
Plant fossils

Redwall Limestone
Marine fossils

Temple Butte Limestone

Muav Limestone
Trilobites

Bright Angel Shale
Trilobites

Tapeats Sandstone

Grand Canyon Supergroup
Single-celled organisms

Vishnu Schist
No fossils

Zoroaster Granite

Permian

Late Carboniferous

Early Carboniferous

Devonian

Cambrian

Precambrian

blocks have been displaced. In a normal fault, one block moves down the fault plane. In a reverse or thrust fault, one block of rock moves up the fault plane relative to the other. In a strike-slip fault, the two blocks move sideways in relation to each other. Erosion of a folded or faulted landscape reveals apparently anomalous rock sequences.

OVER the last 20 years, sedimentologists and stratigraphers, often associated with the search for hydrocarbons, have developed a sophisticated approach to the study of rock sequences. Their work recognizes the importance of unconformities and of sea-level changes. This approach is known as sequence stratigraphy. Young beds positioned over old present a basic stratigraphic sequence. Successive sequences have their own characteristics and, as with individual sedimentary sequences, are often separated by unconformities. Rock layers and sequences are in effect measurements of time, as are unconformities.

Nineteenth-century geologists began piecing together the stratigraphic record, representing the history of Earth, before anybody guessed the timespan represented by this history.

However, rock sequences and unconformities are not true measurements, because they lack definition: they do not indicate how old a rock is, or how much time passed between the deposition of different layers of a sequence. A straight comparison of rock formations indicates only that one is older than another.

Geological science began with the piecemeal identification of rock units such as beds and formations—an example is the Old Red Sandstone of Scotland (identified in the 1790s)—which were merely descriptions of the

INDESTRUCTIBLE FOSSIL LIFE

Fossils are formed from the mineralized remains of ancient plants and animals, embedded in the rock. They are very hard and last for many millions of years. In some rocks the surrounding matrix is softer than the mineralized fossil, resulting in preferential weathering, as in these brachiopods (left), where the 3-D form is perfectly preserved and revealed.

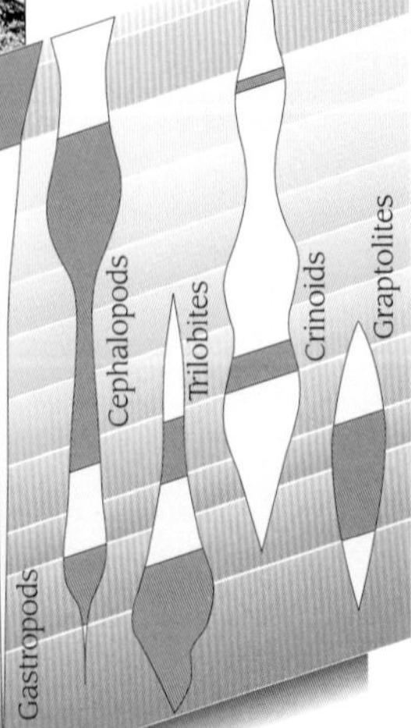

FOSSIL RANGES

The diagram shows the occurrence of common invertebrate fossils. The first four groups still exist today while trilobites, for example, became extinct in the Permian. The width of the shapes corresponds to the abundance of the organisms at that time: the wider the shape, the more abundant was the organism. Dark grey areas indicate important fossils used in zoning (dating a particular rock layer) and correlating rocks from separate geographic locations.

INDEX FOSSILS

A fossil of known age that can be used to date the rocks in which it occurs is known as an index fossil. Most of these fossils existed for a limited, well-defined period of time, were widely distributed, and are easily identified. In-situ index fossils can correlate the dates of rocks in different locations. The presence of two or more index fossils which co-existed can pinpoint a date even more precisely. For example (1), the fossil range of the Ordovician trilobite *Shumardia pusilla* overlaps briefly with that of the gastropod *Cyclonema longstaffae*. A rock suite that contains both species (2) can be dated to the end of the Ordovician, between 442 and 448 million years ago.

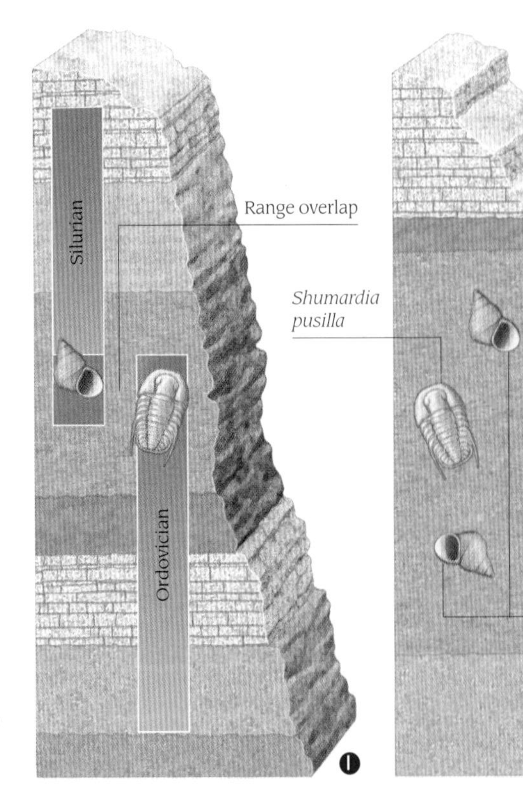

physical characteristics and location of certain rocks, and where they were positioned in relation to other rocks. Most of these formations were readily accessible at the surface, where they were visible as dramatic features in the landscape. Next, geologists worked to uncover formations deep within the Earth and to begin to assign a chronology—a system of relative dating—to rock units according to the fossils found within them.

It was the Dane Nicolaus Steno and the British surveyor William "Strata" Smith who first unlocked the door to relative age dating. Steno applied the law of superposition to undeformed sequences of sedimentary rocks, declaring that upper beds must be younger than lower ones. Smith determined that sedimentary rocks with the same selection of fossils were the same age, and that younger and older beds could be identified by the different types of fossils within them. Thus, relative ages could be assigned and beds of similar age correlated worldwide. Fossils have remained one of the most important tools in geology ever since.

Geologists used this information to piece together a stratigraphic column and work out a geological time scale. This is divided into three eons: the Archean, Proterozoic, and Phanerozoic. The Archean spans the first two billion years of the Earth's history. Fossils from this age are rare, as few Archean sedimentary rocks have been preserved, and the only life forms were primitive single-celled organisms. The Proterozoic lasted from 2.5 billion to 545 million years ago, during which the Earth stabilized and multicellular life began to appear. The Phanerozoic, which spans from 545 million years ago to the present, is subdivided into three eras: the Paleozoic (545–248mya), Mesozoic (248–65mya), and the Cenozoic (65mya–present). Fossils in Phanerozoic rocks are abundant, and the progress of life in this eon is much better understood than the others.

RADIOACTIVE CLOCK

The time taken for half of a radioactive isotope to decay to form another element is called its half-life. The half-life for the decay of uranium-238 through a series of stages to its daughter element lead-206 is 4.5 billion years. By measuring the proportions of U-238 and Pb-206 in a rock, scientists can calculate its age.

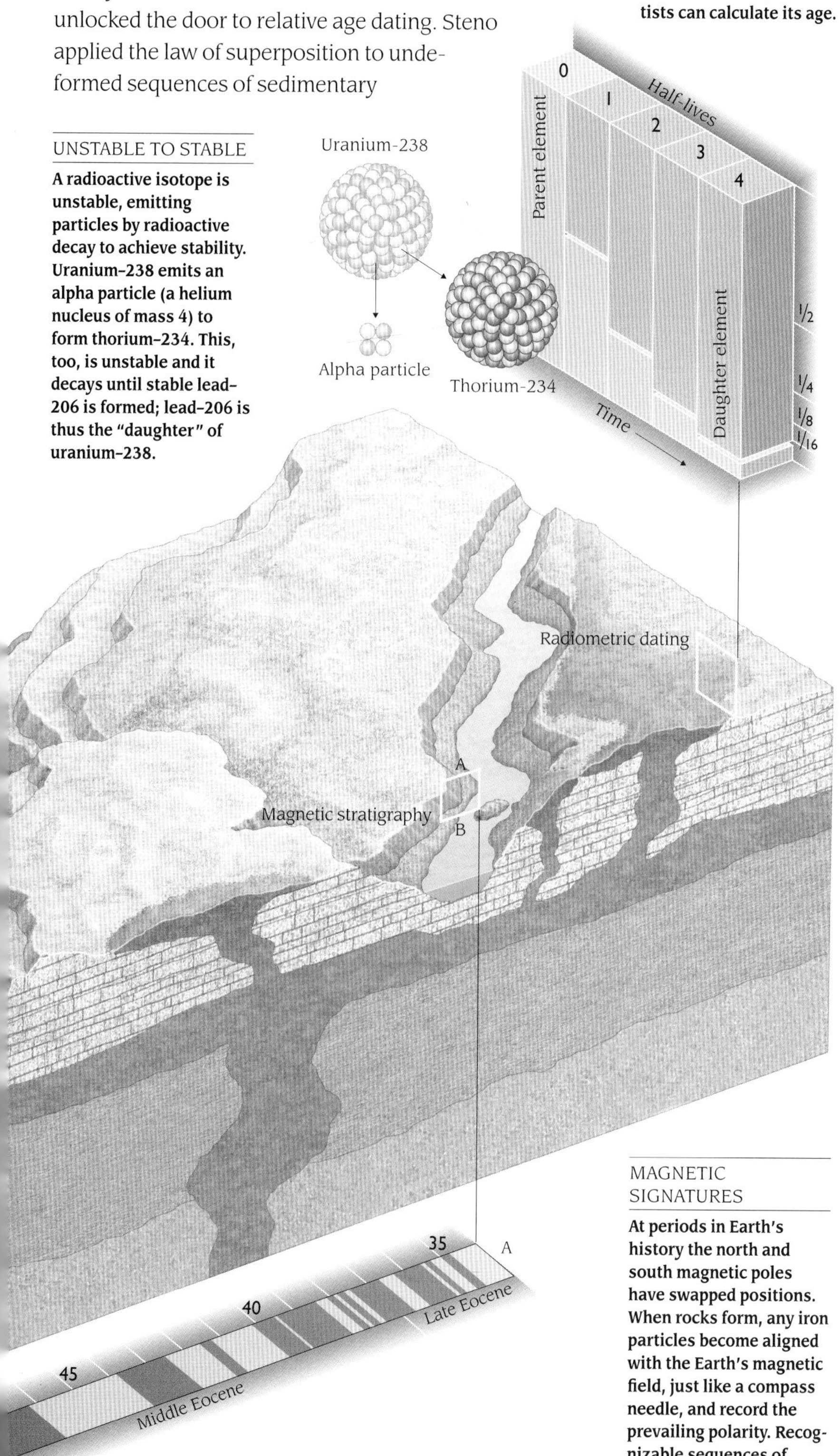

UNSTABLE TO STABLE

A radioactive isotope is unstable, emitting particles by radioactive decay to achieve stability. Uranium-238 emits an alpha particle (a helium nucleus of mass 4) to form thorium-234. This, too, is unstable and it decays until stable lead-206 is formed; lead-206 is thus the "daughter" of uranium-238.

MAGNETIC SIGNATURES

At periods in Earth's history the north and south magnetic poles have swapped positions. When rocks form, any iron particles become aligned with the Earth's magnetic field, just like a compass needle, and record the prevailing polarity. Recognizable sequences of normal polarity (dark tan) and reversed polarity (light) can help date those rocks worldwide.

THIS timescale could only be stated with confidence using absolute dating methods such as radiometric dating, which began with the discovery by Antoine Henri Becquerel of radioactive decay. Radioactive isotopes, such as uranium-235 or thorium-232, decay to produce lead-207 and 208 respectively. In the case of uranium-235 and lead-207 the daughter element, the interval of decay, or half-life, is 0.713 billion years. For thorium-232 to lead-208, the half-life is 14.1 billion years. Over these time intervals one-half of the original nuclei has effectively disintegrated. The ticking of the nuclear clock is constant. Over the period of 713 million years, the accuracy is estimated at ±14 million years. Different radioactive isotopes can be used for ancient or more recent rocks, with carbon-14 (its half-life is just 5730 years) used to date rocks up to 75,000 years old.

Absolute dating of rocks had to wait for the discovery of radioactive isotopes and the development of technology.

Radiometric dating is not the answer to every question raised in the geological time scale. It works well with igneous rocks in which all the constituent minerals are formed at about the same time. But for sedimentary rocks—which are the majority of exposed surface rocks on the planet, and the only ones that contain fossils—radiometric dating has serious limitations. Sedimentary rocks, by their nature, typically contain minerals derived from other rocks of different ages, and the resulting dates can be misleading. Stratigraphers therefore must key the sediments into a regional and temporal framework with interbedded volcanic (igneous) rocks to provide an absolute age for a given sedimentary sequence.

HOW FOSSILS FORM

DEATH, burial, and preservation result in fossils, although the modes of preservation may differ considerably. The first fossils were delicate organisms preserved by chance, such as microfossils trapped in a silica gel or jellyfish rapidly buried in mud or silt. Burial gave protection against scavengers and weather, and the oxygen-free environment stopped decay and inhibited bacterial activity. Fossils became more common when living things developed hard parts. Soft-bodied organisms often vanish without trace, so that they are not well represented in the fossil record.

In most rocks, the mineralized remains of invertebrate and vertebrate organisms are dominant. Their skeletons may have been chitin, calcium carbonate, calcium phosphate, or silica. New minerals crystallize in the existing structure, adding weight but preserving the original detail. Replacement by minerals such as pyrite can produce very beautiful fossils. The minerals involved in these changes are carried in waters that percolate through the enclosing sediment. The changes may occur at different times during a prolonged period of burial.

Plants and graptolites—Paleozoic colonial animals with an organic fibrous skeleton—may be found as black or silvery black films on a fossil rock. The black material is a carbonaceous film, which has survived distillation and reduction of the original tissues. Original shells may be dissolved by acid water passing through the enclosing sediment, leaving a "mold" that may fill with secondary minerals. If the original shell was filled with sediment before death, the internal infilling has traces of the internal structure. These are known as steinkerns.

Organisms that dig or burrow into sediments may leave tracks, trails, or impressions, called trace fossils. Death in peat bogs or in the frozen wastes of a glaciated landscape has resulted in the mummification of whole animals as big as mammoths and humans.

REPLICAS IN ROCK

Many fossils faithfully record the forms of once-living organisms. For example, when a hard-shelled marine mollusk died, perhaps millions of years ago, it would have sunk to the bottom of the sea. Its soft parts—its body—would soon decayed or be eaten by other animals. But the hard shell would survive, slowly becoming covered with sediment, which eventually harden into rock. Later, acidic water may have percolated through the sediment and dissolved the original shell, leaving a three-dimensional "mold" of the mollusk. Mineral solutions could then enter the mold and crystallize, forming an internal cast. If the rock is broken open, it reveals a fossil inside.

FROZEN FISH

This group of fish became "frozen" in stone when they were fossilized around 45 million years ago. Perhaps an exceptionally high tide left them stranded in a pool that rapidly dried up.

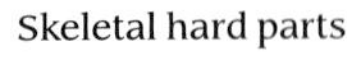

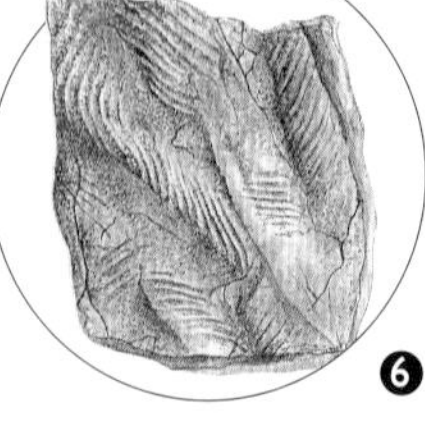

Skeletal hard parts

1 Corals; reef-forming marine animals with calcareous skeletons
2 Radiolarian; microscopic zooplankton with a silica skeleton
3 Bivalve; calcium carbonate shell. Hard parts like this can be preserved unaltered, undergo chemical changes or leave casts
4 Graptolites; branch-like fossils of colonial marine organisms with organic skeletons, usually occurring in black shales
5 Shark's tooth; bones and teeth contain phosphate; they are more resistant to decay than most fossils.

Trace fossils

6 Trace fossil; tracks, impressions, and burrows preserved in sediments

Permineralization

7 Ammonite; its shell has been replaced by iron pyrites
8 Petrified wood; the plant tissues have gradually been replaced by silica
9 Amber; entombs the structure of small organisms trapped within it
10 Carbonized leaves; the plant material has been reduced to a carbonaceous film

BECOMING A FOSSIL

Most vertebrate fossils consist of fossilized bones. The flesh of the dead animal decayed quickly (or was eaten), but before the bones themselves decayed they were covered by sediment. Secondary minerals impregnated the entombed bones as further sediments covered them. Much later, the sedimentary strata were raised back to the surface as a result of regional tectonics. Erosion eventually reveals the fossil, preserved in all its detail. The tracks of walking or crawling animals, and the burrows of worms and mollusks, may also be fossilized. They are known as trace fossils.

CHEMICAL CYCLES

THE MODERN Earth is long past the turmoil that characterized the first few million years of its existence. But it is far from a static world. The slow changes brought about by plate tectonics, erosion, and evolution still continue. Human activities are producing even more rapid alterations. Other changes go in cycles, reflecting the dynamic nature of any perceived equilibrium. Nothing illustrates this point better than carbon, the basic element of life on Earth.

Most of the carbon on Earth exists as carbon dioxide gas. Before the Industrial Revolution, its production was associated with the biological processes of respiration and decay, volcanic emissions, metamorphism, and the formation of limestones and other carbonate rocks. Each of these released carbon dioxide into the atmosphere. Over the long term this was counterbalanced by the loss of carbon dioxide from the atmosphere as a result of photosynthesis, weathering, and the burial of organic matter.

During the Phanerozoic Era (545 mya–recent), fluctuations in the concentrations of carbon dioxide and oxygen changed the climate. This can be documented through the use of stable isotopes, which do not decay spontaneously. Geochemists analyze the isotopic composition of limestones and plot the changes in light or heavy carbon isotopes through time. Fluctuations in the carbon content reflect changes in the composition of seawater, with the rapid burial of organic matter resulting in the depletion in atmospheric and oceanic carbon dioxide of the lighter carbon-12 isotope. This gives rise to a relative increase in the heavier carbon-13, both in the atmosphere and in the limestones deposited in shallow seas. A graph of time against the quantity of a given isotope shows a curve that reflects changes in the carbon dioxide content of the atmosphere. The same is true for oxygen.

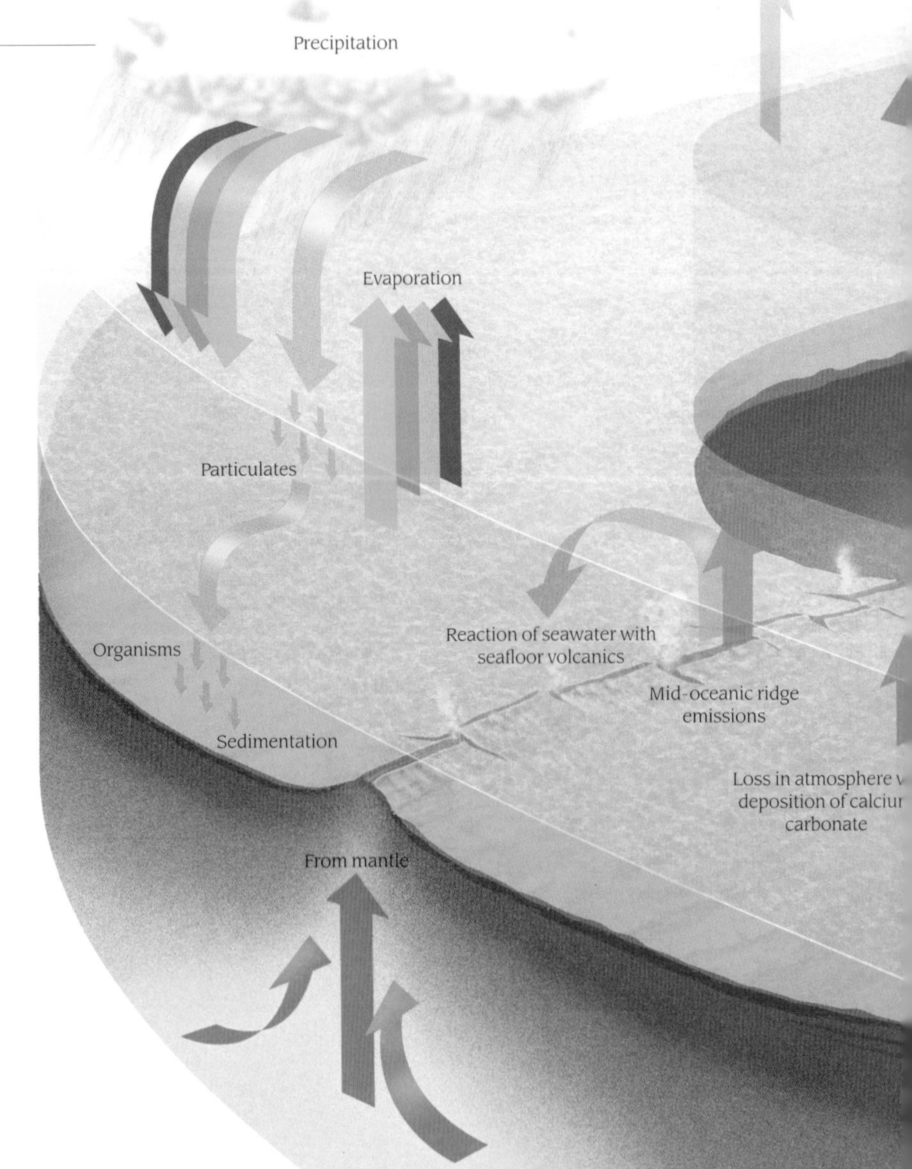

LIMESTONE LANDSCAPE

Rugged limestone outcrops bear evidence of the destructive effects of chemical weathering. The main corrosive factor is rain with dissolved carbon dioxide. This is sufficiently acidic to dissolve some of the calcium carbonate in the limestone, which is carried away in streams and eventually to the sea, thus recycling the calcium.

LIMESTONE CONSTITUENTS

Most carbonate rocks result from the deposition of calcium carbonate in the skeletal debris of marine invertebrates (right). Microscopic protozoans such as foraminiferans take up calcium from seawater to form their shells, which become a major part of the chalky ooze on the sea floor, and eventually limestone.

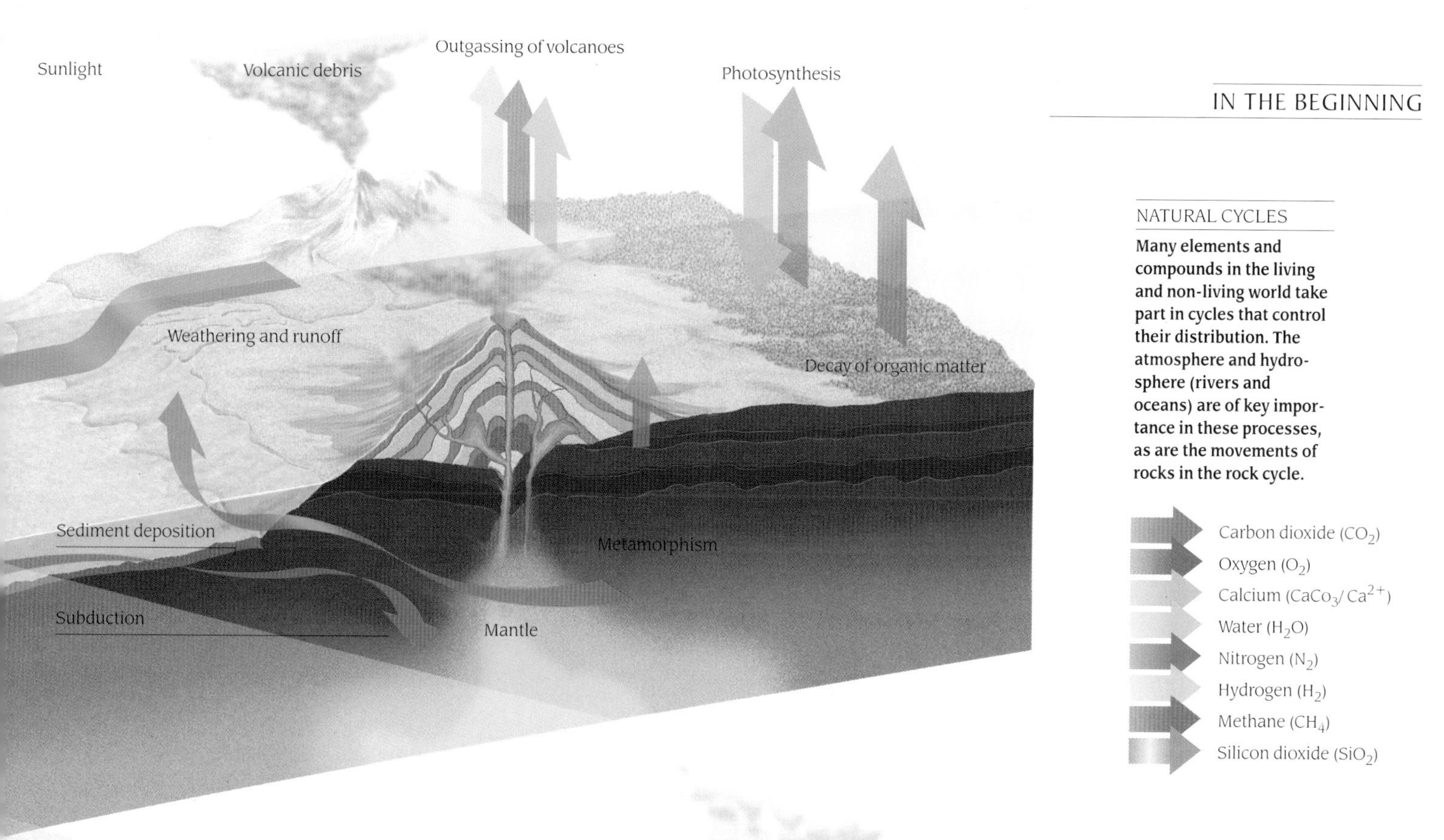

NATURAL CYCLES

Many elements and compounds in the living and non-living world take part in cycles that control their distribution. The atmosphere and hydrosphere (rivers and oceans) are of key importance in these processes, as are the movements of rocks in the rock cycle.

Changes in temperature directly affect other chemical cycles. Increased temperatures lead to increased weathering, and significantly more carbon dioxide in the Earth's atmosphere should result in much more rainfall, chemical weathering, and erosion. This reduces carbon dioxide in the atmosphere and could lead to a new ice age.

Seasonal fluctuations in temperature can be plotted through the study of oxygen isotopes in the skeletons of corals, bivalves, and planktonic microfossils. The ratios of oxygen–18 to oxygen–16 indicate shifts in temperature and seawater salinity. The latter is linked with evaporation, because molecules containing oxygen–16 give rise to isotopically light atmospheric moisture. Rainfall consequently enriches surface waters in lighter isotopes. Since the Late Precambrian, glaciers have lowered sea levels and locked water containing oxygen–16 out of the water cycle. Seawaters rich in oxygen–18 therefore characterize ice ages.

The detailed analysis of such changes reflects the changes in the chemical cycles. The cycles are interdependent, with periods of change followed by episodes in which the natural balance is restored.

THE ORIGIN AND NATURE OF LIFE

LIFE might not have developed at all on a different planet. If the Earth were larger, the increased pull of its gravity would make the atmosphere so dense that sunlight could not get through, and photosynthesis (the process by which plants make their food) would not be possible. If the planet were very small, however, the gases of the atmosphere would escape into space, and there would be no oxygen. (Oxygen is used by most life forms on Earth, but life itself could not have evolved in the presence of oxygen, which destroys delicate material by oxidation.) Nor would there be oceans on the Earth's surface, because the water would evaporate into space. The presence of liquid water, which is crucial for life, also depends on a climate that is neither too hot (as on Venus) nor too cold.

All life is made from the same building blocks. Among the first to appear were amino acids, a group of water-soluble organic compounds containing mostly carbon, oxygen, hydrogen, and nitrogen. Amino acids form proteins, the structural material from which all living matter is shaped.

IN 1953 chemists Stanley Miller and Harold Urey at the University of Chicago devised an experiment using methane, hydrogen, ammonia, and water vapor to simulate atmospheric conditions on Earth during the early Archean. The equivalent of lightning, generated by an electrical discharge, was arced through this gaseous mixture, and eventually formed four amino acids, which are the basic building blocks of proteins. Further experiments along these lines produced carbohydrates and the ingredients for nucleic acids: RNA (ribonucleic acid) and DNA (deoxyribonucleic acid), the chemical messengers of heredity.

Life may have begun accidentally in a storm in the atmosphere above the early Earth, where there was no free oxygen.

An alternate speculation on the origin of life focuses on environments that provided protection from ultraviolet light and oxidation (corrosion by exposure to oxygen), such as geysers and volcanic mud pools, or deep in the ocean. Today's mid-oceanic ridges are characterized by "black smokers," superheated columns of mineral-rich water that gush from vents in the seafloor. The DNA of worms and other heat-tolerant organisms that live near black smokers suggests that life may have formed here.

Somewhere out in deep space, frozen debris from the early solar nebula may still contain evidence of the earliest chemicals of life. If these substances are present in distant space, it is possible that the origin of life on Earth was extraterrestrial. Carbon-bearing meteorites contain compounds similar to amino acids, but the oxygen in the early atmosphere may have killed any life forms transported by meteorites.

FOSSILS show that life on Earth, whatever its source, was established at least 3.5 billion years ago. Cyanobacteria, formerly known as blue-green algae, have been found in the silica-rich rocks (cherts) of western Australia and southern Africa, while rocks from the Baberton Greenstone Belt of South Africa yield microscopic fossils similar to living cyanobacteria. They are examples of the first single-celled organisms. Multicellular organisms appear to have existed at about the same time as cyanobacteria, which suggests that the evolution of life was already underway.

The oldest forms of bacteria could not tolerate the presence of oxygen. Newer forms began to produce oxygen by photosynthesis, and changed the planet.

Cyanobacteria are capable of photosynthesis—using the Sun's energy to make food, releasing oxygen as a byproduct. Organic (carbon-based) compounds discovered in Greenland indicate that this was taking place 3.8 billion years ago. If this is the case, then the earliest components of life must have appeared much earlier. The

KEYWORDS

AMINO ACID
BIG BANG
CARBON
CHERT
CHROMOSOME
CYANOBACTERIA
DNA
EUKARYOTE
EVOLUTION
GENE
NATURAL SELECTION
PHOTOSYNTHESIS
PROKARYOTE
PROTEIN

See Also

THE ORIGIN AND NATURE OF THE EARTH: *The Big Bang, The Early Atmosphere*
THE ARCHEAN: *Black Smokers, First Life Forms, Greenstone Belts, Photosynthesis*
THE PERMIAN: *Mass Extinction*

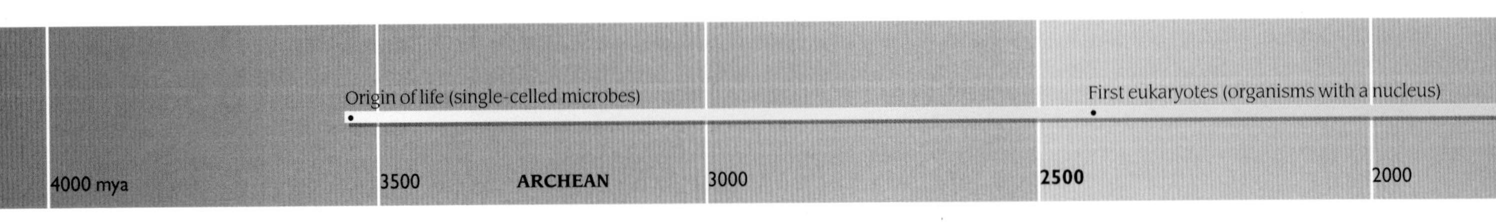

DID LIFE START HERE?

Geysers offer ideal conditions for life to form. They are rich in minerals —the "primordial soup"—with organic material shielded from ultraviolet light and oxidation by siliceous or metallic crusts. Ancient forms of bacteria still exist in low-oxygen environments such as hot springs, mud pools, and in animals' intestines.

A SLOW UNFOLDING

At first, life evolved very slowly. Chemical elements appeared in the Big Bang. Amino acids, RNA, and DNA probably formed a little later. The oldest bacteria existed between 3.5 and 3.8 billion years ago; animals first appeared only about 1 billion years ago. Early life was marine; land plants are about 450 million years old, and land animals are younger. Life accelerated about 545 million years ago in the "Cambrian explosion." Extinctions have wiped out countless species, but new ones soon fill their places.

100 families

most ancient forms of bacteria, called archaeobacteria, cannot survive in the presence of oxygen, and make their food by means other than photosynthesis.

Photosynthesizing bacteria transformed the planet as oxygen accumulated in the atmosphere. Other bacteria were able to use this oxygen to release energy from their food by aerobic respiration. This is a much more efficient way of obtaining energy, and it was these bacteria that began to multiply and diversify. They colonized the young planet, while the archaeobacteria retreated to the depths of oceans, lakes and swamps to live quietly in the mud.

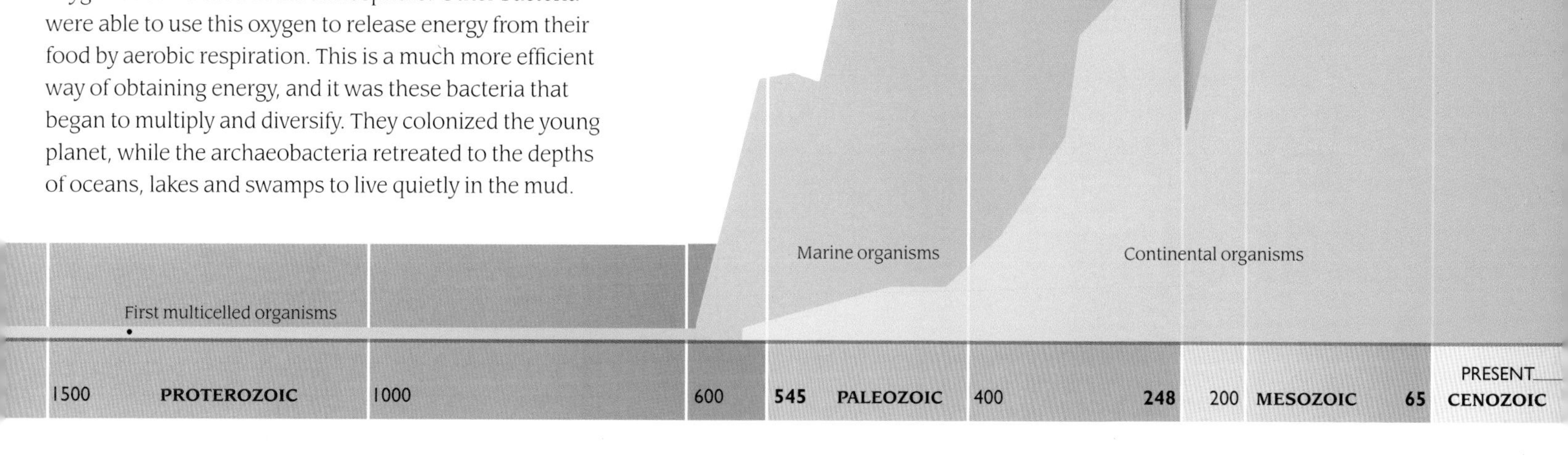

Marine organisms

Continental organisms

First multicelled organisms

1500 **PROTEROZOIC** 1000 600 **545** **PALEOZOIC** 400 **248** 200 **MESOZOIC** **65** PRESENT **CENOZOIC**

BOTH archaeobacteria and eubacteria (which photosynthesize) belong in the taxonomic kingdom of Monera: single-celled organisms that lack a nucleus and whose DNA is not grouped into chromosomes. This kind of cell, called a prokaryote, lacks specialized internal structures that are found in more complex organisms. All other kinds of living thing on Earth today have more structured cells with a nucleus and chromosomes. These organisms are known as eukaryotes, and they gave rise to all of the more complex forms of life.

Bacteria had great potential to evolve: they had DNA, short lives, and many chances to mutate in the ultraviolet light.

These newer life forms differed from the earlier ones in their means of nutrition and reproduction. Photosynthesis and the development of mitochondria (oxygen-burning "turbochargers" in cells) to use energy more efficiently marked the former. Early life forms had reproduced asexually, by simply dividing to make identical copies of themselves. The appearance of specialized sex cells led to a great technical advance here as well. Sexual reproduction, involving two partners that combine their genetic material and hand it on to their offspring, is a much more powerful technique for reproduction. It reshuffles the parents' characteristics into new combinations, making almost endless variations possible within an ever larger breeding group. This is known as a gene pool, and it accounts for the widely varied appearances among humans and other members of a species. Even more significantly, the constant reshuffling of hereditary material leads to long-term change over many generations and improves chances for survival.

MUTATIONS

The fruit fly on the right of this picture has eyes that are smaller and more irregular than normal because of mutation. The original source of all genetic variation, mutations are caused by errors in the replication of DNA molecules, either as a random mistake or by contamination. Most mutations are either neutral or harmful in their effects, though harmful changes are usually weeded out by natural selection. Neutral mutations occur in non-coding parts of the DNA or affect non-functional parts of proteins.

CODING FOR LIFE

DNA is in the form of chromosomes (right): long twisted strands wound around a core of special proteins. Specific segments of DNA, or genes, contain instructions for making proteins or nucleic acids. In sexual reproduction, each parent donates a full set of chromosomes. These pair up two by two (the strands side by side as shown here). If they simply divided, the number of chromosomes would double every generation; instead, by meiosis (reduction division), daughter chromosomes are distributed exactly in new cells so that half come from each parent, increasing genetic variety.

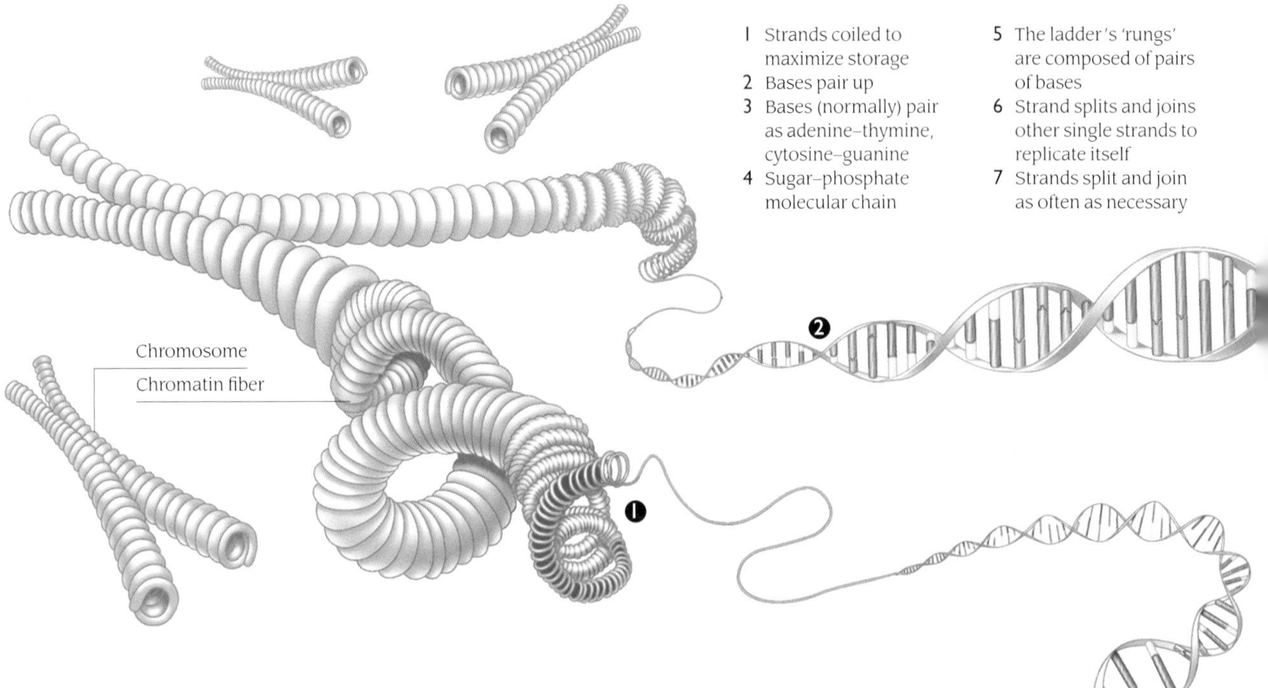

1 Strands coiled to maximize storage
2 Bases pair up
3 Bases (normally) pair as adenine–thymine, cytosine–guanine
4 Sugar–phosphate molecular chain
5 The ladder's 'rungs' are composed of pairs of bases
6 Strand splits and joins other single strands to replicate itself
7 Strands split and join as often as necessary

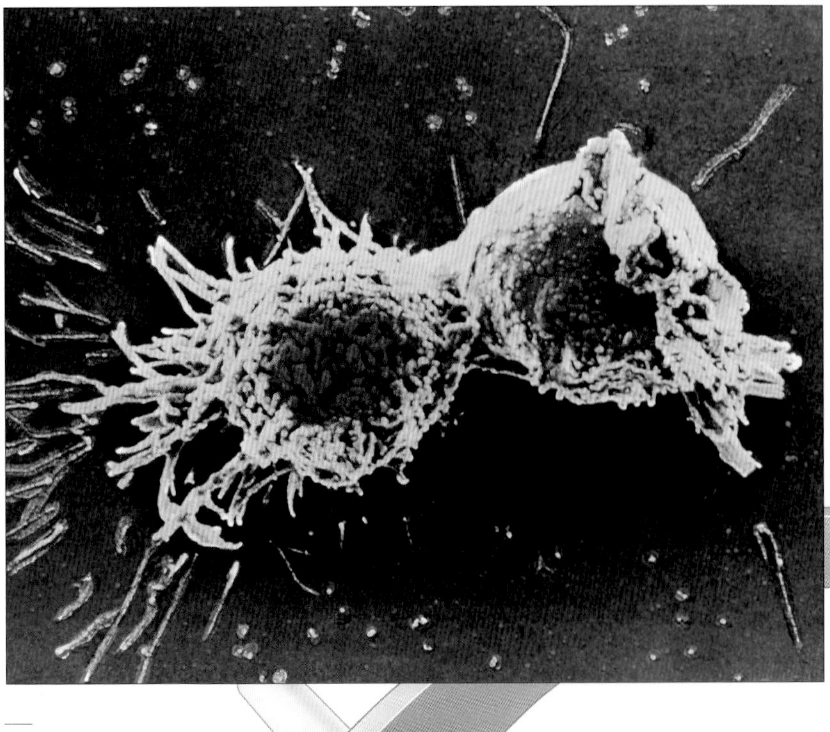

SUPERIOR DIVISION

All reproduction is by cell division (left), but sexual reproduction by meiosis (reduction division) is a huge advantage. One set of chromosomes is inherited from each parent. Each may be beneficial in a different environment, maximizing chances for survival.

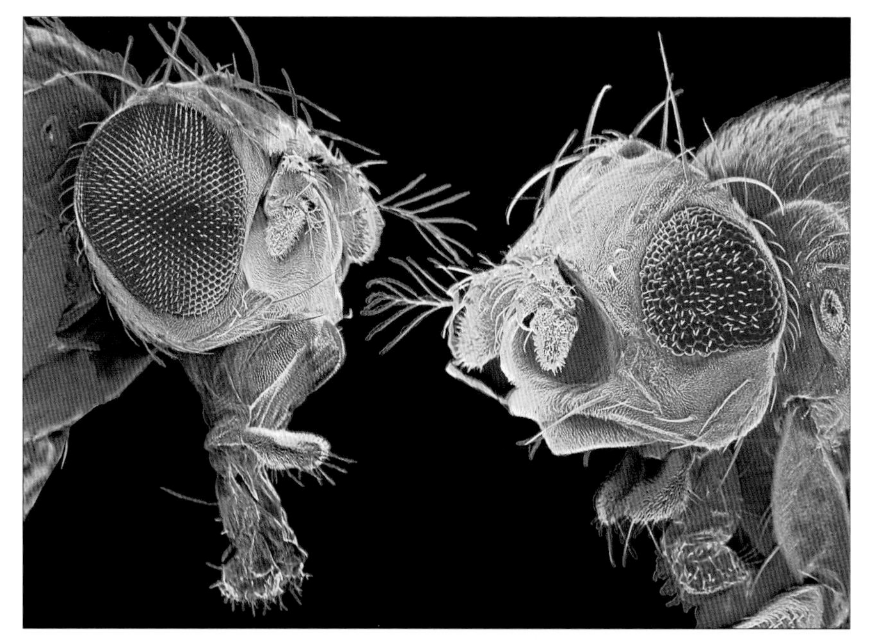

THE ASTONISHING diversity of life forms had long been observed, but it was only in the nineteenth century that scientific investigation of its sources became possible. While experimenting with breeding garden peas, an Austrian biologist and monk named Gregor Mendel discovered the laws of inheritance. His work, published in a natural history journal in Brno in 1865, showed that pieces of information about physical characteristics—now called genes—were passed intact from parent to offspring, with some traits appearing in every generation, and others being masked most of the time. The frequently visible traits are called dominant, and the masked ones recessive. Mendel's findings remained obscure until 1900, when his paper was discovered by researchers studying biological inheritance.

Mendel discovered the laws of inheritance while working with garden peas. His discoveries apply to all living things.

In the first half of the twentieth century, cellular biologists discovered that genes formed strands called chromosomes inside the nucleus of a cell. The strands occur in matching sets, just as genes occur in pairs. Each set contains a basic number of chromosomes. The latter are essentially long threads of DNA molecules, each of which is shaped like a spiral staircase with the steps divided down the middle. The half "steps" alternate, and the combinations within the structure are almost endless. These threads are the mechanisms for transferring genetic information between generations.

DNA's full name is deoxyribonucleic acid, a chemical compound comprised of phosphate, sugar and nitrogen. It probably evolved during the early Archean as a mutation of RNA (ribonucleic acid), a single-strand nucleic acid, which transfers the message of DNA and transports amino acids. RNA is thought to have been present in the most ancient life forms and to have been transformed into DNA at a very early stage.

THE MUTATION of RNA into DNA may have been the first of an almost infinite number of dramatic changes in the structures of life. Mutations occur when an organism is exposed to ultraviolet light, gamma rays, or certain chemicals; they can also happen spontaneously. Mutations cause an individual piece of information to be dropped from the genetic code, to be repeated mistakenly, or appear out of its normal sequence. If the sex cells are affected, this provides further variation in the gene pool, allowing experimentation with yet more diversity. Whether a new life form survives depends to a large extent on its environment. If it does survive, the "mistake" in its genes will be passed on to its offspring.

Genetic coding sometimes takes a sudden detour, giving rise to new material. If this is passed on to new generations, they evolve into new species.

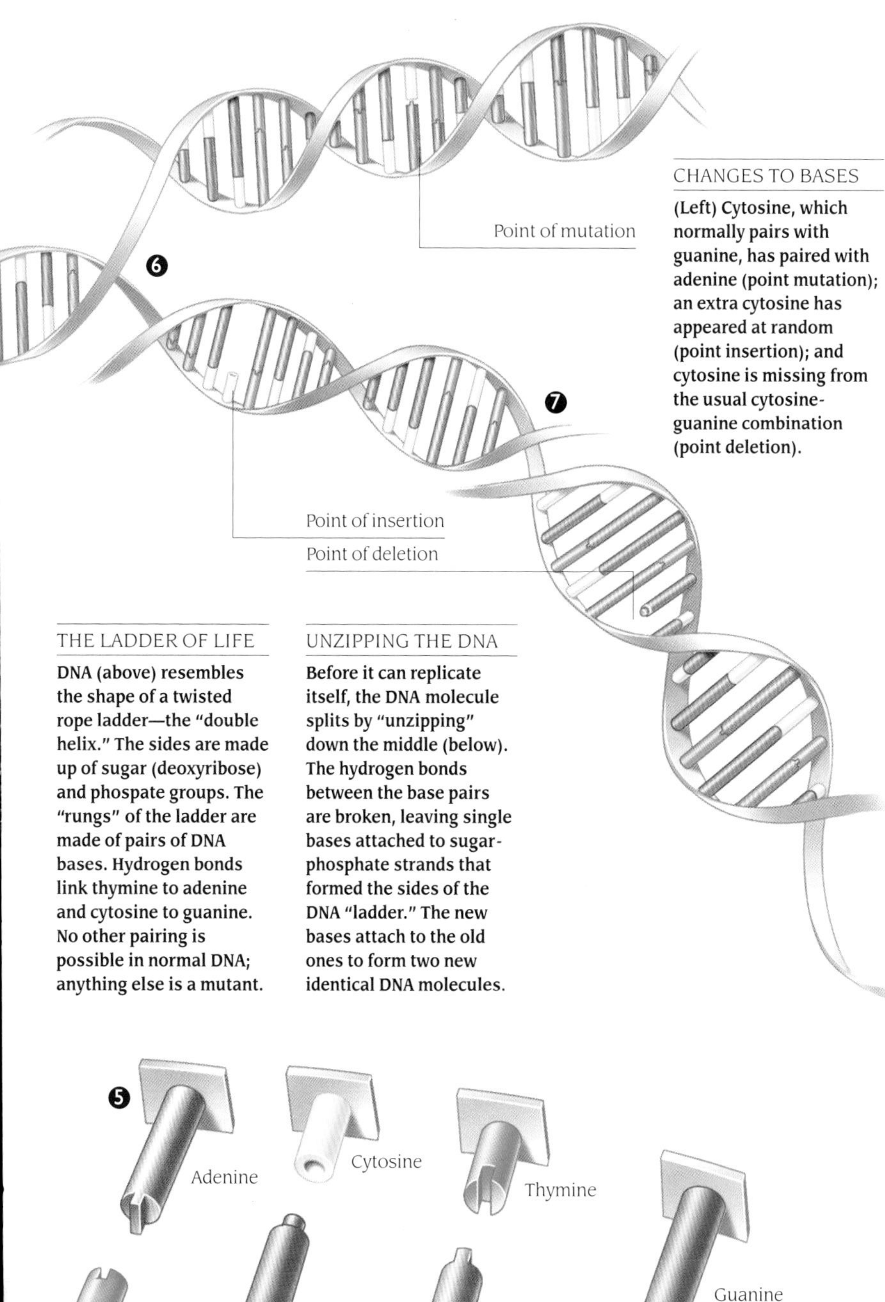

CHANGES TO BASES

(Left) Cytosine, which normally pairs with guanine, has paired with adenine (point mutation); an extra cytosine has appeared at random (point insertion); and cytosine is missing from the usual cytosine-guanine combination (point deletion).

THE LADDER OF LIFE

DNA (above) resembles the shape of a twisted rope ladder—the "double helix." The sides are made up of sugar (deoxyribose) and phospate groups. The "rungs" of the ladder are made of pairs of DNA bases. Hydrogen bonds link thymine to adenine and cytosine to guanine. No other pairing is possible in normal DNA; anything else is a mutant.

UNZIPPING THE DNA

Before it can replicate itself, the DNA molecule splits by "unzipping" down the middle (below). The hydrogen bonds between the base pairs are broken, leaving single bases attached to sugar-phosphate strands that formed the sides of the DNA "ladder." The new bases attach to the old ones to form two new identical DNA molecules.

THE GREAT English naturalist and writer Charles Darwin was the first to identify the origin and adaptation of species to new modes of life and new ecological niches. Darwin's ideas, underpinned by a thorough understanding of contemporary science, were founded on research he undertook during a five-year voyage around the world (1831–1836). The geographic scope of his travels, combined with his training in geology, botany, and zoology, put him in a position to do some highly original thinking. Darwin also collected an impressive range of fossils, which supported his thesis. Gathering the evidence took him five years; interpreting it, and writing up the results, took another 25.

Charles Darwin explained evolution as an interaction of heredity (though he did not know how it worked) and the environment. His was not the first theory of evolution, but it was the first that could be validated by testing.

Darwin's *On the Origin of Species by Natural Selection*, published in 1859, rocked the Victorian world, which believed firmly in life ordained by divine word only a few thousand years earlier. (One European archbishop had calculated that the Creation could be precisely dated to 9:00 a.m. on Sunday, October 23, 4004 BC.) The eminent French naturalist Georges-Louis Leclerc, Comte de Buffon, questioned this timescale, but a substantive theory of evolution had to wait for the work of Darwin, the French naturalist Jean Baptiste Lamarck, and another Briton, Alfred Russel Wallace.

Lamarck was an excellent paleontologist, but his ideas were mostly derided by his contemporaries because he believed that acquired characteristics could be inherited: for example, that longer necks resulting from reaching up for food, or the loss of a limb, would appear in successive generations. However, Lamarck's theories have been revisited in light of the more recent discovery that many newborn animals appear to be equipped with instinctive behavior that allows them to survive in difficult circumstances without parental assistance. This phenomenon raises serious questions as to whether a fully-charged memory bank in a newborn animal is an evolutionary characteristic.

Alfred Russel Wallace was an expert collector and fine naturalist, and he arrived at the same conclusions on evolution by natural selection at approximately the same time that Darwin did. The two published some findings together, but if they had not, someone else would have soon drawn the same conclusions from the same evidence. Nevertheless, it was Darwin's *Origin of Species* that marked the foundation of modern evolutionary science. Opposition to his theories focused on the proposal that humans and apes had common origins, but the story of humans is only a minor detail in the sweeping panorama of life.

If these offspring in turn survive and reproduce, they may eventually become the majority in their population. Thus, tiny transformations in the genetic code of a few individuals can have enormous long-term impact.

A group of genetically similar individuals that can breed only with each other and produce fertile offspring is called a species. In nature, most species produce enough offspring to ensure that some, at least, survive to maturity. Certain individuals have a genetically-based advantage, perhaps speed or strength or other favorable characteristic, which makes them particularly well suited to their environment and improves their own chances of surviving and reproducing. Individuals or species that are not well suited to their environment do not survive for long. If they cannot adapt, they die, or are killed, and are replaced by others that are more robust or flexible. This is the process of organic evolution by natural selection and the "survival of the fittest."

SUCCESSFUL BLUEPRINTS

Sharks are highly successful swimming predators. Their streamlined shape makes them faster than their prey, and they have survived since the Devonian period.

DIVERGENT EVOLUTION

Bats and humans are both mammals, but they have developed very different physical forms and lifestyles through divergence. One clue to their ancestry is in their pentadactyl (five-fingered) limbs. The basic pattern has remained, though humans have evolved hands for grasping, whereas bats have wings for flying. Structures like these are homologous, meaning that they have the same shape but serve different purposes. Birds are only distantly related to bats; however, the structure of their wing is analogous to a bat's, serving the same function. Such anatomical evidence convinced Charles Darwin that completely different species had evolved from common ancestors, and environmental factors played a significant role.

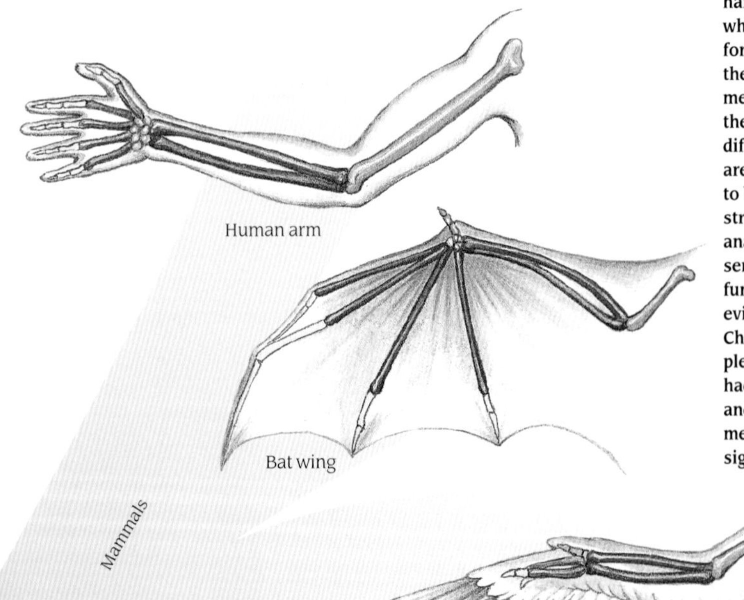

NEW species evolve by four mechanisms of evolution: transformation, speciation, adaptive radiation, and extinction. In transformation, a relatively small population adapts to gradual changes in its environment, eventually giving rise to a group that is different enough from its ancestors to be considered a new species. Sometimes environmental change creates a physical barrier that splits a population, isolating them geographically so that they cannot make contact for mating. Over time, each group will continue to adapt to changes in its local environment. If the changes are sufficiently dramatic or of long enough duration, the groups will become different species; this is called speciation. This process also occurs when part of a group migrates permanently to a distant location.

The evidence for evolution is all around us, in the fossil record and in comparisons of living species whose characteristics prove their common ancestry.

When a new ecological niche opens up, there is a rush to fill the gap. The species that can do so is the one that rapidly increases its number and diversifies to take advantage of the opportunity. Over the course of five to ten million years, what began as a small group of similar organisms becomes a major group of new animals. This is the mechanism of adaptive radiation, which often occurs after the extinction of another group that competed for the same niche. In this way, extinction—the sudden disappearance of a group of organisms, typically after an environmental disaster—is also a force that shapes the evolution of species.

The rise of new species from common ancestors through adaptation to changing environments is an example of divergent evolution. The reverse may also happen: organisms with no modern genetic kinship, which evolved in widely separated geographic locations, often develop similar physical forms in response to their environment. Marsupial mammals, for instance, evolved in geographical isolation on the continent of Australia, and to this day they are found nowhere else. However, the marsupial mole, anteater, and wombat look remarkably like the placental anteaters, moles, and groundhogs found on other continents, suggesting that each of these animals evolved a particular form and lifestyle in response to similar environmental conditions. This is known as convergent evolution.

Related species often evolve in parallel—that is, if they live in similar habitats, they adapt by developing the same characteristics. This is seen in single-toed horses of North and South America, which appeared in the Late Miocene. Both descended from a small five-toed ancestor of the Early Paleocene whose toe structure limited its endurance when running. *Thoatherium,* the horse from Argentina, became extinct; *Equus,* the northern species, became the modern horse.

It is less common for a succession of younger species to be near carbon copies of their ancestors. This phenomenon, known as iterative evolution, seems to occur when there are limits on the species' opportunities to diversify, but their environment has remained secure.

"Laws" or patterns of evolution can be deduced from the fossil record and help to explain why the evolution of a given species progressed as it did. For instance, Haeckel's law states that the development of an individual embryo in a species will reflect the stages of development of its ancestors. Indeed, the early embryos of fish, pigs, rabbits, and humans look alike, with gills

ITERATIVE EVOLUTION

Ammonoids are coiled chambered cephalopods with external shells that first appeared in the Devonian, arising from mollusk ancestors with uniformly coiled shells. Several times during their history, different kinds of irregularly coiled shells appeared—an example of iterative evolution, in which similar shapes and structures are repeated at intervals in different species as a throwback to an original common ancestor. The parent stock of ammonoids were the longest-lived, surviving nearly 400 million years to the Late Cretaceous.

CONVERGENT EVOLUTION

Wolves and woodmice are placental mammals, which are common worldwide except in Australia; the Tasmanian tiger and the mulgara are their Australian marsupial counterparts. They evolved in isolation from each other, but their similar lifestyles are reflected in their very similar appearances.

1 Wolf (genus *Canis*)
2 Thylacine or Tasmanian tiger (*Thylacinus cynocephalus*)
3 Woodmouse (*Apodemus sylvaticus*)
4 Mulgara (*Dasycerus cristicauda*)

TOP OF THE TREE

Panthera leo, the "king of beasts," occupies an ecological niche at the top of the food pyramid, but this mighty, specialized killer had small beginnings. About 55 million years ago a group of unremarkable, tree-dwelling hunters known as viverravines evolved two lineages, cats (Feloidea), and dogs (Canoidea). The cat branch subsequently diversified into four families: mongooses, civets, cats, and hyenas, with the first true cat appearing about 20 million years ago. Further branching of the family tree gave rise to the sabretooths, wildcats, and ocelots. Sabretooth cats dominated the Miocene, hunting large, rhino-like herbivores. The spread of grasslands, together with small, fast prey such as antelopes, gave opportunities for a new family of lightweight, fleet-footed cats, the cheetah-like pantherines. Swift, heavy prey such as zebra, however, were too fast for the sabretooths and too strong for the pantherines. A new niche was waiting to be exploited; 600,000 years ago the big cats, *Panthera*, entered the scene.

Species *Panthera leo* LION

AGenus *Panthera* LION, TIGER, JAGUAR, SNOW LEOPARD

Family *Felidae* LION, CHEETAH, CLOUDED LEOPARD, WILD CAT

Order *Carnivora* LION, WOLF, BEAR, RACCOON, WEASEL, MONGOOSE

Class *Mammalia* LION, ELEPHANT, WHALE, MONKEY, RAT, KANGAROO

Phylum *Chordata* LION, PARROT, CROCODILE, FROG, TUNA, SHARK, SEA SQUIRT

Kingdom *Animalia* LION, OCTOPUS, CRAB, ANT, EARTHWORM, JELLYFISH, AMOEBA

TAXONOMIC PYRAMID

Lions belong to the phylum Chordata (having a backbone or its precursor), class Mammalia (warm-blooded and nurse their young on milk), order Carnivora (meat-eaters), family Felidae (cats with five toes on the front paws and four on the hind paws), and genus *Panthera*.

and tails; and Darwin's contemporary, the biologist Louis Agassiz, noted that he could not distinguish an embryonic mammal from a bird or reptile. However, some simple organisms skip intermediary stages and mature early, so as adults they resemble the larvae of their ancestors—the opposite of what Haeckel's law says.

Another "law"—Williston's—holds that animals with serially arranged structures, such as teeth or legs, will have fewer of such structures as new species evolve, and those structures will take on new, specialized functions. Once a structure is lost or changed it will not reappear in new generations, and once a species has changed—or is wiped out—it is gone forever; this is Dollo's law.

Cope's law states that descendants tend to be larger than their ancestors but to live in smaller populations. Large organisms have the advantage of outgrowing many of their predators, and they are more efficient in their use of food (one large creature consumes less than an equal mass of many small creatures)—but they lose some tolerance of environmental change and are subject to sudden extinction. The workings of Cope's law may be observed in the dramatic demise of the dinosaurs, contrasted with the success of bacteria, which have changed very little in the last four billion years and appear to be virtually extinction-proof.

Extinction may be the result of a sudden catastrophic event such as a meteorite striking the Earth, or it may be gradual. It may be caused by disease, predation, too much competition from other species in the same niche, or a restriction in the environment such as the disappearance of a primary food source. Since the explosion of life at the beginning of the Cambrian period, the origination of new species and extinction have walked side by side. Sometimes species recover from near extinction and go on to diversify and flourish. If they do not, other species will soon appear to take over the vacant niches.

The fact of extinction was accepted before Darwin explained the origin of species. Fossil discoveries had long raised questions about why the animals preserved in this manner were no longer seen on Earth. One explanation was that these animals were still alive and well in remote locations to which no one had ever traveled. By 1786, however, much of the globe had been explored, and it was unlikely that any living thing the size of a fossil mammoth could have remained hidden. This was the argument of the French naturalist Georges Cuvier, who proposed that mammoths must be extinct. The disappearance of species was a less controversial issue than the question of their origins, and Cuvier's argument was quickly accepted.

BETWEEN two and five million species are alive today. Naming and classifying them originated with the power of human speech. Prehistoric peoples would have needed to identify plants and animals: to know which were edible, poisonous, placid, or likely to attack. Dividing living organisms into major groups was first attempted by the ancient Greek polymath Aristotle. He recognized the differences between birds and insects, and was able to classify groups based on similar features. Naming organisms remained rather haphazard until the eighteenth century, when the Swedish botanist Carl von Linné, also known as Carolus Linnaeus, formulated the principle of using two names for a group of animals. The names are Latinized and printed in italic type. Linnaeus introduced a scheme of classification that grouped individual species under a generic name. For example, *Felis* is the generic name for the wild cat, the specific, or species, name of which is *silvestris*. Together with other cats, it is included in the family Felidae. *Felis silvestris* can breed only with another *Felis silvestris*, not with *Felis rufus* (the bobcat) or *Felis concolor* (the mountain lion), or with any of the big cats such as lions and tigers (*Panthera*).

Darwin dealt with the mechanisms by which life forms with new characteristics appear. Identifying and classifying these characteristics belongs to another science, taxonomy.

The system of naming devised by Linnaeus is still the key to modern taxonomy, in which all animals and plants are grouped into five kingdoms (six in some systems). These are divided into a logical system of subordinate groups: phylum, class, order, and family, followed by genus and species, the two designated by Linnaeus. Living things can be readily separated at the larger scale (kingdom, phylum, class, and so on), but characterizing species and subspecies is more challenging.

Linnaeus discovered a natural hierarchy of life; Charles Darwin explained its meaning, which was that all life originated from a common source, and was transformed into its great variety by evolution through natural selection. By following the taxonomic "tree" of any species to its source in one of the five kingdoms, the outline of the evolutionary history of that species is revealed. At the base of the tree are the very primitive life forms from which everything else evolved—the assertion that so upset Darwin's contemporary critics and continues to do so to this day, nearly 150 years later. The branches of the trees represent ever more distant relationships between different groups.

This hierarchy in Nature may be discovered by a method called cladistics, which uses identified primitive and derived shared characteristics to show the links between parent and daughter (sister) taxa. Derived characteristics, such as hair or feathers, which arose only once in evolution, define all the descendants of the organism that first possessed them as a monophyletic (single origin) group or clade. Another new technique with huge potential is molecular sequencing. The sequence of base pairs in DNA and RNA are unique to species. Comparing sequences gives a measure of relationship; the longer the time since two species split in evolutionary history, the more different their DNA and RNA sequences will be.

GALAPAGOS FINCHES

Galápagos finches show how one species evolves into several. The different shapes of their bills reflect their diets. The original species may have come from the South American mainland. As they diversified, they found different food sources: seeds, fruit, insects, cactus. Different species could coexist without competing.

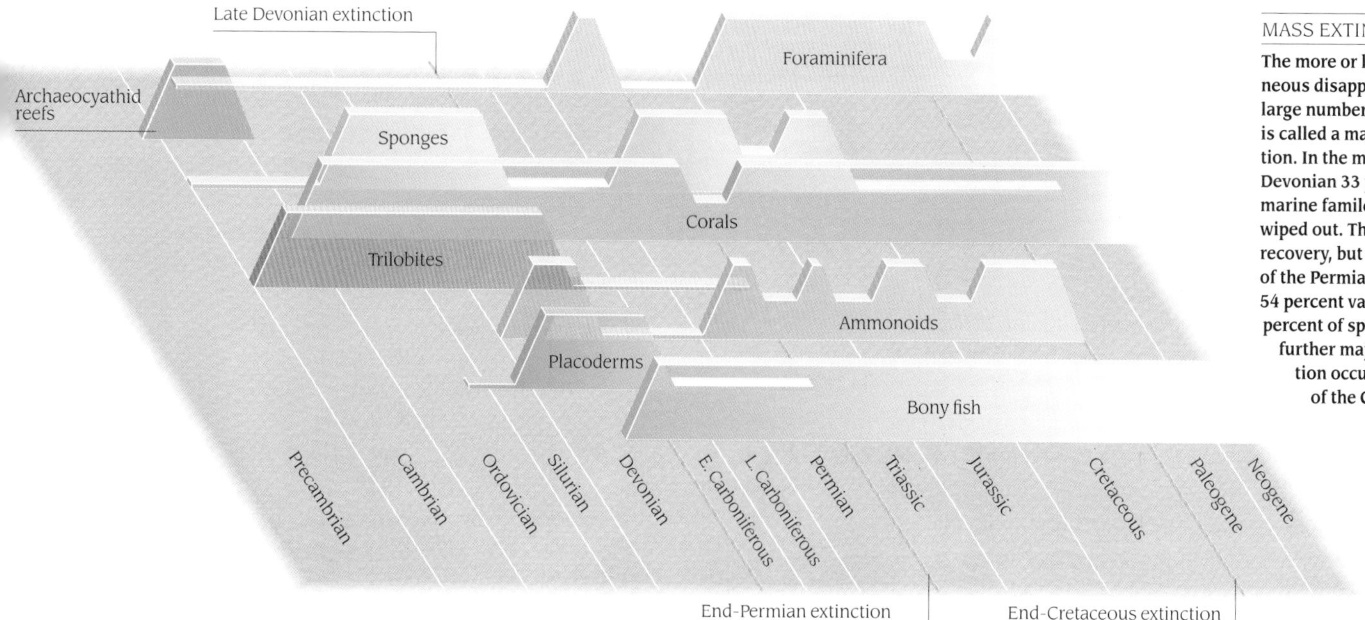

MASS EXTINCTIONS

The more or less simultaneous disappearance of large numbers of species is called a mass extinction. In the middle Late Devonian 33 percent of marine familes were wiped out. There was a recovery, but at the end of the Permian another 54 percent vanished (90 percent of species), and a further major extinction occurred at end of the Cretaceous.

THE FIVE KINGDOMS

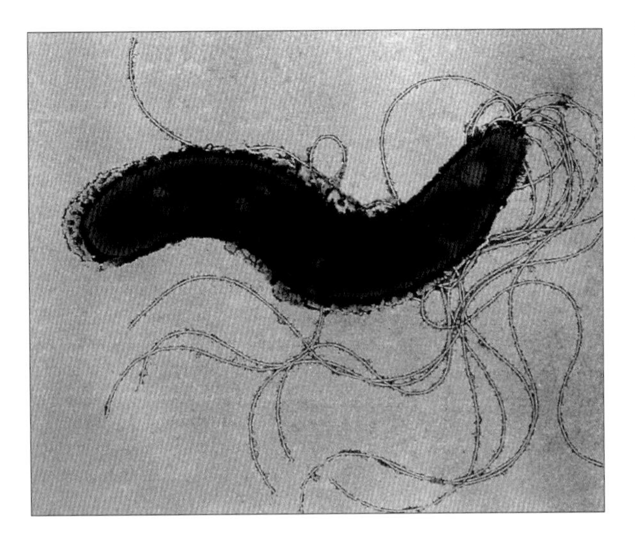

THE SURVIVORS

Bacteria have been a successful form of life on Earth for more than 3.5 billion years and survived the events that wiped out many other species. This *Spirillum* is a eubacteria, of the more recent "true" bacteria. Modern classification systems use two kingdoms for bacteria.

CLASSIFICATION is a shorthand way of noting evolutionary history by separating the levels at which organisms differ from each other. The first and most significant level of difference is between kingdoms. Originally, all organisms were classified as belonging to either the plant or animal kingdoms, but by the late nineteenth century it had become clear that another category was needed for life forms that were neither plant nor animal. The distinguished biologist Lynn Margulis has remarked that many people still dismiss anything else as a "germ."

With further progress in science, a third category was no longer sufficient to classify the rapidly burgeoning list of newly-discovered tiny organisms with strange lifestyles. Eventually, in 1963, a five-kingdom system was proposed based on anatomy, biochemistry, and method of nutrition. The oldest and most primitive kingdom is Monera; the other kingdoms are believed to have evolved from it. Monera contains the prokaryotes, which lack a nucleus and other specialized cell structures. Members of the kingdom Monera have always been the most numerous living things on earth, and they are still flourishing today. Most researchers prefer to divide Monera into two kingdoms, archaebacteria ("ancient bacteria") and eubacteria ("true bacteria"), making six kingdoms.

All other kingdoms contain eukaryotes, which possess nuclei and other specialized structures. The one-celled organisms of Protoctista were the first eukaryotes to evolve. They were followed, in order of appearance, by animals, plants, and fungi, which are classified on the basis of their distinctly different methods for biological processes such as nutrition.

Plants are autotrophs, or food producers: they make their own food by photosynthesis. Many of them are eaten by animals, which are heterotrophs (food consumers), preferring to eat foreign material and digest it to obtain their energy. Finally, fungi are saprophytes, which feed on dead or decaying material; they are decomposers.

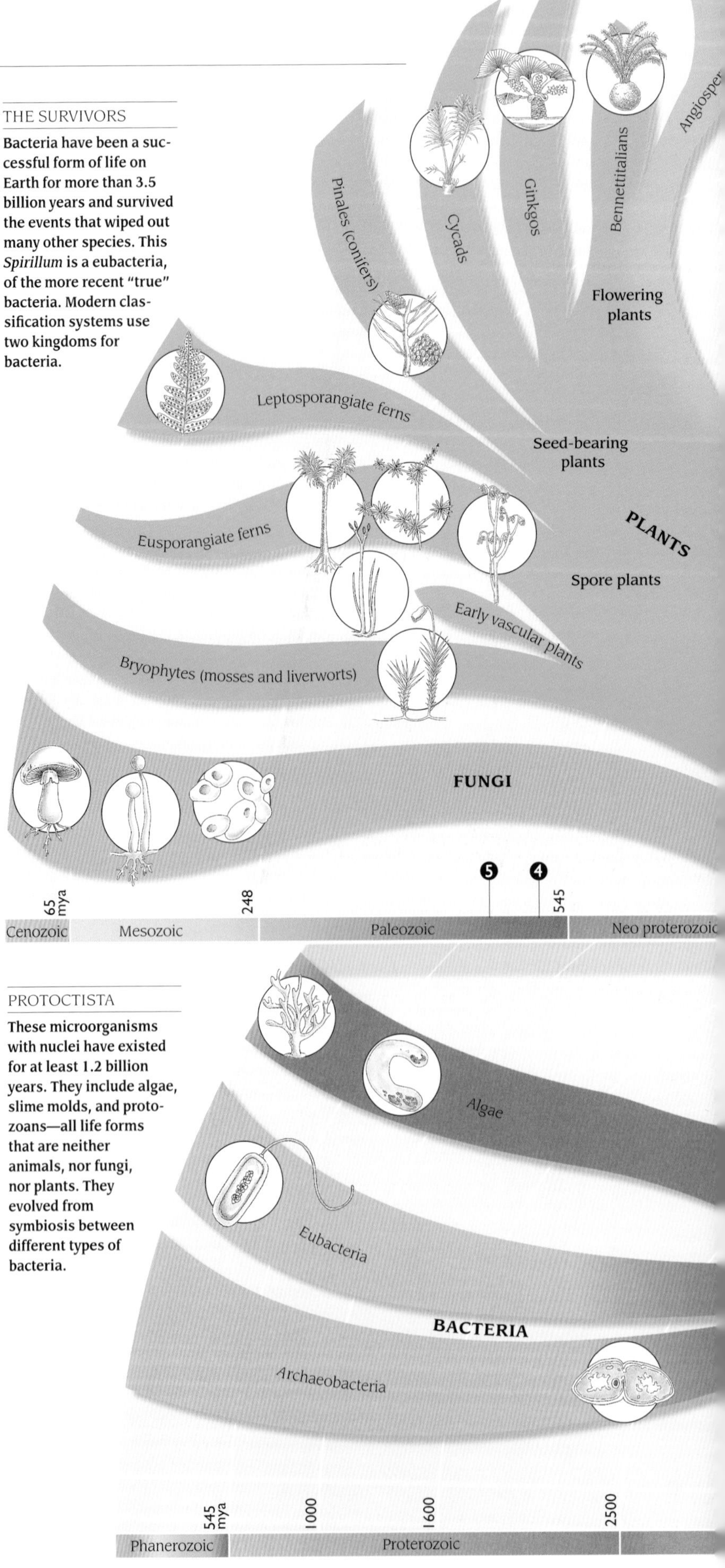

PROTOCTISTA

These microorganisms with nuclei have existed for at least 1.2 billion years. They include algae, slime molds, and protozoans—all life forms that are neither animals, nor fungi, nor plants. They evolved from symbiosis between different types of bacteria.

WHAT IS A PLANT?

Plants develop from embryos, distinguishing them from photosynthesizing bacteria. Most are adapted for life on land. Half a million species are known; there may be double that number. Their fossil record dates back 450 million years.

FUNGI

Fungi evolved from the protoctists and now claim an estimated 1,500,000 species, most of which are unknown. Once considered to be plants, they have more similarities with animals.

THE ANIMAL KINGDOM

Animals may have descended from the Protoctists. Of the 37 phyla of animals, 98 percent are invertebrates. In evolutionary terms, animals are a failure: more than 99 percent of all the species that ever lived have become extinct. There are some notable exceptions; for example, *Lingula*, a species of brachiopod, has survived unchanged from the Cambrian to the present.

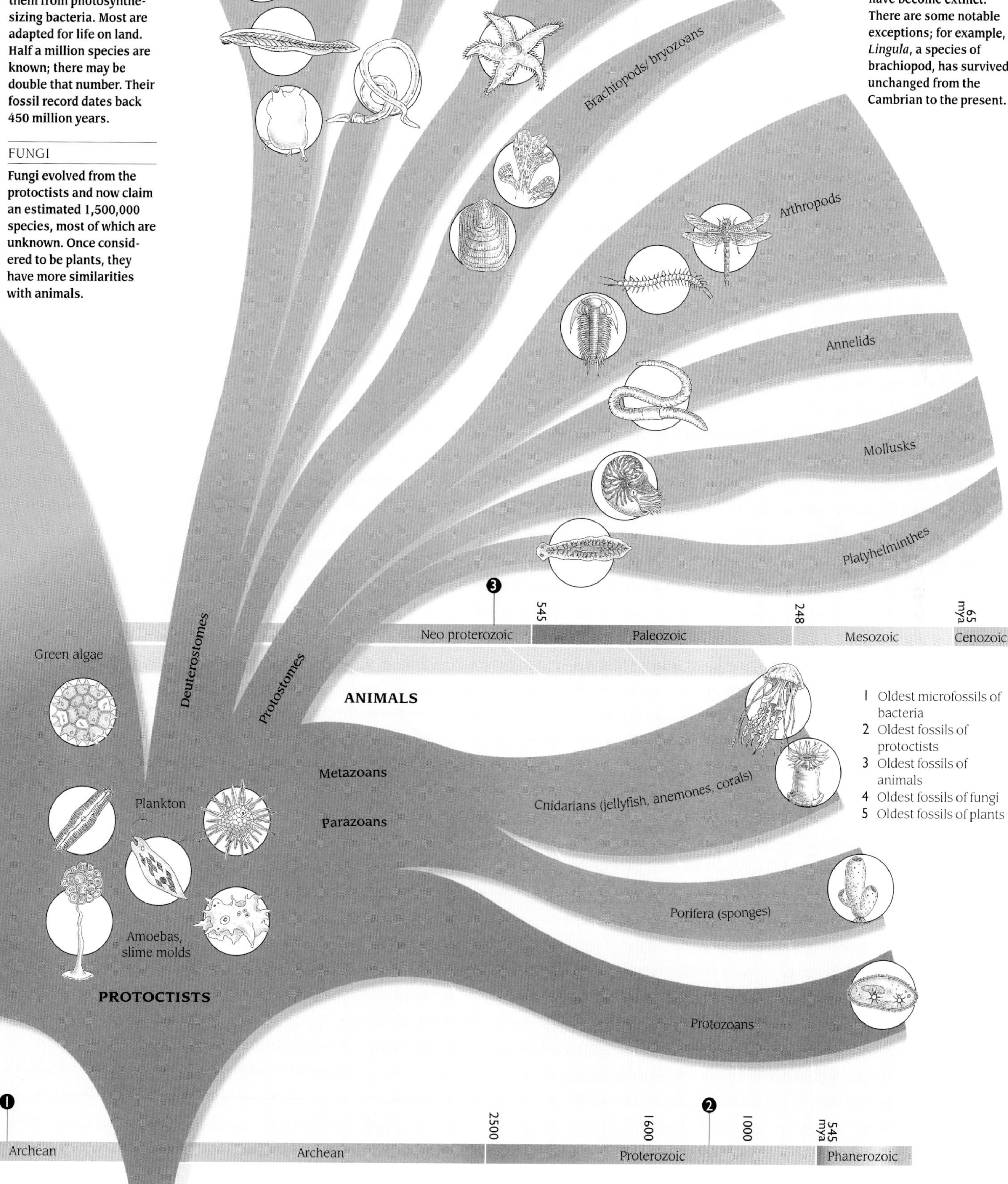

1 Oldest microfossils of bacteria
2 Oldest fossils of protoctists
3 Oldest fossils of animals
4 Oldest fossils of fungi
5 Oldest fossils of plants

THE ARCHEAN

4550 – 2500 MILLION YEARS AGO

THE FIRST two billion years of the Earth's history are encompassed in the Archean eon, representing nearly half of the total span of the Earth's age. The Archean ("ancient age") and the succeeding Proterozoic ("early life") together constitute the span of time known as the Precambrian, which covers 90 percent of the Earth's history. Although this term does not always appear on geological timescales, it is widely used to refer to the time before diverse life forms became established on the planet in the Cambrian. During the Archean, the Earth was transformed from a ball of overheated dust, gas, and cosmic fragments to a full-fledged planet. Its core, mantle, and crust all took shape, and the atmosphere and oceans developed. Numerous microcontinents developed, eroding to produce the first layers of sedimentary rock. Fossils of primitive one-celled organisms—the first evidence of life on Earth—have been found in the Archean rocks that have survived to the present.

MANY scientists depict the early Earth—once it had ceased to be covered by molten magma—with a steam-laden atmosphere and a patchy surface torn by highly active volcanic fissures. Such features persisted during the first part of the Archean eon, known as the Hadean. But dramatic changes were taking place. As the planet cooled, water accumulated and a crust formed. By about 4 billion years ago, the crust was less than 18mi (30km) thick—thinner than today, but comparable in structure and chemistry. Pieces of crust accrued to form the first protocontinents, which floated on the lithosphere. The earliest atmosphere was probably enriched in the noble gases (argon, neon, and so on), but these were swept away by solar winds. They were replaced by an early atmosphere rich in carbon dioxide, nitrogen, and water vapor.

Blistering with volcanoes, pelted by meteorites and asteroids—the first 500 million years on Earth are evocatively named the Hadean eon, from the Greek word for "hell."

KEYWORDS

BANDED IRONSTONE
BIOGENESIS
CARBONATE
CHERT
CRATON
CRUST
EUKARYOTE
FELSIC CRUST
GREENSTONE
MAFIC CRUST
MANTLE
PROKARYOTE
SHIELD
STROMATOLITE
ULTRAMAFIC CRUST

By the Isuan, which spanned about 100 million years from 3.85 to 3.75 billion years ago, granitic platforms were in place on the surface and formed a substrate for layered rocks of volcanic and sedimentary origins. These were subsequently buried to great depths, with heat and deformation resulting in the formation of coarse-grained volcanic gneisses. Rocks from the Isua region of Greenland include the oldest sedimentary rocks known on Earth, about 3.8 billion years old. Ironstones within the Isua sequence indicate that the atmosphere at this time still lacked any significant amount of oxygen.

The continents began to develop approximately 4 billion years ago. Early continental masses, or protocontinents, would have been short-lived, as their constituent rocks were destroyed by convection and early tectonics. They were also considerably smaller than the continents of today: they probably formed from pieces of crust not more than 310 miles (500km) in diameter. By the dawn of the Swazian (3.75–2.8 billion years ago) convection in the mantle had slowed, and scientists suggest that a polygonal pattern of tectonic plates had developed. Spreading of the crust took place radially from a central

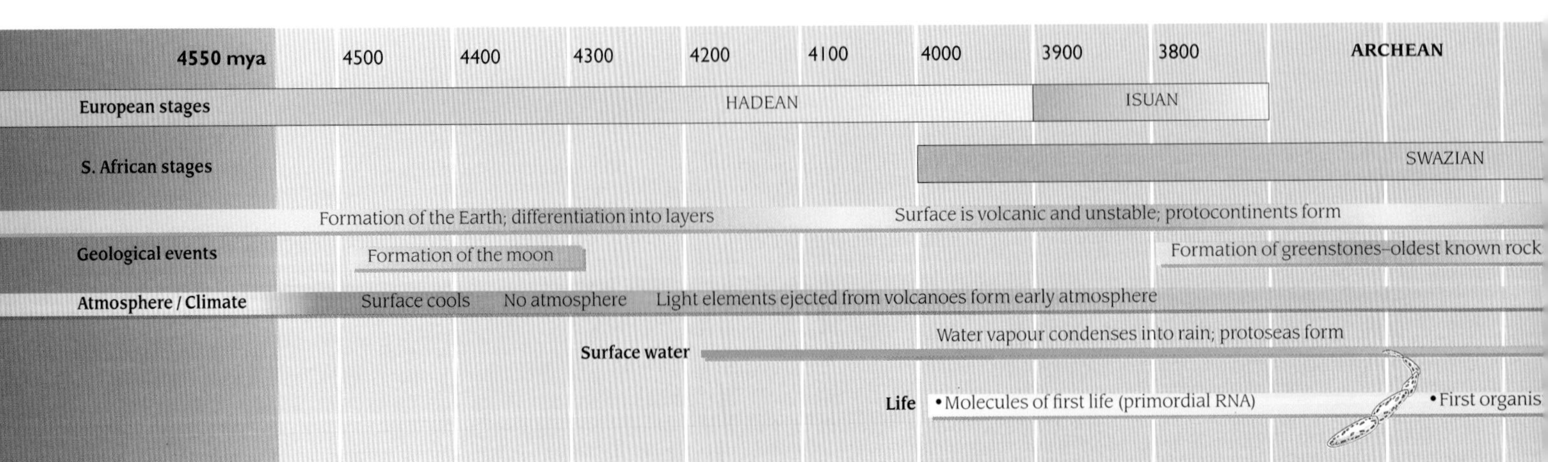

NEW MOON, NEW SKY

Scientists believe that an impactor, roughly the size of Mars and with its own core and mantle, struck the Earth a glancing blow about 4.4 billion years ago. The collision greatly disturbed the mantles of both of the bodies, causing their cores to fuse hours only after impact. The mantle material was blasted in all directions, with a major component moving into Earth's orbit. There it coalesced to form a red ball that was to become the Moon.

More than a billion years passed before this ball finally came to resemble the familiar Moon of today. Initially, a phase of accretion and differentiation took place. The molten Moon gradually became a dead terrestrial body. After the formation of the crust, meteorite craters began to pockmark the lunar landscape. The larger impacts helped to trigger volcanic activity, and basaltic lavas were extruded across the lunar maria between 3.8 and 3.1 million years ago. From Earth, each mare appears dark, the anorthosite highlands being lighter in color and older in age (4.4 billion years).

The crater patterns visible on the Moon were probably formed early in its history. Many are more than 3 billion years old. There are craters within craters, but most would have appeared when meteorite activity in the solar system was at its acme. The dust that covers vast areas of the Moon is a product of early bombardment.

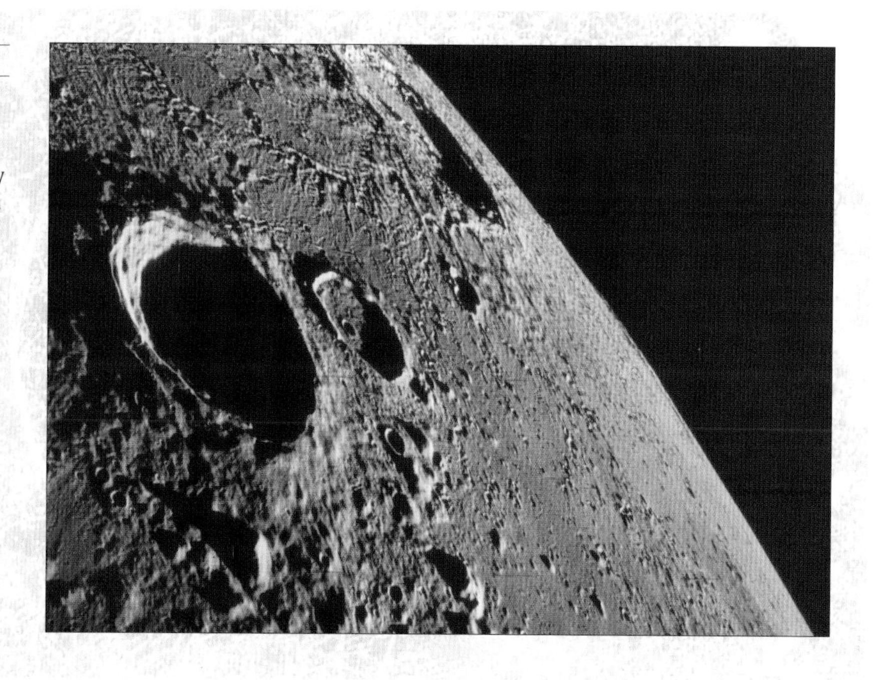

plume, and surface rocks descended into the mantle at the plate margins. The increase in volume of continental crust between 3.5 and 2.5 billion years ago was significant. Scientists estimate that 3.5 billion years ago, this volume may have been only 5 percent of the 15–60mi (25–95km) it is today, whereas a billion years later it may have been 60 percent or more.

By the end of the Randian (2.8–2.5 billion years ago), continental masses were well established. Their margins were continuously destroyed by subducting plates, but stable central areas, known as shields or cratons, survive to this day at the heart of the modern continents.

Erosion of the earliest continents produced the first sedimentary rock. The main sediments deposited were sandstones, gravels, and conglomerates, all types of clastic rocks, rich in silicate minerals, eroded from granites, basalts, and metamorphic rocks. Their erosion also released calcium, which entered the sea through streams and rivers, and gradually built in the ocean, where it precipitated out as limestones and salts. The first limestones were deposited about 2.8 billion years ago. With the increase in atmospheric oxygen and a higher percentage of carbon in the oceans, banded ironstone deposits became less widespread. In the Proterozoic, carbonates were among the major rock types. Their erosion and sedimentation has helped to maintain a chemical balance in the oceans since Archean times.

Gradually the Archean world changed from an inferno to calm landmasses fringed with oceans. Volcanoes still dominated the landscape, and the inland regions of the continents were wastelands. The shores, however, were cloaked with green photosynthesizing organisms which were steadily bringing changes.

Present

Origin of Earth

ARCHEAN

See Also

THE ORIGIN AND NATURE OF EARTH: *Continent, Crust, Lithosphere, Mantle, Plate Tectonics*
THE ORIGIN AND NATURE OF LIFE: *Cyanobacteria, Eukaryote, Prokaryote*
THE PROTEROZOIC: *Algae, Stromatolites*
THE SILURIAN: *Hydrothermal vents*

3400 | 3300 | 3200 | 3100 | 3000 | 2900 | 2800 | 2700 | 2600 | **2500** | **PROTEROZOIC**

RANDIAN

rface covered with volcanoes and lava fields — Early continents form and stabilize

• First free oxygen — Levels of oxygen rise due to increasing photosynthesis

Localized shallow water, tropical seas — Ocean volume 90% of present levels

• Photosynthesizing cyanobacteria, first stromatolites — • First eukaryotes

THE ARCHEAN WORLD

In early Archean times, a huge global ocean covered most of the Earth. The first areas of solid crust consisted of solidified volcanic rocks, which formed over the hot spots that raised them from the mantle below. The existence of these islands was probably short-lived, as the pattern of convection currents in the mantle changed. The hot spots moved away, to form another island somewhere else.

FIRST CONTINENTS

The very first continents were small. As the volcanic rocks that made them up weathered and underwent metamorphism, clays and muds formed, followed by shales, and the small protocontinents fused together in groups to create larger ones. Remnants of these rocks remain today in the shield areas of the modern continents.

LIKE their modern counterparts, the Archean continents were continuously on the move, pushed across the Earth's surface on lithospheric plates, although there is no way of knowing where they moved. Today the only survivors of the Archean landmasses are the shield areas at the centers of all the modern continents. Extensive continental shields (cratons) exist in Canada, Greenland, Scandinavia, Russia, central Asia, China, India, Africa, South America, Australia, and, it is presumed, beneath the ice cap of Antarctica. The largest of these is the Canadian shield, which has been extensively exposed by the action of glaciers in the much more recent Pleistocene. Continental shields provide invaluable data on the evolution of the continents, and allow paleogeographers to construct paleogeographic maps for later times.

Most information about the Precambrian comes from studying continental shields—stable cratons that have not been deformed over time, and that now lie exposed at the surface.

Shield areas have been preserved because they exist at passive, rather than active, continental margins. Through time, extension has resulted in the formation of ocean basins along these margins, with the continents being slowly but continuously pushed apart. The evidence for early island arcs and ocean trenches can also be found within shield complexes, with slices of the ancient sea floor accreted to the continental margins.

Simply assuming that each present-day shield represents a former separate landmass is probably incorrect. But it is likely that several continents moved slowly towards one another, taking 1.5 to 1.8 billion years to amalgamate as the first supercontinent during the Proterozoic.

Some 3.5 billion years ago, during the Swazian eon of the Archean, the protocontinents would have appeared as small islands dotted around a giant ocean. Approximately 3 billion years ago, stable continents of moderate size had developed through the gradual accumulation of igneous, metamorphic, and sedimentary rocks. The two former types—

- Africa and Middle East
- Antarctica
- Australia and New Guinea
- Central Asia
- Europe
- India
- North America
- South America
- Southeast Asia

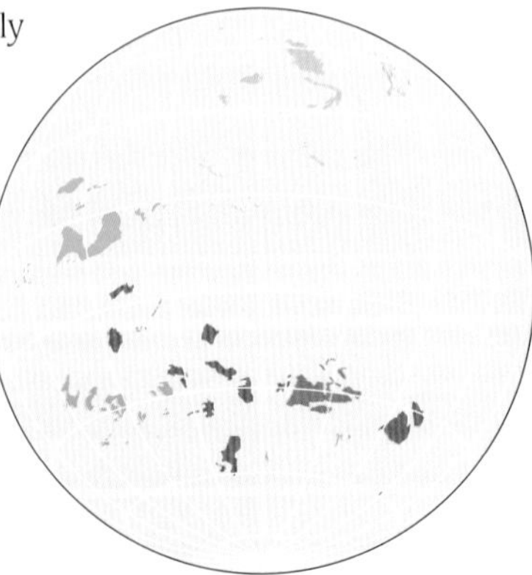

crystalline rock—formed the cores of the new continents, with sediments being deposited in the shallow waters at the continental margins as the cores eroded. Heavy rainfall and huge quantities of volcanic dust and debris would have resulted in the deposition of thick accumulations of volcano-sediments over the margins of the early continents. Significant quantities would also have been blown or transported into the seas and oceans, and deep-water sedimentation is a feature of Archean rock sequences. Snow and ice were probably also present on the higher slopes of giant volcanoes, although the first true ice age did not take place until the Proterozoic.

THE FORMATION of the first continents depended on a decrease in the temperature of the upper mantle, which took place about 4 billion years ago; continental crust is thought to be 4 billion years old. The build-up of the continents must have resulted in a thinning of this mantle, with the shields or cratons floating on an ancient continental lithosphere. This suggests that the lithosphere beneath the shields has remained in the same place for at least 3 billion years. Little or no movement has occurred because the absence of mantle plumes and convection has isolated them from the destructive forces that are operating elsewhere within the mantle.

"Hot spots"—the site of tectonic activity—are much more common in the ocean than on continents, which formed over stable segments of the lithosphere.

ARCHEAN TERRAIN

The Pilbara area in western Australia includes large intrusions of granitic rock from early Archean times. The intrusions have pushed the overlying iron-rich sediments and greenstone belts into large domed structures. The bedding planes of the older rocks are almost vertical, leaving them wrapped around the intrusion.

It is worth repeating that strong convection during the formation of the core and outer layers of the Earth inhibited the formation of stable continents for 400 to 500 million years. There may have been more mantle plumes, or hot spots, with many microplates, each corresponding to a convection cell. This would explain why there were so many microcontinents with intrusions and the seafloor venting of basalts adding to the early crust.

Plumes may also have been associated with the basins that existed behind early island arcs—series of volcanoes that erupted on the continental side of an oceanic trench. If so, the volcano-sedimentary sequences that

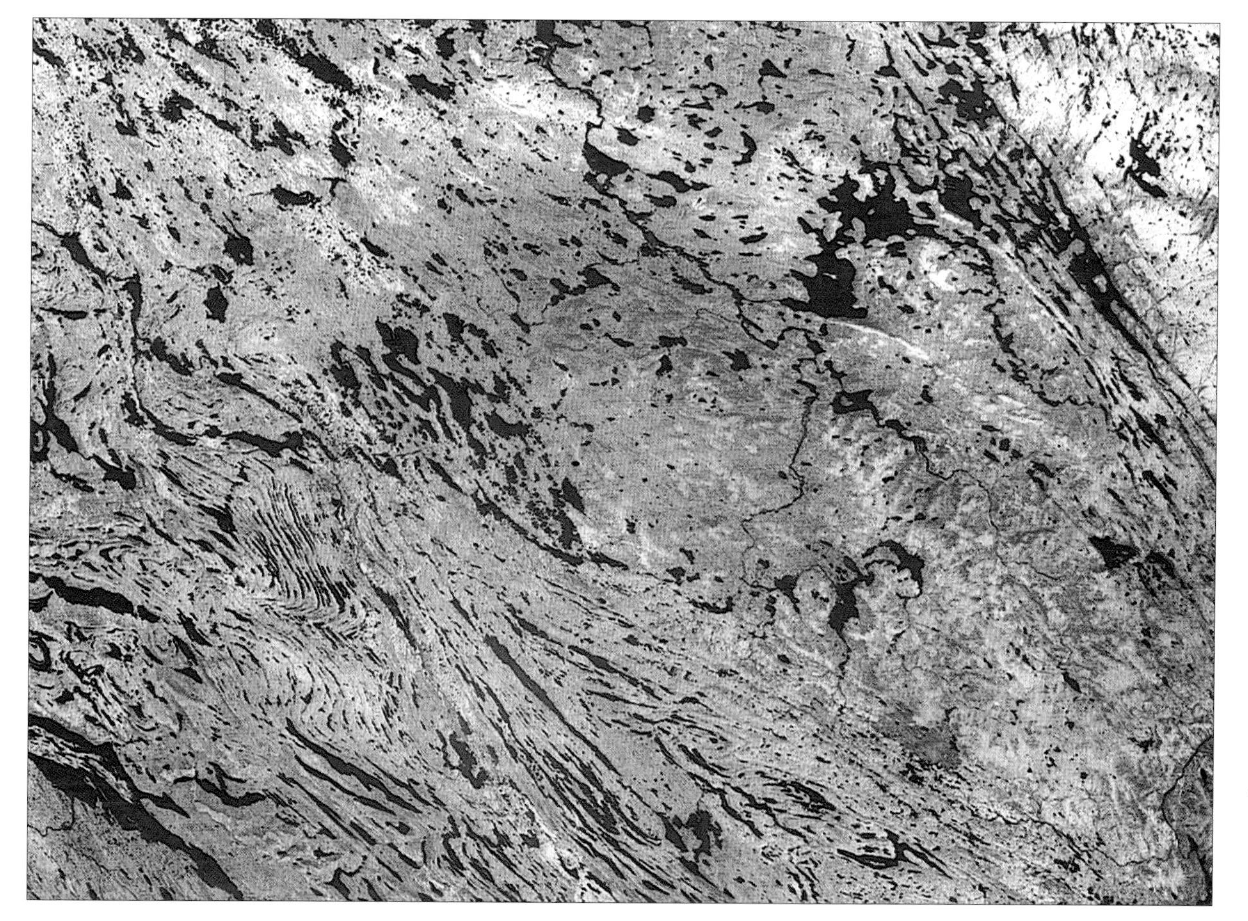

FOLDED SHIELDS

The shield areas of northern Canada consist mainly of granitic rocks from the Archean, and can be seen in the upper right and lower left areas of this satellite photograph of the Labrador fold belt. Such ancient shield rock is very close in composition to continental crust, with a higher proportion of feldspar than ordinary granite. These rocks would have formed deep in the Earth and then been uplifted for long periods before finally being exposed on the continental shields. Between the folds, the layers of rock have been squeezed and folded into convoluted bands. Heat and pressure often caused metamorphism of the sedimentary strata, with gneiss a typical product in these northern shields. Further folding of the gneisses often followed.

THE OLDEST ROCKS

The Earth's oldest known rocks occur within Archean shield areas. The complexly folded gneisses of the Isua region of Greenland (below) were formed volcanically about 3.8 billion years ago and subjected to high-grade metamorphism. Within the gneiss are trough-like bands of greenish rock called greenstone belts. Many of these are rich in metallic minerals and are the basis of local mining industries. There are also conglomerates containing rounded cobbles of even older volcanic rocks, which bear evidence of having been deposited by water. Basic rock from the mantle has in places intruded into fissures in the gneiss to form dikes of younger igneous rock.

accumulated would have been subjected to enormous tectonic forces, resulting in the rocks being added to the emergent continents. These folded rocks are frequently found fringing the granitic cores of the shields or cratons.

THE MINERALS that make up igneous and metamorphic rocks in continental shield areas are themselves indicators of the make-up of early continents and how they developed. Some of the world's oldest rocks, banded ironstones of Greenland and cherts from the Barberton region of South Africa, are examples of shield sediments. Cherts are siliceous rocks that may have a volcanic or biogenic origin (meaning that they may be produced by living organisms), or they may be the products of diagenesis—textural or chemical changes that affect sediments after they have been deposited.

Volcanic activity, sedimentation, and the activity of primitive life forms all contributed to the rocks that make up continental shields.

The Barberton cherts result from volcanic activity, the bacteria entombed in them having existed in the hot water and mud-rich pools of geysers. The banded ironstones were laid down layer upon layer in shallow basin areas on the emergent continental platforms. The siliceous, hematite-rich ironstones are the products of chemical-biogenic processes. Extensive deposits of banded ironstones are also found in North America, western Australia, Russia, and the Ukraine. They are mined locally as iron ore.

The distribution of cherts and banded ironstones around the world provides important clues to the nature of the early Earth. Thick layers of these sediments, several hundred feet thick, accumulated in basins hundreds of miles across. They indicate that the continents were accumulating in a similar manner and that the atmosphere was deficient in oxygen.

The Isua and Barberton rocks also contain clues to the evolution of Archean life. There are no fossils in the Isua ironstones, but traces of carbon suggest that living organisms had originated more than 3.8 billion years ago. In the Barberton cherts, which are dated at 3.5 billion years old, whole single-celled organisms have been preserved, representing evolutionary progress. Bacteria and cyanobacteria (blue-green algae) belong to this category of organism, known as prokaryotes. They developed photosynthesis, which transformed the Earth.

OTHER minerals are indicators of the buildup and composition of the early continents. Minerals characteristic of metamorphic and igneous rocks in shield areas provide data on the depth of burial, the degree of metamorphism, or the composition of the mantle. Metamorphism of rocks within the shields may have been induced by folding and thrust tectonics, or by baking brought about by the intrusion of molten magma from below.

Archean processes left their mark—and a substantial legacy of mineral wealth—in ancient rock.

During early phases of tectonic movement, fractures or faults in the Earth's crust occurred, and these provided

THE ORIGIN OF GREENSTONE BELTS

Greenstone belts formed from repeated deposition of volcanic rocks. Small crustal plates, carried on convection currents (1), stretch and collide and volcanic material from the mantle erupts to the surface (2). Later, the larger currents of plate tectonics transport continental material down into the mantle (3) causing new eruptions. As the continent grows the greenstone sequences are metamorphosed and compressed (4).

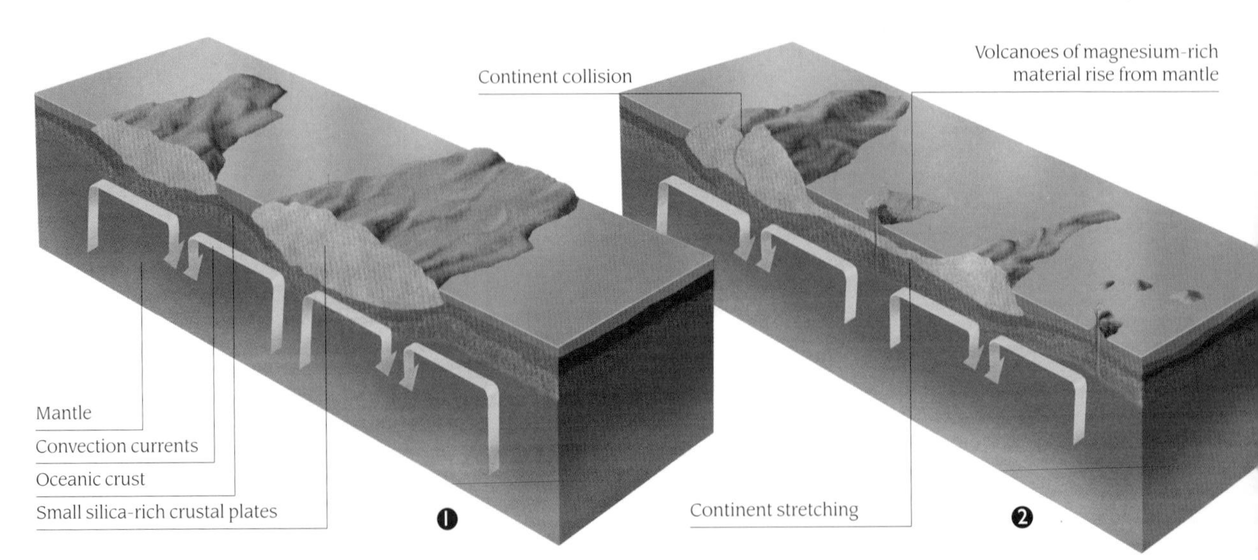

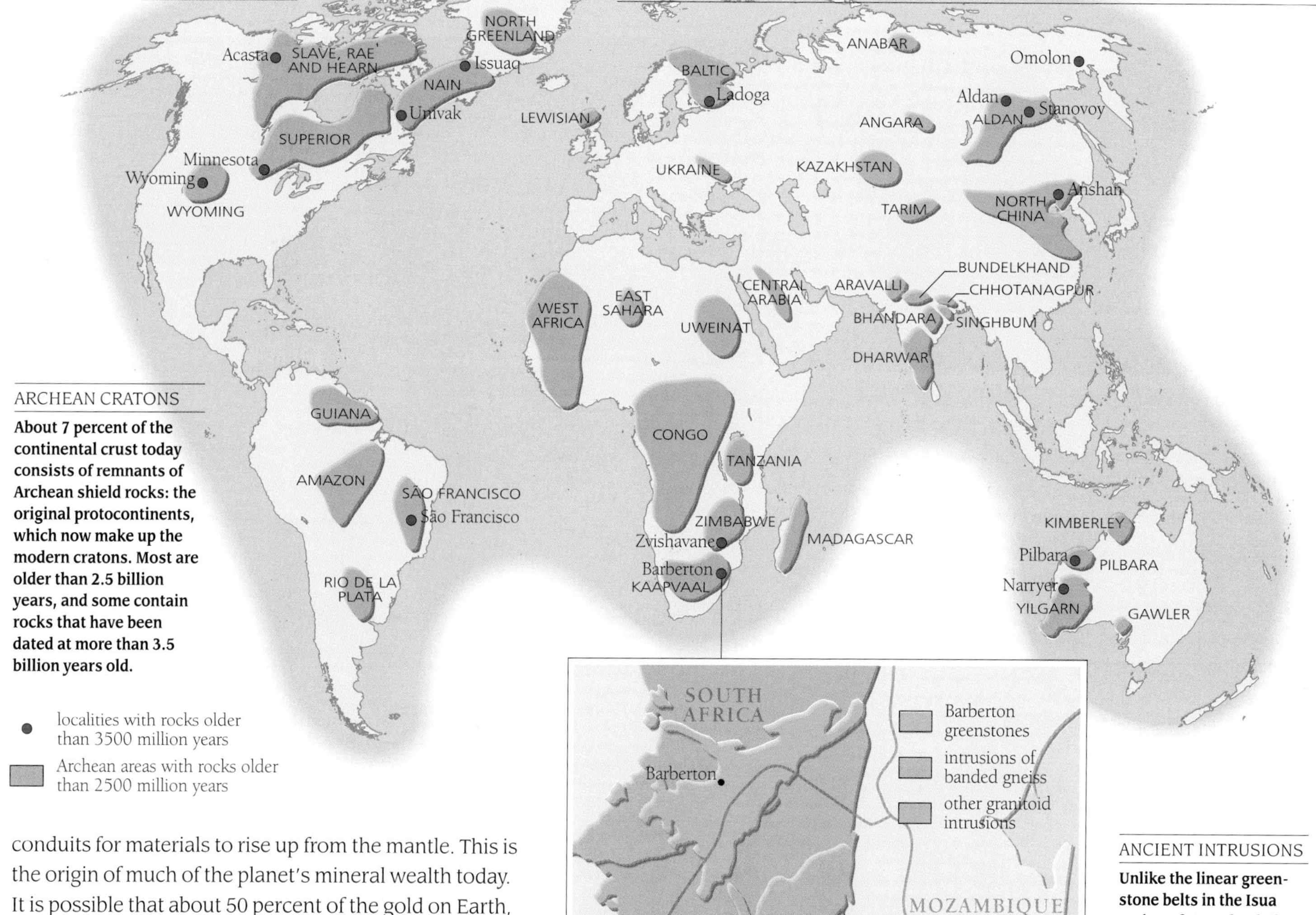

ARCHEAN CRATONS

About 7 percent of the continental crust today consists of remnants of Archean shield rocks: the original protocontinents, which now make up the modern cratons. Most are older than 2.5 billion years, and some contain rocks that have been dated at more than 3.5 billion years old.

● localities with rocks older than 3500 million years

▭ Archean areas with rocks older than 2500 million years

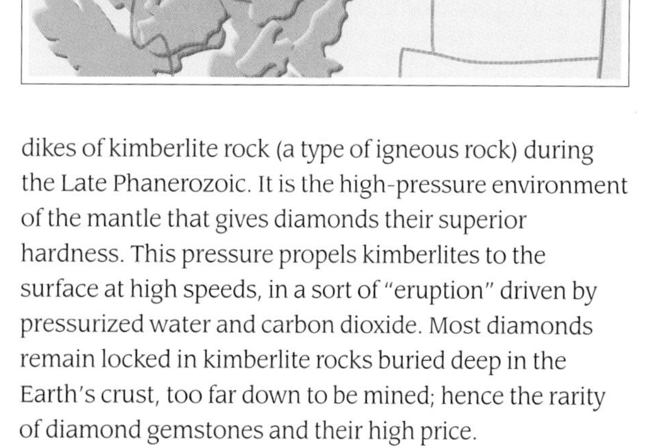

ANCIENT INTRUSIONS

Unlike the linear greenstone belts in the Isua region of Greenland, the greenstones at Barberton in southern Africa have an irregular shape. In the approximately 3.5 billion years since their formation, they have been pushed by earth movements and the intrusion of igneous granitic rocks. The Barberton greenstones have a high concentration of magnesium—an indication of the hotter temperature of the Earth at that time. The slightly younger sediments at the surface contain old placer deposits of gold that has been weathered out of ancient igneous rocks. The best known occur at Witwatersrand, which since they were discovered in 1886 have produced more than half of all the world's gold.

conduits for materials to rise up from the mantle. This is the origin of much of the planet's mineral wealth today. It is possible that about 50 percent of the gold on Earth, for example, is associated with fractured Archean rocks. Gold ore in significant quantities is often produced by hydrothermal processes that deposit metal-rich fluids in magma chambers. These heated fluids can travel great distances through crustal rocks, moving along fractures before cooling and precipitating to form veins of ore. Similar percentages of lead and zinc are also found in Archean shield areas. These minerals were probably flushed onto the sea floor from the upper mantle by seawater circulating through deep fracture systems.

Diamonds, which form under extremely high pressures at 125mi (200km) deep, also originated billions of years ago as a result of Archean processes, but their journey to the surface was delayed until the intrusion of dikes of kimberlite rock (a type of igneous rock) during the Late Phanerozoic. It is the high-pressure environment of the mantle that gives diamonds their superior hardness. This pressure propels kimberlites to the surface at high speeds, in a sort of "eruption" driven by pressurized water and carbon dioxide. Most diamonds remain locked in kimberlite rocks buried deep in the Earth's crust, too far down to be mined; hence the rarity of diamond gemstones and their high price.

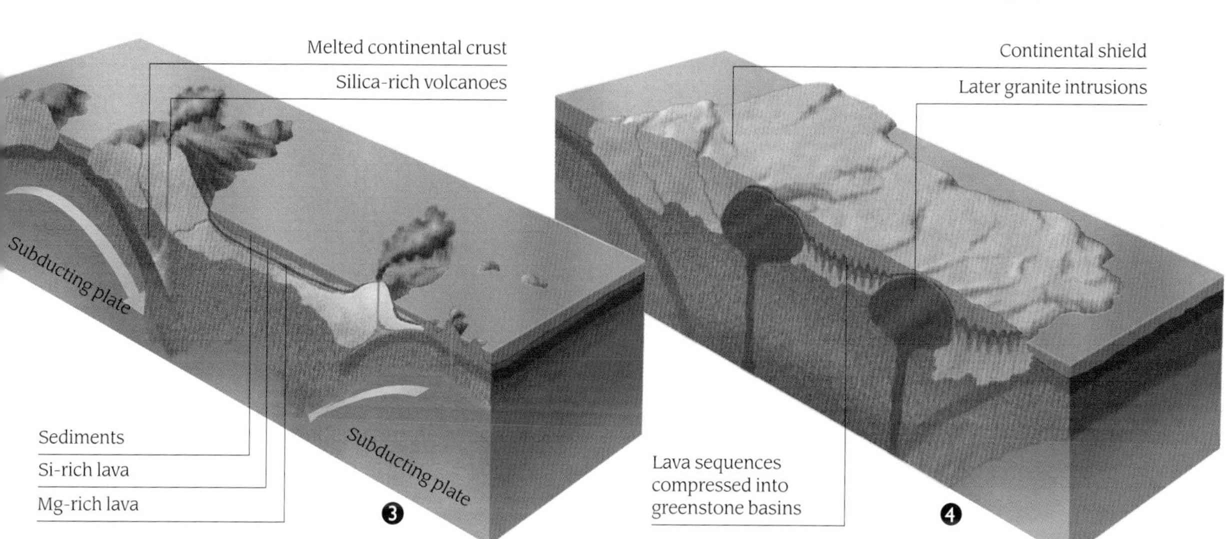

DEEP ocean trenches, submarine canyons, sea mounts, and mountain ranges are the prominent characteristics of the floor of modern ocean basins. In the Atlantic Ocean, the mid-ocean ridge is the major topographical feature; it is an underwater mountain range extending from the Antarctic to the North Pole. It is flanked by older ridges that extend laterally towards the margins of the continents on either side. Between the ridges and the continental slope are broad, flat features called abyssal plains. Their flatness can be attributed to the great thicknesses of sediment that accumulate on the ocean floor. Occasional peaks do occur, but they are probably only the tips of buried volcanoes.

Mid-ocean ridges—long mountain chains under the sea—are the site of continuous generation of new oceanic crust as molten material from the Earth's mantle rises to the surface and is extruded into the water.

The continental slope and platform effectively mark the transition from oceanic floor to continental shelf, and from oceanic to continental crust. The oceanic crust is the older planetary feature, having apparently formed about 4.5 billion years ago as the molten "ocean" of magma covering the Earth began to cool. (The oldest continental rocks identified so far have been dated at just below 4 billion years.) The original oceanic crust would have been composed primarily of komatiite, an ultramafic rock, which forms at extremely high temperatures and is rich in iron and magnesium. Rocks of the modern oceanic crust are predominantly basalt, a magnesium-rich mafic rock that forms at the comparatively cooler temperature of 2000°F (1100°C).

New crust is continuously being created along mid-ocean ridges, adding about 1.35mi^2 (3.5km^2) of new crust to the Earth every year. Slow spreading of this material, as in the Atlantic, creates a stepped valley on either side, whereas the faster-spreading Pacific ocean ridge has no valley.

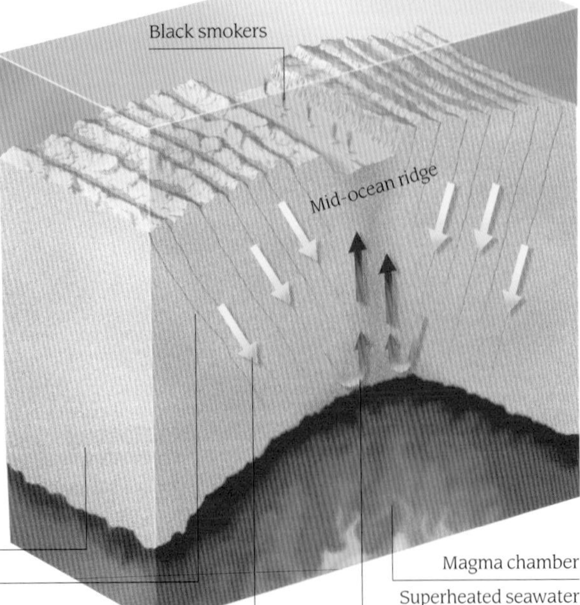

SEAFLOOR SPRINGS

Hot-water mineral springs, or hydrothermal vents, first appeared on the Archean ocean floor along mid-ocean ridges. They still occur when cold seawater seeps down through fractures along the ridge, towards the magma chamber below. The superheated water rises again, dissolving minerals from the rock and emerging at the ridge. The minerals precipitate from the cooled water as a "smoke" of fine particles.

AS NEW crust is generated along mid-ocean ridges, older crust—which is farther away from the ridges, near the margins of continents—disappears in subduction zones as the oceanic plates push under the continents. For this reason, unlike the stable continental shields with their Archean rocks dated at more than 3 billion years old, nothing is left of Archean seafloor. The oldest seafloor present in the oceans of today is 180 million years old—farthest away from the ridge, near the continental margins.

Although oceanic crust appeared before continental, the oldest example today is a mere 180 million years old; the rest has been recycled over and over.

Even during the Archean itself, features of the ocean floor would have been comparatively short-lived due to the vigorous convection of the mantle, which caused

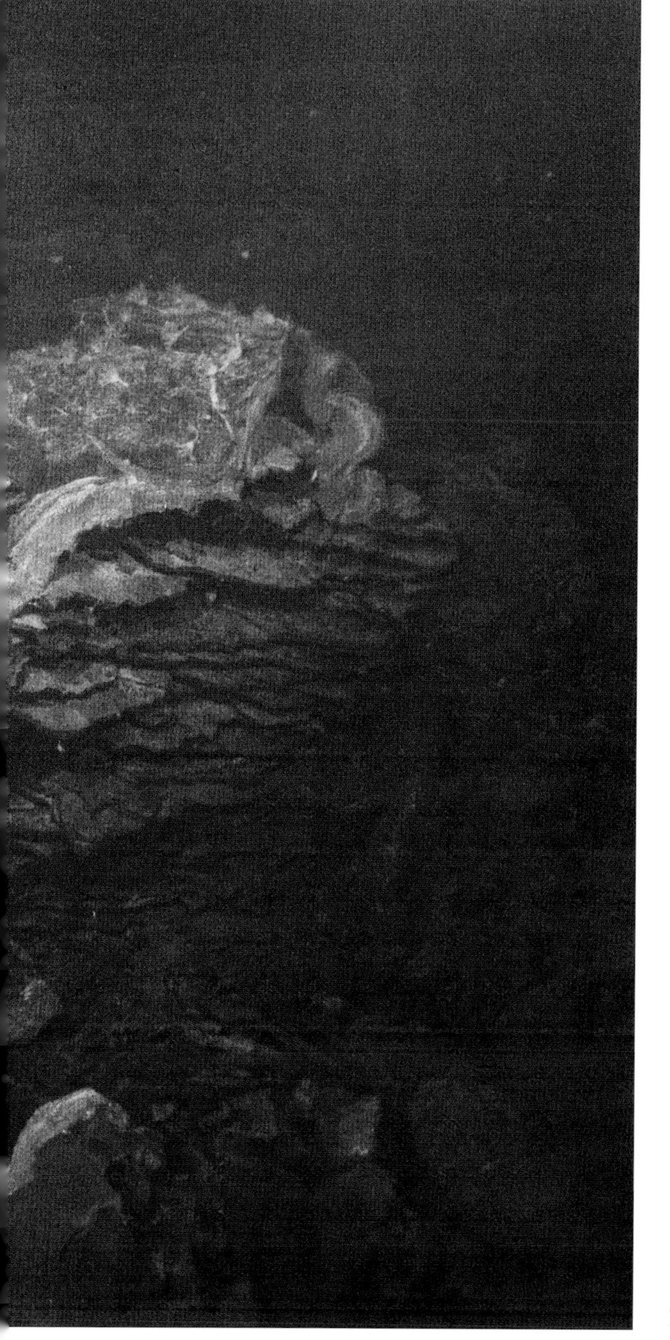

FRACTURES from which sea-floor volcanics emerged were also vents for recycled seawater, which was extremely hot (660°F/350°C) and rich in sulfides. Mineral particles swirling around in the water gave rise to dense clouds of "smoke." The Archean would have had the equivalent of the vents called "black smokers" first found in the late 1970s in the Pacific Ocean. Layered sequences of sulfide-rich deposits of Archean age on Earth today bear witness to the early venting on the seafloor. Modern black smokers are home to communities of worms, mollusks, and fishes. Some scientists believe that during the early Archean they were the source of life itself.

Fissures in the ocean floor gave rise to plumes of superheated water, and clouds of mineral precipitates heralding the first black smokers. From one of these chemical-rich environments the first organic materials of life appeared.

Approximately 3.5 billion years ago the first stromatolites appeared, relics of the activities of cyanobacteria growing in matted "colonies." Fossil stromatolites are more common in Proterozoic than Archean rocks, but it is clear that they enjoyed worldwide distribution by the end of the Archean. The oldest stromatolite fossils have been found in the Pilbara cratonic shield of Australia, and modern stromatolites are still growing in the warm shallow intertidal zone of Shark Bay, Australia.

BLACK SMOKERS

(Left) The plume of dark mineral particles precipitated by hydrothermal vents has earned them the nickname "black smokers." Common among these particles are the sulfides of various metals, such as copper, lead, and zinc. The minerals, precipitated from water superheated to 660°F (350°C) when it meets the cold ocean-bottom water, sometimes form a chimney around the vent. In Archean times primitive bacteria may have metabolized the sulfur in much the same way as modern microorganisms employ oxygen. Some scientists have speculated that sulfur-using bacteria may have been among the original life forms on Earth.

continuous cooling and melting of the crust. Movement and collision of early microplates also destroyed the vast majority of the topographical features produced along the plate boundaries. Also destroyed, if they ever existed, were the magnetic reversal data in rocks that enable the modern seafloor to be dated with accuracy.

Nevertheless, as violent activity gave way to more stable conditions, it is likely that the Archean seafloor had many features in common with today's. No doubt towering submarine volcanic ranges and seamounts were much in evidence, but abyssal plains were also present. Thick accumulations of sediment from the early continents were deposited in these areas, with deep-water sediments constituting a high percentage of Archean deposits. As more crust was generated from within the mantle, the area occupied by the ocean floor would have spread, gradually extending to cover an ever greater proportion of the Earth's surface.

BACTERIAL BUILDERS

A cross section of a fossil stromatolite (right) reveals how it was gradually built up in layers, probably over many years, by the lime-accreting cyanobacteria that constructed it. The structure, in which the carbonates of the original have been converted to silica, is direct proof that these organisms of the shallow seafloors were employing photosynthesis 3.2 billion years ago, although evidence from the Isua greenstone deposits of Greenland dates the origins of photosynthesis more than half a billion years earlier. Stromatolites have survived until this day, and can still be found as stony cushion-shaped structures in Shark Bay, Australia.

1
2
3

THE LANDSCAPE of the Earth during the Archean would have looked very inhospitable to modern eyes. Pools of boiling mud bubbled and steamed, geysers hurled superheated water high in the air, and active volcanoes spewed molten lava and hot ash into the oxygen-starved atmosphere. The only sounds were those of the Earth itself, punctuated by the roar of exploding volcanoes and crashes of thunder as electric storms raged across the landscape to the horizon. Torrential rain poured down from leaden skies. Some nights the clouds parted to reveal the newly formed Moon—still rocky and irregular—on station as Earth's constant companion.

Hot, steamy, barren, and explosive, the Archean landscape was uninviting. Yet it was here that life nonetheless began.

Yet even amid this infernal chaos, early living organisms were beginning to evolve on the fringes of the ancient oceans. Stromatolites formed their characteristic cushions in the warm shallows of the water. Their fossils, known from Archean and Proterozoic rocks, provide evidence that the cyanobacteria (blue-green algae) that formed them as much as 3.2 billion years ago already used photosynthesis as a life-sustaining process.

A different kind of evolution was taking place below the Earth's surface. Areas of solid crust formed the first small protocontinents, some of which have survived to this day as blocks of rock—cratons—embedded in the modern continents. But most of the early attempts at continent formation were dragged down into the molten magma and recycled to the surface by powerful convection currents. Only when this process slowed down did larger, permanent continents begin to build up.

1 Dormant volcano
2 Pyroclastic eruption of parasitic cone
3 Newly formed moon
4 Electric storm
5 Geysers
6 Hot mud pools
7 Stromatolites

THE EVOLUTION OF ALGAE

ALGAE are the most plantlike of the Protoctista. They range in form from tiny single-celled to large multicellular organisms. Most algae live in aquatic habitats or in moist conditions on land; all are capable of photosynthesis. Blue-green algae, also known as cyanobacteria, were among the earliest forms and have a primitive single-celled structure. The first multicelled alga, *Grypania*, appeared in Early Proterozoic times, about 2.2 billion years ago. Algae—and particularly early algae—were soft and lacked skeletal support. The fossil record of these organisms is therefore extremely limited, and a detailed reconstruction of their early evolution is impossible.

First acritarch and then prasinophyte algae followed the appearance of *Grypania* approximately 1.9 billion years ago. Significantly, the acritarchs were to become an important component of late Precambrian and Paleozoic plankton. In contrast, the prasinophytes, which persist today, had motile (free-swimming) as well as fixed modes of life. The fixed, or sedentary, types formed distinct colonies. Acritarchs and prasinophyte species had nucleated cells and were therefore eukaryotic.

The evolution of green algae and red algae can also be traced back into the Proterozoic. Several superfamilies within these groups evolved rigid chalky skeletons for support and protection. During the Phanerozoic, calcareous algae played an important role in the development of reefs. The destruction of the skeletons of shallow-water species produced carbonate sands. Together with cyanobacteria, the algae were vital in the carbon cycle. As cyanobacteria declined in numbers at the end of the Precambrian period, probably because of grazing by gastropods, algae flourished. Planktonic algae, such as coccoliths, diatoms, dinoflagellates, and radiolarians, are used in dating the subdivisions of sedimentary rock sequences. Algal blooms, now associated with nutrient-enriched waters and pollution, probably occurred intermittently during the Phanerozoic. Thus an intricate relationship between global warming, pollution, and extinction exists within the fossil record.

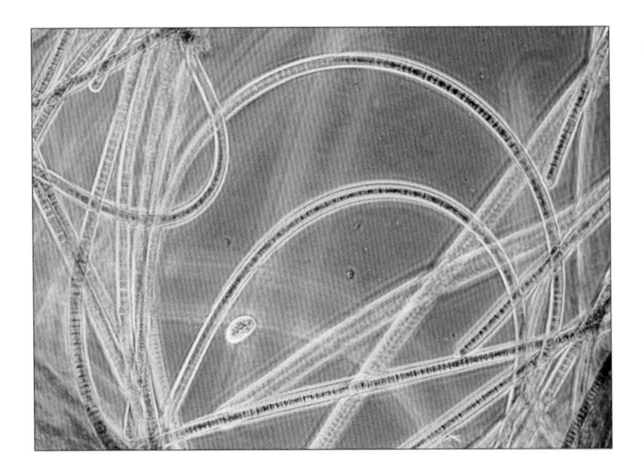

ALGAE LINEAGES

The algae are made up of many unrelated groups, for which numerous taxonomic schemes have been devised. Only the acritarchs, dinoflagellates prasinophytes, coccoliths, diatoms, and the calcareous forms left a useful fossil record, but it is clear that the various lineages evolved independently. Each of the algal groups probably began as a single-celled organism. Some, such as the dinoflagellates, stayed unicellular, though often forming large colonies. Another planktonic group, the acritarchs, with their variety of spherical or many-pointed forms, are some of the oldest and most widely distributed fossil algae.

1 First cyanobacteria (bluegreens); unlike other bacteria they metabolize sunlight (photosynthesis).
2 First multicellular algae. Individual cells were independent and unspecialized in early multicellular forms
3 First acritarchs; spiny forms evolved as the number of predators increased
4 First bangiacean
5 First charaphyte
6 First dinoflagellate
7 First coralline algae
8 Rhodophytes and chlorophytes evolved rigid skeletons
9 First chaetophorales
10 Multicellular algae develop cell connections and specialisms.

ALGAL ANCESTORS

Many cyanobacteria consist of long threads or filaments, although some are single-celled organisms. Today they are allocated a phylum of their own within the kingdom Bacteria. They all contain chlorophyll and are capable of photosynthesis. Some also contain the blue pigment phycocyanin, which gave rise to their earlier name of blue-green algae.

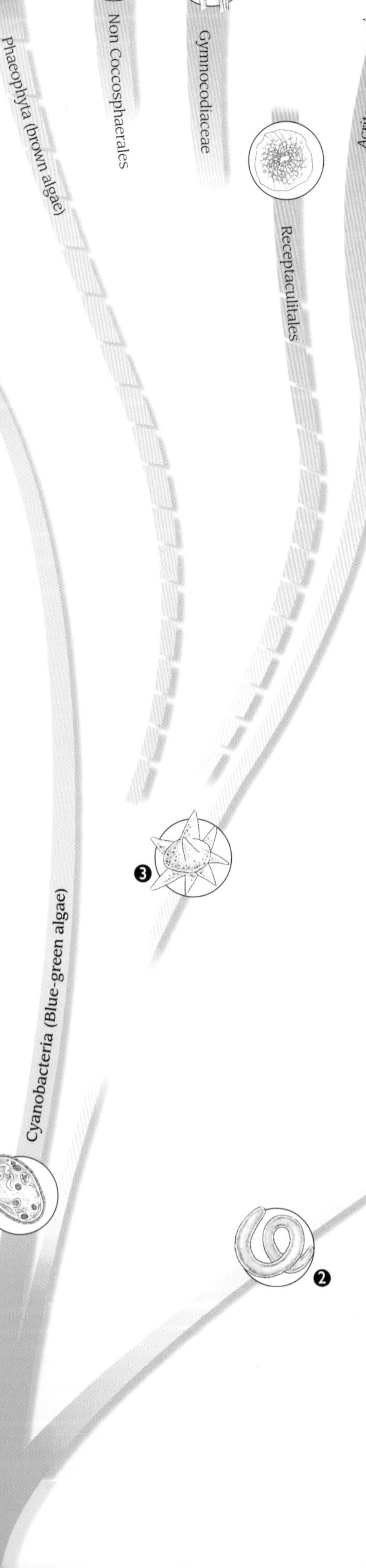

PLANT IMITATORS

Seaweeds are algae but look distinctly plant-like, with broad fronds and a rootlike structure to anchor them; the largest, the kelps, can grow to lengths of 100ft (30m).

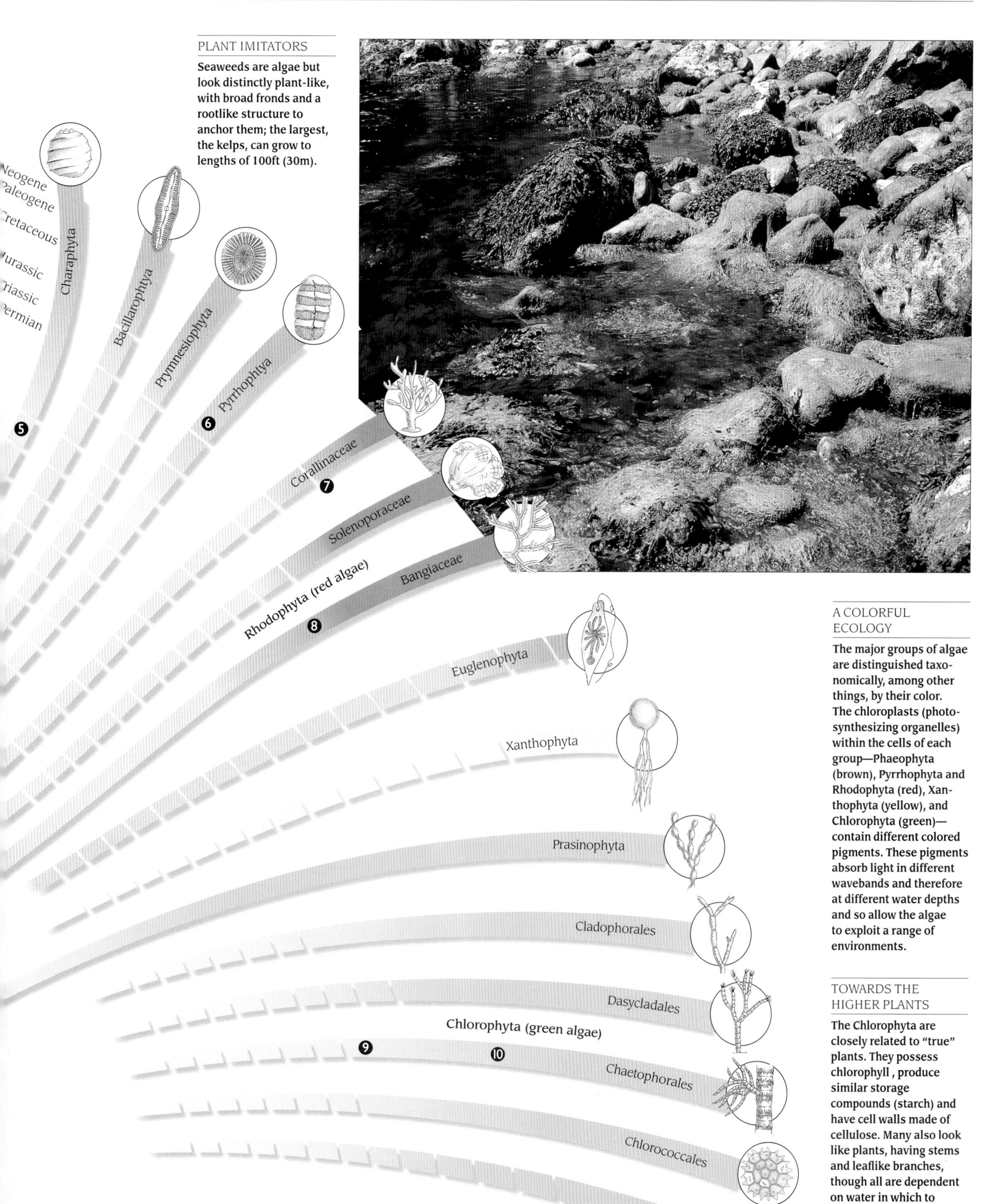

A COLORFUL ECOLOGY

The major groups of algae are distinguished taxonomically, among other things, by their color. The chloroplasts (photosynthesizing organelles) within the cells of each group—Phaeophyta (brown), Pyrrhophyta and Rhodophyta (red), Xanthophyta (yellow), and Chlorophyta (green)—contain different colored pigments. These pigments absorb light in different wavebands and therefore at different water depths and so allow the algae to exploit a range of environments.

TOWARDS THE HIGHER PLANTS

The Chlorophyta are closely related to "true" plants. They possess chlorophyll , produce similar storage compounds (starch) and have cell walls made of cellulose. Many also look like plants, having stems and leaflike branches, though all are dependent on water in which to reproduce. Some produce male and female gametes (sex cells) and undergo sexual reproduction but there is no intermediate embryo stage as in true plants.

THE PROTEROZOIC

2500 – 545 MILLION YEARS AGO

THE PROTEROZOIC is the second and last of the two eons of the Precambrian. During these 2 billion years, which spanned more than 40 percent of the world's total history, many dramatic changes took place on Earth. They culminated in the breakup of the first supercontinent, an ice age, and the first communities of living organisms. The name Proterozoic means "first life," and it is this momentous development that distinguishes the eon from the preceding Archean and the succeeding Phanerozoic ("visible life").

The deep oceans that covered the world in the preceding Archean eon had become wide shallow seas in which layers of sediment hardened into sedimentary rocks. These rocks were less affected by metamorphism but were nevertheless changed by major continent-forming and mountain-building events. Fossils in these same rocks have recorded the evolution of more advanced single-celled organisms and then multicellular animals and plants, some of which have living relatives in modern times.

CHANGES in structural patterns, the composition of granitic rocks, and the deposition of greater thicknesses of limestones, sandstones, and other sedimentary rocks mark the boundary between the Archean and the Proterozoic. The granitic rocks derived from a changing mantle, especially those younger than 2 billion years, can be dated by isotopic methods. These geological tools are extremely useful for dividing the Proterozoic into different stages and series across the world.

The beginning of the Proterozoic eon is marked by changes in the structure, composition, and deposition of rock.

Continental redbeds appeared in the geological record around 1.8 billion years ago. Their presence indicates that atmospheric oxygen had increased significantly and that the oxidation of iron-rich minerals had taken place in subaerial environments. Climate change at this time appears to have been linked with the movement of continents, and evidence for continental glaciation is first documented within the Huronian stage of Proterozoic sequences in North America, about 2.1 billion to 1.65 billion years ago. There is strong evidence that there was regional tectonic activity during the Proterozoic; the Karelian, Wopmay, Grenville, and Gothian orogenies indicate that mountain ranges were being formed intermittently. The Karelian orogeny took place in northern Sweden, the Gothian across Scandinavia, and the Grenville and Wopmay in eastern and northwestern Canada respectively.

Continental masses, larger than those of the Archean, were also present during the early Proterozoic, with continental accretion and a more defined convection system developing. Elongated belts of sediments existed along the continental margins, suggesting that back arc basins were present and that the basins were closing. The term orogenic or mobile belt is used to describe these features.

KEYWORDS

BACK ARC BASIN
CONTINENT
DIAGENESIS
EUKARYOTE
ICE AGE
INVERTEBRATE
LAURENTIA
MOBILE BELT
OROGENY
PROKARYOTE
REDBED
RODINIA

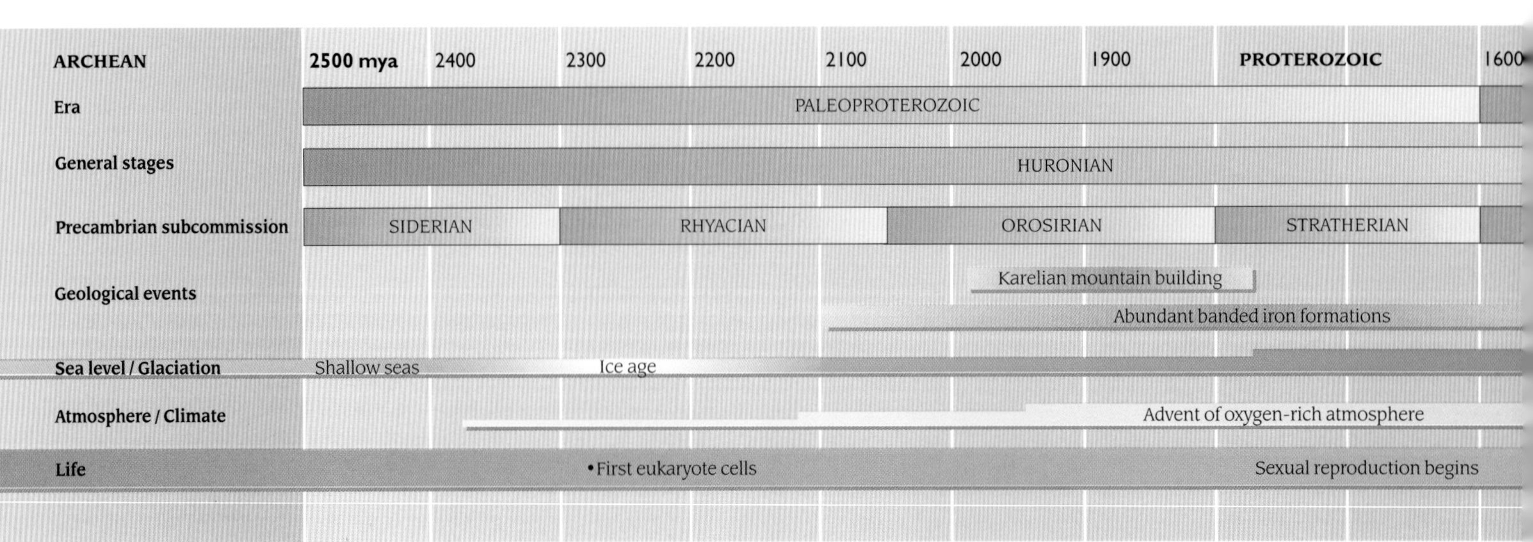

THE ORIGIN OF EUKARYOTES

The oldest known fossils are prokaryotes from about 3.5 billion years ago. These were bacteria and cyanobacteria (blue-green algae), single-celled organisms with no membrane-bound nucleus for their genetic material. The next step up on the evolutionary ladder was occupied by eukaryotes, which also began as one cell but developed a membrane-bound nucleus in which DNA was stored as chromosomes. Eukaryotes also contain other membrane-bound structures called organelles, which may have have evolved by incorporating aerobic (oxygen-using) bacteria into the prokaryote cell, a process called endosymbiosis. At first the bacteria invaded a prokaryote. But instead of being digested, the invaders formed a symbiotic relationship with their host, which provided food and shelter. In time, the bacteria came to depend on the host cell for survival and in return provided the host with energy. These bacteria had become mitochondria, which reproduced within the cell and served the same function in the next generations of the host, which were now eukaryotic animal cells. In a similar way, photosynthetic bacteria became symbionts in eukaryotes, keeping their ability to convert sunlight into food using chlorophyll. They evolved into the chloroplasts in eukaryotic plant cells. In this manner, the plant and animal kingdoms were founded, and the first opportunity for sexual reproduction arose.

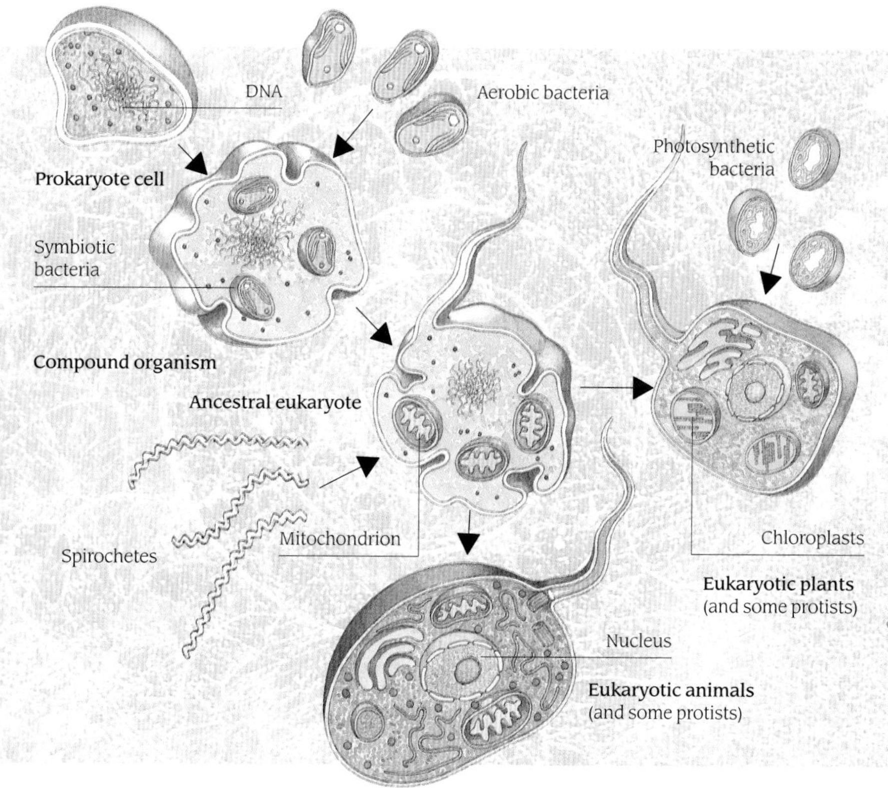

The collision of smaller plates was inevitable, and it is likely that landmasses the equivalent of Siberia, Australasia, Antarctica, and the Americas were slowly drifting towards one another. The vast Canadian Shield constituted the continent of Laurentia during the early Proterozoic. Approximately 1.4 billion years ago, the aggregation of minor continental masses into the supercontinent Rodinia was well advanced. By 1.2 billion years ago, the continents had collided, and mountain-building occurred along the margins of the supercontinent—the Grenville and other orogenies. The formation of continents would have caused dramatic changes to the climate and the movement of water within the oceans. According to data collected throughout the world, four ice ages chilled the planet between 2 billion and 600 million years ago.

With the development of Rodinia and the repeated orogenic episodes, mountain chains became established features of the Neoproterozoic landscape. Intense glaciation and erosion caused by the eventual thaw of the glaciers resulted in the deposition of thick sequences of sediment. Any acceleration in chemical weathering would have given rise to the extraction of carbon dioxide from the atmosphere and, in association with calcium derived from the eroded strata, this would have resulted in the deposition of limestones and other carbonate rocks. As part of the carbon cycle, the carbon dioxide in these sediments was stored for a considerable period of time.

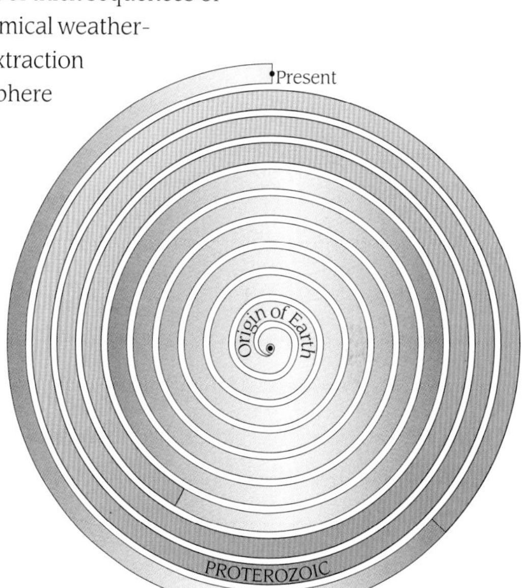

See Also

HOW LIFE EVOLVES: *Prokaryotes, Eukaryotes, Reproduction*
THE ARCHEAN: *First Continents, Cyanobacteria and Earliest Life*
THE CAMBRIAN: *The Cambrian Explosion, Burgess Shale*

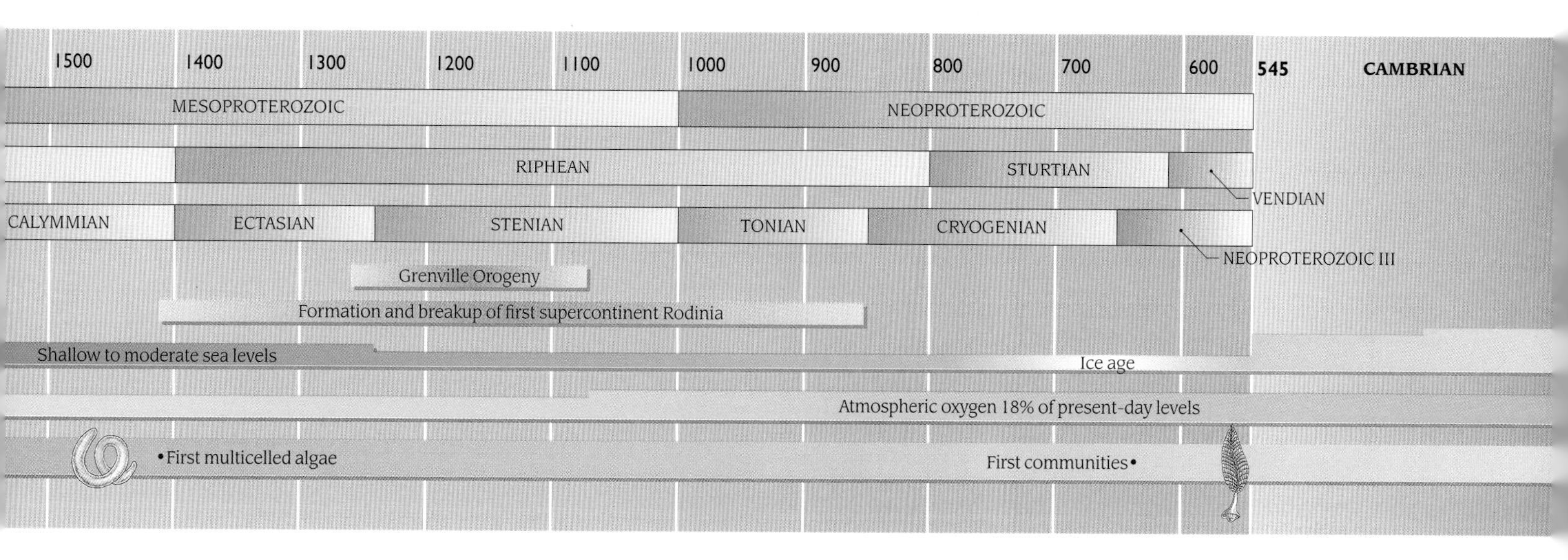

THE FIRST SUPER-CONTINENT

At the end of the Archean the small microcontinents began to come together, giving rise to the supercontinent of Rodinia in the Proterozoic. New crust formed between them by about a billion years ago, creating a large, four-lobed crustal mass, with Laurasia incorporated at the center. Other smaller continents remained to the south. Then the rigid crust that formed Rodinia began to rift apart about 800 million years ago, with a major northwest to southeast divide that later became the Pacific. The whole collection may have drifted south and formed another supercontinent about 550 million years ago, centered on the South Pole.

The transition from the Proterozoic to the Cambrian is marked regionally by unconformities and by shallow-water sedimentation. Over time, a change in sea level would have resulted in the inundation of continental areas. The new species which appeared during the late Proterozoic were now able to exploit a wide variety of new environments.

Shallow to moderately deep seas were characteristic of much of the Proterozoic. Around the edges of the continents were abundant stromatolites in a range of body shapes, indicating diversification in their species and habitats. About 1 billion years ago, however, the numbers of stromatolites began to fall. Environmental change may have been a significant factor in their demise, but it is also possible that the stromatolites had become the prey of primitive grazing animals.

The complex soft-bodied animals that were commonplace in the communities of the late Precambrian did not appear from nowhere. They were the product of the evolution of simple prokaryotes into eukaryotic multicellular organisms during the early Proterozoic, and quickly maximized their diversity by the new method of sexual reproduction. Precursors of sponges, corals, mollusks, and worms probably first appeared around 1 billion to 800 million years ago, although the delicate ancestral forms failed to survive the geological processes of burial and diagenesis (chemical and structural change). The first skeletal fossil, *Cloudina*, is recorded from about 800 million years ago. By 600 million years ago, in the Vendian stage, the first communities of higher organisms had become established in shallow water bordering several continents. Similar organisms occur around the world.

The first communities were somewhat exotic and ephemeral, and the vast majority of their constituent species failed to survive into the Cambrian. The reason

ICE SHEET OVER AFRICA

By the late Proterozoic, Africa and the Middle East were almost completely covered by an ice cap. Widespread glacial deposits indicate that even the regions close to the Equator experienced some glaciation.

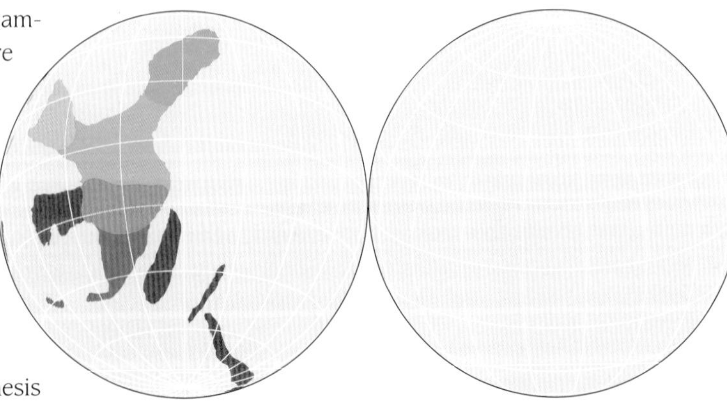

- Africa and the Middle Ea
- Antarctica
- Australia and New Guin
- Central Asia
- Europe
- India
- North America
- South America
- Southeast Asia

for this may be closely linked with glaciation and probable changes in sea level. Carbon isotope data from the late Proterozoic fluctuate over a 250-million-year period (950–600 million years ago). Low isotope ratios and the occurrence of banded ironstone formations in deep-water environments mark the advent of glacial maximums. The descent of cold surface waters rich in oxygen may have triggered the deposition of the banded ironstones. Sooner or later, over 20 to 30 million years, oxygen would have been recycled into the atmosphere. Between 600 and 550 million years ago the rebound of atmospheric oxygen may have been a unique event, stimulating an explosion of life forms that heralded the beginning of the Phanerozoic.

GLACIATIONS at the end of the Proterozoic are associated with the formation of the first supercontinent. The collision of the constituent continental plates took place between 1.4 and 1.1 billion years ago, with Laurentia and Antarctica forming the core of the first great land mass. Periods of intense igneous activity occurred during the early and late Proterozoic, represented by dike swarms and by flood basalts similar to those of the Deccan Plateau in India. These outpourings appear to be linked with the formation and breakup of continents during the Proterozoic. Upwelling of hot mantle rocks and downwelling of cooler rocks from the upper mantle were influential in separating or assembling the constituent blocks. The mechanism is simpler than the processes that affect modern plate tectonic activity, but the movements of the early continents over 200 million years were significant.

Approximately 800 million years ago Rodinia broke in half. Some geologists believe that it formed again about 250 million years later.

Proterozoic continents had cores of Archean rocks. Erosion was effective in stripping these cratonic regions and, gradually, they were covered in younger sedimentary sequences. Stable shield areas characterized the continents of the Proterozoic.

Detailed mapping and sophisticated uranium–lead (U–Pb) dating methods have enabled geologists to analyze the aggregation of Rodinia as well as the individual plates that made up the supercontinent. Laurentia, including Greenland and the Canadian shield, was assembled from five or six microplates over a period of approximately 100 million years. It first appeared 1.5 billion years ago.

Other continents probably appeared over a similar timescale, gradually moving together over a span of 1 billion years. These continents would eventually form the core regions of Gondwanaland, the supercontinent of the southern hemisphere, which did not begin to break apart until the Jurassic and Cretaceous periods of the Mesozoic era. The components of Gondwanaland corresponded to the landmasses that gradually separated and took shape as modern Africa, India, and Antarctica.

Rodinia's assembly was accompanied by a steady movement towards the South Pole. Around 750 to 850 million years ago, the center of the supercontinent lay across the Equator. The positions of the continents changed rapidly during the Late Neoproterozoic, however. The shift is thought to be in the region of 90 degrees, with a movement of 15in (38cm) per year, compared with 3in (7.5cm) in the present. It is almost impossible to register such movement in terms of the modern-day Earth, but one may assume that over a period of 30 million years the continents passed rapidly between polar and equatorial climates.

The movement of the continental masses into an equatorial region would have created vast polar oceans, which may have generated a runaway cooling effect. Some geologists believe that at one stage the Earth may have been a "snowball," with the ancestors of the Cambrian communities hibernating below an ice-

PRECAMBRIAN FOSSILS

A Precambrian formation in Newfoundland is rich in microfossils of single-celled organisms. Cherts and carbonate rocks laid down under water often contain these fossils in abundance, indicating that they probably formed part of the plankton in the ancient oceans. Acritarchs are one of the most common groups, occurring in a wide variety of sedimentary rocks.

covered ocean surface. Volcanic activity may then have generated excessive carbon dioxide, causing a greenhouse effect and melting the ice rapidly. During the Late Proterozoic, these changes in climate were cyclical, with four ice ages recorded over a comparatively short time span. Glacial deposits, such as tillites (lithified tills or boulder clays), and scratches and grooves caused by glacial abrasion are found in many locations in Europe, North America, Africa, and Australasia.

The presence of a supercontinent during the second half of the Proterozoic eon helps to explain the distributions of stromatolite colonies and the earliest communities of complex life forms. When mapped today, these colonies and communities appear to be isolated across the globe, but if their development is plotted along the coastline of a single landmass, then the pattern of their distribution is much more logical. Stromatolites are relatively rare today because they require a warm, shallow-water environment in which to live. The first communities, such as the Ediacara, also occurred in warm-water environments. Elsewhere the continental masses may have been covered with ice, but not for long; global currents and climate conditions fluctuated dramatically.

Normal salinity and oxygen levels are likely to have been requirements, too. The advent of these conditions not only resulted in greater faunal diversity but also in an increase in the size of individual organisms (oxygen being an essential prerequisite for higher life forms) and normal salinity giving rise to a more tolerant life-support system in a marine environment.

Much of the globe would have been covered by a giant ocean during the late Proterozoic. Planktonic organisms, including acritarchs and other eukaryotic cellular material, flourished in the surface waters. Acritarchs may have represented the resting stages for early dinoflagellates. These latter are mainly unicellular algae that exist today in marine or freshwater environments. The resistant nature of the acritarch's outer coat, however, ensures that the cyst stage is present much earlier in the fossil record than the parent organism.

It is likely that the first seaweeds appeared during the late Proterozoic. The fossil remains of *Tawuia* are recorded from the Neoproterozoic. It is the first thallophyte—that is, a plant in which the "body" is not divided into root, stem, and leaves. Patches of green and brown gradually began to cover the barren shorelines.

MULTICELLULAR LIFE

Jellyfish, like many other of the first multicellular organisms, adopted an essentially radial body plan. They are still found throughout the world's oceans and have probably remained abundant throughout the billion years of their existence. Some modern species, like their early ancestors, have brown algae living as permanent symbionts in their mantle tissues.

SOFT BODIES

Fossils of the soft-bodied creatures called *Ediacara* were named after the Ediacara Hills of South Australia, where they were first discovered in a shallow-water formation of the Proterozoic. The fossils are actually casts or molds of the late Proterozoic organisms preserved in reddish-pink sandstone sediments. They have been dated at 640 million years old, but there is still controversy about exactly what the living organisms would have looked like.

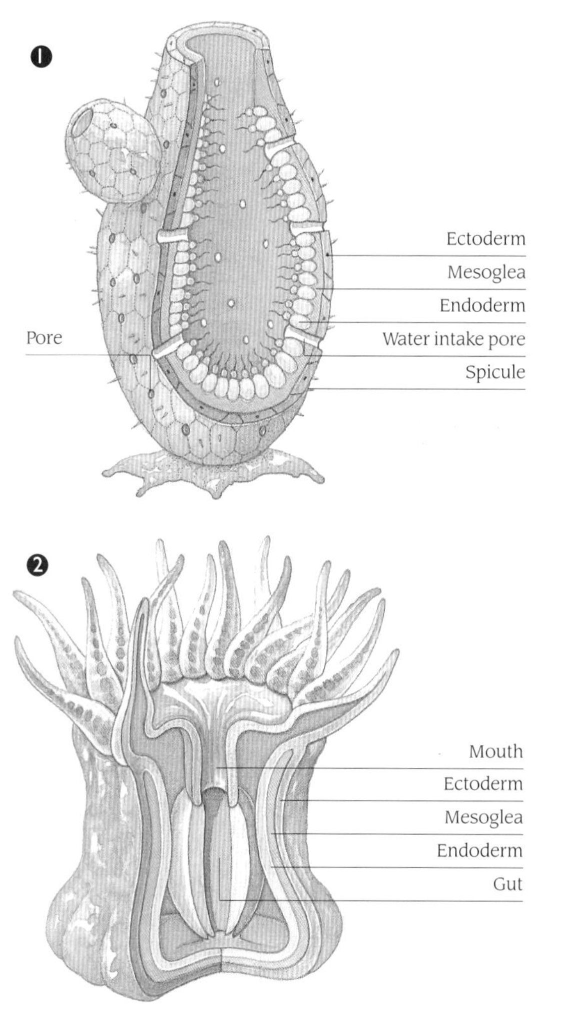

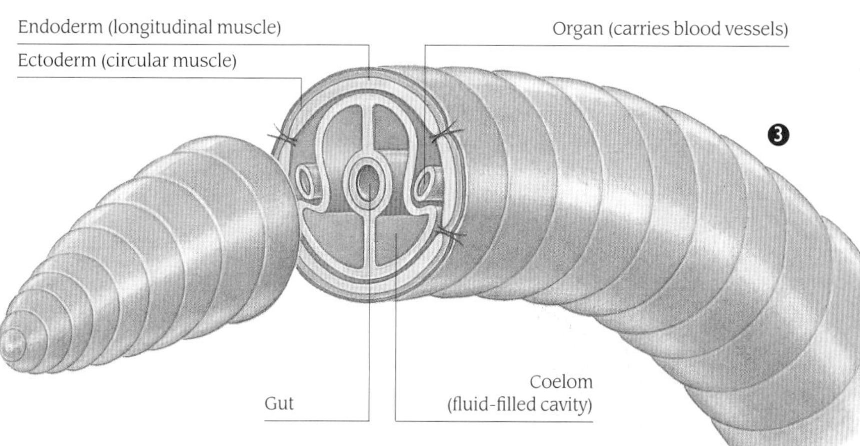

INVERTEBRATE BODY PLANS

Three types of invertebrate organisms (without backbones) represent the increasingly complex body plans that evolved towards the end of the Proterozoic. (1) A sponge belongs to the phylum Porifera, which means "pore bearer." Two layers of cells (endoderm and ectoderm), separated by a gelatinous matrix (mesoglea), surround an inner cavity. The needle-like spicules provide structural support. Seawater enters the cavity through pores, carrying microscopic food particles. (2) A sea anemone—phylum Anthozoa, "flower animal"—has a similar body plan with a proper mouth surrounded by stinging tentacles. It also has a primitive digestive system. In annelid worms (3), phylum Annelida, the segmented body has two outer layers with a fluid-filled body cavity (coelom) containing a central gut. It also has a primitive circulatory system. There are more than 10,000 modern species of annelids.

THE APPEARANCE of the first thallophyte is remarkable in itself, but the true story of the transition to more complex life forms lies in the evolution that took place over a period of several billion years. The initial phase of single-celled organisms may have spanned a period of 500 million years, beginning with the first filamentous fossils of cyanobacteria about 3.4 billion years ago. Within the filaments, the individual cells were effectively identical, each performing a similar function. Their role in the generation of new life forms may have been restricted; the fusion of two prokaryotic cells to form a simple eukaryote was probably much more important. Eukaryotes can generate energy from food through the process of respiration. The earliest forms may have looked like an amoeba with a chloroplast and simple "nucleus" formed by the smaller cells it had in effect devoured.

> ***Evolution was a long slow process until the development of sexual reproduction, about 1.8 billion years ago.***

Within a relatively short period of time, eukaryotes may have given rise to multicellular organisms. Thallophytes such as *Tawuia* would have evolved from a plant-like evolutionary line, with various cells performing specific functions and having a more complex architecture. By this time, about 1.8 billion years ago, the driving force behind the evolution of higher organisms was sexual reproduction. The increase in the gene pool and the greater diversification of life forms provided a perfect launch pad for an explosive rate of evolution during the Late Proterozoic.

According to cell morphologists, sponges and cnidarians have 11 types of specialized cells; in worms and other higher organisms there are 55 types. Complex sponges belonging to the hexactinellid group were present in the Ediacara fauna of Australia. They are represented by *Paleophragmodictya*, which has a supporting skeleton of interlocking spicules, the web-like "skeleton" closely resembling that of modern species. Given this fact, it could be inferred that the pace of evolutionary change took off around 1.4 billion years ago, with at least 11 cell types then present in some animals. Worm-like organisms are known from the Ediacara and older formations. Their presence implies an even more dramatic influence on the process of evolution.

Soft-bodied animals decay easily, and rapid burial in an oxygen-free medium (sediment) is essential for their fossilization. According to the "snowball" theory of Proterozoic climate changes, events related to the freezing of the Earth may have triggered an explosion in metazoan life forms. If the planet did experience such extreme fluctuations, however, the icing and thawing of the oceans would have resulted in corresponding fluctuations in levels of carbon dioxide and in oxygen, causing great stress on the emergent communities. Oases of relatively warm seas around the equatorial landmasses may have provided protected niches in which the the new life forms were able to survive.

PROTEROZOIC fossil communities are rare: only 20 have been found, all dating to about 650 million years ago. The Ediacara fauna of Australia, with a radiometric dating of 640 million years before the present, is the oldest of these. It is exceptional for the beautiful state of preservation of the fossils, and for the rich diversity of the animals and plants represented. Almost identical communities have been discovered in the Mackenzie Mountains of Canada's Northwest Territories, Siberia and in Europe.

Australia's Ediacara fossil community is a rare and beautiful snapshot of life in the Late Proterozoic.

Over 1400 samples of soft-bodied creatures have been collected from outcrops in the Ediacara Hills of South Australia. The fossils are essentially casts and molds, but the definition is very detailed. They are preserved in a fine- to medium-grained reddish-pink sandstone, cross-bedded and ripple-marked—a shallow water sediment, possibly a beach sand. Thin drapes of claystones may indicate seafloor muds or muddy tidal conditions. Lack of predators and a rapid burial allowed the preservation.

Traditional interpretation relates the Ediacara organisms to jellyfish, soft corals, worms, and possibly gastropods. Jellyfish were particularly abundant, with 15 species, and large forms nearly 4.75in (125mm) in diameter. Soft corals represented by *Charniodiscus,* like living pennatulaceans, were seafloor dwellers that fed on microscopic organisms present. Annelid worms and gastropods were free-moving; the annelids *Spriggina* had a well-defined head and 40 body segments.

The paleontologist Adolf Seilacher has described the fossils as flattened and quiltlike, with a hydraulic skeleton and, controversially, proposed that they are unique in evolution and not related to the familiar metazoans at all. He suggested placing the Ediacaran animals in a separate taxonomic category, the Vendobionta. Most paleontologists, however, prefer to see at least some of these creatures as ancestral to modern marine invertebrates.

1 *Spriggina*
2 Sea anemone-type organism
3 Trace of sea anemone
4 *Charniodiscus*
5 Jellyfish
6 Empty trace after *Ernietta*
7 *Dicksonia*
8 Sabellid worm-like organism

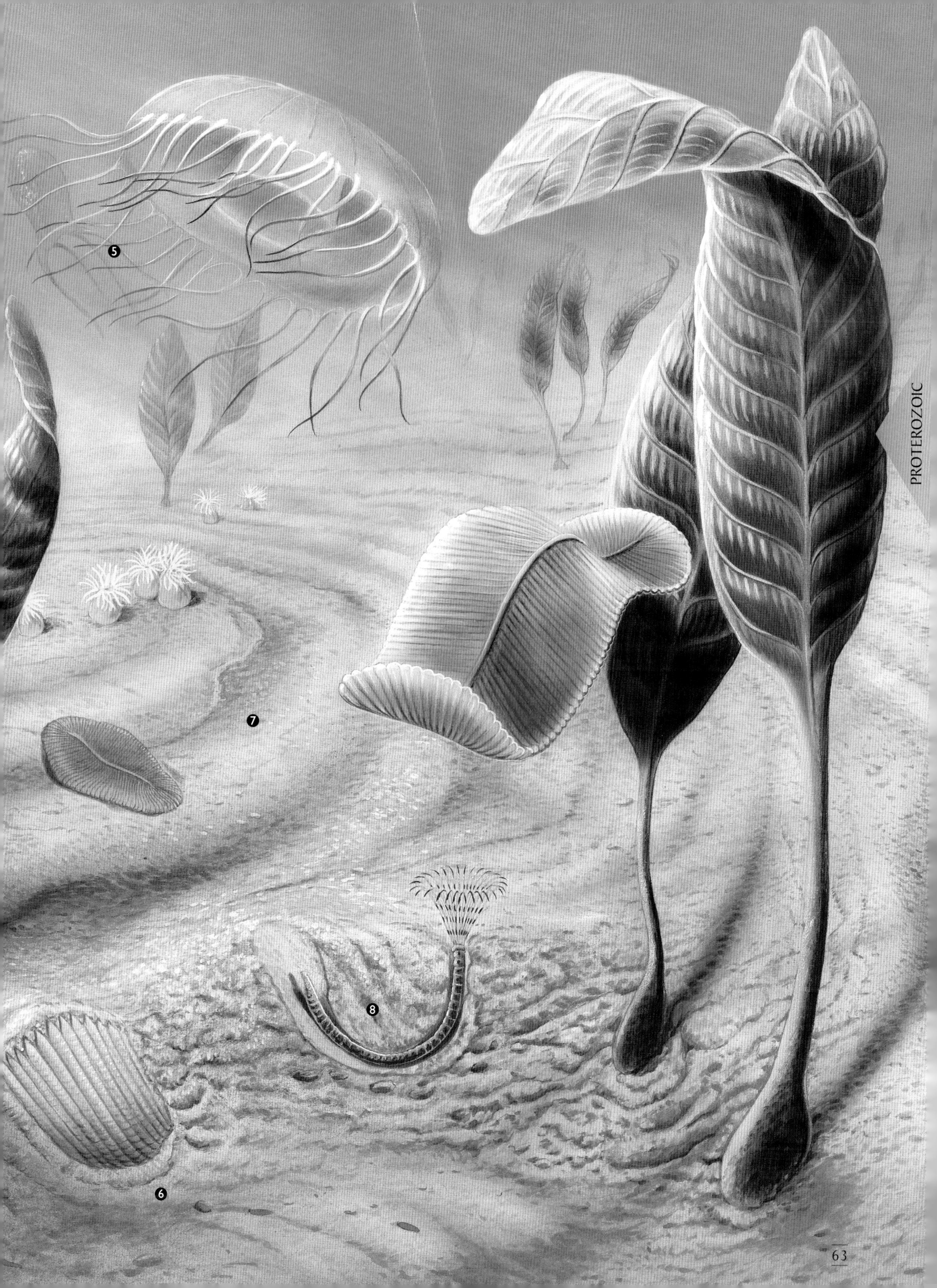
5
7
8
6

THE EVOLUTION OF EARLY INVERTEBRATES

THE EVOLUTION of multicellular organisms is difficult to plot with any certainty. The further back in time one looks, the fewer are the fossil remains of any given organism. This is particularly true of the first multicellular organisms because they were probably very small and lacked any hard parts.

Bacteria and cyanobacteria were among the first organisms to occur in the geological record. It is likely that algae and microscopic organisms belonging to the phylum Protoctista (amoeba-like animals and ciliated algae) coexisted with the early bacteria, but their presence remains undocumented. Eventually, the fusion of two or more cells gave rise to the eukaryote cell, while sexual reproduction and an expansion of the gene pool provided opportunities for adaptation and diversification. The development of an outer membrane to hold the combined material together was essential for cells to fuse.

With the arrival of plants and animals, even simple ones, organisms could be separated into producers and consumers, predators and prey. External stimuli and genetic modification ensured the survival of the fittest, with mutations giving rise to new forms of life.

Sponges are the simplest of all multicellular creatures; they lack the layers of cells or tissues that are characteristic of other invertebrates. The basic building blocks of a sponge were in place with the appearance of the first amoeboid and ciliated cells. All that was needed was a stimulus to bring them together. Cells with whip-like flagella drew water into the sac-like body, and microscopic particles in the water were trapped by individual cells. Once the basic multicelled organism (animal or plant) had appeared, the possibilities for the evolution of more and more complex organisms were almost endless.

Corals, sea anemones, and jellyfish mark the first stage towards the development of specialized tissues, although organisms included in the primitive Vendobionta and Ediacara communities of the Proterozoic may represent intermediate stages. The problem with these communities is that scientists are divided in their opinions of their significance and, in some cases, even as to whether the specimens are actual animals or simply traces of animals.

Tissues in corals and their relatives form inner and outer layers of the body. These animals each have a mouth and a central cavity in which food is digested. The first corals and jellyfish are found in rocks of the Neoproterozoic. It is interesting that they occur at the same time and in the same place as more complicated creatures such as worms and mollusks—animals each with a defined body cavity, or coelom, and a more advanced nervous system. This degree of complexity strongly suggests that the ancestry of the various groups stretches back into the Mesoproterozoic and beyond.

FROM ONE CELL TO MANY

The first living organisms had only one cell as their complete body structure. This had disadvantages, not least of which was that reproduction had to be asexual—by simple division or budding with no exchange of genetic material. The development of eukaryotes and multicellular organisms provided the opportunity for sexual reproduction and the exchange of genetic material between parents. The resulting possibility of mutations greatly accelerated the process of evolution.

1 Archaebacteria, ancestors of all living organisms
2 Photosynthetic bacteria harnessed the energy of sunlight
3 *Grypania,* the oldest multicelled fossil
4 *Tawuia,* thallophyte alga, undifferentiated in stem, root or leaves
5 Parazoans have a two-layer multicellular organization but lack symmetry, differentiated tissues, and organs
6 The extinct vendobionts had radial or bilateral symmetry; they may be the first true metazoans
7 *Cloudina,* the earliest known creature with a mineralized skeleton
8 Alternative patterns of embryonic cell division divide the later metazoans into two groups

FIRST LIFE FORMS

The first forms of life for which there is good fossil evidence are stromatolites. These rock-like mounds were formed by cyanobacteria, which were able to use photosynthesis to build body tissues. They also accreted carbonates, which slowly formed the mound around their colony on the edges of shallow seas.

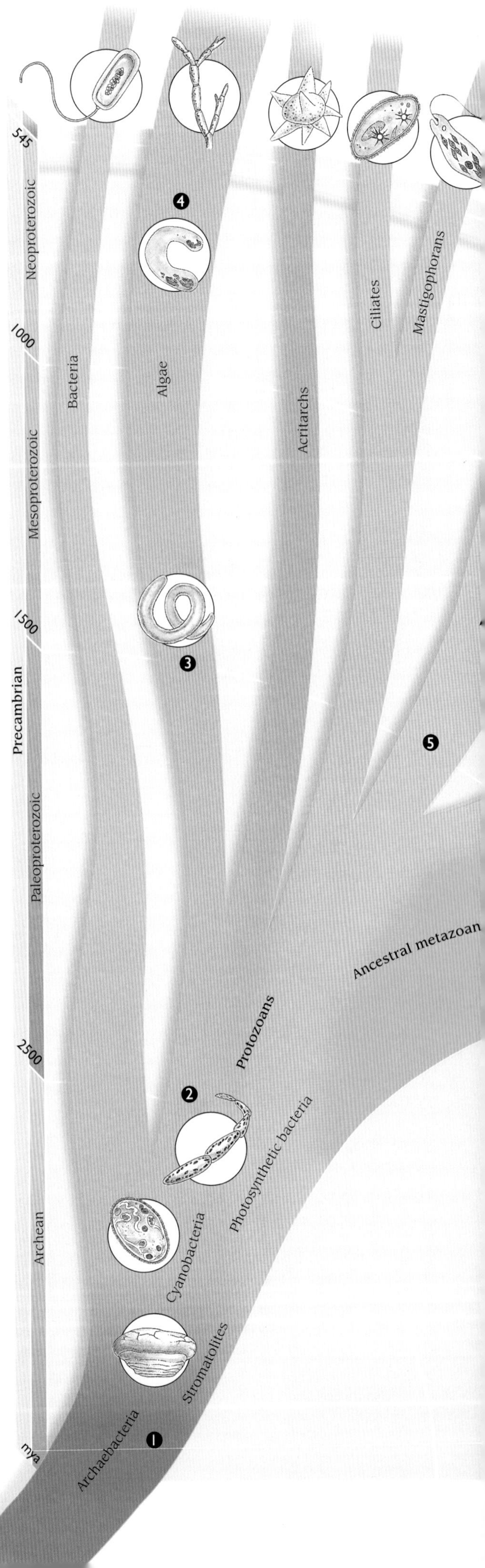

545 Neoproterozoic 1000 Mesoproterozoic 1500 Precambrian Paleoproterozoic 2500 Archean mya

Bacteria, Algae, Acritarchs, Ciliates, Mastigophorans, Ancestral metazoan, Protozoans, Photosynthetic bacteria, Cyanobacteria, Stromatolites, Archaebacteria

THE PARAZOANS

This group is possibly intermediate between the unicellular protozoans and the complex metazoans. The extinct, vase-shaped archaeocyathans had a two-layered body wall perforated by many pores, allowing water and nutrients to circulate. The Porifera, or sponges, have a similar structure. With over 10,000 species, they are still much in evidence in shallow tropical seas. The two phyla diverged more than 1 billion years ago.

THE SIMPLEST METAZOANS

Today's marine cnidarians consist of the anthozoans (corals, hydra, and sea anemones) and the scyphozoans (jellyfish). Some hydra are found in fresh water.

TENTACLE TRAP

A sea anemone, nestling among sponges, waves its purple-tipped tentacles to lure fish into its grasp. A billion years ago there were no fish, and their prey were smaller free-moving creatures.

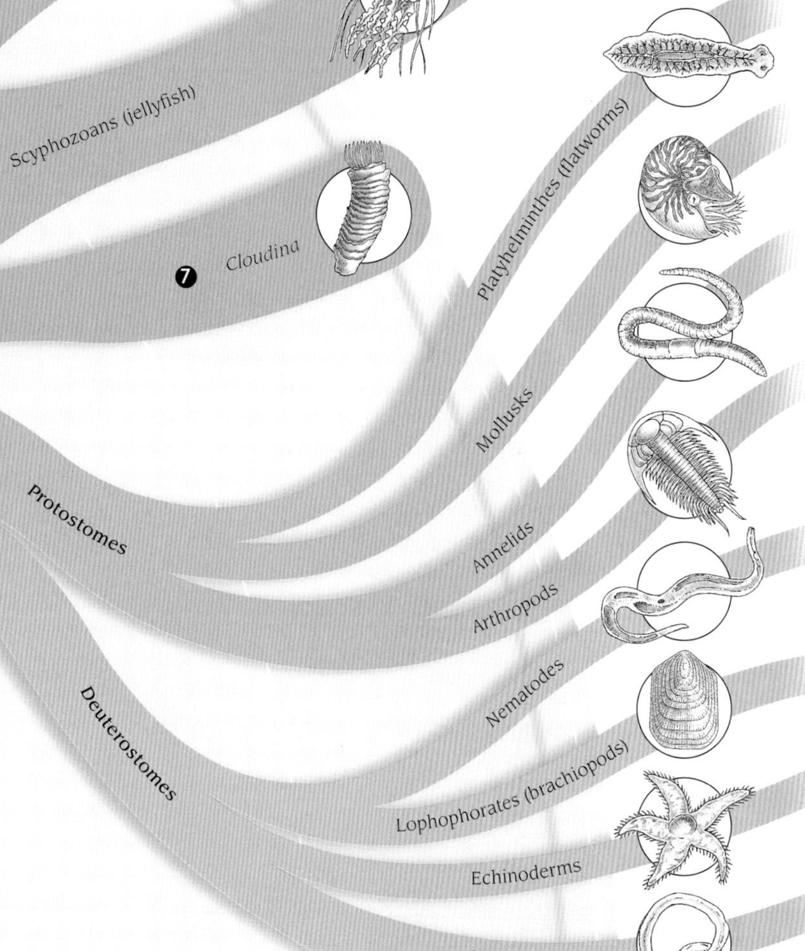

COMPLEX BODY PLANS

As the Proterozoic progressed, animal bodies became increasingly complex, though of relatively few basic types. Body plans tend to be defined in terms of number and type of tissue walls, presence and form of a coelom or body cavity, and patterns of embryonic cell cleavage. There are two main groupings: mollusk-annelid-arthropod and echinoderm-hemichordate-chordate. The evolution of skeletons (external in the case of most invertebrates) was a major advance in protecting and supporting body tissues. The first skeletal fossil, *Cloudina*, appeared during the Neoproterozoic. Its tube-like, calcium carbonate structure probably housed a soft-bodied, polyp-like animal.

VERTEBRATE ANCESTORS

All chordates possess a notochord, a flexible, stiffening rod running the length of the body. The internal skeleton, which made possible the success of the vertebrates, evolved from this.

THE EARLY PALEOZOIC

The explosion of life

PART 2

545–417 MILLION YEARS AGO

PART

2

THE EARLY PALEOZOIC

CAMBRIAN STRATA, with their numerous fossils, are strikingly different from the almost empty underlying beds of the Archean and Proterozoic. For this reason, the entire Precambrian period has been given the name Cryptozoic ("hidden life"), whereas the Cambrian and following periods are assigned to the Phanerozoic ("visible life").

Charles Darwin was one of the first to be struck by this marked difference in the distribution of fossils. In his unsurpassed *The Origin of Species* he wrote: "It is indisputable that before the lowest Cambrian stratum was deposited long periods elapsed, as long as, or probably far longer than, the whole interval from the Cambrian age to the present day, and that during these vast periods the world swarmed with living creatures." This scientific genius could not resign himself to the idea of a sudden appearance of living creatures in all their diversity. According to him, a species very slowly changes into another species by a gradual addition of negligibly small new features. To keep his theory intact, he introduced a hypothetical interval during which life slowly evolved towards its Cambrian composition. We now know that this was definitely not what happened. The sudden proliferation of life during the Cambrian may be fairly described as an "explosion."

With the proliferation of skeletal animals, algae, and cyanobacteria at the beginning of the Phanerozoic, some of the widely distributed rocks, such as carbonates and phosphates, became mostly organic in origin. Even hard rocks, such as lithified carbonates and magmatic basalts, were now bored into by animals, as well as by some algae and fungi.

Living organisms had already become established as the main regulators of the gases in the atmosphere during the Precambrian. During the Phanerozoic, terrestrial plants joined the ranks of the oxygen producers and, together with single-celled planktonic algae, modified the carbon dioxide content of the atmosphere. The development of trees, with their transpiration of water vapor, extended humid conditions over the land areas and made the global climate more even. In turn, rivers, whose banks were stabilized by the growth of vegetation, became more permanent, and the composition of their waters changed. Tree roots penetrated deeper and deeper into the bedrock and markedly affected the rate of erosion by releasing chemical agents in the mountain areas. On the other hand, the accumulation of soil prevented the rapid weathering of bedrock in the lowlands. With the development of active filter-feeders in the oceans, their waters became cleaned faster and faster. Recent marine fauna filter the entire volume of the oceans in six months, and clean its most habitable portion (up to a depth of 1500ft/500m) in as little as 20 days.

THE PHANEROZOIC began with the Paleozoic Era (545 to 248 Mya), which is subdivided in ascending order into the Cambrian, Ordovician, Silurian, Devonian, Carboniferous, and Permian periods. The Cambrian, Ordovician, and Silurian periods represent the Early Paleozoic (545 to 417 Mya). During that interval, life developed mostly in the seas; elaborate terrestrial communities did not appear until the end of the Silurian. Nevertheless, two or three major faunal events occurred then. The first was the Cambrian Explosion, when skeletal animals and calcified cyanobacteria invaded the shallow seas. Unicellular foraminiferans, calcified sponges, mollusks, brachiopods, arthropods, various worms, echinoderms, chordates, and various long-vanished groups of animals appeared and diversified rapidly. The second event was the Great Ordovician Radiation, when the diversity of marine animals tripled. Although no major animal groups except the bryozoans evolved in the Ordovician, a significant replacement occurred within all the groups. Cambrian sponges, mollusks, arthropods, and so on either died out or were displaced into other environments. The makeup of the entire Paleozoic marine biota was intact until the end of the era.

The Cambrian globe consisted of a vast ocean in the north and an aggregate continent in the south, which later split, coalesced, and split again to form the modern continents.

At the beginning of the period, the entire area of the continents was much less than nowadays. Since then, continents have grown up to 15 percent or 6 million sq miles (16 million km^2) because of the enlargement of coastal platforms by the accretion of volcanic arcs and neighboring oceanic basins. A single supercontinent is believed to have existed in the late Precambrian. It was similar in some respects to the Late Paleozoic supercontinent of Pangea, whose existence is beyond dispute. This early supercontinent, called Rodinia, was split by deep rifts and broken into several separate blocks, which started to drift apart. The principal blocks were Laurentia (which eventually became North America), Baltica (which included territories surrounding the present Baltic Sea), Siberia (now part of northern Asia belonging to Russia), Avalonia, and Gondwana. The latter contained the elements that now make up the modern landmasses of South America, Africa, Madagascar, Arabia, India, Antarctica, Australia, and New Zealand. In addition, a number of fragments such as Iberia, southern France, southern Germany, the Middle East, Tibet, Kazakhstan, Tarim, North China, and South China comprised or abutted the periphery of Gondwana. Avalonia was the most peculiar continent. It included eastern Newfoundland, Nova Scotia, Wales, England and some fragments of continental Europe, namely small pieces of northern France, Belgium, and northern Germany. Today Avalonia is completely fragmented and its parts, widely separated by the Atlantic Ocean, have become incorporated into North America and Europe.

The huge Panthalassa Ocean occupied the northern hemisphere, and most of the continents lay in the southern hemisphere. Laurentia and Baltica were divided by the Iapetus Ocean when Rodinia broke apart, and its offshoot, the Tornquist Sea, jutted between Baltica and Avalonia. The Rheic Ocean spread with the drifting of Avalonia away from Gondwana. The Paleoasian Ocean (a southern projection of Panthalassa) washed the Siberian coastline on the west and the Chinese margin of Gondwana on the east.

Andrey Yu. Zhuravlev

THE CAMBRIAN

545 – 490

MILLION YEARS AGO

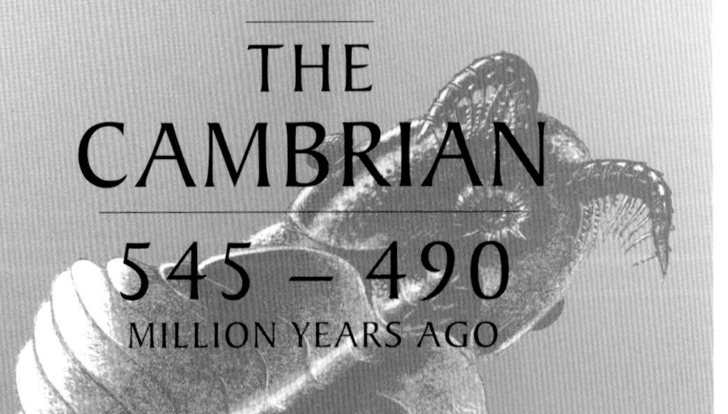

In Wales, whose native name is Cymru (pronounced Kum-ree), one may book a room in the Cambrian Hotel and buy a morning newspaper called the Cambrian News. *It would be interesting to look at the headlines of the* Cambrian News *of about 550 million years ago. They might read:* Is There Any Life on Land? Do Chordates Have a Future? Giant Predator Terrorizes Cambrian Seas. New Fashions in Ready-to-Wear Skeletons. *The last headline was probably the most important, as skeletons marked the turning point known as the "Cambrian explosion."*

Other significant events also occurred in the early Cambrian. Reef-building calcified bacteria were abundant, and animals' lives became more diverse and complicated. Their daily behavior is recorded in trace fossils of burrows, tracks, and trails fixed in sediment. Burrows and holes, like a skeleton, offered protection against predators, which rapidly assumed a significant role in newly diverse ecosystems.

THERE are a number of successions in which Precambrian strata merge smoothly into Cambrian ones. The official Precambrian-Cambrian boundary, which is in Newfoundland, was chosen among these sections. It has a radiometric age of approximately 545 to 550 million years. This part of Newfoundland is a fragment of an ancient continent, Avalonia, which no longer exists. Another fragment of Avalonia is Wales, where British geologists Adam Sedgwick and Roderick Murchison officially defined the Cambrian strata in 1835. They were named for the Cambrian Mountains in northern Wales.

Cambrian strata are distinguished by newly abundant skeletal fossils, but why these should have appeared so suddenly is still contentious.

KEYWORDS

ANOMALOCARIDID — ARCHAEOCYATH — ARTHROPOD — BURGESS SHALE — CAMBRIAN EXPLOSION — GONDWANA — IAPETUS OCEAN — MINERALIZED SKELETON — PELLET CONVEYOR — REEF

At that time, some researchers treated these strata as the lowermost part of the Silurian. The problem was compounded by the nature of Welsh Cambrian rocks, which contain few fossils. It was the French paleontologist Joachim Barrande who described more than 3500 species from the lower Paleozoic of Bohemia and made the Cambrian interval recognizable through primordial fossils, mostly trilobites. By the second half of the 19th century the Cambrian system was widely accepted.

By the middle Early Cambrian all the principal phyla of multicellular animals were already in existence: sponges, cnidarians, ctenophores (comb jellies), cephalorhynchs and annelid worms, arthropods, velvet worm relatives, mollusks, brachiopods, echinoderms, and even chordates. In addition, a number of problematic fossils were unique to the Cambrian (mostly the early part), such as anabaritids, tommotiids, and radiocyaths.

PROTEROZOIC	545 mya	540	535	530	525	CAMBRIAN 520
Series	EARLY/LOWER					
European series	CAERFAI					
Russian-Kazakhian stages	NEMAKIT-DALDYNIAN		TOMMOTIAN	ATDABANIAN	BOTOMIAN	
N. American stages				MONTEZUMAN		DYERAN
Geological events	Rodinia continues to fragment; Pan-African orogeny and collision of East and West Gondwana		Rifting of N. and S. China, Khazakhstan, Mongolia, and other terranes from East Gondwan		Opening of Iapetus Ocean	
Climate / Atmosphere	Increasingly warm; most continents are in equatorial regions with warm, dry conditions; high atmospheric carbon dioxide					
Sea level and chemistry				Rising		ARAGONITE SEAS
Reefs			Archaeocyathan-cyanobacterial reefs			
Animals	• First "shelled" animals appear		Cambrian "explosion" of life		First mass extinction	• Chengjiang community establishe

But why and how did skeletons appear? One theory suggested that early Cambrian life moved into shallow seas, where skeletons evolved as protection from ultraviolet radiation or battering by violent storms. We now know that the first animals appeared on the continental shelves and only spread into the ocean depths and the open seas in the Ordovician, about 490 million years ago.

The appearance of skeletons coincided with the accumulation of phosphorites and the replacement of magnesium carbonates (dolomites) by calcium carbonates (limestones). This led yet other researchers to suggest that marine geochemical changes triggered skeletal mineralization, and the skeleton was perhaps a storage organ or dump for excess phosphate and calcium. However, the mineralization of carbonates and silicates require totally different conditions and could not occur together.

Reefs, which are highly diverse ecosystems, owe their existence to predators; these animals prevent grazers, borers, and competitors from destroying any chain in the complex trophic web, thus maintaining diversity. At first there were no fossils to support theories about Cambrian predators, but the list now includes giant anomalocaridids, cephalorhynchs, arthropods, and other problematic animals.

Separately none of these events could have powerfully influenced the Earth in the Early Cambrian, but tectonic events, geochemical changes, and the appearance of predators all taken together were enough to change the planet forever.

THE LENA COLUMNS

The spectacular Lena Columns (left) spread along 125mi (200km) of the Lena River in southern Sakha–Yakutia (today's Russia). These carbonate structures are one of the major sources of the earliest Cambrian fossils, reefs, and other evidence of events which took place near the boundary between the Precambrian and Cambrian.

THE CAMBRIAN REVOLUTION

The Cambrian period was the first of the Paleozoic era and the entire Phanerozoic eon. It was a time of incomparable biological, ecological, geological, and tectonic changes.

All these species have a common feature: a skeleton, in the guise of siliceous spicules of glass sponges, carbonate mollusks' shells, phosphate brachiopods, and the organic cuticles of annelids and carapaces of some arthropods. Their evolution, with its attendant benefits, may have kick-started Cambrian diversification. Skeletons improve locomotion, as in modern arthropods and vertebrates. They serve as a pedestal for sedentary organisms to filter food from water currents (as do sponges, corals, and bryozoans); they help to grasp and chop up food (claws and teeth) and to breathe (echinoderm plates, insect tracheae). Skeletons are also an effective defense: in an emergency, an animal can simply run away.

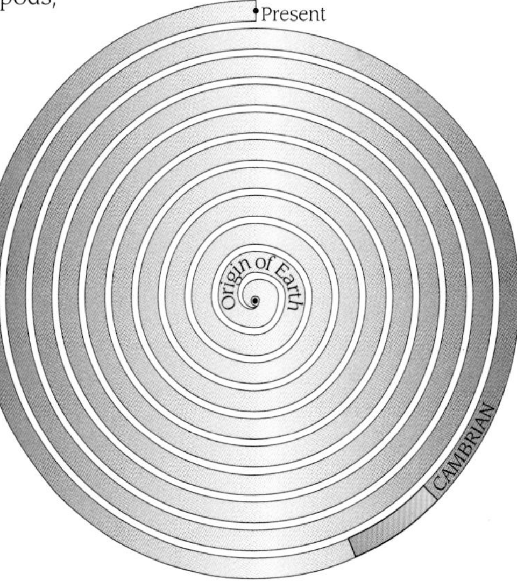

See Also

THE ARCHEAN: *Continent, Crust, Lithosphere, Mantle, Plate Tectonics*
THE ORIGIN AND NATURE OF LIFE: *Cyanobacteria, Eukaryote, Prokaryote*
THE PROTEROZOIC: *Algae, Stromatolites*
THE SILURIAN: *Hydrothermal vents*

CAMBRIAN

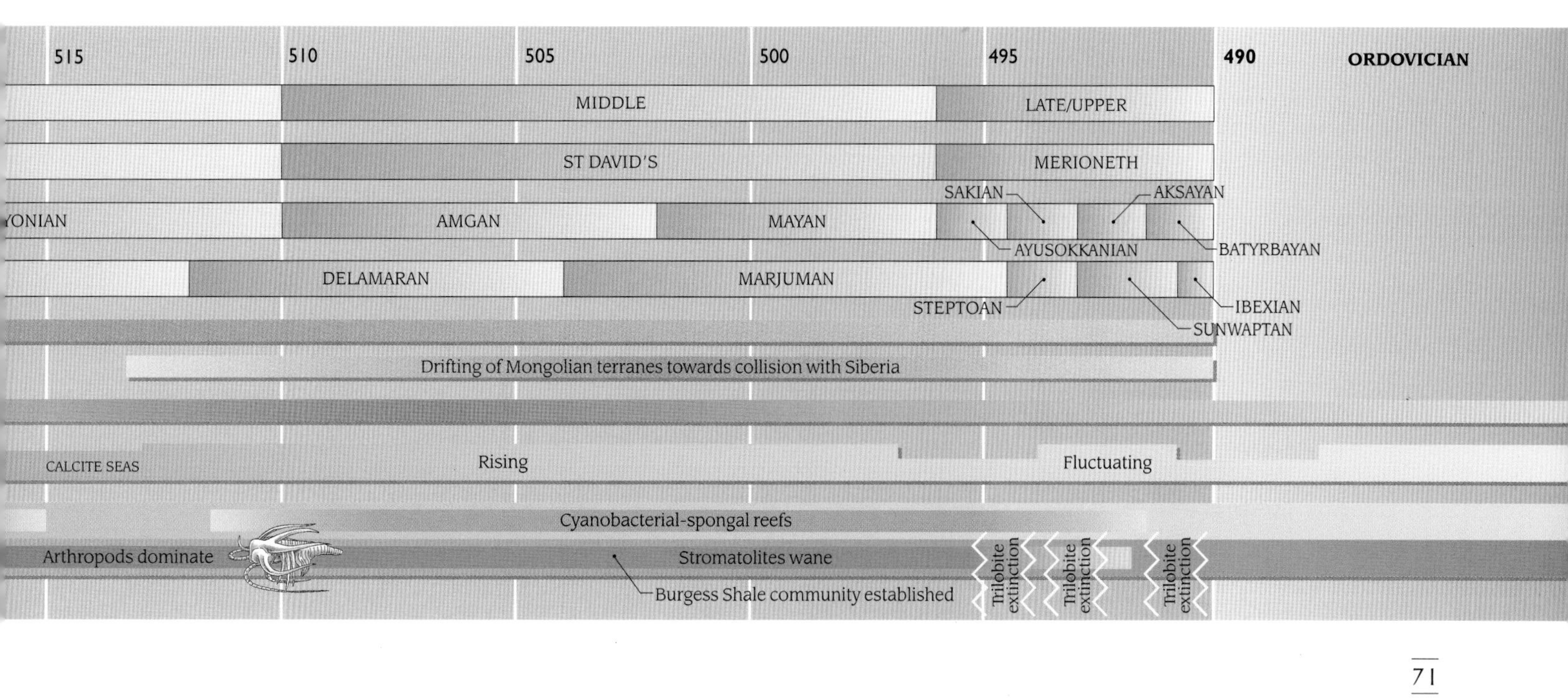

MAPPING ANCIENT CONTINENTS

With the discovery of the magnetic properties of rocks, geologists acquired a very good tool for determining the paleopositions of the continents. The Earth's magnetosphere resembles the field of a huge magnet, with lines of force running from the North Pole to the South Pole. This is why a compass needle allows us to find directions (at least in geographic coordinates). Every iron-containing mineral particle of an elongated shape itself acts as a magnetic needle. Such a particle, falling on the sea bottom and being subsequently "frozen" in the rock, aligns itself with the Earth's magnetic field as it exists at that time. Studying the rocks containing such particles from different continents, scientists can find their original positions (their distances from a pole). The basic assumption for making global reconstructions from paleomagnetic data is that the continents can be treated as rigid bodies and rotated accordingly. Fortunately for most people and unfortunately for scientists, the Earth's magnetic field is axially symmetrical: the scientists have to work out which hemisphere a continent was in, not to mention the longitude. Again, the older the rocks, the less precise is the paleomagnetic data, because rocks undergo metamorphic or magmatic heating, which alters the particle direction and may destroy the rock's magnetic "memory."

CONTINENTAL fragmentation and reassembly at the start of the Cambrian might also have contributed to evolution. Numerous continental fragments separated by shallow seas created barriers to migration, isolating existing species. After a supercontinent breakup, distinctive faunas developed on each new continent. As the sea level rose, shallow epicontinental seas, with their wide variety of substrates, temperature, salinity, and depths, provided new niches. The redistribution of continents across different climatic zones would also partly account for the high number of species, as at present. This hypothesis has some support from the pattern of Cambrian tectonics.

The continents acquired their present outlines less than 200 million years ago, and their ancient outlines are not certain.

Understanding of Cambrian events is limited by inadequate knowledge of the geography and climate. The most ancient data are the least precise because everything has changed, even the hardest rocks. Also, many good indicators of climate, such as land vegetation, did not yet exist in the Cambrian.

Reconstructions of Precambrian and Cambrian continents resemble jigsaw puzzles with hundreds of pieces. It is helpful to compare sediments and faunal successions, because juxtaposed continents have a common history of rifting recorded in their sediments. So, in spite of the difficulties and differences in approaches to reconstruction, scientists agree on the features of the Cambrian globe.

Gondwana revolved around the South Pole and was elevated in the center, with short, broad rivers separated by low hills running to its coast. Australia still retains features of this landscape. The seas of South America and southern Africa faced the South Pole and were filled with cold waters almost without fauna but full of pure siliciclastic sediments. Marine basins more distant from the pole (southern Europe, the Middle East, and southern China) were characterized by more diverse sediments and faunas. Carbonate sedimentation and reef development took place in the seas flooding Australia and Antarctica, which approached the paleoequator.

The finds of similar Cambrian fossil assemblages from Australia in the east to Iberia in the west confirm that Gondwana was a single unit. Huge phosphorite deposits in southern China indicate an east-west orientation for Gondwana, with China on the west.

Laurentia, Baltica, and Siberia were three other large Cambrian continents. Baltica, with its shallow sea and scarce fauna, occupied southern temperate latitudes, while Siberia and Laurentia straddled the paleoequator. Siberia was opposite its current position, and Laurentia was 90° clockwise in relation to modern North America. Only a narrow carbonate belt with reefs extended along its margins. Rises in sea levels during the Middle and Late Cambrian enlarged the area of the seas surrounding Laurentia, and Siberia was almost entirely submerged, allowing wide shallow reefs with very rich fauna to flourish.

Some scientists think that Laurentia split from the Australian-East Antarctic edge of Gondwana in the Neoproterozoic. Others prefer to assign it to a large block that also included Siberia and Baltica, though this is not supported by the distribution of Cambrian fossils and sedimentary rocks.

Avalonia is the most enigmatic Cambrian continent. Some researchers believe that it was a part of Gondwana adjacent to Morocco, but Avalonia's fauna is impoverished and Morocco's is rich. In addition, Moroccan Cambrian strata are full of red carbonates, oolitic limestones, and evaporite traces, which indicate warm, non-polar conditions. It is more likely that Avalonia was an insular continent drifting somewhere at southern latitudes. Florida was even closer to the South Pole than Avalonia was.

EARTH LIKE MARS

The geography of the Early Paleozoic Earth resembled that of Mars of the same time, with its mainly continental southern hemisphere and mostly oceanic northern hemisphere.

NORTH CHINA

GONDWANA

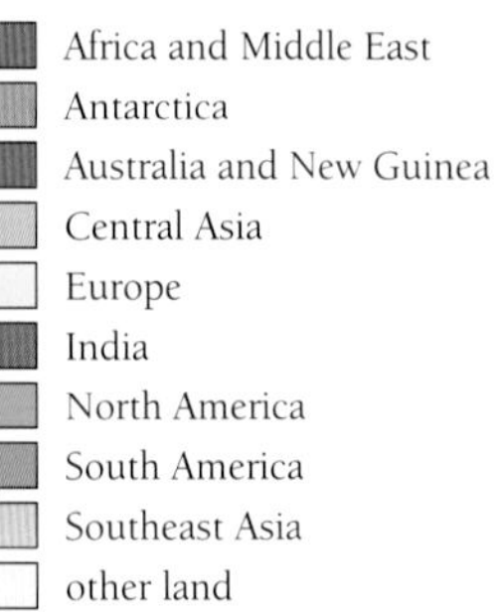

CAMBRIAN

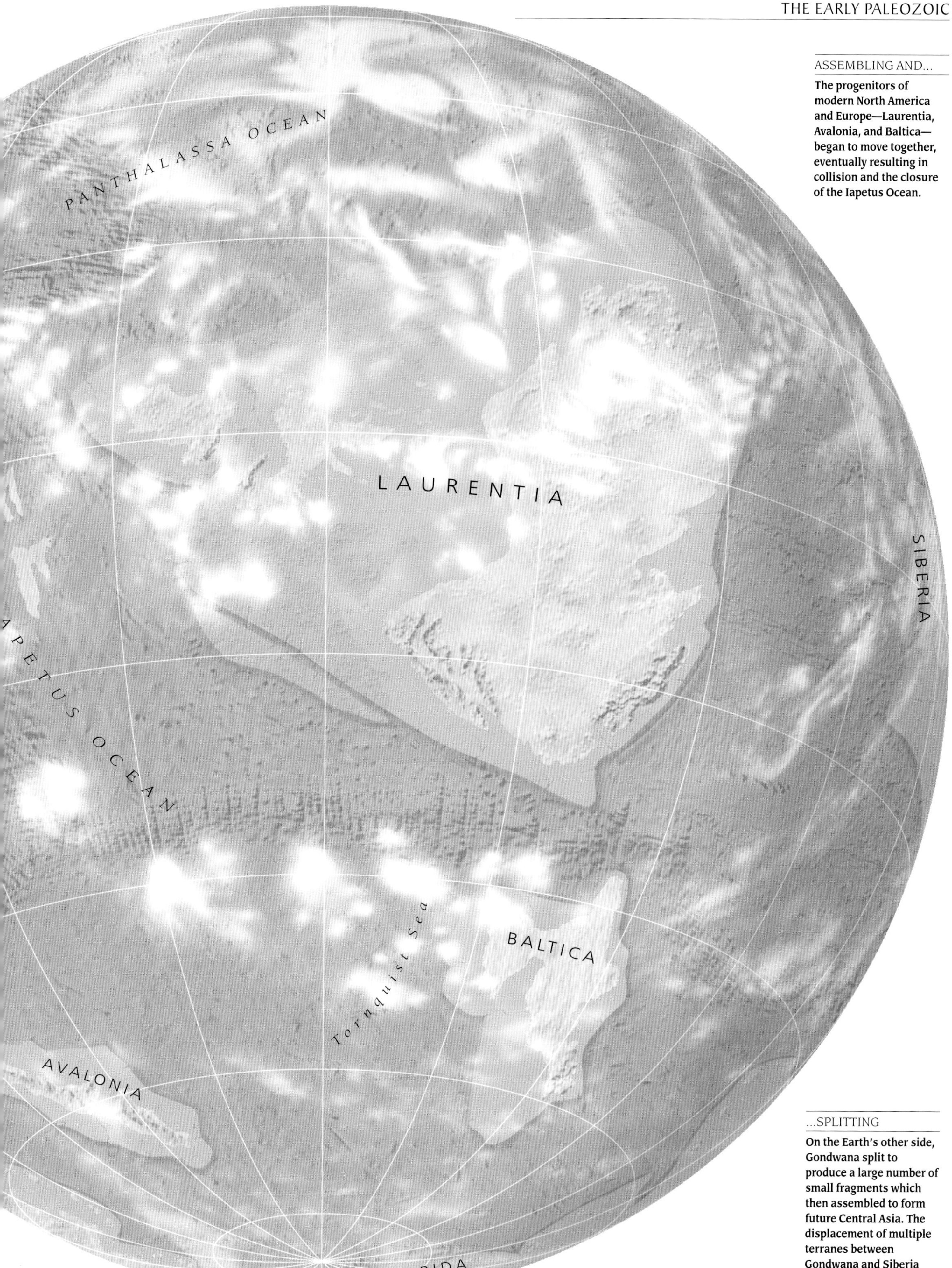

ASSEMBLING AND...

The progenitors of modern North America and Europe—Laurentia, Avalonia, and Baltica—began to move together, eventually resulting in collision and the closure of the Iapetus Ocean.

PANTHALASSA OCEAN
LAURENTIA
SIBERIA
APETUS OCEAN
Tornquist Sea
BALTICA
AVALONIA
FLORIDA

CAMBRIAN

...SPLITTING

On the Earth's other side, Gondwana split to produce a large number of small fragments which then assembled to form future Central Asia. The displacement of multiple terranes between Gondwana and Siberia brought about the closure of the Paleoasian Ocean.

IN ADDITION to the major Cambrian continents, there were a number of continental fragments or terranes. Nowadays a complex array of such terranes comprises the Central Asian Fold System that stretches from the Urals through Kazakhstan and Mongolia to the Sea of Okhotsk, and borders Siberia in the south. This fold system is a dense conglomeration of mountain ridges cut by fresh rivers and covered by high coniferous forests. The oldest ridges have been eroded to low smooth hills and are surrounded by steppes and rocky deserts with ephemeral lakes. Only an experienced geologist can recognize fragments of former ocean basins, island arcs, accretionary wedges, marginal seas, and seamounts. Cambrian rocks here are 26,000 to 40,000 feet (8000 to12,000m) thick.

Continental fragments split off from Gondwana and drifted westwards, eventually colliding with Siberia.

Basalts, gabbros, red hematite jaspers, and certain lavas, forming groups known as ophiolitic rock, are the remains of the late Precambrian–Early Cambrian ocean floor in the Altay Sayan fold belt in northern and central Mongolia. During that remote time, the Paleoasian Ocean was situated within the equatorial belt, between Siberia and the northwestern Gondwana margin composed of Tarim, North China, and South China. By the close of the Precambrian, the margin had been sutured by rifts and a number of terranes had broken from the supercontinent, including Mongolian and Kazakhstan fragments.

The Mongolian terranes started to drift towards Siberia. At the beginning of the Early Cambrian, they still abutted Gondwana, and their shallow epicontinental seas teemed with animals typical of South China. Middle Early Cambrian strata of the same terranes are rich in genuine Siberian fossils: by that time these floating microcontinents had approached Siberia and allowed Siberian animals to migrate into Mongolian seas.

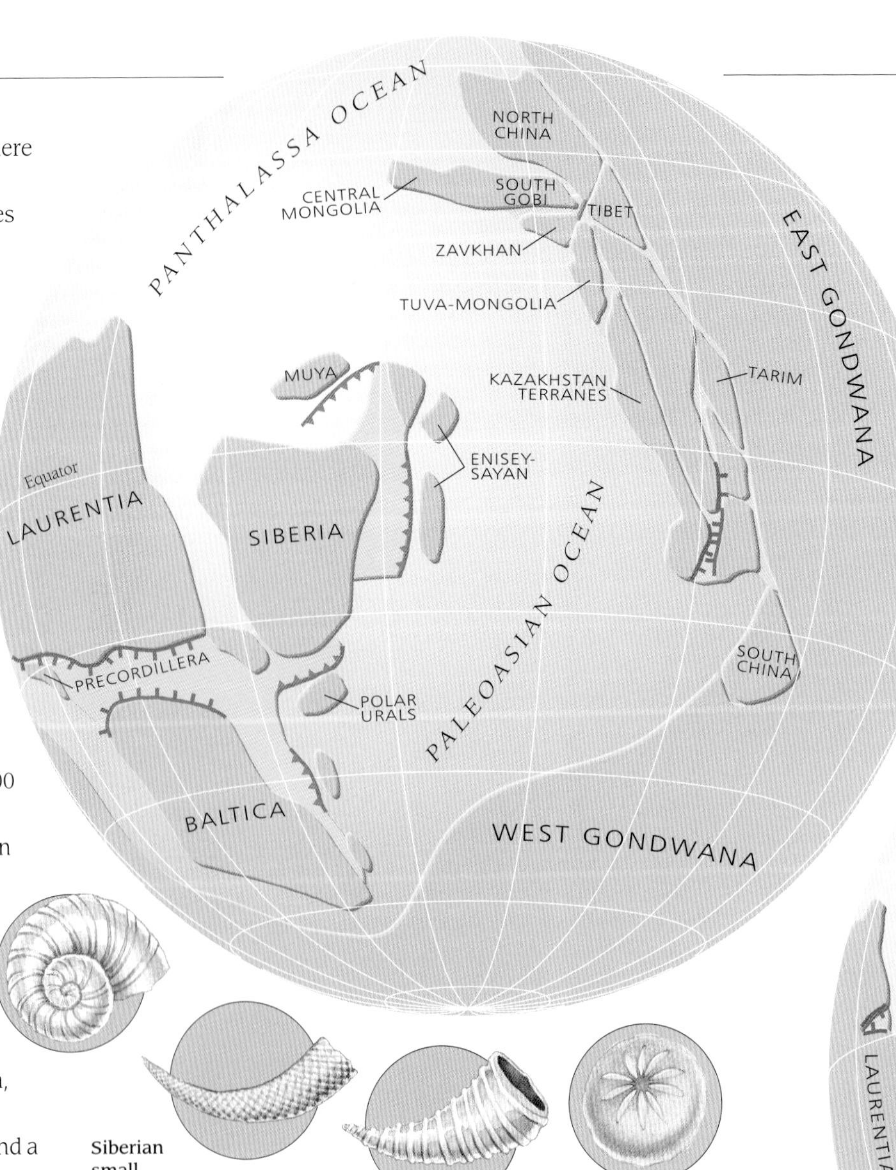

Siberian small shelly fossils

MONGOLIAN ALTITUDES

The Khasagt Khairkhan Ridge, standing more than 9000ft (3000m) above sea level in western Mongolia, consists in part of Early Cambrian archaeocyathan–cyanobacterial reef limestones.

During the late Early to Middle Cambrian, following the terranes' collision with Siberia, island arcs appeared behind them. Some of the islands had highly explosive volcanoes slightly below sea level. Archaeocyathan-cyanobacterial reefs developed on the tops of the volcanoes but were disrupted during subsequent eruptions. Abundant pyroclastic material formed the volcanic cones and moved down to deep-sea depressions.

Between the Middle and Late Cambrian, the collision between island arcs and terranes took place on the periphery of Siberia, changing the whole structure of the Paleoasian Ocean. The major part of the oceanic basin closed as oceanic crust overlapped, and a covering of thick subaerial siliciclastic material buried the marine sediments. This process ended in the Silurian period, when the terranes were finally incorporated into Siberia. As a result, the continent now known as Asia grew by 2 million sq mi (5.3 million sq km).

This is not the only scenario of events leading to the present appearance of Central Asia. An alternative tectonic model suggests that a single giant island arc rifted from a combined Baltica/Siberia hinterland in the Cambrian period. As these two continents converged, the island arc crumpled in the manner of a small car being crushed between two colliding trucks, and this collision

BIG, SMALL, BIG

Continental fragmentation is a recurring theme of Earth's evolution. The assembly and break-up of a single large supercontinent occurring about every 250 million years since the Late Proterozoic is one of the major cycles in a permanent, ongoing transfiguration of the Earth's surface. The supercontinent may split into several large continents or many smaller fragments.

formed the Central Asian Fold System. This suggestion, however, does not fit the testimony of the principal witnesses to these events: the fossils.

Any plate-tectonic event directly affects the diversity of plant and animal life. The diversity of species is regulated by environmental stability, food supply, and local conditions, among other factors. With the fragmentation of large continents and dispersion and isolation of small pieces of it, local conditions become increasingly important in dictating the pace and nature of change. Formerly shared information—the gene pool—becomes disrupted, leading to the appearance of new species in each isolated location; this is the process of speciation. Total species diversity thus increases. So it was that by the mid-Early Cambrian a number of highly localized faunas already existed in Laurentia, Baltica, Siberia, and Gondwana.

RISING sea levels inundated the Cambrian land areas, creating multiple and extensive shallow basins called epeiric seas. These were relatively isolated by deeper basins, and their faunas started to develop and, thus, to diversify independently. During the entire Phanerozoic, the intervals of high sea levels corresponded to times of the most diverse faunas. On the other hand, those were also the intervals of the most intense fragmentation. They had to be, because the two processes were linked by the mechanism of plate tectonics.

High seas flooded low-lying land, isolating local species. At such intervals biodiversity was at its peak.

Continental fragmentation gives birth to new oceans. When an ocean spreads, a mid-ocean ridge lengthens; a volume of the ocean water, which is equal to the volume of the mid-ocean ridge, is displaced onto the land. However, this process, called tectonoeustasy, is not the only cause of the rise in sea level.

An ice cap melting also results in the submergence of continents, or glacioeustasy. Both of these processes operated on the eve of the Cambrian, with tectonoeustasy

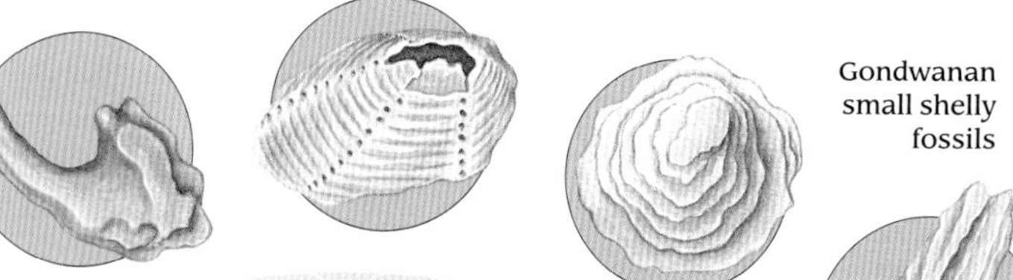

Gondwanan small shelly fossils

MIGRATING TERRANES

During its drifting period a terrane crosses a number of latitudes and migrates between different climatic belts. Organisms inhabiting its seas die out due to changes in the climate and nutrient supply. On the other hand, larvae of animals living in seas in a nearby continent can find the new seafloor appropriate for settlement, and thus organisms expand their geographic range.

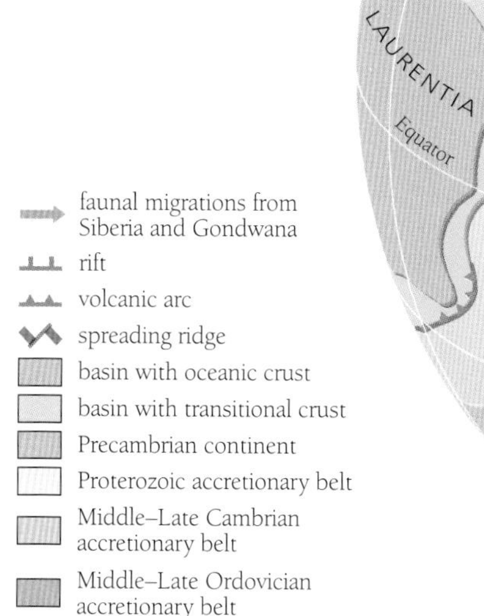

reaching its peak in the middle Early Cambrian, when all the continents, with the exception of the Gondwanan interior and minor island archipelagos, were covered by vast epeiric seas.

That was exactly the time of the maximum of the Cambrian faunal (and floral) diversity. The Cambrian Explosion reached its peak. There were more than 700 genera of animals and over 100 of algae by that time, mostly among groups that did not exist 10 to 15 million years earlier—an extremely short interval on a geological scale. Mammals diversified at a similar rate at the beginning of the Cenozoic era. But mammals are just one class of the phylum Chordata, whereas during the Cambrian Explosion almost all existing phyla radiated explosively.

ON REACHING its peak in the Early Cambrian, species diversification soon declined, rapidly enough to be considered the first global mass extinction event. A number of causes led to the extinction. Continuous rises in sea level brought oxygen-free deeper waters to the continents in which the Cambrian shallow-water fauna could not survive for long. The subsequent fall in sea level, on the other hand, almost dried out the epeiric seas, depriving the established fauna of their habitats. In addition, evolution led to the appearance of animals that were able to modify some of the properties of the water, with profound long-term consequences for the entire planet: the first pellet conveyors had appeared. They collected thin floating particles together and removed them from the water column to the bottom at a speed of over a quarter of a mile (400m) per day.

The Cambrian Explosion was followed by the first mass extinction on a global scale, also linked to changes in sea level. Meanwhile, the clarity of the water was about to change drastically.

Some decades ago, marine scientists decided to investigate ocean current movements by dispersing a brightly-colored plastic dust over the ocean surface. In a few days, they saw nothing but clear water. This was the work of microzooplankton, mostly some groups of tiny crustaceans. These free-swimming animals stuff their stomachs with everything of an appropriate size in the hope of being able to digest it. Some time later, their indigestible waste is bounded by a special membrane excreted at the end of the alimentary channel, forming a pellet that sinks rapidly. Crustaceans need this mechanism to prevent them from feeding on their own waste.

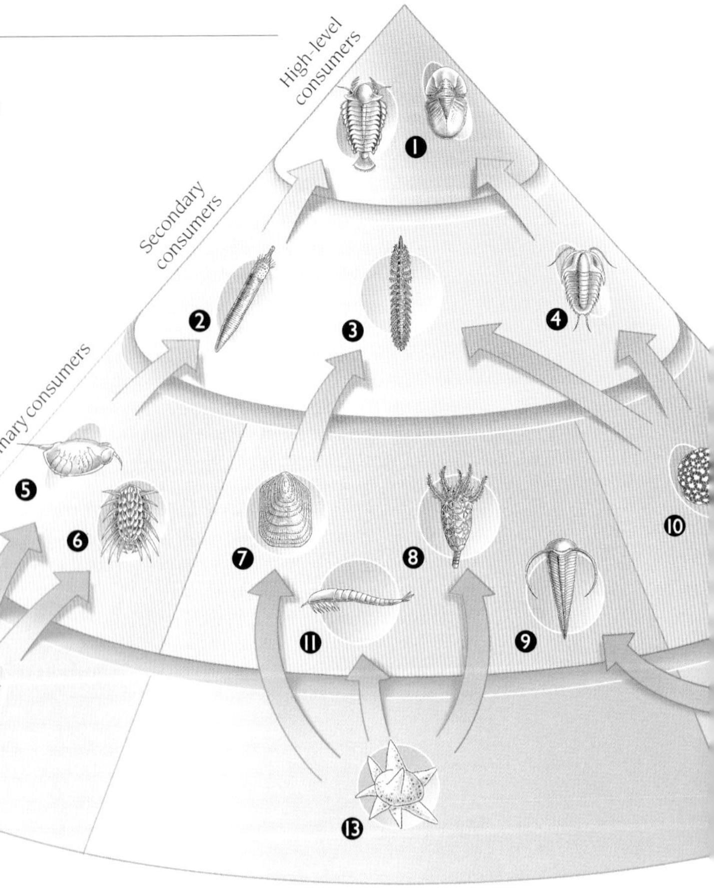

1 Large arthropods
2 Cephalorynchs
3 Tardipolypods
4 Trilobites
5 Bradoriids
6 Wiwaxiids
7 Brachiopods
8 Echinoderms
9 Hyoliths
10 Sponges
11 Zooplankton
12 Benthic algae
13 Phytoplankton
14 Bacterio-plankton

A TROPHIC WEB

A trophic web connects different groups of organisms according to their sources of energy. These are sunlight and chemosynthesis for the producers and organic matter for the consumers, who eat either producers or each other. As a rule, the larger the consumer the higher the level it occupies within a trophic web. Destroyers connect both ends of the web, decomposing dead bodies and transforming the organic matter back into simple sources of energy for producers. Together with parasites they create a three-dimensional trophic web, rather than a simpler pyramid.

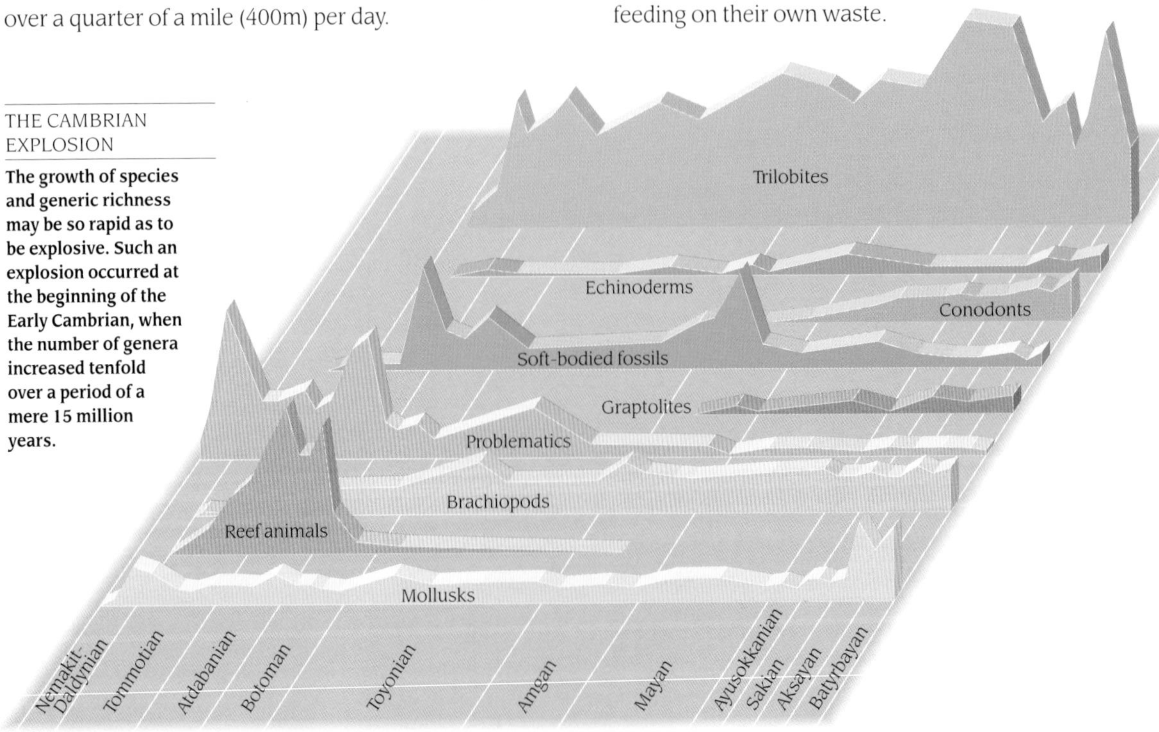

THE CAMBRIAN EXPLOSION

The growth of species and generic richness may be so rapid as to be explosive. Such an explosion occurred at the beginning of the Early Cambrian, when the number of genera increased tenfold over a period of a mere 15 million years.

PLANKTON BLOOM

An increased flux of nutrients creates bacterioplankton, and phytoplankton blooms. These can be visible (greenish area, right, off the coast of Namibia) from space. "Red tides" are dinoflagellate blooms, which produce toxins that kill animals in surrounding waters. Due to the lack of grazers, growing phytoplankton biomass is decomposed by bacteria. Their activity lowers the oxygen level, causing further deaths.

CAMBRIAN

Of course, Cambrian planktonic crustaceans did not eat plastic food; they grazed on the phytoplankton. The rapid descent of the pellets did not allow bacteria and fungi to destroy the organic matter in the water or use a significant amount of oxygen for such destruction. Excess oxygen began to accumulate in the water layer, making it more suitable for a variety of life forms.

At the other end of the pellet conveyor, which was the seabed, detritus-feeders had a permanent source of food. That permitted them to diversify, too. The excess food was mixed with the sediment and buried there, creating a store for animals living under the sediment surface. However, increased activity of burrowers made the sea floor conditions unsuitable for the typical Early Cambrian animals, which were filter-feeders weakly anchored in the soft sediment. Thus, the beginning of the marine "snowfall" in the late Early Cambrian disrupted the habitats of typical Cambrian animals and gave way to the Great Radiation of the Ordovician.

With the introduction of the pellet conveyor, the trophic pyramid was completed. A typical trophic web consists of primary producers(algae, plants, and some bacteria), which produce organic matter by the modification of sunlight or the energy of chemical reactions, and consumers, which eat this organic matter. Consumers are subdivided into several types, including filter-feeders, detritus-feeders, grazers, carnivores, and parasites. The filter-feeders extract the food from water currents moving through their bodies. The detritus-feeders actually digest organic matter buried in the sediment and bacteria living on it (but not the sediment itself). Grazers remove the organic matter cover (such as algal meadows). Parasites prefer to inhabit the body fluids of the host, while ensuring that the host does not actually die. Finally, carnivores eat any other consumers, and decomposers (fungi and some bacteria) break down the remainder of the organic matter to the consistency of simple gases and liquids.

During each period, especially the Cambrian, the consumer types were different. To see a typical Early Cambrian trophic web, the best place to look is at reefs.

TRACE FOSSILS

Deep-sea trace fossils record increasing complexity and regularity of movement as animals looked for food. Meandering (left) became more focused as food patches became restricted. Hexagonal networks (right) were left by burrowers who trapped their food.

REEFS are one of the most complicated and varied ecosystems, supporting more than 5000 species on merely a few square miles of its space. Even for the untrained observer, a reef is a breathtakingly marvelous seascape. Where its features are preserved in fossil form, the beauty and intricacy of the reef's structure may still be seen. Fossil reefs are very hard and resistant bodies of rock. For this reason they retain evidence of the way their constituent organisms interacted as they grew. Rapid marine lithification and subsequent small changes also help. In the Early Cambrian, reef-building organisms were mostly calcified cyanobacteria and sponges.

Calcified cyanobacteria and sponges built the small domed Cambrian reefs—important relics of biodiversity and complex food webs.

These sponges, called archaeocyaths ("ancient cups"), resembled densely perforated cups. The free movement of water through their bodies permitted them to feed on floating bacteria filtered from the water. Although some cylindrical archaeocyaths were up to 3ft (1m) tall and some plate-like ones were over 20in (0.5m) in diameter, most of them were less than half an inch (1cm) in diameter and less than 1in (3cm) tall. As a result, the Cambrian reefs were small domed structures.

Calcified cyanobacteria, individually even smaller than archaeocyaths, developed in colonies and could produce reefs up to several dozen feet in height under suitable conditions. The surface of such reefs is either spotted or covered with an ice-crystal pattern, depending on the shape of building colonies. Calcified cyanobacteria were the principal reef-builders, but they needed archaeocyaths as a firm starting point for further growth.

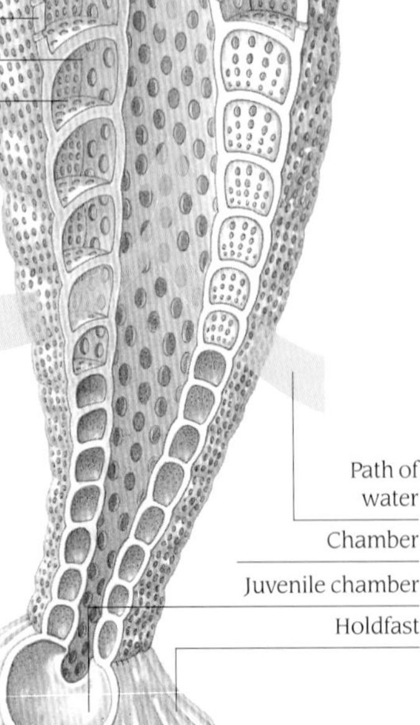
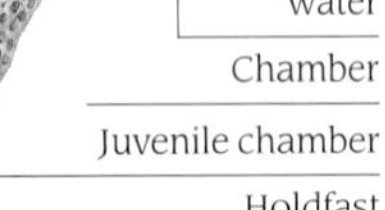

THE CUP-SHAPED ARCHAEOCYATH

Only the calcareous cup-shaped skeleton of archaeocyaths has been preserved. Each resembled an empty ice-cream cone with two walls (outer and inner). The space between the walls (the intervallum) was filled with vertical partitions (the septum). The cup was anchored to the substrate by a laminated holdfast.

TUVINIAN LACE

The former Tuva-Mongolia terrane, which today is fixed along the western banks of lakes Baikal and Khubsugul, has the most spectacular remains of archaeocyaths (above). Geological processes dissolving and precipitating minerals, as well as modern weathering, have revealed the archaeocyath cups in all their antique beauty. It is as if a lacy design covers the rocks. Each cup is under ½in (1.5cm) wide, but still shows vast detail. The double-walled structure can be clearly seen in cross-section.

In addition to the calcified cyanobacteria, representing producers, and the filter-feeding archaeocyaths, Early Cambrian reefs housed a wide variety of organisms. Immobile cup-like echinoderms, without the signs of typical five-sided symmetry, stuck up either on short stout stems or without a stem at all. They caught phytoplankton with their brachioles ("small arms")— articulated, erect, tentacle-like food-gathering appendages covered with calcareous plates. Hyoliths were barely motile, using a pair of crutch-like appendages to orient their conical shells with the main water currents. Under stress, they closed the shell with a well-fitting lid. Some primitive brachiopods and tiny cnidarians hung onto archaeocyathan cups and lodged in the cavities within hardened cyanobacterial colonies. These animals belonged to the filter-feeder group.

In the muddy areas, detritus-feeders left regular traces on and in the sediment. Although the diets of the Cambrian detritus-feeders are well established, nothing is known about the animals themselves.

Both these groups were primary consumers. The secondary consumers were carnivores and included trilobites. However, even a rigid trilobite carapace was no defense against the bites of bigger predators, which included the anomalocaridids. In places, trilobites also fell prey to cephalorhynch worms, which held trilobites in a spiny flexible proboscis and slowly swallowed them.

In this way, a complex trophic web was established from as early as the very beginning of the Cambrian. Carnivores consumed filter- and deposit-feeders, which in turn consumed the primary producers.

Carbonate reefs provide the the only reliable information for investigating Cambrian climates. Both organic and inorganic Early Cambrian mineral carbonates ($CaCO_3$) were precipitated mostly as either aragonite (strontium-rich calcite) or high-magnesium calcite. These minerals dominate over low-magnesium calcite when carbon dioxide levels are low. Because carbon dioxide is the principal gas of the greenhouse effect, one may suggest that the Early Cambrian was rather cool.

The situation probably changed during the Middle Cambrian, when carbonates were predominantly low-magnesium calcite, which happens with the beginning of warmer climatic conditions. The continuous global

METAZOAN REEF

(Below) The oldest reef structures existed as early as the Archean. These were stromatolites built by bacterial communities. The first metazoan reefs appeared almost 3 billion years later, at the end of the Precambrian. They flourished in the Early Cambrian with the appearance of the first sponges able to calcify, whose fragile skeletons were strengthened by calcified cyanobacteria. These archaeocyathan-cyanobacterial reefs occupied the shallow seas of all continents except Baltica and Avalonia, which were too chilly.

Archaeocyaths
1 *Cambrocyathellus*
2 *Okulitchicyathus*
3 *Nochoroicyathus*

Chancelloriids
4 *Chancelloria*

Coralomorphs
5 *Hydroconus*

Radiocyaths
6 *Girphanovella*

7 *Renalcis* (a calcified cyanobacterium)
8 Orthothecimorph hyoliths

warming helped cyanobacteria to surpass the archaeocyaths. Archaeocyathan reefs disappeared forever during that time. Cyanobacterial reefs existed until the middle Cretaceous and sometimes dominated (Early and Middle Ordovician, Late Devonian). Most Early Cambrian animals had become extinct by the end of that epoch.

FEW Cambrian fossils are well enough preserved with reliable information for even a very tentative reconstruction. They were mostly small (less than ⅛in [3mm] long) conical calcareous or phosphatic shells covered with transverse and longitudinal ribs and with a very shallow cavity inside. They were given the common name of "small shelly fossils" and, despite uncertainties as to their affinities, were widely used for a time in the calibration of Early Cambrian rocks. Only with a very vivid imagination could someone sometimes picture such a fossil as an entire animal, whose body was shoehorned into its shell, hanging on a sharp ledge because of the shell's angular shape.

Phosphatic and calcereous shelly fossils, many belonging to unique animals, comprise a peculiar feature of Cambrian strata.

Among the most common Cambrian small shelly fossils are halkieriids. It was recognized that a number of such shells (or sclerites) covered the body of a single animal, creating a scaly casing or scleritome, rather than each shell housing a separate creature. A laborious attempt has been made to count each single sclerite from a rock sample in order to find the exact proportion between different types. As a result, a reconstruction of a spiny cone-like animal with a slug-like foot was made. However, the reality surpassed the expectations. When a complete *Halkieria* scleritome was discovered in Lower Cambrian shales of northern Greenland, the scientific world was shocked: in addition to the spiny scleritome, *Halkieria* bore two shells on its head and tail areas. Called *Halkieria evangelista,* this striking creature included features of such different present animal phyla as brachiopods, mollusks, and annelid worms. Amazingly, even a halkieriid embryo has been found—in Siberia, far away from Greenland. On the eve of the Cambrian, creatures resembling *Halkieria evangelista* generated many animal branches that then filled the seas and lands of the Earth.

Another bizarre animal is *Hallucigenia,* which was first reconstructed (nightmarishly) as standing on seven pairs of stilts, with a row of seven soft appendages on its back. Further finds of remains of Cambrian animals finally helped to clarify their true appearance. Chinese relatives of *Hallucigenia* have indicated that the stilts were actually top spines, while the appendages were (paired) telescopic legs. These allowed the animal to squeeze easily through dense algal meadows or burrow in search of prey.

CAMBRIAN DAGUERREOTYPES

The flattened organic remains of creatures such as the mysterious 1in (2.5cm) *Hallucigenia*, above, on Burgess Shale plates resemble the early black-and-white photographs known as "daguerrotypes." Despite their venerable age, their superior state of preservation records the shape and form of different animals long before the arrival of photography.

Modern velvet worms look similar. They possess telescopic legs and a rigid spiny covering, but belong to a terrestrial fauna. The only difference is in the underside position of the mouth bearing the horned jaws. *Hallucigenia* and its relatives had terminal jawless mouths and were given the name tardipolypods ("slow many legs").

The first predators were the anomalocaridids. They may have been the fearsome hunters that left bite marks on trilobite carapaces with their stout spiny mouth limbs. Modeling of these interesting animals proved that they could bite with their unusual teeth and swim using their powerful side flaps and tail lobe. Close analysis of arched predation scars indicates that the trilobites were bitten predominantly from the right side. Such a preponderance shows clearly that anomalocaridids possessed an elaborated brain and behavioral asymmetry like many other members of the carnivore group. One-sided behavior is typical of hominids, which are among the largest and most advanced modern predators.

THE STRUCTURE and affinities of many Cambrian organisms like *Halkieria*, *Hallucigenia* and *Anomalocaris* have been clarified because of the discoveries of Lagerstätten ("place of deposits"). This German term has come to mean any unique locality that houses unusually well-preserved fossils, because this is the term used to describe the German limestone deposits that contain the fossilized soft tissues of ammonites, belemnites, ichthyosaurs, and the famous half-bird, half-reptile *Archaeopteryx*.

The Burgess Shale "Lagerstätte" is a fossil treasury of over 120 genera of Cambrian animals.

The history of the Cambrian Lagerstätten began in Canada in 1909 with an expedition to British Columbia by the eminent American geologist Charles Walcott. On repeated trips he collected more than 40,000 specimens from a Cambrian outcrop about 200ft (60m) long and at most 8ft (2.5m) thick on the slopes of Mt. Stephen. He named the formation the Burgess Shale, and his books, describing the many creatures he discovered and identified, became the bible of generations of paleontologists.

In the late 1960s, a Canadian–British team started to re-evaluate the Burgess Shale. They added a number of new specimens, re-described and reinterpreted many of the old fossils, and outlined the principal features of the Cambrian Lagerstätten. Using their observations, paleontologists soon discovered new Lagerstätten in many areas of the United States and Canada, in Australia, China, Spain, Greenland, and Siberia. Each of these locations bears its own unique fossils. Sirius Passet in Greenland produced *Halkieria evangelista*; the Australian Emu Bay Shale diverse anomalocaridids; and Chinese Chengjiang the earliest chordates, some unusual arthropods, and tardipolypods.

ANOMALOCARIS, THE JIGSAW PUZZLE

Not so long ago, reconstructions of the Cambrian seascape were scattered with preposterous-looking animals. *Hallucigenia* minced along the substrate on seven pairs of stilts with a row of flexible appendages waving gently on its back. Jellyfish-like *Peytoia* floated above the sea floor like the Cheshire Cat's disembodied grin. *Anomalocaris* hid between large sponges (*Laggania*) and algae, showing only its fleshy shrimp-like tail.

In other illustrations, *Anomalocaris* either looked like a huge millipede with multiple pairs of stout jointed legs, or was reduced to a single pair of claws surrounding the mouth of a large arthropod called *Sidneyia*. Its true appearance became apparent when the former jellyfish *Peytoia*, sponge *Laggania*, and *Anomalocaris* itself, were eventually assembled together as one creature. The complete animal had a huge head with a pair of compound eyes on its upper surface, and a rounded mouth with two spiny appendages below and flat sharp teeth inside. The body bore side flaps and a fanlike tail (forked in some anomalocaridids). A row of paired soft appendages resembling those of velvet worms have been suggested, joined to the lower body surface. The intact anomalocaridid bodies were up to 2ft (60cm) long, but some individual mouth parts hint that among them were animals over 3ft (1m) long.

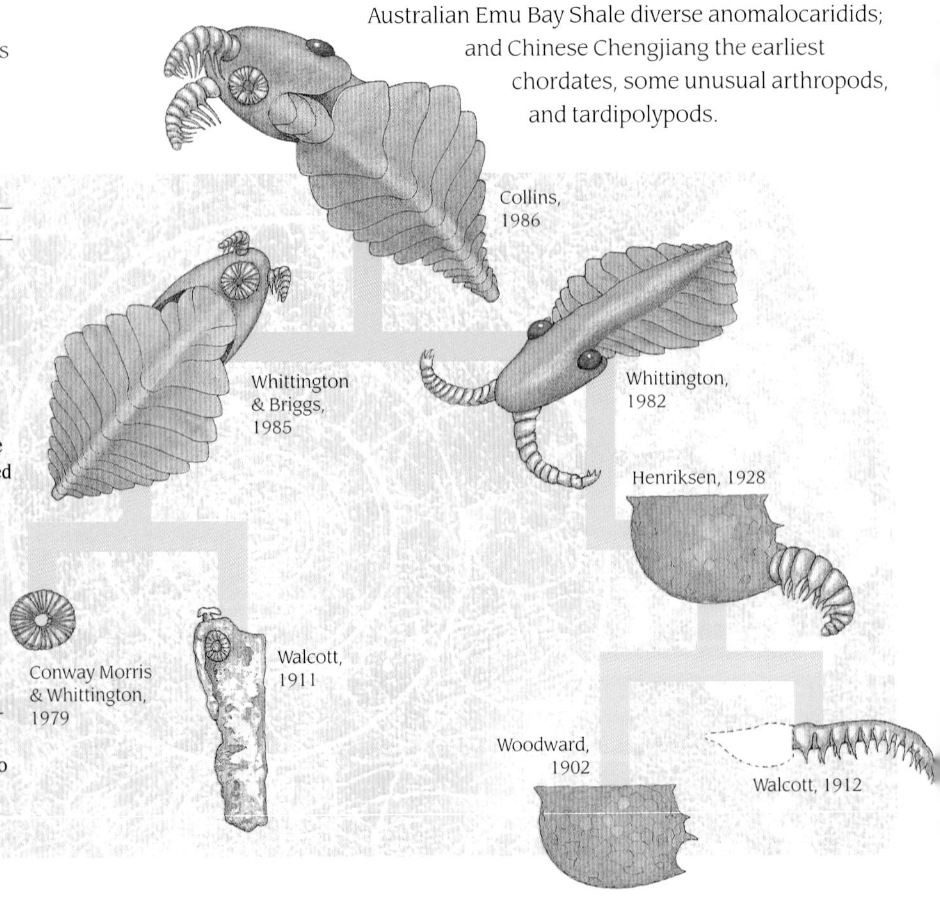

Collins, 1986
Whittington & Briggs, 1985
Whittington, 1982
Henriksen, 1928
Conway Morris & Whittington, 1979
Walcott, 1911
Woodward, 1902
Walcott, 1912

ALMOST all Lagerstätten are restricted to a relatively narrow interval from the middle Early Cambrian to late Middle Cambrian. Only a few are known from the remainder of the entire Paleozoic, when a window to the ancient seascapes was opened briefly and then firmly closed again. The nature of this phenomenon is not well understood. Perhaps such localities were established with the appearance of lightly skeletonized animals and disappeared when active burrowers settled in deeper water environments, having disturbed forever the rest of the organisms buried there.

Buried by a mudslide from a submarine cliff, the Burgess Shale fauna were preserved in great detail; the lack of oxygen prevented their decomposition.

The Burgess Shale locality is still the undisputed queen of all Cambrian Lagerstätten. Among Burgess Shale fossils, there are about 130 species of cyanobacteria, algae, sponges, cnidarians, ctenophores, cephalorhynch and annelid worms, anomalocaridids, tardipolypods, arthropods, brachiopods, mollusks, hyoliths, coeloscleritophorans (*Halkieria* relatives), echinoderms, hemichordates, chordates, and some animals whose affinities are still unknown. Their preservation allows reliable reconstructions of the life style, dietary habits, interactions, and trophic webs of the Cambrian biota.

Once upon an early Middle Cambrian time, the Burgess Shale was a thin, relatively deep-water marine sediment that had accumulated near a high escarpment. Mud currents captured organisms living on the seafloor, and buried them in the deep-water sediment. Tightly packed by fine particles, and isolated from oxygen and the activities of scavengers, the remains were preserved undisturbed and some soft tissues became "mummified" through mineralization processes. Of course, even under such favorable conditions, not everything is perfectly preserved. "Soft-bodied" animals actually possessed resistant cuticles, rarely fossilized, coating its surface and alimentary canal. As a result, the exoskeleton and digestive tract are the features to be preserved.

THE BURGESS SHALE

The Early Middle Cambrian Burgess Shale forms part of the high mountain peaks of eastern British Columbia (Canada) around Lake Louise and the town of Field. During the Cambrian Period it was a northern marginal sea of Laurentia. The shales have been quarried since the early 20th century. Several books have been written based on data from material dug out of the Burgess Shale, though paleontologists continue to contest each other's theories about the structure and attributes of the wonderful Cambrian animals.

1
3
4
5
6
7
9
10
11
4

IF IT were possible to look back at the Burgess Shale as it once was, it might have looked something like this. A huge variety of creatures, including 40 different kinds of arthropods, inhabited the site. A flattened *Odontogriphus* floated, using its lophophore (bristle-bearing) loop for a food search rather than for filtration as brachiopods do. Giant anomalocaridids,nearly 2ft (60cm) long, propelled themselves through the water by means of lateral flaps, ready to seize their prey in their spiny graspers. *Hallucigenia*—finally the right way up—scavenged on the seafloor. Cephalorhynch worms such as *Ottoia*, resembling modern priapulids, waited in their U-shaped burrows for careless trilobites and hyoliths. Spiny wiwaxiids, the descendants of *Halkieria*, grazed with their pointed teeth. Sponges rose above the seafloor, making a tower for the tardipolypod *Aysheaia* to climb at a slow pace with its telescopic appendages and, possibly, stop for feeding. The large arthropod *Sidneyia* held tightly onto a smaller creature in its claws. Other arthropods swam, walked, and scurried around, while the wormlike *Pikaia* swam along in a series of S-shaped bends, rather like snakes do, using the paired, segmented muscle blocks and stiffening notochord that are unique to vertebrates.

Arthropods dominated the Burgess Shale environment.

1 *Anomalocaris* (an anomalocaridid)
2 *Marrella* (an arthropod)
3 *Aysheaia* (a tardipolypod)
4 *Olenoides* (a trilobite)
5 *Wiwaxia* (a halkieriid)
6 *Habelia* (an arthropod)
7 *Dinomischus* (no known phylum)
8 *Odaraia* (an arthropod)
9 *Ottoia* (a cephalorhynch worm)
10 *Sidneyia* (an arthropod)
11 *Pikaia* (a chordate)
12 *Odontogriphus* (a lophophorate)
13 *Hallucigenia* (a tardipolypod)
14 *Nectocaris* (an arthropod)

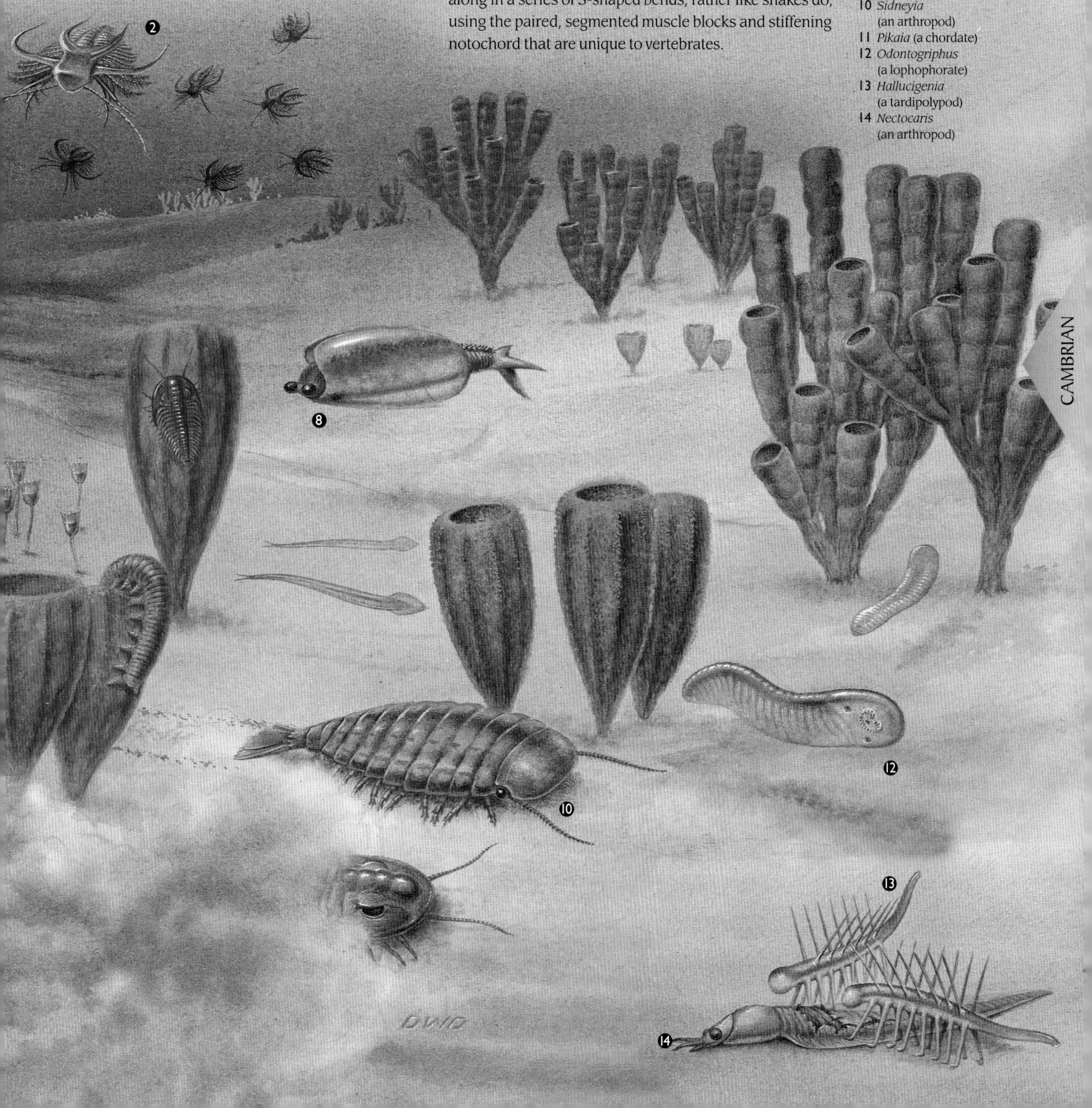

2

8

10

12

13

14

DWD

EVOLUTIONARY TRENDS

The principal trends in arthropod evolution were the improvement of skeletal shading during molting, and the modification and differentiation of body segments and appendages. The basic modern arthropod forms had been established by the Jurassic.

1 Skeletal cuticulization and mineralization
2 Differentiation of segments and appendages
3 Appearance of peritrophic membrane and pellet conveyor
4 Adaptation to a terrestrial life by the development of tracheae for breathing and Malphighian tubes for urination
5 Adaptation to wood and cellulose consumption
6 Acquisition of flight
7 Appearance of sessile filtrators
8 Adaptation to green mass consumption
9 Plant pollination
10 Further increase of types of appendages

CAMBRIAN ARTHROPODS

The most primitive arthropods, the Marrellomorpha, are known from the Cambrian to the Devonian. They were characterized by multiple and simple limbs consisting of a lower walking branch and an upper feathery gill. Already the oldest Cambrian Lagerstätten, Sirius Passet of Greenland and Chengjiang of China, were yielding plentiful and highly diverse arthropods.

CHELICERATES

Chelicerates have six pairs of limbs. The first pair, called chelicerae, are grasping and jawlike. The second pair are usually clawlike, while other pairs are mostly used as walking legs.

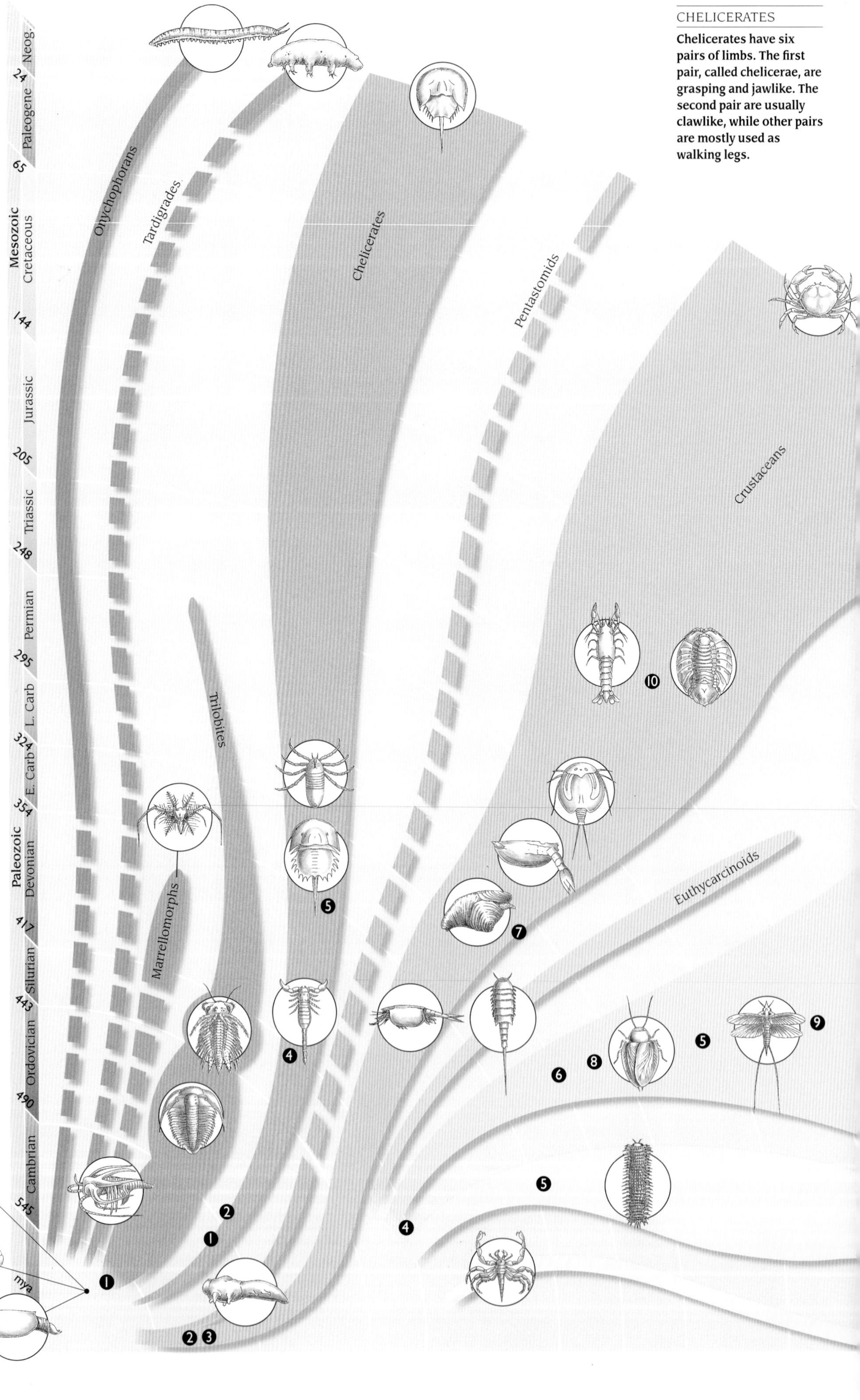

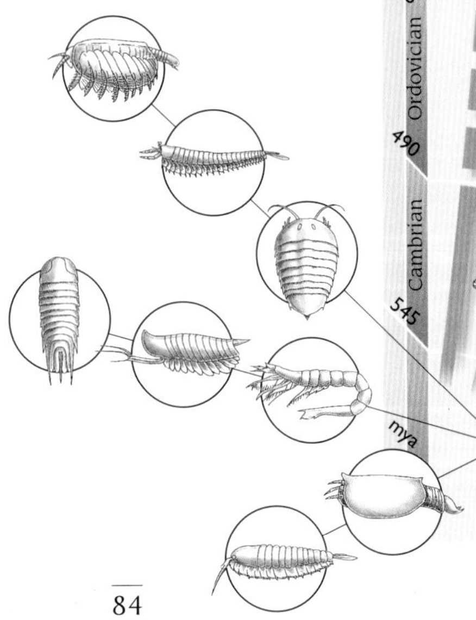

THE EVOLUTION OF ARTHROPODS

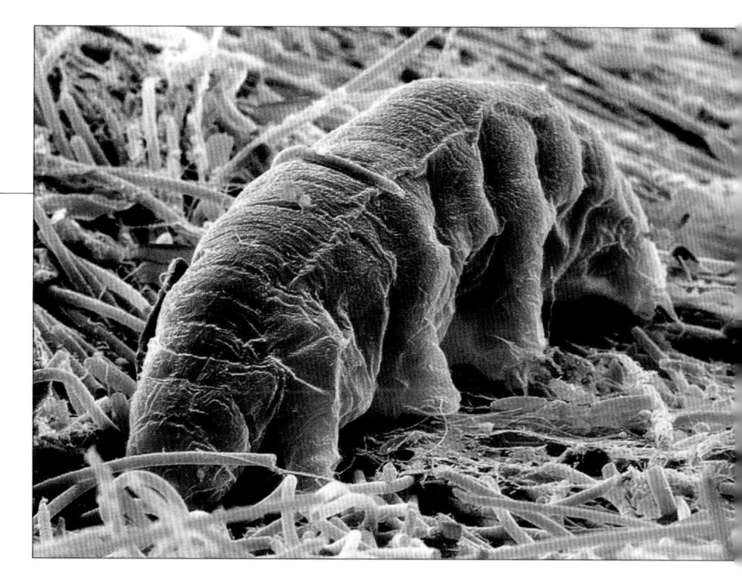

TYPICAL arthropods are distinguished by their segmented bodies and jointed appendages. The body is usually subdivided into a head, a thorax, and an abdomen. The entire body and limbs are covered with a rigid external skeleton made of chitin, a polysaccharide. In some arthropods the exoskeleton is hardened by mineralization; it is calcified in trilobites, ostracodes (seed shrimps), and barnacles, and phosphatized in some extinct bivalved arthropods. As a result, these groups are better preserved in the fossil record. However, the presence of an exoskeleton was an impediment to arthropod evolution by restricting their maximum possible size. Also, in order to grow, arthropods have to shed the exoskeleton periodically by molting, rendering them almost totally defenseless.

Since the middle Early Cambrian, arthropods have dominated the marine world from shallow to deep waters. The earliest terrestrial communities were also full of these articulated creatures. At present, the number of insects (from 1 to 10 million species by different estimations), which comprise just one of the three main living arthropod groups, greatly exceeds the number of all other species on the planet.

Arthropods' role in the Earth's ecosystems is as important as their number. From the Cambrian through to the Silurian, arthropods were the dominant marine predators, crowning the food web. At the same time, tiny crustaceans (krill) were at the base of this web, being the primary consumers cropping the phytoplankton. They were the link between the almost-invisible but highly productive world of unicellular life and the world of large animals. During the Mesozoic, marine crustaceans were among the major burrowers, digging 6ft- (2m-) long shafts and, thus, aerating the deep sediment and making it habitable. In terrestrial ecosystems, arthropods play the leading role in both the planting and the harvesting of plants—the primary producers. Some fly and beetle larvae (such as scarabs) consume dung-heaps of large vertebrates and, thus, manure the soil. Honey bees, butterflies, and some other insects pollinate flowers and improve their reproduction. (The fossil record shows that Permian insects had pollen grains in their stomachs). Termites possess symbiotic flagellate protozoans, which in turn bear symbiotic bacteria, and together constitute a powerful factory for the conversion of the wood and cellulose into digestible organic matter. Giant millipedes and their extinct relatives, the arthropleurids, probably did the same work in the Carboniferous coal forests.

ARTHROPOD RELATIVES

The Tardigrada (water bears, right) are segmented invertebrates which range from 0.002–0.05in (0.05–1.2mm) and lack respiratory and circulatory organs. Other relatives of early arthropods are the Onychophora (velvet worms), and Pycnogonida (sea spiders).

CRUSTACEA

Crustaceans are almost exclusively aquatic gill-breathing animals with a bivalved carapace. They include ostracodes, barnacles, malacostracans (e.g., lobsters, crabs, shrimps), and some other groups. By the Early Cambrian there were already many groups of crustaceans. Recently, the crustacean affinity with the parasitic invertebrates called pentastomes (tongue worms) has been established.

INSECTS

Arthropods are subdivided into four major groups: Crustacea, Chelicerata, Uniramia, and Trilobita. Insects belong to the Uniramia, which also includes millipedes, centipedes, and some minor groups of myriapod-like animals. Recently, a separate origin for some of these terrestrial arthropods has been suggested and insects have been linked to the crustaceans.

Insects

Myriapods

Pycnogonids

Phylum
PROTOSTOMIA (arthropods, mollusks, brachiopods, bryozoans, various worms)
ANNELIDA (segmented worms)
ONYCHOPHORA (velvet worms)
MYRIAPODA (millipedes, etc)
MARRELLOMORPHA
TRILOBITA
CHELICERATA (scorpions, spiders, horseshoe crabs)
CRUSTACEA (crabs, shrimps, etc)
INSECTA
extinct

ARTHROPOD CLASSIFICATION

Conventionally, arthropods were thought to be descended from annelid worms. However, the recent comparison of DNA sequences has indicated that they are closer to cephalorhynch worms than to annelids. The intermediate cephalorhynch-arthropod features of some Cambrian fossils—anomalocaridids and tardipolypods—fit very closely to the molecular data.

CAMBRIAN

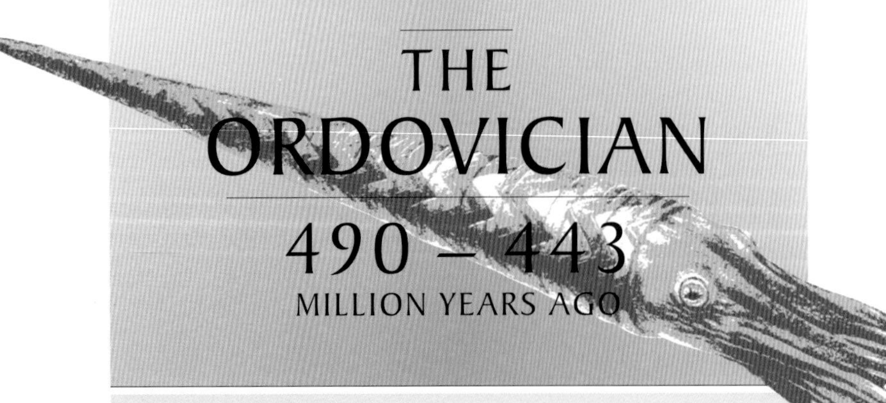

THE ORDOVICIAN

490 – 443 MILLION YEARS AGO

THE ORDOVICIAN was a period of high sea levels with lows of short duration, suggesting fluctuations caused by melting polar ice. By the Middle Ordovician, little land remained exposed. The maximum sea level was about 320 to 740ft (100 to 225m) higher than it is today, a high point exceeded only in the Cretaceous, and then only slightly.

The Iapetus Ocean was beginning to close, consuming most of the newly formed ocean crust as Baltica and Siberia edged closer to Laurentia. The Taconic orogeny added new territory and a great mountain range along the eastern margin of Laurentia. Shallow seas occupied extensive areas at low latitudes, facilitating a warm and humid climate around the globe, though southern Gondwana was covered by ice. Along with sea level and the composition of seawater, atmospheric chemistry also changed. Concentrations of carbon dioxide rose sharply at the beginning of the period, contributing to "greenhouse" global warming, before dropping markedly again at the end.

RECOGNITION of the Ordovician began with a controversy between two leading British geologists, Roderick Murchison and Adam Sedgwick. After years of investigation in and around Wales, Murchison published an article in 1835 in which he outlined the Silurian system with its two major subdivisions, the Upper and Lower Silurian. In the same year, Sedgwick and Murchison had drawn attention to the Cambrian-to-Silurian succession in a paper published jointly. However, because one of the authors described mainly rocks while the other preferred to characterize fossils, they soon caused the lowermost Silurian to overlap with the uppermost Cambrian. Furthermore, Murchison claimed that there are no strictly Cambrian fossils. His popularity and official position caused most of his peers to accept his division of the Silurian.

Now allocated to the second period of the Phanerozoic, the Ordovician was ratified last.

KEYWORDS

CARBONATE
CEPHALOPOD
EXOTIC TERRANE
HIRNANTIAN ICE AGE
NAUTILOID
PALEOZOIC FAUNA
PRECORDILLERA
SHELL BED
SILICICLASTIC
TACONIC OROGENY
TRILOBITE

Once again, Joachim Barrande's descriptions of fossils helped to solve the problem. He was able to separate the Lower Silurian faunas from the Upper Silurian and primordial (Cambrian) ones. In 1879, using Barrande's data and his own observations from Scotland and Wales, Charles Lapworth validated both the Cambrian and Silurian systems but renamed the former Lower Silurian as the Ordovician system, after the ancient tribe of the Ordovices from North Wales.

One of the most remarkable things about the Ordovician is the complete turnover of marine faunas that occurred. The Cambrian fauna was almost eliminated and a more active, powerful, and diverse Paleozoic fauna replaced it. This was the fauna that was to reign over

CAMBRIAN	490 mya	485	480	475	470	465
Series	EARLY/LOWER					
Series	TREMADOC		ARENIG			
European stages	CRESSAGIAN	MIGNEINTIAN	MORIDUNIAN	WHITIANDIAN	FENNIAN	
N. American stages	IBEXIAN			WHITEROCKIAN		
Geological events	Narrowing of Iapetus Ocean and widening of Rheic Ocean; Terrane drift of Argentinian Precordillera; Avalonia rifts from Gondwana and drifts northwards					
Climate / Atmosphere	Warming and increase in carbon dioxide levels					
Sea level	Rising					
Plants / Animals	Stromatolites decline	• First corals, bryozoans, and stromatoporoids				

marine ecosystems for the next 250 million years, in spite of a mass extinction at the end of the Ordovician.

The Ordovician radiation was marked by a fourfold increase in global diversity of marine animal genera. More than 4500 genera are known from the Ordovician, representing 12 percent of all marine animal genera from the entire Phanerozoic. These were articulate brachiopods, echinoderms, foraminiferans, and bivalve mollusks, which became much more abundant than in the Cambrian; cephalopods, graptolites, chitinozoans, and conodonts appeared at the very end of the preceding period; stromatoporoid sponges, corals, bryozoans, ostracodes, and some water scorpions and jawless fishes. The diversification of the later, typical Paleozoic groups really started in the early Ordovician and culminated in the later part of the period. Trilobites, the representatives of the Cambrian fauna, re-expanded, and inarticulate brachiopods also re-diversified.

In addition to the diversity, the overall biomass grew in size. This trend is exemplified by an increase in the concentrations of shells, which become a significant part of the stratigraphic record at the base of the Middle Ordovician. Three-foot (one meter) thick shell beds appeared, whereas Cambrian and the lowermost Ordovician shell beds were lens-shaped and thin. Across the Lower to Middle Ordovician boundary, shell concentrations changed composition, with articulate brachiopods and ostracodes replacing trilobites. (Articulate brachiopods had jointed half-shells with hinge teeth and complementary sockets.) Individuals also became larger: Cambrian mollusks and brachiopods rarely achieved sizes of one inch (3cm) and were commonly only one tenth that size, but Ordovician brachiopods grew up to 3in (8cm) wide, gastropods to 8in (20cm) in diameter, and some cephalopods up to 2.6 ft (800cm) long. Large calcified sponges and corals scaled up their reefs proportionally.

Whole new realms, such as pelagic waters and the bottom of the ocean, were pioneered by these new life forms. Larger individuals provided more room under their shells and branches for smaller species. The space both above and below the water-sediment interface became finely subdivided into tiers of subcommunities, reducing competition for food between different tiers. Vast epicontinental seas and carbonate shelves yielded a variety of communities, prompting a geographical differentiation to the global diversity.

Another source of diversity in Ordovician communities was the distribution of distinct local faunas. Numerous groups of trilobites, brachiopods, acritarchs, conodonts, chitinozoans, graptolites, and others were distinct because of their adaptations to different substrates, depths, oxygen level, pressures, and temperatures in the water they occupied.

ROCK STRATA

Newfoundland's Green Point is the official boundary between the Cambrian and Ordovician. It is one of the few areas not affected by falling sea levels at the time.

Present

Origin of Earth

ORDOVICIAN

See Also

THE CAMBRIAN: *Archaeocyath Reefs, Arthropods, Predators*
THE SILURIAN: *Limestones and Coral Reefs, Hydrothems*
THE PLEISTOCENE: *Ice Age Cycles*

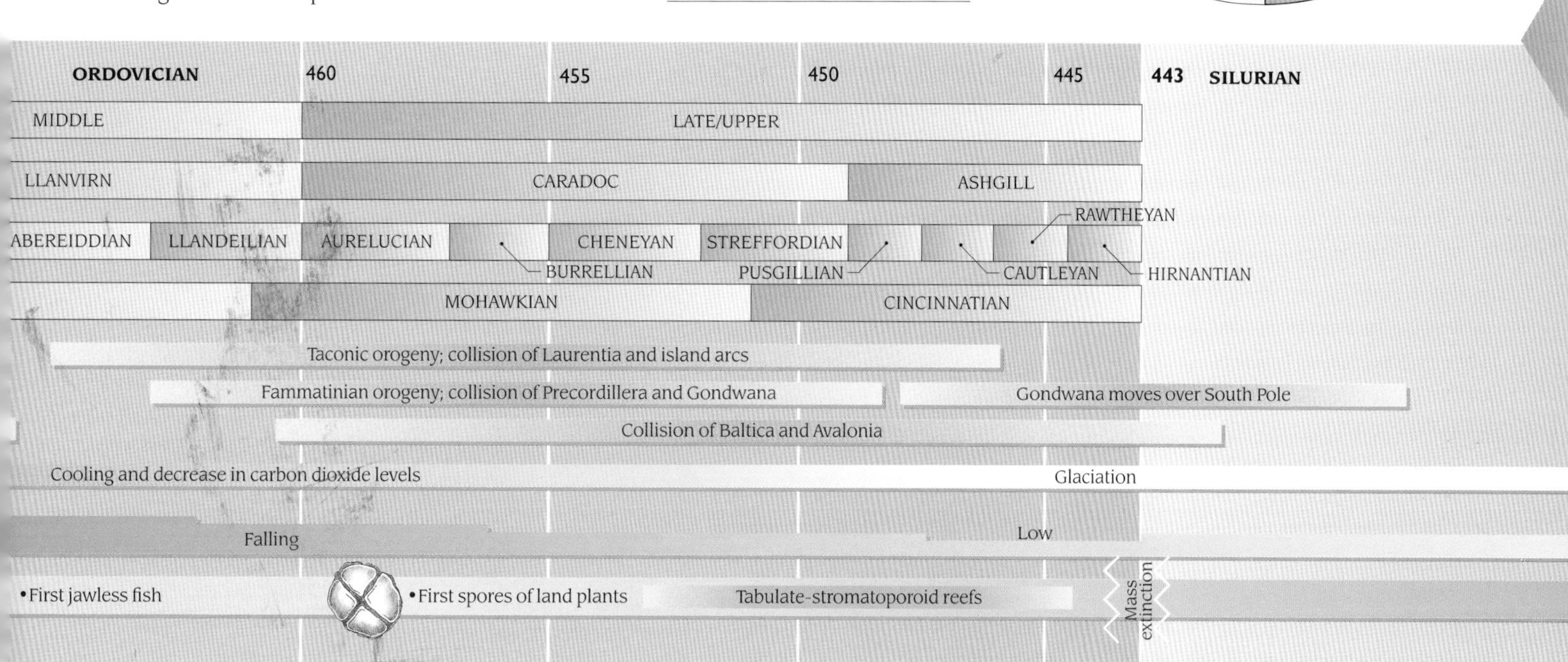

ORDOVICIAN paleogeography has been tentatively reconstructed in the following way. In the northern hemisphere, above 40°N latitude, there was an ocean, the Panthalassa. Consequently the northern oceanic currents circulated in a closed loop, and little opportunity existed for mixing cooler northern hemisphere waters with the relatively warmer waters of the tropics. The separate continents of Laurentia, Siberia, North China, and Kazakhstan lay within the tropics. Evaporites (rock salt and calcium carbonate) were common there.

Warm, shallow seas covered much of the Ordovician globe. Laurentia and Siberia were now about 600 miles apart, and the Iapetus Ocean was closing.

Carbonates were also widespread in Laurentia and Siberia, where there was intense reef-building. These continents were probably less than 630 miles (1000km) apart, and were characterized by similar warm, shallow water in which cephalopods, trilobites, brachiopods, graptolites, and conodonts dwelled. A southern equatorial current may have flowed from Siberia to Laurentia, but because of the presence of island arcs, it would probably have been diverted south.

Kazakhstan was subdivided into two microcontinents bounded to the east by volcanic arcs. A system of great carbonate seamounts, similar to the modern Bahama Bank, existed along the continental slope.

Oceans of different depth and width separated the North China, South China, and Tarim (western China) plates. Chinese Ordovician faunas have a close affinity with those of Laurentia, Siberia, and Australia, because China occupied an intermediate position between them.

Gondwana dominated the southern hemisphere. Its lobes extended northwards into the tropics, including parts of modern Chile and Argentina, as well as a huge landmass containing much of modern Antarctica, Australia, the Middle East, and some of China. A temperate climate zone, aided by proximity to the Iapetus Ocean, prevailed from South America in the west to Baltica in the east. Cool-climate siliciclastics and cold-water fauna were widespread, including occurrences in central and southern Europe, north Africa and adjacent Arabia, and much of South America. The South Pole was located in northwestern Africa.

Baltica lay in the mid-latitudes of the southern hemisphere and drifted slowly north, possibly rotating as it went. Baltica was home to distinct trilobite, brachiopod, graptolite, and conodont faunas, clearly showing that it was an isolated continent. It constituted the eastern margin of the temperate and subtropical zones in the Iapetus Ocean.

Avalonia moved north from Gondwana, from about 60°S latitude to about 30°S. Relatively rapid plate movements in this region hint at significant subduction at plate margins involving Avalonia and Baltica.

The overall tectonic situation altered dramatically in the Middle Ordovician when regional tension changed to regional compression. During these collisions, some basins were transformed into high mountain ranges, and new volcanic arcs appeared as a result of the subduction of the oceanic crust. Several island arcs outlined the eastern margin of Laurentia, some of them located far out in the Iapetus Ocean. A progressive collision of these arcs with Laurentia started in the north during the Early Ordovician and ended with the early Late Ordovician Taconic orogeny in New England.

Processes like those that led to the Taconic orogeny may be observed nowadays in Japan and the surrounding region. An oceanic plate composed of heavier rocks is subducted—drawn down below a continental plate. Volcanic mountains are split and submerged piece by piece into deep-ocean marginal trenches, as Japan's Mt. Kashima is currently disappearing into the Japan Trench. A collision of continental and oceanic plates (the latter bearing island arcs) occurs, releasing great energy through volcanic explosions, but does not create high mountains like the collision of two continents.

THE POSITIONS of large landmasses may be guessed more or less accurately, but the occurrences of smaller terranes represent a puzzle. One fragment was the Argentinian Precordillera in South America. It contains fossils and strata that appear alien in South America but fit perfectly with the Appalachian Mountains in North America. The eastern Precordillera bears thick Cambrian to Ordovician limestones, in contrast to the mainly siliciclastic or volcanic rocks of the surrounding Ordovician basins. Carbonate rocks and strong affinities among trilobites,

Argentina's Precordillera is a puzzling fragment: it has more fossils in common with the Appalachians than with the rest of South America.

GLOBAL FEATURES

Ordovician Earth inherited the features of the Cambrian globe. Panthalassa dominated the northern hemisphere while continents were concentrated in the southern hemisphere.

PANTHALASSA OCEAN

GONDWANA

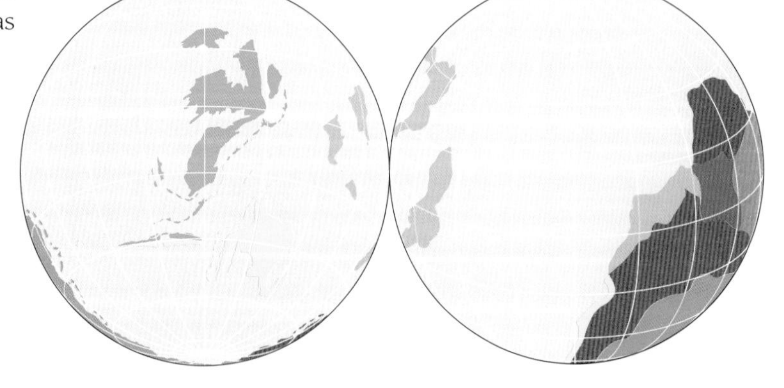

ORDOVICIAN

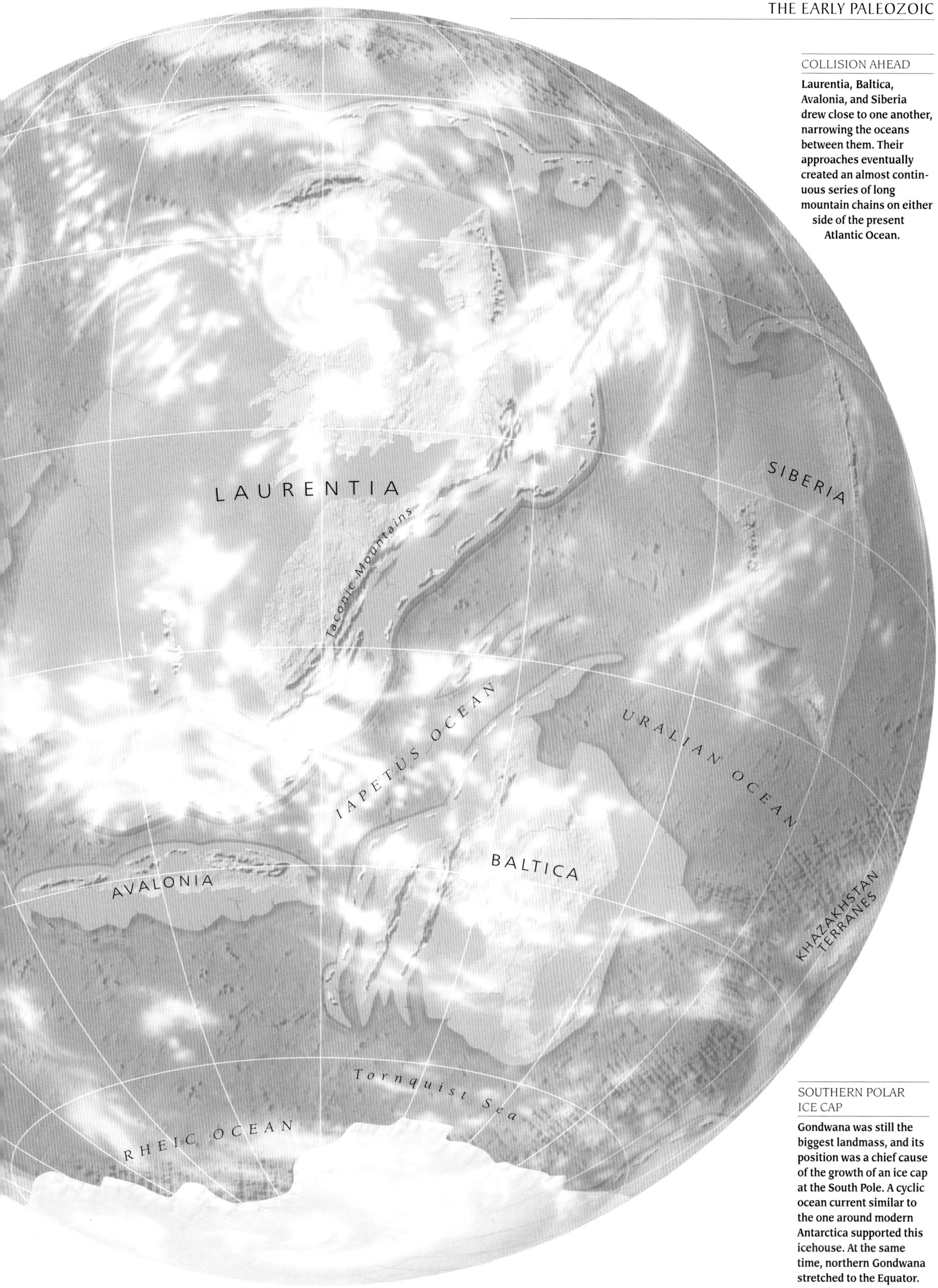

COLLISION AHEAD

Laurentia, Baltica, Avalonia, and Siberia drew close to one another, narrowing the oceans between them. Their approaches eventually created an almost continuous series of long mountain chains on either side of the present Atlantic Ocean.

SOUTHERN POLAR ICE CAP

Gondwana was still the biggest landmass, and its position was a chief cause of the growth of an ice cap at the South Pole. A cyclic ocean current similar to the one around modern Antarctica supported this icehouse. At the same time, northern Gondwana stretched to the Equator.

GREAT SMOKY MOUNTAINS

The Great Smoky Mountains are part of the Appalachians in the eastern United States. The constituent rocks represent sediments that accumulated in back-arc basins of the Iapetus Ocean. The vertical position of once-horizontal layers indicates a major island-arc collision with Laurentia, pushing seabeds onto the continent and turning them through 90 degrees.

conodonts, and brachiopods reveal a close relationship between the Precordillera and the Appalachians.

In 1984 the American geologist Gerald Bond and his colleagues suggested that the Precordillera could be a terrane related to Laurentia. Two principal mechanisms have been invoked to explain the connection. According to the first, the Precordillera and the Appalachians were one continuous mountain chain cut off during the Ordovician separation of Laurentia and Gondwana. In the second model, the Precordillera was an elongated fragment which was torn off the ancestral Gulf of Mexico (known as the Ouachita embayment) and became amalgamated to Gondwana, causing the mid-Ordovician Fammatinian orogeny in Argentina. If the Precordillera was the southward continuation of the Appalachians, the post-Cambrian faunas of both regions should be identical. However, the available evidence indicates that the Precordilleran fauna became differentiated from the Appalachian in the Early Ordovician, though some typical Appalachian genera persisted.

A wide variety of data confirms the location of the Precordillera at a low latitude during the Cambrian. As the Cambrian strata give way to the Ordovician, the rich faunas of stromatoporoid and sponge reefs, as well as the abundance and diversity of sponges, in the limestones still indicate warm, shallow waters at low latitudes and an abundance of warm-water conodonts.

Beginning with the late Early Ordovician, black shales and dark deepwater mudstones overlapped the shallow-water carbonates. These sediments together with cold-water conodonts imply cooling. From the Middle Ordovician, cold-water faunal assemblages became more and more important. By the Late Ordovician, most of the Precordilleran successions are siliciclastic, related to falls in sea level, although the presence of abnormal cold-water carbonates in the Central Precordillera is significant. In contrast to low-latitude carbonates, these were devoid of oolites and large thick-shelled organisms, verifying their development in temperate mid-latitude waters. The associated fauna was adapted to temperate shelf carbonates. An increasing proportion of high-latitude Baltic genera contributes to the overall picture, suggesting that the Precordillera terrane had been displaced.

By the early Late Ordovician, a few Precordilleran brachiopods and trilobites had clear affinities with Gondwanan fauna. Their relative scarcity does not preclude the possibility that the Precordillera was still isolated from Gondwana but connected to the open sea to the west. By the latest Ordovician, temperate fauna of the Precordillera had become virtually indistinguishable from other regions of Gondwana's margins. During the Hirnantian Stage, the Precordillera and its basement to the east were partly covered by ice. Bedded layers of sea ice with dropstones (stones carried and dropped by icebergs) and massive tillite deposits are preserved on its periphery. (Tillites are rocks formed by the consolidation of glacial till, a mixture of clay, sand, gravel, and boulders deposited by a glacier). Just above them, shales and siltstones include shell pavements with a typical *Hirnantia* (brachiopod) assemblage, characteristic of a cold Gondwanan province. By that time the Precordillera was joined to Gondwana, later to become part of South America.

OLENELLID TRILOBITE

***Olenellus thompsoni*, a trilobite, is found in the North American Applachians, in southwest Scotland, Spitsbergen (Norway), and the Argentinian Precordillera. Its wide flat body indicates that it had well-developed gills to allow it to breathe in the deep oxygen-poor waters that covered these places.**

ORDOVICIAN rocks of the northern hemisphere tell quite a different story. Here, large continental masses started to draw closer, giving rise to a series of orogenies that is now western Europe and eastern North America. In place of the Atlantic Ocean there was a single long mountain chain. Its origin in the island arcs of the Iapetus (some of which accreted to Laurentia in the Taconic orogeny) is confirmed by the presence of ophiolites, relics of the ocean floor, in the northern Appalachians, and less definitely in Britain and Scandinavia. Ophiolites represent a relatively rare example of heavy crust squeezed out of the ocean floor, and are an important indicator of a former ocean and thus of a collision.

Both ophiolites and conodonts—rocks and fossils—tell the story of the Taconic orogeny, which laid the foundations of the Appalachian range.

Certain species of conodonts had died out during the Taconic orogeny. During such an event, younger ocean basalts involved in the orogeny became the source of some rare earth elements and certain of their isotopes (such as a light isotope of neodymium). The isotopes tend to accumulate in phosphatic skeletal fragments of dead organisms. Thus, a paleontological search of the conodonts allows geochemists to detect isotopes that are evidence of a very old orogeny.

Other fossils and isotopes help to explain these events. Belemnites, the skeletal fragments of a close relative of the squid, resemble a thermometer in shape. Perhaps this prompted a scientist to measure the temperatures of long-disappeared seas with belemnites and with other calcareous fossils. It is possible because some organisms build skeletons in equilibrium with the environment, preserving the isotope ratio of the seawater.

When polar ice caps are formed, the ice contains mainly a lighter isotope of oxygen (oxygen–16), while the reminder of the seawater becomes relatively enriched with its heavier isotope (oxygen–18). Organisms absorb the heavier isotope from the seawater. Ordovician seas, in contrast to Cambrian ones, were teeming with articulated brachiopods whose tough calcareous shells are perfect for studying isotopes. Following the oxygen isotope ratio curve across the record of Ordovician fossil brachiopods shows that a general lowering of global temperatures took place. The time of the lowest temperatures coincides with the Late Ordovician glaciation, which was one of the biggest in the entire Phanerozoic.

The development of the southern ice cover over a significant part of Gondwana coincided with the intensification of the Taconic orogeny because the chemical weathering of freshly exposed rocks served as an important sink for carbon dioxide gas stored in the atmosphere. When mountains grow higher, the greenhouse effect becomes less. Of course, orogeny was not the only cause of global cooling. The peculiar geography of the Late Ordovician could have significantly increased the global ocean heat transport towards the pole in the southern hemisphere. If the pole was located close to the edge of a supercontinent (such as Gondwana), glaciation would exist even with a higher level of carbon dioxide in the atmosphere. The ocean's thermal inertia would prevent summer temperatures from rising above freezing, which is a major prerequisite for glaciation.

Use of carbon dioxide by organisms in the formation of both carbonates and dark shales may have influenced

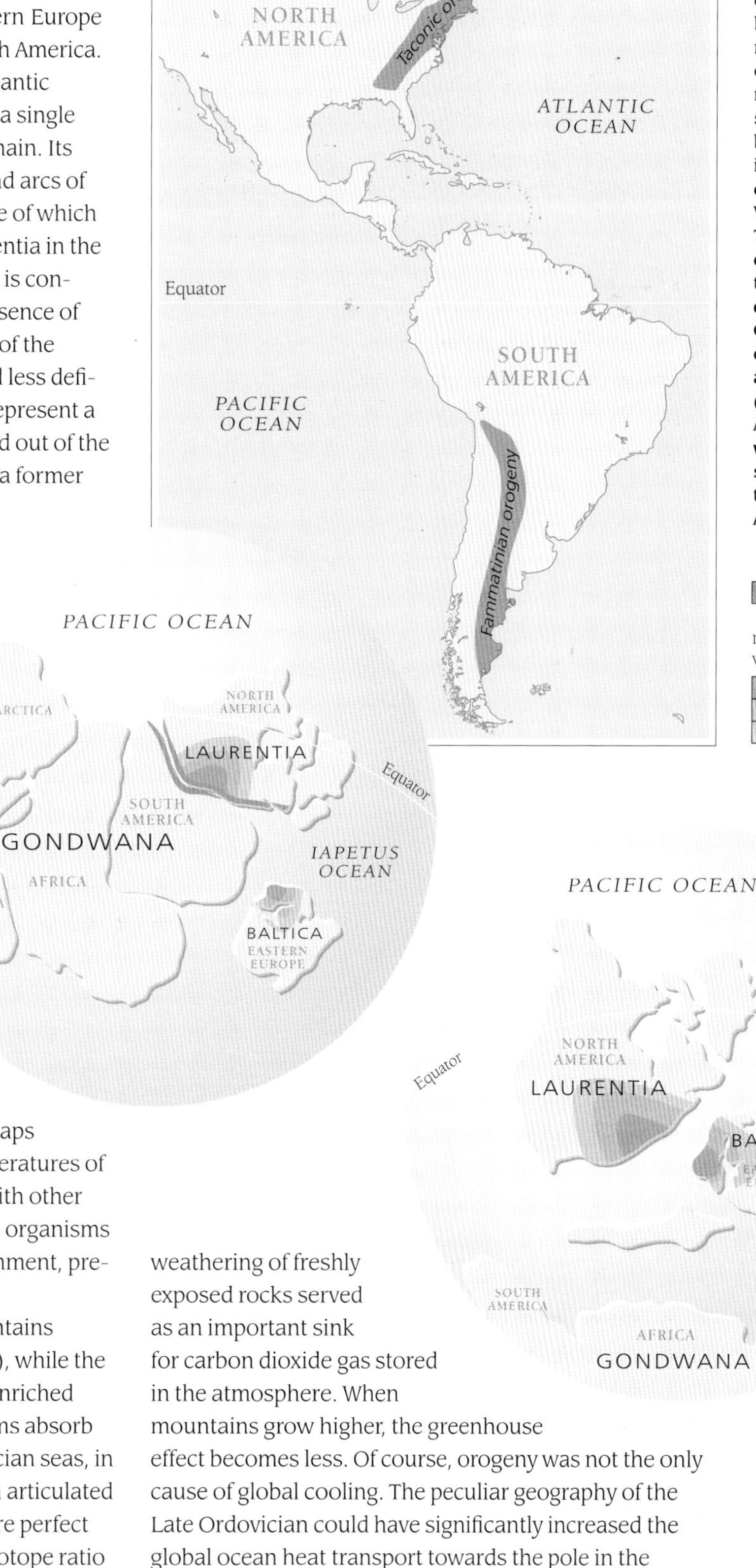

ANOTHER TECTONIC PUZZLE

If continents are close enough, the narrow ocean barriers between them do not prevent the dispersal of animals. In such cases rock formations from a single short interval can be more help than fossils in determining a continent's paleoposition. Volcanic activity from the Taconic orogeny spread a quantity of ash whose thickness decreased with distance from the source. Comparing ash falls, it is clear that North America and Europe were involved (lower globe) but South America (upper globe) was not, despite the stratigraphic affinities of the Precordillera with the Appalachians.

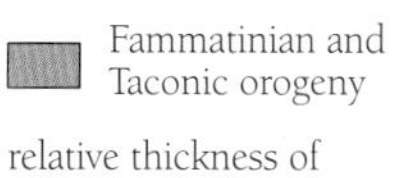

relative thickness of volcanic ash deposits
thick
thin

ORDOVICIAN

a reduction in the atmospheric carbon dioxide concentration. It is also possible that an extreme orbital position of the Earth could have played a part in causing an ice age. These were sufficient causes for a global ice age.

THE GREATEST extent of the ice age, during the Hirnantian Stage of the latest Ordovician, coincided with one of the greatest mass extinction events in all of the Phanerozoic. Many genera were wiped out, reducing diversity to the level that had existed at the beginning of the Ordovician radiation. The overlying Lower Silurian strata contain a relatively impoverished, low-diversity, relatively dwarfed, recovery fauna.

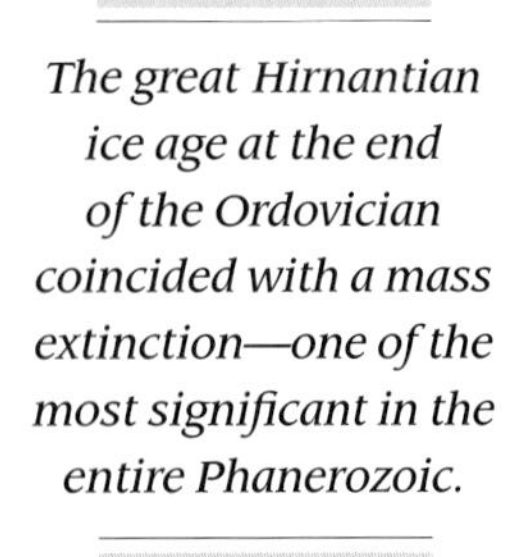

The great Hirnantian ice age at the end of the Ordovician coincided with a mass extinction—one of the most significant in the entire Phanerozoic.

Small (less than 5-8 mm) brachiopods dominated; sporadically abundant were small highly conical gastropods, nautiloids, and trilobites. Matchstick size and leafy bryozoans existed, but did not form reefs. Large modular and solitary corals were absent.

The extinction occurred in two phases, separated by about 0.5–1 million years. The first phase, at the beginning of the Hirnantian, was approximately contemporary with the onset of a major Gondwanan glaciation, and the consequent start of a substantial fall in sea level. The second phase coincided with the end of glaciation and a rise in sea level as the ice melted.

It is too simplistic to attribute the mass extinction solely to the widespread glaciation of the time. Prior to the glaciation, the Ordovician biota was evidently already undergoing considerable environmental pressures due to changes in sea level, ocean temperature, and circulation. Climate change, which resulted from the growth and subsequent melting of the Gondwana ice sheets, may have provided a critical force for extinction. Increasing specialization of the fauna, making them adapted to a narrower range of conditions, constituted another cause of the extinction event. While generalists continued to thrive among some animal groups, specialists were eliminated. Reef ecosystems, which were full of specialists, were particularly vulnerable to the icehouse conditions.

ROCK AND CLIMATE

Glacial rocks speak of cold high latitudes; reef carbonates require a warm climate in which to accumulate. The worldwide distribution of these different rocks shows the influence of climate zones on sedimentation and the dispersal of organisms throughout the entire Phanerozoic.

limit of glacial cover
limit of rocks related to the glaciation
arid area
stromatoporid reef
relative positions of modern land

SIBERIA
SOUTHEAST ASIA
NORTHEAST ASIA
EUROPE
INDIA
AUSTRALIA
AFRICA
NORTH AMERICA
SOUTH AMERICA
ANTARCTICA

AROUND much of the world, Early Ordovician reefs appear to have been relatively simple in biological terms, with complexity increasing in younger reefs. There were a variety of Early and Middle Early Ordovician reef communities, each dominated, at least in terms of biomass, by a particular organism. Which species came to dominate was controlled by the same environmental factors that affect reef communities today: temperature, sunlight, degree of oxygenation, the presence and strength of currents, the frequency of environmental disturbance, the amount of terrigenous input, nutrient and food availability, the nature of the substrate, turbidity, and the interactions between the organisms themselves.

Reefs grew in the shallow waters, 30 to 60 feet deep, that had flooded the surface of many Ordovician continents.

Despite the long time interval and wide geographic separation, all of the Early and Middle Ordovician reefs

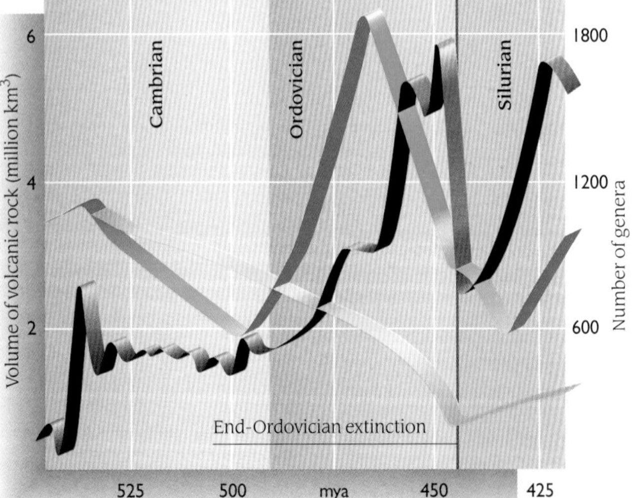

VOLCANOES, ISOTOPES, ANIMALS

A single volcanic eruption sends large volumes of ash tens of miles into the air. The ash dissipates in the upper atmosphere and reflects part of the solar radiation back into space, lowering temperatures by several degrees over large areas of Earth for several years. Major eruptions in the Middle to Late Ordovician (green) could have caused a sharp drop in temperature which was recorded in oxygen isotopes (yellow) and speeded up glaciation and mass extinctions (purple).

AFRICAN FREEZE

Central Africa in the Late Ordovician looked much like this Antarctic scene. That ancient landscape is recorded in tillites and other glacial rocks found in Africa today, as well as its generally flat terrain. Glaciers smooth out even the hardest rocks, while a significant mass of ice causes an overall depression of the crust beneath. Sizeable areas of modern Antarctica actually rest below sea level due to the extreme weight pressing down on the continent.

were remarkably similar. Most of these accumulations probably formed as scattered patch reefs in the vast epeiric seas deeply indented into continental interiors. Such predominantly inshore reefs were especially abundant in Laurentia, Siberia, and North China.

The reefs were associated with moderately to well-sorted sediments with occasional high-energy deposits, which indicate conditions below or near the normal wave base but above average storm wave base. For similar conditions on most modern continental shelves, 30ft (10m) is the maximum depth significantly affected by surface waves, although storm waves can reach up to 330ft (100m) or more. The greatest damage to reef organisms during storms occurs in the upper 80ft (25m). To accumulate the large numbers of sponge fragments seen in Ordovician mounds, the sponge–cyanobacterial reefs had to develop in fairly shallow waters, about 33 to 65ft (10 to 20m). Such depths closely match the distribution of modern filter-feeding sponges.

All the reefs were small accumulations ranging in height from less than one and a half to six feet (0.5 to 1.8m) and were 3 to 23ft (1to 7m) in diameter, although some mounds did reach larger sizes. The most important constituents were lithistid demosponges (comprising about 50 percent of the mound volume), soanitids, and calcified cyanobacteria, although the accumulations were also partly bound by stromatoporoids or bryozoans by the Middle Ordovician and by various other organisms.

Two principal groups of the Ordovician reef communities had Cambrian forerunners. Lithistid sponges resembled archaeocyaths, which were their predecessors and relatives, but lithistid skeletons were built of needle-like siliceous elements (spicules) rather than of calcareous

GREENHOUSE, ICEHOUSE

Ordovician climate fluctuations from icehouse (1) to greenhouse and back are recorded in Antarctic ice cores. High levels of carbon dioxide (CO_2) in the atmosphere create a "greenhouse effect" increasing temperatures and, in turn, the amount of CO_2 taken up from the oceans. The polar ice caps melt and sea levels rise, encroaching on the continents and depositing carbonates. Global warming increases the chemical weathering of rocks and the carbonate compensation depth (CCD) falls as carbonate ions accumulate and CO_2 solubility reduces. However, the accelerated weathering extracts CO_2 from the atmosphere and the process eventually reverses, resulting in icehouse conditions: low sea levels, rising CCD, increased runoff, and high nutrient availability and plankton productivity.

granules. Spicules fused together formed a rigid, almost lithified framework that was resistant to moderate wave action. Soanitids (*Calathium*) were possible descendants of the largest Cambrian reef-builders, the radiocyaths. Their branching skeletons consisted of numerous multi-rayed rosettes, with each opposite pair connected by a shaft. The external rosettes, being joined together, formed the outer wall of the skeleton, and the internal rosettes were fused to the inner wall.

Soanitids had four rays per rosette and were covered with rhombic plates. In receptaculitids, evolved from soanitids and extinct in the Permian, the plates finally became one continuous cover. These organisms did not leave any living representatives, and neither their construction nor the complete constriction of body openings has been repeated among modern organisms. Even so the porous radiocyaths and soanitids may be compared in lifestyle to sponges, though their diet is not known.

WITH THE disappearance of tiny but diverse Early Cambrian reef animals, calcified cyanobacteria took over reef communities during the Middle and Late Cambrian. A new epoch of reef-building by skeletal animals started in the Ordovician. These reefs, generally larger than the Cambrian mounds, better exhibit a community succession. This means that a former community modified the environment (hardening the substrate, creating a rigid framework, and so on), making them more suitable for its successors— and, often, less suitable for itself.

Reefs modify their environment, facilitating the development of more diverse communities to succeed themselves.

An ecological succession began with a stabilization phase, when a crinoid (sea lily) colony forced the substrate to become stabilized. During the colonization stage, a reef usually formed from mutually interconnected lithistids and soanitids on top of lithified crinoid granules. As the reef grew, its surface and position higher in the water provided a suitable habitat for a number of species, and the community as a whole became diverse.

At this stage of diversification, calcified cyanobacteria joined the community by binding the resulting framework. If the cyanobacteria suppressed other organisms, the domination stage set in. It was not a problem for them because the reef rose towards sea level as it grew, and other species were unable to live in the higher-energy and fluctuating environments at these very shallow depths. This stage is called domination because only a few organisms were present, but these were significantly abundant to dominate the community. Finally, the mounds were cut by channels, buried by sediments, and the whole cycle began all over again.

The Ordovician radiation resulted in increasing complexity in reefs as well as other habitats. Like many other natural communities, reefs consist of interactive organisms. Species with similar appearance can easily replace each other, making communities volatile associations. Later Middle Ordovician and Paleozoic reef associations contained a variety of new reef-constructing organisms, such as bryozoans, various heavily calcified sponges (stromatoporoids and chaetetids), and corals. Because of the addition of framework organisms, the accumulations turned out to be quite different from the sponge-microbial fabrics characteristic of the older reefs, but that is more of a Silurian story.

Many other creatures, including cephalopods, brachiopods, echinoderms, gastropods (such as the nearly sessile snail *Maclurites*), chitons, and trilobites, were also present in varying abundance among Ordovician reefs.

GROUND FLOOR

Brachiopod-crinoid settlements (left) commonly formed the base of a reef. After death, crinoid skeletons of multiple calcite elements turned into a vast area of grainy substrate. These skeletal elements lithified rapidly, while more durable brachiopod shells made a good base by themselves.

ORDOVICIAN REEF

Ordovician reefs (below) retained some of the Cambrian reef-builders (calcified cyanobacteria). Further members of this guild were added, such as corals, which proliferated from that time although modern corals belong to a different group. Crinoids and brachiopods became widespread reef-dwellers during the Paleozoic but declined in the Mesozoic.

1 Thromboliths (bacterially precipitated mud)
2 *Renalcis* (a calcified cyanobacterium)
3 *Lichenaria* (a tabulate coral)
4 Trilobite
5 Crinoids
6 Orthid brachiopods
7 *Maclurites* (a sessile gastropod)
8 Homoiostelean echinoderms

COMPARING the trilobite with various recent creatures, one may suggest that its closest relatives are among arthropods (shrimps, lobsters, as well as insects, spiders, millipedes, and other similar creatures). A more precise comparison would reveal that shrimps and lobsters look very much alike. A lobster can be divided into a head shield with eyes, a segmented thorax and a tail, which is similar in outline to a trilobite pygidium. In comparison with the trilobite, a lobster also possesses multi-jointed antennae, claws, several pairs of walking limbs, and compound mandibles.

At first glance, trilobites look very "brainy," but this is just a huge stomach.

Trilobites buried alive by marine mudslides make the perfect specimens. This is the type of fossil that allows the study of features not preserved under the usual conditions of fossilization. It appears that a living trilobite had a pair of multi-jointed antennae, which transferred signals to a receptor organ (brain), and several pairs of forked appendages as well as the intestine. Thus, the trilobite could crawl along the seafloor using the lower branches of its limbs. Upper limb branches consisted of long, flattened filaments, closely spaced and forming fan-like sheets. These fans served like gills for respiration.

The trilobite had neither claws nor mandibles. This suggests that it probably held and ground up its prey by means of stout spines that covered the lower edge of the inner joints of its walking limbs. These joints were quite suitable for passing the prey, already chopped up into the consistency of digestible food, into the mouth. The mouth was located under the cephalon and opened rearwards. On a trilobite that had eaten just before its sudden death, an alimentary tract was preserved, which ran back from the stomach to the posterior tip of the pygidial axis.

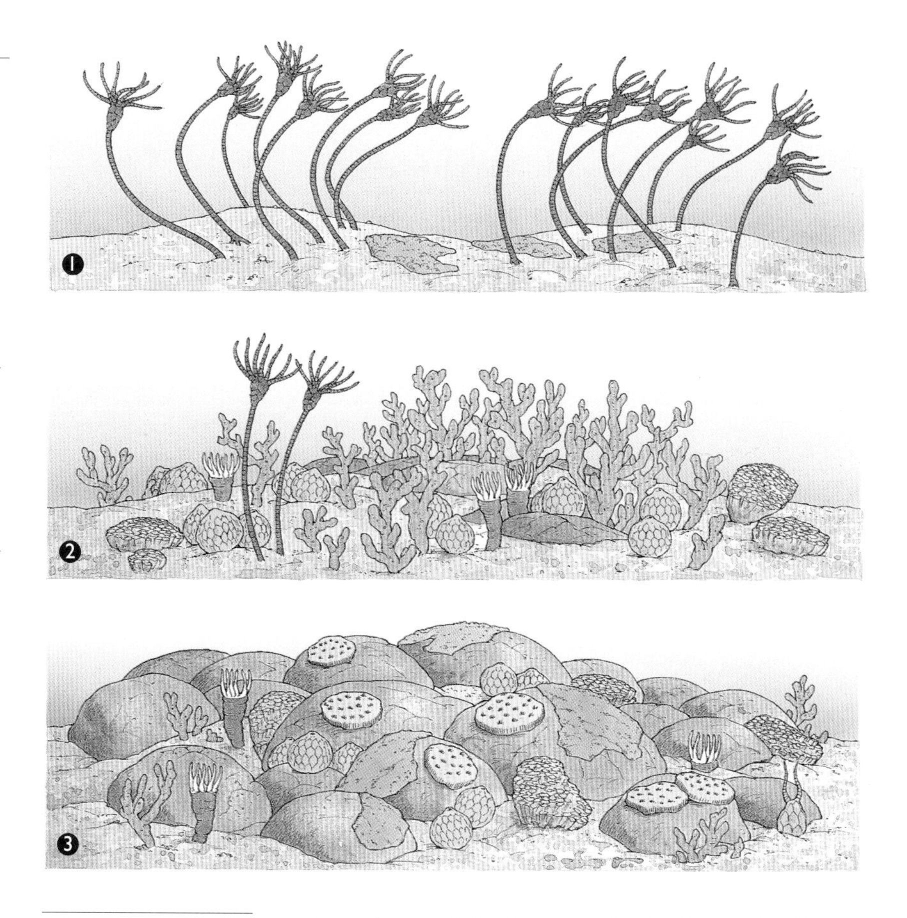

REEF FORMATION

Reef-building (above) often began with a settlement of crinoids (1), who anchored themselves to the seabed. Their disarticulated skeletons helped stabilize the substrate. During colonization, ball-like receptaculitids, bryozoans, and solitary corals created the first framework (2). Various builders, including large corals, contributed to the diversification stage(3).

Some modern lobsters, when closing their claws, develop a force of 800 newtons, enough to chop off your toe. Lobsters need such power to extract the soft bodies of mussels and snails from their thick rigid shells. Because such potential meals close a valve or lid as tightly as possible having noticed any moving shadow, it is necessary to use a powerful claw as a tin opener.

Such shells were hard to crack for trilobites. In any case, both bivalves and snails were so small during the times of trilobite proliferation that it was much easier to stuff the stomach with them, together with the mud with which they were covered, and then to lie down in the mud, out of harm's way, and to digest them slowly. That is why the trilobite stomach was so big and remarkable.

Trilobites did not overlook any other larger animals (lacking hard coverings) that burrowed in the loose sediment. The surface of such a sediment, if it has hardened rapidly, preserves trace fossils, allowing the entire hunting drama to be reconstructed more than 450 million years later. Here something worm-like, pulsating its fragile body, burrowed a straight cylindrical tunnel in the sediment. Intersecting the worm burrow, a trilobite moved, touching the seafloor with its antennae and leaving a track. The trilobite overtook the worm, loosening the mud with the sharp edge of its cephalic shield; squeezed the worm (which was not yet aware of the danger) with its spiny joints; and began to pass it, still alive, into its mouth. The imprint of the entire trilobite rested on the excavated sediment. Mud slowly settled down on it.

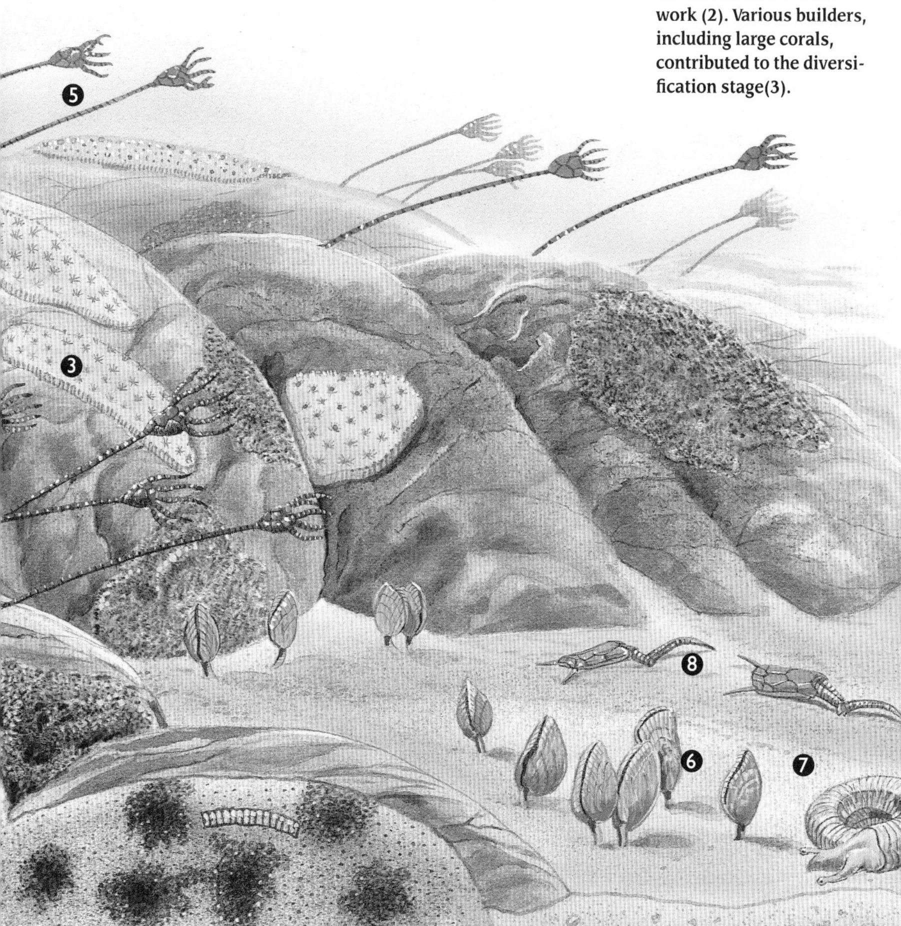

TRILOBITE organs of sight differed from those of a lobster. They resembled very much the bulging, compound, multi-lensed eyes of a dragonfly. Each lens (altogether there were up to 15,000 lenses in each eye) was a magnifying glass. As the trilobite grew, the eyes became bigger, and the field of view gradually changed from 30 to 90 degrees. The lenses were arranged in such a way that they brought light to a focus at about the same relative distance below them within the trilobite's head where, probably, there was a light-sensitive organ. The resolving power of a lens in an eye of an advanced trilobite has been estimated as ten times that of a modern frog's eye. In addition, adjacent lenses within one eye could have been used to provide stereoscopic vision. In the way that a highway patrol's radar sends out a series of pulses to detect the speed of a moving car, a pair of adjacent lenses covering a particular area of the visual field could pick up the same passing object on their respective retinas. As it moved towards or away from the trilobite, its distance could be inferred by comparison of images in adjacent retinas at any one time, while its sideways movement could be detected by comparison at successive time intervals (like the radar). Such a complicated organ could have operated in this way if the neural relay system was as advanced as the lenses. Of course, trilobite eyes were well adapted for a life under dim light intensities, and some of them could see even in almost total darkness.

Trilobite eyes had up to 15,000 lenses each, and some were extended on stalks, periscope-like.

TRILOBITE EYE

Trilobites had compound eyes with as many as 15,000 lenses in each eye. Sharp vision gave them an advantage in hunting and defense that allowed them to survive longer than the dinosaurs did.

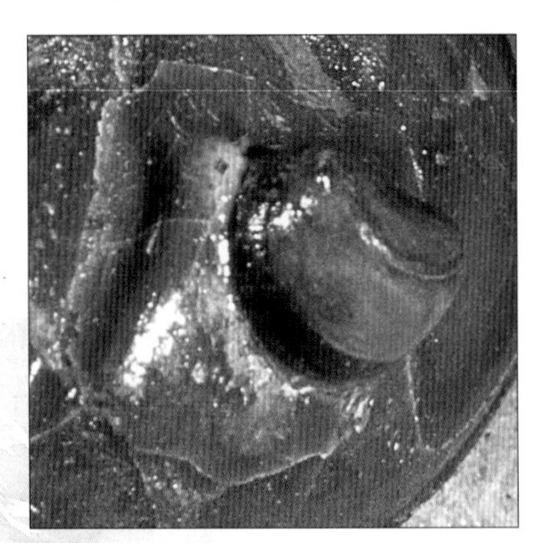

If sensitive antennae helped trilobites to look for prey, the eyes, on the other hand, helped them to avoid becoming the prey of something else. Some trilobite eyes extended into a rigid stalk. The animal itself could be completely concealed under the mud, leaving only its periscope-style eyes exposed. Other trilobites used their eyes to calculate when to roll themselves up into a solid, spiny ball which hardly any predator could unravel.

The eyes were especially important during the time of molting. The rigid external skeleton had to be shed from time to time to allow growth, leaving the soft parts totally exposed. During molting the old exoskeleton split along prominent sutures on the surface. From 80 to 90 percent of mortality among arthropods occurs during molting, when they become easy prey, or when they get stuck inside their own skeletons because they have grown too quickly. Different trilobite species had to shed their skins from 8 to 30 times during their lifetimes, and each time it became one or two thoracic segments longer.

Sometimes a surface of the same slab of several dozen square feet contains up to several hundred similar-sized trilobite exoskeletons. An analogy with living marine arthropods can be drawn, suggesting that such clusters could have resulted from a behavioral pattern of a species gathering at a particular stage, including seasonal molting and reproduction.

TRILOBITE LIFESTYLES

Despite their simple design, trilobites were adapted to different lifestyles. Those with light skeletons floated in the water column, hunting for smaller zooplankton. Benthic trilobites crawled along the seafloor, eating whatever they could find. Others, with powerful spiny appendages, hunted burrowing worms. Some reef trilobites became genuine reef-dwellers. They became nearly-immobile filter-feeders anchored to the substrate.

Pelagic (free-swimming) trilobites
1 *Carolinites*
2 *Irvingella*

Benthic (bottom-dwelling) trilobites
3 *Isotelus*
4 *Triarthrus*
5 *Acaste*
6 *Conocoryphe*
7 Fully rolled up trilobite (*Acaste*)
8 Shed skin with a trail in the sand

TRILOBITE LOCATIONS

Trilobites and other animals were separated by oceans into distinct provinces. The warm reef-filled onshore seas of paleoequatorial Laurentia and Siberia had spiny bathyurids. Equally warm but remote seas of Gondwana (South China, Australia, Argentina) were colonized by asaphids, dalmanitoids, and leiostegiids. Other asaphids occupied Baltica. High-latitude Gondwana (southern Europe, Arabia) was dominated by large calymenoids and dalmanitoids. Only the deepwater olenids had a wide distribution in both Laurentia and Gondwana.

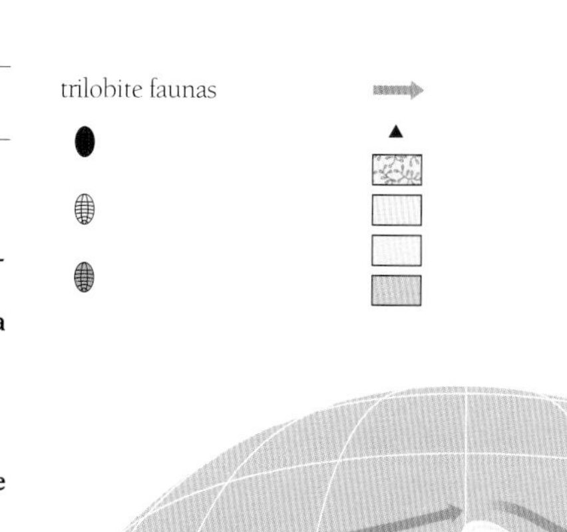

The improving eyesight and ability to molt on time because of advances in the ability to shake off the old tight skeleton became essential features in the long history of trilobites, which had originated in the Early Cambrian and died out on the eve of the Permian mass extinction. These features allowed them to escape carnivores such as the cephalopods.

CEPHALOPODS ("head-foot") are not the most common mollusks in the modern seas. Bivalves and gastropods ("stomach-foot") both are much more diverse. In addition, they have settled down in rivers, lakes and ponds, and snails have crept out onto the land. On the whole, mollusks are the second largest phylum after arthropods. Cephalopods, however, are the most advanced mollusks. They are even called "the primates of the seas." (The primates of the land are humans, apes, and monkeys.) The most primitive mollusks are aplacophorans ("none-plate-carrying"), which never had a shell (modern slugs have lost their shells secondarily); polyplacophorans ("many-plate-carrying"), and univalved monoplacophorans ("one-plate-carrying").

Cephalopods, sometimes called the "primates of the seas," were among the most important predators in Paleozoic seas.

TRILOBITE MOLTING

Trilobites had several steps to shedding their exoskeleton. (1) Inverting the body against the seabed and contracting the muscles. (2, 3) Hunching up the shield produced a break around the cephalon. (4) Extra sutures on the cephalon allowed the old one to be split into several small pieces to make the head easier to free. This was a major evolutionary advance.

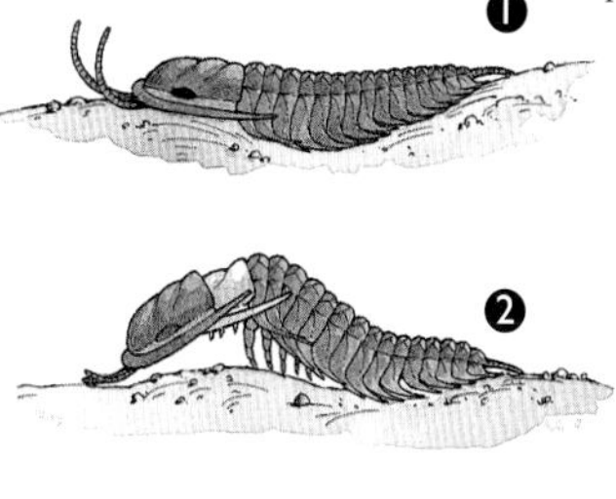

Diminutive monoplacophoran-like animals are usually considered to be the ancestors of the cephalopods. They originated in the Late Cambrian, and their first remains are known from the continental fragment that became northern China. At the very end of the Cambrian, cephalopods had already achieved their first peak of diversity, but those lived mostly in the same region and had relatively small and simple shells of a few inches long. The principal expansion of cephalopods began in the Ordovician, and since then they have steadily grown in variety and abundance.

Two features ensured the success and proliferation of cephalopods: a chambered shell that functioned as a well-constructed hydrostatic apparatus to provide effective buoyancy; and a jet propulsion system, making them very dynamic animals. In addition, relatively well-developed brains and big eyes connected to the brain were important hunting equipment for these mostly carnivorous swimmers. It is worth noting that mammalian and cephalopod eyes are strikingly similar in structure, with a cornea, lens, and retina, but the two have a very different origin from each other.

A mollusk also possesses a mantle, a fold in the body wall that lines the shell and secretes the calcium carbonate of which it is made. A basically conical shell is subdivided by septa into multiple chambers. Each septum has an opening, and a tubular outgrowth of the rear body side (siphuncle) continues through the entire shell. With the siphuncle, an animal regulates the gas/liquid pressure in the chambers, allowing it alternately to rise to the ocean surface and to submerge again like a submarine. It is no coincidence that the last living cephalopod with an external shell has the same name as the submarine commanded by Captain Nemo in Jules Verne's classic novel *Twenty Thousand Leagues Under the Sea*: the name of *Nautilus* (meaning "small ship").

A NAUTILOID body occupies the last or living chamber, from which protrude a head bearing large eyes and sharp beaklike jaws; numerous tentacles; and a funnel. By energetic contractions of muscles of the abdominal part of the mantle, an animal creates jet propulsion in the funnel and moves rapidly backward. Most Cambrian nautiloids were probably bottom-dwelling detritus feeders and scavengers crawling along the seafloor. A chambered buoyant shell was used as a float to minimize the effort of movement.

Ordovician nautiloids owed their expansion to the range of designs for their shells, which grew up to 20ft (6m) long in some groups.

The success of the Ordovician nautiloids lay in a different use for the shell. Actinoceratids and endoceratids had long straight shells with a series of chambers and calcareous fillings on the bottom that allowed them to move horizontally above the seafloor. Some endoceratids achieved a length more than 20ft (6m), although the shell stayed narrow at less than 12in (30cm) in diameter. Where orthoceratids and endoceratids had ten tentacles each, oncoceratids possessed numerous stout tentacles like the modern nautilus. Their shells were thick-walled and sculptured, with chambers that were too small. Such heavy animals were merely slow crawlers able to hang suspended in the water for only a short time.

Ascoceratids changed the shape of their shell several times during their life cycle. A juvenile ascoceratid had a tiny wide-conical shell and floated in the sea as if on a parachute. As new chambers were added, the shell became stick-like and the animal went down to the sea floor. The later chambers had a flattened shape, and the entire shell turned into a swollen thin-walled bag with a stick-like extension. This dead end finally broke off, and the ascoceratid was able to float again.

Following the Ordovician, nautiloids and their relatives the ammonites (which flourished from the Devonian until the Cretaceous) acquired hard jaws and coiled flat-spiral shells, better shaped for more active swimming. Mesozoic cephalopods evolved into animals with extremely streamlined shells, but even this was a great handicap not suffered by those that had no heavy external coverings at all. These were Triassic–Cretaceous belemnites, squids (their descendants), cuttlefishes, and octopuses. The latter might have evolved from the ammonites. Squids are capable of bursts of speed up to about 50ft (15m) per second (equivalent to 30 knots) and, having reached it, can jump up to 150ft (45m), whereas the nautilus, retaining its external shell, swims at a maximum speed of 9in (0.25m) per second (0.5 knots). About five species of nautiloids are left in modern seas, and there are hundreds of cephalopod species with an inner shell that has proven a successful design.

Ordovician mature nautiloids slowly voyaged in their bright colorful shells across shallow (less than 655ft/200m deep) epeiric seas. They were far from good swimmers, unable to maneuver quickly, and they hunted weak or dead soft animals that posed no great

CEPHALOPOD SCENE

Early cephalopods were not good swimmers. Ascoceratids in their can-shaped shells floated passively in the water column. Streamlined orthoceratids were better swimmers but remained bottom-dwellers for a long time. Flat cone-shaped actinoceratids could glide over the seafloor like modern flatfish. It was only the coiled shell, which appeared in the Middle Ordovician, that allowed cephalopods to leave the bottom. The last living nautiloids still retain this shape.

1 *Heracloceras* (barrandeocerid nautiloid)
2 *Orthonybyoceras* (actinoceratid)
3 *Mandaloceras* (discosorid nautiloid)
4 *Gonioceras* (actinoceratid)

ORDOVICIAN

NAUTILOID LIMESTONE

In the Ordovician and Silurian, orthoconic nautiloids were so common that their remains formed dense shell beds; this one (left) is in Sardinia. Ancient cephalopods swam in large schools.

ESTONIOCERAS

***Estonioceras* (above), from the Middle Ordovician of Estonia, is a tarphyceratid, the first cephalopod with a coiled shell. Its small diameter (1.5in/ 4cm) and flattened shape reveal its adaptation for swimming. Coiling made the shell stronger; evenly arranged chambers made it more buoyant.**

challenge to a hunter. Such prey was readily available in the Leningrad Region (near modern St. Petersburg) of the Silurian Plateau of modern Russia. During the Ordovician this was a coastal part of the Tornquist Sea, which covered Baltica, with the Leningrad Region in its shallowest area. The deeper part of the sea opened into the Iapetus Ocean.

THE LENINGRAD Region is located on the Silurian Plateau, a geological structure built of Ordovician limestones whose name recalls the heated debates about the validity of the Ordovician period. The plateau stretches from the Lake Ladoga in the east across the Leningrad Region to northern Estonia. These fossil-rich strata, cut by numerous rivers and exposed along the Baltic Sea coast, have been under close study since the eighteenth century. Pioneering geological descriptions and excellent woodcut prints of the area were made in the 1830s. Later, a number of Russian geologists and biologists with Western roots continued the quest. Edward Eichwald published the first full treatise on fossils of European Russia, with more than 2000 illustrations, including many Ordovician species. Christian-Heinrich Pander discovered very important fossils, the conodonts, which are now ascribed to the chordates. At the beginning of the 20th century, Roman Hecker, the father of paleoecology, described the extremely strange echinoderms of this area.

In the warm shallow seas of the Iapetus Ocean over Baltica flourished a diverse community of bottom-dwelling fauna.

The fossils of the Silurian Plateau are as unique as their investigators. Echinoderms comprised the core of the Middle Ordovician shallow-water, bottom-dwelling, storm-dominated community in the Leningrad Region which, together with Baltica, was located in southern temperate latitudes. In the early days of investigation a German paleontologist gave a reconstruction of an Ordovician echinoderm in which he joined together representatives of three different classes. Those were sea lilies, ophiocistioids, and several eocrinoids. Another early attempted reconstruction shows a starfish with prominent bulbous spines (bolboporites).

Nowadays there are five classes or principal body plans of echinoderms: starfish, sea urchins, sea lilies, sea cucumbers, and brittle stars. In the Ordovician there were about 12 classes in the Leningrad Region alone. A free-living homoiostelean ("equal column"), which has sometimes been suggested as a vertebrate ancestor, had a flattened asymmetrical body with two tail-like appendages in front and behind. It is almost impossible to guess whether it used these to burrow in the substrate or hang onto the sea lilies. An ophiocistioid ("snake and bubble") possibly walked on its arms like a modern brittle star, but another reconstruction suggests an opposite downward orientation of its cup-like body with a whorl of filter-feeding arms.

A bolboporite ("porous tuber") was reinterpreted as a small eocrinoid (0.5in/1.5 cm tall). It had a single brachiole and a solid tuber-like calyx. Typical eocrinoids ("dawn" and "lily-like") staggered on long stems and caught food carried by water currents using a bunch of long brachioles. However, some of them looked like upright cucumbers planted on sturdy holdfasts.

Sea lilies had even longer stalks and arms. Each arm carried a groove along which tube feet passed food particles to the mouth. A paracrinoid ("near lily-like") had a lens-shaped calyx with arms sticking out like porcupine quills. A rhombiferan ("rhombi-bearing") resembled a water tower with a reservoir of riveted diamond-shaped plates. It was given the common name of "crystal apple."

All these stemmed echinoderms occupied different tiers above the seabed to filter the food from currents at different levels. The unicellular planktonic alga *Gloeocapsomorpha* was especially abundant and comprised the bulk of the organic matter in an oil shale (kukersite) deposited in a nearby deeper basin of Estonia.

THE EVOLUTION OF NAUTILOIDS

In the Ordovician and Silurian, cephalopods produced an immense variety of shell types, some of which were more than 12ft (3.6m) long and over 6600lb (3000kg) in weight. Straight (orthoconic) conches were common among early Paleozoic cephalopods such as nautiloids, having developed from the Cambrian endogastric form (lower figure). Nautiloids with exogastric shells became coiled into a flattened spiral (upper figure). Coiling made the shell stronger with the imbrication (overlapping) of shell layers, while the even spacing of the chambers improved buoyancy. These nautiloids had a distinct evolutionary advantage and proliferated rapidly. After the end-Ordovician and Late Devonian mass extinctions, almost all cephalopod lines had a coiled shell.

In the Mesozoic, the dominant shelly ammonoids rarely repeated other shell types. Some of them built something resembling a small whorl, saxophone, or ball of rope, but such forms did not exist for long. In the Mesozoic, with the appearance of large carnivorous marine reptiles and sharks, even the most streamlined fast-swimming ammonoids were not safe. Belemnites and their descendants lost the heavy outer shell but gained the advantage of superior maneuverability to become the cephalopods of new Cenozoic times.

ORDOVICIAN

1 Treptoceras (a nautiloid)
2 Crinoids
3 Thallograptus (a graptolite)
4 Rhipidocystis (an eocrinoid)
5 Neorhipidocystis (an eocrinoid)
6 Conodonts
7 Pachydictya (a bryozoan)
8 Asaphus (a trilobite)
9 Cuneatopora (a bryozoan)
10 Dittopora (a bryozoan)
11 Cryptocrinites (an eocrinoid)
ORDOVICIAN

12 *Bockia* (an eocrinoid)
13 *Simankovicrinus* (an eocrinoid)
14 *Paraconularia* (a conulariid)
15 *Bolboporites* (a bolboporite)
16 *Volchovia* (an ophiocistioid)
17 *Treptoceras* (a nautiloid)
18 *Siphonotreta* (a brachiopod)
19 *Clathrospira* (a gastropod)
20 *Monticulipora* (a bryozoan)
21 *Cheirurus* (a trilobite)
22 *Heckericystis* (a homoiostelean)
23 *Echinoencrinitos* (a rhombiferan)

BECAUSE echinoderm skeletons consist of multiple calcareous plates, after death a huge amount of debris accumulates. The debris, being rapidly lithified, enlarges an area of hard ground suitable for settlement by new echinoderms, which in turn produce more debris, which upon becoming lithified enlarge an area, and so on. Some scientists have even suggested that a small settlement of echinoderms would be enough to start a great expansion of the community.

The Middle Ordovician hard ground of the Leningrad Region in Baltica had one of the richest Early Paleozoic marine communities.

Finally, a diverse community of echinoderms and dome-like branching bryozoans, articulate brachiopods, receptaculitids, graptolites, and conulariids became established and proliferated. Conulariids had a four-sided, elongate pyramidal skeleton with arched transverse ridges crossing each side and a four-lobed lid folding in the manner of origami. They existed from the Ordovician until the Triassic and had some similarities with some scyphozoan medusas. Conulariids, as well as other filter-feeding sessile organisms, preferred to attach themselves to the ground and to other organisms. Dendritic graptolites entangled empty nautiloid shells, to which rhombiferans and craniid brachiopods also adhered. Sea lilies twined together and were encrusted by bryozoans. Even living nautiloids' shells were used as substrate on which bryozoans and cornulits (small wormlike animals living in calcareous conical tubes) settled. Unknown large borers drilled the hard ground—the first significant living destroyers of hard rocks.

Detritivorous ostracodes and trilobites, and carnivorous-jawed polychaete worms, crawled among those inhabitants of the hard-ground community. Above the substrate, tiny sharp-toothed conodonts and long narrow endoceratids plied in the search for not-too-fast food. Scars found on the fossils of trilobites and brachiopods may be the marks left by unsuccessful hunters.

THE EVOLUTION OF TRILOBITES

PEOPLE have known about trilobites for centuries. Some Native American tribes from Utah, where Cambrian and Ordovician trilobite fossils are very common, call them "little water bugs that live in a stone house" and once valued them as amulets. In the vicinity of the capital of the Czech Republic, these Lower Paleozoic fossils are so widespread that, from the Middle Ages, local bakers used impressions of large trilobites as molds in which to bake ornamental honeycakes. Now trilobites are among the most desirable trophies for fossil collectors. Their abundance and rapid rate of evolution have made them ideal tools for geologists.

There are trilobites among the fossils picked up during the study of Paleozoic strata, especially from the Cambrian and Ordovician. So what can we see on a trilobite? It is stony, of course. If this fossil is treated with a drop of hydrochloric acid, the drop becomes cloudy, fizzes, and bubbles like good champagne. The reaction indicates that the trilobite remains consist of calcite (calcium carbonate), a mineral that is easily dissolved by an acid with the release of carbon dioxide.

The length of adult trilobites varied from a fraction of an inch up to 2ft (1mm to 70cm) in different species. With a specimen of about 4in (10cm), its principal features can be seen without a microscope. The trilobite has a flattened bilaterally symmetrical body: its right side is a mirror image of its left. It slightly resembles the back half of a coat of mail, consisting of several jointed convex shields, as if the mail has been split along the shoulders by a sharp sword. An elongated oval in outline, the trilobite's coat is subdivided by clear sutures into three parts: a horseshoe-shaped head shield usually bearing eyes (the cephalon); a subtriangular tail never bearing eyes (the pygidium), and a body shield (the thorax). Multiple similar segments form the thorax. (Segmenta-

354 Devonian 417 Silurian 443 Paleozoic Ordovician 490 Cambrian 545 Precambrian mya

TRILOBITE LINEAGES

With the appearance of cephalon sutures facilitating molting, trilobite evolution became a theme rather than an innovation. Trilobites never gained the diversity of structure and function as other arthropod groups. Their few modifications were miniaturization; increasing and reducing thoracic segments; smoothing of the carapace's narrowing (mostly in swimming forms) and widening of the body; and the development of compound eyes.

1 The fused cephalon and pygidium, trilobation (division into lobes), eyes on carapace, and skeleton calcification are present
2 Dorsal ecdysal sutures and eye ridges developed
3 Free cheeks sutured from glabella entry
4 Miniaturization and reduction of thoracic segments
5 Enrollment appeared
6 The most advanced eye type evolved

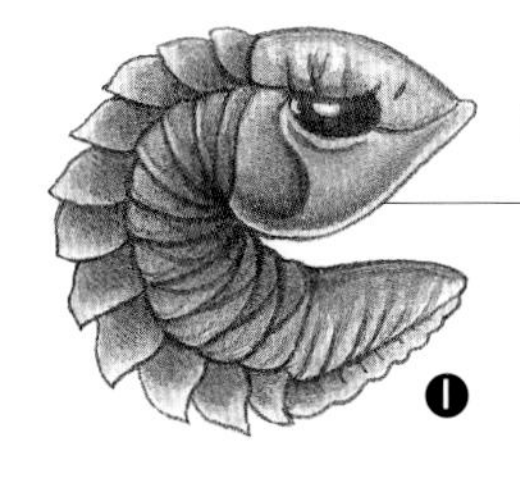

Cephalon and
pygidium
meet to
form a seal

ROLLING UP

Cephalopods and jawed fishes, which appeared in the Silurian, became dangerous predators of trilobites. In order to defend its more vulnerable bottom side, a trilobite could roll itself up into a ball to be protected by its rigid upper shields. The cephalon and pygidium were nearly the same size for an easy fit, with a groove along the anterior of the cephalon into which the posterior of the pygidium could slot (1); the two parts then locked together (2). Projecting spines made the trilobite too big for a predator's mouth.

TRILOBITE ANATOMY

A trilobite skeleton is made up of the head (cephalon), body (thorax), and tail (pygidium) shields. Both cephalon and pygidium consist of several fused segments. The three major head lobes are central bulbous glabella and flat-sided cheeks bounded by facial sutures. Compound eyes are located on the sides of the glabella. The thorax includes two to 40 almost identical segments. Trilobite remains from Lagerstätten show a mouth, three pairs of legs, and the bases of two articulated antennae on the underside of the head. The mouth leads to a large stomach located under the glabella and tapers into an intestine running back to the pygidium. Each segment has a pair of jointed legs and each leg is divided into a lower walking branch and an upper branch with feathery gills. Exceptional specimens show a simple heart above and behind the stomach, and segmented vessels running through the axial lobe.

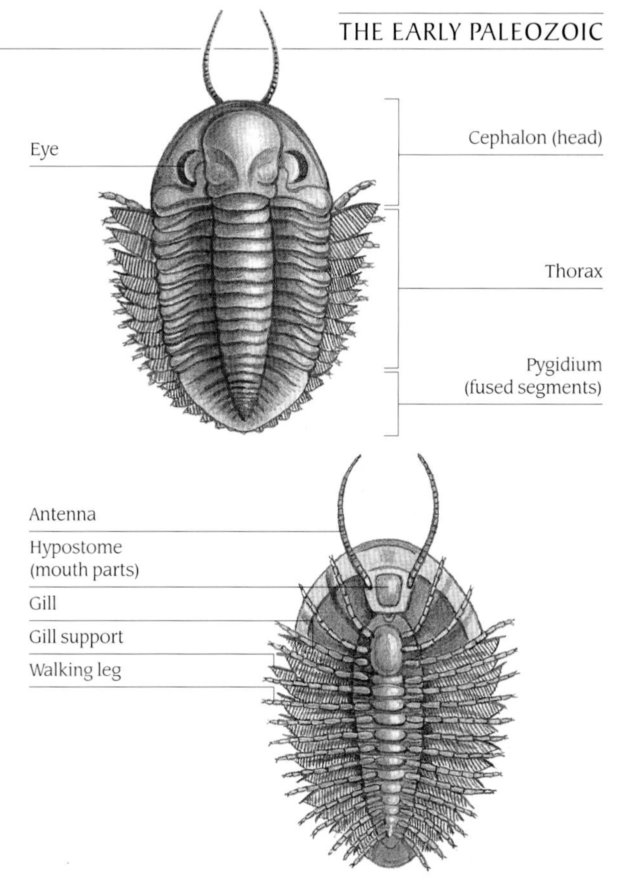

tion may also be seen in thin human bodies. However, in this case, segmentation is produced internally by the ribs; the trilobite, possessing an exoskeleton, has external segmentation. In the seventeenth century these differences were not yet understood, and trilobites were ascribed to the vertebrates as having true ribs.)

In addition, the entire trilobite exoskeleton is divided by shallow infolds, evident externally as furrows. Two longitudinal axial furrows running from the cephalon to the pygidium separate the convex central axis from the parts at the sides. Thus, the entire exoskeleton is three-bladed, giving the name to an entire class of animals (the Trilobita). Spines—hollow projections from the exoskeleton—may occur in any of its parts.

Soon after their appearance, trilobites spread across the world and became the principal animals in many marine communities in terms of both diversity and abundance. There were both pelagic and benthic, burrowing and crawling, carnivorous and suspension-feeding trilobites in the seas. A strange group of Cambrian–Ordovician trilobites lacked the eyes and all but two thoracic segments; they were extremely small and had an almost indistinguishable cephalon and pygidium. It is possible that these enigmatic creatures were bivalves, floating in the water and filtering food with their spiny limbs.

Ptychopariids

Phacopids

Lichids

Odontopleurids

Proetids

Early Carboniferous 324 Late Carboniferous 295 Permian

THE SILURIAN

443 – 417 MILLION YEARS AGO

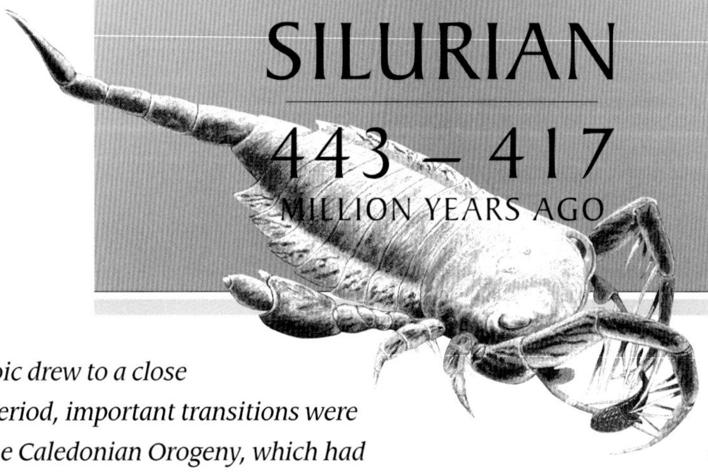

As the Early Paleozoic drew to a close with the Silurian period, important transitions were taking place on Earth. The Caledonian Orogeny, which had begun in the Ordovician, united Laurentia with Baltica and Avalonia as a great northern continent that was to remain largely intact until the Cretaceous. This event left its mark in visibly similar mountain ranges along the edges of what later became North America and northwest Europe. Glaciers withdrew from the poles, the sea level rose, and gentle, equable climate conditions spread over the globe. Marine invertebrate animals continued to thrive while significant new life forms appeared. Among these were diverse jawless vertebrates and the first fishes. Plants and invertebrate animals had become permanently established on land by the end of the period. Stromatoporoid reefs grew to great size.

THE SILURIAN System is the third subdivision of the transitional "old greywacke" beds of Paleozoic rocks. It was identified by Roderick Murchison, a prominent early British geologist, along the border between England and Wales. In 1835 Murchison published an article in the *London and Edinburgh Philosophical Magazine* in which he named the system after an ancient tribe, the Silures, who once inhabited the region. He defined the Lower Silurian, which was later renamed the Ordovician System, and the Upper Silurian, which was also reclassified and became the Silurian as it is defined today. Because of its distinctive fossil content, the system was soon recognized in many other countries of Europe, in America, and across the Himalayas. Charles Darwin reported finding Silurian fossils in the Falkland Islands, though they later proved to be Devonian.

Distinctive fossils made the Silurian strata easy to identify in most locations, and its dates were quickly accepted.

KEYWORDS

CALEDONIAN OROGENY
CHORDATE
CONODONT
CORAL
GRAPTOLITE
GREYWACKE
HYDROTHERM
STROMATOPOROID
URALIAN OCEAN
VASCULAR PLANT
VERTEBRATE

The Earth in Silurian times was a hospitable planet without polar ice caps, with gentle climates, high sea levels, and a widespread, nearly global fauna unaffected by mass extinctions. Perhaps it was the favorable climate that triggered the first radiation of the vertebrates, the most adaptable animals. By the end of the Silurian, all the major groups of jawless vertebrates and fishes existed. Another great milestone also occurred during the period: the appearance of the first land-living communities.

Sea level rose rapidly at the start of the early Silurian as ice caps melted and then fell again at the end of the period to its lowest since the Cambrian. The Late Silurian widening of lowlands, perhaps, forced previously aquatic plants and animals to adapt to unpleasantly (for them) dry conditions. The average global temperatures were

SILURIAN

ORDOVICIAN	443 mya – 440 – 435 – SILURIAN – 430
Series	EARLY/LOWER
Series	LLANDOVERY
European stages	RHUDDANIAN; AERONIAN
N. American stages	MEDINAN
Geological events	Closure of the Iapetus Ocean; The Rheic Ocean begins to close as Gondwana moves west
Climate	Deglaciation; Decreasing levels of carbon dioxide
Sea level	Moderate; Rising
Plants	
Animals	•First hydrothermal community; •Radiation of graptolites; •Radiation of jawless fish

some 7° to 9°F (4° to 5°C) higher and the pole-to-equator temperature gradient was not as extreme as nowadays. That fact, together with the absence of permanent polar ice caps, prevented the mixing of ocean waters and caused stratification into a relatively oxygenated upper layer and an anoxic lower layer. Black shales were deposited in abundance in deeper basins.

The environment of the early Paleozoic oceans may be analyzed by the use of phosphatic fossils and rare earth elements accumulated in the sea sediments. In modern marine sediments deposited under oxygenating conditions, the rare earth element cerium is very rare compared with the related elements lanthanum and neodymium. In contrast, biogenic apatite (a phosphate mineral) of the early Paleozoic is enriched with cerium. This anomaly confirms that the early Paleozoic oceans, including the Silurian, were predominantly anoxic. During major upheavals the oxygen-deficient waters shifted onto continental shelves and into epeiric seas, resulting in local extinctions and the expansion of black shale deposits. Graptolites flourished in the deep anoxic waters, and they evolved rapidly.

Soon after the recognition of the Silurian the German geologist Albert Oppel showed that Jurassic rocks in southern Germany could be divided into 33 zones based on pelagic ammonites. According to him, exploring the vertical range of each species allowed such precise subdivisions to be made. In 1878 Charles Lapworth applied this method to the graptolites of the early Paleozoic and found that many species had very short time ranges. He used this new zoning to map the contorted stratigraphy of southern Scotland. Current studies of these extinct animals have revealed that an even finer subdivision is possible, although Lapworth's pioneering zonation is still valid today over a century later. The duration of a zone ranges from 0.44 million up to 1.43 million years—a mere blink of an eye in geological time. These short segments allow a close-up look at the succession of biological and geological events throughout the Silurian.

THE CALEDONIAN mountains of Scotland, northwestern Ireland, and Scandinavia were forced up when Avalonia and Baltica collided with Laurentia. When the temperature rose slightly, global volcanic activity and the mountain-building between Laurentia and Baltica slowed down. Magma intruded into what were to become the Highlands and absorbed sedimentary strata, turning them into coarse-grained igneous rocks known as granites. Their coarse crystalline texture resulted from the slow cooling of molten material many miles below the Earth's surface. In eastern England the Caledonian orogeny is concealed beneath the flatlands of East Anglia. In contrast, the Scottish Highlands have retained features of the Caledonian landscape because of the granite's resistance to erosion, despite millions years of rain pouring down on the stone and a multitude of glaciers passing over.

Continuing tectonic activity after the Silurian buried some strata, while others remained grandly exposed.

These continents straddled the paleo-equator with their centers of mass in the southern hemisphere. During the Silurian, Laurentia was stationary at the equator, whereas Avalonia and Baltica continued their northward movement with a sideways drift of approximately 3 to 4 inches (8 to 10cm) per year until the continents met and fused.

See Also

THE ORIGIN AND NATURE OF THE EARTH: *The evolving atmosphere*
THE EARLY CARBONIFEROUS: *Acadian–Caledonian orogeny; life on land*
THE PERMIAN: *New Red Sandstone*

SILURIAN HEAT

The Silurian Period was one of the warmest in Earth's history. The collision of several continents produced a huge amount of heat through tectonic–volcanic processes. Magma from the collision events was intruded and extruded as the world's first high mountain chains, including the Caledonians of northern Europe. The average global temperatures increased, melting the polar ice caps. In the ocean depths, settlements of worms basked in the hydrothermal heat. Warmth also facilitated the development of coral-stromatoporoid reefs, floating hemichordate graptolites, and swimming true chordates, both jawless and jawed fishes.

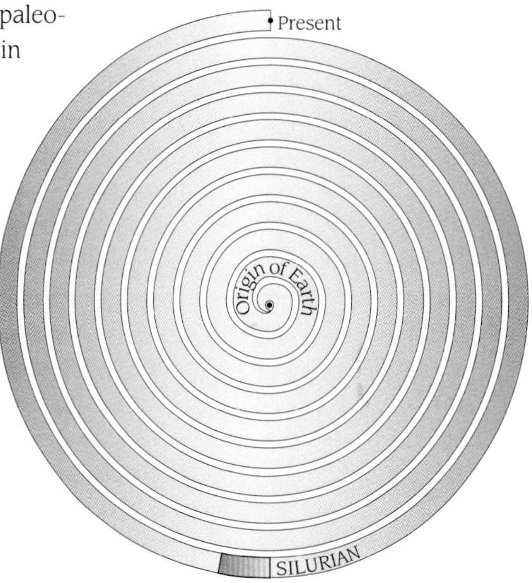

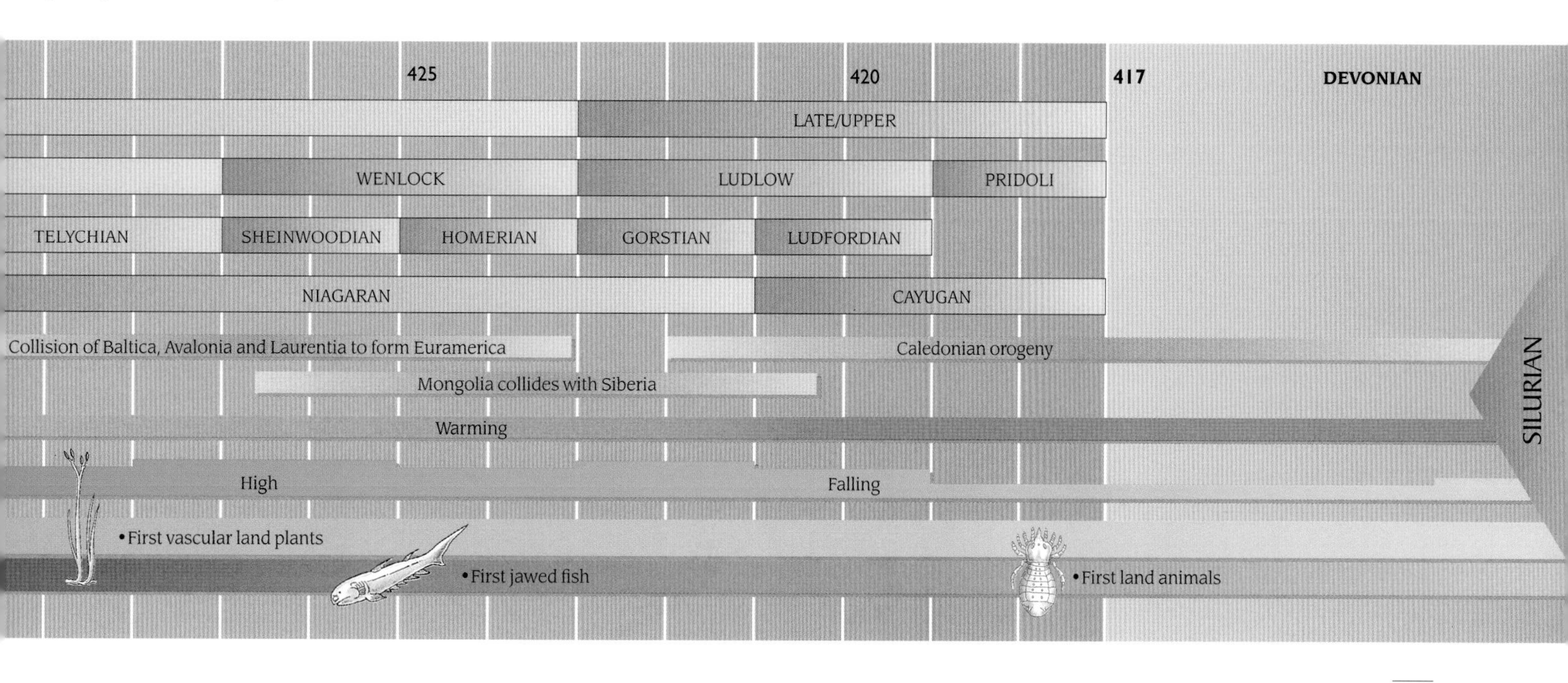

THE CALEDONIANS

The Caledonians (left) were some of the first high mountains, and gave their name to the entire epoch of orogeny (mountain-building) which extended from the Silurian to the Early Devonian.

The Iapetus Ocean was closed and the remainder of its basin transformed in sheetlike nappes of the present Europe and North America. The ocean's northern part disappeared first. The southern parts, between Avalonia and Laurentia, were preserved until the middle Devonian.

THE PRESENT border between England and Scotland is a monument to the collision of Avalonia and Laurentia, with the composition of rocks between England's Midland Valley and Lake District to Scotland's Southern Uplands showing increasing similarity until they correspond perfectly in Middle Silurian strata. In Scotland, a structural feature called the Iapetus Suture divides the ancient Laurentian terranes from the Avalonian ones. The igneous rocks of northwestern Ireland and Scotland are covered with greywackes composed of immature sandstones and redbeds consisting of pebbles, cobbles, and boulders which formed with active mountain building and erosion.

The collision between Avalonia and Laurentia has been recorded in rocks on the border between England and Scotland.

Geochemical analysis helps to determine the provenance of rocks in the Caledonides and the pattern of events related to their formation. For instance, the ratio of alumina to silica provides an indication of the relative proportion of quartz to clay, whereas the ratio of potassium oxide to sodium oxide serves as a measure of the potassium feldspar and clay against plagioclase content, and so on.

Feldspar is a common aluminum- and silica-rich, white transparent mineral, and the potassium content is typical of such rocks as granites. It accounts for 60 percent of the entire lithosphere. Plagioclase (meaning "oblique split") is another aluminum- and silica-rich mineral but of a dark color and with a significant sodium content. Both minerals and chemical compounds vary with rock sources, which in turn depend on the local magmatic processes and volcanism resulting from plate tectonic movement. Even so, the geochemical fingerprints of the vanished Ordovician to Silurian active northern margin of the Iapetus Ocean can be distinguished among the Caledonian greywackes.

During the Silurian Period, the northern hemisphere was still oceanic, and Gondwana occupied the southern hemisphere with the South Pole in either its South American or central African part. The Paleothetis Ocean divided Gondwana from Laurentia and Baltica. In the Mesozoic this ocean would reappear between the rapidly disintegrating Gondwana and growing Asia, former parts of the short-lived supercontinent of Pangea.

Carbonate rocks of Silurian age were laid down in Laurentia, northeastern Avalonia (England), Baltica, equatorial Gondwana (Chinese fragments), Siberia, and the Kazakhstan terrane. The last two of these continental blocks gradually drifted into the northern temperate belt. Falling sea levels isolated the wide Siberian seas from the open ocean, and dolomites with evaporites were precipitated there.

The vast continent of Gondwana, with its contrasting relief of high mountains and deep canyons, still affected sedimentation in its epeiric seas, where large deposits of siliciclastics accumulated (Arabia, northern Africa, South America, and Australia). The Bohemian Massif completely detached itself from Gondwana and left the vicinity. By Late Silurian times it had reached latitudes at approximately 20°S abutting the southern margin of Baltica. Carbonates with minor siliciclastics were formed on this terrane.

Faunas were evenly distributed in the Silurian marine basins. Most shallow seas were inhabited by corals, stromatoporoids (massive calcified sponges), receptaculitids, calcified algae, bryozoans, crinoids and other echinoderms, trilobites, brachiopods, bivalves, and gastropods. Only the fauna of the temperate southern seas were impoverished, although two groups, the conodonts and conulariids, had many forms.

Africa and Middle East
Antarctica
Australia and New Guinea
Central Asia
Europe
India
North America
South America
Southeast Asia
other land

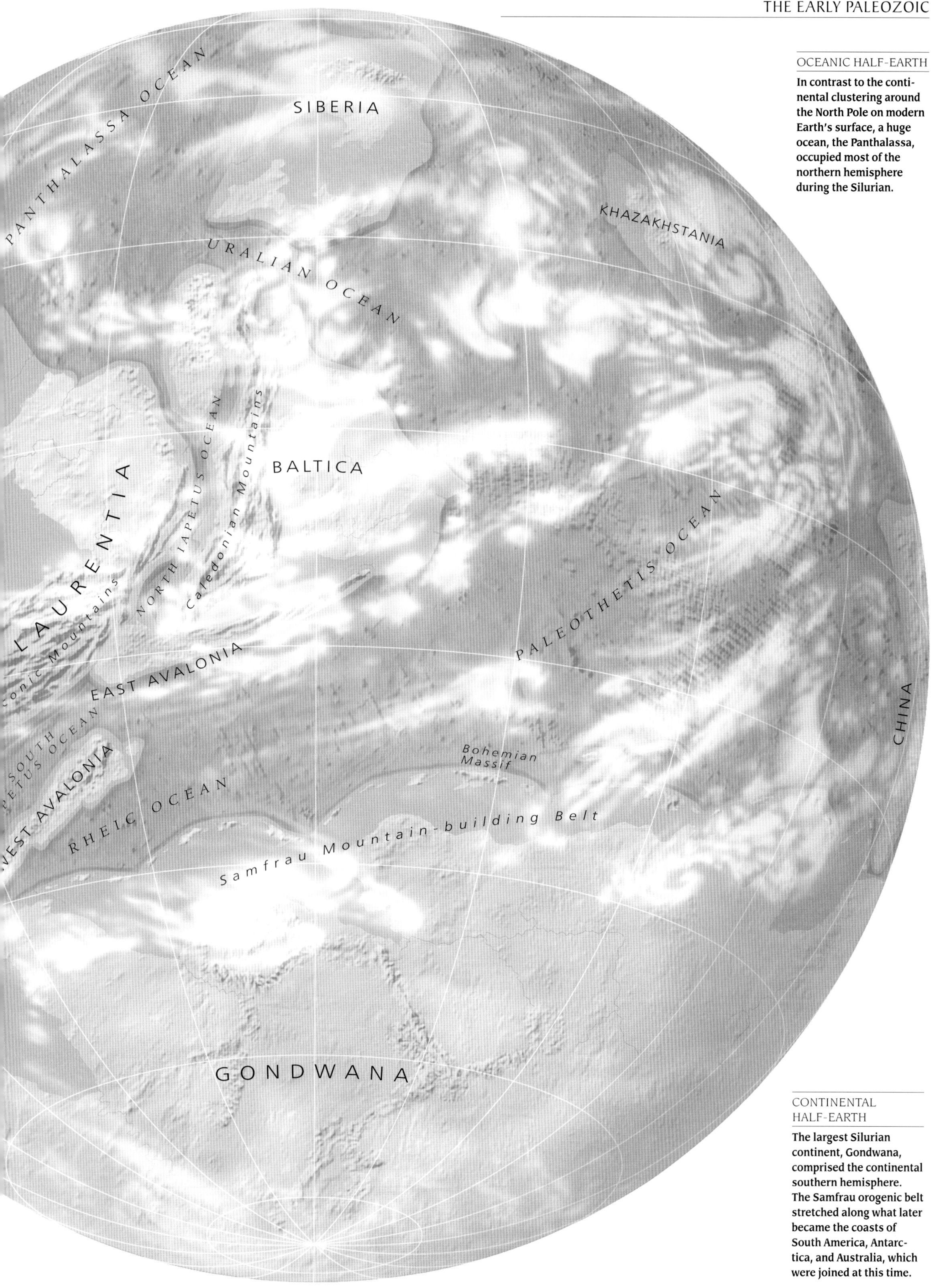

OCEANIC HALF-EARTH

In contrast to the continental clustering around the North Pole on modern Earth's surface, a huge ocean, the Panthalassa, occupied most of the northern hemisphere during the Silurian.

CONTINENTAL HALF-EARTH

The largest Silurian continent, Gondwana, comprised the continental southern hemisphere. The Samfrau orogenic belt stretched along what later became the coasts of South America, Antarctica, and Australia, which were joined at this time.

BLACK SMOKERS

The discovery of black smokers, named for the clouds of murky-looking matter that stream from their chimneylike structures, caused a sensation in geology and biology at the end of the 20th century. This revelation from the very floor of the deep sea strongly challenged the established thinking about the origin of sulfur ores and the sources of life energy. The ores are formed as a significant contribution of sulfur-metabolizing bacteria, while a huge biomass of worms and clams exists on the seafloor at depths previously thought unsuitable for any life.

WHILE the Iapetus Ocean was closing, the Uralian Ocean, between the newly composite Euramerica and Siberia, was at its mature stage. The Ural Mountains themselves are thought to be a suture dividing Europe and Asia. An ore belt more than 1240 miles (2000km) long extends along their length; copper has been intensively mined here for centuries, and ore concentrations are particularly high in the southern region. The deposits have preserved their original appearance of knolls with steep slopes. The knolls are built of pyrites, which are minerals consisting of sulfur and metals (copper, iron, zinc). In 1979 the Russian tectonist Lev Zonenshain proposed a startling hypothesis: that the pyrite knolls were fossil "black smokers," hydrothermal vents at hot spots along mid-ocean ridges on the seafloor, and that the entire copper ore belt of the Urals marked an extinct submarine volcanic chain. The oldest of these knolls discovered to date has been shown to be of Silurian age.

The Ural Mountains, suturing Europe and Asia, mark a major seaway of the Early Paleozoic, now closed.

Mid-ocean ridges are mountain chains with some peaks even higher than Mount Everest. Deep rift valleys split the ridges along their full length. Hot magma (about 2200°F/1200°C) rises up where the water meets the seafloor, bringing with it heat from the Earth's interior; seawater percolates down the crevices and fractures to meet the magma 10 miles (15km) under the seafloor. There the water is heated to 500–850°F (250–450°C) and enriched with a cocktail of dissolved minerals, especially metals, sulfates, and sulfides, and spewed out—a "black smoker," so colored by the sulfide particles.

Several temperature belts exist around a hydrothermal vent, each with its own community of organisms. The wormlike vestimentiferan *Riftia* lives in a white tube up to 10ft (3m) long from which a fringe of scarlet tentacles sticks out. A specific organ, each gram of which contains up to 10 billion symbiotic bacteria, occupies most of its body cavity. These worms inhabit the periphery at temperatures of about 75°F (23°C), while some like it even hotter. The polychaete *Alvinella* (named for the submersible research vessel *Alvin*) survives at temperatures of between 100° and 175°F (40° and 80°C) and seems to prefer to keep bacteria on its outer surface.

1 2 3 4

UNSHAKEABLE EXISTENCE

The Silurian fossil black smokers revealed that a very similar ecosystem of sulfur bacteria, (1) vestimentiferans, (2) tubicolous polychaetes, (3) large clams, and (4) tiny snails passed unchanged through 400 million years of the Earth's history, despite permanently moving across zones.

The entire hydrothermal ecosystem is united by a single reservoir of heat that is involved in chemosynthesis, rather than light energy and photosynthesis as in the majority of Earth's ecosystems. Another general adaptation of hydrothermal animals is the susceptibility to infrared rays. Because of this "sixth sense," they can find their way home even after being swept away by strong currents—but the flash of a camera can permanently blind them. Before the first black smoker was discovered, nobody thought that such diverse and abundant communities might exist and survive the high pressure, lack of light, constant cold, and extremely reduced supply of nutrients. However, sulfur bacteria convert the energy of hydrogen sulfide decomposition into the source of life for the entire hydrothermal ecosystem, creating oases with shoe-sized clams and tangled masses of vestimentiferans at depths of more than 8200ft (2500m). Other organisms either contain symbiotic sulfur bacteria (bivalves, vestimentiferans, annelids) or feed on such animals (crabs, shrimps, fishes).

Following the discovery of black smokers on the Pacific seafloor by a Franco-American expedition in 1977, the search for ancient black smokers and their faunas became intense. Strikingly, the finds made of fossil Silurian communities highly resemble the present high-temperature vent communities. There are long vertical vestimentiferan tubes; short, mostly horizontal tubes that may be annelids; bivalves; and univalve mollusk shells. Relics of symbiotic bacteria have been detected with a powerful microscope. The only difference lies in the presence of articulated brachiopods. These were common members of Paleozoic communities, including those of hot vents and cold seeps.

All the fossils are strongly pyritized and are restricted to sulfide rings whose shapes suggest that of modern chimneys. Thus, the deep marine chemosynthetic ecosystem is among the oldest on Earth.

RIFT VALLEYS

Along deep-sea rift valleys (above), melted magma flows into the water and is solidified there as "pillows" of basaltic lava. The constant repetition of such spreading leads to old basaltic flows being replaced with new ones, while the heat from the magma provides the energy with which black smokers produce their metallic ores.

THE URALS

The Ural Mountains (main map) form the eastern boundary of the Baltic Platform, replacing what was once the Uralian Ocean. The ocean itself was over 1200mi (2000km) wide and framed by island arcs and marginal seas. The basin opened in the Ordovician and closed in the Permian after a collision between Euramerica and Siberia.

COPPER FIELDS

The rich copper ores of the Urals (inset), as well as arsenic, zinc, iron, and sulfur, began as a result of bacterial communities around black smokers.

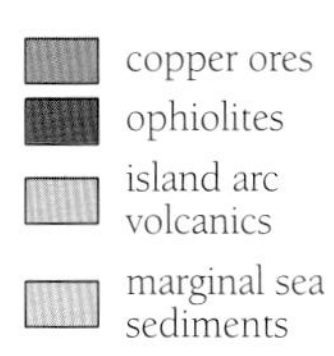

ALTHOUGH Silurian graptolites were chiefly pelagic animals, their lifestyle was connected with relatively deep oxygen-poor waters. Their fossils are particularly abundant in paper-thin shales and look like scribbles scratched on the rock surface. (The word graptolite means "written stone" in Greek.) Graptolites were initially classified among mineral dendrites. Later, they were affiliated to marine plants, cephalopods, coelenterates, and bryozoans before finally being placed close to pterobranchs ("wing and gill"), which themselves are fairly advanced animals belonging to the hemichordates.

Common since the Ordovician, graptolites built colonies of interdependent individuals. They perfected their design during the Silurian.

Partly by comparing them with living pterobranchs and partly by studying the fossils themselves (especially the astonishingly well-preserved graptolites from Ordovician cherts of Poland), the appearance and lifestyle of graptolites have been reconstructed. They were exclusively colonial animals a few inches long. A colony is a collection of zooids, each of which is an individual but lacks the ability to live separately. Graptolite zooids possess a common organic, proteinaceous dendritic casing, which is called the rhabdosome ("stick body"). A tubelike slender soft-tissue outgrowth, the stolon, connects the chain of zooids and runs through the entire rhabdosome. The stolon served to spread the colony, budding off new zooids and communicating with them.

The newly formed zooid had to perforate the tube wall to emerge from the parental tube. Each zooid built and occupied its own balcony-like chamber of about .002in to .08in (0.05 to 2mm) in diameter and, probably, had a collar surrounding its mouth and extending into pairs of hollow arms bearing ciliated tentacles. If a colony was broken during a storm, some fragments could survive and regenerate into a completely new colony. There were males and females among zooids of many sessile graptolites. As a colony grew, older males were probably eliminated and females transformed into hermaphroditic zooids. Graptolites evidently became able to reproduce sexually when budding adversely affected the hydrodynamic properties of the rhabdosome.

Such properties were very important for graptolites grazing on floating plankton. Tenants of the same rhabdosome may have been able to synchronize the flapping of their tentacles for a coordinated movement, although they were too small to get a significant advantage after such an action. This is why graptolites modified their common house, the rhabdosome. Simple modeling of planktonic graptolites demonstrates that a variety of shapes of rhabdosome creates a spiraling motion, which has been observed and recorded using video cameras. Many aspects of rhabdosome morphology have a measurable hydrodynamic function. This implies that hydrodynamic effects were a major control on the evolution of different graptolite types. For instance, a spiral-conical colony could rotate in the water column, slowly sinking and rounding up the plankton as in a trawl net. A propeller-like rhabdosome did the same job even better, because it created vortices, enabling the colony to draw in the plankton. The best-designed rhabdosomes appeared in the Silurian. They looked like a multiple loop with a loose end hanging down. Such a loop could move up and down depending on either inflation by or the release of gas bubbles

THE GRAPTOLITE ANIMAL

The graptolite was a colony of nearly identical individuals connected by common tissue. Each individual possessed a collar of tentacles to feed itself and its close relatives. The entire colony was encased in an organic outer skeleton.

Mouth

Extended zooid

Feeding tentacles (lophophore)

Colonial casing (rhabdosome)

Living chamber (theca)

Retracted zooid

Growth lines

ROCK WRITING

Graptolites usually occur as flattened carbonized fossils (below). Their "scribbles" record a significant part of the history of marine sedimentary rocks from the Middle Cambrian to the Early Carboniferous. Their diversity resulted from their rapid evolution while their wide distribution, due to their planktonic habit, makes them important zone fossils.

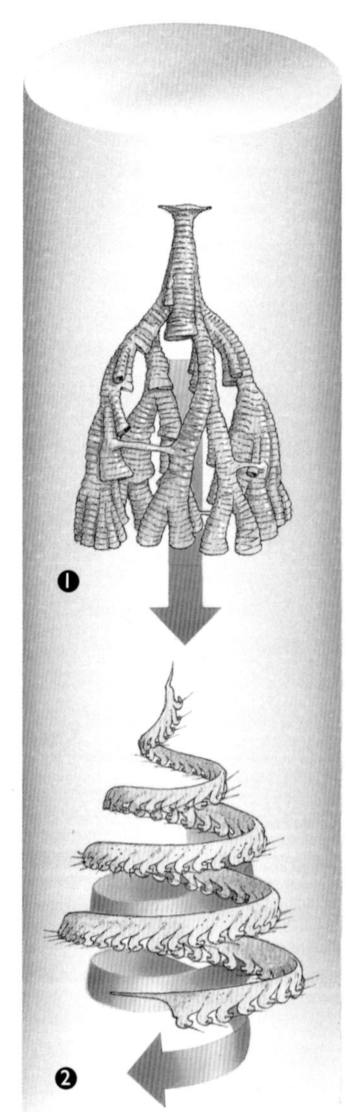

FEEDING STRATEGIES

The varied forms of graptolite colonies gave them a wide range of feeding strategies. The earliest forms were anchored to the seafloor, depending on currents to bring food particles. After the Early Ordovician, most graptolites became planktonic (free-floating) filter-filters. They moved through the water column like trawl or drift nets, each zooid feeding from its immediate vicinity. Among the conical types (left) the spiral motion of *Monograptus turriculatus* (2) provided a more efficient harvesting mechanism than the linear motion of *Rhabdinopora* (1).

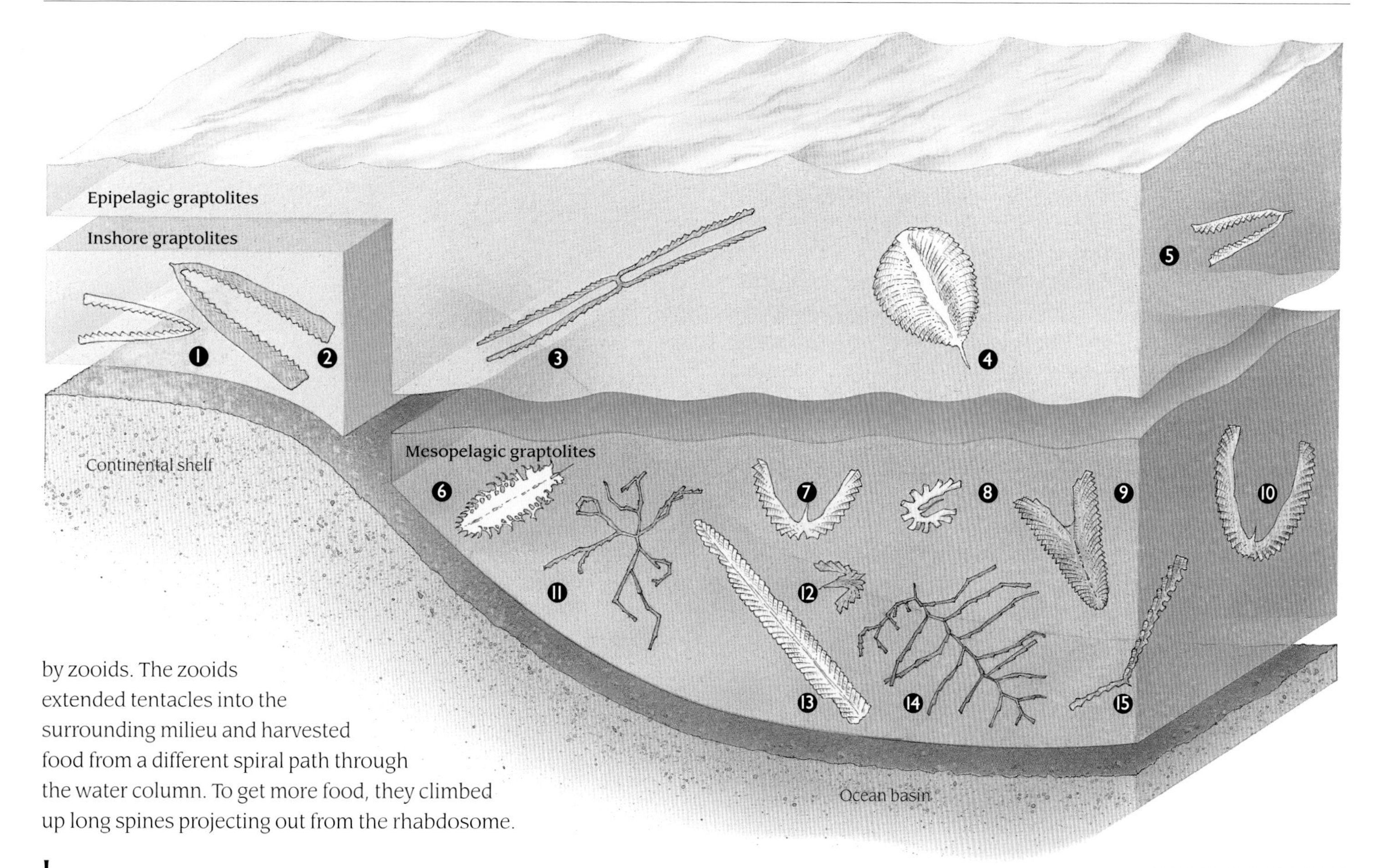

by zooids. The zooids extended tentacles into the surrounding milieu and harvested food from a different spiral path through the water column. To get more food, they climbed up long spines projecting out from the rhabdosome.

INITIAL radiation among planktonic graptolites took place at the beginning of the Ordovician. It was at this time that formerly substrate-bound (sessile) graptolites developed the capacity to float. Planktonic graptolites did not live continuously throughout the oceans during their Ordovician to Early Devonian existence. (Sessile dendroid graptolites survived until the Early Carboniferous.) They flourished in certain limited areas and during those specific times where and when upwelling took place. Under such conditions, extending from certain continental margins, oxygen-poor but nitrogen-rich waters come to the surface. Owing to the nutrient transport to the photic zone, essential primary productivity and high densities of phytoplankton and zooplankton characterize relatively narrow bands of upwelling waters in modern oceans. Bacteria, another important food resource, proliferate at the margins of the oxygen-depleted water that develops beneath upwelling waters due to the oxygen consumption by the decomposition of abundant organic matter. Because of the nutrients, the oxygen content and plankton composition vary vertically in the upwelling and underlying waters, and a vertical differentiation of grazing species occurs.

Graptolites flourished in specific areas at times when there was upwelling of nutrient-rich, oxygen-poor waters from the deep.

This differentiation resulted in a wide variety of planktonic graptolites. Epipelagic graptolite species within the platform-margin habitat probably fed on nannoplankton in the photic zone, and mesopelagic graptolites feeding on bacteria lived on the edge of the oxygen-minimum zone. Physical properties of the water, such as density difference and vertical flows, further facilitated the buoyancy of graptolites. The oxygen-depleted zone was extended in the stratified (unmixed) ocean, and with the return of warm seas graptolites diversified. Their high evolutionary rates and planktonic mode of life, which allowed their wide distribution, make graptolites an important tool for high-precision biostratigraphy of the Ordovician and Silurian.

FORMATION of a colony is merely a special case of the use of building modules. Modular organisms are united by a common living place. However, if zooids in colonies (such as graptolites) do not observe the customary boundaries and freely exchange their secretions, individuals can be isolated by skeletal barriers in many modular organisms. Both the modular structure and the colonial aspect in its strict sense were important for the development of reef-building, but during the Paleozoic it was mostly modular organisms that were responsible for forming the reefs.

Silurian corals and calcified sponges (long mistaken for corals) built reefs the size of today's Great Barrier Reef off Australia.

Two principal groups of Early Paleozoic reef-builders achieved their prime in the Silurian and Devonian periods, when reefs the size of today's Great Barrier Reef encircled the planet's warm seas. These were corals and calcified sponges. They are common in

GRAPTOLITE DIVERSITY

A riot of graptolite forms originated from different modes of colony growth. The curving, looping, bending, and interconnecting of the rhabdosome produced a variety of graptolite colonies. It also enabled them to use the entire water column, in accordance with water density and dominant currents, and to graze on phytoplankton or planktonic bacteria. The variety of twisted skeletons is reflected in their tongue-twisting scientific names.

Graptolites

1 *Pendeograptus*
2 *Didymograptus*
3 *Tetragraptus*
4 *Cardiograptus*
5 *Aulograptus*
6 *Glossograptus*
7 *Pseudoisograptus*
8 *Psilograptus*
9 *Oncograptus*
10 *Isograptus*
11 *Clonograptus*
12 *Dicellograptus*
13 *Glyptograptus*
14 *Sigmagraptus*
15 *Tylograptus*

SILURIAN CORALS

Silurian corals (right) belonged to two principal groups which became extinct by the end of the Paleozoic. Tetracorals (rugosans) were mostly solitary, branching forms covered with a thick outer skeletal layer. Tabulates were tall modular colonies of numerous zooids (coralites) forming massive chainlike, fan-like skeletons. Inside the coralite skeleton were multiple vertical partitions and the same number of tentacles which stretched out to catch food.

WENLOCKIAN REEF COMMUNITY IN ENGLAND

Carnivorous corals, filter-feeding branching bryozoans, and massive stromatoporoids formed the core of a typical Silurian reef community. They were encrusted by low (brachiopods) and high (crinoids) suspension-feeders. Deposit-feeding trilobites scratched the seafloor and hunting cephalopods floated in the water column.

modern reefs, but are either represented by taxonomically different groups or reduced in their ability to be the chief builders.

In fact, generations of paleontologists mistook Paleozoic calcified sponges, the stromatoporoids, and the chaetetids, for corals. Stromatoporoids ("porous blanket") are mostly dome-shaped heavily calcified structures resembling a pile of pancakes with starlike grooves (astrorhizae) on the upper surface. Chaetetids ("hairlike") consist of a bunch of hair-thin vertical tubes. Both are quite big, nearly 2ft (0.6m) in diameter, so that their appearance is much more coral-like than spongelike. Most sponge skeletons consist of tiny spicules and look very fragile, although bath sponges are soft.

Even when modern representatives of stromatoporoids and chaetetids were caught in the Caribbean in the early 20th century, they were viewed with great skepticism. Only since the late 1960s has it been confirmed that such skeletons belong to sponges and consist of both calcareous layers and siliceous spicules. A thin but tough organic envelope prevents the spicules from being dissolved in the calcareous mass when the sponge is alive. After its death, microscopic cavities are left instead of spicules. The astrorhizae themselves show the affinity of such skeletons with sponges, because these structures are designed exclusively for filter-feeding.

Other Paleozoic creatures such as tabulates have been wrongly ascribed to sponges. Tabulates have close-set horizontal skeletal partitions in corallite tubes. Examples

with fossilized living tissue have been collected in Silurian rocks of Quebec, proving that these honeycomb-like skeletons had contained a corallite with 12 tentacles in each tube. Thus, Ordovician–Permian tabulates were distinct from Ordovician–Permian tetracorals (rugosans) as well as from modern hexacorals (scleractinians) and octocorals (sea pens). The stromatoporoids, chaetetids, and tabulates were large modular forms, while rugosans tended to be branched. Many rugosans were solitary corals bearing a thick outer cover with coarse concentric wrinkles. Solitary rugosans inhabited loose substrates and were not active reef-builders.

Tabulate corals
1 *Favosites*
2 *Heliolites*
3 *Halysites*

Tetracoral
4 *Streptelasma*

Articulate brachiopods
5 *Leptaena*
6 *Atrypa*

7 Bryozoan *Hallopora*
8 Stromatoporoid sponge *Actinostroma*
9 Crinoid
10 Orthoceratid nautiloid cephalopod
11 Trilobite *Dalmanites*

THE GREATEST SILURIAN REEF

During the Wenlock, reefs became widespread and abundant in Mid-Continental Laurentia (today's North America), especially in the Hudson and Michigan basins. The Michigan reef was one of the broadest and longest reef tracts of Earth's history. It covered an area of about 680mi (1100km) across, and a continuous buildup occupied about 300,000sq mi (800,000 sq km). Several tens of thousands of individual reefs developed there on carbonate shoals and along the margin of the intervening basin.

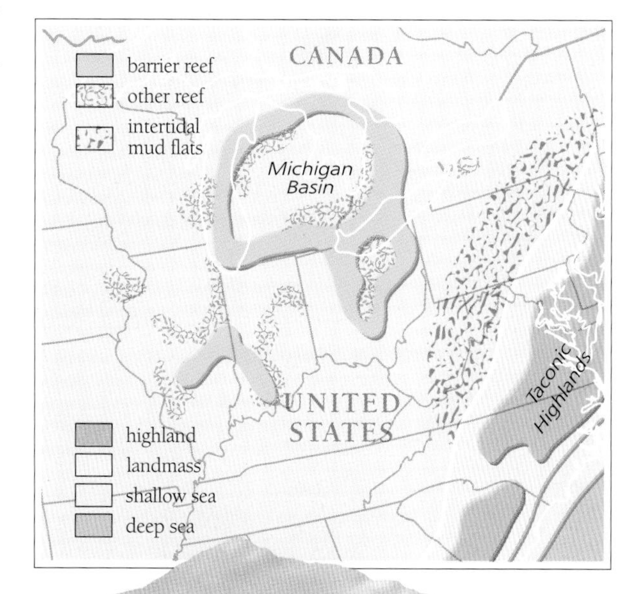

Evaporites

Algal layers

Stromatolite mounds

Pinnacle reefs

Stromatoporoid/coral patch reefs

REEFS OF A RESTRICTED SEA

A restricted inland basin (above), surrounded by desert, accommodated diverse reef types: nearshore stromatolite mounds precipitated by bacterial communities; coral-stromatoporoid edifices patching shallow marine areas (patch reefs); and ring-like atolls. While shallow-water reefs grew quite slowly on the edge of the basin, corals had to build their vertical framework more rapidly to keep their settlement at the same depth while the seafloor rapidly subsided. These reefs resembled columns or pillars.

IT WAS modular calcified sponges and corals that made the reefs. Although their individual modules were small, the modular organization allowed them to grow far beyond the limited size possible for an individual. Also, because they are less aggressive than their solitary relatives are, modular forms are much more compatible with each other. Thus, they mutually bind to each other to create a rigid framework resistant to deterioration (such as by wave action), which is the core of the reef. Finally, owing to the rapid growth of reef-builders, framework reefs reach tremendous sizes. They are strengthened by marine cementation infilling the reminder of the cavities.

Reef-building depended on a certain level of cooperation between individual animals, allowing the reefs to reach huge proportions.

The entire reef community, which includes encrusters and destroyers (borers and scrapers) in addition to builders, comprises a whole carbonate factory, producing a multitude of rock particles. A hardened framework turns it into a boundstone. Well-sorted fragments of reef organisms deposited by strong currents or by storm action convert into grainstone. A calcareous mud, precipitated under calm conditions, alters it into mudstone, and so on. Each carbonate fabric is thus a fairly good indicator of the conditions under which a certain rock was formed. Particular lateral and vertical successions of distinct lithologies compose facies, such as fore-reef, lagoonal, and others. The entire pattern provides information, allowing the reconstruction of the history of a basin and prediction of the distribution of possible accumulations of oil, gas, and some other deposits.

Evolutionary radiation of coral and stromatoporoid faunas and innovations in associated rock types produced distinguishable fore-reef, reef, back-reef, and lagoonal facies. Middle Silurian coral-stromatoporoid reef tracts were larger in extent than modern reef tracts, and were concentrated in subtropical and equatorial climatic belts. In addition, in terms of volume, calcified cyanobacteria and algae were important in many warm-water Silurian reefs, and siliceous spicular sponges formed mounds under temperate and deep-water conditions. A wide array of reefs luxuriated, from nearshore fringing structures and lagoonal patch-reefs to shelf-margin barriers, atolls, and deep-slope mounds. The biggest reefs influenced contemporaneous regional sediments and local climates, being barriers to the circulation of water in the basin.

ALL THE reefs teemed with a variety of organisms, including jawless fishes. Jawless fishes look much more fishlike than jellyfishes, crayfishes, and even starfishes. Nonetheless, the word "fish" in the name "jawless fish" means only that such an animal lived in the water. Being chordates and craniates, with a skull and a brain, they differ from all other craniates, including cartilaginous and bony fishes, in their absence of jaws. They also lack pelvic bones and, in most cases, have no paired fins. In the language of biological taxonomy, jawless fishes comprise a sister group to all other gnathostomes (vertebrate animals with jaws), namely other fishes, amphibians, reptiles, birds, and mammals.

Jawless fish are rare today, but numerous species flourished in salt and fresh water alike during the Silurian period.

Only a few dozen species of jawless fishes exist in modern seas and fresh waters; they are hagfishes and lampreys. These scaleless, wormlike animals possess a round mouth like a suction cup and exist as voracious parasites. Hagfishes have been fixed in the fossil record from Carboniferous rocks when they probably already

THE CONODONT CONUNDRUM

In 1856 the Russian scientist Christian-Heinrich Pander described tiny phosphatic teeth from Ordovician strata of the Baltic Sea coast. Called conodonts, they have been classified as plants, mollusks, annelids, arrow worms, and other invertebrates. In fact, they are craniate chordates. At first, conodonts were arranged into several different groups according to shape. Later, fishlike conodont-containing bodies were discovered in the Carboniferous of the United States, although these proved to be no relation; the conodont-bearing organ was redefined as being a conodont-digesting one.

The conodont animal itself was discovered in a collection of Carboniferous fossils at the Institute of Geological Sciences in Edinburgh, Scotland. Regular series of V-shaped muscle blocks, a laterally flattened body, an unequal tail fin, and a multi-chambered assemblage in the head all fit a jawless craniate. Large-eyed Silurian conodonts from South Africa had teeth that were used to crush and shear food: shoals of these fast, sharp-eyed animals could have terrorized Early Paleozoic seas.

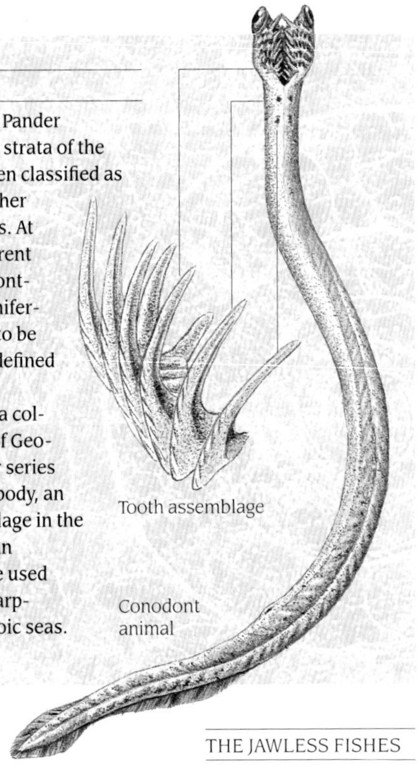

Tooth assemblage

Conodont animal

lived on other fishes, boring into them with a strong sharp-toothed tongue.

However, it was in the Silurian and Devonian periods that a large variety of jawless fishes (agnathans) swam in the world's seas, lagoons, and lakes. They were totally different from modern agnathans in their external appearance and were covered with shields. Judging by their external rather than internal skeleton, fossil agnathans were affiliated with arthropods for a long time.

The earliest known representatives of the group were Middle Ordovician astraspids and arandaspids (Aranda is a tribe in Australia, near which such remains have been found). Arandaspids, inhabiting the shallow seas of Gondwana, had an elongated head protected by a large bony nodular shield, a tail bearing rod-shaped scales arranged in chevrons on each flange, and a tail fin. Their mouth was armed only with thin rows of small bony plates, which formed a scoop for scraping the muddy sea bottom. Long series of gill openings were situated on each side of the body. The eyes were at the very front of the head, flanking a pair of nostrils. On the top of the head was a duplex "third eye." In modern lampreys, such an eye has an underdeveloped crystalline lens which can detect the shadow of possible danger falling on it.

Astraspids lived in the sea basins around Laurentia and Siberia. They were distinguished from the arandaspids by the presence of eight large gill apertures, eyes at the side of the head, a single "third eye," and scales on the tail, which were much wider and diamond-shaped. Both arandaspids and astraspids had carapaces made up of a kind of bone that lacked cells, and no paired fins. A lateral line system was well developed over the head and body; such a system in fishes gives awareness of balance and senses vibrations in the surrounding water produced by other moving organisms (either welcome prey or unwelcome predators).

These craniates had no clear relationship with other groups. Their nearest relatives were the heterostracans.

THE JAWLESS FISHES

Jawless fishes (agnathans) existed over 500 million years ago. Their oldest remains are found among Early Cambrian Chengjiang fauna in China. Shortly after the Ordovician they achieved a rampant disparity and were covered with bony armor: shields, tubercles, and thick scales. Their fossils have been used in dating Silurian and Devonian sedimentary rocks. The heterostracans, thelodonts, galeaspids, and osteostracans underwent major radiation during the Silurian. Anaspids also appeared during this time. Only bare-skinned lampreys and hagfishes survived beyond the Devonian.

AFTER deglaciation in Early Silurian times, armored jawless fishes with jawbreaking names suddenly bloomed in the sea: heterostracans ("different plates"), osteostracans ("bony plates"), thelodonts ("nipple tooth"), anaspids ("shieldless"), and galeaspids ("lamprey shields"). Of these, Silurian–Devonian thelodonts and anaspids were relatively widespread; heterostracans and osteostracans inhabited the northern hemisphere; and galeaspids were restricted to Chinese seas. The heterostracan mouth was lined by a few small oral plates, which did not abut the beak-like rostrum but probably scooped up food. The snout, or rostral region, had a broad notch beneath it. The entire tapering head was covered with a single plate and large diamond-shaped scales protected the body. The tail was paddle-shaped.

Armor developed in the first fishes and fishlike animals, which were defenseless creatures. Some later became predators.

Most osteostracans had a horseshoe-shaped head shield with a pair of shoulder fins attached. The body was protected by rows of small, elongated scales, and there were fins on the back. The tail pointed upwards, as in modern jawed fishes, with a large lower web and a peculiar

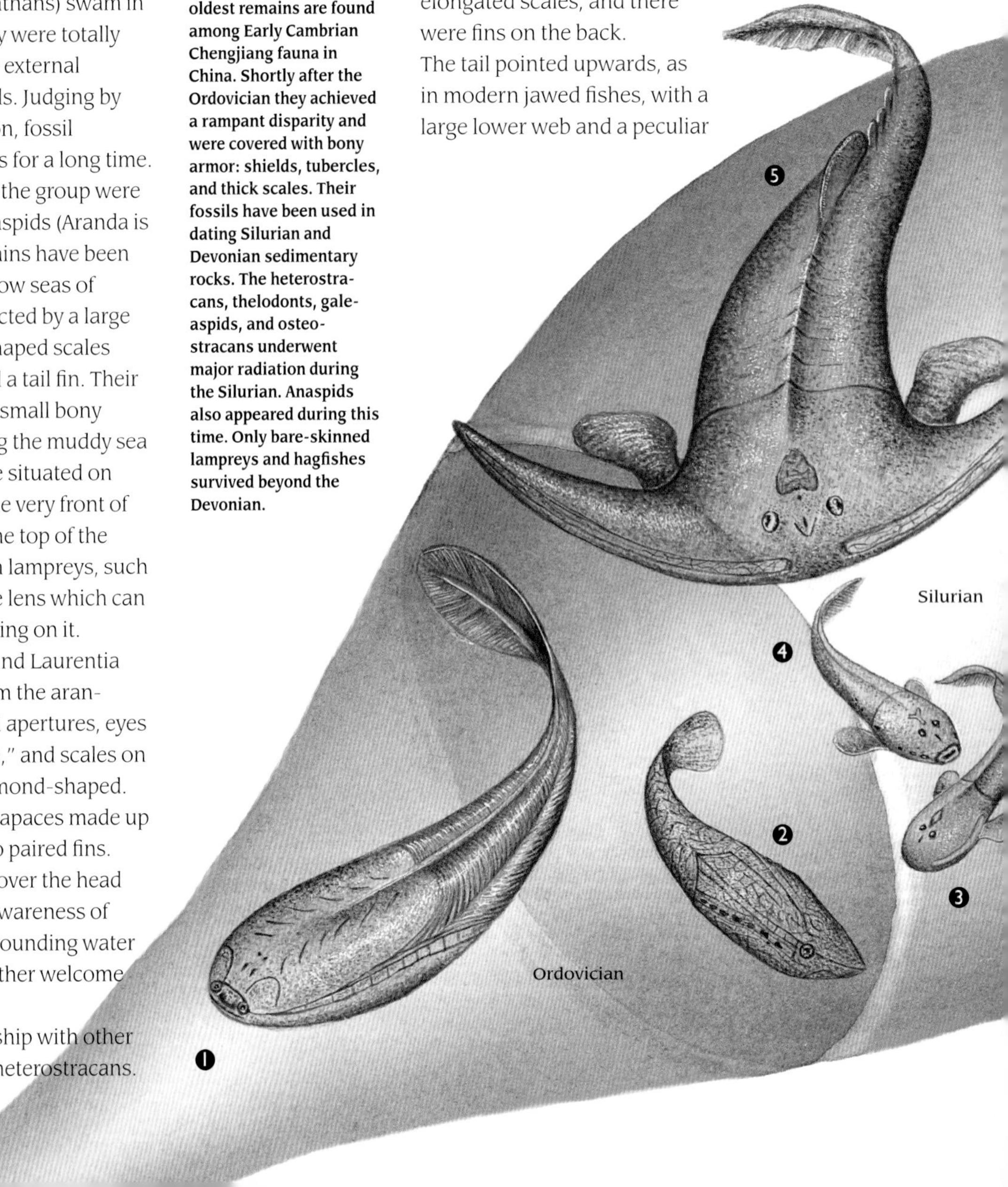

5

Silurian

4

2

3

Ordovician

1

SILURIAN

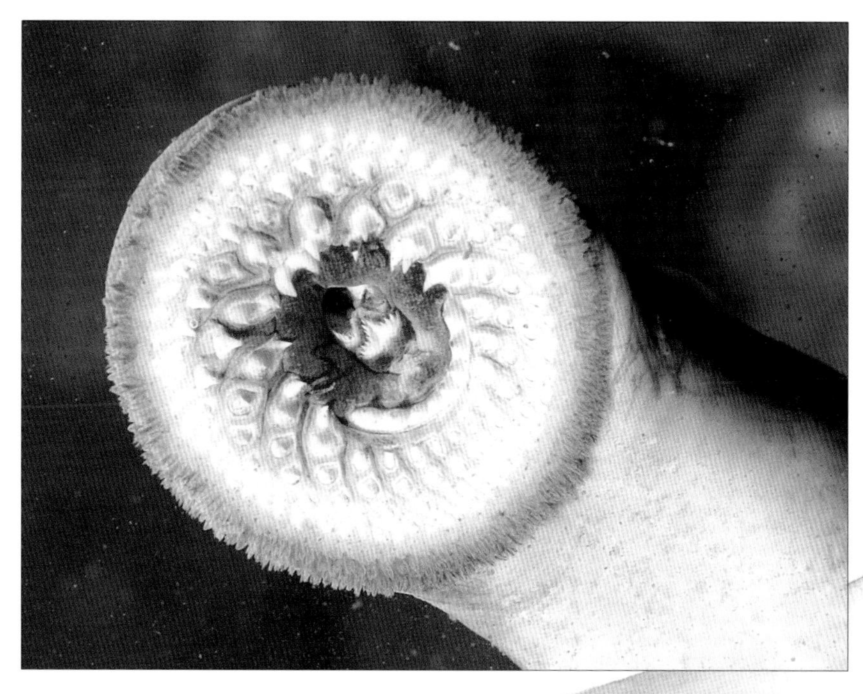

horizontal lobe along the bottom of the tail fin. They were characterized by strange lengthways and sideways shallow depressions on the upper surface of the head shield. These shields were covered with small, loose polygonal bony plates and connected to the labyrinth cavity by branched canals. These may have housed either well-developed electric sensory organs or other kinds of sensors linked with the lateral line system.

Thelodonts were deep-bodied fishes entirely covered with tiny scales superficially resembling some scales of modern sharks. They had paired flaps over the gill openings, broadly separated eyes, a forked tail, and developed dorsal and anal fins.

Slender, laterally compressed anaspids, with large eyes at the side of the head, were possibly close relatives of modern lampreys and were not armored like other agnathans. Their heads bore minute scales and a few large plates, and their bodies were wrapped with elongated scales arranged in chevrons. Their tails were tilted downwards, with a prominent upper fin lobe. The mouth was rounded, with upper and lower lips furnished with plates that could bite vertically against each other, much in the same way as typical jaws.

RING OF TEETH

The lamprey's mouth, with its ring of teeth, is a formidable weapon. It attaches itself to its prey and rasps away, enabling it to suck its blood.

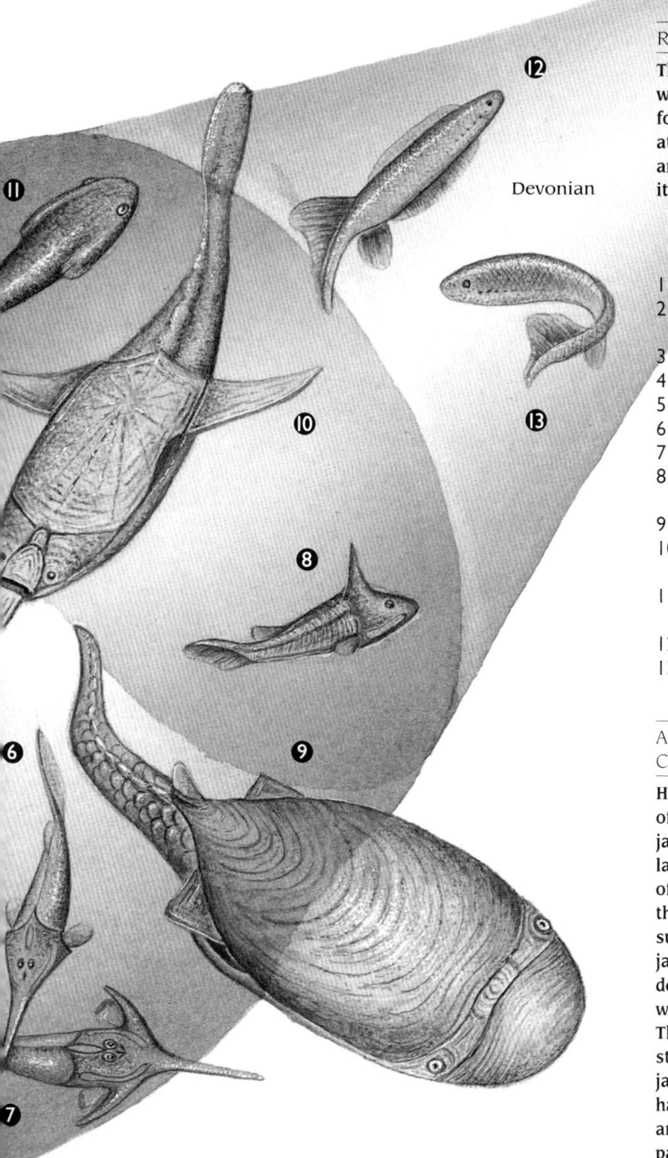

Astraspids/ Arandaspids
1 *Sacabambaspis*
2 *Astraspis*
Osteostracans
3 *Norseaspis*
4 *Gustavaspis*
5 *Parameteroraspis*
6 *Belonaspis*
7 *Boreaspis*
8 *Machairaspis*
Heterostracans
9 *Zascinaspis*
10 *Doryaspis*
Thelodont
11 *Turinia*
Anaspids
12 *Endeiolepis*
13 *Euphanerops*

UNLIKE the dazzling diversity of Silurian reefs, lagoons and other marine communities, the land is thought to have been almost barren during most of the period. It is quite probable that the first land microbial crusts existed as early as the Archean Eon, as indicated by some odd mineralizations. Later on, bacteria and lichens (symbiotic associations of green algae or cyanobacteria and fungi) decorated the land. By the beginning of the Cambrian Period, seaweeds acquired a tough envelope and spring-like organs. The envelope permitted them to be exposed to the air without the support of the water, while the twisted spring could be

Plants at last moved onto land in the Silurian, bringing small patches of green to what had been a barren wasteland.

AGNATHAN CLASSIFICATION

Hagfishes have features of the most primitive jawless fishes. Whether lampreys were ancestors of anaspids or vice versa, the reason for their survival is unclear. Other jawless fishes evolved by developing heavy armor, which protected them. Thelodonts and osteostracans may be related to jawed fishes; thelodonts had scales (like sharks) and osteostracans had paired fins and ossified bones round the eye. This is related to the development of the lower jaw in fish embryos.

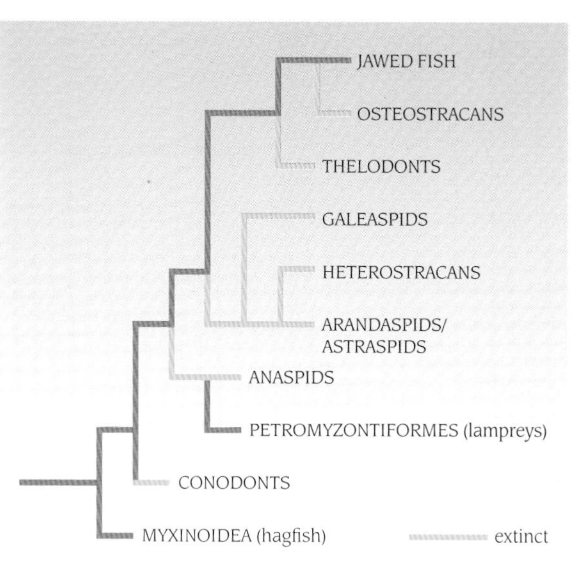

SILURIAN

uncoiled to throw out spores, but only in the air. Life forms developed on the virgin lands step by step. On the Late Cambrian sandy beaches of New York, unknown animals left traces whose size and pattern resembled imprints of the tires of a solitary biker. In temporarily emergent Ordovician conditions, non-marine arthropod tracks were made on soft volcanic rocks in the English Lake District. The animals that left the traces followed a more-or-less straight course on several paired legs.

The earliest generally accepted fossil evidence for land plants is from Middle Ordovician non-marine and nearshore marine deposits, and consists of isolated spores and plant fragments. A spore is a cell that enables asexual reproduction. It has a cell wall tolerant of drying up and exposure to ultraviolet radiation. The acquisition of spores was an important first step in the colonization of land by plants. Ordovician spores are interpreted as coming from land plants. They are similar in size and gross shape to the spores of land plants and their wall structure is akin to that of spores from primitive extant land plants. Besides, identical spores occur from what definitely were land plants from somewhat later deposits. Their small size and abundance suggest that they were produced in vast numbers, blown by the wind, and thus capable of long-distance dispersal. The earliest and simplest spores corresponded to moss-like or, more

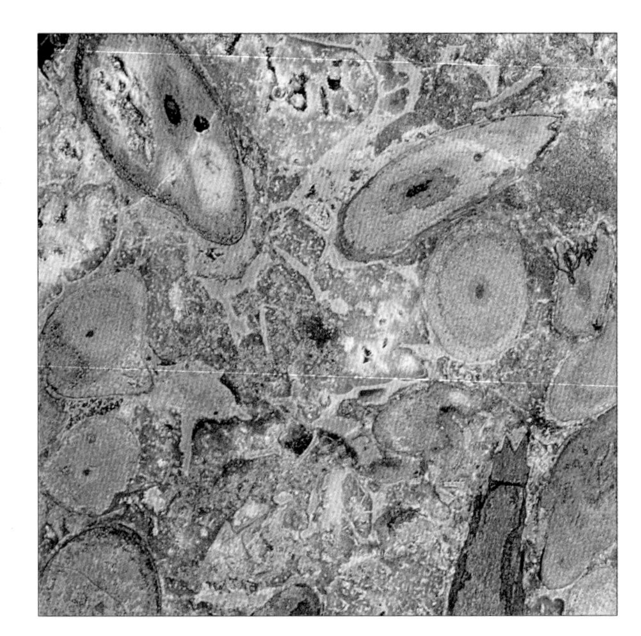

DEVELOPMENT OF VASCULAR PLANTS

Until recently all plants found in the Silurian and Early Devonian were ascribed to a group called the psilophytes. Now a huge variety of forms is recognized among the earliest land plants, including zosterophyllophytes, trimerophytes, and rhyniophytes, which can now be subdivided into several evolutionary lines (below left). This shows that adaptation to life on land has occurred independently in many plant groups.

accurately, liverwort-like plants. Late in the Ordovician, spores appeared in forms that are produced primarily by ferns today.

The exact affinities of the earliest land plants remain unclear. However, it seems likely that they evolved from some green algae. The Ordovician plant fragments might derive from nematophytes ("thread plants"), which were enigmatic land plants of unknown affinity with tubular anatomy and a waxy covering. The nematophytes could have been pathogens or decomposers, and were related to fungi or perhaps even lichens.

The oldest actual plants are first recognized in the fossil record from the Early Silurian, roughly 30 million years after the appearance of land-plant spores. They became progressively more abundant during the period. Early terrestrial assemblages of northern Europe, Bolivia, Australia, and northwestern China include clubmosses, related early plants (zosterophylls and rhyniophytoids), and various remains of uncertain affinity, such as *Salopella*.

FOSSIL PLANTS

The best-preserved fossil plants are known from silicified peat deposits such as the Rhynie Chert. Plant tissues and cells can be seen in thin sections of the rock put under a microscope (above).

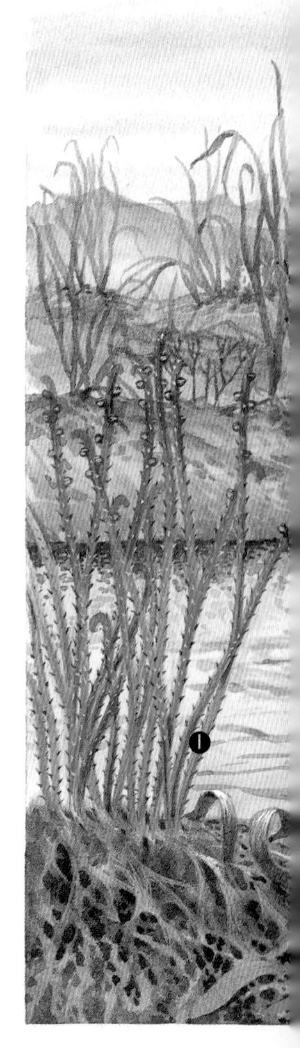

DURING the Late Silurian, Gondwana and Laurentia were dominated by rhyniophytoids, which united the plants resembling *Rhynia* from the Rhynie Chert of Scotland. These were branching plants with globular to kidney-shaped sporangia (spore-generating organs) terminating the branches. Zosterophyllophytes ("garland leaves"), with spiny protrusions and multiple sporangia directly attached to less regularly branching axes, indeed looked like garlands. *Salopella* had an axis resembling that of a clubmoss. None of these plants was differentiated into stem, leaves, and roots.

Early plants were small and grew in moist lowlands. Height was linked to their expansion into drier conditions.

These early plants varied but were generally small organisms (usually less than 4in/10cm tall) restricted to moist lowlands. Nonetheless, they were indisputably vascular plants. This means that they had tubular tissues of elongated cylindrical cells for conducting water and other nutrients from organs attached to a water-containing substrate to the very top of the plant. This was the innovation that had allowed land plants to become taller and in due course to expand over drier sites. Small skin pores (stomata) in the waxy covering permitted gas exchange and controlled transpiration and water vapor loss. Finally, symbiosis with bacteria that could fix nitrogen directly from the air made vascular plants almost independent of the nutrient supply.

The advent of land plants had a major impact on the atmosphere and other aspects of the global environment through creating leaf litter, modifying nutrient flow and drainage, and helping soil form. The evolution of roots is thought to have been important in reducing atmospheric carbon dioxide concentrations through increased weathering of minerals brought about by mechanical disruption and soil acidification.The part played by vascular plants in evapotransportation, for which no counterpart exists in other organisms, has a primary influence on rainfall, average temperature, and atmospheric circulation. From the latest Silurian onwards, the evolution of terrestrial ecosystems (and even marine ecosystems) has been driven by plants, and principally by vascular plants.

A TYPICAL Silurian landscape would have existed in Saaremaa, a low-lying island of modern Estonia that is now overgrown with heath. The island is made up exclusively of Silurian dolomites and limestones, which have been quarried for building stone throughout the Baltic Sea countries since the 13th century. At the beginning of the Late Silurian (Ludlow), a regression was developing, and Saaremaa was on the edge of a warm shallow open marine basin surrounded by a smooth lowland.

In the warm shallow waters near modern Estonia lived the first known community in which chordates played a major part.

From north to south in today's coordinates, the basin was subdivided into several belts. On the north there was a semi-isolated lagoon, where dolomitic muds were precipitated and stromatolites developed in the arid climate. Farther south, on the peritidal shoal above the normal wave base, well-washed oolitic sands were deposited between reefs, and sessile benthic (stromatoporoids, tabulates, and calcified algae) as well as some motile animals (ostracodes, polychaetes, and jawless fishes) proliferated. The shoal passed to the subtidal open shelf . Mudstones accumulated there below the normal wave base and sustained a variety of stromatoporoids,

AUSTRALIAN GARDENS

Whereas the early rhyniophyte floras were typical of Euroamerican deposits, Late Silurian vegetation of Australia at the same time was farther ahead in its evolution. *Baragwanathia* had leaves and root-like organs. It formed a thick, 7ft (2m) high growth on unstable floodplain in warm, dry climates, along with an early zosterophyllophyte, *Sawdonia*, and another problematic plant, *Salopella*. *Sawdonia* had kidney-like sporangia attached to axes, while *Salopella* had small, round terminal sporangia.

1 *Sawdonia* (zosterophyllophyte)
2 *Buthotrepis* (thallophyte algae)
3 *Salopella* (uncertain affinity)
4 *Baragwanathia* (possible clubmoss)
5 Myriapod
6 Track of eurypterid

SILURIAN

tabulates, brachiopods, ostracodes, conodonts, rare trilobites, and various fishes. The shelf plunged into the slope, where, under calm muddy conditions, vagrant trilobites and ostracodes as well as some brachiopods dominated, and chitinozoans, graptolites, and thelodonts floated in the water column. Where the slope ended in a deep depression, bituminous black siltstones formed and the remains of nektonic (free-swimming) acanthodians, conodonts, and graptolites sank to the bottom.

The community that inhabited the brackish lagoon is of particular interest. Brachiopods and relatively large (up to 0.2in/0.5cm long) ostracodes flourished. Brachiopods were filter-feeders, and ostracodes gathered food from the detritus. Filter-feeding bivalve crustaceans, the phyllocarids ("leaf legs"), achieved 2in (5cm) in length. A few bryozoans, stromatoporoids, and tabulates clung to slightly hardened areas of the floor. Scolecodonts, the remains of annelid jaws, indicate the presence of carnivorous polychaetes, and dense trace fossils reveal the activity of detritus feeders.

THE OSTEOSTRACANS of Saaremaa were small—2 to 4in (5 to 10cm) long. They exhibited very different lifestyles. Some of them with olive-shaped heads, locked up in a heavy shield and lacking paired fins, could swim only slowly by moving the tail. They spent most of the time hidden in the mud and passively feeding on detritus. Others were more active swimmers because of a lightly-built skeleton and well-developed muscles on a long flexible tail, whose large lower lobe provided a means of uplift. In addition, the head shield was well profiled for gliding, while water jets ejected from downward-pointing slits helped it to leave the bottom. Paired shoulder fins improved maneuverability. Smeared with slime excreted from cavities on the shield, the fish slid above the seafloor seeking out small soft invertebrates.

Many predators, from polychaete worms to giant eurypterids, inhabited the Saaremaa basin

1 *Lingula* (brachiopod)
2 *Tremataspis* (osteostracan)
3 *Myxopterus* (eurypterid)
4 *Thyestes* (osteostracan)
5 *Ceratiocaris* (phyllocarid crustacean)
6 *Nostolepis* (acanthodian)
7 *Phlebolepis* (thelodont)

1

2

SILURIAN

Both anaspids and thelodonts were free-swimming planktivores that kept themselves well above the sea floor. Round in cross-section, thelodonts had the typical swimmer's camouflage of light bottom and dark top, and had a wide, forward-facing mouth. Powerful fins and tiny longitudinal grooves in the scales improved streamlining. Laterally-placed eyes and a well-developed lateral line system gave them information equally from above and from below. It was especially important in the presence of such pitiless predators as 7ft (2m) eurypterids (water scorpions) and acanthodians with well-developed jaws and big conical or triangular teeth. Slender acanthodians looked like small, spiny sharks and had a long bony spine in front of each fin except for the tail lobe.

The trophic web was simple, consisting almost exclusively of seafloor scavengers and benthic and nektonic predators, though the chain of predators was long. In spite of the supremacy of the eurypterids, the Silurian lagoons of Estonia yielded the first community shaped significantly by the presence of chordates.

❼

❻

❸

❹

❺

THE EVOLUTION OF CHORDATES

THE FIRST CHORDATE

The lancelet, or amphioxus, *Branchiostoma* (left) bears some resemblance to *Cathaymyrus*, found in the Lower Cambrian Chengjiang locality of China, and thought to be ancestral to all vertebrate (backboned) animals.

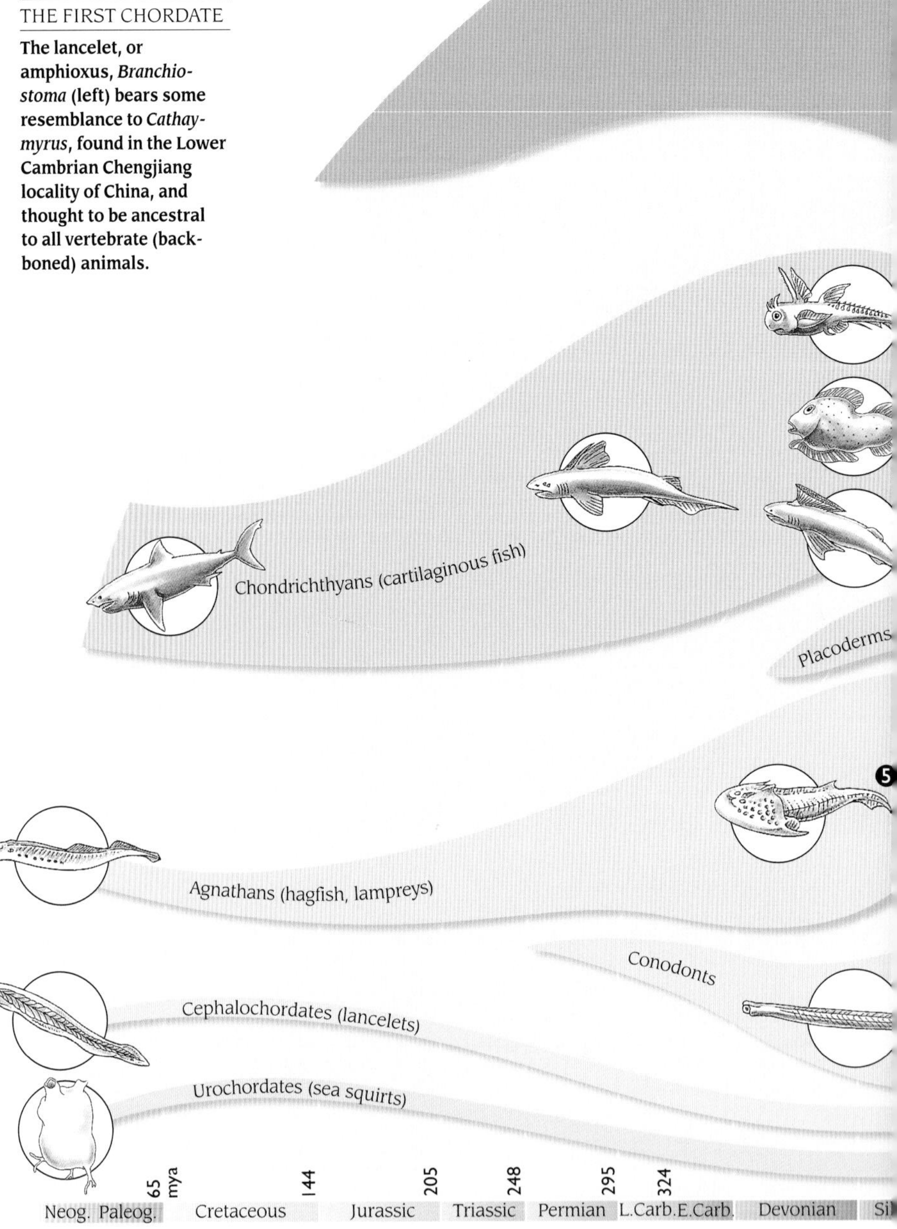

In THE Cambrian seas a tiny, soft, defenseless creature survived the trilobites, anomalocaridids, and other large carnivores. This maggot-like animal, with slits running on each side of its head and a fibrous bar extending through its body, was the ancestor of all chordates.

Chordates, hemichordates (pterobranchs and graptolites), and echinoderms are all deuterostomes ("other mouth"), meaning that the mouth of the adult develops in a different place from the primary opening of the embryo. In chordates the nerve cord runs along the spinal side of the body, whereas in the majority of bilaterally symmetrical animals it follows the ventral (belly) side. Chordates also have a notochord, a cartilaginous rod that serves as an anticompression device against which the muscles can work. It persists in lancelets and lampreys but occurs only in swimming larvae of sea squirts. In vertebrates the notochord grows into a spinal column. The third feature of chordates is the presence of gill slits in the throat at some stage of life.

Lancelet-like and agnathan-like chordates inhabited Cambrian seas. In the Ordovician they were joined by the true agnathans (jawless fishes), which remained widespread in the Silurian and Devonian. A number of so-called vertebrate scales and spines, still reported occasionally as being present in Cambrian rocks, are merely the remains of arthropods and cephalorhynch worms.

Jawed vertebrates might have existed in the Ordovician, although the evidence is not conclusive. The first vertebrates were onshore marine organisms because their kidneys could not function in fresh water. In the Silurian, chordates spread into brackish and freshwater basins. The transition between brackish to full marine conditions may have stimulated the high rates of vertebrate evolution. The jaws were developed from the skeletal arches of front gills and may originally have been used for holding food rather than for chewing.

FEW STEPS—MANY WAYS

The key evolution of chordates began in the Early Cambrian, but it was not a single straightforward process. Early Cambrian chordates had features of urochordates, cephalochordates, and even agnathans, but at different levels of development. Some fishes developed lobe fins and lunglike organs, and mammal and bird features developed in reptiles. These parallels show how evolution has its own patterns and directions.

During the Silurian, many groups of jawed craniates first appeared, such as primitive sharks (chondrichthyans), placoderms (armored fishes), and perhaps bony fishes (osteichthyans). The last named were the earliest known representatives of a large group including modern herring and cod. The acanthodians are possibly related to bony fishes. They became extinct in early Permian times, but most modern aquatic vertebrate taxa seem to have been present by the end of the Silurian.

At the end of the Devonian, the tetrapods, descendants of lobe-finned fishes, made the first steps onto land. It was only a few steps away from the croaking, feathered, hairy world of higher vertebrates. At the close of the Early Paleozoic, chordates, once prey, became the most powerful predators. The Late Paleozoic had begun.

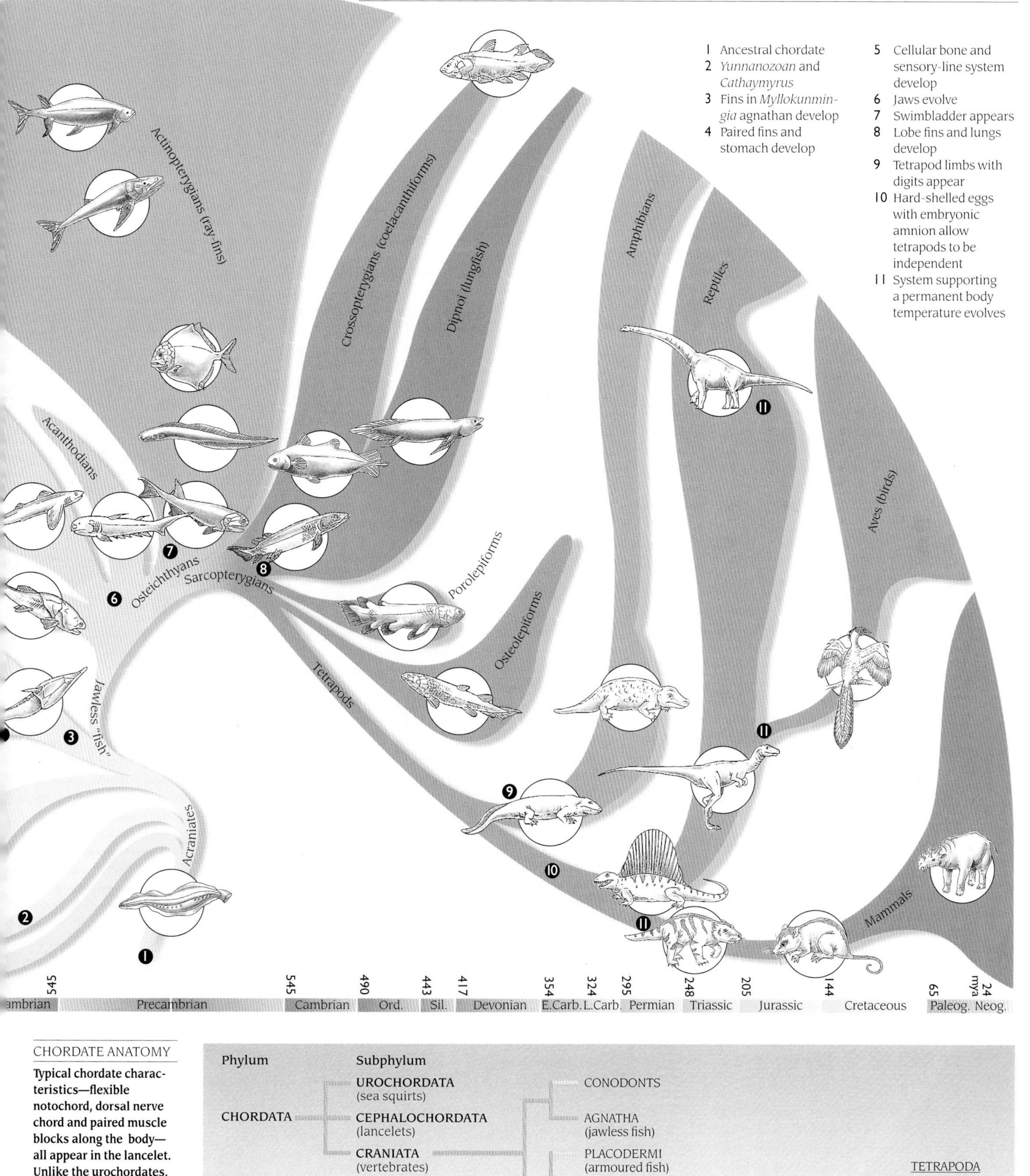

CHORDATE ANATOMY

Typical chordate characteristics—flexible notochord, dorsal nerve chord and paired muscle blocks along the body—all appear in the lancelet. Unlike the urochordates, the other basal group of living chordates, the notochord extends into the head, hence the name cephalochordate.

Nerve chord
Notochord
Gills
Mouth

Phylum: CHORDATA

Subphylum:
- **UROCHORDATA** (sea squirts)
- **CEPHALOCHORDATA** (lancelets)
- **CRANIATA** (vertebrates)
  - CONODONTS
  - AGNATHA (jawless fish)
  - PLACODERMI (armoured fish)
  - ACANTHODII (primitive jawed fishes)
  - CHONDRICHTHYES (cartilaginous fish)
  - OSTEICHTHYES (bony fish)
    - ACTINOPTERYGII (ray fins)
    - SARCOPTERYGII (lobe fins)
      - TETRAPODA
        - MAMMALIA (mammals)
        - AVES (birds)
        - REPTILIA (reptiles)
        - AMPHIBIA (amphibians)

extinct

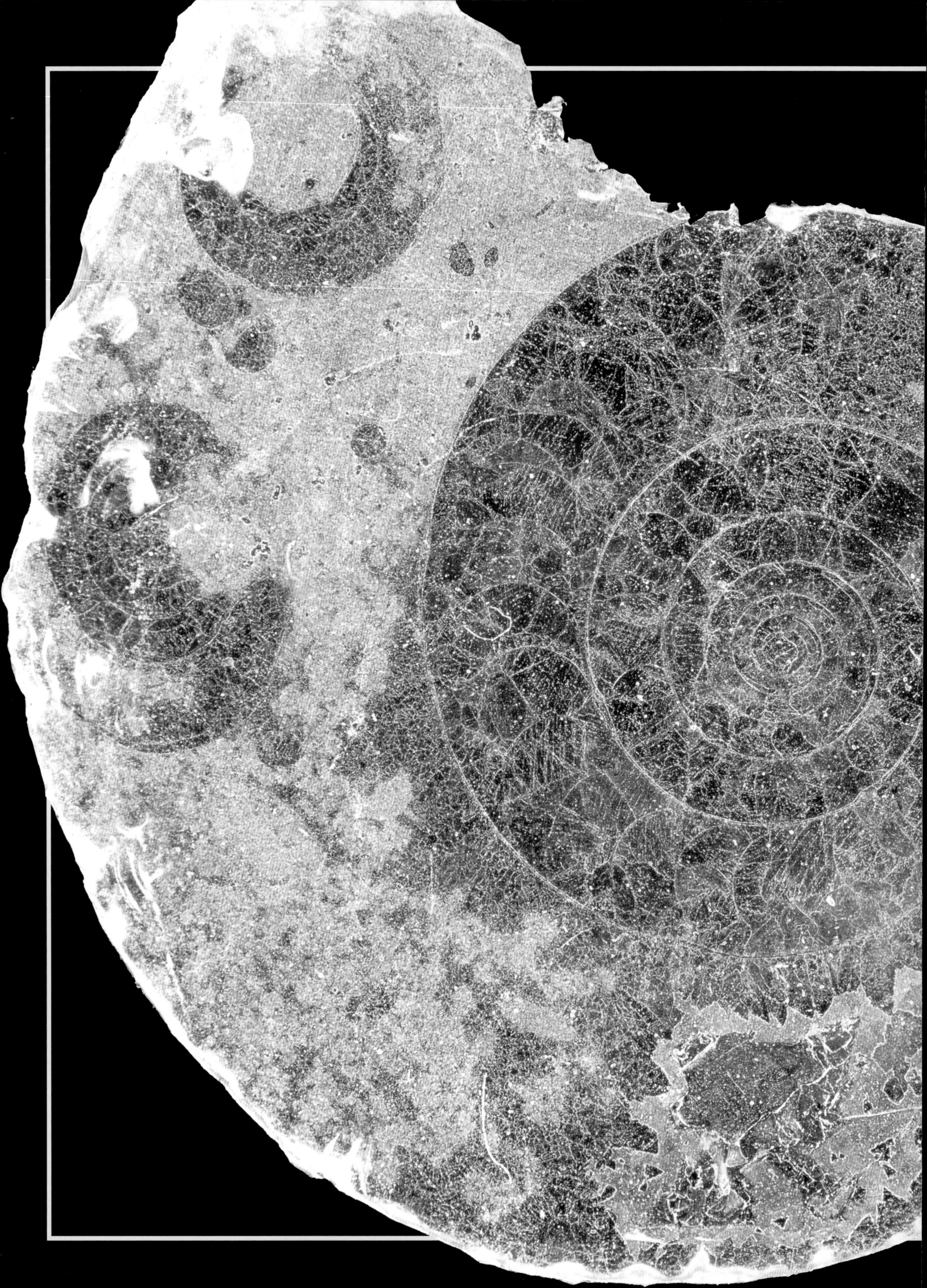

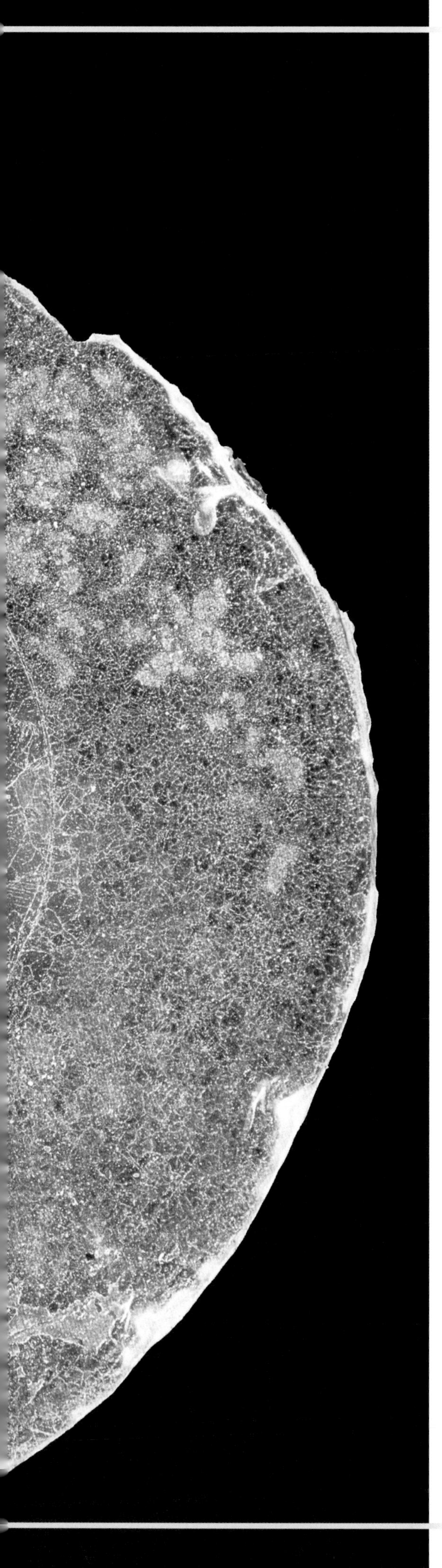

THE LATE PALEOZOIC

Life moves onto the land

PART 3

417–248 MILLION YEARS AGO

THE VIEW of Earth from space in the early part of the Late Paleozoic would have been reminiscent of the present day, with a scatter of continents separated by oceans. As the Late Paleozoic progressed, the gradual motion of the earth's lithospheric plates caused these continents to drift, collide, and fuse, forming bigger continents, uniting with other combined continents, and accreting island arcs and micro-continents as they went, until, at the very end of the Paleozoic, all the fragments of continental rock had aggregated into a single landmass—the supercontinent of Pangea.

It is tempting to think of Pangea as a body consisting of all of the modern continents sitting snugly together, ready to drift apart again when the geological time was right. The many paleogeographic reconstructions that superimpose the outlines of the modern continents on the ancient geography might suggest this but, in fact, the truth is much more complex. Pangea consisted of several cratonic shields of ancient Precambrian rock—the tectonically stable heart of any continent, surrounded and separated from one another by fold mountains. The age of these mountain chains varied; some were old and eroded even then, such as the oldest parts of the Appalachians, representing the suture lines of collisions that had taken place hundreds of millions of years previously. Others, such as the ancestral Ural mountains, were high and jagged, the results of recent collisions.

Occasionally there is some indication in the geography of Pangea of the lines that would, in the future, split apart, bringing about the familiar shapes of the continents we know today. When continents collide, grinding up big mountain ranges between them, thrusting down roots of metamorphosed sedimentary rock deep into the lower crust and mantle, these rugged boundaries tend to be less rigid than the cratons at each side, despite their greater thickness. When a supercontinent finally fractures, as a result of mantle heating and expansion beneath it, it is usually along these boundaries. Over the history of the Earth there have been several cycles of supercontinent aggregation and break-up of which Pangea was the latest.

Significant among the Late Paleozoic events was the collision between Laurentia (North America) and Baltica (northern Europe). Fifty million years later, when Pangea began to split up, one of the major rifts appeared along this very line, forming the Atlantic Ocean.

The climates of the Late Paleozoic reflect this gradual accumulation of a single landmass. While the continents were still drifting, the conditions on their land surfaces were tempered by the proximity of the oceans. Once they had all fused together there is evidence for the extremes of climate associated with vast landmasses. The oceans are very slow to warm up during hot weather, but once they have done so they hold on to their heat for a long time, even as the air temperatures cool. This delayed response has a moderating effect on the climate in coastal areas, but far inland it is a different story. By Permian times, when Pangea was complete, the center of the continent would have been uninhabitable because of the intense heat and aridity. At the same time, the southerly extreme of the supercontinent lay over the pole and the region suffered an ice age, as glacial deposits found today in Africa testify.

Other features of this past environment are also reflected in the rocks. The continents, whether single islands or one supercontinent, were not all dry land. Even today, large areas of continent are covered by shallow seas; the Seychelles and Mauritius in the Indian Ocean are the mountain peaks of a continental mass that was once part of Africa, and the Bahamas protrude above a huge underwater plateau that, geologically, is part of North America. So, in Late Paleozoic times, many of the continental fragments that combined to form Pangea would have been continental shelves and submerged plateaus. Vast beds of limestone were formed at this time from the deposition of fine particles in warm shallow water. Continental collision produced huge mountain ranges that immediately began to erode, spreading sandbanks and mud over the shallow seas. Terrestrial deposits consist of river sediments washed down from these new mountains and desert deposits formed in the hot interior. Sediments laid down in hot, dry conditions tend to have a red tinge to them, caused by the oxidation of iron in their particles. Red sandstones are the typical continental rocks of these times.

By the end of the Late Paleozoic, the supercontinent Pangea extended from pole to pole. The Earth had also begun to change color as forests spread across the globe.

AS THE LATE PALEOZOIC progressed the land began to change color, from the grays, reds and yellows of bare rock and sediment, to the greens of vegetation. This was when life left the water and became fully established on dry land. Although there are tracks of animals crawling along beaches dating from the Cambrian period, and evidence for the presence of land-living algae as far back as the Precambrian, it was not until the late Paleozoic that land life really took hold.

By Silurian times, at the end of the Early Paleozoic, there were primitive land plants, mere photosynthetic stalks growing at the edges of shallow waters. In the Devonian, the stems of these first vascular plants were branching and producing leaves. By the Late Devonian there were forests: lycopods, horsetails, and seed ferns grew on sandbanks. These became the vast swamps and coal-forests of the Carboniferous, with their giant club-mosses. On higher ground the first conifers grew.

In the wake of the plants came the animals. Insects and other arthropods were first to colonize the land; the first insects were wingless, but by the Late Carboniferous they had taken to the air. Vertebrates also made the transition from one element to another. The Devonian radiation of fish produced a group known as lobe-finned fish, with strongly muscled paired fins and rudimentary lungs. These evolved into tetrapods—the earliest amphibians—which were ancestral to all land vertebrates. Amphibians, however, still needed to lay their eggs in water. With the arrival of reptiles and shelled eggs, animal life became truly established on land.

Dougal Dixon

THE DEVONIAN

417 – 354 MILLION YEARS AGO

EARTH and all of its inhabitants began to undergo some remarkable changes during the Devonian period. Vertebrate animals were coming to the fore. Fish, which so far had only been small, jawless, tadpole-like animals, developed into a wide range of types and began to leave the water. Land plants, too, were spreading, establishing the world's first forests and causing dramatic changes to the atmosphere. And the movements of the continents were bringing the landmasses together, producing even larger supercontinents. Along the joins between them, huge mountain ranges grew. These shed sediment in both directions, forming distinctive redbeds that stretched from the Catskill Delta of New York all the way to western Russia. The remains of these beds are seen today in the sequences known as the Old Red Sandstone.

THE DEVONIAN marks the beginning of Late Paleozoic times. During this period, much of the landmass that now constitutes North America and Europe lay in arid zones above and below the Equator, and so accumulated sands that became known as the "Old Red Sandstones." By Carboniferous times these landmasses had drifted north, giving rise to climate changes that nourished the tropical forests, swamps, and river deltas that produced the rock sequences known as coal measures. Later, with the dawn of the Permian, they continued to drift north, laying down a new sequence, known as the "New Red Sandstones."

William "Strata" Smith, the English surveyor, had defined Scotland's Old Red Sandstones as early as the 1790s.

KEYWORDS

ACADIAN OROGENY
BASIN
CALEDONIAN OROGENY
CRATON
EVAPORITE
LAURENTIA
MOLASSE
OLD RED SANDSTONE
REDBED
TERRANE
VASCULAR PLANT

None of this was particularly relevant to the scientists who first studied and named the Devonian. It was the great geological figures of the early nineteenth century who identified the system: the British geologists Roderick Impey Murchison and Adam Sedgwick, whose most significant work had already been done in recognizing the systems of the lower Paleozoic. They named the Devonian system in 1839 for a sequence of marine beds in Devon on the southwest coast of England. A publication by William Lonsdale, an expert on corals, in the following year suggested that this sequence came between the Silurian system of Murchison and the already-identified Carboniferous, and that it may have been the same age as the great sequences of red sandstones found to the north. The sequence of marine rocks for which the period was originally named was rather limited. Lonsdale's theory is significant because it was

SILURIAN 417 mya	415	410	405	400 DEVONIAN
Series	EARLY/LOWER			
European stages	LOCHKOVIAN	PRAGHIAN		EMSIAN
N. American stages	ULSTERIAN			
Geological events	Caledonian orogeny; collision of Baltica with Laurentia in Scandinavia and eastern Greenland; Continued fragmentation of Gondwanaland			
Climate				
Atmosphere			Marked decline in levels of CO_2	
Sea level				Increasing
Plant life	Primitive vascular plants on land		Spore-bearing plants become more common on land	
Animal life	• First ammonoid		• First insects	Rapid diversification of jawed fish

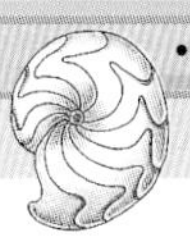

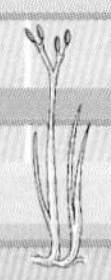

the first time that it had become known that two different types of rock—marine and desert sandstone—could have been formed at the same time, and that the world was not the same all over in any geological period.

Although a familiar part of the landscapes of southern Wales, central and northern Scotland, and the southern tip of Ireland, the Old Red Sandstones had been largely ignored by the leading geologists of the time. The general opinion was that they were a localized, fossil-lacking part of the Silurian or Carboniferous.

MUCH of the Old Red Sandstone continent lay in tropical latitudes to the north and south of the Equator. In Devonian times, as today, trade winds brought moisture north and south to the Equator, where the wet air was heated and rose, dropping its water into the soils and rivers below and encouraging greenhouse-like growing conditions. The dried air at high altitude drifted north and south, cooled, and descended again at about the latitudes of the tropics, producing hot, dry climates with occasional bursts of heavy seasonal rainfall.

Climates were warm throughout the Devonian, particularly on the northern continent; as the new landmass grew its interior became drier.

The continents of Laurentia (modern North America) and Baltica (modern Scandinavia), which had been moving inexorably towards one another for the whole of the Paleozoic so far, at last collided at the end of the Silurian. The Iapetus Ocean between, with its shelf seas, its deep trenches, its chains of volcanic islands, was crushed out of existence. In its place, marking the boundary between the two great continents, arose a towering mountain chain, the Acadian-Caledonian range. Before the theories of plate tectonics and continental drift were accepted, geologists believed that these mountains had once stretched across the Atlantic, and that the intervening portion had been eroded away.

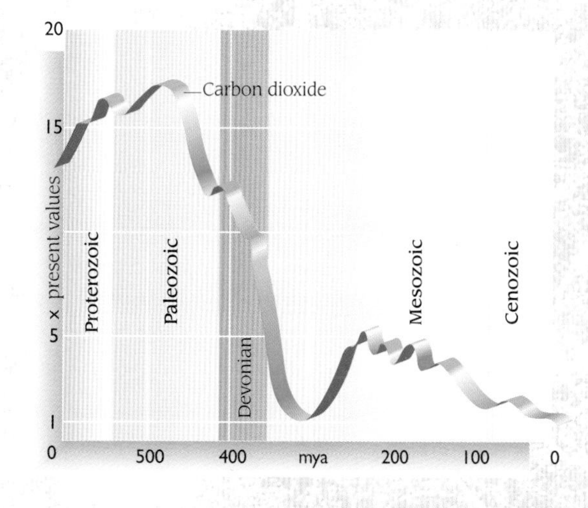

A FRESHENING ATMOSPHERE

The rapid development of plant life in the Devonian meant that much of the carbon dioxide in the atmosphere became converted to oxygen. The proportion of oxygen increased throughout the period, reaching the level at which animals with lungs could live on land. However, carbon dioxide increased again during the Mesozoic before dropping back to modern levels.

The collision of continents has been a repeated occurrence throughout the Earth's history and has been a major factor in shaping the geographies of the various periods. It is still happening today. Australia is closing with south-east Asia and will collide within the next 50 million years. Africa is already on the point of collision with Europe—the contorted, volcanic, earthquake-riven tangle of the Mediterranean and the stretched-out string

Present
Origin of Earth
DEVONIAN

See Also

THE ORIGIN AND NATURE OF THE EARTH: *The Evolving Atmosphere*
THE EARLY CARBONIFEROUS: *Acadian-Caledonian Orogeny; Life on Land*
THE PERMIAN: *New Red Sandstone*

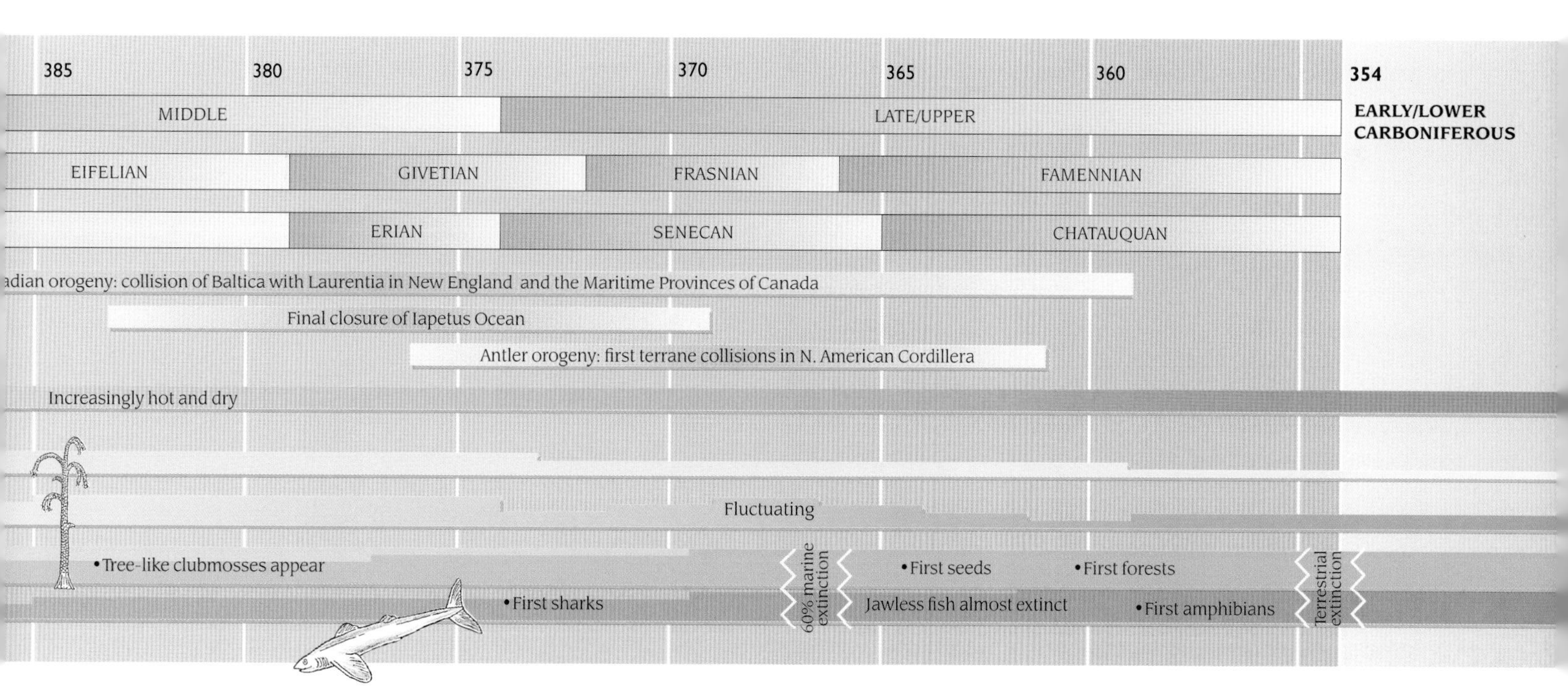

of drying puddles that are the Black Sea, the Caspian Sea, and the Aral Sea are the remains of the ocean area that once existed between. The next few million years will see the Mediterranean region thrust up into a new mountain chain as high as the Himalayas.

FOR 200 million years the Iapetus Ocean, between Baltica and Laurentia, had been collecting sediment. The shallows at the sides had built up coastal sands. Limestones had been deposited by marginal reefs that had abounded with life: corals, crinoids, and trilobites. The deeps had produced layer upon layer of thick black muds, with the embedded remains of graptolites and other drifting creatures. Chains of volcanic islands had erupted parallel to the opposing coastlines, fed by molten material squeezed upwards as the Earth's structure crumpled between the approaching continents, and had spread out their molten rock as deposits of pillow lavas on the seafloor. Now all these sediments were crushed into one another and compacted, ground down into the Earth's mantle and forced up into jagged highlands that reached to Everest heights and stretched for thousands of miles, marking a boundary.

Fragments of ancient seafloor and other marine artifacts still surface today in Scotland and New England, marking the boundary of the great collision.

Modern North America formed the western part of the Old Red Sandstone Continent. The ragged stumps of the Acadian-Caledonian mountain range are still to be seen in the northern part of the Appalachians. Meanwhile, the western coastline lay close to a subduction zone, and built up an active volcanic island arc offshore. The sea poured in at this edge, flooding the coastal plains and further inland as the sea level rose. Further deformation, known as the Antler Orogeny, built mountains where the modern Rocky Mountains now lie.

Off the northeastern edge of the Old Red Sandstone continent stretched a sea area, the dimensions of which are unknown. Beyond it lay the Siberian continent. In times to come this oceanic area would close and eventually form the Ural Mountains as the continents fused.

At the southern edge of the Old Red Sandstone continent, in the region of modern Germany, there was deep water not far offshore. This seems to be evidence of ocean trenches between the Old Red Sandstone continent and Gondwana to the south, showing that they were approaching each other. The sea surged and receded, producing deep-water, shallow-water, and dry land sediments in turn. An extension of these sediments formed the deposits in southern England for which the Devonian was named.

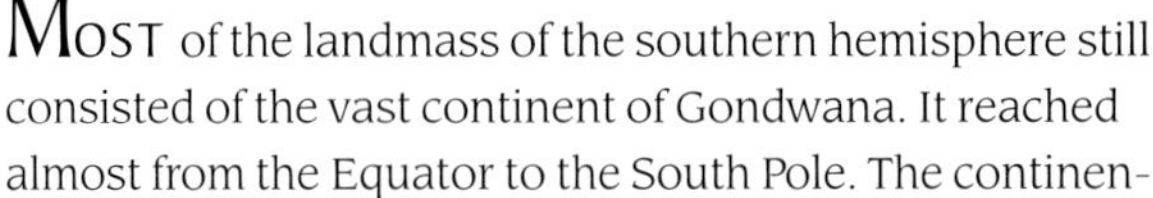

MOST of the landmass of the southern hemisphere still consisted of the vast continent of Gondwana. It reached almost from the Equator to the South Pole. The continental fragments that now represent South America, Africa, India, Antarctica, and Australia were still part of a single landmass. Cratons, formed in the earliest parts of the Archean, made up the core of this great continent; they were separated by tracts of worn-down mountains, exposed to the roots. This ancient landscape had been eroded flat, with the only mountainous regions existing along limited areas of coastline where the continent's edge coincided with an active plate margin. Parts of the modern northern Andes and eastern Australia—the Tasman Belt—show signs of such Devonian activity. In places Gondwana was so flat and low that it was encroached by shallow seas creeping in from the edge. Northern Australia, closest to the Equator, had shelf seas complete with barrier reefs. Other regions of shallow seas were found in the region of the South Pole. There were no barrier reefs here; fossil finds indicate a fauna adapted to very cold conditions. These were little affected by the mass extinctions that wiped out huge numbers of tropical species twice during the Devonian.

Gondwana, huge and flat, dominated the southern hemisphere, but it was beginning to fracture. Its climate ranged from tropical to polar.

The rest of the continents were scattered across the northern ocean. To the east and northeast of the Old Red Sandstone Continent, Siberia and Kazakhstania were approaching one another, and both were closing on the Old Red Sandstone Continent. Only the landmass of China was well separated from all others.

Laurentia and Baltica did not have a simple head-on collision and fusion. As they closed in on each other there would have been fragments of crustal material and landmasses between them. A similar movement is taking place today as Australia moves towards Asia. Between them are loops of volcanic island chains like Sumatra and Java, formed along the subduction zone of the Java Trench. There are also chunks of continental fragments, such as Borneo and New Guinea. There was not a clear

NORTHERN FUSION

Baltica (Europe) and Laurentia (North America) joined as the North American and Eurasian plates moved closer together, eventually forming Laurasia.

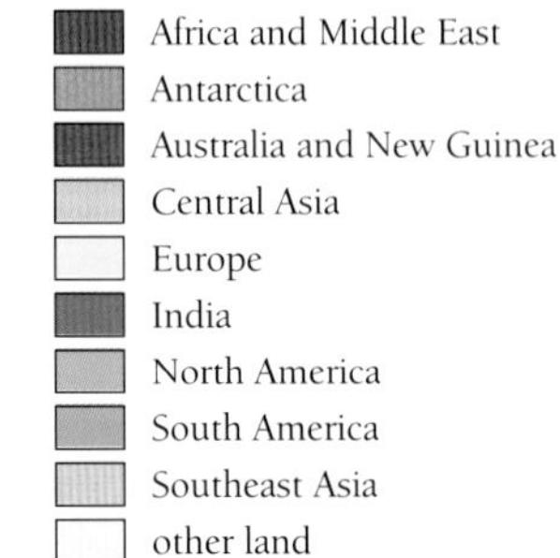

THE FLOATING FRAGMENTS

Fragments of Asia lay scattered across the northern hemisphere. Eventually they would unite, with India, as one great landmass.

SOUTHERN SUPERCONTINENT

Although fragmentation was taking place along its northern margin, Gondwana was still the largest of the continents, containing South America, Africa, Australia, India, and Antarctica. It extended from almost the Equator to the South Pole.

DEVONIAN ROCK

The "Devil's Chimney" on the coast of Pembrokeshire, Wales, is an example of Late Devonian sandstone. The rock formed by alternating marine and continental deposition as the sea advanced and retreated and sediment was shed by the Caledonian mountains. Vertical tilting of the beds occurred when the Old Red Sandstone continent collided with Gondwana during Late Carboniferous times.

OLD RED SANDSTONE CONTINENT

The heart of the continent was the Acadian-Caledonian mountain range, consisting of lower Paleozoic rocks that had been crushed and moved as individual blocks, or terranes. Old Red Sandstone deposits were spread over the craton of the surrounding continent. Rivers carried more sediment out towards the continent's edge.

boundary between the two continents, but a jigsaw of discrete rock masses, each piled upon the next. The impact had a sideways motion as well, shearing the blocks against one another. Island chains and peninsulas in the modern Mediterranean are twisted into great S shapes by a similar sideways movement of Africa against the southern edge of Europe.

The resulting mountain range consisted, then, of elongated masses of different types of rock, separated from one another by faults that were the focus of earthquakes for long after the continents had fused. Indeed, earth tremors are often felt along these faults today. Molten material from the roots of the mountains forced its way up along zones of weakness generated by the faults, and burst out as volcanoes in the heart of the mountain chain. Long valleys, bounded by the faults, carried huge lakes of water that eventually dried up in the hot, dry climate, leaving behind their own record of lake sediments and an abundance of freshwater fish that gave the Devonian one of its nicknames: "The Age of Fishes."

MODERN Scotland's Highlands, along with the Welsh mountains, the mountains of Norway, and the northern part of the Appalachians, are what remains of the huge mountain system that formed during the collision of Laurentia and Baltica. There the Cambrian, Ordovician, and Silurian rocks are twisted, baked, sheared, and thrust over one another and now form the majestic Trossachs, Grampians, and Cairngorms of Northern Scotland, and the gentle hills of the Southern Uplands. They are worn down to such depths that the cores of metamorphic rock are exposed, as well as the granitic masses that show where molten material gathered in the heart of the mountains. It was in this dramatic landscape that the early geologist

Today's Scottish Highlands and the North American Appalachians are the worn stumps of the Acadian-Caledonian mountains.

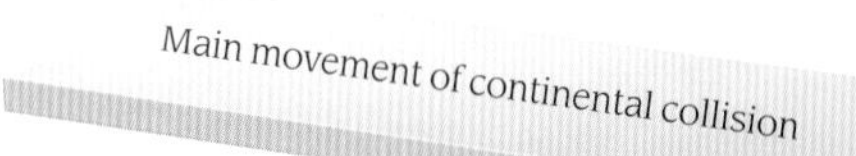

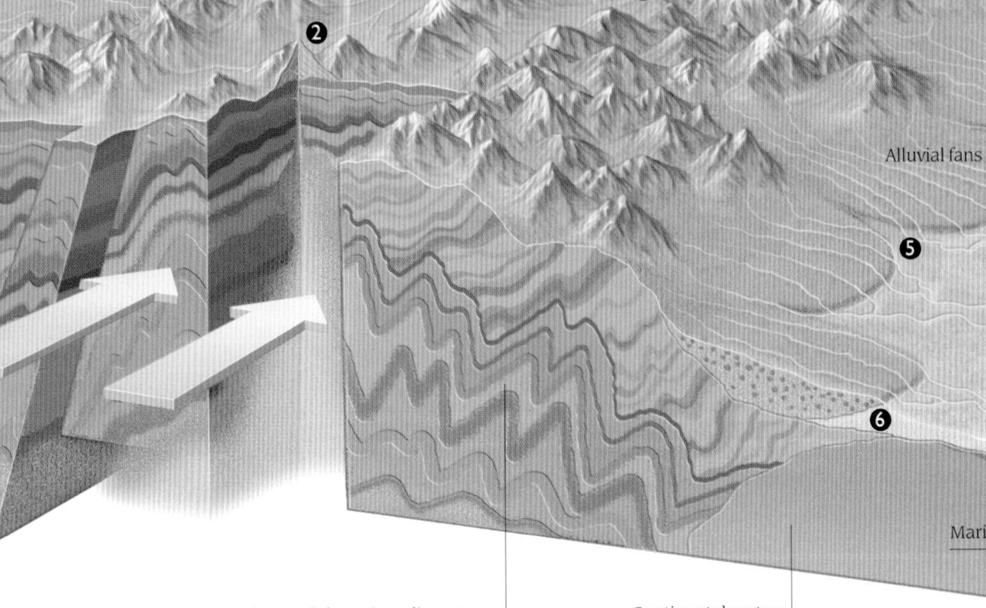

James Hutton recognized and described the principle of stratigraphic unconformity.

A physical geology map of Scotland shows that it is crossed by a number of southwest-to-northeast trending faults that cut the whole area into slices. One fault separates the Outer Hebrides from the mainland. The Moine Thrust crumples the whole of the northwest coast. The Great Glen Fault cuts off the northernmost part of the country along a string of lochs. The Highland Boundary Fault and the Southern Upland Fault form the sides of the Central Valley. A vaguely demarcated faulted area called the Iapetus Suture (after the vanished ocean) divides Scotland from England.

All these faults define distinct masses of rock, which geologists refer to as "exotic terranes." They consist of masses of old seafloor, volcanic island chains, chunks of continental material—anything caught up between the closing continents—which all ground past one another during the formation of the Old Red Sandstone Continent during the Caledonian Orogeny.

As the mountains were thrust upwards they were continuously worn away by the beating of the weather and the destructive pull of gravity. Broad fault-bounded valleys filled up with the debris washed down the gullies and wadis from the rising mountains. Rivers carried eroded rocky matter out on to the plains of the surrounding continent, where there was little plant cover or soil to hold the new material in place. There the finest parts were picked up and blown about by winds, the iron minerals in them oxidizing to a red color in the warm, dry conditions. The sand that accumulated on these valleys and plains eventually solidified to form characteristic red sandstone beds—the Old Red Sandstone.

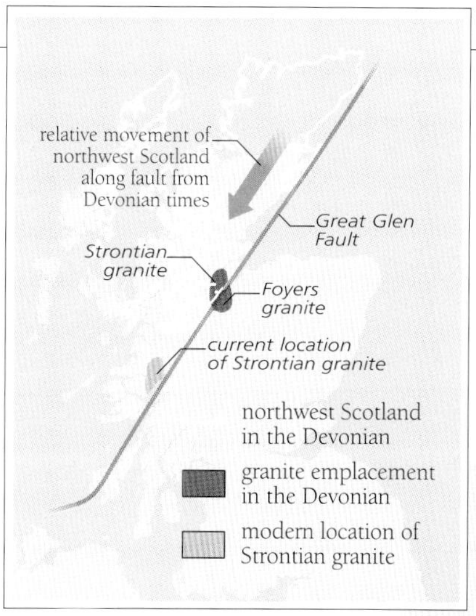

Because this formation was so clearly a desert formation, it was originally thought to have no connection with the marine sandstone of Devon for which the period was named. In fact, the two kinds of Devonian sandstone were formed at the same time from the same mountain debris, but Devon was periodically inundated with seawater, whereas Scotland, located farther inland on the great continental landmass, was not.

THE GREAT GLEN FAULT

Along the Great Glen Fault in Scotland, an emplacement of granite at Foyers on the south side of the fault can be correlated with one at Strontian on the north. These two masses are now 65mi (105km) apart, showing the distance that the fault has moved since the granite was emplaced. Earthquakes still occur in Scotland today, acting along the old Devonian fault lines. The other fault-bounded terranes show similar movement.

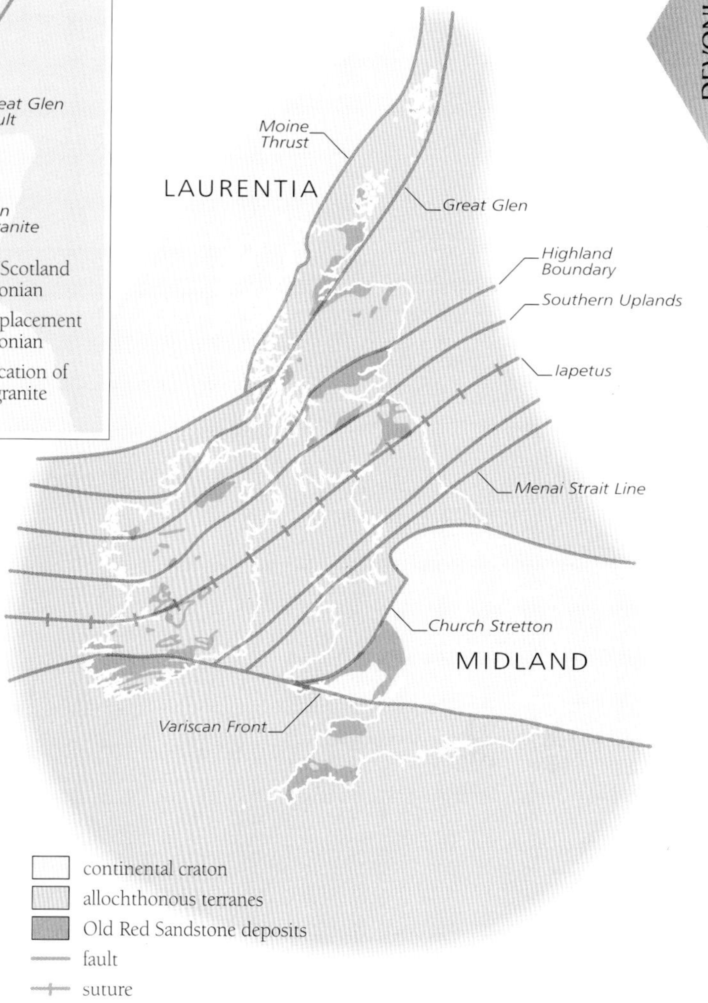

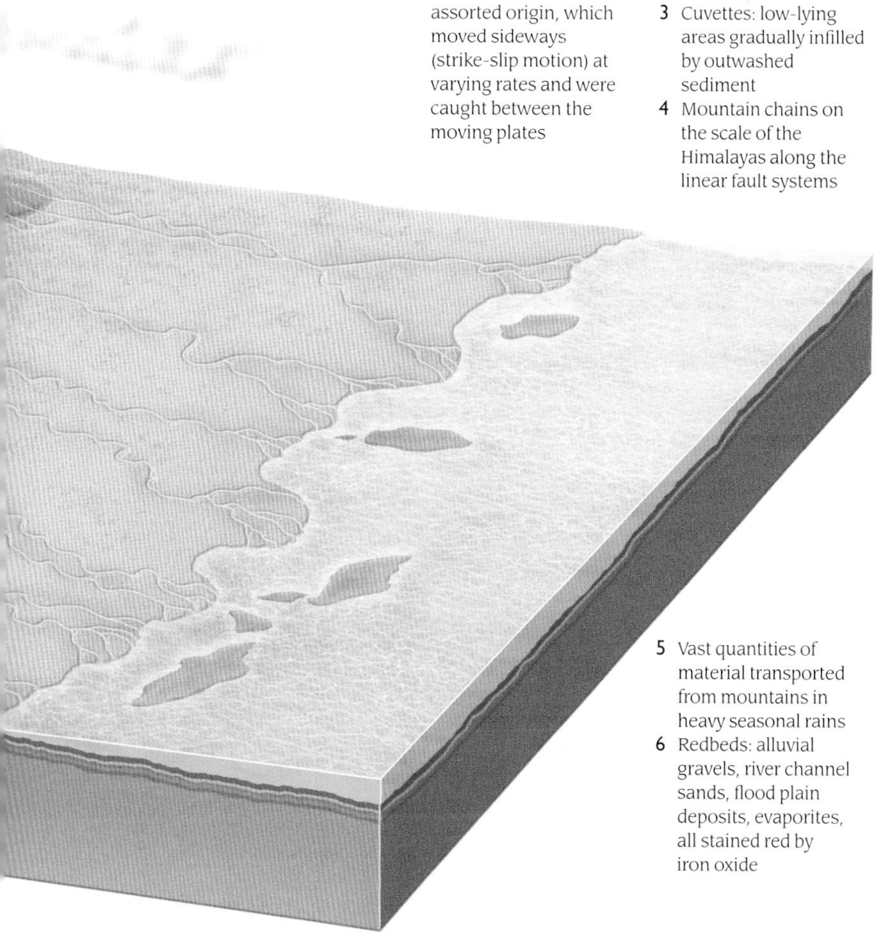

1 Exotic terranes: fault-bounded blocks of assorted origin, which moved sideways (strike-slip motion) at varying rates and were caught between the moving plates
2 Volcanoes associated with major fault lines
3 Cuvettes: low-lying areas gradually infilled by outwashed sediment
4 Mountain chains on the scale of the Himalayas along the linear fault systems
5 Vast quantities of material transported from mountains in heavy seasonal rains
6 Redbeds: alluvial gravels, river channel sands, flood plain deposits, evaporites, all stained red by iron oxide

THE OLD Red Sandstone consists of all kinds of river-deposited and wind-deposited material. Close to the mountains it was predominantly largish chunks of rock that had been broken from the outcrops. These chunks were irregular and jagged, showing that they had not traveled very far and had not had their rough corners worn off. Such deposits were formed in alluvial fans, where debris was washed from the mountains by flash floods during rainy seasons and dropped on the surrounding lowlands as soon as the current slackened. Eventually these deposits were transformed into

The Old Red Sandstone was not all red, and not all of it was sandstone.

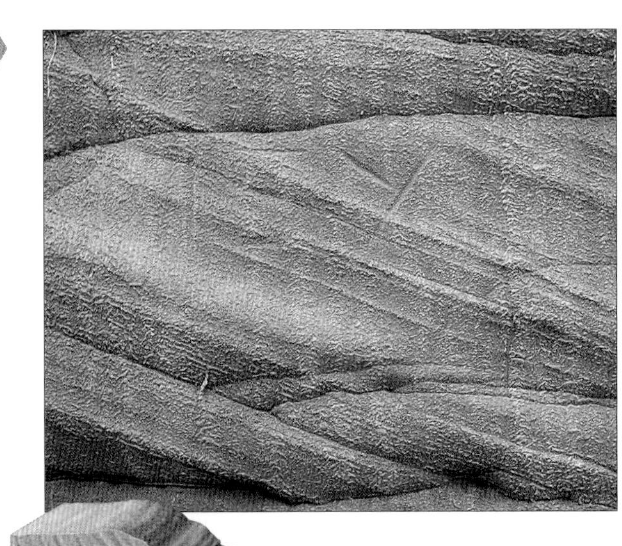

CROSS-BEDDING

The inclined planes of this Devonian sandstone (left) show cross-bedding caused by strong currents of water affecting their deposition. In a typical delta, where a flowing river drops its sediment load on reaching deeper water, there are horizontal or very gently shelving topset beds. At the delta front are inclined foreset beds, and gently sloping bottomset beds meet the sea floor in front of the delta. Current flows in the direction of the down-sloping strata.

coarse-grained sedimentary rocks geologists call breccia. This rock is mixed liberally with river silts and sands in the redbeds of Scotland.

Further out on the Old Red Sandstone Continent's plain—for example, in Wales—most of the sediments consisted of sands that were river-deposited. The beds have a characteristic S-shaped cross-section, a sign of river currents. In Europe the orientation of these current beds show that the rivers flowed from the Caledonian Mountains in the northwest. The individual rivers appear to have been up to 40mi (64km) apart. The rocks formed from these beds are interspersed with horizontal beds of silt laid down by the rivers in flood. These may show mud-cracks and roots, indicating that they dried out and supported a vegetation. They also contain limestone bands called "calcrete" or "kunkar," formed in hot arid regions with seasonal rainfall. When the occasional moisture evaporates from the soil, it draws up dissolved calcite and deposits it close to the surface. Closer to the sea, the rocks consist of river-deposited sandstones interspersed with intertidal beds. Nearer the southern edge of the continent, the sediments that were deposited on land, in rivers, and in lakes gave way to coastal and marine beds, as in Devon, producing partly marine rock.

OLD RED SANDSTONE SEQUENCES

Near the mountains the Old Red Sandstone consisted of deposits left by alluvial fans. Out on the plains they were mostly river deposits and windblown sands; it is difficult to tell their derivation because the mountain hinterland is so jumbled. Deposits close to the sea were interbedded with marine sediments.

Alluvial fans

River flood plains

Coastal deposition

River sands with current bedding

Conglomerate/breccia

River silts

Siltstone with mud cracks

Calcrete (limestone formed in dry soil)

Sea mudstone

ALTHOUGH this sequence of deposits was originally identified in Britain and Europe, it is also found in North America, formed from material washed from the other side of the mountain chain. The Catskill Wedge of the northeastern United States consists of a sequence of redbeds that grades westwards through sandstones, into siltstones and finally marine shales and limestones. In the east, in Pennsylvania, closest to the old mountains, it is some 9000ft (2740m) thick, thinning to a few hundred feet of marine deposits in Kentucky and western Ohio.

The Catskill Wedge is sometimes called the Catskill Delta. This is misleading as it suggests a river running into the sea, when in fact most of the deposition was on land.

The sheer thickness of the Old Red Sandstone attests to the heavy erosion of the mountain hinterland. Deposition in central New York has been calculated from 23ft (7m) per million years at the beginning of the Devonian, rising to 230ft (70m) by the end of the period. While mountains are rising they are being subjected to continuous erosion, so that much of the material is worn from them before they attain their greatest height. To account for the volume of sediment, there was even a suggestion that a vast continent, now submerged, existed in the region of the Atlantic Ocean. Geologists have often tried to calculate the height of the Acadian–Caledonian range by measuring the volume of Old Red Sandstone derived from them, and have reached unrealistic conclusions.

The interior fault-bounded basins of the mountains had their own depositional systems, sometimes known as "cuvettes." Lower than the surrounding mountains,

they had their own rivers draining into them and developed vast areas of freshwater lakes. These lakes collected thin layers of fine sand, which eventually solidified to form flagstones, often used for roofing and floors, and to make fences. More important as a building material, however, is the granite. Formed as molten material cooled in the heart of the mountain system, and exposed after 400 million years of erosion, this coarse-grained igneous rock is ideal for large-scale stone masonry.

The Devonian, with its large tracts of shallow reefs in western North America, also provided a significant amount of oil on that continent. The reefs of the shallow waters of Devonian North America began as a growth of stick-shaped corals on the muddy seafloor. Flat spreading corals then used these as a foundation and spread over them. Finally, the stromatoporoids—sponge-like animals with dense skeletons—took over from the corals as the main producers of reefs, building up towering ridges. In the quiet waters behind these ridges, carbonate mud gathered, contributing to the widespread Devonian limestone formations of the continent.

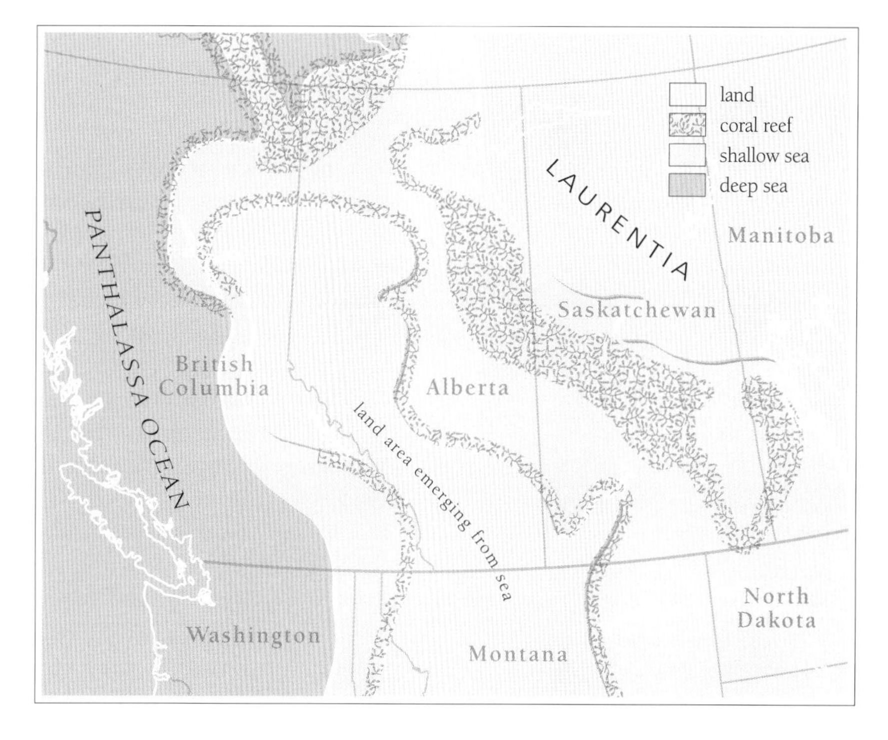

SHALLOW SEAS AND REEF-BUILDING

Along the western edge of North America, in the far west of the Old Red Sandstone Continent, the sea came in through the island arcs that bordered the ocean, between the hilly islands that resulted from the Antler Orogeny. It spread out across the low-lying interior of the continent, forming a broad shallow shelf sea. Along the edges of this shelf sea, limestone reefs built up in the basins of the North American craton. During the Devonian, this region was in a tropical climate, and the reefs were almost completely wiped out by the mass extinction at the end of the period.

TOWERING REEFS

Devonian reefs, such as this exposed reef in the Rocky Mountains of Canada, stretched 600ft (200m) from the seafloor to sea level. Unlike modern tropical reefs, which consist mainly of algae and corals, Devonian reefs were built by tabulate corals and sponges. Later these reefs accumulated more organic debris, eventually giving rise to oil fields.

SIMPLE plants had been taking the air since Silurian times. The first non-aquatic plants grew a photosynthetic stem above the shallow freshwater in which they lived. The earliest detailed flora of the Devonian consisted of plants that were similar to this basic plan. In a cuvette that lay where northeast Scotland now lies, a lake supported a growth of such simple plants. A mat of roots and stems intertwined on the lake bed and erected stems with spore-bearing bodies on the end. These stems contained vascular tissue, a plumbing system that carried water to the photosynthetic parts of the plant, and food back again.

For the first time, parts of the Earth's surface were turning green.

Evidence for this comes from a particularly well-preserved fossil plant community at Rhynie in Scotland in which the organic material was slowly replaced—molecule by molecule—by silica, preserving the cellular structure of the plants in a natural glass called chert. The source of this silica may have been hot water from the volcanic mountains that ringed the cuvette.

New plants began to stabilize the land with their roots, encouraging the development of more sophisticated vegetation as soil replaced sand. Later plants became very complex, with strong tree-like trunks giving support, specialized leaves manufacturing the food, and cone-like structures producing the spores. The rivers of the Catskill Wedge in New York state supported forests of such plants, especially in a site called Gilboa, foreshadowing the great forests of the Carboniferous to come. The distribution of stomata—the pores on a plant's surface where gases are exchanged—on well-preserved fossil plants suggests that at the beginning of the Devonian there was ten times as much carbon dioxide in the air as there is today. Then, as all the new plants underwent photosynthesis, the oxygen level rose and the carbon dioxide level fell to something more like they are now. From then on, land plants became the major producer of oxygen—twice as much as phytoplankton in the ocean, though land takes up only one third of the Earth's surface.

Only slightly less dramatic changes were occurring in the water with an explosion of new forms of swimming

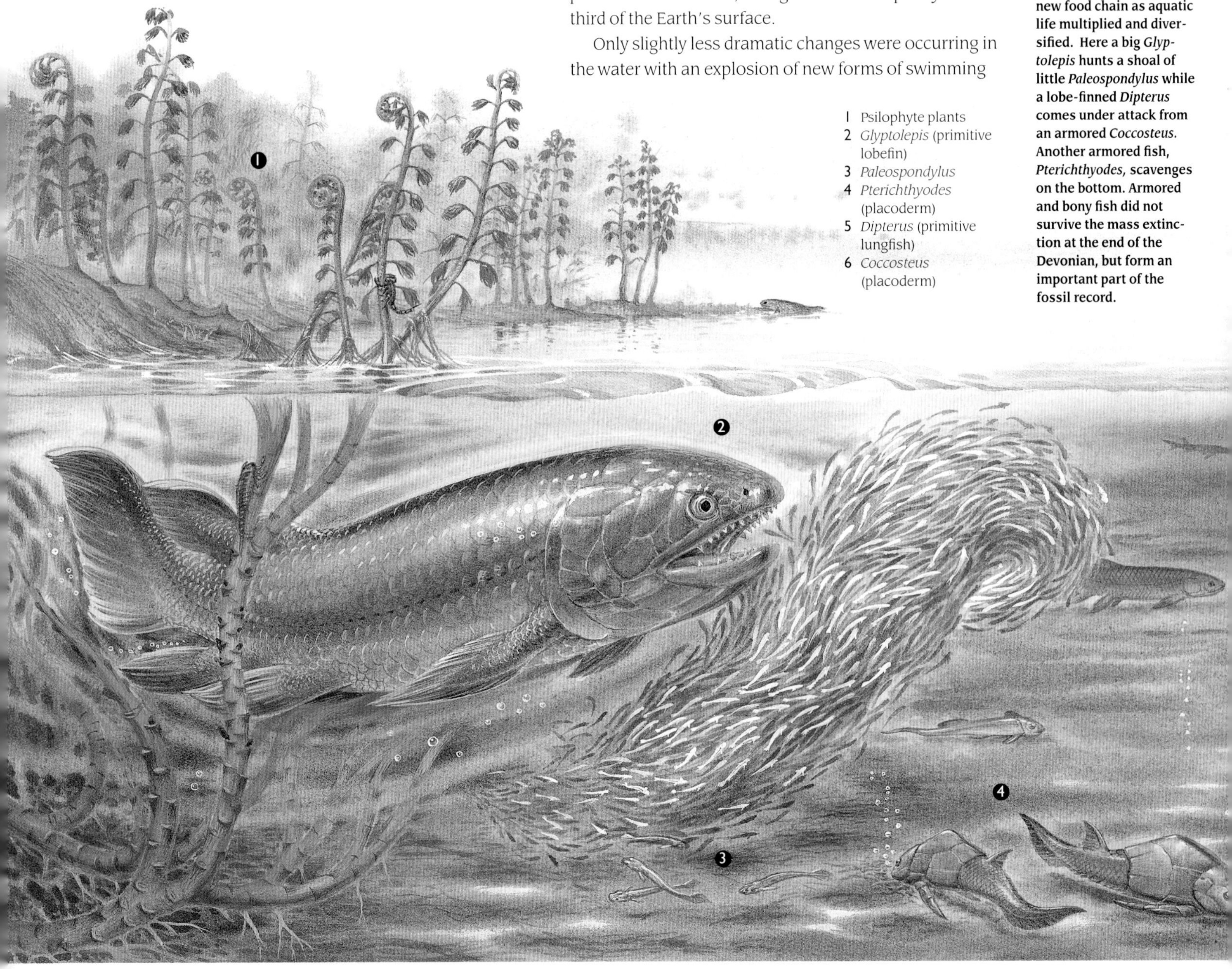

FISH LAKES

Most surviving fish fossils from the Devonian come from fresh water. The cuvettes in the mountains of the Old Red Sandstone continent had lakes that supported vast numbers of fish, many of them predators, reflecting the new food chain as aquatic life multiplied and diversified. Here a big *Glyptolepis* hunts a shoal of little *Paleospondylus* while a lobe-finned *Dipterus* comes under attack from an armored *Coccosteus.* Another armored fish, *Pterichthyodes,* scavenges on the bottom. Armored and bony fish did not survive the mass extinction at the end of the Devonian, but form an important part of the fossil record.

1 Psilophyte plants
2 *Glyptolepis* (primitive lobefin)
3 *Paleospondylus*
4 *Pterichthyodes* (placoderm)
5 *Dipterus* (primitive lungfish)
6 *Coccosteus* (placoderm)

animals. Ammonoids, a kind of mollusk that could live in deep or shallow water, evolved from nautiloids in the early Devonian. Another group of arthropods, the predatory eurypterids, ranged in size from 8 inches (20cm) to over 7ft (2m). Eurypterids were among the first animals that could live in brackish or fresh water. They gave rise to the first true (freshwater) scorpions, which eventually evolved into land spiders and mites.

Fish were the first vertebrates (animals with backbones), and until the end of the Devonian, the only ones. Fish that existed in Ordovician and Silurian times were small jawless creatures. By the Devonian the jawless forms were still around, but now so were primitive sharks, spiny acanthodians, armored placoderms, actinopterygians that were the forerunners of today's bony fish, and lung-bearing lobe-finned sarcopterygians—the ancestors of amphibians and so of all land animals.

It is not clear when fish first moved into freshwater environments, but the transition had been made by the time of the Silurian, when most fish fossils document freshwater species. By the Devonian, thousands of types of fish were living in fresh water all over the world, as discoveries of lake deposits attest, but the most famous occurrence is in Scotland, at Achanarras and Dura Den. Gogo and Canowindra in Australia are similar sites that show that occurrences were not restricted to the Old Red Sandstone Continent. A fossil Devonian fish was brought back from Antarctica by an early explorer, Captain Scott.

The famous fish beds indicate that large numbers of fish died together. Perhaps the freshwater lakes dried out or perhaps the water suddenly became toxic. It is likely that the lake fish lived in the well-oxygenated waters by the shore, but now and then the toxic stagnant water of the deep lake surged upwards and made even the shallows uninhabitable. By the end of the Devonian, armored fish and bony fish, like the shallow-water corals and other tropical species earlier in the period, had become extinct. This mass extinction seems to have been fairly gradual, lasting several million years. This makes it unlikely that the extinction was due to a catastrophic event such as a meteor impact rather than a climatic or environmental change. An ice age was just beginning in the southern hemisphere; the cooling global climate associated with this, and a reduction of shallow water habitats, may have been responsible.

CORAL CALENDARS

In the 1960s the British paleontologist Colin Scrutton found that corals produce a layer of skeleton every day, and that this layer is a different thickness at different times of the lunar month and different times of the year. With this information he studied fossil Devonian corals and found that the lunar month was longer in Devonian times than today—30 days as opposed to 28. More remarkable was the fact that there were between 385 and 405 days in the Devonian year. Extending this study, Scrutton found that the Cambrian year was about 428 days long. The length of the year is constant; this means that the Earth is slowing its rate of spin, becoming 0.0016 of a second slower every century. Astronomers had already predicted this because of the tidal drag of the oceans across the planet, but it was Scrutton's work that provided the conclusive proof.

TOWARDS the end of the Devonian a side-shoot of the sarcopterygian fish, the rhipidistians, took the next step towards life on land. Over the next 10 million years, with continuous small changes, they evolved into the amphibians. All jawed fish already had lungs, and the crossopterygians were adapted to fresh water. Their two pairs of flexible, muscular fins became longer and stronger, and developed hinge joints and toes, enabling them to walk. Stronger joints between the vertebrae of the spine gave support to replace the missing cushion of water. Finally, adults lost their gills.

The transition from fish to amphibian took about 10 million years—a relatively swift transition on the scale of evolution.

The best-known of the Devonian amphibians were *Ichthyostega* and *Acanthostega* from Greenland. They were found in sediments that showed that they lived in meandering rivers in ephemeral forests—almost in the northern extension of the Catskill Wedge. *Ichthyostega* had eight toes on its hind feet; the five-toed pattern had not yet evolved.

There were other Devonian amphibians as well. Fragments of a land-dweller called *Hynerpeton* have been found in Pennsylvania. Five-toed footprints on land have been found in Scotland, Greenland, Canada, Australia, Russia, the Paraná Basin in Brazil, and Ireland. Land life had come to stay.

AN EARLY DEVONIAN VASCULAR PLANT

The minute cell structure of some plants is preserved in fossils from the Rhynie Chert of Scotland. This plant is a member of the genus *Rhynia*, a land plant of the early Devonian. It had a supporting stem with two kinds of tubes: one for carrying water and nutrients to make food, the other to distribute the food. *Rhynia* was displaced by larger plants with stronger stems and more vascular tissue for more efficient food production.

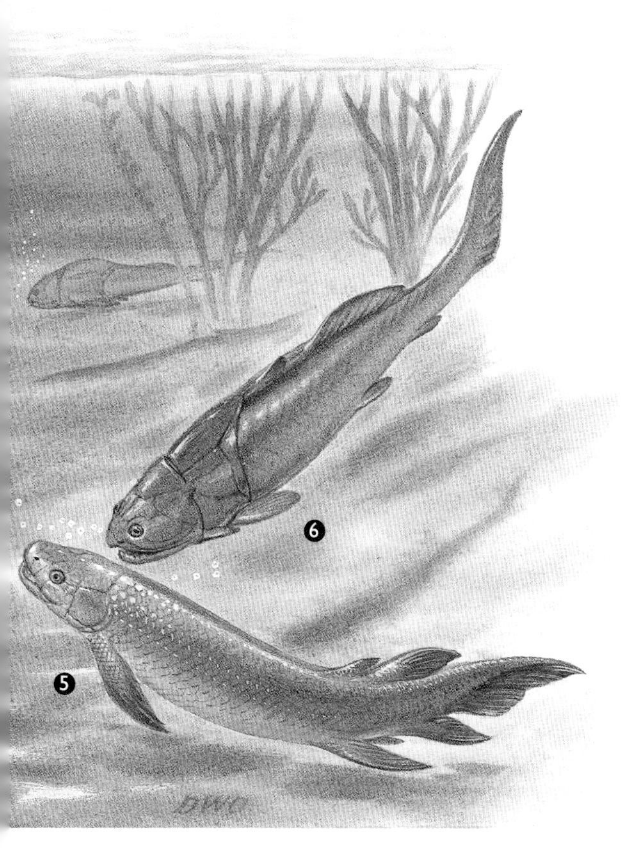

THE AIR over the Late Devonian Old Red Sandstone continent was hot and dry, but it was perfectly breathable—probably for the first time in the Earth's history. A vast extent of dried mud stretched to the horizon, where a distant range of mountains rose above the haze of suspended dust.

By the end of the Devonian, the air on Earth was fit to support animal life.

The mud, deposited when a nearby river last flooded in the rainy season, had cracked into polygonal slabs, curled at the edges. The dusty concave surfaces of the polygons were tinged green with microscopic algae and showed concentric patterns of white salt where moisture had evaporated. Red dust filled the cracks and little dunes of red sand, driven by the wind, crept across the flatness.

In the other direction the aspect would have been very different. A river meandered across the landscape, its route picked up by a winding band of greenery in which the open water sparkled in the sunlight. Vegetation flourished along the banks of the river, along its backwaters and oxbows, wherever the soil was saturated and the roots could reach the moisture.

Stands of tall trees were clearly visible, their trunks silhouetted against the water, their crowns standing out against the dusty plain beyond. They were the curly-branched *Protolepidodendron*, an early member of the lycopsids or clubmosses, and the tree ferns *Aneurophyton* and *Archaeopteris*. Their heavy trunks rose from an undergrowth of indistinct low-growing plants.

A closer look would have shown the undergrowth plants to be very spindly, giving an ethereal wispy appearance to the ground cover. The plants had few leaves as

1 *Asteroxylon* (a lycopsid—clubmoss)
2 Mosses and liverworts (non-vascular plants)
3 *Duisbergia*
4 *Rhynia* (an early tracheophyte—vascular plant)
5 *Aneurophyton* (tree fern)
6 *Protolepidodendron* (a lycopsid—clubmoss)
7 *Calamophyton* (early horsetail)
8 Lungfish
9 *Pseudoscorpion* (an aquatic arthropod)

yet, and those that they did have were very small and narrow. A rooty tangle of lycopsid stems spread in a mat from the edge of the water, their sporebodies waving at the ends of vertical stems. Branched stems of *Calamophyton* stood knee-high here and there, their segmented shoots showing them to be primitive relatives of the horsetails. Elegant fern-like fronds also carpeted the ground wherever there was enough moisture.

Apart from the singing of the wind, blowing the distant sand, and the rustle of leaves jostling one another, there was silence everywhere; the land seemed uninhabited. Then a loud *Plop*! Out on the placid surface of the still backwater, a run of ringed ripples was spreading out from a momentary disturbance below. Something had risen from the water, taken the air, and submerged itself once more. There was animal life here after all.

5

6

7

8

9

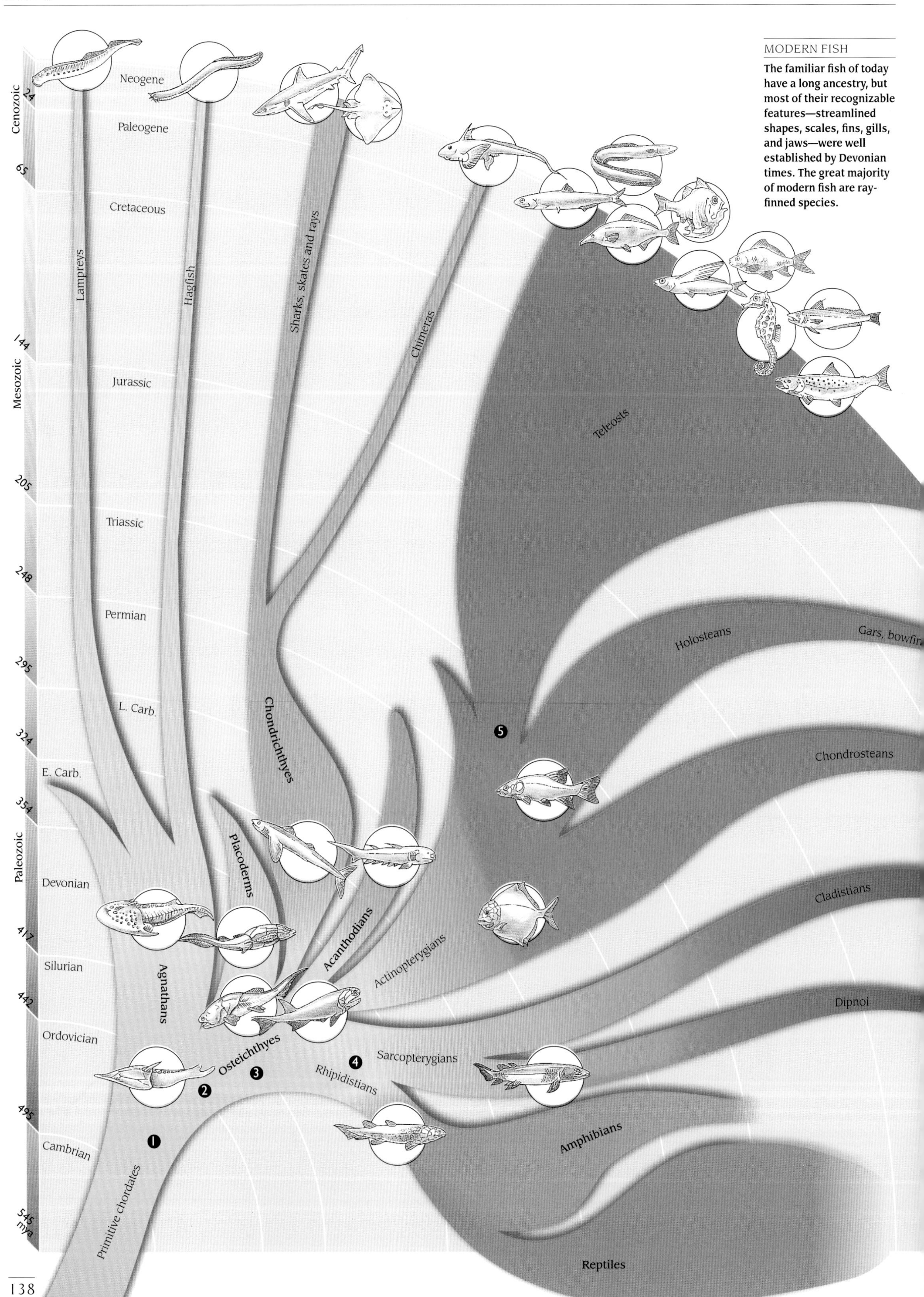

MODERN FISH

The familiar fish of today have a long ancestry, but most of their recognizable features—streamlined shapes, scales, fins, gills, and jaws—were well established by Devonian times. The great majority of modern fish are ray-finned species.

THE EVOLUTION OF FISH

FISH is a broad and unscientific term covering an enormous range of creatures no more related to one another than they are to humans. A fish is simply a swimming vertebrate animal that breathes by means of gills and has no legs for terrestrial locomotion.

Like other vertebrates, fish evolved from small segmented worm-like creatures with nervous systems supported by a beam of gristle that ran the length of their bodies. Segmentation meant that eventually this beam became divided into individual units that evolved into the vertebrae of the backbone. The nerve center, or brain, became encased in a skeletal box, which eventually evolved into the skull. Gills were supported on skeletal units that grew in pairs on the segments close to the head; the foremost of these evolved into jaws. Paired skeletal structures on segments down the length of the body became ribs, fins, and eventually hips, shoulders, and legs. This very sketchy view of the evolution of the vertebrates can be seen in the development of the fish, particularly in Devonian times.

The earliest creatures classified as fish were the agnathans, or the jawless fish, which first appeared in the Ordovician period. The parasitic lamprey and hagfish are the closest modern equivalents. They evolved paired fins, and eventually the hinged jaw developed. The placoderms, a group of armored fish of the Devonian, were the first to show this feature. At about this time the sharks and rays evolved. They were so well adapted to their habitat that they have remained almost unchanged until the present day. All these creatures had skeletons that were made of cartilage—a gristle-like substance. The development of sturdier bone was the next big evolutionary step.

Bony fish have two main groups: lobe-finned fish and ray-finned fish. In the former, two pairs of fins developed into muscular structures, packed with bone and with the fin itself forming a kind of a fringe around the edge. The same structure can be seen in modern lungfish and coelacanths. Some even developed a lung, which enabled them to breathe air, and the muscular lobe-fins enabled them to crawl—the first steps towards living on land.

The vast majority of modern fish are ray-fins, which underwent a dramatic radiation during the Carboniferous. Their fins consist of a fan-like arrangement of struts without a muscular base. Their other major feature is a swim bladder, which probably evolved from a primitive lung sometime around the Triassic period. The swim bladder is used to adjust buoyancy during swimming. This organ makes the modern bony ray-finned fish a perfectly-adapted creature for the water environment.

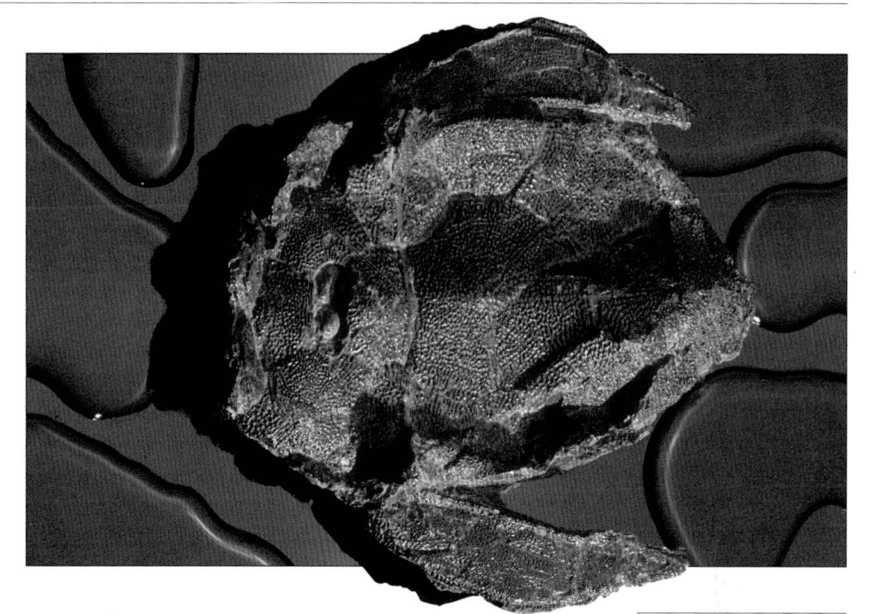

FOSSIL FISH

The placoderms were a group of fish that were restricted to Devonian and Early Carboniferous times. Their heads and the fronts of their bodies were covered in interlocking armor. These shields are seen in spectacular fossils from freshwater deposits found in the Old Red Sandstone.

FISH EVOLUTION

The Devonian is called the "Age of Fishes" because most major groups had evolved by then. The modern groups—ray-fins and lobe-finned bony fish—were all present. Lobe-fins gave rise to the first amphibians and thus to today's land animals.

1 First animals with a backbone appear
2 Hinged jaws and paired fins develop
3 Lungs develop, probably among freshwater fish in times of drought
4 Fleshy lobed fins develop, the forerunner of the tetrapod limb
5 Fully pouting jaws of teleosts (modern fish) develop; swim bladder evolves

FISH CLASSIFICATION

The various types of fish are classified on the development of jaws, the kind of skeleton (cartilage or bone), the types of scales and armor, and the structure of their fins. Three classes (Agnatha, Chondrichthyes, and Osteichthyes) have survived into the present time, of which the Osteichthyes (bony fish) are the predominant group.

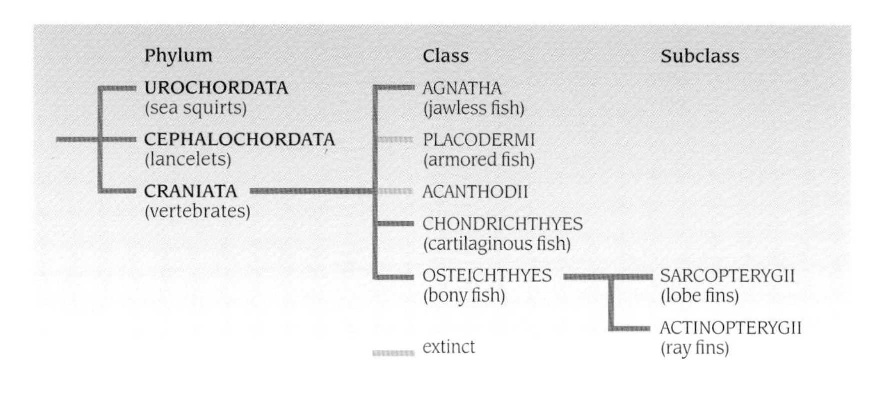

THE EARLY CARBONIFEROUS

354 – 324 MILLION YEARS AGO

GLOBAL changes that took place in the Devonian period were carried over into the Carboniferous. Continents continued to drift together, inching towards the assembly of the single supercontinent of Pangea. The sea level stood high, flooding low-lying areas of the northern continent of Laurentia and leaving its mark in vast limestone deposits. These were important in distinguishing the Early Carboniferous from the Late, which is strongly associated with the formation of coal. Trees with woody trunks appeared and grew up to 100ft (30m) tall. Output from the photosynthesis of newly abundant plants increased the proportion of oxygen in the atmosphere, encouraging the growth of more plants. The Earth became home to diversifying amphibians and insects, and possibly the first reptiles appeared. Extensive swamps and bogs from the Early Carboniferous were the forerunners of the great coal forests of that characterized the Late Carboniferous.

THE TERM Carboniferous ("carbon-bearing") was first used by two British geologists, William Conybeare and William Phillips, to describe strata in the north of England that contained beds of coal. Conybeare, like many other figures in the history of early geology, was a clergyman by profession; Phillips was a printer. They introduced the term Carboniferous in their *Outlines of the Geology of England and Wales*, published in 1822. This book soon became the standard reference volume on the stratigraphic succession in Britain, preceding Charles Lyell's classic *Principles of Geology* by eight years. Building on the findings of other geologists who preceded them, Conybeare and Phillips were the first to name and date a formal geological period.

Completely different rock suites were deposited at the beginning and end of the Carboniferous. When geologists recognized this, they divided the period.

KEYWORDS

AMPHIBIAN
ANTLER OROGENY
CARBONATE
COAL
CRINOID
ECHINODERM
GONDWANA
KARST
LIMESTONE
PANTHALASSA
REEF
SHELF SEA
TETRAPOD

As they defined it, the Carboniferous included the system that is now recognized as the Devonian, which was given a separate designation by two other geologists in 1839. They recognized that the lower layers of the system—above the sandstone of what later became the Devonian—were characterized by a high proportion of limestone, whereas it was the upper layers that contained most of the coal for which they named the Carboniferous. In the opinion of Conybeare and Phillips, these two types of sequence, "though entitled to the character of distinct formations, are yet so intimately connected, both geographically and geologically, that it is impossible to separate their consideration."

Their contemporaries soon disagreed. The Belgian geologist J.J. d'Omalius d'Halloy proposed a division of the system into Lower (limestone) and Upper (coal), and in Europe they have been recognized ever since. There are considerable regional differences, however, especially in the upper part of the Carboniferous strata. The lower limits are approximately the same in all locations.

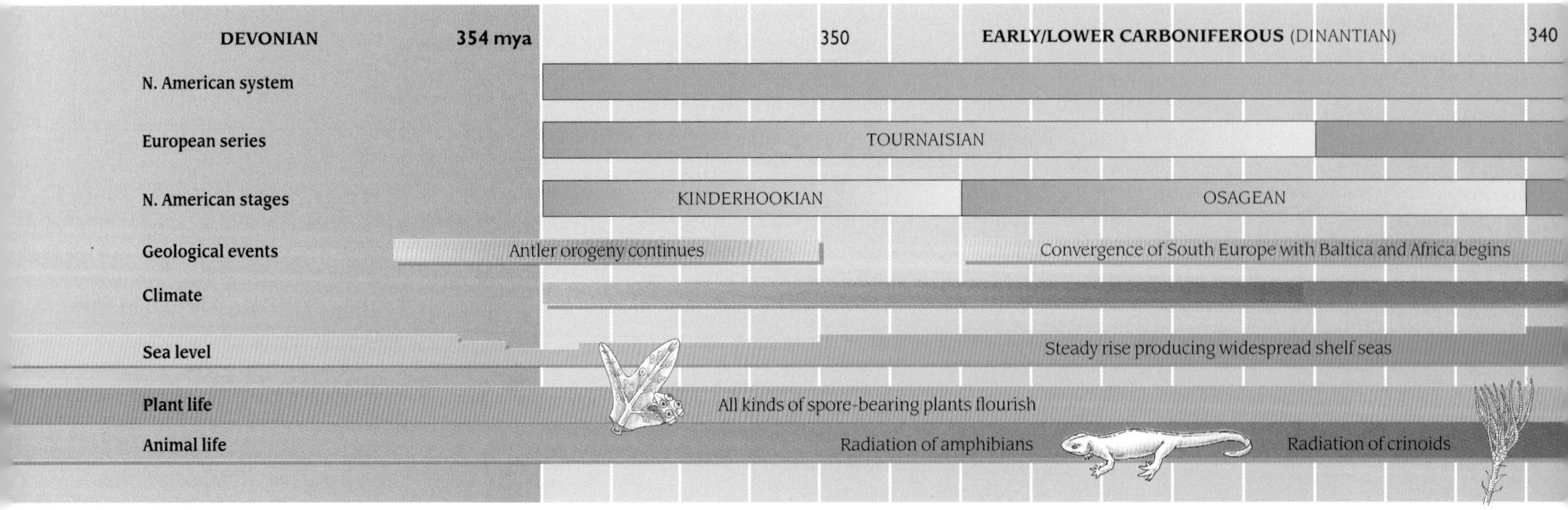

DEVONIAN	354 mya	350	EARLY/LOWER CARBONIFEROUS (DINANTIAN)	340
N. American system				
European series	TOURNAISIAN			
N. American stages	KINDERHOOKIAN		OSAGEAN	
Geological events	Antler orogeny continues		Convergence of South Europe with Baltica and Africa begins	
Climate				
Sea level			Steady rise producing widespread shelf seas	
Plant life	All kinds of spore-bearing plants flourish			
Animal life		Radiation of amphibians	Radiation of crinoids	

American geologists found that the division of Carboniferous rock strata in North America was even more extreme. In 1839—the same year that the Devonian was split off from the original system proposed by Conybeare and Phillips—the American D.D. Owen used the term "Upper Carboniferous" to describe coal measures at the top of the sequence and "Subcarboniferous" for the limestone below. His study was conducted along the upper valley of the Mississippi River, and in 1870 the term Mississippian was adopted in place of the Subcarboniferous, because the limestone formations were particularly well exposed in the river valley. The Upper Carboniferous became known as the Pennsylvanian for the state's characteristic coal-rich strata. Both names were widely used long before they were officially adopted by the United States Geological Survey in 1953.

THE EARLY Carboniferous Period can be regarded as the "age of limestone." There are a number of reasons for this. Rising sea levels meant that much of the low-lying northern continent of Laurasia was covered by shallow seas. Large areas of shelf sea such as these are too far away from land to contain much river-borne sediment; the sea sediments consist mostly of materials deposited from salts dissolved in the seawater, or from the skeletons of animals and plants that live there. Calcite—calcium carbonate, the mineral of limestone—is an important component of the dissolved material in seawater and also forms in the hard parts of many sea animals, and so this was the principal mineral to be deposited. The climate was still tropical over most areas at the time, and there was always a great deal of evaporation from the sea surface, concentrating dissolved materials and encouraging their precipitation.

The Carboniferous was named for its plant life, but the vast limestone deposits of the early period came from marine species that flourished in the warm shelf seas.

The southern continent of Gondwana was less inundated by the sea, and Carboniferous limestone is scarce.

Plant life became abundant during the Early Carboniferous as established Late Devonian flora continued to colonize the land, spreading to form dense plant cover. These flora rapidly diversified in a sort of evolutionary experiment, producing a few species that left traces of coal in the Lower Carboniferous strata. It was the most successful species that later made up the great coal swamps. The first tall trees with woody trunks grew up during this time; some of them reached 100ft (30m) in height. Seed plants also made their first appearance.

There was a very high oxygen content to the Early Carboniferous atmosphere: 35 percent as opposed to 20 or 21 percent nowadays. This was due to the presence of extensive forests that were now flourishing along the coastal areas. The bulk of these trees was made from lignin, a newly-evolved organic substance that gave plant cells their strength; it makes up about 20 to 25 percent of the wood of trees. In modern times, a dead tree normally rots away through various biochemical processes that use atmospheric oxygen to break down the organic materials. During the Early Carboniferous, lignin was too new a substance to have evolved a biochemical process for degradation, so the oxygen that would have been used to degrade the lignin was left in the atmosphere.

Forest fires were common, judging by the quantity of charcoal found in Lower Carboniferous freshwater deposits—particularly in Scotland. This is understandable given the high level of oxygen in the atmosphere, which would have supported combustion. This may also account for the build-up of non-decaying woody material in sediments that led to the formation of coal strata.

See Also

SILURIAN: *Hydrothermal vents*
DEVONIAN: *Sandstone; Life on Land*
LATE CARBONIFEROUS: *Coal; Insects*

THE AGE OF LIMESTONE

The Early Carboniferous (or Mississippian) seems to have been a fairly quiet time geologically—a lull between episodes of mountain-building. During this interval, plant and animal life became firmly established on land, assisted by predominantly tropical conditions. Low-lying continents in the northern hemisphere were inundated by warm shallow seas, leaving substantial deposits of carbonates that formed massive beds of limestone. Change was underway, however, as Laurentia and Gondwana edged closer to collision. As they did, early Paleozoic marine conditions gave way to the steamy swamps of the Late Carboniferous and finally to dry land in the Permian.

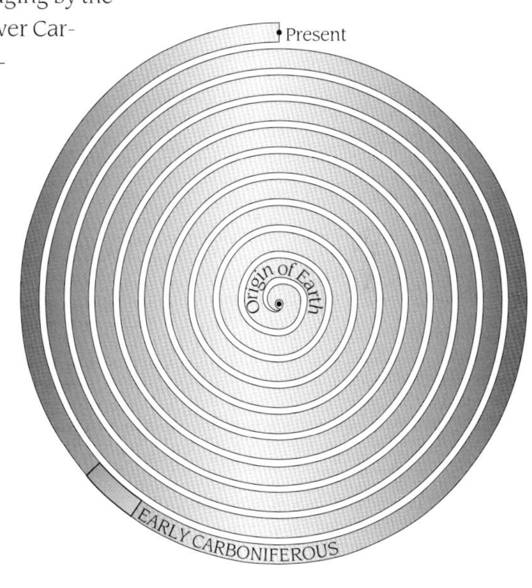

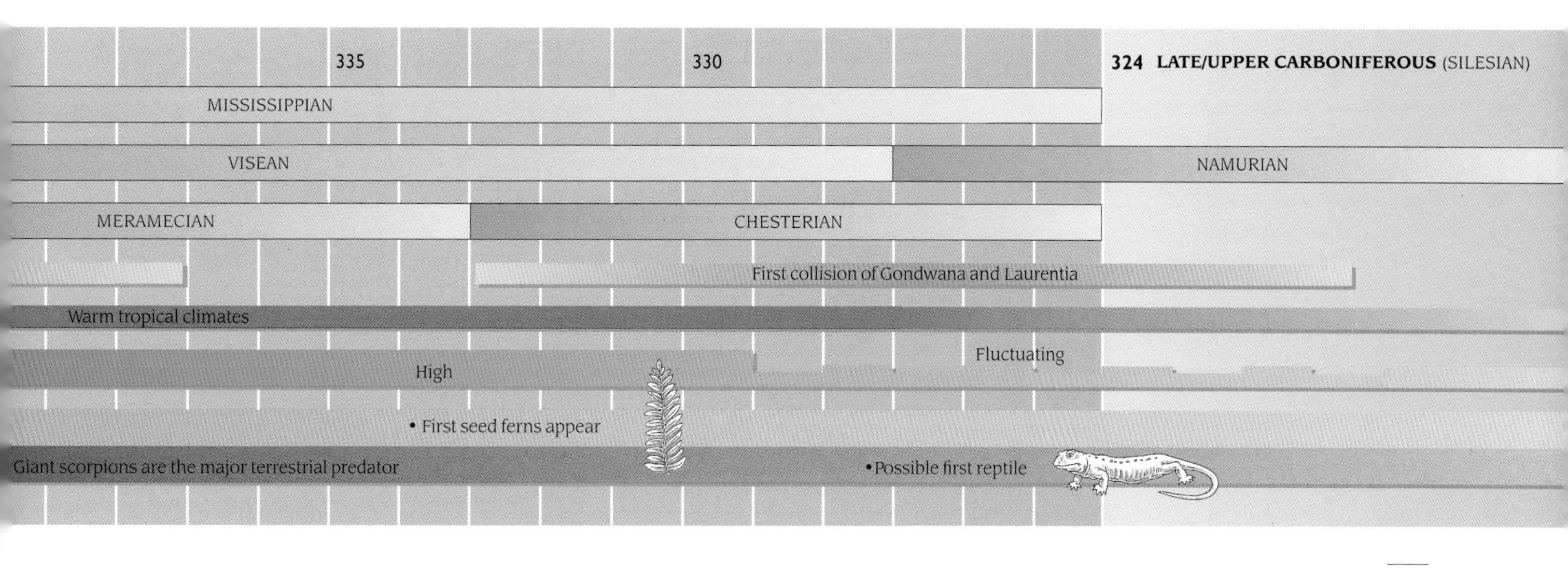

THE CONTINENTS were still moving towards one another in Early Carboniferous times. The Old Red Sandstone Continent, formed in the Devonian and consisting of North America and northern Europe—was approaching the southern supercontinent of Gondwana. Deep ocean trenches developed off its southern coast. The gap between Laurentia and Kazakhstania was also closing, and the narrowing ocean between them, known to geologists as the Pleionic, must have been very much like the Mediterranean Sea is today. Along this join the Ural Mountains of Eurasia eventually formed. Long before their collision became apparent on land, the closing continents would have twisted and contorted the ocean bed, crushing it into a chaotic pattern of deeps and rises, dotted with volcanic islands and sheared and twisted peninsulas, and the whole area split by earthquakes.

The ocean of Panthalassa, which covered all the planet not covered by continent, would have had all the oceanic features of active ocean ridges, abyssal plains, islands, but none of these have been preserved.

Along the western edge of Laurentia, where it faced the vast ocean of Panthalassa, there was an active fold mountain belt similar to the West Coast ranges of modern North America. For the duration of the Early Paleozoic this had been a passive margin—the edge of a continent where there was no movement of tectonic plates. Then, in the late Devonian, a subduction zone developed here, swallowing the oceanic plate below the edge of the continent.

This activity produced an oceanic trench and crumpled up sediment and rock at the rim of the continent. Parts of the descending ocean plate melted, causing molten material to rise through the overlying continental plate and burst through, forming an arc of volcanic islands. The pressure of the uplift here was transferred across the western part of the continent and sheared up the basal rocks further inland into another parallel mountain range: the Antler Mountains, whose remains are found in modern Nevada and Idaho. Their creation is known as the Antler Orogeny.

Another active tectonic area lay along the western edge of Gondwana (which became South America), also facing the Panthalassa ocean. The result here was a mountain range similar to the modern Andes, and lying in about the same position. Many of the sedimentary rocks of the modern Andes were formed from material washed off the continent and gathered in the deep ocean just offshore at this time. The mountain range in the late Paleozoic and most of the Mesozoic formed part of the Samfrau range, from an acronym for South America, Africa, Antarctica, and Australia. This mountain range was just forming in late Paleozoic times.

Fold mountain belts are caused by the impact between moving plates. The destruction of oceanic plates at plate margins must be balanced by the generation of new plate material in oceanic ridges. Such ridges are known throughout the modern oceans, but evidence for their existence in past geological times is slim. This is because the ocean floor sediments and structures are eventually destroyed in subduction zones, usually within a few tens of millions of years of their formation, leaving no trace of their former existence. However, in Lower Carboniferous rocks there is some indication that this action was taking place. In modern oceans the ridges are volcanically active and produce mineral-rich hydrothermal vents known as "black smokers". Bacteria feed on the chemicals, small invertebrates feed on the bacteria, crustaceans feed on these, and giant vestimentiferan tube worms wave their feeding tentacles at the top of the food pyramid.In two locations, in Ireland and Newfoundland, there are Lower Carboniferous fossils of tubeworms in association with the kinds of sulfur-based minerals found in black smokers. These tubeworms are not as large or spectacular as the modern type, but they represent the first fossil evidence for black smoker faunas.

Despite all this mountain-building activity, the tectonics of the world were fairly quiet through the Early Carboniferous. The sea level rose worldwide, broadening the continental shelves of Gondwana and flooding across the low-lying areas of Laurentia. North America, between the Antler Mountains and the Acadian–Caledonian Mountains, and south of the old craton of the Canadian Shield, was covered by a shallow limy sea called the Kaskasia Sea. To the east of the Acadian-Caledonian Mountains, shallow water also covered most of Europe.

A contemporary river—the Michigan River— flowed into the inland sea of North America from the Canadian Shield; it was one of the few river systems large enough to cross the distance and join the sea. The Michigan River delta built up from molasse deposits shed by the Appalachian Mountains in the northeast, and continued to grow into the Late Carboniferous. When the sea began to recede, the river carried more sediment southward to the exposed continental margin, which gradually

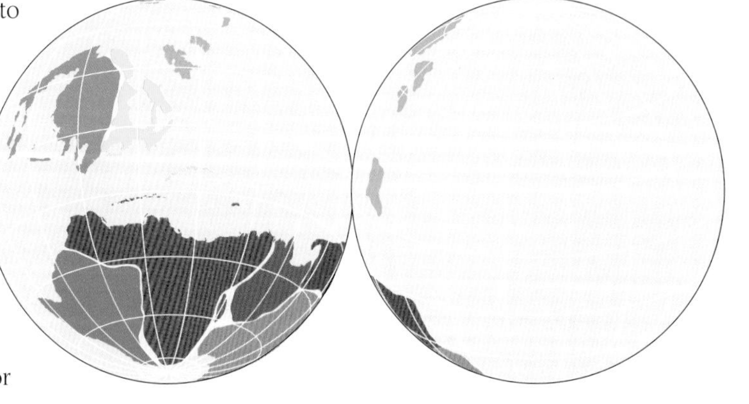

Africa and Middle East
Antarctica
Australia and New Guinea
Central Asia
Europe
India
North America
South America
Southeast Asia
other land

FLOODED NORTHERN HEMISPHERE

The most significant feature of Early Carboniferous geography was the extensive shelf seas inundating the northern continent of Laurentia. Limestone beds up to 2300ft (700m) thick were laid down as a result.

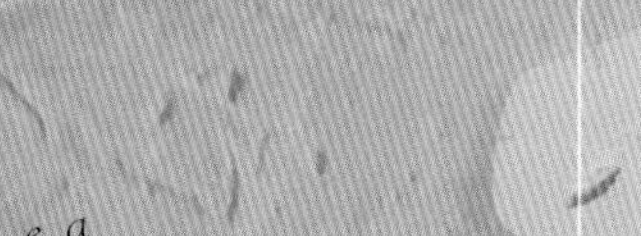
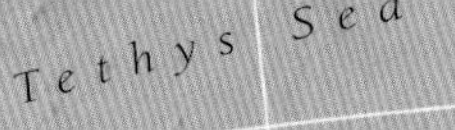

TECTONIC ACTIVITY

The Samfrau orogenic belt, which lay along the southern margin of Gondwana, was the site of active subduction of the oceanic lithosphere. The first uplift of the Andes was one result.

accreted a new southern portion of landmass corresponding to the states around the modern Gulf of Mexico (except Florida, which had been part of Gondwana). The Michigan River system eventually expanded into the southern part of the continent to become the Mississippi, which also ran all the way to the sea.

MOST limestone is made from the remains of once-living organisms. The process by which the layers build up can be observed today in the coral reefs of the tropical oceans. The fringing reefs, barrier reefs, and atolls of the Indian and Pacific Oceans are areas in which limestone is currently being deposited. Corals build up the reefs themselves, each generation of corals constructing their shells upon the dead remains of the previous generation, while loose shelly material is deposited in the lagoons cut off by the reefs. These occurrences of reef limestone, however, are comparatively restricted in their extent.

Coral reefs in modern tropical seas are still producing limestone, but they do not occupy a large enough area to match the huge sheet limestone formations of the Carboniferous.

Shelf limestones form from dispersed fragments rather than growing where reef-building organisms live. These are being deposited off the coast of Florida and around the Bahamas in modern times, but they are nothing compared with the continent-wide deposition of shelf limestones that took place during the Early Carboniferous. At its greatest extent, the Kaskasia shelf sea of North America washed over nearly all of the Canadian Shield. Perhaps the most spectacular evidence of this vast sea is the outcrop of the Redwall Limestone in the Grand Canyon of the Colorado River. At 574ft (175m) thick, it comprises one of the major vertical cliffs in the Grand Canyon. Its red color is derived from iron minerals percolating down from overlying Permian deposits; the natural color of the rock is white to gray, typical of limestone. The Redwall outcrop can be traced as an escarpment across the Nevada Desert, where it is known as the Monte Cristo Limestone.

The shelly substance that provides the raw material for the formation of limestone varies from place to place, and, more importantly, from time to time. Gastropod shells, bivalve shells, brachiopod shells, and corals have all contributed to the formation of limestone at one time or another, but most Early Carboniferous limestone was made up of the remains of a group of animals called crinoids that lived in the shelf seas. They were echinoderms, related to the starfish and the sea urchins. The best way to describe a crinoid is as a starfish on a stalk. It was a filter-feeder, with a bunch of feeding arms spread out around a cup-shaped body. This was connected to the sea bed by a cylindrical column, with a holdfast at the bottom that kept it in place. Its structure consisted of calcite plates—disc-shaped in the column, hexagonal in the cup, and small and irregular in the arms. All these elements became disassociated in death and piled up on the seafloor. The polished surface of a Lower Carboniferous limestone may show masses of these disc-shaped columnals, all packed in together.

A MODERN KARST

Clints and grikes are the obvious features of a modern karst surface such as here at Malham Cove in northern England. The landscape is very dry, as the rain seeps away into the rock, and so there is not much vegetation and little development of soil. In a close-up view, the circular plates of the crinoid stems—part of the animals that formed the limestone—can be seen in the stone itself.

CRINOID FOSSIL

A form of echinoderm, crinoids were a major component of Lower Carboniferous limestone. There were more than 400 kinds in the Mississippi Valley alone.

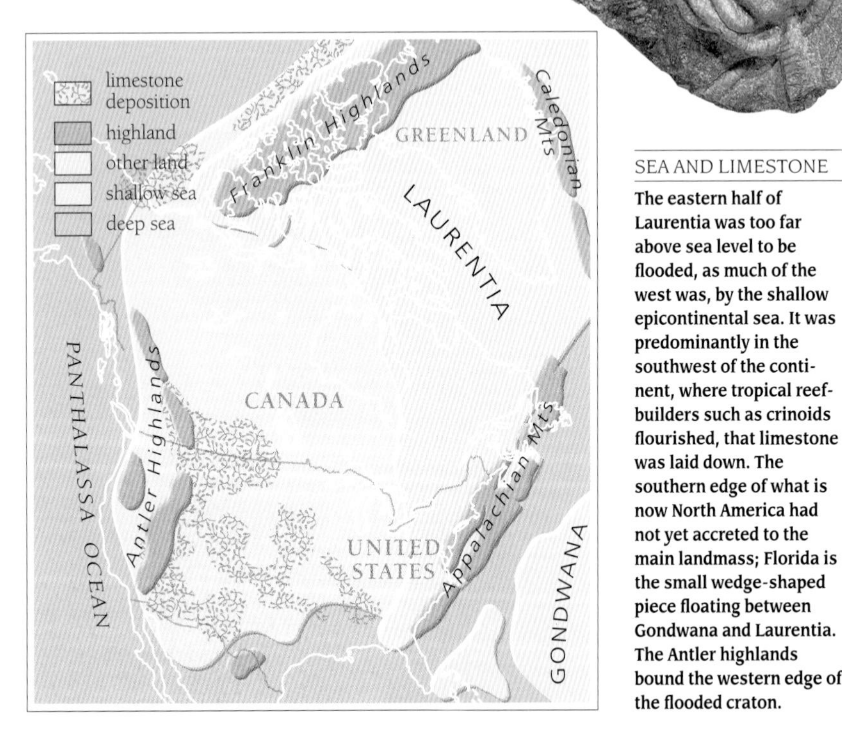

SEA AND LIMESTONE

The eastern half of Laurentia was too far above sea level to be flooded, as much of the west was, by the shallow epicontinental sea. It was predominantly in the southwest of the continent, where tropical reef-builders such as crinoids flourished, that limestone was laid down. The southern edge of what is now North America had not yet accreted to the main landmass; Florida is the small wedge-shaped piece floating between Gondwana and Laurentia. The Antler highlands bound the western edge of the flooded craton.

IN MOST sedimentary rocks the particles in a bed lie there, separate from one another, for perhaps millions of years. The weight of overlying beds can compact these and reduce the spaces between. However, it is only when groundwater deposits minerals between the grains, cementing them together into a solid mass, that a sediment becomes a true sedimentary rock. Limestone normally does not have to wait for this process to happen; once the grains of calcite are deposited, they may become cemented to one another almost immediately.

Limestone is unique among sedimentary rocks in that it can solidify almost as soon as it is deposited, forming this "instant rock."

A bed of limestone laid down and solidified at the bottom of a broad shelf sea can emerge at the surface as a solid bed of rock if the sea level falls slightly. Because limestone is particularly vulnerable to chemical weathering, the exposed bed immediately begins to break down. Rainwater dissolves carbon dioxide from the atmosphere and so becomes a weak acid. The acid reacts with the calcite of the limestone and dissolves it. The action takes place most readily along zones of weakness, for example along bedding planes and joints where the solid substance of the rock has cracked. Long straight hollows are weathered out of solid limestone. The geologist's term for such a fissure is a grike; the intervening rock is left as a series of pinnacles and blocks, called clints. This kind of surface is known as karst after the area in Slovenia and Croatia where it was first identified. Below the surface the acid groundwater erodes systems of caves, again along the lines of weakness, and also along the water table. In these caves the dissolved calcite can be re-deposited as the classic stalactites and stalagmites so typical of limestone caverns.

Another subtle change in sea level can then submerge the eroded surface and flood all this. More limestone is deposited, first in the hollows and gullies and the caves, and then as beds on top. Many of the extensive limestones laid down in the Early Carboniferous show successive series of depositions over karst surfaces like this. During the karst erosion, non-calcite minerals are washed from the decaying limestone and gather in the bottoms of grikes, or are washed into cave systems and deposited there. Old karst structures in the Redwall Limestone have yielded both copper and uranium ores, which have been mined since the nineteenth century.

LIMESTONE FORMATION

Reefs begin to develop as corals or other reef-building organisms anchor themselves to hard rock outcrops on the seafloor (1). Once mature, the reefs cut off quiet shallow lagoons in which limy material settles and forms beds of limestone (2). If the sea level drops, the newly-formed limestone is exposed and becomes eroded forming a karst landscape with caves and gullies (3). After the next rise in sea level, more limestone is deposited, burying the eroded surface and preserving the erosional features (4).

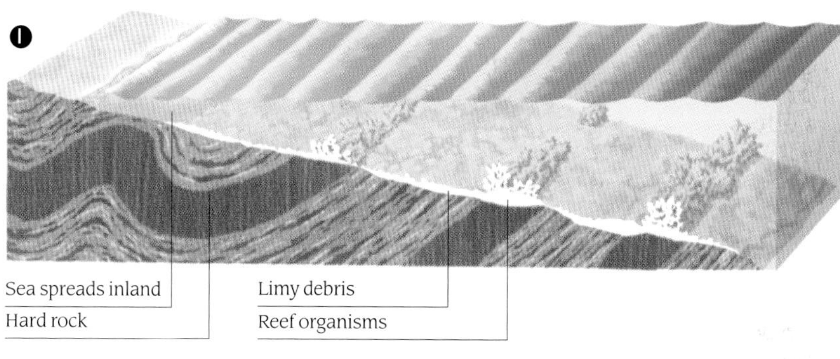

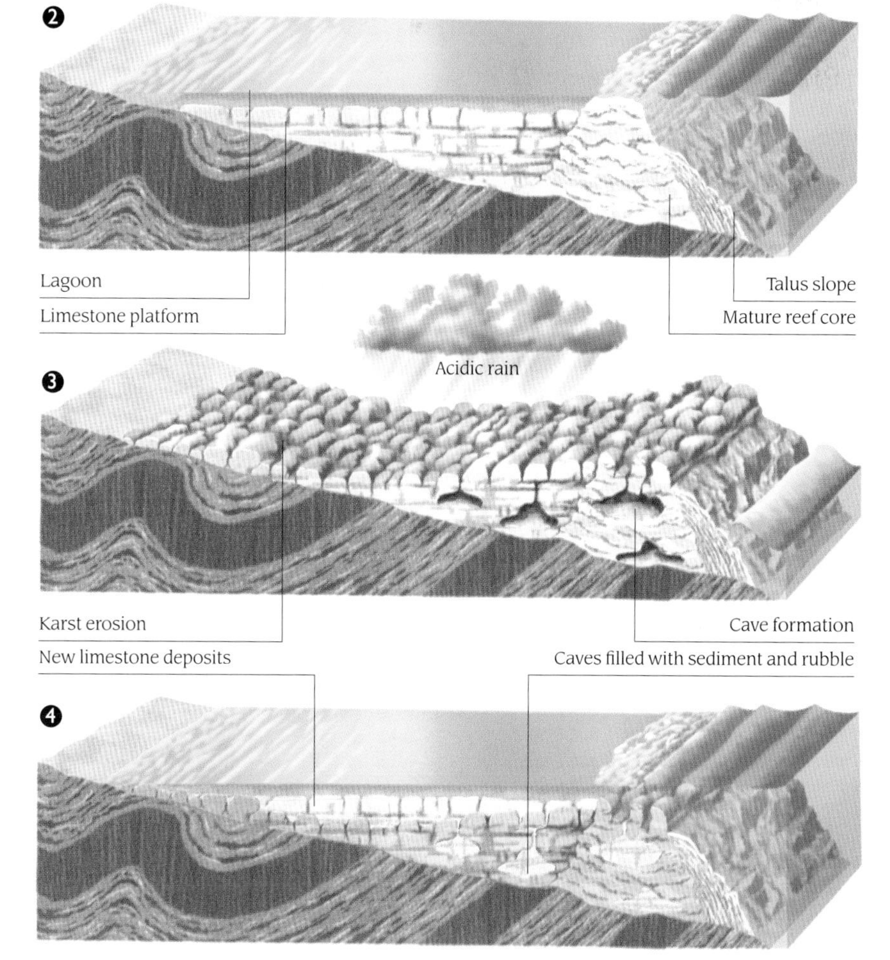

EAST KIRKTON FAUNA

338 million years ago, the East Kirkton quarry was a tiny lake, possibly a volcanic hot spring. The fossils are the remains of animals that were washed into the mineral-rich water and quickly preserved in the limestone.

1 *Hibbertopterus* (eurypterid)
2 *Sphenopteris* (seed fern)
3 Millipede
4 *Balanerpeton* (temnospondyl)
5 Lycopod (clubmoss)
6 *Eldeceeon* (anthracosaur)
7 *Ophiderpeton* (aïstopod)
8 *Pulmonoscorpius* (giant scorpion)
9 Opilionid (harvestman)

LAND plants were spreading across the continents. On the flanks of the eroding mountain chains, fed by the river systems, grew forests, consisting mostly of ferns and clubmosses. Although they had changed little from the Devonian flora, Early Carboniferous plants were much more varied than those of the Late Carboniferous, but they contained few coal-producing species. Lycopods, which became the giants of the Late Carboniferous coal forests, were merely small spore-bearing swamp plants.

The modest amount of coal found in the Lower Carboniferous is due to the types of plant that grew.

Land animals were developing too. Scorpions first emerged as land animals in the Early Carboniferous, growing up to 1.5ft (0.5m) long. The high proportion of oxygen in the atmosphere meant that land-living arthropods could grow to much greater sizes than they can today. The East Kirkton site in Scotland—an out-of-use lime quarry with Lower Carboniferous deposits dating to 338 million years ago—has produced scorpion fossils that illustrate this, with breathing mechanisms and their mode of feeding that show these were land animals, and eyes that indicate that they hunted by daylight. These must have been the greatest land predators of the Early Carboniferous, hunting through the undergrowth for other land-living arthropods. Towards the end of the period, however, scorpions lost their niche to another predator and became the small (but still deadly) nocturnal creatures that are familiar today.

Eurypterids (which had first appeared in the Silurian) continued to proliferate in the Early Carboniferous, when they became land animals, colonizing the undergrowth of what is now Scotland. Some eurypterids were giants; the partial remains of one fossilized head-shield show a head some 2ft (60cm) across, with eyes the size of plums. However, the mouth-parts of these creatures were adapted to sieving small animals from the water. These were not the predators that replaced the scorpions.

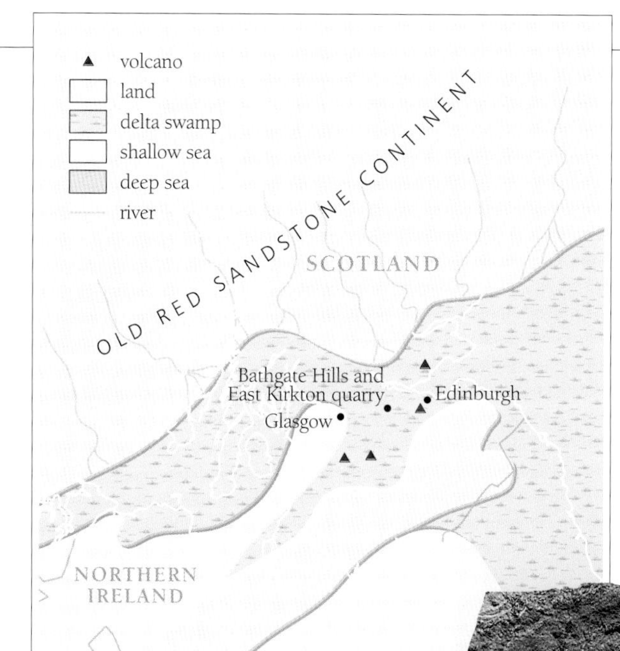

SCOTLAND'S EARLY CARBONIFEROUS

Debris from the slopes of the Caledonian mountains spread into the shallow seas, building up delta swamps. Local volcanoes may have poisoned the swamp and killed the inhabitants, which were preserved in limestone. Fossils were found in a disused quarry at East Kirkton in the 1980s.

THE BATHGATE BEAST

The early temnospondyl *Balanerpeton* is a fossil preserved as "part and counterpart." An impression of the fossil can be seen on each side of the split rock slab.

Amphibians had already evolved late in the Devonian, when they still retained many of their ancestral fishlike features. Now, however, they were branching out into all kinds of habitats and lifestyles. The basic amphibian shape is somewhat salamander-like, or newt-like, with a long moist body, four legs with typically five fingers and toes, strong teeth for eating live prey, and a long tail adapted for swimming. *Balanerpeton*, a 1.5ft (0.5m)long amphibian found at East Kirkton, possessed these features. Its contemporaries showed all kinds of developments from this basic shape. Some, the aïstopods, abandoned their newly-evolved limbs and lived like snakes or eels in moist undergrowth and shallow water. Others developed very much stronger limbs and spent most of their life on land. Predominantly land-living amphibians can be distinguished from aquatic ones by their lack of fishlike characteristics such as traces of gills and the lateral-line sensory system on the skull.

Vertebrates were the major success story of the Early Carboniferous. Terrestrial invertebrates—the scorpions and the eurypterids—had already developed large sizes because of the oxygen content of the atmosphere; oxygen was able to reach the tissues through the pores easily, an advantage to a creature without lungs. However, an animal with an external shell can contain only so much internal muscle; above a certain size, the armor is too heavy to be moved. Amphibians, with their internal skeleton, had no such constraints, and a land-living amphibian could grow much bigger than a land-living arthropod. Big scorpions, the main predators at the beginning of the Carboniferous, became the prey of even bigger amphibians as the period progressed, and vertebrates became the rulers.

It was about this time that some of the more specialized amphibians evolved into the reptiles, dispensing with the need to spend a larval stage in water by laying their eggs on land. In effect, they produced their own little ponds within waterproof membranes in which their larval stages could develop. Evidence of this ability is very difficult to determine from fossil specimens—which are rare in any case—and so the precise point of transition from amphibian to reptile has not been identified.

One of the most significant finds of a transitional form is *Eldeceeon rolfei*, also discovered at the quarry in East Kirkton. This land-living reptiliomorph ("reptile form"), which grew to about 14in (35cm) in length, is the earliest example known of its kind. *Eldeceeon* is believed to be a primitive anthracosaur, possibly a sister group of better-known suborders such as the Seymouriamorpha.

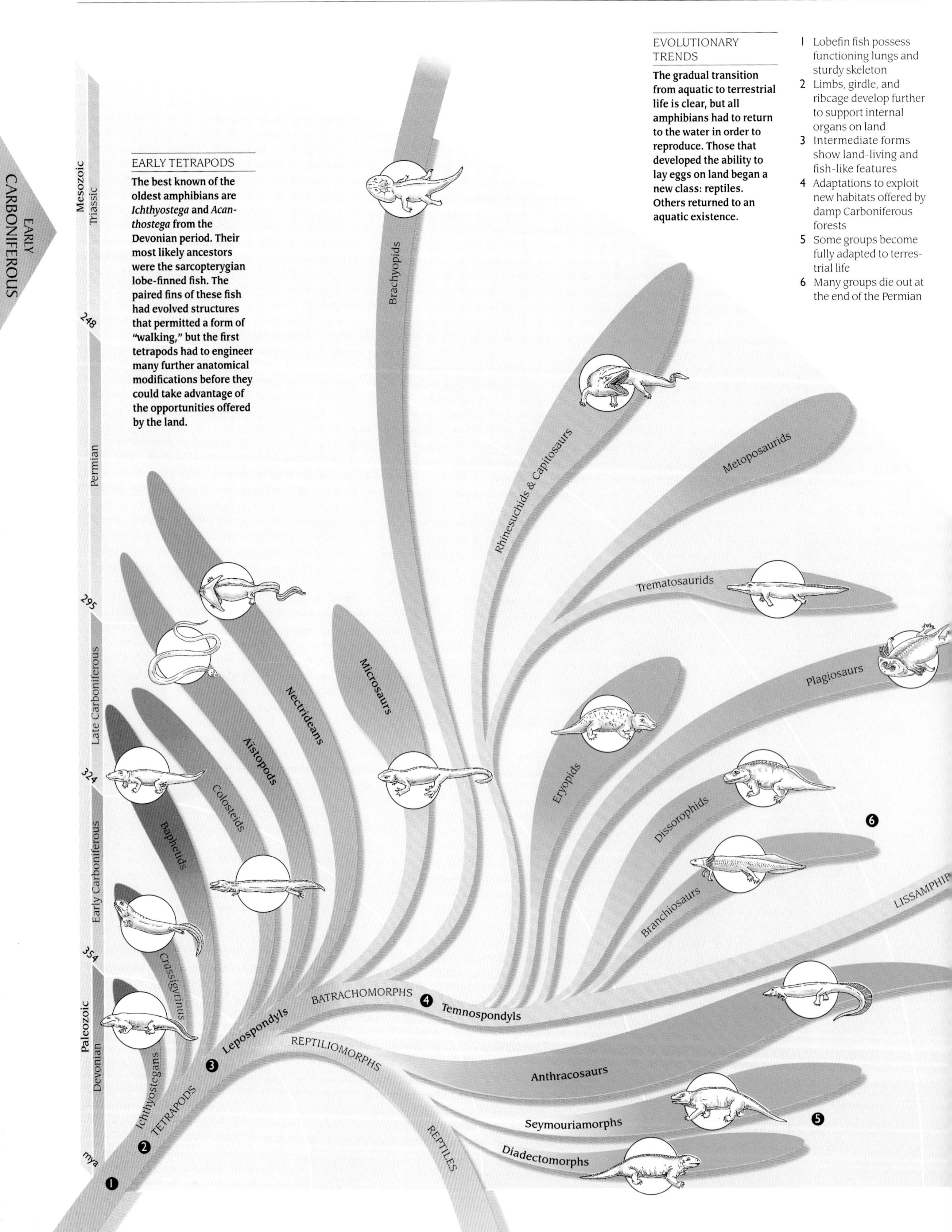

EARLY TETRAPODS

The best known of the oldest amphibians are *Ichthyostega* and *Acanthostega* from the Devonian period. Their most likely ancestors were the sarcopterygian lobe-finned fish. The paired fins of these fish had evolved structures that permitted a form of "walking," but the first tetrapods had to engineer many further anatomical modifications before they could take advantage of the opportunities offered by the land.

EVOLUTIONARY TRENDS

The gradual transition from aquatic to terrestrial life is clear, but all amphibians had to return to the water in order to reproduce. Those that developed the ability to lay eggs on land began a new class: reptiles. Others returned to an aquatic existence.

1 Lobefin fish possess functioning lungs and sturdy skeleton
2 Limbs, girdle, and ribcage develop further to support internal organs on land
3 Intermediate forms show land-living and fish-like features
4 Adaptations to exploit new habitats offered by damp Carboniferous forests
5 Some groups become fully adapted to terrestrial life
6 Many groups die out at the end of the Permian

THE EVOLUTION OF AMPHIBIANS

AMPHIBIAN VARIETY

The smooth newt (right) has four fingers. Early tetrapods experimented with a varying number of digits—up to eight—before evolving the more familiar five-toed arrangement.

AMPHIBIANS are four-footed vertebrates that have an aquatic larval stage but spend much of their life on land. They are the most primitive living members of the tetrapods, a group that consists of all four-footed vertebrates except fish. This designation can be confusing, since it includes birds, whales, and snakes. Nevertheless, all are descended from a four-footed ancestor and share a common pattern of limb bones. "Tetrapod" is also a useful term for describing the diverse early fossil forms before the characteristics of amphibians and reptiles emerged. Rocks from the Late Devonian period have a whole range of animals that seem to be part fish and part amphibian. Tetrapod features include four legs and a strong ribcage to support lungs for breathing out of the water. The fish features include the arrangement of bones in the skull and the presence of a fin on the tail. Six or seven fossil animals share these features to some degree. Another unusual characteristic is the inconsistent number of fingers and toes. Six, eight, or even more toes were tried out before five became the norm for tetrapods; modern amphibians have four fingers and five toes.

With the dawn of the Carboniferous period, amphibians became established in a number of different lines. They colonized all of the continents and were the dominant predators of their time, which lasted about 80 million years. Unlike modern amphibians, many were enormous, up to 13ft (4m) in length. As other classes of animals appeared, the amphibians faced competition and predation, and the survivors were smaller species that could exploit less conspicuous niches.

Gaps in the fossil record make the history of amphibian evolution contentious. The classic approach is to divide them labyrinthodonts (named for the convoluted tooth enamel passed on by their sarcopterygian ancestors, and comprising the orders Ichthyostegalia, Temnospondyli, and Anthracosauria), lepospondyls (small, mostly aquatic early forms), and modern amphibians. This system has now been rejected in favor of a split between batrachomorphs and reptiliomorphs; the latter eventually led to reptiles.

Modern amphibians form three groups: frogs and toads, newts and salamanders, and caecilians. The first frogs appeared in the Triassic while the oldest newt and caecilian fossils are Jurassic. There is no firm evidence for how the modern groups evolved, but most experts believe they are descended from the temnospondyls.

Tertiary
65
Cretaceous
144
Jurassic
205
Triassic
Gymnophiones (caecilians)
Anurans (frogs, toads)
Urodeles (newts, salamanders)

AMPHIBIAN GROUPS

Traditionally, the hugely diverse ancient amphibians were divided into three groups: labyrinthodonts (mostly large), lepospondyls (mostly small), and lissamphibia (modern amphibians). This classification has now been revised, and the various orders are grouped in two broad lineages: batrachomorphs, or "true " amphibians; and reptiliomorphs, which include the ancestors of the reptiles. Some of these were terrestrial; others, such as the anthracosaurs, readapted to an aquatic lifestyle. The temnospondyls, a large order containing several terrestrial forms, dominated the Carboniferous; the later groups were aquatic.

MODERN AMPHIBIANS

Living amphibians are assumed to be closely related, on the basis of the structure of their tiny teeth. They probably arose within the temnospondyl group during the Triassic period. Today there are only about 4000 species, the smallest class of living vertebrates.

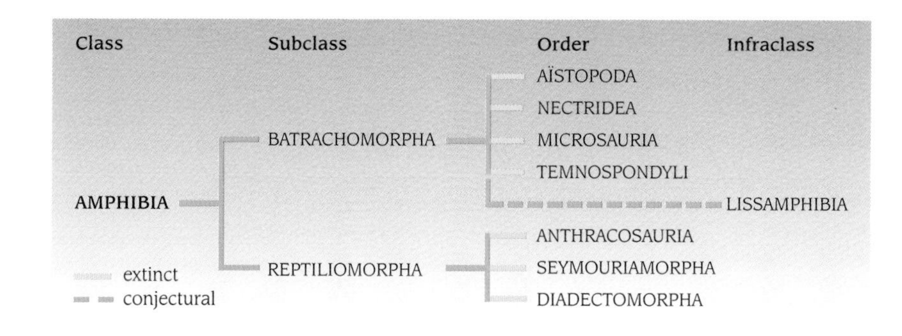

THE LATE CARBONIFEROUS

324 – 295 MILLION YEARS AGO

As the Early Carboniferous gave way to the Late Carboniferous, the Iapetus Ocean continued to close, bringing Laurentia and Gondwana ever closer together to form the supercontinent of Pangea. The areas where the continents merged enjoyed tropical to subtropical climates, while more southerly locations were cool, and an ice cap began to spread across the bottom of Gondwana.

Rapid changes were taking place on the land. The swamps that had remained as sea levels dropped grew into luxuriant forests. Distinctive northern flora stretched across a huge portion of Laurentia (in what is now northern Europe and eastern North America), while different southern flora spread across Gondwana (especially South America, Africa, and Australia). Both types of vegetation produced coal beds that gave the period its name. These forests provided a welcoming environment for new fauna, particularly for insects and reptiles, which rapidly began to expand and diversify.

Of all the periods of the Earth's history, the Late Carboniferous has probably had the most profound effect on human technological development.The characteristics of carbon-bearing rocks, and consequently the period in which those rocks were formed, were first recognized in England and Wales. The Carboniferous system was the first of the geological systems to be identified as a specific suite of rocks, and so it was the first to be scientifically named. (The distinctive red sandstone of the Devonian had been informally recognized as early as the 1790s, but it was not named until later.) This is hardly surprising, since the Carboniferous system was the source of the deposits of coal and much of the iron that fueled the Industrial Revolution. Records show that coal was worked in England in the time of the Romans. However, the discovery of flint tools in coal seams in the English county of Derbyshire suggests that earlier peoples also extracted it. Marco Polo, the Venetian traveler who began his journeys to the Far East in 1271, mentions in his journals that the Chinese used a black stone as fuel—something he had never seen at home.

Geology had long been driven by theology. With the coming of the Industrial Revolution, it became driven by technology.

Eighteenth- and nineteenth-century Europe and North America underwent vast economic changes as communities abandoned their traditional agricultural way of life and turned to manufacturing and industry. Technological advances made at this time, particularly in physics and metallurgy—such as the ability to manufacture fine-quality steel, which led to the development of railways, and the production of steam-powered machines and mills—were all based largely on the ability

KEYWORDS

ACADIAN OROGENY

ALLEGHENIAN OROGENY

ANGARALAND

APPALACHIAN OROGENY

CLUBMOSS

CYCLOTHEM

EXOPTERYGOTE

HERCYNIAN OROGENY

IAPETUS OCEAN

PANGEA

EARLY CARBONIFEROUS	**324 mya**	320 **LATE CARBONIFEROUS** (DINANTIAN) 315
N. American system	PENNSYLVANIAN	
European stages	NAMURIAN	
N. American stages	MORROWAN	
Geological events		
Climate	Warm,tropical climates	
Sea level	Moderate, rapidly fluctuating	
Plant life	Spore-bearing plants	Coal-swamp forests widespread
Animal life	• First fixed-wing insects	• First pelycosaurs

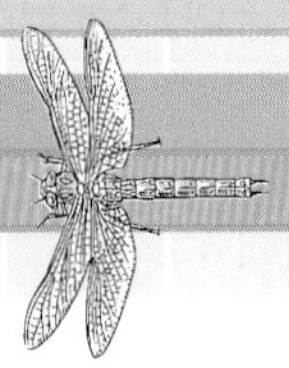

to exploit the available fuel and raw materials. Coal and iron were the most important, and they were both found in the same rock system of England and Wales.

Naturally, because of the economic importance and abundance of coal and iron, the geologists of the time paid great attention to their occurrence, and to the conditions under which they formed. The openings of massive quarries and mines led to the discovery of a great wealth of fossils, especially of plant life, prompting people to imagine what conditions on Earth had been like at the time the fossils were formed. It is, however, only the Upper Carboniferous system that is significantly coal-bearing, because before Late Carboniferous time the plant species that produced the most significant coal deposits had not yet evolved.

In North America, the Carboniferous is regarded as two distinct systems: the Mississippian, with its predominance of shallow water limestones, is equivalent to the Lower Carboniferous in Europe; and the Pennsylvanian, with its massive deltaic deposits and coal measures, is equivalent to the Upper Carboniferous. These American system names, taken from the two regions with their characteristic strata, were a matter of informal usage until 1953, when they were formally adopted by the United States Geological Survey. The boundary between the two systems is marked by one of the largest unconformities in the world, left by the inland sea, the Kaskasia, as it retreated from the North American continent late in the Early Carboniferous. Marine and nonmarine sediment alternates repeatedly in the strata of the Pennsylvanian system, with coal deposits being greater towards the highlands in the east (in modern Pennsylvania and West Virginia), mixed sediments in the center, and limestone predominating to the west.

The division between the Mississippian and Pennsylvanian does not correspond exactly to the division between the Dinantian and Silesian subdivisions that make up the Lower and Upper Carboniferous of Europe, but it is close enough for a general discussion of global conditions throughout the period.

DURING the Late Carboniferous, changing conditions on Earth made possible a new variety of plant life. The shallow tropical seas that covered much of the world throughout the Early Carboniferous slowly receded, leaving behind coastal swamps and deltas. Mineral-rich soil and mud, left behind by the waters, and the warm climate in the northern hemisphere produced a favorable environment for the growth of new kinds of plants and animals. Dense forests like the Louisiana swamplands of modern North America stretched across entire continents.

Coal and other fossil fuels took millions of years to form. It has taken a mere fraction of that time to use them up.

For tens of millions of years these swamp forests absorbed the energy of the sun and fixed the carbon dioxide of the atmosphere. When the plants died, they toppled into the stagnant water they grew in. The slightly acidic water preserved them, not perfectly intact, but enough so that they did not decay. For hundreds of millions of years, this plant matter remained untouched. Underground, with pressure from the rocks above, the energy became increasingly concentrated as the plant material turned to coal.

Only since the 1700s has the energy of the prehistoric sun and carbon dioxide been extracted and exploited intensively in the service of modern industry. It has taken less than 300 years to use up most of these millions of

See Also

THE ORIGIN AND NATURE OF THE EARTH: *The Evolving Atmosphere*
THE DEVONIAN: *Acadian–Caledonian Orogeny; Old Red Sandstone; Life On Land*
THE PERMIAN: *New Red Sandstone*

HALFWAY THROUGH THE LATE PALEOZOIC

By the Late Carboniferous (Pennsylvanian) period, the supercontinent of Pangea was nearly complete. New plants and animals spread quickly over the land, enjoying warm climates—though the southern part of Gondwana was drifting over the South Pole, and an ice sheet spread from what is now South America all the way to Australia.

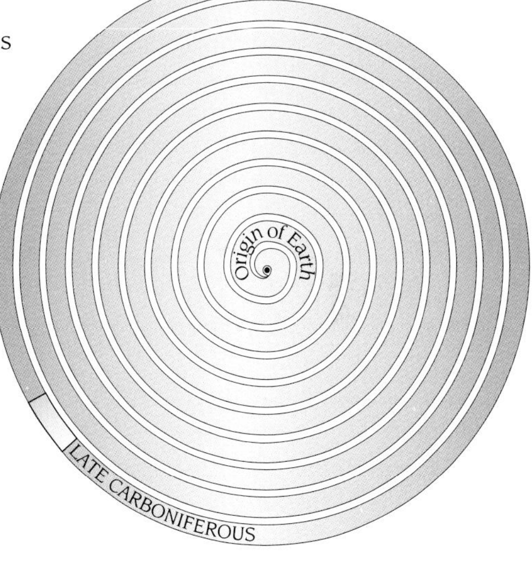

310 | 305 | 300 | 295 | PERMIAN

WESTPHALIAN | STEPHANIAN

ATOKAN/DERRYAN | DESMOINIAN | MISSOURIAN | VIRGILIAN

Alleghenian mountain building

Principal coal-bearing strata, Europe

Principal coal-bearing strata, N. America

Glaciation, southern hemisphere

Gradually rising

• First conifers

• First diapsid reptiles

• First hinged-wing insects

years' worth of stored energy and return huge quantities of carbon dioxide to the atmosphere. Although it is entirely possible, it is difficult to imagine the Earth returning to conditions that would allow such forests to return and the energy to be replaced.

THE GRADUAL but massive movement of tectonic plates beneath the Earth's surface had brought most of the continents together by the Late Carboniferous period. In the northern hemisphere, the Iapetus Ocean, which had separated Laurentia and Baltica since Precambrian times, had slowly closed up during the Devonian. As it did so, Laurentia and Baltica collided, forming the Old Red Sandstone Continent Hundreds of millions of years would elapse before the Iapetus reopened and grew into the Atlantic Ocean, once again separating Europe from North America. An alternative name for the Iapetus Ocean, reflecting its destiny, is the proto-Atlantic.

The collision and suturing of the great landmasses of Gondwana and Laurentia-Baltica began the formation of a new supercontinent.

Meanwhile, in the southern hemisphere, the vast supercontinent of Gondwana was approaching Laurentia from the south with a clockwise rotation, so that its eastern part (India, Australia, and Antarctica) moved south and its western part (South america and Africa) moved north. (Although it later broke up to form these modern continents Gondwana had always been a coherent single landmass, as far as geologists have been able to tell, rather than individual continents joined along obvious suture lines as some paleogeographic maps seem to suggest.)

Between Gondwana and Laurentia, the Tethys Ocean became smaller and smaller as the two giant continents converged. The only areas that were yet to be incorporated into this northern supercontinent were those of Angaraland and China, but their days as island continents were numbered. Angaraland was closing in on the eastern edge of Laurentia—the ocean between them becoming narrower all the time, and its sediments beginning to crumple up as Europe and Asia fused.

With this, the great supercontinent of Pangea was almost complete. It occupied one side of the globe. The other side consisted almost entirely of water: Panthalassa, the great world ocean, which stretched across 300° of longitude.

The orientation of the continents was markedly different from the present time. The North Pole was covered by water and the South Pole was located in Africa or South America. The equator bisected Laurentia along what is today a north-south axis, so that it ran straight up through the middle of Canada and northern Europe. This explains why the fossils of tropical plants and amphibians have been found buried in layers of ice at the Arctic Circle in modern times. Even in the heart of Antarctica there are fossils of Late Carboniferous plants, showing that nearly all of Gondwana was capable of supporting vegetation, at least in the earlier, warmer part of the period.

THE CLIMATE of the Late Carboniferous world was probably similar to that of today, varying from arctic conditions near the poles to tropical rain forest close to the equator, with seasonal temperate climates between. Because of the sheer size of Pangea, parts of the interior were arid, far from the reach of the shallow seas that flooded the edges. The Late Carboniferous South Pole, deep in the heart of Gondwana, had an ice cap, with glaciers extending as far north as 30° south latitude; this is a subtropical latitude in modern times. The evidence for this ice cap is found in South America, South Africa, India, and Australia, as well as modern Antarctica—all part of the supercontinent Gondwana. The ice lasted into the Permian period.

In spite of tropical conditions along the Equator, there was so much glaciation that the Late Carboniferous may be considered an ice age.

There is no direct evidence for an ice cap at the North Pole at that time. This may be due to the fact that the North Pole was all ocean and so there would be nothing useful preserved. However, there is some indication of floating icebergs in northern Siberia.

It was probably the presence of the Gondwana ice cap that helped to trigger coal-forming conditions.With a great temperature gradient between the pole and the Equator, there must have been a very vigorous circulation of the atmosphere through convection currents. This would have meant that moisture was carried from ocean areas over land by northeasterly trade winds, and then dropped as rain when the moist winds met mountains. This was the source of rivers and the swamps where coal formed.

PANTHALASSA OCEAN

A VIEW OF EARTH

From space, the Earth of the Upper Carboniferous period would have appeared like this. The continents were clumped around the South Pole, leaving the rest of the globe as ocean with just a few massed islands in the northern hemisphere.

JUNGLE AND ICECAPS

The concentration of the landmasses in the region of the South Pole created an extensive ice cap, contrasting with the development of thick tropical forest along the equator, with a steep temperature gradient between the two.

Although it took millions of years, the collision of the continents was a violent event. The impact of Laurentia and Baltica had squeezed up mountains between them, thrusting and grinding them upwards into a mighty range. This mountain-building event is called the Acadian Orogeny in North America, where the remains are found in the northern Appalachians, and the Caledonian Orogeny in Europe, where the modern results are the mountains of Wales, the Highlands of Scotland, and the fjord coastline of Scandinavia. As Gondwana approached Laurentia, offshore island arcs generated by the subduction of the plates became welded to the edge of Laurentia, compacted into the coastal mountain range that had already been generated by the Ordovician Taconic Orogeny. The impact of Gondwana's collision with the northern continents produced mountains along much of Europe, in an event called the Hercynian Orogeny. The resulting mountains are almost all gone now, eroded over hundreds of millions of years, but their granite cores can still be seen in the moors of southwest England and the craggy coastline of northern France.

The impact of two colliding continents had far-reaching effects. On the western coast of Laurentia, the embryonic Rocky Mountains showed themselves as island chains, crumpled up and shot through with volcanoes as the continental plate of Laurentia pushed westwards against the oceanic plate of Panthalassa. Much of the mountainous interior of modern North America was formed at this time.

MOUNTAIN BUILDING

The Allegheny River cuts through the ancient sediments of the once rugged Appalachian Mountains (right). The region can be divided into discrete geological provinces (left) with an east-to-west gradient of intense to weak deformation following repeated episodes of mountain building. The Alleghenian orogeny saw the highly metamorphosed rocks of the Piedmont Province pushed great distances over the underlying strata; inland the older rocks of the Valley and Ridge underwent folding and thrusting. The Blue Ridge Province is an uplifted slice of Precambrian basement.

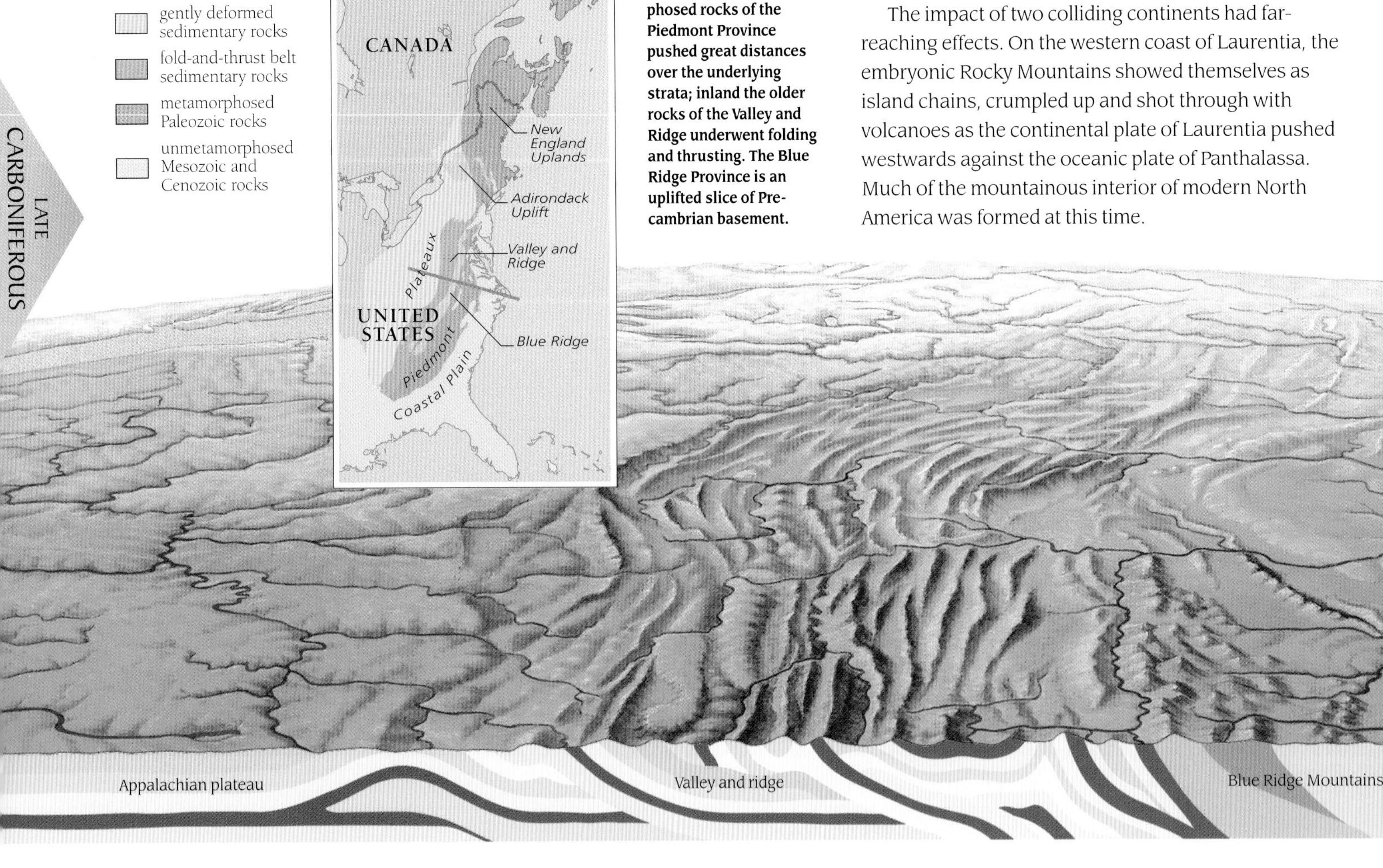

THE STORY OF THE APPALACHIANS

The Appalachians began in Cambrian times, when Avalonia (now northern Europe) approached Laurentia (North America) (1). They collided, fusing during the Taconic Orogeny (2). Meanwhile, the corner of Gondwana was approaching. This collided in the Late Carboniferous, thrusting up the intervening ocean sediments in the Alleghenian Orogeny (3). During the Mesozoic Era the peaks were gradually leveled by erosion and the Atlantic Ocean opened up to the east.

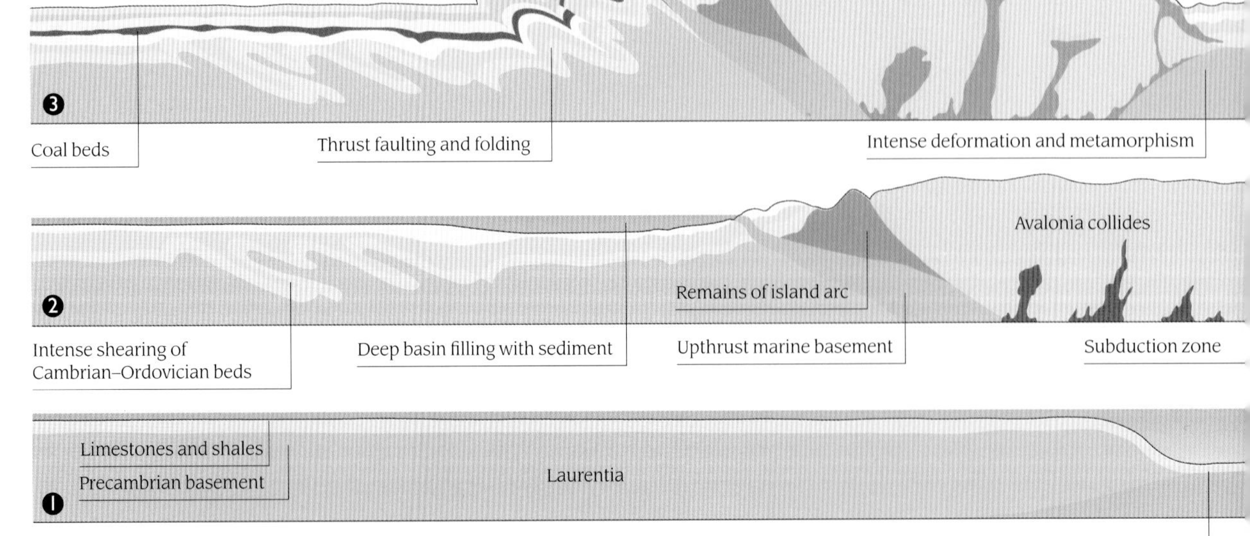

AS GONDWANA slowly advanced and finally crashed into Laurentia, even larger mounds of sediment from the ocean floor were crumpled up between them. This event, known as the Alleghenian Orogeny, regenerated the Acadian and Caledonian mountains, producing an enormous new range on top of their remains. These newer mountains still exist today as the southern Appalachians in North America and much of the Atlas Mountains in North Africa. Although insignificant today compared with the Rockies and no longer volcanically active, the original Appalachians would have rivaled the height and extent of the modern Himalayas, which were formed much later. The present Appalachian Mountains are merely the worn stumps of this once-huge system, eroded away over millions of years.

The collision of the Laurentia with, first, Baltica and then Gondwana pushed up vast mountain ranges along the suture lines.

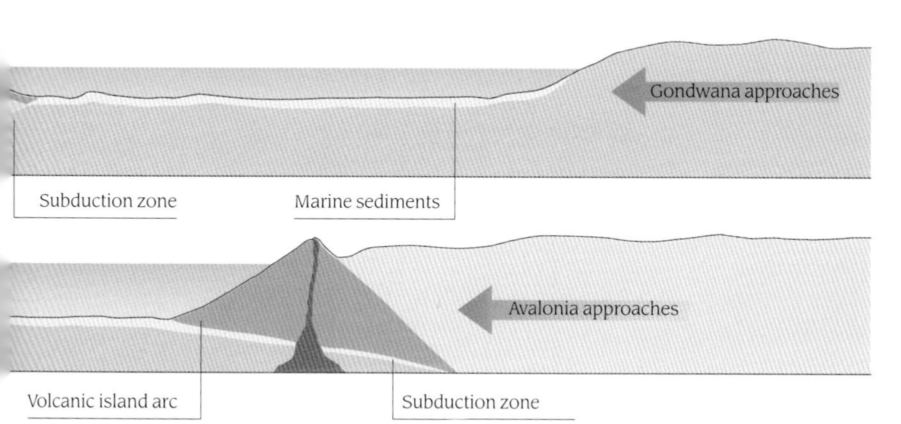

Mountain-building took place at the edge of continents. The interiors of the Late Carboniferous continents were generally flat. Areas such as this, lying close to sea level, are flooded by the sea at regular intervals. This is one of the reasons why most of the rocks formed in Late Carboniferous times consist of distinct sequences in which shallow sea sediments alternate with sediments representing river deposition and forested dry land. Along the eastern highlands of the Appalachians the rocks are mostly continental sandstone and shale. At the edge of the Appalachian belt, about half the rocks are marine in origin, and farther west they are predominantly marine limestone, sandstone, and shale. This pattern corresponds to the sloping of land from mountains down to lowlands and shallow flooded plains.

IT WAS a time of extensive erosion. All the newly arisen mountains were rotted down by the wind and the rain, which split open the exposed rocks and broke them apart. The resulting rubble was trundled down the mountain streams, breaking and shattering as it went. The rivers of the foothills carried the ground-up sand and silt on towards the sea, depositing it where the currents slowed. The great warm shallow seas that covered much of North America and the other continents in Early Carboniferous times became

Great deltas formed from tons of river-borne sediments.

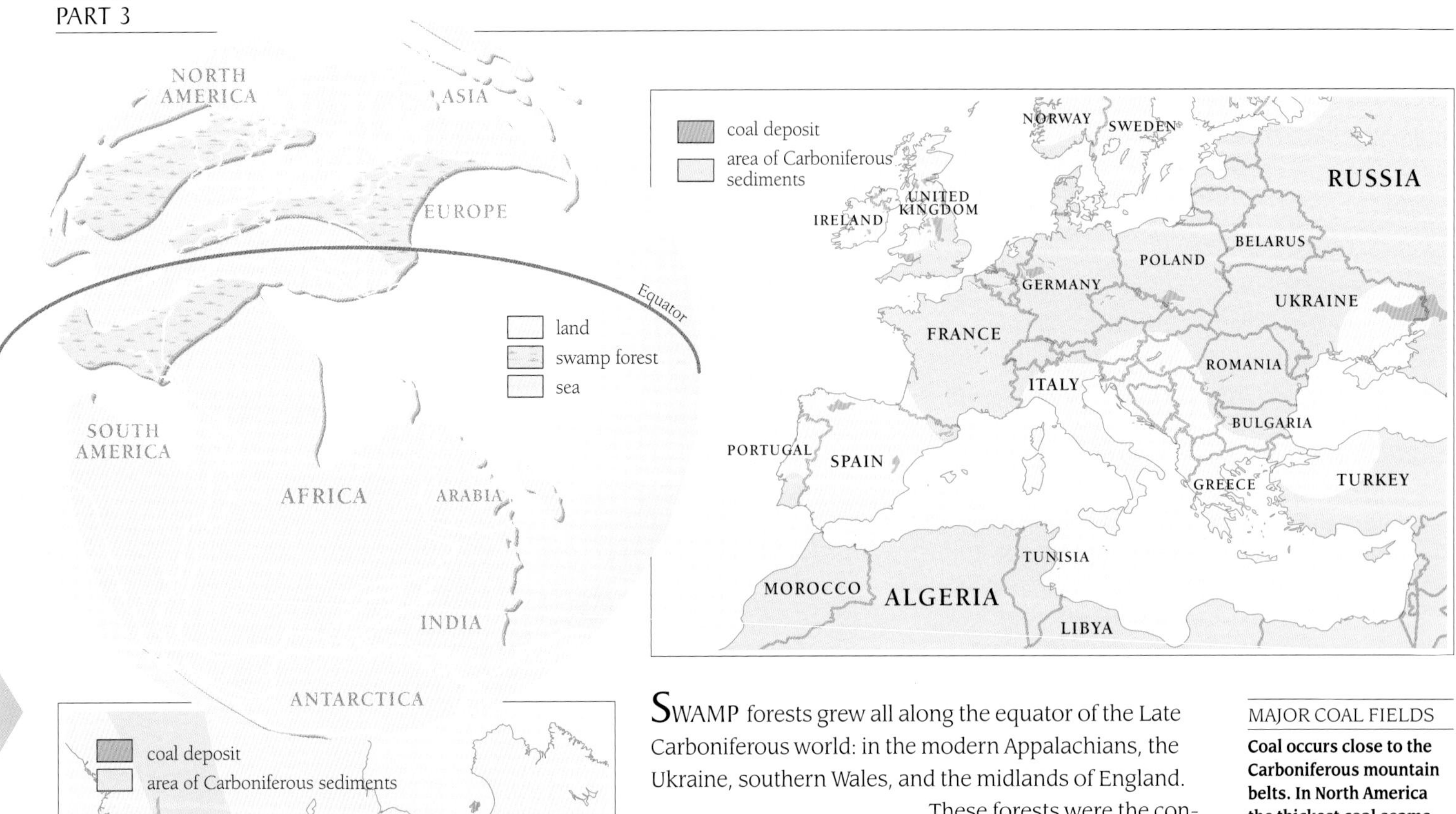

encroached upon by deltas that crept gradually outwards from the decaying mountainous landmasses as more and more sandy material was deposited. The coastlines on each side of the Caledonian and Taconic mountain chain must have looked very much like the delta landscape of modern Bangladesh. But today's deltas have large human populations, while those of the Late Carboniferous were the habitats of fish, amphibians, and insects.

The constant rain of the equatorial climate would have fed mountain streams, building up to great rivers, so there was always a vigorous circulation of water as well as a great deal of suspended mud and sand being washed along. The biggest rivers of the modern world carry about 700 million tons of sediment a year, most of which is deposited in deltas and river plains. The great rivers of the Late Carboniferous would have carried similar amounts, filling in the shallow continental seas and building up the delta fronts rapidly. This low-lying land, rich in mineral deposits and flooded by warm tropical water, provided ideal conditions for the growth of plants along the margins of North America and northern Europe.

SWAMP forests grew all along the equator of the Late Carboniferous world: in the modern Appalachians, the Ukraine, southern Wales, and the midlands of England. These forests were the contemporary equivalents of today's tropical rainforests, like those of the rivers Amazon, Congo, and Mekong. These great areas of swamp forest seem to have appeared in central Europe first, triggered no doubt by the approach of Gondwana to Baltica. The narrowing ocean space in between was soon filled with sediment washed from the rising Hercynian mountains, producing thick sequences of sandstone known as millstone grit. This built up a river plain on which the first of the coal forests grew at the very beginning of the Late Carboniferous. It was not until a few million years later that the forests spread out to North America in one direction and the Ukraine in the other. The coal forests did not appear farther east in China until about halfway through the Late Carboniferous, and deposits found in the central and western United States and Australia date from the Permian period.

Coal seams were first discovered in cool northern climates, but the great swamp forests that formed them were originally tropical.

There was no variation in climate from one part of the year to another in these regions, judging by the lack of growth rings in the fossil trees of the time. The shapes of the fossil leaves were evolved to shed large quantities of water quickly, indicating that rainfall was heavy.

The lifespan of the swamp forest determined the depth of the coal seam it eventually produced. Transient sandbanks only produced thin coal seams, while swamps and backwaters that remained stable for a long period of time accumulated the great masses of vegetable material that produced thick coal beds. It takes a 33-foot (10-meter) thickness of dead plants to produce one meter bed of bituminous coal. This mass of plant material would take up to 7000 years to accumulate. Although most coal

MAJOR COAL FIELDS

Coal occurs close to the Carboniferous mountain belts. In North America the thickest coal seams border the Appalachians; in Europe they lie between the Caledonians in the north and the Hercynians in the south.

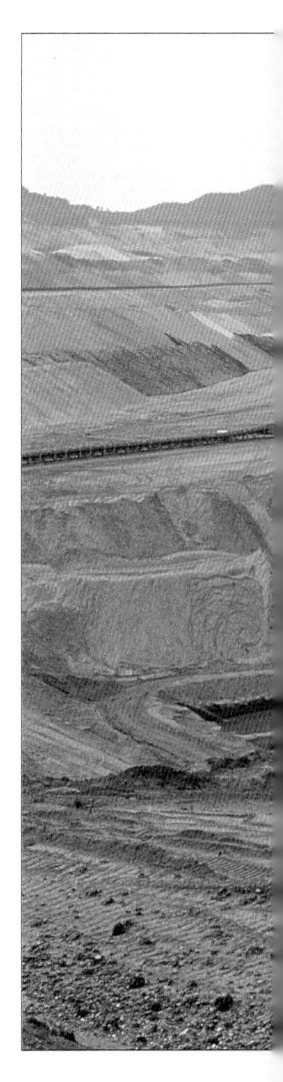

was formed at the bases of standing vegetation, there were other types forming at the same time. Fine plant material, such as spores, drifted along in water and was deposited to form the compact fine-grained "cannel" coal. Algal debris building up underwater produced the similar "boghead" coal.

COAL is classed by its grade: the amount of carbon it contains relative to its other constituents. In general, the older and more compressed a coal, the higher its grade. Peat, widely available across the world, can be regarded as the lowest grade of coal, consisting merely of compressed plant matter. Its development is quite recent in geological terms and its carbon content is about 55 percent, which is little different from wood's 50 percent. Lignite or brown coal, found mostly in Tertiary deposits, and of economic importance in eastern Europe, contains about 73 percent. Bituminous coal, such as most of that from the Late Carboniferous, contains about 84 percent. The highest grade of coal is anthracite, with a carbon content of over 93 percent. Anthracite is a result of the metamorphism of bituminous coal. Some Late Carboniferous coal deposits have been metamorphosed and occur as anthracite, particularly in Wales.

Most high-quality coal comes from the Late Carboniferous. Coal beds laid down in later periods produce an inferior grade.

Coal is not the only economic product of the Upper Carboniferous system. Large quantities of petroleum are frequently found. Many of the swamp sandstones also contain a fair proportion of iron, concentrated at the time of the rock's formation, and sometimes this is found in rich nodules of oxide. Iron was one of the most important metals of the Industrial Revolution, and much of the iron that was found in Britain came from Carboniferous deposits. A rock that contains between 20 and 40 percent of iron in a compound from which it can be extracted can be regarded as an economic ore. Nowadays, mining the original deposits is uneconomical, but smelting may be carried on at old sites using ores from different locations, maybe from a different rock system.

Other metals are concentrated in ores during mountain-building. Hot underground water rises to the surface rich in dissolved metallic salts. Under certain conditions these salts are converted to insoluble minerals and precipitated. When the Appalachians formed, limestone from the Lower Carboniferous was penetrated by the hot hydrothermal waters, rich in dissolved salts of lead, zinc, and copper. They reacted with organic matter in the limestone and precipitated out as insoluble sulfides. Ores of lead, zinc, and copper formed in this way are now mined along the Mississippi River.

MINING FOR COAL

Modern techniques of open-cast or strip mining remove vast quantities of coal-bearing rocks along with their overburden. This mine near Cologne, Germany, extends for over 3mi (5km). Traditional methods concentrate on thick seams and remove only the coal itself.

❶

THE FORMATION of coal was due to the constantly-changing conditions of the swamps. As the sea level rose and fell, the coastlines of the deltas advanced and retreated. Such changes are reflected in the alternation of sediments that formed at that time. The bulk of each delta consisted of sand, laid down in three beds: the topset, the foreset and the bottom set. When the current changed, the topset was washed away along with the topmost part of the forest. Only the bottom part of the foreset beds and the bottom set beds remained. Upon the fresh erosion surface at the top, another delta tongue was built out and this in its turn was eroded and then built upon.

Cycles of deposition, colonization, and inundation created the anaerobic conditions needed for the formation of coal.

❷

Eventually, when all this was turned to rock, the result was a thick bed of sandstone showing a distinctive curved structure, instantly recognizable to any geologist as the result of river deposition. Frequently the bed of sand arose above the surface of the water. On this exposed sandbank plants took root and grew. The roots spread into the sand, drawing from it whatever nourishment they could find. The plant growth may have been long-lived or it may have been quite temporary, but eventually it would have been killed as the water arose to cover it. In the rock record this is preserved above the sandstone as a layer of clay or of bleached sandstone that is full of carbonized roots. Immediately above this "seat clay" or "seat earth," sometimes called ganister by mining engineers, lies the remains of the vegetation: the coal itself.

❸

Sooner or later the coal-producing vegetation was flooded, usually as the land sank and the sea came in. Mud was deposited on top of the dead vegetable matter, and eventually this mud turned to shale. If the flooding persisted and the sea became deeper, limy deposits would build up, giving beds of limestone. The thickness of these beds is not a good indication of the speed of

A CYCLOTHEM

Rocks are laid down in typical sequence as a delta front advances and retreats. This is an idealized example; in reality, stages may be incomplete or missing.

10 Shale formed as delta muds encroach again
9 More marine limestone
8 Marine shale—formed in deep water
7 Marine limestone formed as sea floods in
6 Muds and shales from flooding by sea
5 Coal
4 Seat earth or clay, containing roots
3 Limestone nodules
2 Silts and muds from levees and floods
1 Sandstone deposited by river channels

Delta
Sea
Delta

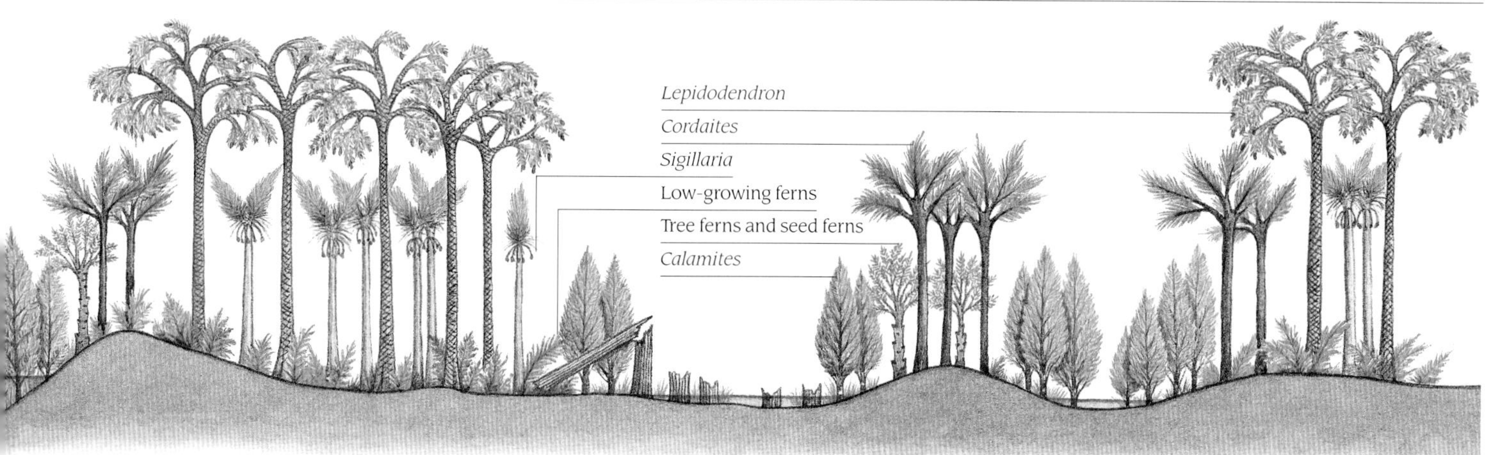

accumulation of the sediment: it would have taken 7000 years to produce enough vegetable material to give 3ft (1m) of coal. But it would have taken only five years to deposit enough mud to form a similar bed of shale.

Then things would change, and the delta would build back out, starting with a bed of river-deposited sandstone once again. This sedimentary sequence or cyclothem—sandstone, seat-earth, coal, limestone, and back to sandstone again—is never found just once but is repeated many times, showing the instability of local conditions. Although such a sequence consists largely of sandstone and shale, with very little coal, it seems likely that the coal-producing conditions prevailed. It is estimated that the forest areas of the Late Carboniferous world were flooded for only about 2 percent of the time.

The sequence described is an ideal one. Often some of the elements are missing. Perhaps the sandbank never reached the surface of the water and so no coal is present. Perhaps the return of the sea was too short to allow limestone to develop. In any case, the Upper Carboniferous systems of North America and Europe show thousands of such cycles.

4

THE COAL SWAMP

The distribution of plants in a coal swamp was not random. In the wettest portions, in the creeks and flooded backwaters, the horsetail *Calamites* grew. More intermittently, flooded areas supported the giant club mosses like *Lepidodendron* and *Sigillaria*. On the driest parts, the crests of levees, there grew the primitive conifer *Cordaites* amidst an undergrowth of ferns and seed ferns.

A SHIFTING ENVIRONMENT

As river-borne sediments built up, delta fronts crept out over the sea (1). Islands and levees formed in advance of the main delta area. At its height, the delta consisted of a constantly shifting pattern of meandering streams and backwaters (2). When sea levels rose, floodwaters drowned the forest and began the process of laying down marine sediment. (3). Eventually, a new delta would form (4).

LATE CARBONIFEROUS

THE TREES of the Carboniferous forests were mostly similar to the trees of Devonian times. The most typical were giant lycopsids: clubmosses 98ft (30m) high. Modern clubmosses are insignificant little things, strings of scale-like leaves, easily overlooked in the long grass of moorland. Their ancient relatives, however, were the giant redwoods of their day. Their trunks, 7 to 10ft (several meters) in diameter, rose into the canopy of branches. All the branching was dichotomous: that is, it forked into two equal branches, each of which forked in two again and so on. The root system did the same, as seen in fossils in coal-bearing rocks. The branching underground part of the stem, which held the roots, is known as *Stigmaria* and is recognized by spiral marks from which the roots grew. The trunk itself had a pattern of scars from which the simple leaves sprang. *Lepidodendron* had a diamond-shaped arrangement, while *Sigillaria* had the leaf scars in vertical rows. Reproductive cones of these trees are known as *Paleostrobus*. Because fossils are mostly found scattered, it is almost impossible to tell which cone came from what tree, or whether it had been attached to the tip of a branch or hung beneath.

Few people now would notice a clubmoss if they saw one. But the giant clubmosses of the Late Carboniferous dominated the forest.

Around the edge of a modern swamp are beds of horsetails, primitive plants related to ferns. They rarely grow to more than 3ft (1m) in height, but the Late Carboniferous version reached 33ft (10m). They were called *Calamites,* and there were several different species. The ribbed stems are often found as fossils. Ferns existed too, and probably formed much of the undergrowth. These were all primitive plants that reproduced by means of spores rather than by seeds. There were ancestors of modern conifers too, with thick woody trunks and long thin leaves. These are called *Cordaites.*

All these different plants lived in different parts of the forest: the *Calamites* growing in the water, the *Sigillaria* and *Lepidodendron* along the river banks, the ferns and seed ferns forming the undergrowth, and the *Cordaites* growing on the drier upper land. Horsetails that were adapted as creepers hung from the trunks and branches.

LATE
CARBONIFEROUS
1
2
3
4
5
6
7

IN THE forest, vegetation was so dense that it would have been difficult to see more than a few paces ahead. The air was hot and humid, smelling of rotting plants. The banks of the rivers, built up into levees by mud and silt, were the only places where it was dry enough to walk. This was where most of the plants grew. Between their great size and rampant growth—even the undergrowth was head-high—it would have been almost impossible to get through.

Insects, not birds, were the first conquerors of the air.

There were no birds; the air was the dominion of the first flying insects. Other insects (and other arthropods) crawled, scuttled, and jumped on the forest floor. The spiders and scorpions of the time were almost indistinguishable from those today, while other arthropods were completely different. The dragonfly *Meganeura* with its 2.3-foot (70cm) wingspan was as big as a parrot. The water and the undergrowth crawled with amphibians the size of alligators. In the mud down the side of the levee, between the thick horsetails, could be found enormous amphibians like salamanders, 6.5ft (2m) long. *Hylonomus,* the first reptiles, lived in these forests too. They were small lizard-like animals which probably spent most of their time hiding from the larger amphibians.

The swamp forest stretched on: 600mi (1000km) of green impenetrability. Somewhere beyond it lay the sandbanks and islands of the delta mouths, where the swamp met the great inland sea.

1 *Psaronius* (tree fern)
2 *Aphthoroblattina* (cockroach)
3 *Keraterpeton* (amphibian)
4 *Lepidodendron* (a lycopsid—clubmoss)
5 *Eogyrinus* (amphibian)
6 *Hylonomus* (reptile)
7 *Ophiderpeton* (snake-like amphibian)
8 *Calamites* (horsetail)
9 *Annularia* (creeping horsetail)
10 *Sigillaria* (lycopsid)
11 *Meganeura* (dragonfly)
12 *Arthropleura* (giant centipede)

8

11

9

10

12

LATE CARBONIFEROUS

THE EVOLUTION OF INSECTS

LATE CARBONIFEROUS

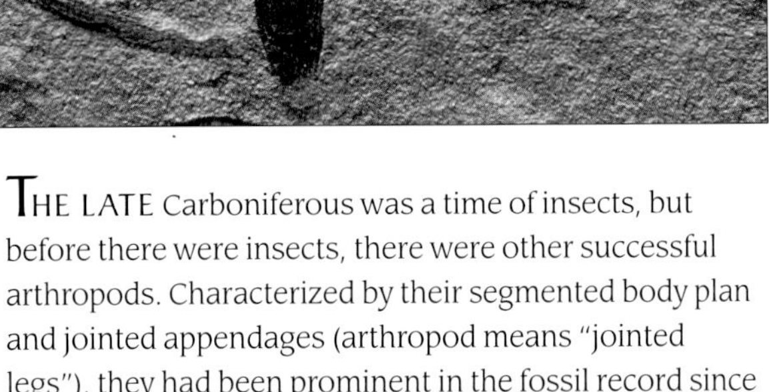

INSECT FOSSILS

This carbonized image of a beetle (order Coleoptera) preserves a wealth of detail. Many of the Upper Carboniferous insects have been identified mainly by means of of their fossilized wings. Wings are made of a very tough natural substance, chitin, to withstand the stress of flight. Occasionally, if the preservation is good enough, the color and pattern of the wings may be visible.

THE LATE Carboniferous was a time of insects, but before there were insects, there were other successful arthropods. Characterized by their segmented body plan and jointed appendages (arthropod means "jointed legs"), they had been prominent in the fossil record since Cambrian times. The earliest were marine species such as the trilobites and their relatives and were among the most active of the early Paleozoic invertebrate fauna.

Arthropods were also among the pioneers of land-dwelling animals. In the Lake District of England are 450-million-year-old footprints from an arthropod that resembled a millipede and walked on dry land. Huge footprints from a scorpion-like animal are known from 420-million-year-old Silurian shoreline deposits of Australia. However, it was with the evolution of insects that arthropods finally became at home on land.

During the early Silurian period, a transitional form—part millipede, part insect—appeared in Western Australia. The body had three segments (head, thorax, and abdomen) like an insect's, but it had 11 pairs of legs rather than the insect's three. This creature is called an euthycarcinoid. The oldest true insect (or hexapod, "six legs") is possibly a springtail from the Devonian Rhynie Chert of northeast Scotland.

By the Late Carboniferous, with the development of flight, insects had come into their own. The most primitive insects, such as springtails, are flightless and do not go through a larval stage; an immature insect resembles an adult. Flight evolved with the development of dragonfly-like forms, such as *Meganeura*, with fixed wings that could not be folded back over their bodies. Later Carboniferous insects like the cockroach *Aphthoroblattina* had more sophisticated folding wings. Now insects began to undergo metamorphosis, changing from larvae to pupae to breeding adults.

Much later, the appearance of pollen-producing flowering plants during the Mesozoic Era created a niche for the social insects, ants and bees.

EVOLUTIONARY TRENDS

Insects, representing 90 percent of all arthropods, are the most successful group of animals that has ever existed; today there are more than a million species. Insects and myriapods, the Uniramia, developed as a separate line during the Silurian and specialized rapidly. Their ancestry is unclear but it is possible that they share their origins with the annelid worms. The Onychophora (velvet worms) provide some evidence of this, having characteristics of both groups, though recent DNA analysis has suggested a closer relationship to cephalorynch worms. The Uniramia show several evolutionary trends typical of all arthropods: gradual reduction in the number of limbs, fusion of some segments, and the specialization of others for particular functions.

1 Unsegmented ancestor
2 Segmentation first appears
3 Hard exoskeleton develops, plus legs and antennae
4 Legs become jointed
5 First segment fuses with head; limbs on first four segments reduced to mouthparts
6 Legs of last three segments develop into male genitalia; other abdominal legs lost
7 Wings develop from leg-shields

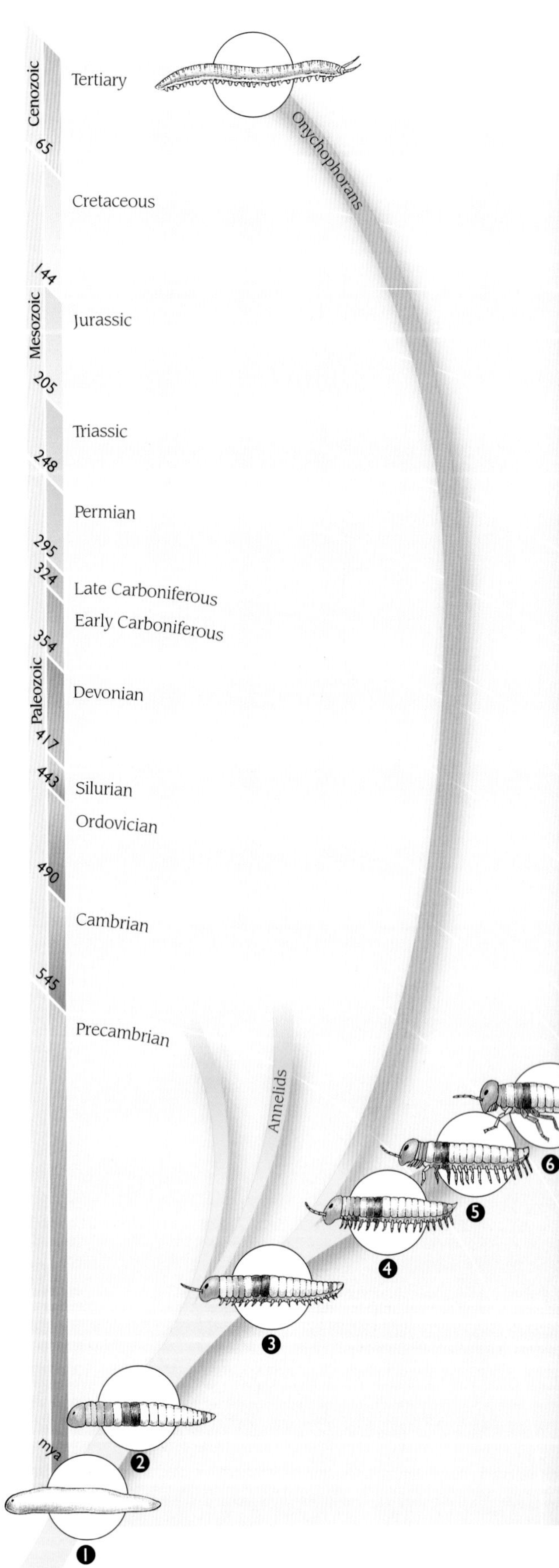

LATE CARBONIFEROUS

WINGED INSECTS

The most remarkable feature of the evolution of insects is the power of flight. During the Late Carboniferous, modified leg-shields, useful for gliding, eventually developed into wings. The majority of winged insects, called "endopterygotes," have a larval stage that is distinctly different from the adult. They undergo a complete transformation to the adult form, entering a dormant pupal stage while the larval structures are transformed into adult characteristics. The other group, known as "exopterygotes," hatch from the egg as miniature adults, but without wings or reproductive apparatus.

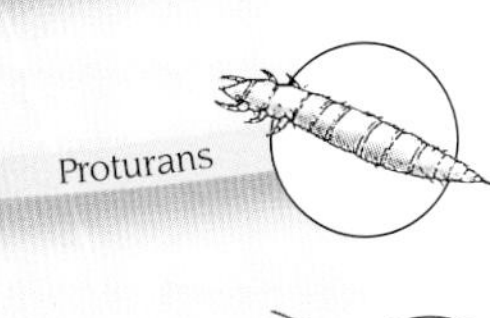

WINGLESS INSECTS

There is a subclass of insects which includes the bristletails (Thysanura), whose ancestors never developed wings—as distinct from insects such as fleas which have at some time lost theirs.

CLASSIFICATION

The phylum Uniramia is one of the three modern arthropod phyla. It has two major groups: the myriapods and the hexapods (insects plus the classes Collembolla, Protura, and Diplura).

Phylum	Class
ANNELIDA	
ONYCHOPHORA	
UNIRAMIA	INSECTS
	MYRIAPODS
CHELICERATA	SCORPIONS
	SPIDERS, MITES ETC
	SEA SPIDERS
	HORSESHOE CRABS
	SEA SCORPIONS
CRUSTACEA	
TRILOBITA	

extinct
conjectural

THE PERMIAN

295 – 248 MILLION YEARS AGO

The Permian period is the last in the Paleozoic era. The ancient continents were continuing to move towards one another and the fusion of the supercontinent Pangea was almost complete. It was a time of widespread mountain-building and volcanic activity. Throughout the period the landmasses rose and the sea retreated from the continents, producing fewer and fewer marine sediments. The characteristic rocks of the system are terrestrial sandstones, so similar to the Old Red Sandstones of Devonian times that they are referred to as the New Red Sandstones. These continue right through the succeeding Triassic system, causing some confusion as early geologists attempted to fit these new pieces of the stratigraphic puzzle. However, the Permian and the Triassic could hardly be more different biologically. The end of the Permian period is marked by an unprecedented rapid change of plant and animal life, caused by the greatest extinction in history, which wiped out 96 percent of all species. This episode, some 250 million years ago, heralded the end of the Paleozoic era.

MINING engineers and geologists of the early Industrial Revolution, especially in England, observed that coal-bearing Carboniferous rocks were overlain by a sequence of barren red-colored sandstones and, in the northeast of the country, by a bed of magnesium-rich limestone. The red sandstones contained evidence of sedimentation by sand dunes, alluvial fans, and rivers, all very similar to the Old Red Sandstone sequences of the Devonian period. By the 1820s the term New Red Sandstone had become applied to the Permian beds to distinguish them from the older deposits.

Two layers of red sandstone, from the Devonian and the Permian, make a "sandwich" of the Carboniferous strata.

In 1833 the great geologist Charles Lyell was the first to use the term New Red Sandstone as an official geological designation. His definition encompassed all rocks that lay between the Carboniferous and the lower Jurassic, known as the Lias at that time. In Germany the redbeds that succeeded the Carboniferous were known as the Rotliegendes and the magnesium-rich limestone as the Zechstein (a local quarryman's name meaning "hard stone"). The latter had been economically important for centuries because it contained a significant source rock for copper. Only a year after Lyell's classification was published, the German geologist F.A. von Alberti introduced the term "Trias" for the uppermost part of the New Red Sandstone, based on the fact that sections of it could be divided into three distinct sequences, at least in Germany. This is the basis for the now-acknowledged Triassic period that follows the Permian.

Classification of these sequences was difficult compared with some of those, such as the Devonian and Carboniferous, that had already been identified. The beds of the New Red Sandstone are the products of terrigenous deposition or sedimentation in continental salt

KEYWORDS

- BIOGEOGRAPHIC PROVINCE
- CONTINENTAL DRIFT
- GLOSSOPTERIS
- ICE AGE
- MAMMAL
- MORAINE
- THERAPSID
- TILLITE
- VARVE

LATE CARBONIFEROUS	295 mya – 290 – 285 – 280 – PERMIAN – 275
Series	LOWER PERMIAN (ROTLIEGENDES)
European stages	ASSELIAN; SAKMARIAN
N. American stages	WOLFCAMPIAN
Geological events	Alleghenian orogeny continues; Gondwana fuses to Laurussia (Laurasia + Baltica)
Climate	Permo-Carboniferous glaciation ends
Sea level	Fluctuating
Plants	Clubmosses dominant
Animals	Land animals spread, including pelycosaurs (mammal-like reptiles); Sponge/ bryozoan reefs

lakes and shallow gulfs. There is not a fossiliferous marine bed among them, and it is only the ever-changing and widespread fossils of marine beds that allow for strict correlation between one area and another. Another problem was the fact that throughout western Europe there seemed to be a depositional break between the Carboniferous strata and the New Red Sandstone, and there was no indication of how much of the geological succession here might be missing.

IN 1840–1841 Roderick Murchison, who co-named the Devonian, toured Imperial Russia along with the French paleontologist Edouard de Verneuil and the Latvian scientist Alexandr Keyserling. He was mindful of the stratigraphic break above the Carboniferous in western Europe and hoped that he could find a continuous sequence from Carboniferous rocks right up through the New Red Sandstone. With the help of local geologists, this is precisely what he did. The discovery was made in the vast tract of post-Carboniferous deposits that spread from the Barents Sea in the north almost to the Kazakhstan border in the south, between the Volga River in the west and the Ural Mountains in the east. In a letter to the Society of Naturalists of Moscow in October 1841, Murchison named this rock system the Permian; its namesake was the nearby industrial city of Perm, founded 60 years earlier. The Carboniferous rocks here graded up into a marine sequence, which was followed by New Red Sandstone beds very similar to those in western Europe.

The Permian is still sometimes merged with the Triassic, although one falls in the Paleozoic and the other in the Mesozoic.

Permian rocks were first identified in North America in a large area between the Mississippi and Colorado rivers by geologist J. Mancou in 1853, and he recognized the similarities between these rocks and those of Europe.

Many popular-level books still use the New Red Sandstone as a catch-all division encompassing the Permian and the Triassic periods, despite the fact that they are separated by a major time scale boundary: that between the Paleozoic and Mesozoic eras. From the point of view of the general geography the two are somewhat similar, with desert conditions spreading everywhere over united continents.

It is generally agreed now that the Permian can be divided into two sections, the lower being by far the greater and represented in Europe by the Rotliegendes, and the upper represented by the Zechstein. From time to time scientists have argued for a three-part division. However, even in China the Permian is known as the *Erdie*, which translates as "two-fold," although the division is not quite the same as in Europe. The lower part is again by far the greater, and the upper part is even smaller than the Zechstein.

The shortage of marine beds has always been a difficulty for establishing an exact correlation of the Permian. This has been a particular problem for the later stages of the Permian, when the gradual retreat of the sea left greater and greater land areas, and produced fewer and fewer marine deposits. The actual transition from Permian beds to Triassic beds has only been identified in one place, in the Salt Mountains of Pakistan. Even the lower parts of the sequence in Murchison's work have at times been regarded as Carboniferous rather than as Permian. In fact the US Geological Survey did not officially acknowledge the period until 1941, and the lowermost part of the American Permian succession—the Wolfcampian—was not accepted as officially Permian until 1951 because of ongoing controversy in what was then the Soviet Union as to precisely where the boundary of the Permian lay.

See Also

THE DEVONIAN: *Old Red Sandstone*
THE EARLY CARBONIFEROUS: *Coal, Pangea*
THE LATE CARBONIFEROUS: *Alleghenian Orogeny, Ice Age*

TURNING POINT

The Permian's position at the very end of the Paleozoic era makes it very significant in the history of the Earth. At the beginning of the Paleozoic life was confined to the sea. As the era progressed the land gradually became colonized, first by plants, then by invertebrates, and finally by vertebrates. By the end of the period, life on land was fully established, and the Mesozoic era was beginning.

PERMIAN

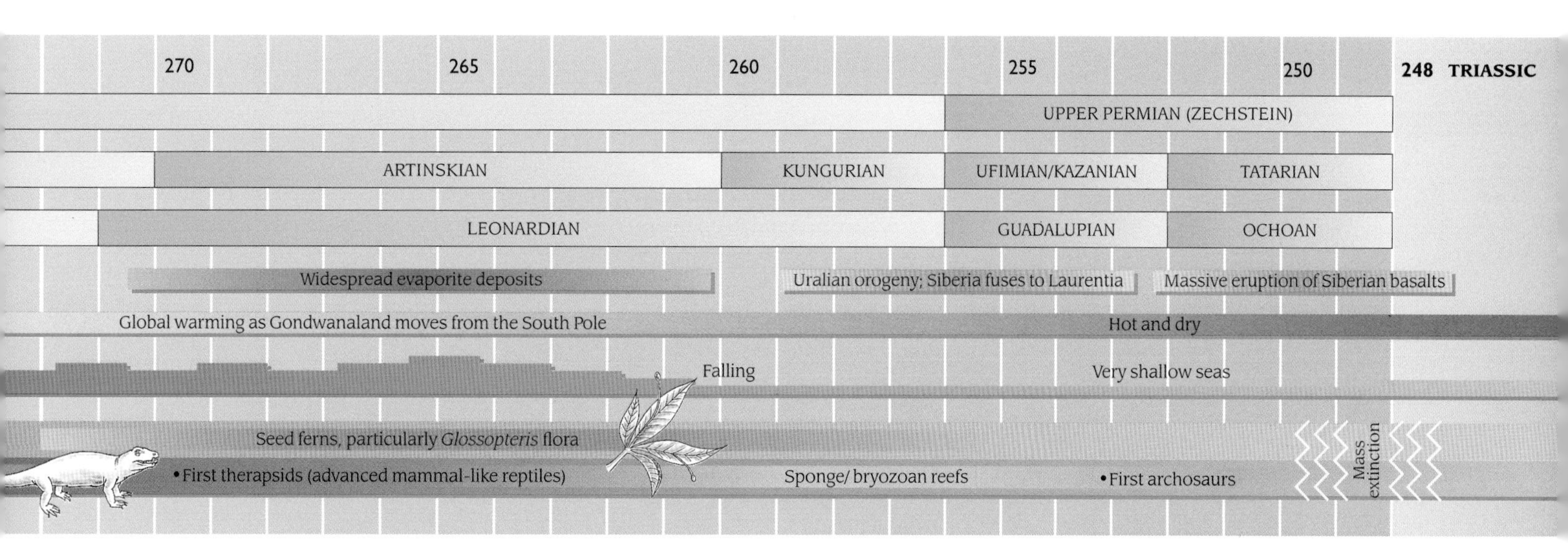

INHOSPITABLE HEART OF THE SOUTHERN CONTINENT

There are few fossils of living things dating from the Permian sediments of Gondwana. This absence is due to the huge size of the supercontinent, and the extreme climatic conditions that must have existed there. The center was thousands of miles away from the moderating influence of the ocean. Computer models show that the heart of the southern continent could not have received more than an average of 0.08in (2mm) of rain in a year. In summer the highest temperatures on Gondwana may have reached 113°F (45°C), while between seasons in some places the temperature difference could have been varied between 77°F (25°C) in summer and −77°F (−25°C) in winter. Some winter temperatures could have been below −86°F (−30°C). In such harsh conditions it would have been difficult for life on land to thrive, and any that did must have migrated between places where the seasonal changes were less stressful.

THE SUPERCONTINENT of Pangea was creeping towards its completion in Permian times. Only the landmass of China and some of the nearby islands had yet to be incorporated. Gondwana had finally closed with Laurentia and a vast range of mountains, the Alleghenies, was uplifted between them. The Alleghenian orogeny was the final stage of the formation of the Appalachians, the deformation being superimposed on that produced by the fusion of Laurentia and Baltica during the Devonian. This mountain-building episode produced the Hercynian mountains across what is now southern Europe, and the Mauritanides in what is now North Africa, swallowing up the intervening ocean in the process.

The new supercontinent of Pangea stretched from pole to pole. Only parts of eastern Asia floated in isolation off the coast of Siberia.

In the north the old Caledonian Mountains were beginning to wear down, and in at least one place the northern seas broke through, spreading a shallow gulf over the area that is now occupied by the North Sea, northeast England, and Germany. The slow evaporation of this shallow sea produced the limestone deposits and the beds of evaporite minerals that constitute the Zechstein formation. Around this inlet the erosion of the new and the old mountains produced the alluvial fans and sandy deposits that ultimately formed the New Red Sandstone of the Permian.

Farther east, the eastern margin of the Old Red Sandstone continent finally collided with Angaraland (Kazakhstania and Siberia), forming the landmass of Laurasia. The ancestral Urals were building up along the join. Although the continental masses had fused, there was still a sea area along the edge of the new mountains, like the Persian Gulf today—a shallow shelf sea forming as the Arabian and Iranian plates fuse, with the Zagros mountains on the Iranian side. It was this Permian sea, the Uralian Seaway, that provided the marine deposits identified by Roderick Murchison. As the Permian continued, this sea eventually dried up and its place was taken by deserts, salt lakes, and alluvial fans spilling down from the mountains, forming more New Red Sandstone. Enormous expanses of lava flows spread to the east of the Urals at the end of the period, generated by the last pulses of the mountain-building process.

Out to the west, the far coast of the Laurasian landmass continued to be marked by arcs of volcanic islands trapping shallow basins between them and the continent itself. The continued subduction of the oceanic plate continuously produced new volcanoes and islands. The mountains of the Antler orogeny, begun in the Carboniferous, were still present in the region of modern Nevada and Idaho. As the volcanic arc of Sonomia docked, towards the end of the period, this uplift was renewed in the Sonoma orogeny. The rest of the North American continent to the south of the Canadian shield varied between redbeds and shallow shelf seas. This succession in the low-lying region in the southwest produced the desert-formed Coconino sandstones followed by massive Kaibab limestone, both of which are significant horizons in the rim of the Grand Canyon.

WHERE Laurasia and Gondwana met, in southern Texas, stretched the Ouachita mountains formed by their impact. This mighty range represented the western section of the Allegheny range; its most imposing remains can be seen today in the state of Arkansas. Immediately to the north of the range lay the Midland and Delaware basins, two deeps in the shelf sea, formed by the same process that raised the Ouachita mountains. Towards the end of the Permian these basins filled in with sediment and disappeared underground. They were rediscovered in the twentieth-century search for oil.

The uplift of the Ouachita mountains created the Delaware and Midland Basins, now important sources of petroleum.

A WARMING CLIMATE

The Permian was a time of climatic contrasts, ranging from glaciation as Gondwana moved over the South Pole to increasingly hot, arid conditions in the continental interiors of northern Pangea.

SIBERIA
ANGARALAND
Ural Mountains
KHAZAKHSTANIA
Uralian Seaway
AURASIA
Caledonian Mountains
Hercynian Mountains
TETHYS OCEAN
PANTHALASSA OCEAN
Alleghenian Mountains
G E A
GONDWANA

LINING UP

During the Permian almost all the continental areas had assembled on one side of the globe forming a U-shaped supercontinent. However, several microcontinents, including Tibet and Malaysia, began to rift from Gondwana and migrate north.

PERMIAN

NEW MOUNTAINS, NEW OCEANS

As the ocean between Siberia and Baltica finally closed, the resulting collision formed the Ural mountains. Meanwhile, to the southeast, the Tethys Ocean became defined.

GLACIAL DEBRIS

The Dwyka Tillite of South Africa is typical of the features left by the Permo-Carboniferous glaciation. A rock surface polished by ice is partially covered by glacier-borne debris (till) that has been consolidated into a solid rock (tillite) over 250 million years. Scratches show how the ice moved.

Gondwana, the southern continent, was still a single undivided landmass consisting of South America, Africa, Australia, India, and Antarctica. It was so huge that moist winds from the sea rarely reached its center. Vegetation was present but was confined to areas close to the edges. The build-up of dry heat in the center meant that the climate was extreme. Continental glaciers that covered parts of South America, Africa, India, and Australia during the late Carboniferous were dwindling. They had disappeared altogether by the end of the Permian period.

PERMIAN GLACIATION

A typical landscape of an ice age. The landforms associated with an ice age, such as drumlins, eskers, and kames, have not been preserved from the Permo-Carboniferous glaciation. These structures are made of loose till, such as gravel and sand, or soft material like clay, and are easily eroded away. Most of the preserved till from the Permo-Carboniferous glaciation is in the form of "drop stones"— rocky debris that has fallen from the undersides of floating glaciers and built up on the seafloor, where it has subsequently been covered with sediment and incorporated into the stratigraphic succession.

MANY of the landforms associated with ice ages are easily eroded, but before the Permian glaciers gradually melted as the climate heated up, they left their mark.

The discovery of glacial deposits in India in 1856 puzzled geologists. These deposits provided evidence that led to modern understanding of the Earth's movements.

Geologists surveying the region of Orissa in northern India in 1856 came across a startling find: a boulder bed that must have been left by glaciers in late Carboniferous times. Such ice-deposited structures were already very familiar to geologists from their studies of the Pleistocene Ice Age. The perplexing thing was that this glacial deposit was found in the tropical subcontinent of India, a location and climate not obviously associated with ice.

Called the Talchir tillite, the Indian glacial bed was not officially identified until the survey was completed in 1874. The deposits consist of beds of boulders, grooved and faceted by the force of ice, dumped in a random manner. A bed such as this deposited by the Pleistocene ice sheets would be called a till; solidified and cemented into a rock, it becomes a tillite. These overlie striated surfaces—bedrock that has been scratched and gouged by the passage of glaciers over them. Such surfaces have also been found in Rajasthan and Madhya Pradesh.

Closer to the Himalayas, rocks show the presence of layers of alternating texture, called varves. As a glacier begins to melt during the summer and freezes again in

4 Meltwater lakes; thin layers of sediment are deposited and preserved as patterns in the rock—varves
5 Moraine features erode and do not survive from Permo-Carboniferous times

1 Striations are marks left by embedded rocky material grinding against rock surface; these show the direction of glacier movement
2 Glacier tongue transports debris; as the ice melts, it falls to the sea bed
3 Icebergs carry rocky material for long distances

THE EVIDENCE FOR CONTINENTAL DRIFT

Patterns of glaciation have been used to show that Africa was positioned over the south pole for much of the Carboniferous. As Gondwana broke up and drifted north, Antarctica drifted south, where it has remained. Continent-sized ice sheets were associated with each location that drifted near the pole.

Fossils provide important evidence that the continents were not stationary through time. An important Permian plant was the seed fern *Glossopteris*, whose fossils are found in Africa, India, South America, Australia, and Antarctica. The modern continents are far apart and have distinct floras. Some conventional 19th-century thinkers suggested that *Glossopteris* seeds must have blown across the ocean to fertilize each continent. The German geophysicist Alfred Wegener believed that only the joining of landmasses could account for the near-global range of *Glossopteris*, and proposed the theory of continental drift.

Other evidence comes from the animal kingdom. *Mesosaurus* was an early Permian swimming reptile that lived only in freshwater lakes. Its fossils have been found in Africa and South America, which would not have been possible if the Atlantic Ocean had always separated them. Also, in 1969 a fossil of the Triassic mammal-like reptile *Lystrosaurus*, whose remains had previously been found only in India and Africa, was unearthed in Antarctica, linking that landmass to the supercontinent of Gondwana.

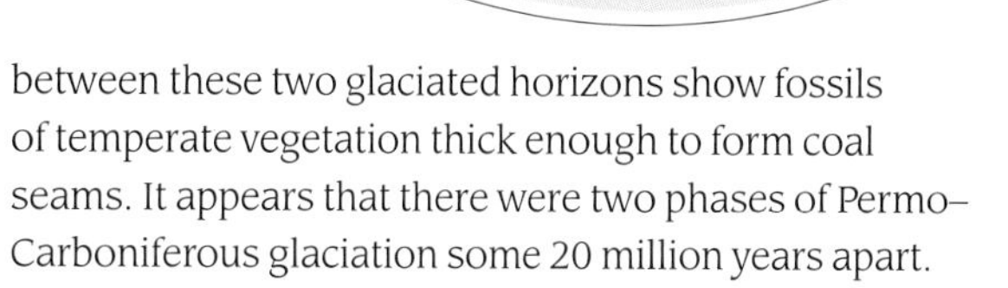

the winter, the meltwater flowing from it carries different grades of rocky material. This material builds up in lake bottoms as distinct annual layers (varves). The direction taken by the ice, as deduced from the erosional structures, suggested to the early surveyors that the source of the glaciation lay somewhere out in the Indian Ocean.

Similar beds were identified in South Australia in 1859, in South Africa in 1870, and in Brazil in1888. The glacial beds of Africa—the Dwyka tillite—are the most extensive. They occur in scattered exposures from Transvaal to the Cape of Good Hope, and from Namibia to Natal, with other occurrences as far north as Kenya. Again most of the evidence is from rock surfaces scraped and grooved by the passage of glaciers. There are even ancient U-shaped valleys filled with tillite. Some of the erratic boulders found in the tillite had evidently been transported over great distances from the north.

Another Carboniferous glacial bed was discovered in Brazil in 1888. These tillites and others found in South America were deposited by glaciers traveling from the east, from the Atlantic Ocean.

The evidence from Australia is more patchy. Several regions of glaciation are known, some of which date from late Carboniferous and others from the middle Permian. The lower Permian rocks between these two glaciated horizons show fossils of temperate vegetation thick enough to form coal seams. It appears that there were two phases of Permo–Carboniferous glaciation some 20 million years apart.

These discoveries suggested that the Earth must once have been covered in ice from the poles to the equator. However, further investigation showed that the Indian bed was the same age as tropical desert and forest beds in Europe, meaning that the ice sheet could not have been global. (The only Carboniferous glacial bed found in the far north lies in Alaska and is formed from mountain glaciers.) Geologists therefore concluded that the ice had covered only the southern hemisphere, spreading from the oceans to the continents and reaching north beyond the equator to central Africa and India.

Alfred Wegener's theory of continental drift challenged this view early in the 1900s. He built on the work of other geologists who had speculated on the existence of a supercontinent called Gondwanaland. It consisted of India, Africa, and South America, whose geological and biological similarities had already been noted by Charles Darwin and Alfred Russel Wallace. Wegener proposed that Gondwanaland had once been joined to Laurasia in the single supercontinent of Pangea. In either case, the ice would have been contiguous, and the Permo–Carboniferous glaciation would have been much more modest, affecting only the landmasses near the south pole.

All evidence pointed to this. In 1960, boulders in the youngest tillites of Brazil were proven to have originated in Namibia; Africa and South America must have been joined. Once this was known, there was no more mystery about the distribution of ice sheets.

5

6 Erratic rock is recognizable from a particular location; it is useful in determining the direction of flow
7 Till—beds of rocky material, eventually solidified to tillite

MESOSAURUS FOSSIL

This freshwater reptile has been found on both sides of the Atlantic, showing that South America and Africa were joined when it was alive.

ANOTHER line of evidence that indicated the former existence of Pangea was the distribution of land animals of Permian times. The spread of the drier conditions marked by the extensive red beds opened the way for the development of animals that could live full time on land.

Different continents usually have distinctly different land animals, but specimens of the same fossils are found in Permian rocks on all of the continents.

There were still many amphibians around, mostly in the wetter parts of the globe. One development was the evolution of amphibians with tough armored skin that were at home on dry land after their larval stage. Some of the larger Permian amphibians are difficult to tell from reptiles by the evidence of their fossils alone.

Reptiles began to flourish in the Permian. All modern lines developed from the small lizard-like forms of the Carboniferous; other kinds of reptiles became extinct.

Permian plants were essentially a continuation of the kind of flora that produced the coal swamps of the Late Carboniferous. Low-growing and bushy vegetation was supplied by the seed ferns, particularly the *Glossopteris* type characteristic of Gondwana. Giant horsetails still grew by the water. The gigantic clubmosses were still the significant trees, along with the cordaites, the early relatives of the conifers. True conifers evolved during the Permian, as did the cycads and ginkgoes, all of which were better adapted to dry conditions than the more primitive types. Although the dominant landscape was dry and lifeless, there were still local areas that were thickly vegetated. Significant coal seams were formed at this time, particularly on Gondwana after the ice ages; modern Zimbabwe and South Africa have mining industries based on Permian coal deposits. Due to the widely varying conditions, each region had a markedly different characteristic flora. The concept of such biogeographic provinces was first suggested by the two pioneering naturalists Charles Darwin and Alfred Russel Wallace.

ONE OF the most spectacular sequences of Permian marine fossils comes from the Delaware and Midland Basins of Texas. Sandstones, limestones, and shales built up in the deep waters of these basins while reefs grew around the edges.

At the edge of the Permian basins of Texas, life flourished in tropical reefs; more than 350 species have been identified.

Behind the reefs, between them and the desert landmass, lay shallow lagoons giving rise to thin deposits of limestones that gradually gave way to red-beds. In the early part of the Permian these basins were shallow and connected to the sea in the west. Water, rich in oxygen, was constantly flowing in and circulated to the basin floors. Later in the period the basins deepened; oxygenated water could not reach the depths, which became lifeless. The only fossils

THE CAPITAN REEF: A PERMIAN FOSSIL

The Capitan Limestone of Texas is part of the Permian reef structure. After 255 million years the landscape still reflects the topography of the time of formation. The low-lying area to the south, in the foreground, represents the depths of the Delaware Basin. The talus slope below the cliff is formed from the debris that broke from the reef rim and tumbled into the depths. The cliff is the reef itself, and the plateau behind is the remains of the lagoon.

THE LIVING REEF

The 465mi (750km) of reef that fringed the Permian basins of west Texas and New Mexico were mainly composed of colonies of bryozoans, spiny brachiopods, calcareous sponges, and green algae. The outcrop detail (below) shows calcareous sponges (lighter tan) hung upside down, attached to a frond-like bryozoan. Layers of encrusting organisms and microbial "goo" helped bind the framework of the reef together. Calcareous sediment (gray) and botryoidal aragonite, a marine cement, filled in the gaps.

1 Brachiopods
2 Glass sponges
3 Crinoids
4 Green algae
5 Ammonite
6 Spiny brachiopod
7 Calcareous sponge
8 Gastropod
9 Bryozoan
10 Stromatoporoid

known from the beds formed at this time are those of swimming and floating creatures—cephalopods and radiolarians—that lived in the oxygenated upper waters and sank to the stagnant floor when they died. Towards the end of the Permian the connection to the shallow western seas was cut off, and the basins filled with sediment. The water eventually evaporated, leaving great deposits of gypsum and salt baking in the desert sun.

When they existed, the Texas reefs must have been spectacular structures. Formed from the buildup of corals, sponges, and bryozoans, they arose 2000ft (600m) from the floor of the basin to the water's surface. Nearby lagoons were only a few feet deep. The chemical composition of the reef limestone has been altered to dolomite, and this process has destroyed much of the fine structure of the reef itself. However, many of the shells of the reef organisms have been replaced by tough silica, turning them into durable fossils.

Limestone structures produce such excellent oil traps that these reef complexes form the basis for the extensive modern petroleum industry of the Permian Basin of Texas. Their economic significance means that they have been thoroughly mapped and studied as geologists and engineers plot how to extract the trapped fuel.

THE DELAWARE BASIN

By the Permian period the seas that had covered much of the North American continent were in retreat but in west Texas marine conditions still prevailed. A shallow seaway lay between the Ancestral Rockies and the young Ouachita Mountains. Events associated with the uplift of these ranges created three deep basins only connected with the open ocean to the west by a narrow channel. They lay close to the Permian equator, in the rain shadow of the equatorial trade winds. The arid climate made for extensive evaporation and sediment deposition. The Midland Basin gradually became filled in but the reefs surrounding the Delaware Basin grew rapidly upwards, eventually rising above a basin 2000ft (600m) deep.

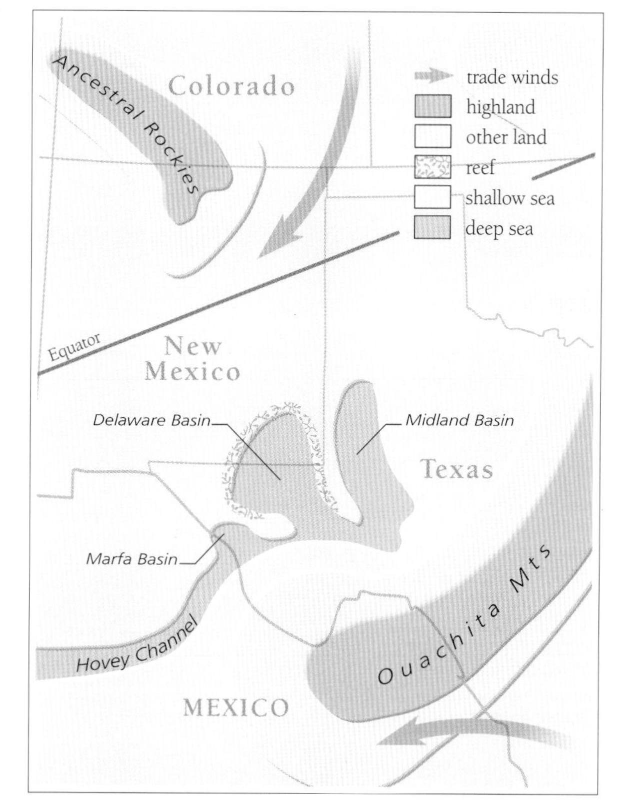

FOSSILS from the Permian are so dramatically different from those of the succeeding Triassic that it was recognized that the end of the Permian also marked the end of the Paleozoic era. This change was due to an unprecedented mass extinction of both plants and animals. The reason for this catastrophe has never become clear. It does not seem to have been an abrupt event; there is evidence that some groups of animals were declining for about ten million years in the late Permian, although many of the reef faunas appear to have thrived right up to the end.

At the end of the Permian, 96 percent of the world's species perished. This was the greatest mass extinction in the entire history of the Earth.

The continued falling of the sea level and the restriction of sea areas may have had an influence. There was a great deal of volcanic activity at the end of the period, particularly around the Ural Mountains, and this may have had an effect on the composition of the atmosphere—a factor implicated in other mass extinctions during Earth's history.

extent of Siberian traps
relative movement of Laurasia and Angaraland
highland
other land
shallow sea
deep sea
SIBERIA
ANGARALAND
Ural Mountains
KAZAKHSTAN
LAURASIA
sea area closing
EUROPE
TETHYS OCEAN

Short term cooling
Volcanic ash in atmosphere
EXTINCTION
Release of carbon dioxide
Eruption of Siberian traps
Flora and fauna poisoned by fluorine
Increase in carbon dioxide
Oceanic anoxia and global warming
EXTINCTION
Permian seafloor

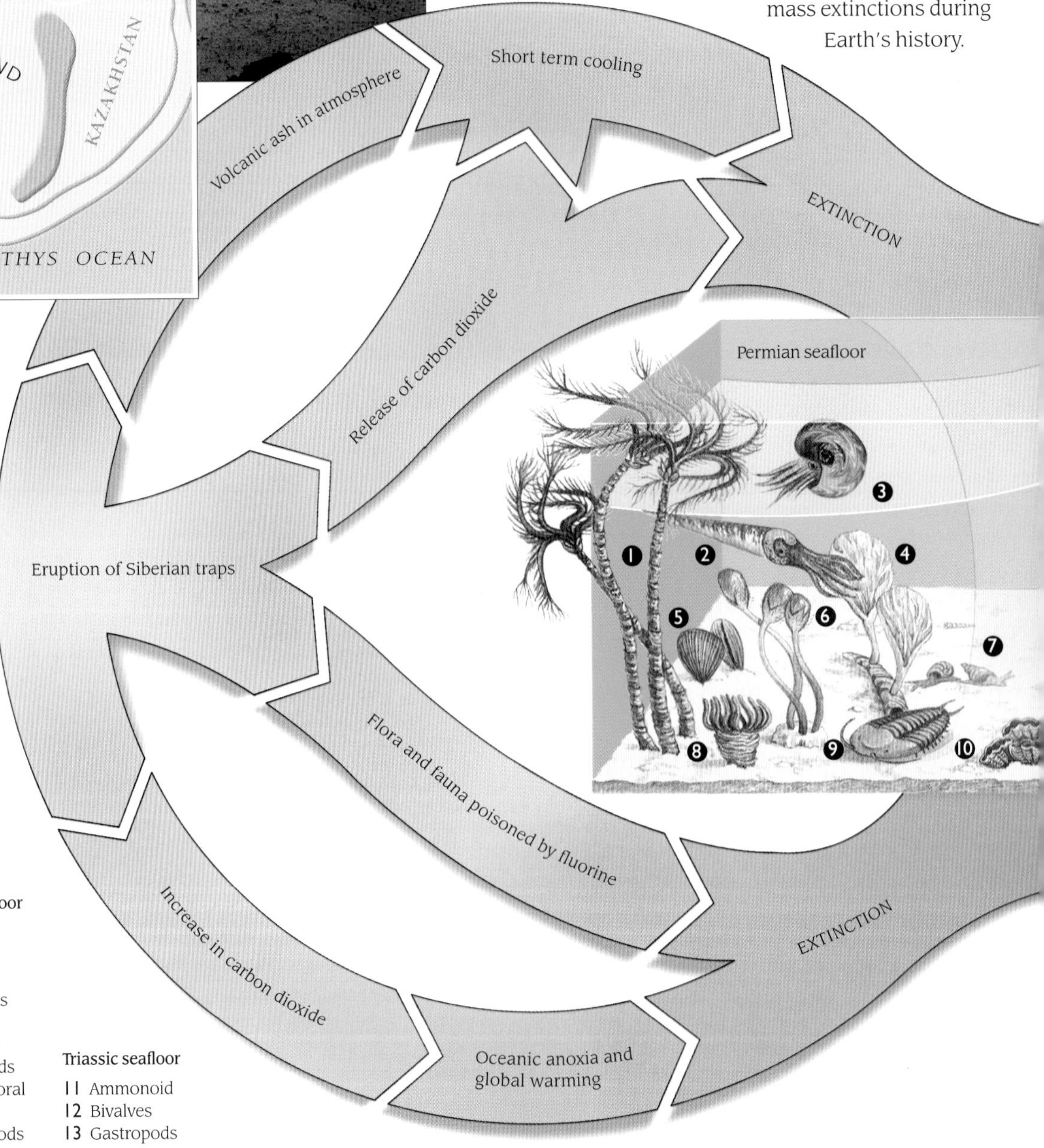

SIBERIAN TRAPS

A trap is a stair-like structure usually formed from successive flows of basalt, a runny lava that covers wide areas before solidifying. In the Late Permian a trap developed to the east of the Urals. This level of volcanic activity must have affected the composition of the atmosphere.

MARINE EXTINCTIONS

Permian fossil strata show typical Paleozoic animals. Crinoids, solitary corals, brachiopods, and bryozoans are common. The cephalopods are mostly straight-shelled nautiloids or tightly-coiled goniatites. The last of the trilobites crawl on the sea bed. Just before the Triassic boundary, the sea bed is almost empty, with only the gastropods and the bivalves present in any great numbers. Surviving cephalopods soon evolved into the ammonites.

Permian seafloor

1 Crinoids
2 Nautiloid
3 Goniatite
4 Bryozoans
5 Bivalves
6 Blastoids
7 Gastropods
8 Rugose coral
9 Trilobite
10 Brachiopods

Triassic seafloor

11 Ammonoid
12 Bivalves
13 Gastropods

CAUSES OF EXTINCTION

An extinction event is a complex network of interrelated factors. Intense volcanic eruptions could cause the climate to cool and poisonous gases to spread. Either of these environmental disruptions can wipe out whole species of plants and the animals that feed on them; in turn, the animals that prey on the plant-eaters also become extinct. A fall in sea level reduces the areas of the shelf seas and hence the diversity of marine habitats. It also leads to the release of carbon dioxide trapped in sediments, and to the oxidation of coal and other carbon-rich sediments that had previously been underwater. On land, the expansion of continental areas also causes a more extreme climate, to which the animals and plants may not adjust.

Studies show that the carbon dioxide content increased and the oxygen content decreased at the Permian–Triassic boundary. The lowering sea levels exposed beds of coal that had been building up on land earlier. This coal oxidized in the air, converting the atmospheric oxygen to carbon dioxide, causing a "greenhouse effect."

Before the Permian extinctions, marine life had been a rich and complex system of interlocking food chains for 100 million years. A typical Paleozoic seafloor contained a number of characteristic organisms. Many were stationary filter-feeders, eating the tiny organisms and scraps of organic matter swept towards them by the water current, and so were anchored to the sea bottom. There were more filter-feeders in those times than there are today. There were the crinoids or sea lilies, and blastoids; both stalked relatives of the starfish. The corals were mostly solitary corals, like sea anemones in shelly cups. Bryozoans, looking like encrusting mosses, spread over the sea bed. The shellfish were nearly all brachiopods, a group of animals that resembled modern bivalves but in fact proved to be no relation at all.

The extinction wiped these out, with only one type of crinoid and no blastoids at all surviving. The solitary corals disappeared, as did three quarters of the the bryozoans. Nearly all the 160 known species of brachiopods died out, with only a handful surviving to the present day. Their places were taken in the Triassic by the more familiar bivalves that had somehow managed to survive the extinction with the loss of only about ten percent of their number. The free-swimming predators of the Paleozoic were mostly cephalopods – shelly relatives of the octopus and squid. They became almost completely extinct but those that survived actually developed very quickly into the ammonite group that was to become such an important part of the Mesozoic marine fauna.

On land, the most significant group of Permian animals were the mammal-like reptiles such as the therapsids. These died out almost completely, although a few survived into the Triassic. Their continued evolution eventually produced all of today's mammal life.

Plants were equally affected by the Permian extinctions. The great conifer forests of Laurasia disappeared, not to return for another 5 million years. By comparison, in the extinctions that wiped out the dinosaurs at the end of the Cretaceous, the affected plants took only 10,000 to 100,000 years to return. Again, the relative scarcity of Permian fossils has made this subject difficult for geologists and paleobotanists to study in depth, but it is thought that it took 500,000 years for the northern forests to grow back. Conifers in Australia and Antarctica fared better due to their location near the pole.

Ecological instability
Habitat diversity reduced
EXTINCTION
Increase in climatic seasonality
Triassic seafloor
11
12
13
Sea level fall
Oxidation of organic carbon
Increase in carbon dioxide
EXTINCTION
Oceanic anoxia and global warming
Increase in carbon dioxide
Gas hydrate released

DECLINE IN OXYGEN

The ratio of carbon-13 to carbon-12 in the atmosphere declined abruptly in the Late Permian as changes in ocean circulation released carbon dioxide from deep-sea organic sediments rich in carbon-12. This led to "greenhouse" warming and anoxic marine conditions that are now revealed in uplifted rock sequences.

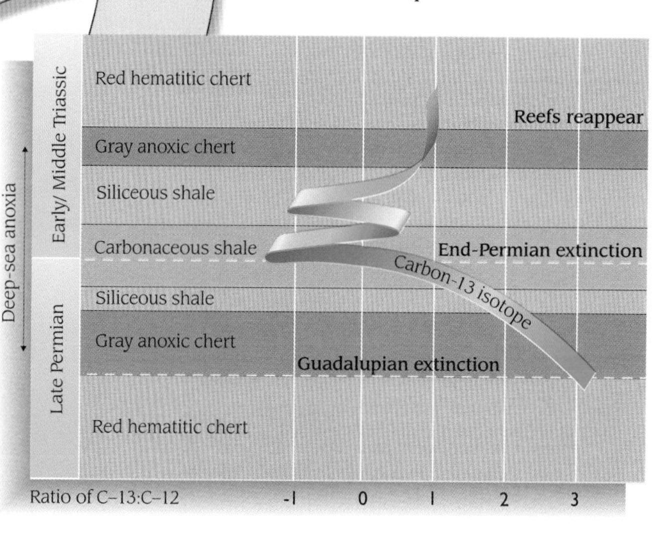

PERMIAN

THIS might have been the harsh landscape of the western foothills of the newly-risen Ural Mountains at the end of the Permian period—a landscape that would look, at first, as lifeless as anything from the Precambrian era. The sun would have scorched down on the hot rubbly slope of alluvial fans that reached from the eroding mountains down to the flatness of the salty desert plain. A hot dry wind sent specks of sand hissing against stony surfaces. Against the the white dazzle of the salt flats that stretched into the distance, the rusty-colored sand and pebbles looked dark.

At the end of the Permian period, plant and animal life were at the transition point between the Paleozoic and the Mesozoic.

It was not a particularly inviting environment. But there was life here. Nearby, at the foot of the slope, water that trickled down from the mountains, or welled up from the aquifers, gathered in temporary pools before leaching away into the distant salt lakes and evaporating. Where the water was able to stand for any length of time, an oasis developed. The oasis, consisting as it did of horsetail thickets and stands of tall straight lycopods and *Cordaites*, would have resembled a throwback to the great coal swamps that flourished in the Late Carboniferous period. At the feet of the trunks a bushy undergrowth of ferns spread wherever the ground was moist enough. However, there were more advanced plants here too. Away from the water grew the more recently evolved trees, such as conifers and ginkgos, that had developed a greater tolerance of the dry conditions.

Animal life could be detected by movement in the ferny thicket just before the appearance, perhaps, of a number of massive creatures that lumbered out of the shade. The big browsers of these oasis plants were the pareiasaurs—distant relatives of the turtles, though they hardly looked like it. Bodies like elephants, stumpy legs that stuck out sideways like reptiles', small spiky heads—these were *Scutosaurus*, among the biggest of the pareiasaur family. A small group of them might have emerged from the ferns and moved out along the slope towards the next stand of trees.

Then, from the shadow of a sand-blasted rock, appeared a sinister shape, a cross between a crocodile and a saber-toothed tiger. The mammal-like gorgonopsid *Sauroctonus*, feeling hungry, sniffed the air for prey. It caught the distant movement of the pareiasaur group. It cast its hunting eye on the harmless herbivores, lowered its forequarters into a posture of stealth, and stalked down the slope toward them.

1 *Sigillaria*, a lycopod
2 *Cordaites*
3 Ferns
4 *Walchia* (a conifer)
5 Horsetails
6 A ginkgo
7 *Scutosaurus*, a big pareiasaur
8 Skeleton of *Elginia*, a small pareiasaur
9 *Sauroctonus*, a gorgonopsid

PERMIAN

9

8

1
2
3
4
5
6
7

THE EVOLUTION OF MAMMAL-LIKE REPTILES

BEFORE the appearance of dinosaurs, another group of reptiles ruled the Earth. These evolved in late Carboniferous times, and by the Permian they had spread into every ecological niche in every part of the world. They were the mammal-like reptiles, which originally took the form of small undistinguished lizard-like animals. Before the end of the Carboniferous they had evolved into an order called the Pelycosauria.

A number of pelycosaur suborders appeared. Some of these retained the basic lizard-shape but others developed tall spines on their backbones which, in some families, supported a spectacular sail. *Dimetrodon* is perhaps the best-known of these pelycosaurs, and it was a meat-eater. *Edaphosaurus* was a plant-eating type. The sails may have been used for signaling to one another, but it is very likely that they also functioned as heat exchangers. In the morning, while the air was still cold, the pelycosaur could turn and hold its sail at right angles to the rays of the rising sun. The sail, being full of blood vessels, would absorb the heat and warm the animal's blood. In this way a pelycosaur could become active and go hunting early. In the heat of the desert midday, it could hold its sail into the wind and cool itself off. This primitive temperature regulation system anticipated the more sophisticated systems of these animals' descendants.

Pelycosaurs died out well before the late Permian extinction, but by then they had given rise to a more advanced mammal-like reptile order, the Therapsida. The several suborders of therapsids encompassed the basic lizard-like forms, burrowing mole-like forms, heavy browsing rhinoceros-like forms, and active hunting wolf-like forms. These flourished in the Permian times but, like the rest of the animal kingdom, were badly hit by the end-Permian extinction.

The two therapsid suborders that survived into the Triassic and flourished until the end were the dicynodonts, browsing animals that ranged from rabbit-like beasts to hippopotamus-like beasts, and the hunting wolf-like cynodonts. As time went on the latter became more and more mammal-like, developing a dentition consisting of specialized nipping, killing, and shearing

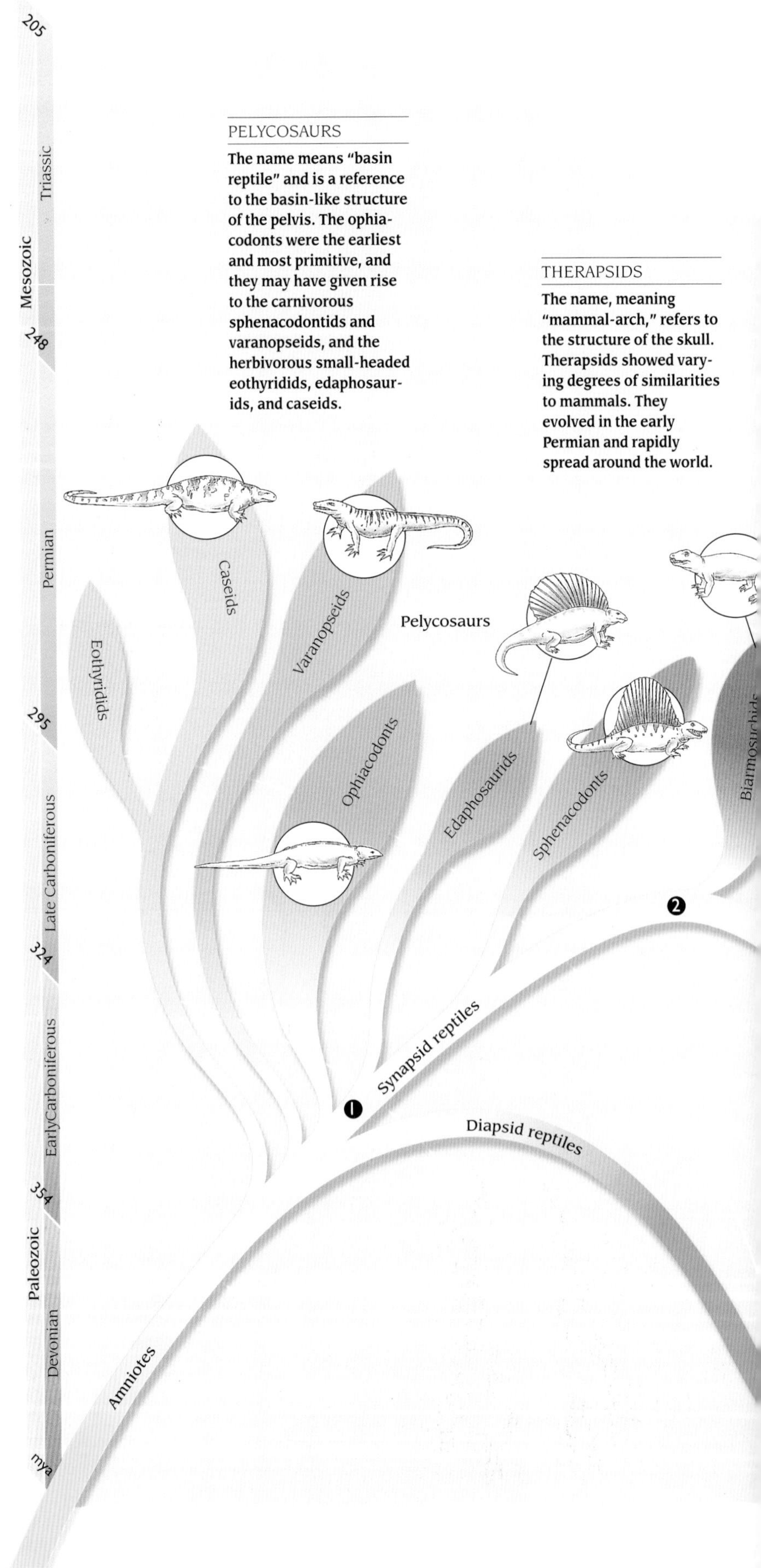

PELYCOSAURS

The name means "basin reptile" and is a reference to the basin-like structure of the pelvis. The ophiacodonts were the earliest and most primitive, and they may have given rise to the carnivorous sphenacodontids and varanopseids, and the herbivorous small-headed eothyridids, edaphosaurids, and caseids.

THERAPSIDS

The name, meaning "mammal-arch," refers to the structure of the skull. Therapsids showed varying degrees of similarities to mammals. They evolved in the early Permian and rapidly spread around the world.

SUCCESS STORY

The mammal-like reptiles are the first great success among reptiles. They developed early, spread quickly into all kinds of habitats around the world, and survived for 70 million years. When they finally died out, they left descendants that eventually evolved into the most important animal group on Earth.

1 Specialized teeth appear for slashing, chewing, shearing
2 Limb girdles appear that allow for a semi-erect posture
3 Mammal-like jaw musculature develops
4 Mammal-like tooth arrangement and palate develop
5 Furry skin probably evolves to support warm-bloodedness

PLACERIAS FOSSIL

This dicynodont has a body and legs like those of a herbivorous mammal, and a very specialized browsing tooth structure.

teeth, producing hairy coats, suckling their young, and adopting mammal-like stances on straight legs. Eventually they gave rise to the mammals themselves, a group that continued after the ancestral species had all died out.

At what point did cynodonts cease to be reptiles and become true mammals? The jaw mechanism is the crucial factor. A reptile's jawbone consists of several bones, the biggest being the dentary (lower jaw) that holds the teeth. A mammal's jawbone has only the dentary, and the other bones have been incorporated in the structure of the ear: the hammer, anvil, and stirrup that form the chain of little bones between the eardrum and the inner ear. This defines the true mammal.

Class	Subclass	Order
	ANAPSIDA (no temporal openings)	PELYCOSAURIA (sail-backed mammal-like reptiles)
REPTILIA	SYNAPSIDA (one temporal opening)	
	DIAPSIDA (two temporal openings)	THERAPSIDA (mammal-like reptiles with differentiated teeth, mammals)
	extinct	

PERMIAN

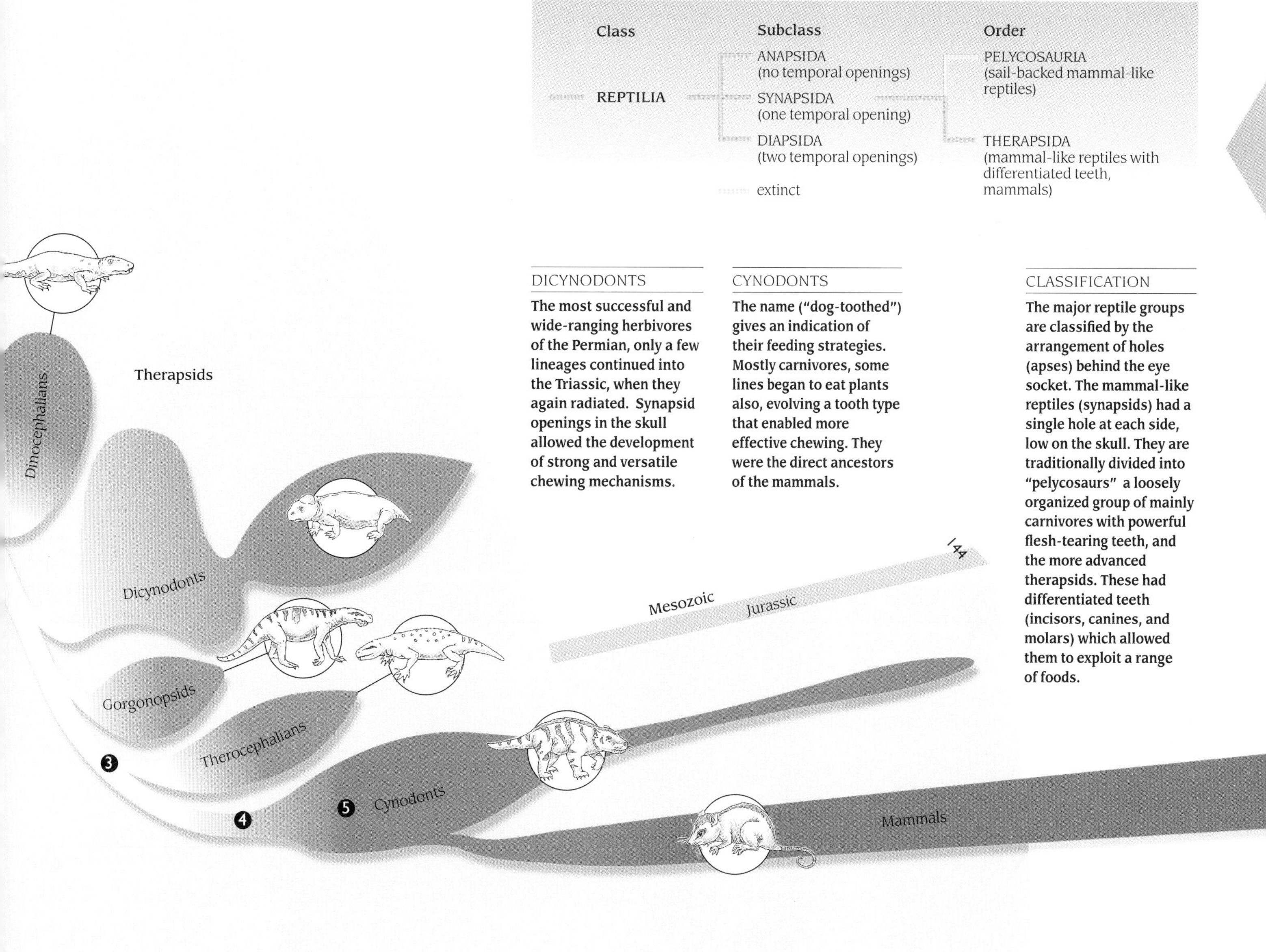

DICYNODONTS

The most successful and wide-ranging herbivores of the Permian, only a few lineages continued into the Triassic, when they again radiated. Synapsid openings in the skull allowed the development of strong and versatile chewing mechanisms.

CYNODONTS

The name ("dog-toothed") gives an indication of their feeding strategies. Mostly carnivores, some lines began to eat plants also, evolving a tooth type that enabled more effective chewing. They were the direct ancestors of the mammals.

CLASSIFICATION

The major reptile groups are classified by the arrangement of holes (apses) behind the eye socket. The mammal-like reptiles (synapsids) had a single hole at each side, low on the skull. They are traditionally divided into "pelycosaurs" a loosely organized group of mainly carnivores with powerful flesh-tearing teeth, and the more advanced therapsids. These had differentiated teeth (incisors, canines, and molars) which allowed them to exploit a range of foods.

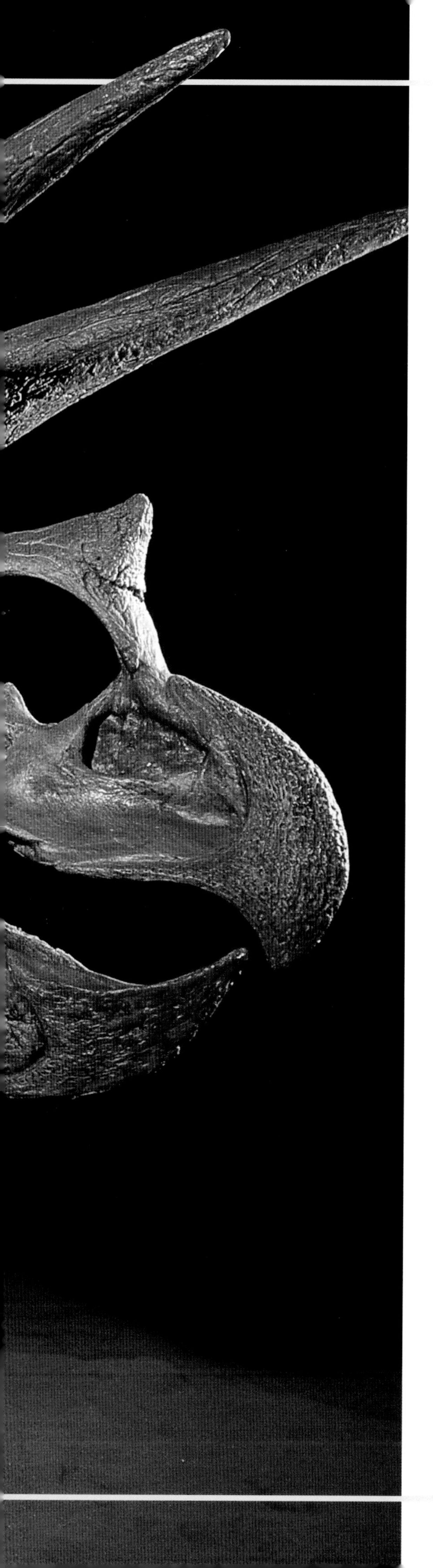

Reptiles take over the Earth

PART 4

THE MESOZOIC

248–65 MILLION YEARS AGO

MIDDLE LIFE is the meaning of the name "Mesozoic." It signifies the transition, marked by the Permian mass extinctions, between the distinctly ancient life forms of the Paleozoic and the more advanced (though not fully modern) forms that followed. During the 160 million years of the Mesozoic, both plants and animals underwent dramatic transformations. In the seas, modern reef-building corals and large marine reptiles made their first appearance; on land, gymnosperm (seed-bearing) plants continued to predominate, and birds and mammals were now present. But the animals most widely associated with the Mesozoic were the dinosaurs.

These extraordinary creatures have exerted a fascination on the general public since their first fossils were discovered in the nineteenth century; and so the Mesozoic is the ancient era that receives more attention in the popular press than any other. But it is typical to see dinosaurs represented in an anachronistic jumble: a late Cretaceous carnivorous dinosaur from North America attacking a late Jurassic herbivorous dinosaur from East Africa, out of scale with each other, knee-deep in a vegetation that never existed anywhere, against a background of jagged mountain peaks and bubbling volcanoes. This is the paleontological equivalent of showing a historic battle with a Sherman tank in desert camouflage fighting a Sumerian war chariot on a glacier, and the chariot is pulled by horses 33ft (10m) tall.

Like every other plant and animal that ever existed, dinosaurs lived in distinctive communities in their own specific ecosystems. The geography of the time dictated where they lived and what their lifestyles were.

EARLY IN THE ERA, during the Triassic period, all the land areas of the globe formed the single supercontinent Pangea. This was so vast that most of it was far from the modifying influence of the ocean, and temperatures over the interior soon began to rise. The heat was sufficient to melt glaciers in Africa, Australia, South America, and India, all of which had been clustered around the South Pole. The general warming trend was assisted by the presence of shallow, epicontinental seas, and the Triassic climate of inland regions became extreme. Only around the edges of the continent could life exist comfortably. With the unification of the landmass, animal populations from disparate regions became widespread: the small meat-eating dinosaurs that lived in what was to become Arizona were the same as those living in what was to become Zimbabwe; long-necked plant-eaters in South Africa were similar to those in Germany. Plants, too, migrated with relative freedom. Great forests grew up, which were dominated by cycads, conifers, and ginkgoes—all gymnosperms (seed-bearing plants).

Pangea began to break up during the Jurassic period. Rift valleys spread across the vast landmass as the Earth's tectonic plates moved the continents, and newly-forming oceans insinuated themselves between individual land components. These spreading oceans, and the shallow seas that lapped over the low-lying areas, brought more equable climates to most of the world. Most of the landmass was still joined together, however, and so it is possible to find similar fossil faunas from the Jurassic as far apart as Tanzania and Wyoming. It was in the more isolated places that different populations of flora and fauna began to appear.

The dinosaurs had a marked effect on their environment. Plant-eaters of the Triassic period fed on conifer trees, and these conifers soon developed knife-like leaves to minimize the damage done by them. In the Jurassic the biggest of the plant-eaters migrated seasonally from one feeding ground to another, wherever food was to be found, and the meat-eaters followed them. Upland woodlands of the Cretaceous period supported the duckbilled dinosaurs and armoured dinosaurs, cropping the newly evolved herbivorous undergrowth, while on the lowlands, where archaic conifers and cycad-like plants still existed, the horned dinosaurs and the last of the long-necks thrived.

As the landmasses of the Mesozoic moved together and then apart, forming and then disbanding the super-continent of Pangea, the mighty dinosaurs appeared, radiated, and dominated—before a catastrophe of global proportions caused them to vanish forever.

DURING THE CRETACEOUS period, most of the modern continents became individual landmasses, and plant and animal life began to reflect this. Long-necked plant-eating dinosaurs continued to thrive in South America because it was an island, whereas in North America, the duckbilled dinosaurs were taking over from the long-necks. However, North America was still attached to Asia and so there were similar populations of duckbills and armored dinosaurs in these places. Then there were small islands, where animals tended to develop small body sizes in response to the limited quantities of food found there. Dwarf armored dinosaurs and tiny duckbilled forms existed in the archipelagos that stretched across the shallow seas of the eastern European region at that time.

The breakup of Pangea brought about the recognizable shapes of the modern continents and developed the oceans in the patterns that are familiar to modern geographers. On the other side of the world from these scattering continents, the super-ocean of Panthalassa gradually shrank as the surrounding continents impinged upon it. It was developing into the Pacific Ocean, the greatest of the modern oceans, with its tectonically active "Ring of Fire" marking the borders of plate margins.

The Mesozoic ended as it began, with a mass extinction that marked a transition to an era of distinctly new life forms: the hitherto insignificant mammals. Undisputed predators of their world—or, in the case of the plant-eaters, often too large to be preyed upon—the dinosaurs did not survive into the modern age. It is possible that they had already begun to decline, but they disappeared after a catastrophic event at the end of the Cretaceous. Buried under the sediment of millions of years, they would only be discovered and remembered by those other great predators who eventually succeeded them: humans.

Dougal Dixon

THE TRIASSIC

248 – 205 MILLION YEARS AGO

*P*ANGEA *dominated every aspect of global conditions in the 40-million-year span of the Triassic. A hot, arid climate over much of the great landmass combined with the continued erosion of the Paleozoic mountain chains to deposit more red sandstone. Although sea levels were generally low, the supercontinent was almost cut in half along an east-west axis by a huge embayment called the Tethys Seaway. Around the edges of the water the climate was more temperate, creating marginal regions with enough rainfall to produce vigorous forests along river banks.*

The animal and plant life of the Triassic represents a slow recovery from the mass extinction at the end of the Permian. Clubmosses and seed ferns, which dominated the late Paleozoic, were giving way to conifers. Bivalves succeeded brachiopods as the most important shellfish, and ammonites spread into the niches vacated by extinct types of cephalopods. It was the dawn of the Age of Reptiles, an astonishingly diverse group. Mammal-like reptiles declined, but other forms flourished, grew huge, and became the most fearsome predators of the age—as were their descendants, the dinosaurs.

THE TRIASSIC is the first of the three periods of the Mesozoic. The term "Trias" was first used by Friedrich August von Alberti in 1834 to refer to the three distinct sequences of rocks that outcrop in Germany. Before that it was merely the upper part of the New Red Sandstone. Alberti's threefold sequence had been known in medieval times because of the rocks' importance as sources of building stones and of chemicals such as salt and gypsum. The sequences were the Bunter sandstones, the Muschelkalk limestones, and the Keuper marls. Since this classification was based on localized rocks derived from sediments laid down in deserts with only the Muschelkalk (the "mussel chalk") containing any fossils, this classification has never been useful in worldwide stratigraphy.

The terrigenous rocks of the Triassic are similar to those of the Permian, and the two are often grouped together as the New Red Sandstone.

KEYWORDS

AMMONITE
BRECCIA
CONIFER
DINOSAUR
DREIKANTER
DUNE
MAMMAL
MANICOUAGAN EVENT
NEW RED SANDSTONE
REPTILE
SEED FERN
TETHYS SEAWAY

Low sea levels at the end of the Permian caused a lack of marine successions—hence the difficulties of correlating rocks worldwide, and the problems with the precise dating of the Permian–Triassic boundary. The sea level remained low during the transition to the Triassic period but it began to rise steadily from then on. Towards the end of the Triassic the sea level was generally high, with a great deal of fluctuation. As a consequence there are few marine rocks from early in the system but many more later. Correlation improves as the period progresses.

Despite the parochialism of Alberti's "Trias," the term "Triassic" appeared in Charles Lyell's stratigraphic classification of 1872, and has been used ever since, although "Trias" is still used informally by geologists.

Marine Triassic rocks are known from the Alps of southern Austria and the first attempt at a classification of the period based on marine fossils was done there between the 1860s and 1920s. However, the stages of the Triassic based on these rocks proved to be just as problematic, as the successions were incomplete and the

PERMIAN	248 mya	245	240	235	TRIASSIC	230
Period	EARLY/LOWER (BUNTSANDSTEIN)		MIDDLE (MUSCHELKALK)			
General stages	INDUAN	OLENEKIAN	ANISIAN		LADINIAN	
	Pangea — fusion of all continents					
Geological events		Docking of Sonoma terrane with western N. America			Deposition of sands and evapori	
Climate	Tropical aridity; Pangea has strongly seasonal climate					
Sea level						
Plant life	Lycopods dominant		Dicroidium seed ferns dominant			
Animal life	Slow recovery after the end-Permian extinction	• First hexacorals		Development of all kinds of reptiles		

precise stratigraphic position of the individual beds was difficult to establish because of the intense deformation in the mountain range. Since then, far better marine Triassic exposures have been found in Siberia, in China, and in the western mountains of North America. A detailed Triassic stratigraphy is now known, based on the rapidly-evolving ammonite fauna of the oceans.

THE BURGEONING ammonite population was an indirect result of the extinction of other swimming animals during the end-Permian extinction. The nautiloids and goniatites were the main swimming cephalopods up until then, but suddenly they became extinct. Their surviving relatives, the ammonites, were able to diversify into all the niches left by them. It was not only the ammonites that spread everywhere; the base of the Triassic series is marked by the beginning of the slow recovery from the end-Permian extinction. All kinds of animals had become extinct, and the survivors were beginning to take over.

The Triassic begins with the recovery of life from one mass extinction, and ends with its disruption in another.

Some of these were forerunners of the mammals, as preserved in rocks of the Karoo basin at the southern tip of Africa. From late Carboniferous times to the late Triassic, this basin was gradually filled with deposits from swamps, lakes, and rivers in which a huge range of fossils were preserved: thousands of plant, fish, amphibian, and reptile species, as well as mammal-like reptiles, which evolved into mammals late in the Triassic. The Karoo sequence, which consisted mostly of silt and sandstone, was buried under more than 3300ft (1000m) of basalt lava at the end of the Triassic and was only discovered in the 1840s. This great outflow of lava may have been caused by rifting as Africa began to pull away from South America and Antarctica, eventually breaking up the ancient continent of Gondwanaland.

Another extinction marks the end of the Triassic period. At this time about 33 percent of sea animals, 32 percent of land-based vertebrates, and 90 percent of the land plant species died out. The nature of the plant species that survived and the anatomy of their leaves suggest that temperatures suddenly rose at the end of the period and instigated the biological disruption. This was probably due to some kind of greenhouse effect but the exact cause is still a mystery. Some experts have suggested that a meteorite impact at this time, similar to the one that may have taken place at the end of the Cretaceous period, was responsible. One discovery in support of this theory is the presence of shocked quartzes—crystals that show the stresses and deformations produced by a meteorite impact—at the top of the Triassic system in the Tuscany region of Italy. However, there is no evidence of a buildup of iridium at the time, and such a buildup in the rocks is often taken as a sign of a meteorite impact.

A large meteorite did strike the Earth late in the Triassic. An impact known as the Manicouagan Event left one of the world's largest known meteorite craters in Quebec, Canada. This crater has a diameter of about 62mi (100km) and was probably caused by a lump of rock about 37mi (60km) in diameter. It has proved difficult to find an exact date for the Manicouagan Event, but it seems to have taken place about ten million years before the end of the Triassic. This episode may have initiated the collapse of fauna, as towards the end of the period different groups of animals were becoming extinct at different times, and different groups were coming into being at different times. Instead of a single extinction there may have been a gradual turnover of animals for about 20 million years.

Other evidence suggests, however, that the extinction was sudden, taking less than a

See Also

THE PERMIAN: *New Red Sandstone, Mammal-like Reptiles, Mass Extinction*
THE JURASSIC: *Ammonites, Dinosaurs, Pangea Begins to Break Up*
THE PALEOGENE: *Mammals*

INTO THE MESOZOIC

The Mesozoic era began with the Triassic period. During the 43 million years of the Triassic, from 248 to 205 million years ago, the supercontinent of Pangea was finally completed, and the sea level gradually rose from the low level it occupied at the very end of the Permian. These environmental changes had a profound impact on life. Both terrestrial and marine ecosystems became more complex and diverse, with many new life forms appearing: swimming and flying reptiles, turtles and frogs, coniferous plants, and ginkgoes.

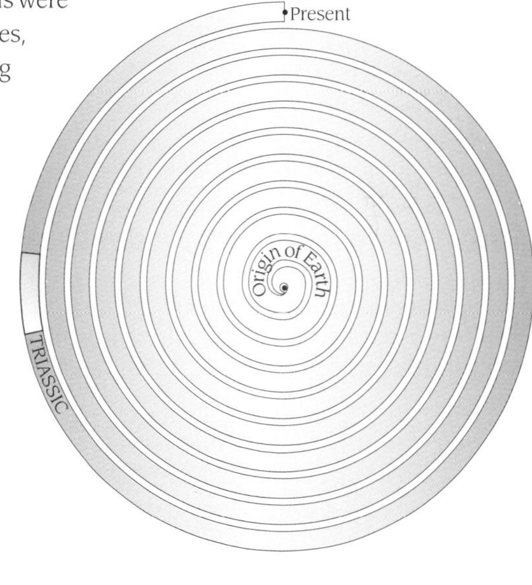

TRIASSIC

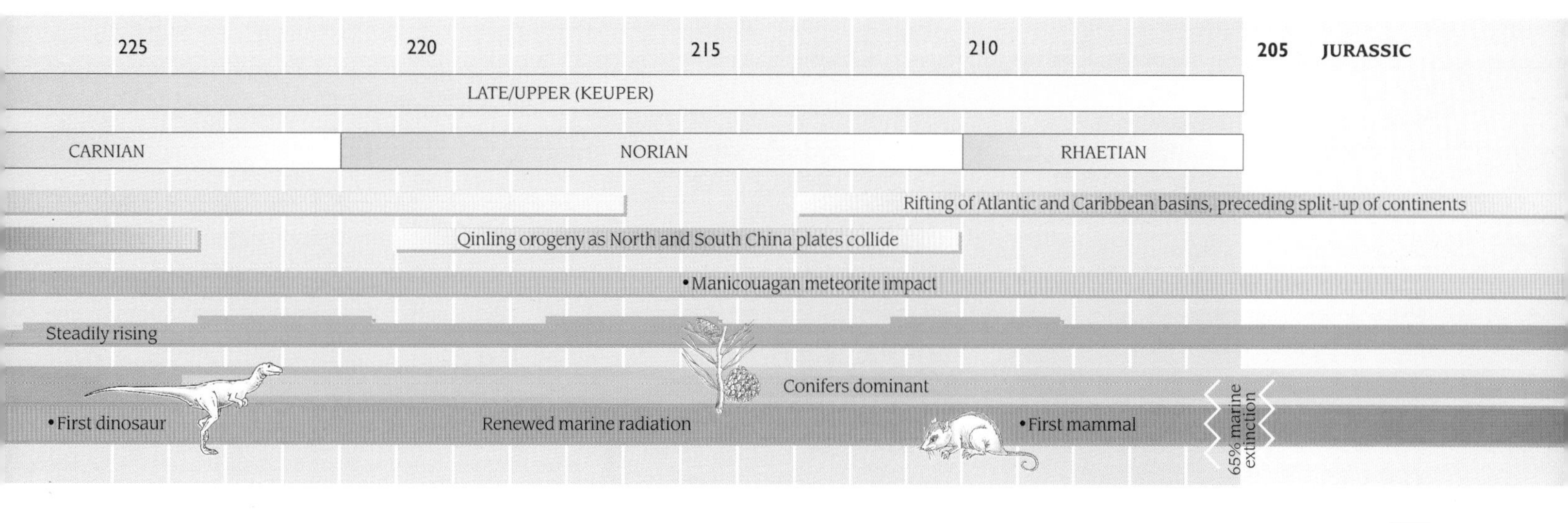

million years, and that the Manicouagan Event was far too early to have had an influence. A smaller mass extinction did take place about 15 million years before the end of the period, and perhaps the Manicouagan Event triggered this one.

AT THE START of the Triassic Pangea was a single, huge continent—an situation that is proven by the occurrence of fossils of Triassic land-living vertebrate animals in all today's separate continents. However, this was a very temporary state. No sooner was the supercontinent assembled than it began to fragment again. Even before parts of eastern Asia had accreted to the mainland, the edges of Gondwana were beginning to break up and drift northwards. The actual splitting apart of Pangea would not really begin until the following Jurassic period, but by the end of the Triassic the indications of future fracturing began to make themselves apparent. Cracks and faults began to shear their way through new mountain ranges and ancient cratons. In the Appalachians, rift valleys appeared and began to fill with sediment and salt lakes. Another rift valley subsided through the Caledonian mountains in the north, between what is now Scotland and Norway, producing the structure that is now the basis for the oilfields of the North Sea. A long narrow sea, resembling the modern Red Sea in structure and appearance, cracked into Gondwana, beginning to separate the continental pieces that later became India and Africa.

If the Paleozoic is characterized by the continental collisions that led to the formation of Pangea, the Mesozoic was the era when it began to break up.

Pangea was shaped like a giant C, with a northern section consisting of Laurasia and the eastern segments, and a southern section consisting of Gondwana. Their fusion, during the Permian, formed a giant seam, with the entire east coast of modern North America and parts of southern Europe attached to South America and northern Africa. Scattered islands in the region of what is now China and the Far East had joined with the eastern edge of Laurasia, to complete the supercontinent. All land was concentrated on one side of the globe, the other side consisting of the Panthalassa Ocean. Between the northern and southern regions lay a huge bay known as the Tethys Seaway. This body of water had a modifying effect on the geography and the climate of the landmasses. The ocean current in the Tethys would have been counterclockwise, as gauged from the Coriolis forces generated by the turning of the Earth. Warm waters would have swept northwest along the north coast of Gondwana bringing moist winds to that edge of the continent, and returning cooler waters would have lowered the temperatures of the southern edge of Laurasia. The interiors, isolated by mountains, were desert-like: arid, hot in the summer, but bitterly cold in the winter. Both northern and southern hemispheres probably had monsoon seasons.

The broad continental shelf that spread along the northern margin of the Tethys produced the shelf sediments that are today found in Italy and Poland, and those in the Austrian Alps that contained the marine fossils from which the Triassic was dated. Eastwards from the area now occupied by the Black Sea were subduction zones with ocean trenches and island arcs, showing that the Tethys was evolving. To the west, where Europe, North America, and Africa met, the Tethys narrowed and shallowed, and became confined between mountains. There it formed inlets and gulfs that dried out, depositing limestone and evaporites.

ALONG the join between Laurasia and Gondwana, the recently-formed Appalachian mountains still stood high but were beginning to erode, forming intermontane basins that filled up with red sandstone. Rivers flowed west from these mountains, across southwestern North America, to reach the ocean on the west coast. The Ural Mountains of Eurasia were still new in the Triassic, and there continued to be the eruption of basaltic lavas to the east of them.

The mountain chains that had been so vast at the end of the Permian were beginning to erode into gentle Triassic hills.

The Gondwana part of Pangea remained more or less the same as it had been, with desert areas in the center and active mountains along the western and southern edges. Here and there were areas of inland drainage, into which the seasonal rivers flowed producing intermittent inland (playa) lakes, something like Lake Eyre in modern Australia.

As in Permian times, Gondwana was so large that rainfall rarely reached the center. Harsh desert conditions existed here. Seasonal rainfall would have occurred

THE RED EARTH

During the Triassic, most of Pangea was so far from the sea that little moisture could reach it. Oxidation of iron minerals in the soil gave the soil a distinctive red color.

PANTHALASSA OCEAN

Africa and Middle East
Antarctica
Australia and New Guinea
Central Asia
Europe
India
North America
South America
Southeast Asia
other land

TRIASSIC

NORTHERN CONTINENTS

Northern Pangea was still dominated by mountains where Europe, North America, and Africa met, but these were now eroding, and rift basins were beginning to appear.

Ural Mountains
LAURASIA
Caledonian Mts
Manicouagan Crater
palachian Mountains
TETHYS SEAWAY
PANTHALASSA OCEAN
PANGEA
GONDWANA

SOUTHERN CONTINENTS

Gondwana remained a stable continent: ancient cratons surrounded by old mountains, with fresh fold mountains around its southern edges.

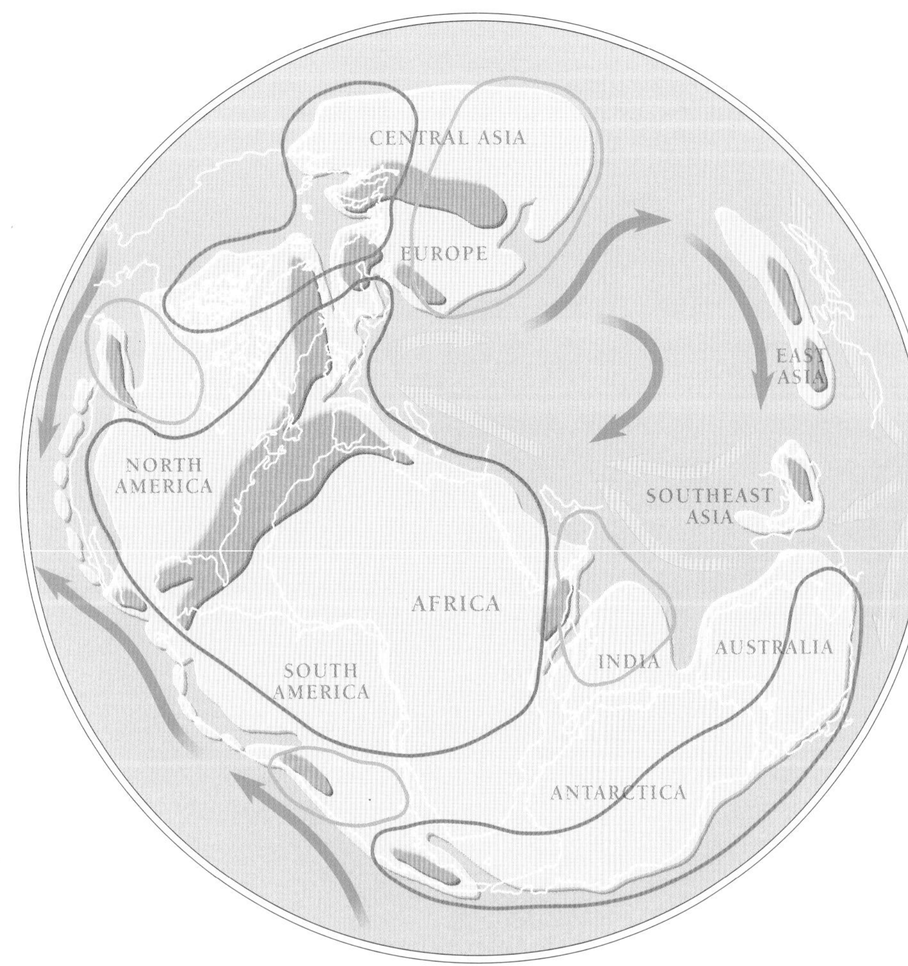

climate
- dry
- seasonal rain
- cool, rainy
- warm ocean current
- cold ocean current
- highland
- other land

around the edges, depending on whether or not the prevailing winds blew in from the oceans. At this time there was water at both poles: an epicontinental sea covered the North Pole, while the South Pole lay in the Panthalassa Ocean.

THE VAST desert landscapes of the Triassic period produced very distinctive deposits and characteristic rock types. Perhaps the most familiar type of desert feature is the sand dune. In fact, sandy surfaces cover only about 20 percent of modern deserts, the rest being gravel plains or naked rocky outcrops. In the Late Permian and Early Triassic the proportions would probably have been similar. In dry areas the wind can pick up a small grain of sand and fling it about, wearing away rocky outcrops and generating more sand in the process. Bare rocks can become polished and sculpted into strange shapes, with most erosion taking place close to the ground where sand particles bounce along in the wind. The sand itself may accumulate in certain places, and be pushed over long distances across the desert in wind-driven dunes. Sandstones formed from desert sands often show that the original beds were laid down as sand dunes.

A microscopic view of a Triassic sandstone will show that it is made up of spherical grains, showing that it was deposited by winds in a desert environment. The grains may have a coating of red iron oxide around them.

Dune bedding consists of curved beds, concave upwards, representing subsequent slip faces of ancient sand dunes. The orientation of these dune-bedded sandstones can be used to work out the prevailing winds of the time. Those found in the Permian–Triassic strata of central England and Scotland indicate that the prevailing winds were blowing from the east during the Permian and Triassic, while those found in the southwestern United States show that the prevailing winds were blowing from the northwest.

Gravel desert is largely the result of deposition by alluvial fans. As mountains erode, broken fragments are washed down towards low-lying lands by temporary or permanent streams. In desert areas the streams flow only during the wet season, and usually only for short distances. Broken eroded material accumulates where the streams reach flat lands and disappear. Heavy material is normally deposited first while the finer fragments are carried further, building up in broad low semi-circular accumulations along the edges of the slopes as alluvial fans. Alluvial fans dating from ancient times can be identified in desert sandstones by their alternating beds of coarse rubble and fine fragments. The individual grains are jagged and angular, showing that they have not traveled very far. Rock formed from large angular fragments is called "breccia."

Out on the open plains of a desert, large fragments of rock are abraded by the sand grains whirled around by the force of the relentless wind, and the sand grains themselves are worn into rounded shapes. Anyone who picks up a rock lying on the surface of a desert plain will see that its windward side has been "polished" in this manner until it is smooth. Often the erosion on such a rock is so great that it becomes unbalanced on the exposed side and topples over, exposing another side to the force of the wind and sand. Such a rock has three polished sides and is called a "dreikanter"; its presence is another indicator of desert conditions.

PATTERN OF RAINFALL

The present-day movement of the atmosphere—a result of the Earth's rotation and of the distribution of land and sea—can be used to construct the likely climatic pattern of the Earth in Triassic times. Most of Pangea had a year-round dry climate, its edges had seasonal rainfall, and the high latitudes were cool and rainy. This ties in with the geographical occurrence of climate-sensitive Triassic deposits such as redbeds and evaporites. The flora followed a similar distribution, with the heat-adapted plants of the Euramerican flora in the moister areas, Siberian flora in the far north, and Gondwanan flora in the far south.

FOOTPRINT MYSTERY

One crucial piece of information that cannot always be determined from a fossil footprint is what kind of animal left it. Because of this, paleoichnologists (who specialize in fossil footprints, or ichnites) give scientific names to the footprints themselves. *Chirotherium*, a common Triassic fossil footprint, is a series of sets of hand-like prints: big prints from the hind foot, and small prints from the fore. The strange thing about the arrangement of toes is that the little digit that looks as if a thumb is on the outside of the track rather than on the inside. It probably was not a thumb but a specialized fifth finger.

Because the author of the *Chirotherium* tracks was not known, this led to all kinds of speculation and fanciful reconstructions of what the creature might have been like, including one model arranged as if it had walked cross-legged. When the skeleton of the rauisuchid reptile *Ticinosuchus* was subsequently found in Triassic rocks in Switzerland, its small fifth toe fitted the "thumb print" well. It is now widely believed by paleoichnologists that this animal was the one that produced the *Chirotherium* prints.

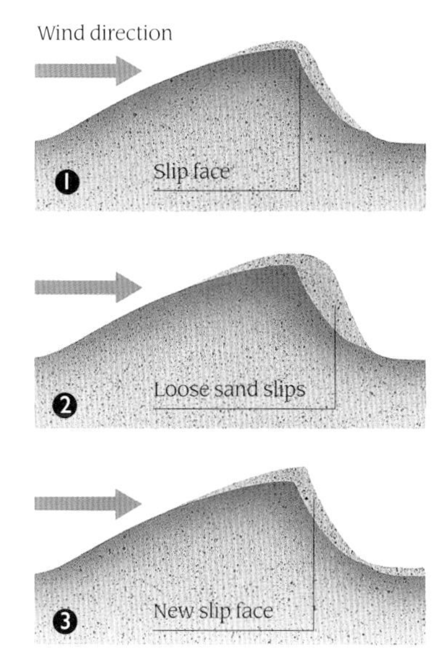

DUNE FORMATION

Wind drives loose sand across the desert surface as dunes. The movement of dunes is distinctive. Each grain exposed on the windward face is blown up the slope (1) and will drop into the lee, or the "slip" face (2). As this happens to every grain, the dune advances in the direction of the wind (3). Trains of dunes follow one another across the desert in the direction of the wind. Often this takes place on different scales, with huge dunes like low hills having smaller dunes across them, criss-crossed by ripple marks.

IN DESERT areas that are subject to periodic flooding and drying, the muddy material that gathers on the beds of playa lakes dries out in the arid season. The layers of dried mud shrink and crack into polygonal slabs. The next layer of mud to be deposited will fill the cracks, forming casts of them. When such patterns of mud cracks are found in the rocks, either as actual cracks or as casts of the infilling material, they can be used to interpret the climatic conditions of the time. When the rains come after a dry period, the pits formed by the raindrops in the dried mud are sometimes preserved in the rock.

The soil on the surface of Mars must resemble that on much of the Earth in Triassic times. The photographs sent back by the Mariner and subsequent probes showed red sand and dreikanters.

Playa lakes dry out from time to time, and any dissolved mineral that has been washed down into the lake's waters will concentrate, precipitate, and accumulate on the dry lake bottom. Beds of halite, anhydrite, and gypsum are often found in desert sandstones.

TRIASSIC

DUNE DESERTS

Sedimentary rocks are usually formed underwater, where sediment settles. Landscapes are mostly areas of erosion rather than deposition. Dune deserts are different, accumulating masses of sandy material which may become buried and solidify into desert sandstone. Such a sandstone often has the slope of the dunes preserved within the bedding.

A valuable source of information from red sandstone beds is provided by the tracks of animals that lived in the area. The study of fossil footprints is called "ichnology" and is a very important part of paleontology. Fossil footprints are much more common than the fossils of the actual animals—each animal will produce millions of footprints in its life but only one skeleton—and are much more useful as an interpretation of lifestyles and environmental conditions. Footprints reveal whether animals traveled singly or in herds, at what time of year they were active, whether they moved slowly or quickly. Unfortunately, they do not usually tell which animal made them. The first dinosaur tracks known were discovered in Triassic sandstones in the American state of Connecticut in 1802. Because of their appearance they were originally thought to have been formed by very large birds from ancient times; dinosaurs had not yet been identified.

THE FLORA of the Permian had largely been similar to that of the coal forests, consisting of ferns, seed ferns, and giant lycopods, the clubmosses. Plant life across Gondwana particularly had been dominated by the seed fern *Glossopteris*. At the end of the Permian the *Glossopteris* flora died out. With it went the majority of the herbivorous mammal-like reptiles that fed on it. Afterwards, in the Early Triassic, the clubmosses became briefly dominant once more and then, in Middle Triassic times, a long-lived flora based on the seed fern *Dicroidium* developed. It was the dicynodonts, the herbivorous group of mammal-like reptiles, that suffered at the end of the Permian, while the cynodonts, the carnivorous group, survived and flourished into the Triassic. When the *Dicroidium* flora appeared, the cynodonts gave rise to herbivorous types that exploited it. At the same time another group of herbivorous reptiles appeared, flourishing on the newly-developed floras. These were the rhynchosaurs, a group of pig-like reptiles with broad jaws carrying tusks that could dig up the plants and chopping teeth that could break them down. Rhynchosaur remains are found in Triassic rocks all over the world, from Brazil to Great Britain, from Tanzania to North America.

Although the landforms and geographies had not changed much since the Permian period, the plant and animal life developed spectacularly.

Towards the end of the Triassic the *Dicroidium* flora died out and was replaced by the first flora dominated by conifers, which had been around since late Paleozoic times. Now, in the Early Mesozoic, they became the most significant type of plant.

The plants that abounded at the end of the Triassic were those with with few stomata, the pores in the leaves that are used for gas exchange. This suggests a high carbon dioxide level in the atmosphere, which in turn indicates a warmer climate. Another trend was that broad-leaved plants like seed ferns were replaced by narrow-leaved plants like conifers. The narrow leaves were better for temperature control in hot conditions.

This turnaround in the plant life had a profound influence on the animal life. The rhynchosaurs and the

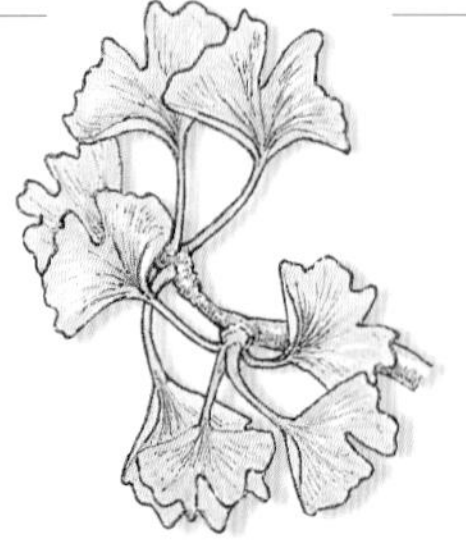

FLORAL SUCCESSION

The rapid turnaround of plant types during the Triassic eventually produced a flora that remained stable for much of the Mesozoic. The modern ginkgo tree (above) is almost unchanged from the ginkgoes that existed in Triassic times. The most famous occurrence of Triassic plants, from the Petrified Forest in Arizona (below), consists of the fossilized trunks of monkeypuzzle trees that are very similar to modern types.

3
5
1
4

MARINE REPTILES

Many new types of marine reptiles evolved in Triassic times, including those adapted to living along the shores of the Tethys. The long neck of *Tanystropheus* (background) may have been used for dipping for shellfish in rock pools. The long jaws and sharp teeth of *Nothosaurus* (foreground and skeleton below) were ideal for catching swimming fish.

plant-eating mammal-like reptiles—the most important animal groups on Earth—suddenly became extinct, probably because they were denied their *Dicroidium* diet. It was time for a new kind of animal to come to the fore.

IT WAS the meat-eating mammal-like reptiles, rather than the plant-eating types, that survived the Permian–Triassic extinction. As well as giving rise to a new group of plant-eating mammal-like reptiles at the beginning of the Triassic they also gave rise to a new group of animals at the end, just before their own extinction. These were the mammals themselves. However important the mammals were in the overall scheme of the evolution of life on earth, they were not the animals that were to dominate the land for the next 160 million years. Another group of meat-eaters had also survived the Permian–Triassic mass extinction. These were the crocodile-like thecodonts, which began to flourish at the end of the Triassic. These animals eventually became the dinosaurs.

Once ocean life had re-established itself after the end-Permian extinction, there was plenty of food in the sea to inspire the evolution of sea-living reptiles from land-living ancestors.

The evolution of the dinosaurs and of the conifers seems to have gone hand in hand. The first plant-eating dinosaurs were the prosauropods, long-necked animals that could reach up into trees and scrape the needles from the branches. The conifer trees that were most common at that time were the araucarias, the modern monkeypuzzles. These had broad needles with edges like knives—probably a defense against over-browsing by the tall dinosaurs.

In other fields the reptiles really took over in the Triassic. As well as being the principal plant-eating and meat-eating land animals, they also returned to the sea. Only a few tens of millions of years after they had evolved into land-living animals they had recolonized the realm of their ancestry. All kinds of different reptile lineages produced water-living forms. The placodonts were walrus-like reptiles that plucked shellfish from the sea bed and crushed them in broad flat teeth. Some placodonts were even covered with armor, which made them look somewhat like the turtles that also evolved a little later in the Triassic. The nothosaurs were long-necked, needle-toothed, fish-eating creatures that were the forerunners of the plesiosaurs to come. The most well-adapted sea reptiles of all, the dolphin-shaped ichthyosaurs, also evolved at the end of the Triassic.

Reptiles even took to the air. A number of unrelated lizard-like animals, such as *Icarosaurus* and *Coelurosauravus*, independently developed gliding mechanisms that consisted of webs of skin stretched across projections from the ribcage. These primitive kite-like animals were quite short-lived, but there were more successful flying animals around. Towards the end of the period the first pterosaurs appeared. These had wings attached to the forelimbs and were able to fly with a sophisticated flapping action. So successful were these that they were the animals that dominated the skies for the remainder of the entire Mesozoic era.

THE FIRST MAMMAL

The earliest mammals evolved at the end of the Triassic. They were tiny shrew-like or opossum-like animals such as *Megazostrodon* (left). For the next 160 million years they would scamper insignificantly around the feet of the great dinosaurs.

Fossils of the Molento and Lower Elliot Formations of South Africa reveal the life of the Triassic.

SILVERY braided streams wound down the almost imperceptible slope from the distant mountains, merging with one another and diverging once more, producing a broad plain of muddy islands and sandbars. Along the edges of sluggish waters stood thick reedbeds of horsetails, with thickets of fan-leaved ferns on the emerged spits themselves. On the more permanent stream banks were forests of conifers and ginkgoes, with a ferny undergrowth. Desert landscape lay beyond the narrow belts of riparian forest.

This area of Gondwana in the Late Triassic period is now known as the Karoo region of South Africa. Now it is high plateau land. Then, however, it was a sedimentary depression, where streams ran down from the mountains along the southern edge of the supercontinent, laying down typical redbeds as they went, and eventually gathering and evaporating in a region of inland drainage in the heart of the southern continent.

Where vegetation grew, animals flourished. It was the beginning of the age of dinosaurs, and for the next 160 million years these animals dominated the land. However, the dinosaurs here were far from dominant. Soft mud at the edge of a stream trapped unwary drinkers such as an early prosauropod, a forerunner of the big plant-eaters. Its noises of panic alerted hunters and scavengers in the forest. Cruising through the shallow waters came the broad shapes of 13ft (4m) long amphibians, the main aquatic predators of the Late Paleozoic. On land the first hunter at this scene was a rauisuchid, like a land-living crocodile, also to disappear soon. A group of small theropod dinosaurs trotted up behind to scavenge. Their descendants would do the hunting in later times.

1 *Ginkgophytopsis* (fan-leaved fern)
2 *Equisteum* (horsetail)
3 *Rissikia* (coniferous tree)
4 Capitosaurid amphibian
5 *Euskelosaurus* (prosauropod dinosaur)
6 *Basutodon* (rauisuchid reptile)
7 Theropod dinosaur
8 Dragonflies

6

7

8

2

TRIASSIC

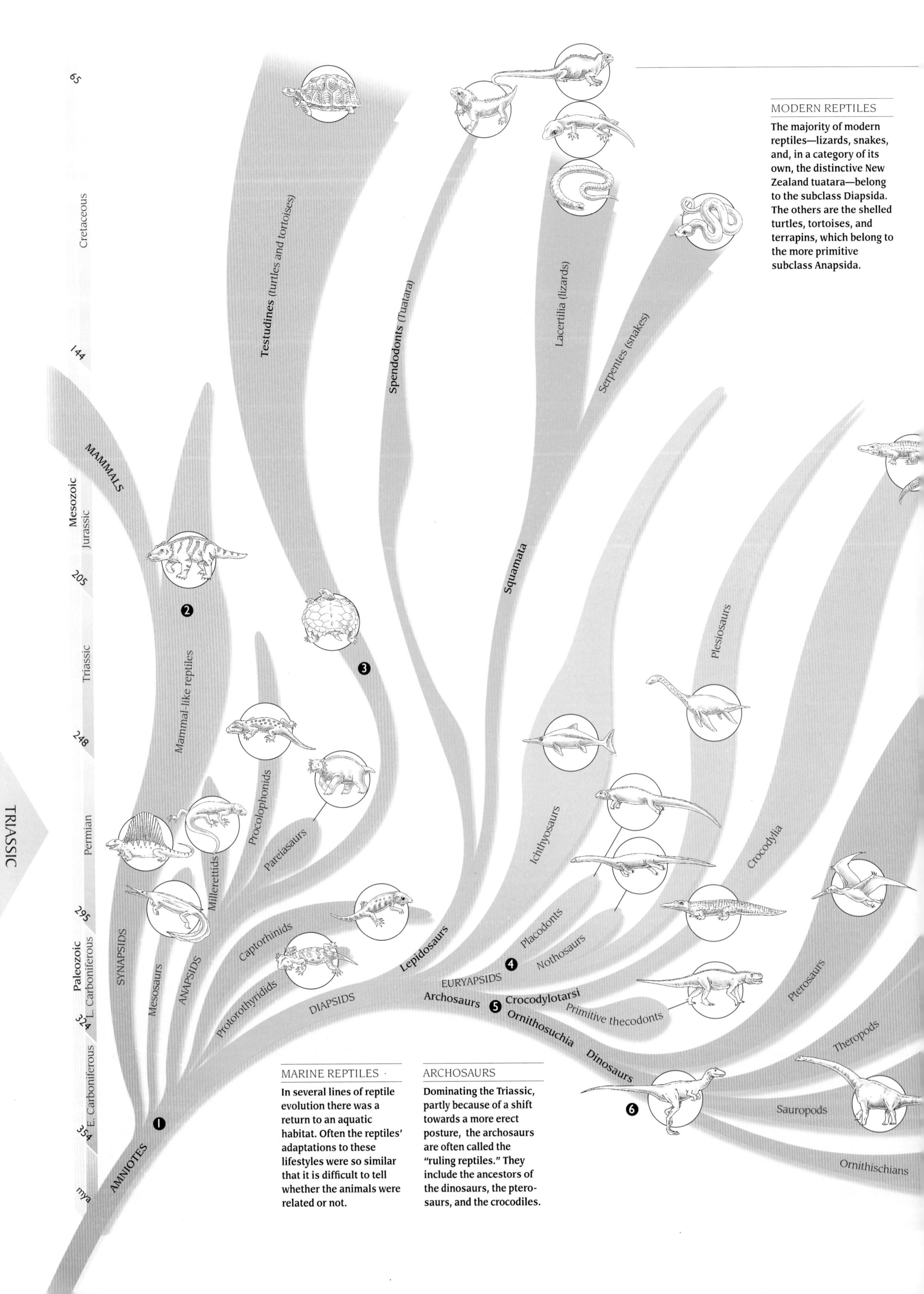

MODERN REPTILES

The majority of modern reptiles—lizards, snakes, and, in a category of its own, the distinctive New Zealand tuatara—belong to the subclass Diapsida. The others are the shelled turtles, tortoises, and terrapins, which belong to the more primitive subclass Anapsida.

MARINE REPTILES

In several lines of reptile evolution there was a return to an aquatic habitat. Often the reptiles' adaptations to these lifestyles were so similar that it is difficult to tell whether the animals were related or not.

ARCHOSAURS

Dominating the Triassic, partly because of a shift towards a more erect posture, the archosaurs are often called the "ruling reptiles." They include the ancestors of the dinosaurs, the pterosaurs, and the crocodiles.

THE EVOLUTION OF REPTILES

REPTILES evolved from the amphibians in the late Carboniferous and dominated the Earth for more than 220 million years, producing some of the most amazing life-forms the planet has ever hosted. They evolved to take advantage of the newly forested land habitats of the Carboniferous, with its new range of insects and plants on which to feed. Amphibians had begun the transition, but reptiles were the first vertebrate group to live entirely on dry land without going through an aquatic larval stage.

They achieved this by the evolution of the waterproof egg, in which the individual young developed in its own self-contained pond. The aquatic larval stage of the amphibian was a mechanical necessity, because a tiny embryo would not have been able to support itself on land. Even if it could have, the ratio of its surface area to its mass would have meant that it dried out too easily. This restriction was also overcome by having early development take place in a protective egg.

The reptile egg contains a yolk that nourishes the embryo, and a structure called the allantois that stores the waste products. The living matter is surrounded by a membrane called the amnion that contains the fluid but enables gases to pass through. A hard or leathery outer shell provides protection. These important developments distinguished reptile eggs from those of fish and amphibians, and gave the reptiles a great advantage in reproductive ability.

WHOSE BABY?

It is almost impossible to match a fossilized egg to the animal that laid it. This egg was probably laid by a sauropod dinosaur, but it has also been suggested that it was laid by a big emu-like bird that lived at the time. It has its own species name, *Dughioolithus siruguei*.

Reptiles are coldblooded, although warmbloodedness developed independently in distinct evolutionary lines among their descendants. They have dry scaly skin and their eggs are fertilized inside the females. After they appeared, about 320 million years ago, they quickly developed into the main groups that are known from the fossil record. The criterion used for the classification of the reptiles is the structure of the skull, specifically the arrangement of gaps between the bones immediately behind the eyes. The functional significance of these gaps involves the attachment of the jaw muscles.

Despite the fact that the reptiles were able to live on land, many groups returned to an aquatic existence from time to time in their evolutionary history. In at least one group, the ichthyosaurs, this even involved abandoning the egg stage and adopting a reproductive system in which the young were born alive. This is an advantage in northern climates in which eggs are vulnerable to the cold, and most modern reptiles bear their young live.

REPTILE GROUPS

The most primitive subclass is the Anapsida, which have no skull openings (apses) behind the eyes. Turtles and tortoises are the only survivors of the Permian anapsids. The Synapsida encompass the mammal-like reptiles, such as pelycosaurs, which gave rise to the mammals at the end of the Triassic. The Euryapsida, which are now extinct, contain most of the successful marine reptiles of the Mesozoic: placodonts, ichthyosaurs, and, possibly, the long-necked nothosaurs and plesiosaurs. The Diapsida include the dinosaurs and most of the modern reptiles.

AVES (birds)

REPTILE SKULLS

Arched recesses (apses) behind the eye sockets distinguish the four subclasses of reptiles. The Anapsida have no apse; Synapsida have one large central one; Diapsida have two apses; and Euryapsida have an upper one.

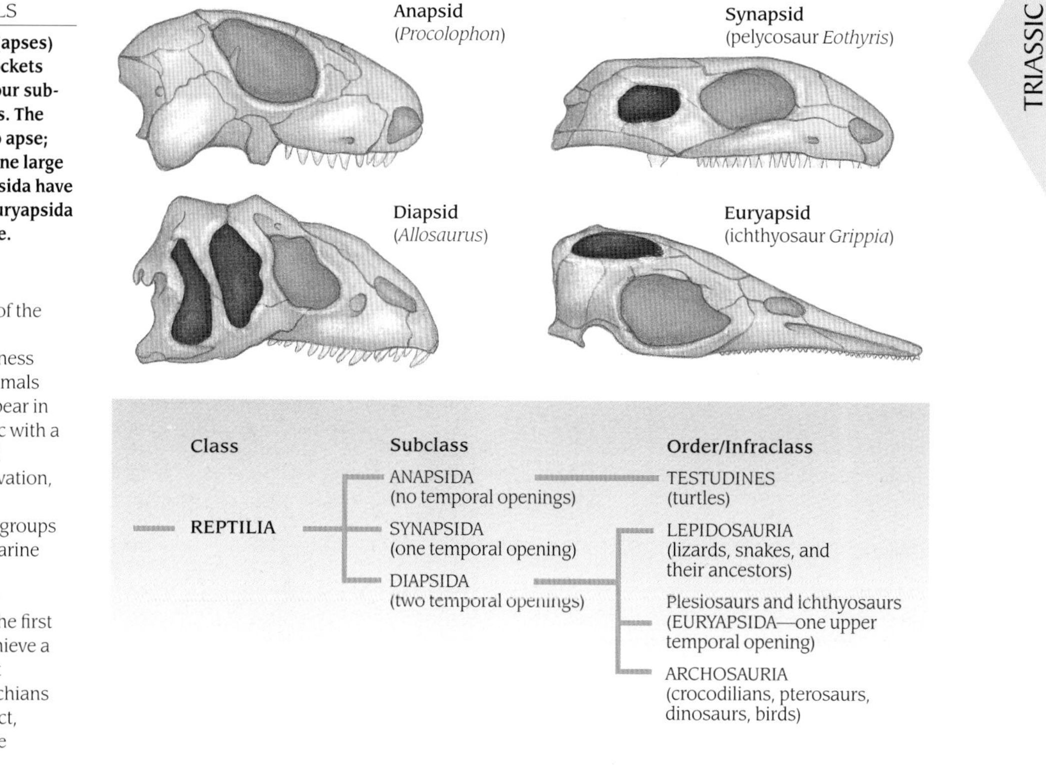

1 Development of the amniote egg
2 Warmbloodedness presages mammals
3 The turtles appear in the late Triassic with a successful and enduring innovation, the shell
4 The euryapsid groups readapt to a marine environment
5 The archosaur *Euparkeria* is the first tetrapod to achieve a semi-erect gait
6 The ornithosuchians develop an erect, bipedal posture

THE JURASSIC

205 – 144 MILLION YEARS AGO

BY THE beginning of the Mesozoic Era the great supercontinent of Pangea had finally been completed, but it soon began to rift apart again, breaking into the landmasses that would ultimately become today's continents. This did not occur all at once but in stages over 150 million years or so, continuing into the succeeding Cenozoic era. As the continents separated, water poured into the gaps between them, forming the familiar oceans of today's globe. The sea level stood high during the Jurassic, producing humid tropical climates over the formerly arid interiors of even the large continents. Most fossils of the time are of marine animals such as ammonites. They were so abundant and rapidly-evolving that they are the basis for the correlation of the entire Jurassic period. In the minds of the public, however, the Mesozoic is firmly established as the heyday of the Age of Reptiles. In marine deposits there are the remains of huge swimming reptiles such as plesiosaurs and ichthyosaurs. Land deposits of the Jurassic are much less common, but they contain the most fascinating reptile fossils of all: those of that incomparable and diverse group, the dinosaurs.

THE JURASSIC system, although it was not known as such at the time, was the first stratigraphic unit to be mapped and classified in a formal manner. The English engineer William "Strata" Smith, when building waterways across southern and central England in the late eighteenth century, noticed that the different layers of rocks that his workers were excavating contained different sets of fossils. Smith realized that a particular assemblage of fossils could be used to identify a particular layer of rock wherever it occurred. He was the first to use what was to become known as the "law of faunal succession"—a concept that has remained one of the cornerstones of stratigraphy to the present day. Between 1797 and 1815 Smith compiled the first geological maps, in which the outcrops of different ages were identified by different colors. It was the Jurassic system in which all this work was done.

As early as 1795, the distinctive limestone of the Jurassic was recognized in both Switzerland and Britain.

KEYWORDS

ANGIOSPERM
ARCHAEOPTERYX
CALCITE
FOOD WEB
GRABEN
GYMNOSPERM
HEXACORAL
OOLITE
ORNITHISCHIAN
RIFT VALLEY
SAURISCHIAN
TELEOST
TRIPLE JUNCTION

The name "Jurassic" comes from the Jura Mountains in northern Switzerland. Alexander von Humboldt published the term "Jura Kalkstein" (limestone from the Jura) in 1795 to describe the limestones outcropping there. The rock sequences which were to become classified as the Jurassic were originally known by quarrymen as the "lias" and the "oolite." The lias—a Dorset quarryman's term meaning "layers"—consisted of alternating beds of limestone and clay, and this outcropped across England from the south to the northeast. The oolite consisted of thickly-bedded limestones, excellent for use in building, which lay above the lias sweeping across the country along the same line; it was referred to as the Oolitic Series by Conybeare and Phillips, who identified and named the Carboniferous system. In 1829 the French geologist Alexandre de Brogniart used "Terraines

JURASSIC

TRIASSIC	205 mya	200	195 JURASSIC 190	185	180	
Series	LOWER (LIAS)					
General stages	HETTANGIAN	SINEMURIAN	PLIENSBACHIAN	TOARCIAN	AALENIAN	BAJOCI
Geological events	Pangea begins to break up		Extensive rift valleys between N. America and Africa and Gulf of Mexico			
Climate	Continuing warm					
Sea level	Shallow, rising					
Plant life	Seed ferns disappear			Gymnosperm flora dominate, especially cycads		
Animal life	Hexacorals form large reefs	•First frog	Heyday of the dinosaurs		•Last therapsids	

SEAFLOOR SPREADING AND THE AGE OF OCEANS

The oldest part of the ocean floor has been dated at approximately 180 million years, which means that it formed during the Jurassic Period. Deep-sea research during the 1960s showed that ocean crust is youngest at ocean ridges, where new crust is continuously formed at a rate of about 1.3 square miles (3.5 square kilometers) per year, and oldest near the margins of continents. This is the result of seafloor spreading, a mechanism proposed by the American geologist Harry Hess in 1960. The ocean floor is produced at ocean ridges where molten lava wells up from the Earth's mantle. The lava reacts with the cold ocean water to form ocean crust, which is denser and heavier than continental crust. A conveyor-belt action carries the crust sideways away from the ridge towards the continents on either side. As it cools, it contracts and sinks, forming deep basins layered with sediment. This theory explains the relatively young age of the ocean floor, and why there are mountains under the sea. The great Mid-Atlantic Ridge, which follows a curving north-south axis from the Arctic to the Antarctic, began to form in the Early Jurassic when the supercontinent of Pangea began to "unzip" along a fracture between present-day North America and Africa—where the ancient continents of Laurentia and Gondwana had joined.

Jurassiques" for outcrops of oolite in mainland Europe. Within Europe the Jurassic system falls easily into three distinct parts recognized today: the Lias, equivalent to the Lias of the early quarrymen; the Dogger, equivalent to the lower part of the Oolite; and the Malm, which is essentially the middle and upper Oolite. Between 1842 and 1849, other European geologists identified eleven stages within these parts, each named after the place where it was first identified.

The great wealth and variation of marine fossils in the whole Jurassic system meant that its biostratigraphy was quickly established. The zone fossils that were used were mostly ammonites, that evolved quickly and spread over wide areas. Each stage in the Jurassic succession has several ammonite zones, and the zonation that is widely used today was established in 1946.

The beginning of the period is marked by the recovery from the end-Triassic mass extinction. A narrow sequence called the Rhaetic lies at the boundary. Throughout the history of geology there has been debate as to whether the Rhaetic is Triassic or Jurassic, but current classifications put it at the end of the Triassic. Similarly, there is a debate about the Purbeck that lies at the top of the Jurassic strata in Dorset in southern England. It is a freshwater deposit and so has no marine fossils to allow a proper zonation, but it is currently believed to straddle the boundary between Jurassic and Cretaceous.

THE JURASSIC period was characterized by a gentle tropical climate over much of the globe, allowing an ever greater diversity of life to flourish both on land and in the water. New species of phytoplankton appeared in the oceans, and new corals formed extensive tropical reefs. It was an age of gymnosperms—seed plants that lack flowers—such as cycads, conifers, and ginkgoes. Seed ferns died out during the period, but the new gymnosperms continued to flourish into the present.

Dinosaurs, though among the largest and most successful, were not the only significant species in the Jurassic; others were plankton, pine trees, frogs, birds, and the first of the modern fishes.

In addition, the Jurassic was the Age of Reptiles, when these varied animals dominated the land, the water, and even the air. The first frogs and modern fish made an appearance, along with the first known bird, *Archaeopteryx*, a curious specimen that shared many features with its famous relatives, the dinosaurs.

THE MIDDLE OF THE MESOZOIC ERA

The Jurassic period, in the middle of the Mesozoic era, lasted approximately 61 million years—twice as long as the Triassic period, but not quite as long as the Cretaceous. The Jurassic was a time of particularly swift change effected by the breakup of the single supercontinent Pangea and the resulting development of new oceans across the globe. During this time the stage was set for the physical geography of the world to assume its present contours. Familiar life forms, from birds and frogs to pine trees, also appeared for the first time.

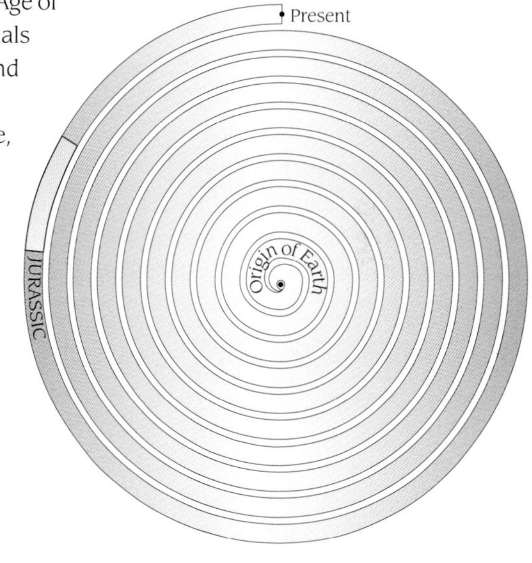

See Also

THE ORIGIN AND NATURE OF EARTH: *Crust, Mantle, Oceans*
THE ARCHEAN: *The Precambrian Seafloor*
THE EARLY CARBONIFEROUS: *Limestone*
THE TRIASSIC: *Pangea, Reptiles*
THE CRETACEOUS: *Dinosaurs and Mass Extinction, Flowering Plants*

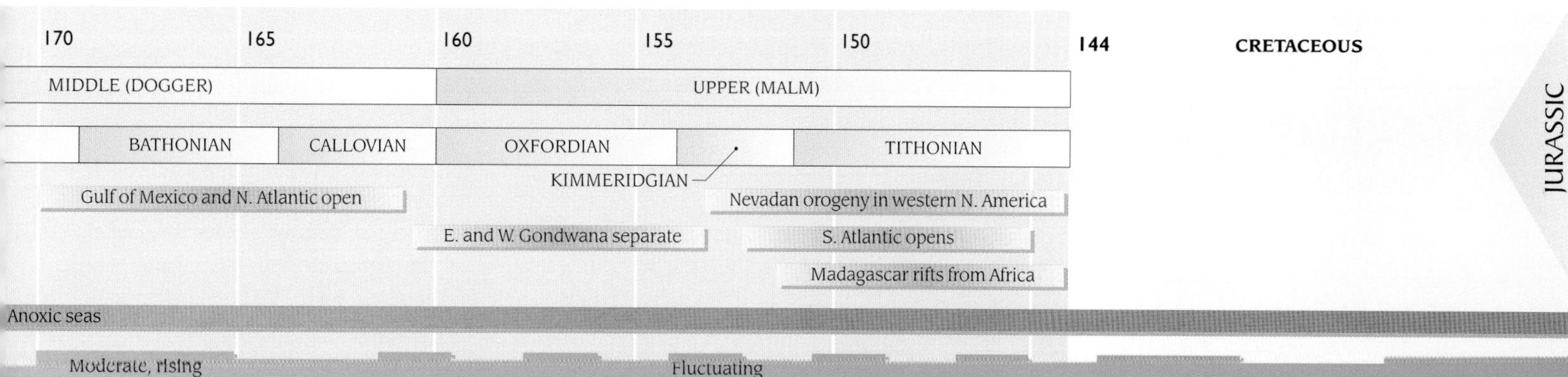

PANGEA was beginning to break up. Even by the end of the Triassic the strains were beginning to be felt as the forces that had brought the continents together now began to move them apart. Rift valleys appeared, crisscrossing the supercontinent—the first manifestation of its disassembly into the modern continents. By the Early Jurassic the great landmass began to split up in zigzag cracks stretching out across the continent. These were frequently the sites of great outpourings of lava as molten rock welled up from the mantle. More than 3300ft (1000m) of basalt lava buried the Triassic Karoo landscape of South Africa.

Pangea was a supercontinent for only a few million years. The Atlantic and Caribbean basins began to open as the landmasses rifted.

The most significant rifts had appeared between what is now North America, South America, and Africa, separating these landmasses, and running the length of the division between South America and Africa. Other rifts had developed along the line that was to separate Africa from India and Australia, and along the site of the present North Atlantic between North America and Europe. Greenland also began to separate from North America, though this process was never completed as for the other continents. Paleomagnetic data suggests that during the Jurassic the continents also came to occupy approximately the same latitudes that they do now, though their orientation has shifted slightly in some cases.

Rising sea levels, possibly caused by the melting of glaciers as the southern continents began to drift away from the pole, meant that the rifts between the continents were becoming flooded, with water reaching in from their ends and slowly expanding to form infant oceans: the Atlantic, Pacific, and Indian Oceans, three of the largest oceans of today. The geography of the Earth was beginning to take on the shape familiar from maps.

The Tethys Seaway spread into the rift that was separating North America from Africa, as if driving a wedge between the continents. This rift valley flooded periodically, drying out to deposit thick beds of salt and evaporites in the area that now constitutes the Gulf of Mexico. The west coast of North America, meanwhile, was also being flooded from the north and west.

Many of the low-lying areas at each side of the Tethys became submerged, covering the limestone landscapes of southern Europe, and chains of low-lying islands spread across this shallow sea area. Northern Europe also became flooded. The waters swept in from the north through the rift between Greenland and Scandinavia. Permian and Triassic redbeds were becoming replaced by the shales and limestones of the Lias, or lower Jurassic.

Gondwana was still Gondwana in spite of the rifts that were beginning to wrench it apart. Although North America and Africa were slowly separating along the seam where they had joined— the old Hercynian mountains—Africa and South America were to remain fused until the succeeding Cretaceous period. Along Gondwana's northeastern edge, two large continental fragments, which later became part of the Himalayan mountain chain, had broken away and were beginning to drift north across the broad eastern end of the Tethys.

Out to the west, chains of islands and continental fragments were converging upon the subduction zone that still lay along the western margin of Pangea. Eventually these would combine with the coastal mountain ranges and produce the "exotic terranes" (blocks of lithosphere sutured to the much larger mass of an existing continent) that are so much a part of the region today, particularly along the west coast of North America, where dozens of such terranes constitute almost three-quarters of the Cordillera of the west. This process, sometimes called the Nevada orogeny, is thought to have added as much as 185mi (300km) to the west coast of North America.

THE LOCALIZED seas, in addition to the great Tethys Seaway, brought moist temperate climates deep into the hearts of the Jurassic continents. Even the great desert-like expanses of the interiors of landmasses, which had remained barren for millions of years, were beginning to turn green as plants expanded into the hospitable environment and diversified there. The many fossil floras that are known from Early Jurassic times suggest that the whole world basked in a subtropical climate. There was not much difference in temperature between the equator and the poles, as indicated by the discovery at very high latitudes of Jurassic fossil ferns that would not have survived in a climate where there was a regular occurrence of frost.

Moisture from the new oceans had a profound effect on the Jurassic continents. Plant life, long confined to the coastal margins, now spread inland and flourished rapidly.

THE AMERICAS SEPARATE

As the Gulf of Mexico opened, the land link between the Americas was broken. The shallow Sundance Sea covered the southwest part of North America.

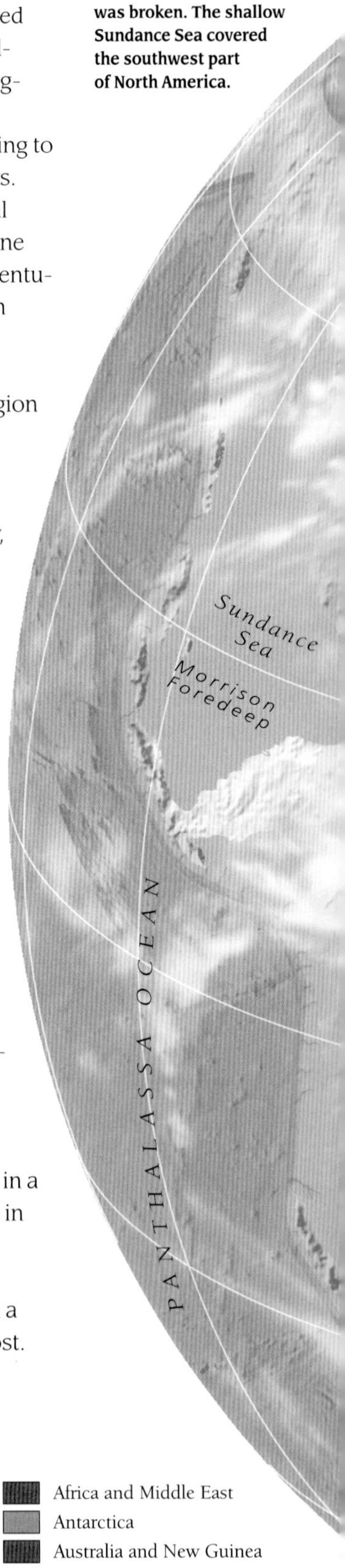

Sundance Sea
Morrison Foredeep
PANTHALASSA OCEAN

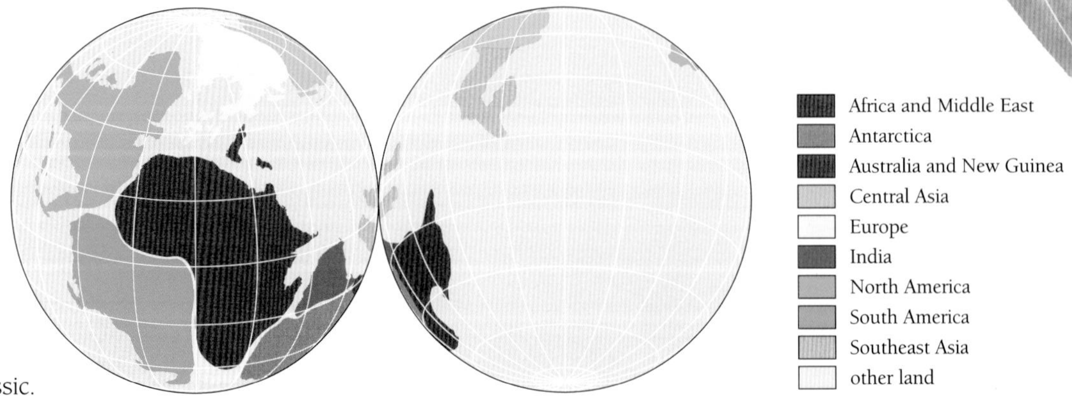

Africa and Middle East
Antarctica
Australia and New Guinea
Central Asia
Europe
India
North America
South America
Southeast Asia
other land

JURASSIC

CRACKUP

Jagged rift valleys criss-crossed Pangea, fracturing the land from one triple junction to another. For the first time, the shapes of the modern continents are visible on the face of Gondwana.

A GREENING EARTH

From space the Early Jurassic Earth would have been a fertile-looking planet as vegetation encroached upon formerly arid areas. The great rifts would probably have been visible as block mountain ranges and valleys. There were no ice caps.

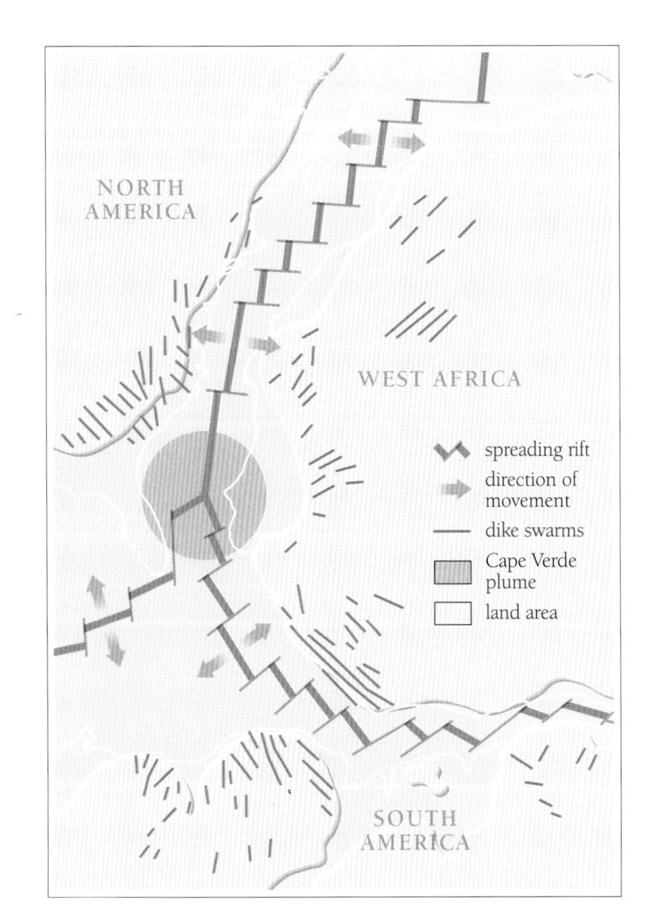

WHEREVER rifts developed across Pangea, the shape of the cracks was the result of the turmoil beneath the crust that was pulling the surface apart. Convection currents in the mantle pushed up against the crust, raising it and cracking its surface. The cracks would typically radiate out from a point, and usually form three major branches, each representing a boundary between lithospheric plates. As the process continued, two of these rifts would develop into a continuous wide valley, which would merge with those from neighboring triple junctions and form the continuous zigzagging rifting zones across the continent. The third rift remained only partially developed; geologists refer to this as a "failed rift." Two of the world's longest rivers, the Amazon and the Mississippi, flow through valleys formed by such dead-end rifts where Pangea came apart.

Rifting in Pangea developed along the junctions where plates had collided to bring the supercontinent together. Modern rifting can be seen where the Arabian peninsula is pulling away from Africa.

The modern Red Sea is a flooded rift valley—an ocean in its infancy. At its northern end it splits at a triple junction, forming two smaller rifts, the Gulf of Suez and

A TRIPLE JUNCTION

(Left) Upwelling in the mantle (the Cape Verde plume) produced a rise in the land surface that resulted in three rifts, separating North America from Africa and South America, and separating South America from Africa. As well as the major rifts, igneous dikes radiated from this point too. Seawater flooded these new valleys, to form the young Atlantic Ocean.

RIFT VALLEY

(Below) The elongated depression called a rift valley is bounded by fault scarps on one side. The smaller faults on the other, downwarped side, are covered by sediments and do not produce landscape features. A cross-section of a modern day rift valley shows the same features as those of the rifts that began to tear Gondwana apart at the beginning of the Jurassic.

Triple-point fracture

Dikes

A MANTLE PLUME

A plume of hot material rising from the molten interior of the Earth initially produces a dome the crust (detail above). A radial fracture pattern develops, with the two main cracks spreading to become a rift valley.

1 Rising magma lifting the continental surface
2 Major faults, allowing slices of the crust to slip downwards
3 Downwarped crust producing the valley
4 Minor faults covered by sediment
5 Valley sediment, mostly brought in from the down-warped side
6 Ephemeral lake, from intermittent flooding

the Gulf of Aqaba. The latter continues north through the Jordan Valley and through the Dead Sea as an active rift valley, leaving the Gulf of Suez as a failed rift. At its southern end the Red Sea meets another triple junction, with the Gulf of Aden as the active rift and the northern end of the East African Rift Valley, stretching south from Afar in Ethiopia, as the less important feature.

UNTIL the 1980s it was assumed that rift valleys were relatively simple structures caused by tensions pulling continents apart, with faults at right angles to the direction of stretching appearing where the tension is greatest. Blocks of the continent subside between these parallel faults, producing a geological structure called a graben, which is visible on the surface as a rift valley. More recent studies, particularly in the Great Rift Valley of East Africa, have shown that rifting is not as symmetrical as previously thought. Faults do form, but only on one side of the line of tension. Material on the other side warps down against these faults, causing the depression that is seen as a valley.

Rift valleys are visible above the structures called grabens, where one continental block has subsided against another.

Most of the sediment that gathers on the floor of a rift valley comes from the downwarped side, brought by rivers. On the faulted side, sediments are brought in by short streams that cut gorges through the faulted escarpments. Most drainage on the rift side of the valley takes place away from the valley, often producing large rivers running parallel to the valley's course. In the modern world the Nile flows in a direction parallel to the Red Sea rift, but is beyond the flanking block mountains.

This configuration—faults along one side and downwarping along the other—can run along the rift valley for tens or hundreds of miles. Then the structure reverses, with the faults and the downwarping exchanging sides. This cuts the rift valley into a number of basins, each of which may contain a lake. Because most rift valleys occur in the middle of continents, the lakes evaporate, so that the dissolved minerals brought down by the streams and rivers are precipitated on the valley floors.

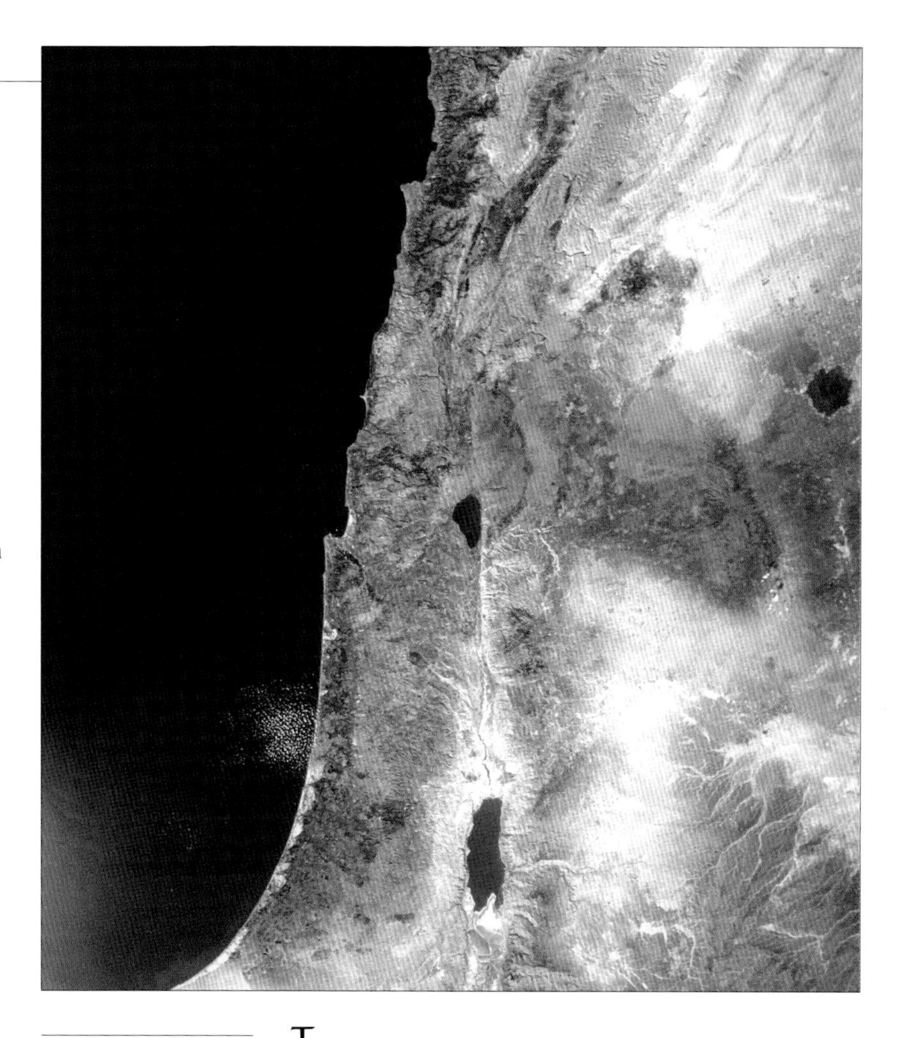

A MODERN RIFT VALLEY

The rift valley of the Jordan River cuts from north to south through the Middle East, meeting the Red Sea in a triple junction at its southern end. The Dead Sea and Lake Tiberias are ephemeral salt lakes in areas of inland drainage.

THE SEAWARD end of a rift valley may be open to flooding from the ocean, again giving rise to deposits of minerals produced by evaporation. There are many evaporite beds dating from the Late Triassic and Early Jurassic that are associated with rift valleys. When minerals are deposited by evaporation in lakes or restricted gulfs, precipitation takes place in the same sequence. When seawater is drying out the first minerals to precipitate are the calcite salts, the carbonates, which give rise to limestone beds. Calcite salts are all removed by the time the water shrinks to 15 percent of its original volume. At about 20 percent

Rift valley volcanoes have a range of compositions, because the lava has absorbed some of the local continental rock before erupting.

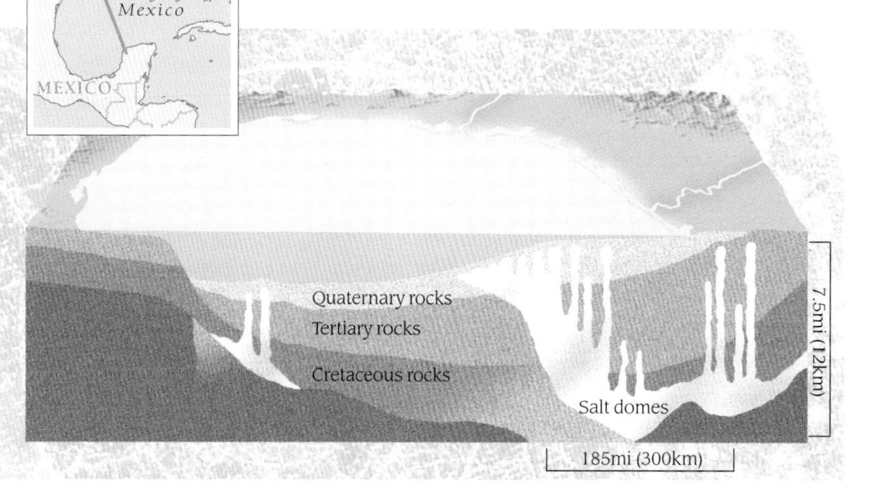

DEPOSITS FROM ANCIENT GULFS

One arm of the developing Atlantic rift separated North America from South America. It was flooded sporadically by the newly emerging Pacific Ocean. The evaporation of this restricted gulf produced a vast deposit of rock salt that now underlies the Gulf of Mexico. Subsequent deposits covered the salt bed and buried it deeply. Eventually the salt, being much less dense than the overlying sedimentary rocks, began to rise through them, producing tall dome-like structures called diapirs or salt domes. As each dome pushed its way upwards it dragged up the strata in the immediate vicinity and produced pockets of sedimentary rocks that made ideal oil traps. Petroleum that developed at a later date migrated to these traps and concentrated there, forming the basis for today's oil industry.

ZONE FOSSILS

Clusters of ammonites (below) found together tend to be of the same species, reflecting its abundance at that time. Younger or older beds will have ammonites of a different species, since they evolved so quickly. This rapid turnover of species is the basis of biostratigraphy: dating beds by their fossils. Ammonites were very sensitive to variations in water conditions, with different groups living in different habitats. As well as being a key index fossil, they are important indicators of ancient marine environments.

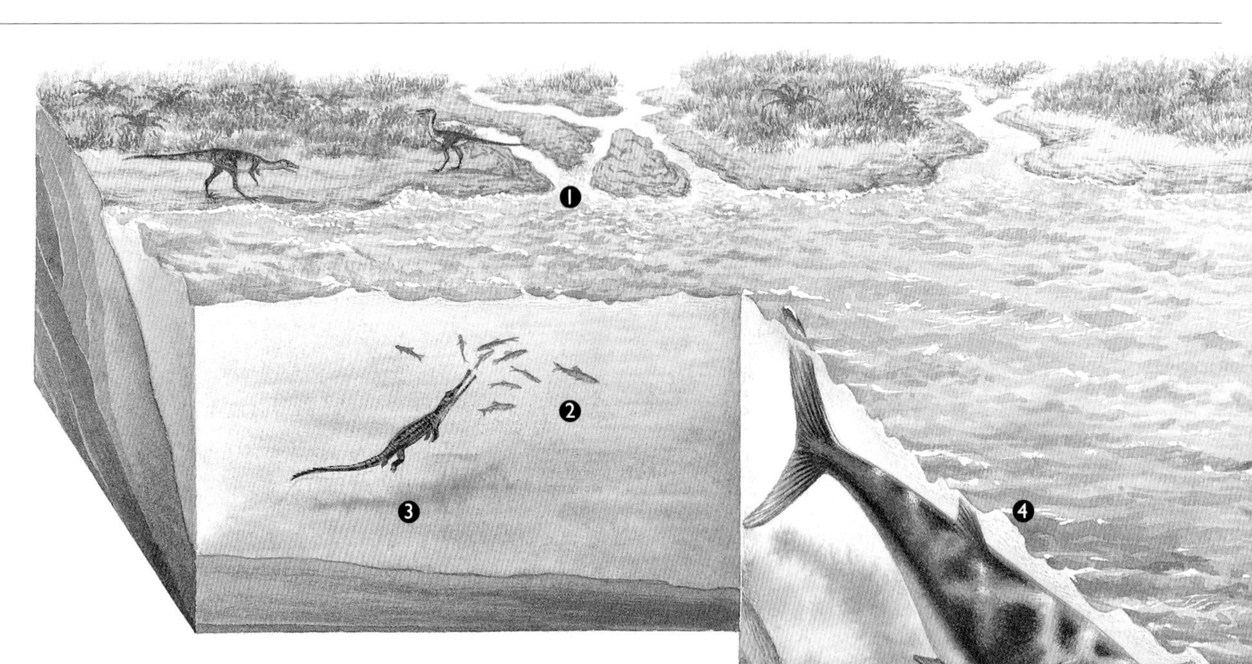

TETHYS WEB

The food chains of the Jurassic seas (right) were as complex as they are today. Plants absorb sunlight to make their own food. Microscopic plants and plant debris are eaten by tiny animals, which are eaten by bigger animals. The biggest meat-eaters are at the end of this chain. However, even the big meat-eaters die; their bodies decompose and provide more nourishment, giving rise to all kinds of other interactions along the chain. At this level of complex interweaving, the system is more properly described as a web.

of its original volume, calcium sulphate begins to precipitate. By far the biggest component of dissolved mineral is sodium chloride, or common salt, which begins to precipitate at 10 percent of the original volume and produces the bulk of the precipitated material.

With all the plate tectonic activity that causes rifting, it is not surprising that volcanoes are another landscape feature often associated with rift valleys. Molten material rises from the base of the Earth's crust and the top of the mantle, through the lines of weakness provided by the faults, and erupt at the surface close to the faulted edge of the valley. The volcanoes are of the basaltic variety, with the molten material derived from the mantle and having a relatively low proportion of silica.

However, on its way up through the crust the molten material tends to absorb the minerals of the rocks through which it passes, resulting in surface volcanoes that may produce a wide range of lava types when they erupt. The East African Rift Valley today even has volcanoes that erupt lava that is so full of the mineral calcite that it forms the igneous equivalent of a limestone—a highly unusual occurrence.

THE SHALLOW shelf seas that bordered the Tethys in Jurassic times were full of life. Among the vertebrates were bivalves, gastropods, and cephalopods. The cephalopods had developed into a variety of shelled forms, the most important of which were the coiled ammonites and the straight squid-like belemnites, both abundant as fossils in Jurassic rocks. The ammonites ranged from button-sized to some as large as the wheels of trucks, and varied greatly in their ornamentation of ribs, ridges, and spines. Starfish and their relatives were common, as they had been in Paleozoic times. Crinoids, or sea lilies, were much more abundant than they are today, growing wherever there was a hard substratum, or clinging to driftwood.

The invertebrates on the floor of the Jurassic shelf sea would have been remarkably similar to those that we find today.

Modern fishes, the teleosts, came to the fore in the Jurassic. They were distinguished from more primitive bony fish by the lighter scales and the hinging of the jaws.

1 Nutrients washed into the sea from land
2 Fish (*Pholidophorus, Leptolepis*)
3 *Steneosaurus* (sea crocodile)
4 *Leedsichthys* (giant fish)
5 *Metriorhynchus* (sea crocodile)
6 *Rhamphorhynchus* (pterosaur)
7 *Cryptoclidus* (long-necked plesiosaur)

The sharks, which had first appeared in the Devonian, showed no sign of fading out (and still have not today).

Marine reptiles were the most spectacular of the sea animals. There were seagoing crocodiles, such as the long eel-like *Teleosaurus* and fin-tailed *Metriorhynchus*. The plesiosaurs were large paddle-finned animals and came in two types: long-necked forms like *Cryptoclidus*, and big-headed short-necked pliosaurs like *Liopleurodon*, which grew as long as 40ft (12m). The most highly adapted marine reptiles were the ichthyosaurs, such as *Opthalmosaurus*, which had evolved in the Triassic. Shaped like like sharks or dolphins, they bore live young and are well known from fossils in the Holzmaden region of Germany, which may have been a favored place to give birth. Also found among Jurassic marine fossils are those of the flying pterosaurs.

As IN the seas of today, each member of the Jurassic marine community depended on the others for its existence, forming a linked feeding chain known as a food web.The basis of any food web is the input of energy, usually from the sun. The lands surrounding the Tethys and its offshoots were low and vegetated. The warm moist climates ensured that many plants grew, and plant debris was constantly washed into the seas by rivers and streams. This vegetable matter, along with phytoplankton that grew in the sunnier levels of the sea itself, provided the raw material that kept the food webs active.

As life expanded and diversified, complex food webs had evolved in which each living thing was dependent on the presence of others.

In the uppermost layers the phytoplankton—microscopic floating plant life—were fed upon by tiny animals, mostly larval fish and invertebrates. These, in turn, were eaten by fish and ammonites. Some of the fish that ate the tiny animals were huge, like today's whale sharks, which eat only the smallest animals despite being the largest fish in the oceans. The smaller fish were eaten by larger fish and by the fishing pterosaurs. At the top of the food web were the big carnivores, like plesiosaurs, pliosaurs, and sea crocodiles.

Near the bottom of the sea most of the nutrition came from material washed in from land and drifting down from the surface waters, and also the carcasses of big animals that sank to the bottom. Shellfish, such as the oyster *Gryphaea*, thrived on the suspended food matter. Belemnites and skate-like fish fed on the shellfish and were preyed upon by sharks and ichthyosaurs.

More species lived at the surface than on the seabed. This may have been because the seabed had little variety and only a few kinds of animals could live there, or it may have been because there was little oxygen near the seafloor. The large number of islands prevented bottom currents, which in turn limited the circulation at depth.

8 *Liopleurodon* (short-necked pliosaur)
9 *Peloneustes* (short-necked pliosaur)
10 Ammonites (cephalopods)
11 *Ophthalmosaurus* (ichthyosaur)
12 Belemnites (cephalopods)
13 *Gryphaea* (bivalve)
14 *Hybodus* (shark)
15 *Spathobatis* (ray-like fish)

FOSSIL ICHTHYOSAUR

An ichthyosaur fossil from Holzmaden shows the dolphin shape of the body as a dark silhouette of carbon. Its internal organs, including in some specimens of unborn young, can be seen, as can the fleshy fins on the tail and on the back.

DINOSAUR COUNTRY

Apatosaurs (brontosaurs) of the Jurassic period left their tracks in mud in this Colorado section of the Morrison Formation (right) Herds of them followed seasonal supplies of food and water.

1 Channel-fill deposits: current-bedded sands and conglomerates, sometimes with dinosaur remains
2 Overbank deposits on levee—sand and silt from times of flood
3 Channel-fill deposits from old river course
4 Crevasse splay where levee was breached; occasional footprints
5 Mud bed from spring-fed pond, often with plant fossils
6 Caliche—layers of limestone deposited by calcite-rich water migrating upwards
7 Limestone and salt from salt lakes; may contain fossils of small animals poisoned by the water
8 Soil layers from the plain are indistinctly bedded due to constant trampling

THE JURASSIC period is widely regarded as the heyday of the dinosaurs, and the site that probably above all others established this connection was the Morrison Formation of the western United States. Named after the small town of Morrison in Colorado, the formation consists of a vast sweep of shales, siltstones, sandstones, limestones, and conglomerates, between 100 and 900ft (30 to 275m) thick, all laid down in Late Jurassic rivers and lakes, stretching north–south from Montana to New Mexico, and east–west from Nebraska to Idaho. Each single meter of rock in the Morrison formation represents a period of time about ten times the length of human history.

More than 70 dinosaur species were found in the Morrison Formation, a huge Jurassic deposit that took 6 million years to be laid down and now stretches across ten states.

The Morrison formation, and its spectacular fossil contents, first came to the attention of science and the world in 1877 when schoolteacher Arthur Lakes found a dinosaur vertebra embedded in the rock close to Morrison, near Denver. It had only been about 30 years since anyone first proposed that such a creature as a dinosaur had ever existed, and most of the discoveries of dinosaur fossils had been made in Britain and Europe. The only American specimens known at that time had been found all the way across the country in New Jersey.

JURASSIC

Lakes' discovery sparked off an incident known as the "Bone Wars." Two distinguished American paleontologists, Othniel Charles Marsh and Edward Drinker Cope, tried to outdo each other in their attempts to unearth dinosaur fossils. For decades the two sent rival teams across the Morrison Formation, following leads, bribing guides, and sabotaging each other's excavations to find more and more bones and bring them back to the museums and universities of the East Coast. The positive outcome of their rivalry was that Marsh and Cope discovered 136 new dinosaurs by the end of the century.

IN LATE Jurassic times the Sundance Sea, a shallow continental sea that had spread southwards across the Midwest, began to dry up and retreat north. In its place a vast apron of sediment was deposited from a shifting network of rivers and streams that ran down from the ancestral Rocky Mountains to the west. The apron of sediment formed a low-lying plain. This was the Morrison Formation environment. The climate was semi-arid, but the plain was so flat and lay so close to sea level that it remained soggy for most of the year. Huge lakes shrank and swelled seasonally; rivers flooded, spreading silt, and dwindled to trickles. Trees and other permanent vegetation were confined to the river banks, and there were sand dunes in between. Over this landscape dinosaurs roamed. Sauropods, big long-necked plant-eaters such as *Diplodocus* and *Brachiosaurus*, fed from the trees and undergrowth and migrated in herds to and from seasonal feeding grounds. Plated *Stegosaurus* browsed the thickets. Smaller plant-eaters like *Camptosaurus* and *Dryosaurus* left footprints across drying lake beds. Meat-eating *Allosaurus* and the smaller *Ornitholestes* harassed them. Pterosaurs flew overhead.

The Morrison Formation was laid down on a plain by the evaporation of the Sundance Sea in the Late Jurassic.

RETREAT OF THE SUNDANCE SEA

Early in the Jurassic the Sundance Sea spread over most of midwestern North America, the southern portion forming the Morrison Foredeep. As the period went on, the sea retreated northwards, to be replaced by the vast plain of river and lake deposits that became the Morrison Formation. This material was eroded from the new Rocky Mountains that were rising in the west, and gradually covered an area of approximately 386,000 square miles (1 million square kilometers).

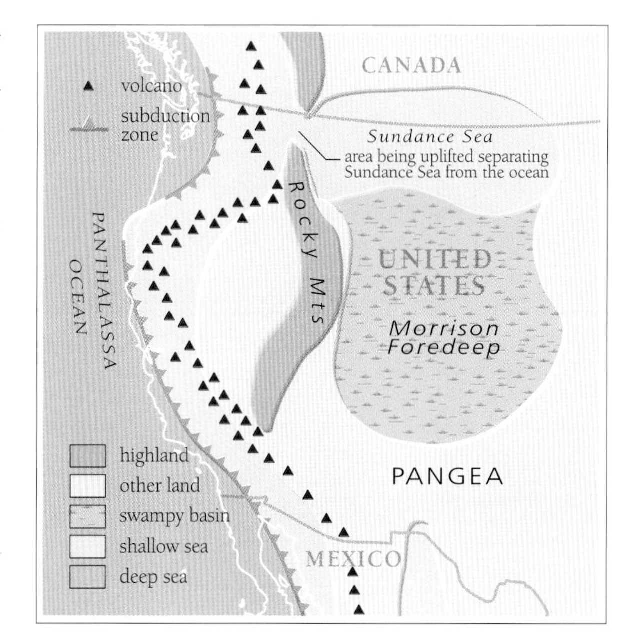

SEASONAL CHANGE

The Morrison landscape in the wet season (below, far left) had rivers that were so full that the water surface was higher than the plain, the water being held back by levees. Floods were frequent, and there were many springs. Some lakes were salty and poisonous. The landscape supported thickets of ferns but the only forests were by the rivers. In the dry season (below)the rivers dried to trickles and clouds of dust drifted; lakes dried out and sand dunes appeared. The land was trampled by herds of migrating dinosaurs.

PERHAPS the most important vertebrate fossil ever found was that of an extraordinary creature—part bird, part dinosaur—discovered in a limestone quarry in Bavaria in 1861. Called *Archaeopteryx,* it was found just as the implications of Charles Darwin's *On the Origin of Species* were beginning to be felt publicly. *Archaeopteryx,* as a transitional form between the reptiles and the birds, provided a beautiful example of the process of evolution. It was seen to be such a peerless example because of its mode of fossilization: the limestone in which it was preserved was so fine that the impression of the

***Limestone from the Upper Jurassic strata of southern Germany has preserved one of the most important vertebrate fossil finds:* Archaeopteryx.**

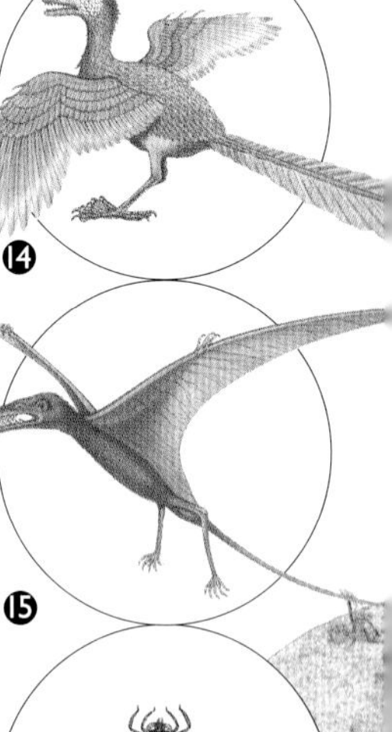

ARCHAEOPTERYX– THE PROVEN LINK

(Above) The link between the dinosaurs and birds was first discovered in the Solnhöfen limestone in 1860 as a single feather. The first almost complete skeleton was found the next year.

THE SOLNHOFEN LAGOON

Fine-grained limestone was laid down in the quiet waters of a lagoon, cut off from the turbulent Tethys by barrier reefs of coral built on old sponge reefs. The constant evaporation precipitated very fine limy particles and concentrated the salt, producing a poisonous water layer near the bed that killed anything that entered it.

BOTTOM-DWELLER?

The "sea lily" *Saccocoma*, above, is an extraordinarily common Solnhöfen fossil. There are clues that it adapted to the salty sediments and flourished.

1 Open Tethys Sea
2 Living sponge reefs
3 Sponge reefs die as they are raised. Corals grow upon them
4 Barrier reefs absorb energy of waves
5 Coral debris broken from reef by storms
6 Fine limy debris washed into lagoon

17

10 Jellyfish
11 Fish
12 Horseshoe crab
13 *Saccocoma* (crinoid)
14 *Archaeopteryx*
15 *Rhamphorhyncus*
16 Dragonfly
17 *Compsognathus*

Animals that enter the lagoon die and sink to the bottom where the fine sediment preserves them

7
8
9

7 Constant evaporation from surface produces a poisonous hypersaline water level on the lagoon bed
8 Limy debris settles to form fine-grained lithographic limestone
9 Older limestone platform faulted and raised towards the north

minutest structure of the feathers showed up. The limestone was the upper Jurassic limestone, known as the Solnhöfen Limestone, which outcrops in an area of about 43x28mi (70x30km) between Nuremberg and Munich in what is now southern Germany. At its thickest there are about 310ft (95m) of the Solnhöfen limestone, representing half a million years of deposition. The beds are flat and thin, and so fine grained that they record the tiniest impressions etched upon them. This also makes them the ideal medium for preserving the finest structure of fossil animals. Occurrences such as the Solnhöfen limestone in which the animal life is preserved in such detail are known by the German term *lagerstaten*.

Although *Archaeopteryx* was the most famous fossil from the site—there have been eight specimens found, including a single feather—it is not the only one. Many pterosaurs have been found, the structure of their wing membranes preserved in detail. There are also skeletons of small dinosaurs, lizards and lizard-like sphenodonts. Of the invertebrate fossils the most common is the free-floating sea lily *Saccocoma*, which resembles a brittle star. There are also jellyfish, dragonflies, and horseshoe crabs preserved at the ends of their trails, and the earliest octopus known. Despite this richness, the number of fossil finds is very low, and they have only been discovered because of the extent of the quarrying.

THE SPONGE REEF

The remains of the great reef that once spread across the north of the Tethys can nowadays be found in outcrops in central and southern Spain, southeastern France, southwest Germany, Switzerland, central Poland, and throughout Romania as far as the Black Sea. Many kinds of sponge grew in the reef, all feeding on microscopic food particles.

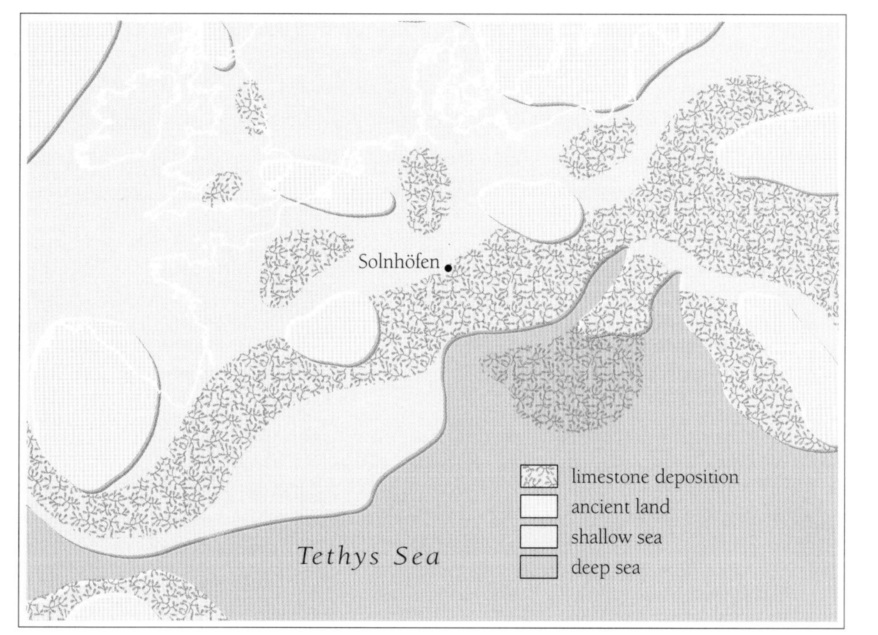

THE SOLNHÖFEN limestone was laid down in tropical seas and produced by reef-building hexacorals, a group that first appeared in the Triassic. Early hexacorals built mounds that grew only to modest proportions, about 10ft (3m) high; these involved only a few species. By the end of the Jurassic several dozen hexacoral species were producing reefs, and these grew much larger. An enormous sponge reef spread across the northern part of the Tethys, stretching from Spain to Romania and Poland, a distance of about 1800mi (2900km)—half as long again as the modern Great Barrier Reef off northeast Australia. The Tethys sponge reef grew in deeper water than modern coral reefs, around 500ft (150m) deep, and was anchored on the gentle slope of the continental shelf.

A huge reef stretched across much of the northern Tethys. By the end of the Jurassic its quiet lagoons turned deadly from high concentrations of lime and salt—ideal for preserving fossils.

The faulting that produced the block mountains and rift valleys over the whole Jurassic world also raised and lowered the seabed. When the reef came too close to the surface, the conditions were wrong for the sponge growth, and long sections of it died. The dead skeletal portions were used as the foundation on which corals built later reefs closer to the surface. These reefs cut off the quiet lagoons in which Solnhöfen limestone was gently deposited. The concentration of lime and salt in these lagoons became so high that the water was toxic, killing any animal or plant that came into contact with it.

BIOGEOGRAPHIC PROVINCES OF THE JURASSIC

The imminent separation of Pangea into today's continents was beginning to show in the different types of life that existed in different parts of the Jurassic world. Long before anybody suspected that the continents had moved, marine life in Europe was classified into northern or southern provinces, now known as the Boreal and the Tethyan. The presence of corals indicates that the Tethys province, in the south, was tropical, and the Boreal, where there were no reefs, was not. Ammonite species of the Jurassic also vary between the two marine provinces.

Plant fossils show that Asian flora was becoming distinct. Gondwana, beyond the rift that ran down the eastern margin of Africa, had a characteristic flora, although North America still shared many species with northern Africa, despite the opening of the Atlantic. There was also a great similarity in animal life between east Africa and North America: the same dinosaurs roamed both lands. This represents a paradox, since animal life could be expected to diverge quickly in separated areas. There is still much work to be done on this aspect of paleogeography.

In a mere 35 million years from the beginning of the Jurassic, the breakup of Pangea had proceeded so quickly that new continents and oceans were already evident.

THE BREAKUP of Pangea took a total of 150 million years to complete, but from the Early to the Late Jurassic—a span of about 35 million years—separate continents had split away, with new oceans forming between them. North America was slowly rotating anticlockwise and ripping away from the rest of the landmass. The North Atlantic ocean appeared, separating North America from Gondwana in the south and from Europe in the east. This was no shallow shelf sea, but an actual ocean, with an abyssal plain between continental shelves and a volcanic oceanic ridge that generated the new ocean crust material. It opened up a seaway that connected the Tethys all the way to the western edge of the continent and the ocean of Panthalassa. The immediate impact of this continuous seaway was that the prevailing winds generated by the convection currents of the atmosphere and the turning of the Earth would have driven a permanent ocean current that flowed westwards in the tropical latitudes all around the world. The waters warmed in this equatorial seaway would then have circulated north and south along the shores of the fragmenting Pangea. The effect on the climate would have been that warm moist air would have penetrated much deeper into the continents than previously, and that the worldwide climate would have become much less extreme.

Gondwana itself was well on its way to becoming a number of separate continents. The rift valleys that crisscrossed the supercontinent were now much wider and more extensive. What was to become Africa was almost completely separated from the other fragments that were to become Antarctica and India. The two large sections of continental crust that had detached themselves from the region of northern India were migrating northwards across the Tethys, drawn by the subduction zone at the southern edge of Asia. Spreading ridges would have appeared between these and Gondwana.

Likewise, landmasses were continuing to accrete along the west coast of North America. These were made up from island chains and small continental fragments that had been scattered across Panthalassa, and were now drawn towards the continent by the subduction zones along the western margin. There were evidently several subduction zones along this continental margin, and deep sea sediments were piling up against the continent, crushed in by the approaching landmasses. This pressure produced local mountain-building in the region of Nevada, giving rise to a small orogeny that persisted into Cretaceous times.

Away to the east, parts of what was to become China were beginning to break away from the northern landmass.

The Tethys was still part ocean and part shallow shelf, the northwestern region remaining as a shelf sea with a number of large low-lying islands. Late in the Jurassic there was a rise in sea level, making the shallow seas over Europe particularly extensive. This rise was a result of the appearance of the new oceanic ridges which raised themselves from the ocean floors, displacing water across the surrounding low-lying lands. The seas flooded into northern Asia to the east of the Urals, forming the broad Ob Basin in that area. Then, towards the end of the period, the sea levels fell once more. The Sundance Sea, which had covered much of the American Midwest, retreated north and was replaced by vast dryland deposits. Northern Europe's shallow-water marine beds were replaced by deltas and estuaries.

PANTHALASSA OCEAN

Ancestral Rockies

Ancestral Andes

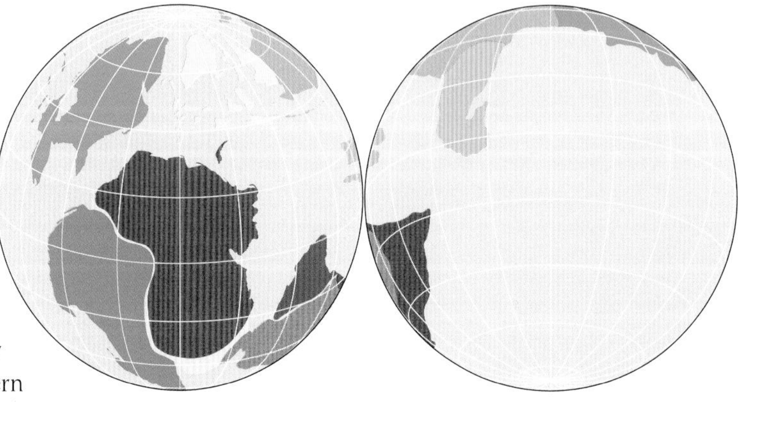

Africa and Middle East
Antarctica
Australia and New Guinea
Central Asia
Europe
India
North America
South America
Southeast Asia
other land

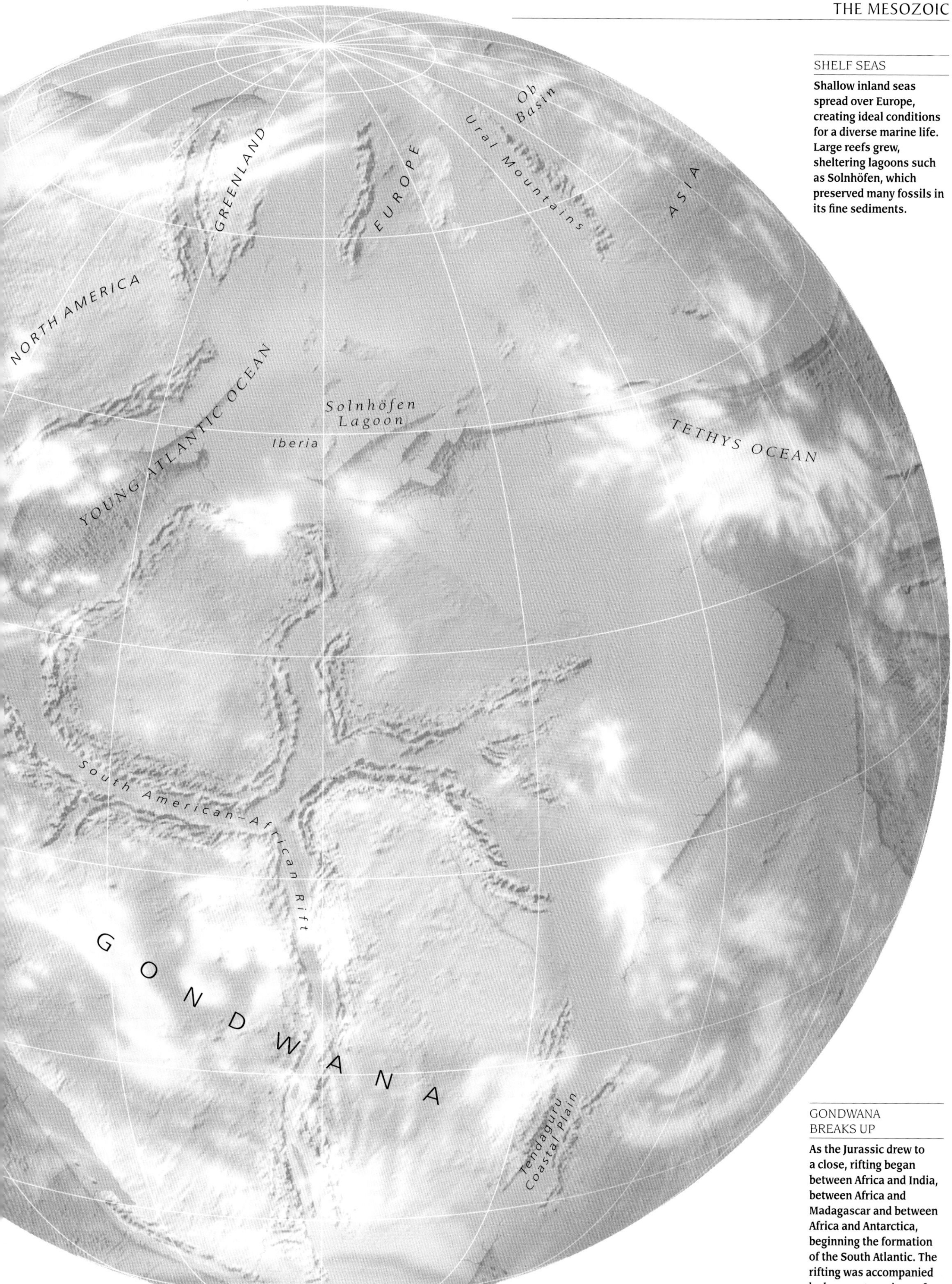

SHELF SEAS

Shallow inland seas spread over Europe, creating ideal conditions for a diverse marine life. Large reefs grew, sheltering lagoons such as Solnhöfen, which preserved many fossils in its fine sediments.

GONDWANA BREAKS UP

As the Jurassic drew to a close, rifting began between Africa and India, between Africa and Madagascar and between Africa and Antarctica, beginning the formation of the South Atlantic. The rifting was accompanied by huge outpourings of basaltic lava.

1
2
3
4
6
7
8

At the edge of the widening ocean that was splitting Gondwana in Late Jurassic times, separating what was to become East Africa from what would be India, lay the Tendaguru coastal plain. The sea lapped gently on the shoreline; there was no high surf here because the ocean was still quite narrow and the waves did not have enough distance in which to build up to great heights. Additionally, a wooded offshore sandbar kept back what waves there were. Inland, the horizon was formed by scarp slopes, the edges of the fault blocks generated by the rifting of the continent. The slopes dropped steeply away into a steamy Jurassic tropical forest of cycads, tree ferns, and conifers, whose resemblance to evergreens would have made this ancient landscape appear more familiar to modern eyes.

The Late Jurassic animal life of East Africa was very similar to that of North America, from which Africa had only recently separated.

Along the shoreline the high tide left tidemarks of seaweed and of vegetation brought down by rivers, as well as remains of the animals of the sea. The coiled shells of ammonites lay half buried in the sand, as did the elongated shapes of belemnites. Here and there were larger relics. Clouded in flies and the scrabbling shapes of pterosaurs, the corpse of a plated stegosaur lay rotting. It had been lying there long enough to attract scavengers. Small theropod dinosaurs gathered to tear at the decaying flesh, but these were themselves disturbed by a bigger, fiercer theropod charging down the sand, scattering the smaller animals, to claim the meat for itself.

Oblivious to this carnivorous frenzy nearby, the big plant-eaters of the forest placidly munched their way through the vegetation as they stood in the dappled sunlight at the head of the beach. Long-necked sauropods, towering over the crowd, browsed peacefully side by side in the thickets of conifers and ferns. At about halfway through the Age of Dinosaurs, these animals were already at their most successful.

1 *Stachyotaxus* (conifer)
2 Ferns
3 *Brachiosaurus* (sauropod dinosaur)
4 *Dicraeosaurus* (sauropod dinosaur)
5 Plesiosaur
6 *Nillssonia* (cycad relative)
7 *Ceratosaurus* (theropod dinosaur)
8 *Elaphrosaurus* (theropod dinosaur)
9 *Kentrosaurus* (stegosaur)
10 *Rhamphorhynchus* (pterosaur)
11 Ammonites
12 Belemnites

5

9

10

11

12

JURASSIC

THE EVOLUTION OF AMMONITES

THE MOLLUSKS are among the most important of animals to the paleontologist. Their hard shells mean that they fossilize easily, and their evolutionary history is well-known since Cambrian times. The main groups are the bivalves, the gastropods, and the cephalopods. This last class may seem odd to be a major fossil group, since the modern forms—squids and octopi—are mostly without shells, but the fossil record of cephalopods shows a vast range of shell types and structure.

The evolutionary development of the shell was essentially a continuing experiment in buoyancy. A cephalopod shell was divided into chambers. As the animal grew, it lengthened the shell and laid down a new wall or septum behind it to form a new chamber. The chambers were filled with gas, introduced from a porous tube (siphuncle) that ran the entire length of the shell, and this gas was used to adjust the buoyancy of the animal in the water. Other motion would have been provided by jet propulsion, as in the modern squid and octopus.

By the Mesozoic one group of cephalopods, the ammonites, became among the most common animals in the ocean, and as such they are now the most important to the stratigrapher. They were wide-ranging, with each species spreading over entire sea areas, and they evolved quickly, with species replacing one another every million years or so. This means that when a recognizable species is found in a rock stratum, that stratum can be dated accurately, and the fact that ammonites were so wide-ranging means that such evidence can be used to date rocks that are far apart. The whole of the Mesozoic has been divided into time zones based on the species of ammonites that are found in particular levels.

Identification of individual ammonite species is based on the shapes of the shells, the ornamentation of the shells, and the patterns made by the suture lines. Although they are interior structures, the suture lines can easily be seen, because in a fossil ammonite the outer shell has usually been worn away over time.

LIVING FOSSIL

The ammonites are long extinct, but their anatomy may be compared to that of their only surviving modern relative, the nautilus (left). The structure of the eyes, the digestive system, and the arrangement of tentacles—all of these being parts that do not fossilize—were probably very similar.

EVOLUTIONARY TRENDS

The most primitive shelled cephalopods had straight shells, divided internally into chambers by simple walls or septae. As evolution progressed, coiled shells and more elaborate septae became more common.

1 Earliest types, with straight (orthocone) shell; buoyancy is achieved by counter-balancing
2 Curved shell (cyrtocone) develops; skeletal modifications allow a horizontal position to be maintained
3 Coiled shell becomes standard, allowing for greater control
4 The ammonites become the major group of cephalopods
5 Belemnites evolve an internal skeleton
6 Shell becomes reduced or disappears altogether

CLASSIFICATION

There are several orders of cephalopods today, but it is the extinct orders with their shells that are important in the evolutionary record.

Phylum	Class	Order
MOLLUSCA	AMPHINEURA (chitons)	NAUTILOIDA (simple sutures)
	SCAPHOPODA (tusk shells)	AMMONOIDA (complex sutures)
	GASTROPODA (snails, slugs)	BELEMNOIDA (belemnites)
	BIVALVIA (clams,mussels)	SEPIOIDEA (cuttlefishes)
	CEPHALOPODA	TEUTHOIDA (squids)
		OCTOPODA (octopi)

extinct

AMMONITE FORMS

Ammonite shells varied according to their habitat and life modes. The basic shape was a serpenticone, (1) wound so that the coils just touched one another. Other main types were oxycones (2), very flat in section; cadicones (3), with the coils partially enveloping one another; and, altogether different in style, the rather bizarre heteromorphs (4, 5).

SUTURE PATTERNS

The septa that divide the ammonoid shell into a succession of chambers meet the outside wall along a line called a suture. The shape of this line is very important in identifying a fossil cephalopod. Early nautiloid types had simple, broad undulations (1). Goniatites had sutures that formed a jagged pattern, (2). Ceratitic sutures (3) were more complex, while ammonites themselves had very ornate, frilled, and fluted patterns (4).

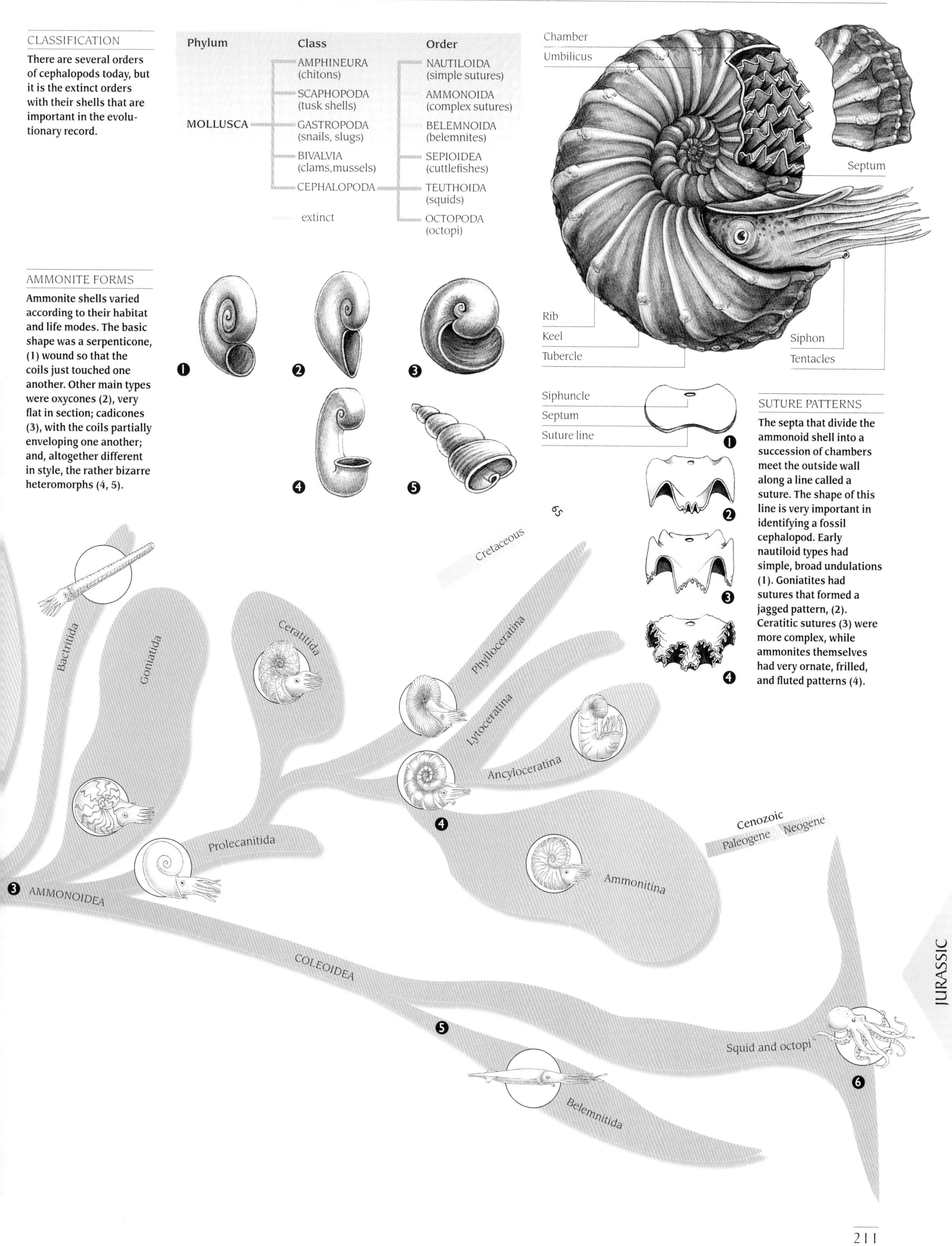

THE EVOLUTION OF DINOSAURS

CLAWS

The popular image of a dinosaur is of a fierce killer, and the bones of the big meat-eaters like *Allosaurus* live up to the stereotype. Their hands would have been big enough to grasp a man's head, and were armed with three long claws, strong enough to disembowel the biggest of the animals at the time. However, it is also easy to see its resemblance to the delicate claw of a modern bird, the dinosaurs' ultimate descendant.

DINOSAURS are the most famous and spectacular of all extinct animals, but they have an undeserved reputation as evolutionary failures. They developed about 225 million years ago and survived until the end of the Cretaceous period 65 million years ago. No group of animals that survived for 160 million years can be called failures.

Along with the mammals that survived them, dinosaurs evolved in Late Triassic times, and swiftly spread to every continent and occupied every environment, taking up a wide variety of lifestyles. Their ancestors were in the diapsid line of reptiles, and their contemporary relatives were the crocodiles and the flying reptiles or pterosaurs. Their modern descendants are the birds, to which they were so closely related that some scientists insist that birds do not form their own group but must be regarded as dinosaurs themselves.

Constant research means that modern appreciation of dinosaurs is changing continually. It was once thought that they were slow-moving coldblooded animals like the rest of the reptiles. Then in the mid-1960s it was proposed that they were warmblooded just like their descendants, the birds. The debate continues, and it is even possible that they were somewhere between the two, with meat-eaters on the warmblooded end of the scale and the big plant-eaters on the coldblooded end.

Another scientific dispute is about the reasons for their sudden extinction 65 million years ago. The most popular theory involves the impact of an asteroid or a comet with the Earth. However, widespread volcanic eruptions, like those that produced the Deccan Traps of India at the end of the Cretaceous period, could have had a similar effect. Other scientists believe that the demise of the dinosaurs was a much slower and more gradual event, possibly due to changing climates or sea levels, and that it is only the nature of the fossil record that makes it appear sudden. Or all these factors may have come together in a fatal combination.

EVOLUTIONARY TRENDS

At the start of the Triassic the dinosaurs split off from the archosaurs—the reptile lineage that also included the pterosaurs (flying reptiles) and the ancestors of the crocodiles. There were two major dinosaur groups, defined by the arrangement of their hip bones. The "lizard-hips" (Saurischia), which comprised the meat-eating theropods and the giant plant-eating sauropods, dominated the Jurassic, while the "bird-hipped" group (Ornithischia), made up of the two-footed plant eaters and all the horned and armored types, became the most common herbivores of the Cretaceous period.

1 Oldest true dinosaurs were bipedal "lizard-hipped" carnivores
2 The theropods were the major predators of the Mesozoic
3 The sauropods were giant herbivores which fermented their food in their guts; they developed weight-bearing features and elongated necks
4 The ornithopods were highly successful plant-eaters; unlike the sauropods, they chewed their food
5 Horned and armored types became widespread

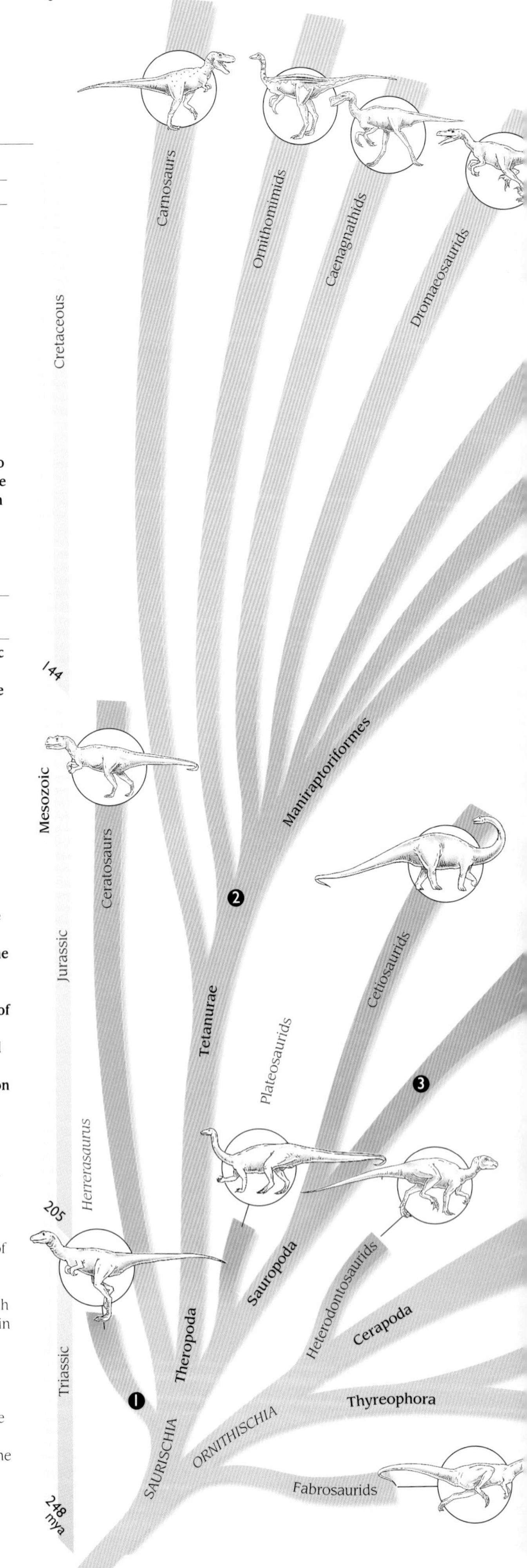

LIVING DINOSAURS

At some point in their evolution the theropod dinosaurs, by the development of feathers for temperature control and of wings for flying, evolved into birds.

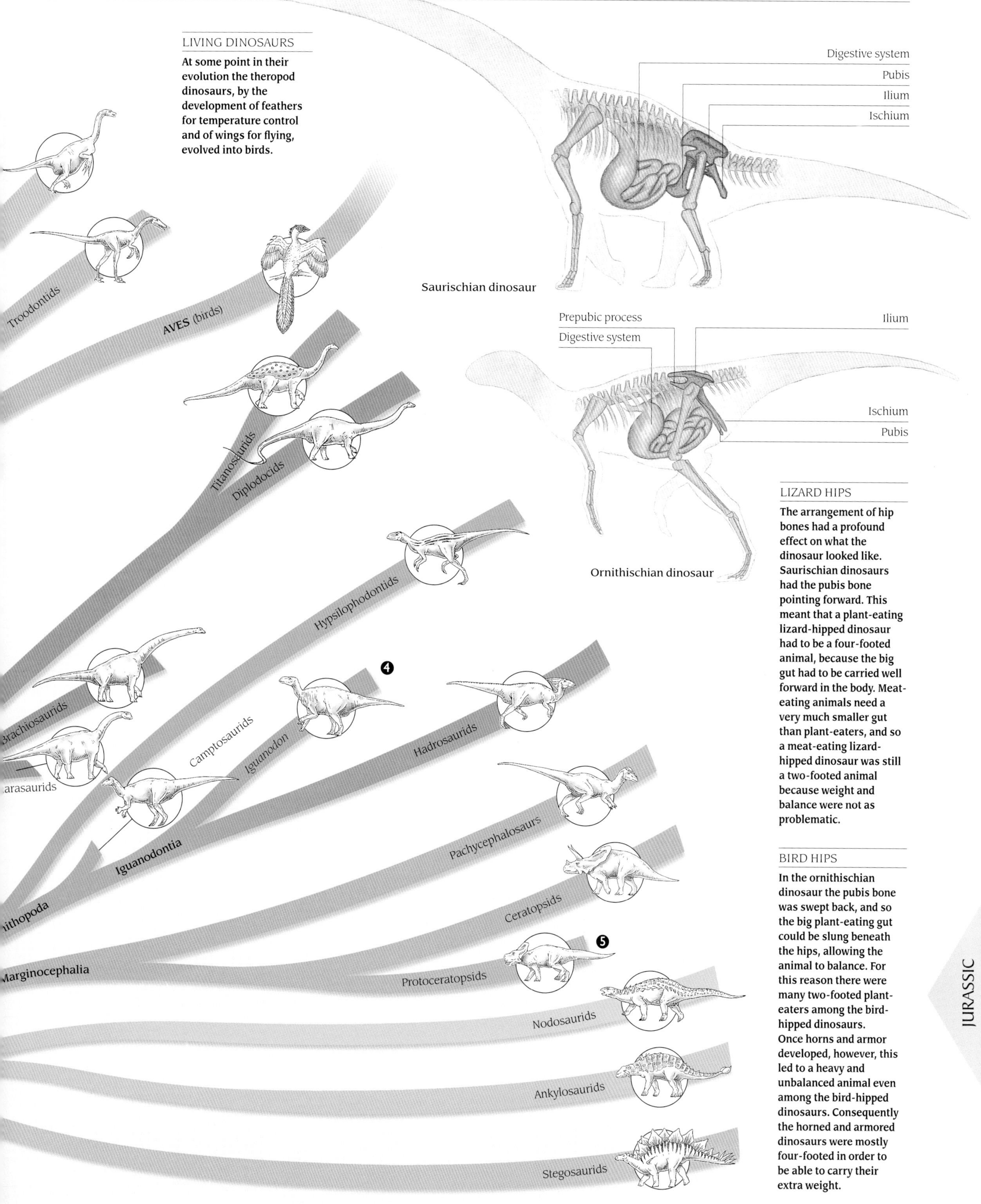

LIZARD HIPS

The arrangement of hip bones had a profound effect on what the dinosaur looked like. Saurischian dinosaurs had the pubis bone pointing forward. This meant that a plant-eating lizard-hipped dinosaur had to be a four-footed animal, because the big gut had to be carried well forward in the body. Meat-eating animals need a very much smaller gut than plant-eaters, and so a meat-eating lizard-hipped dinosaur was still a two-footed animal because weight and balance were not as problematic.

BIRD HIPS

In the ornithischian dinosaur the pubis bone was swept back, and so the big plant-eating gut could be slung beneath the hips, allowing the animal to balance. For this reason there were many two-footed plant-eaters among the bird-hipped dinosaurs. Once horns and armor developed, however, this led to a heavy and unbalanced animal even among the bird-hipped dinosaurs. Consequently the horned and armored dinosaurs were mostly four-footed in order to be able to carry their extra weight.

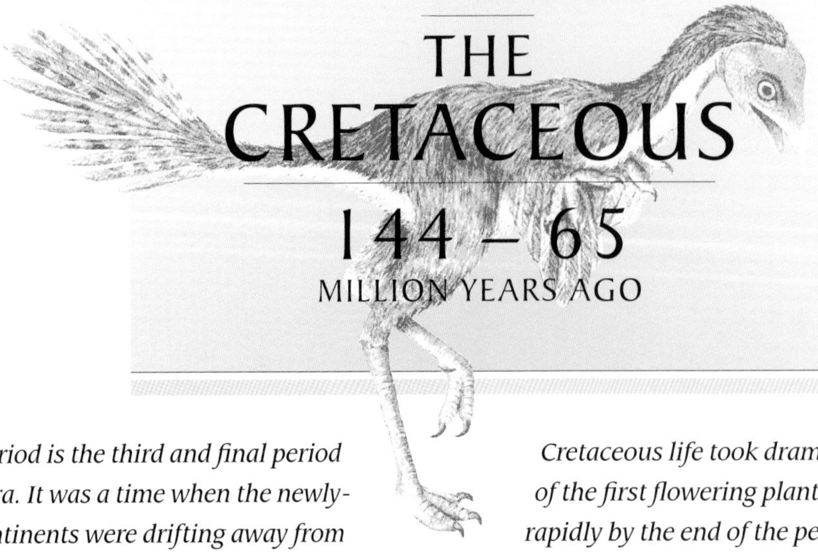

THE CRETACEOUS

144 – 65 MILLION YEARS AGO

THE Cretaceous period is the third and final period of the Mesozoic era. It was a time when the newly-formed individual continents were drifting away from one another after the breakup of the supercontinent of Pangea. The sea level rose continuously and finally flooded these various landmasses to an extent not seen in any other geological period, laying down great deposits of limestone, including the chalk for which the period is named. Climates everywhere around the globe were warm and mostly humid.

Cretaceous life took dramatic new turns, with the evolution of the first flowering plants, which diversified and spread rapidly by the end of the period. In the animal kingdom, snakes and mammals now existed, though they were small and insignificant figures in landscapes dominated by dinosaurs and huge reptiles. The geographical separation of landmasses began to have a marked effect on the evolution of new species, but the most dramatic event of all was the mass extinction that finished off the dinosaurs and their era.

BY THE beginning of the Cretaceous the supercontinent of Pangea had broken up completely, and the individual continents were scattering, giving rise to different environmental conditions on each. Cretaceous rock sequences are therefore more difficult than their predecessors to correlate between one continent to another. The name Cretaceous is derived from the Latin for "chalk," and the term was first proposed in 1822 by J.B.J. Omalius d'Holloy when he was commissioned to compile geological maps of France. *Terrain Crétacé* was the name he gave to the chalk deposits of the Paris Basin. At the same time, the rocks of the Cretaceous were being studied in England. By then the surveyor William "Strata" Smith, who had established the system of identifying particular beds of rocks by their fossils, had already recognized four distinct sequences between the Portland stone (the top of the Jurassic) and what he called the Lower Clay, which was eventually designated the bottom of the Tertiary. Smith's sequences were called "brick earth," "micaceous clay," "brown chalk," and "white chalk." In 1822 William Conybeare and William Phillips, who also named the Carboniferous, combined Smith's four sequences into the Lower and Upper Cretaceous. This division is still accepted internationally today.

With the breakup of Pangea, the individual continents developed different climates and correspondingly different rock sequences.

The identification of the twelve regional stages of the Cretaceous dates back to a period between the 1840s and the 1870s, when a number of European geologists analyzed the Cretaceous rocks of France, Belgium, the Netherlands, and Switzerland. The stage names are derived from typical Cretaceous sites in these countries.

KEYWORDS

ALULA
ANGIOSPERM
CHALK
CORDILLERA
GYMNOSPERM
EXOTIC TERRANE
HOT SPOT
MARSUPIAL
OVULE
PLACENTAL
PYGOSTYLE
SEVIER OROGENY

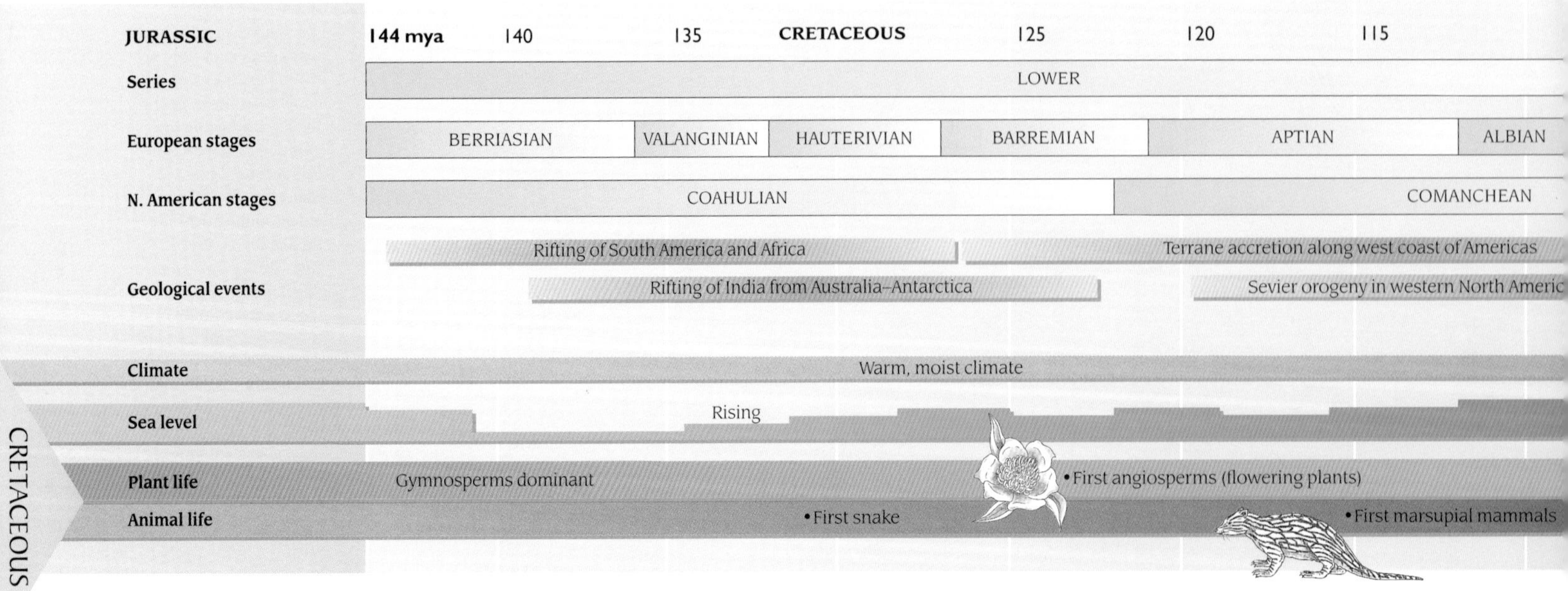

JURASSIC	144 mya	140	135	CRETACEOUS	125	120	115
Series	LOWER						
European stages	BERRIASIAN	VALANGINIAN	HAUTERIVIAN	BARREMIAN	APTIAN		ALBIAN
N. American stages	COAHULIAN				COMANCHEAN		
Geological events	Rifting of South America and Africa; Rifting of India from Australia–Antarctica				Terrane accretion along west coast of Americas; Sevier orogeny in western North Americ		
Climate	Warm, moist climate						
Sea level	Rising						
Plant life	Gymnosperms dominant				• First angiosperms (flowering plants)		
Animal life	• First snake				• First marsupial mammals		

The Lower Cretaceous rocks of Europe consist mostly of soft clays, muds, and sandstones; the Upper Cretaceous consists essentially of the chalk deposits. The classification into these dozen stages has stood until the present, although in 1983 there was a move to divide the longest stage, the Albian, into three.

The European stages are not easy to distinguish in North America. The Cretaceous system on that continent is divided into the Coahulian and Comanchean, which together constitute the Lower Cretaceous, and the Gulfian of the Upper Cretaceous.

The layers of the Cretaceous system are near the top of the stratigraphic column, and have been less deformed by metamorphism and erosion than the older sequences below. For this reason, Cretaceous sediments and fossils—unlike those of the preceding Triassic and Jurassic—are abundant on all continents and in the deep sea. Documenting the events of the Cretaceous has also been aided by the presence of flowering plants, which appeared during the period and whose sensitivity to environmental conditions makes them excellent recorders of climate changes.

DURING the Cretaceous there was still a continuous seaway around the world in tropical latitudes, although this was now becoming restricted by island arcs between North and South America. Sea level stood higher than at any other time in the Phanerozoic. At its height the water surface may have been 1150 to 2130 feet (350 to 650m) higher than it is now. As a result, only 18 percent of the globe's surface was dry land, compared with 28 percent today. The reason for this was probably the fact that the new oceanic ridges, which were driving the spreading motion of the continents, were rising from the ocean floors and displacing the waters upwards and outwards. The spreading action at this time was much faster than usual and so the oceanic ridges were probably higher and more voluminous than they are at the present time.

High sea levels during the Cretaceous left vast deposits of limestone and chalk, and also of organic matter, especially around what is now the Persian Gulf.

The westward-flowing current in the tropics continuously heated the ocean water, which brought warm climates to higher latitudes as it circulated. The Cretaceous climate was probably the warmest of any during the entire Phanerozoic eon. Tropical conditions existed as far north as 45° north latitude, and both poles were temperate. There was little difference in temperature between the poles and equatorial regions: seawater was about 86°F (30°C) all year round at the equator and no lower than about 57°F (14°C) at the poles. Temperatures on the ocean floor were about 63°F (17°C) —much warmer than today's—and so there was little vertical convection between the layers of water.

This was one of the factors leading to the accumulation of undecaying organic material that gave rise to oil deposits. Over 50 percent of the world's known oil reserves originated in Cretaceous rocks, and three quarters of this are found around the Persian Gulf. Most of the remainder is found around the Gulf of Mexico and off the north coast of South America. The oil was formed from organic matter that collected in deep still waters. The Jurassic salt domes of the Gulf of Mexico rose up though Cretaceous rocks, twisting them and producing pockets in which the oil could gather and accumulate in significant quantities. Coal measures, relics of the hot

See Also

THE ORIGIN AND NATURE OF THE EARTH: *Ocean Crust, Seafloor Spreading*
THE TRIASSIC: *Formation of Pangea, Reptiles, Early Dinosaurs*
THE JURASSIC: *Dinosaurs*

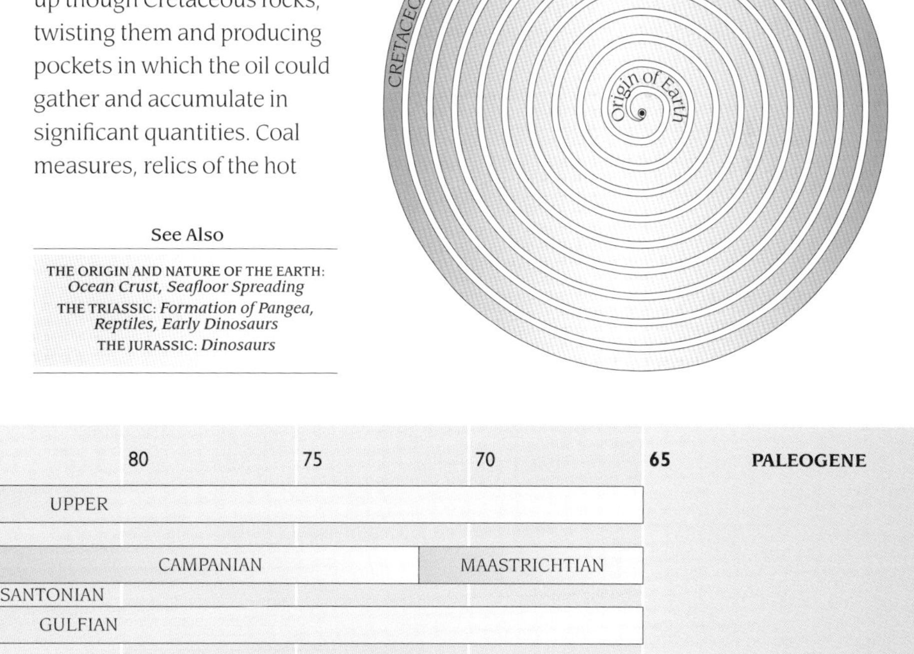

THE END OF THE MESOZOIC ERA

The Cretaceous is known as the "Age of Chalk" for the widespread deposition of this very fine limestone in the second half of the period. During the Cretaceous the continents of the former supercontinent of Pangea became separated into the individual landmasses familiar today. As the modern era approached, species of animals and plants on each continent became increasingly different from those elsewhere. The end of the Cretaceous was marked by the mass extinction of the dinosaurs, which had been dominant for two thirds of the Mesozoic.

100 95 90 85 80 75 70 65 PALEOGENE
UPPER
CENOMANIAN | TURONIAN | CONIACIAN | SANTONIAN | CAMPANIAN | MAASTRICHTIAN
GULFIAN
Initial rifting of Australia–Antarctica
India begins to move rapidly north
Laramide orogeny uplifts central Rockies
Widespread deposition of chalk
Turkey and Iran–Afghanistan collide with Eurasia
High seas; many continental regions flooded
Radiation of angiosperms
Radiation of calcareous plankton
• First placental mammals
Radiation of teleost fishes
Mass extinction
CRETACEOUS

humid climate and the abundant vegetation, were laid down on every continent except Antarctica. The western United States, Canada, Nigeria, and eastern Asia all have extensive deposits of Cretaceous coal—largely lignite, which is of a lower grade than the coal from Carboniferous deposits. Many of the ore minerals of the Rockies formed at this time in the intense mountain-building process that was taking place along the western edge of North America.

ONE OF the most significant developments of the Cretaceous was the change in global geography as the former components of Pangea continued to drift apart. Although the process had begun earlier in the Mesozoic, at the beginning of the Cretaceous the continents were not as widely spaced as today. The southern continent of Gondwanaland, in fact, was still closely grouped at the start of the period. The individual continents of Africa and South America gradually emerged, and the Indian peninsula detached itself and crossed the ocean to join the completely separate continent of Asia. Only Australia and Antarctica in the far south remained joined.

As the Cretaceous progressed, the individual continents, which had been in close formation, began to drift apart.

New oceans were forming, and older ones were expanding as the distances between the continents increased. The southern Atlantic, the Caribbean, and the Gulf of Mexico all opened during the Early Cretaceous. In their infancy the modern oceans were much narrower than they are today. The Atlantic Ocean zigzagged from pole to pole, separating the Americas from Europe and Africa. Its continental shelves were broader on the western edge than on the eastern, because of the way the original rift valley faulted. The symmetrical spreading of the Atlantic at each side of its mid-ocean ridge gave rise to correspondingly symmetrical patterns of hot-spot volcanic island chains, forming in the same manner as the modern Hawaiian chain.

Further east, the Indian Ocean was forming from the Tethys. The continental masses that had separated from the northern margin of Gondwana and had drifted across the Tethys had now fused with the southern edge of the Asian landmass, producing the first wrinkles that were eventually to become the Himalayas. The subduction process that had drawn them across the ocean was continuing. Once the continental fragments had collided, a new trench and island arc, signifying a new subduction zone that resumed the process, appeared offshore. The main portion of India had broken free and was following these two continental masses, shedding fragments behind it that were to become Madagascar and the Mascarene submarine plateau that now peeps above the Indian Ocean as the Seychelles Islands.

Vast shelf seas spread over the lower-lying areas of the continents. In Asia, immediately east of the Urals, the Ob Basin reached southwards from the northern ocean. Further west, almost the whole of western Europe was flooded, the shelf seas of the northern ocean meeting those of the Tethys to produce a vast island-studded shelf on which the great Cretaceous chalk deposits accumulated. One dry landmass here was the area that was to become Spain and Portugal. The opening of the North Atlantic wrenched this landmass from the edge of the continent, and the relative movement between the African and European continents twisted it around in a counter-clockwise direction, opening up the Bay of Biscay and crumpling up the Pyrenees along the join.

On the North American landmass the Mowry Sea reached southwards, separating the newly rising Rocky Mountains from the Canadian Shield. As the period went on, this inland sea met the newly emerged Gulf of Mexico, producing a continuous north–south interior seaway that divided the continent into two land areas. To the east lay the Appalachian mega-island, which had generally low relief and was not tectonically active during the late Cretaceous. To the west was the cordillera formed by the continuously rising Rocky Mountains, whose buildup

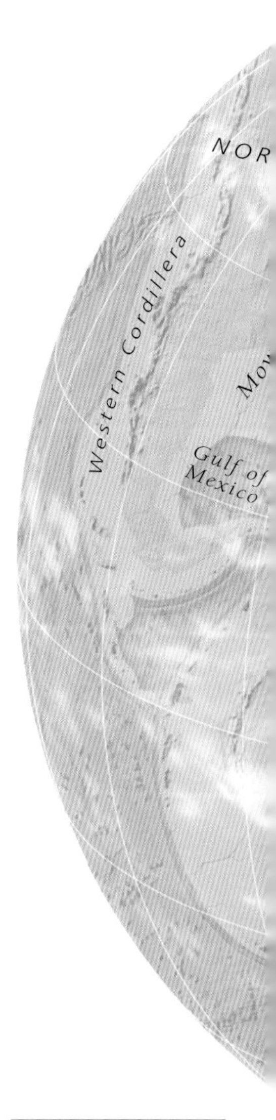

CHALK SEAS OF EUROPE

Chalk accumulated on the beds of the European shelf seas producing the deposits that stand out as white cliffs (above left) in Dorset on the south coast of England. Other famous chalk deposits are found in Kansas, deposited in the North American interior seaway, and in Tennessee and Alabama, deposited on the shelf between the seaway and the new Gulf of Mexico.

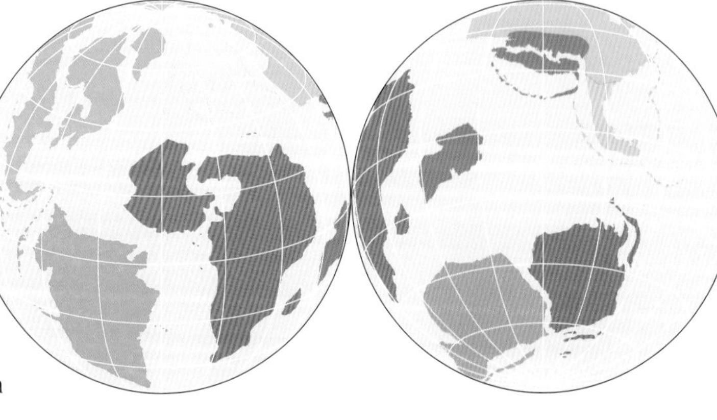

Africa and Middle East
Antarctica
Australia and New Guin
Central Asia
Europe
India
North America
South America
Southeast Asia
other land

during this time is referred to as the Sevier orogeny. The cordillera stretched far into the north and swung west across the site of the modern Bering Strait, connecting this part of the North American continent with northern Asia. To the south, the cordillera was broken in the region of Mexico, and an open seaway lay between its end and the northern tip of South America. Island arcs between the Americas were the forerunners of the land bridge that eventually became Central America.

The open seaway between the Americas meant that the equatorial current was still active, bringing mild climates to every part of the globe. In the North Atlantic, the precursor of the Gulf Stream ocean current existed, demonstrated by the fact that modern Newfoundland and Ireland have fossils of animals that could only have originated in tropical waters.

North America was not the only continent divided by shallow sea. West Africa was isolated by a trans-Sahara seaway and the Iullemmeden Basin, which followed the line of the failed rifts produced by the splitting of Pangea.

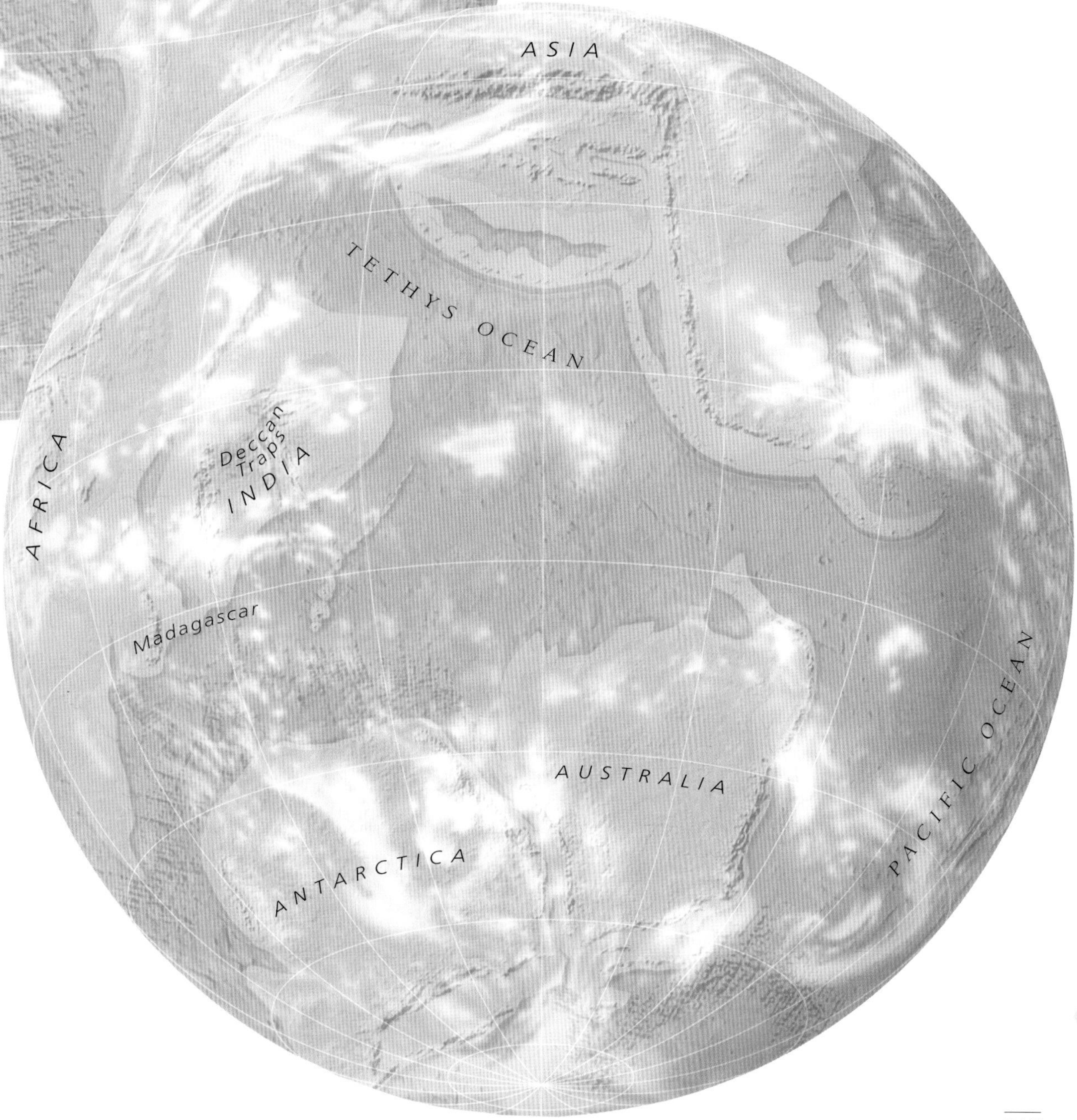

THE CRETACEOUS WORLD

From space the late Cretaceous world would have appeared as a world of oceans. The individual continents had by now broken away from the old Pangea landmass. Only Australia and Antarctica were still joined, and these continents were moving away from each other. Panthalassa, the vast world ocean that had occupied the area not covered by the landmass of Pangea, no longer existed. On each side it was being encroached by the spreading continents, shrinking, and becoming the modern Pacific. Seafloor spreading here was pushing Central and South America together.

THE QUESTION OF AUSTRALIA

Of all the continents, only Australia does not seem to have been affected by spreading shelf seas in the late Cretaceous. Early in the period it did have significant inland seas, but in late Cretaceous times, when all other continents were flooded, almost all of Australia remained dry. Only small areas in the west and northwest had epicontinental seas. The continent must have stood higher than sea level, estimated at 1150 to 2130ft (350 to 650m) higher than today. This is possibly because a complex subduction zone west of the continent caused the landmass in late Cretaceous times to override an ocean plate that had not been completely destroyed. Southwest of modern Australia lie the Kerguelen and Broken Ridge submarine plateaus. These were originally one piece of continental crust but are now split by the Indian Ocean Ridge. Plant remains found by deep sea drilling show that this area was partly above water, probably due to the same phenomenon.

OCEANS are born in rift valleys where continents are splitting apart. Convection currents within the mantle send material up towards the surface and spread it outwards in the mantle's top layers. This activity raises the lithosphere on the surface, along with any continental crust that lies there. Tensions are set up, causing the continent to stretch, form faults, and finally split, with a rift valley forming along the line of weakness. This was the situation that prevailed across the continents in Jurassic times. Sometimes rifts seem to have taken place in areas where continental masses have already come together. An excellent example is in the North Atlantic, where the rift follows the line of the Appalachian and Caledonian mountains formed as the northern continents fused in Devonian and Carboniferous times. Perhaps the distortion of the crust produced by the mountain-building produces a zone of weakness that can be exploited by the tectonic convection currents.

Oceans fill the gaps where continents break and move apart. Pangea's many rifts provided many opportunities for this.

Once the rift valley is established, its development into an ocean is easy to trace. The first sediments to accumulate are freshwater and subaerial, as swamps and freshwater ponds form along the rift valley floor. After about 20 million years the valley has widened to about 60 to 125mi (100 to 200km), and seawater begins to flood in. Salt deposits are formed at this time as the seawater gathers in lakes and evaporates.

After another 20 million years or so the two continental parts actually break apart and new ocean crust begins to form between them. The opposing flanks of the original rift valley are now 125 to 370mi (200 to 600 km) apart, with open seawater between them. By around 30 to 50 million years later the raised continental areas at each side begin to subside and continental

Rift valley
Magma chamber
Dikes
Continent stretched by tectonic action
1
Sea floods into rift valley
Shearing of continental crust
2
3
Narrow continental shelf (Africa)
Solidified chamber material
Dikes
Lava flows
Broad continental shelf (N. America)
Sediment

THE GROWTH OF OCEANS

Oceanic crust is formed by new material welling up from the mantle and solidifying. As the surface continental crust bulges up and cracks into a rift valley a magma chamber of molten rock gathers beneath (1). From this chamber molten material pushes upwards through cracks, sometimes solidifying there as dikes and sometimes erupting at the surface as volcanoes. As the two areas of continental crust pull apart, the igneous matter builds out between, formed from lava flows on the surface, solidified dikes below that and rock that cooled and crystallized at the edge of the magma chamber at depth. This is the structure of the oceanic crust (2). The edges of the continents subside, faulted on one side and downwarped on the other, to form the continental shelves (3).

ISLAND CHAINS

As new ocean crust moves away from the ridge that created it a volcano that appears above a hot spot (1) is carried away from its origin. This volcano dies and begins to erode while another active volcano appears in the fresh crust above the hot spot (2). The eventual result is a string of progressively older volcanic islands stretching away from an active hot spot volcano (3).

A MODERN HOT SPOT

Hawaii's eight main islands stretch in a line extending for 430 miles (700km) across the Pacific. Beyond that, the chain is completely submerged. Only the Big Island (upper right) is currently active,though there are new vents on the seafloor further east.

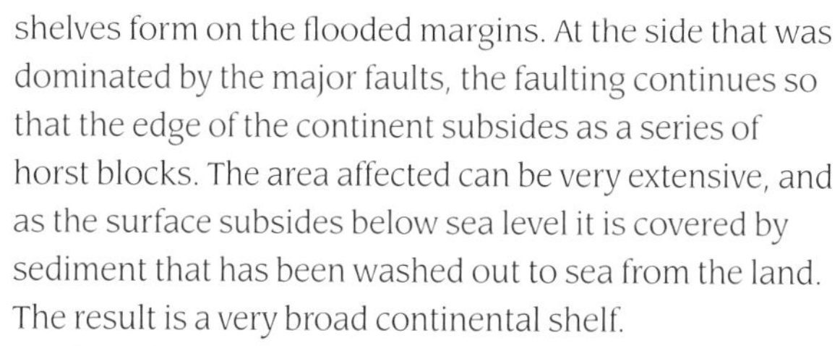

shelves form on the flooded margins. At the side that was dominated by the major faults, the faulting continues so that the edge of the continent subsides as a series of horst blocks. The area affected can be very extensive, and as the surface subsides below sea level it is covered by sediment that has been washed out to sea from the land. The result is a very broad continental shelf.

The other side, the downwarped side, subsides more abruptly and a narrow continental shelf results. A permanent oceanic ridge develops: a line along which new ocean crust material is emplaced by the constant rising of igneous material from the Earth's mantle, forming dikes and volcanoes. If there are no subduction areas at the ocean margin, this ridge runs up the center of the ocean, and the ocean structure is symmetrical at each side. The area is now a full-fledged ocean.

This is precisely what was happening in various areas around the Cretaceous globe. The Atlantic Ocean had widened to maturity from its Jurassic rift valleys, with broad continental shelves on the American side and narrow shelves on the African. Similarly, the gap between India and Africa had widened into the mature ocean stage (the Indian Ocean), as had that between Africa and Antarctica. The boundary between Australia and Antarctica, however, was still a rift valley.

Throughout all the activity associated with the breakup of Gondwana, Africa seems to have remained motionless while the other continents split away from it. All the other continents have been moving apart and have fold mountains along their leading edges. Even today Africa has no extensive chain of fold mountains, but the Great Rift Valley shows that the continent is dispersing.

ELSEWHERE, molten material was rising from the mantle, through the crust and out on to the surface. Volcanoes formed over these "hot spots." Continuously moving tectonic plates carried the volcanoes away from the center of activity and they became dormant. As this activity continued, new volcanoes would appear alongside the extinct ones. The result of this was a chain of progressively older and more eroded volcanoes stretching away from the hot spot in the direction of plate movement. The modern Pacific Ocean is full of such volcano chains, and there were many throughout the oceans of the Cretaceous. The symmetrical opening of the Atlantic Ocean meant that such chains were arranged in a pattern at each side of the spreading ridge. The Walfisch Ridge and Rio Grande Plateau in the South Atlantic are remains of chains that grew in the Cretaceous, and modern Tristan da Cunha is the active volcano that currently stands over their mutual hot spot.

"Hot spots" continued where molten lava emerged from the mantle, leaving a trail of live and extinct volcanoes.

OCEANIC crust has a relatively simple structure: a layer of lava, over a layer made of vertical dikes, over a layer of material crystallized from a magma chamber. Continental crust is extremely complex. It consists of ancient metamorphosed rocks shot through with igneous material and topped by sedimentary rocks. The sedimentary rocks in the center of a continent may remain relatively undisturbed, but at the edge of the continent, close to a subduction zone, they are twisted, sheared, faulted, broken, and completely mangled. It is at the edges of continents that new continental crust is created.

Continental crust is produced in subduction zones at the margins of continents, and is much more complex than oceanic crust.

As the oceanic plate slides beneath the continental plate, any sediment that it carries is scraped up and mixed with sedimentary material being deposited at the continent's edge. This mixture is then sheared and faulted by the movement between the two plates, and the result is a fold mountain range at the edge of the continent. Often this activity takes place offshore, producing a chain of elongated islands parallel to the coast. This can be seen today in Vancouver Island off the west coast of North America, and the Chilean Archipelago on the west side of the southern tip of South America. These islands are all temporary features and will eventually become welded to the main part of the continent.

As the oceanic plate descends into the mantle, it begins to melt, and the molten material rises through the overlying continental plate. There it gathers in magma chambers below the surface, from which volcanoes erupt in the coastal mountain range. The magma in these chambers has a different composition from that in the magma chambers below the oceanic ridges, since the melt derives from molten crustal material rather than from the mantle. Volcanic eruptions in these areas, such as Mount Saint Helens and Mount Cotopaxi, are more violent than those of oceanic sites, such as in Iceland and Hawaii. If the subduction zone is some distance from the continent, then the volcanoes erupt as island arcs close to the ocean trenches that mark the subduction zones.

EDGE OF A CONTINENT

A fold mountain range at a destructive plate margin is not made up simply of oceanic sediments and continental rocks all crumpled up together. Instead, it consists of pieces of islands and scattered landmasses that have been brought across the ocean by subduction and welded to the edge of a continent. Such pieces are all different from one another and give rise to completely different suites of rock in the mountain range. These pieces are known as exotic terranes.

1 Isolated continental mass moving towards the continent
2 Old volcanic island chain (destined to become exotic terranes)
3 Fold mountain islands, incorporating ocean sediment and exotic terranes
4 Coastal fold mountains
5 Crustal melting produces magma chambers and volcanoes
6 Old magma chambers cooled and eroded to batholiths
7 Old exotic terrane
8 Ancient metamorphosed basement
9 Alluvial plain

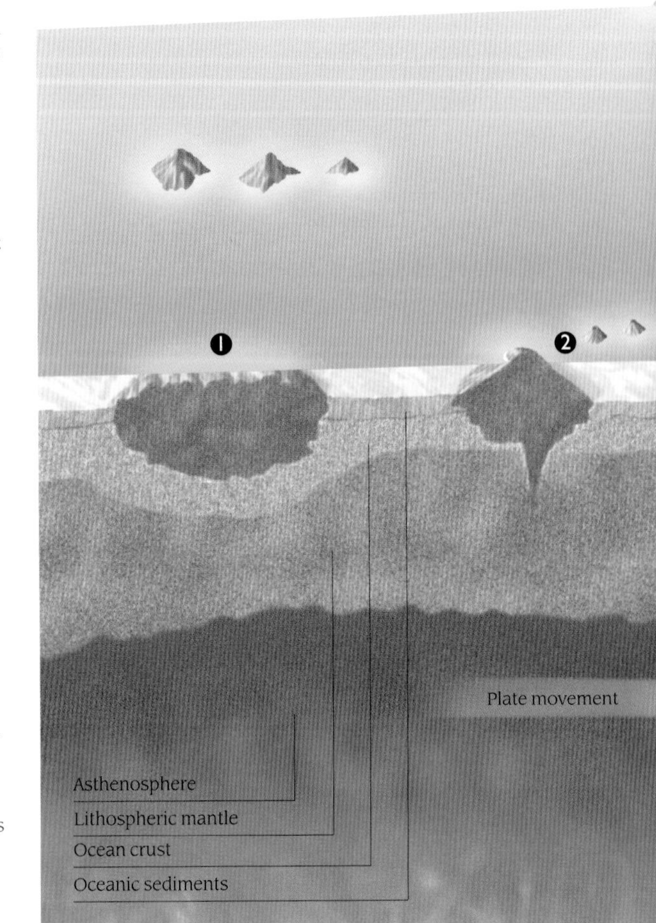

CONTINENTS as well as oceans were growing during the Cretaceous period. An important part of the process, especially on the west coast of North America, was the accretion of exotic terranes, sometimes called "allochthonous terranes," "suspect terranes," "displaced terranes," or even "alien terranes." These are patches of crustal material that have been brought to the continent and welded there. They consist of old island chains, continental fragments, and pieces of submarine plateau with a former existence somewhere in the ocean. As the oceanic lithosphere moves towards a continent and is destroyed in the subduction zone, these masses are brought to the

The Western Cordillera, made up of exotic terranes, added a large western margin to North America. Before its accretion the edge of the continent lay where Nevada and Arizona now are.

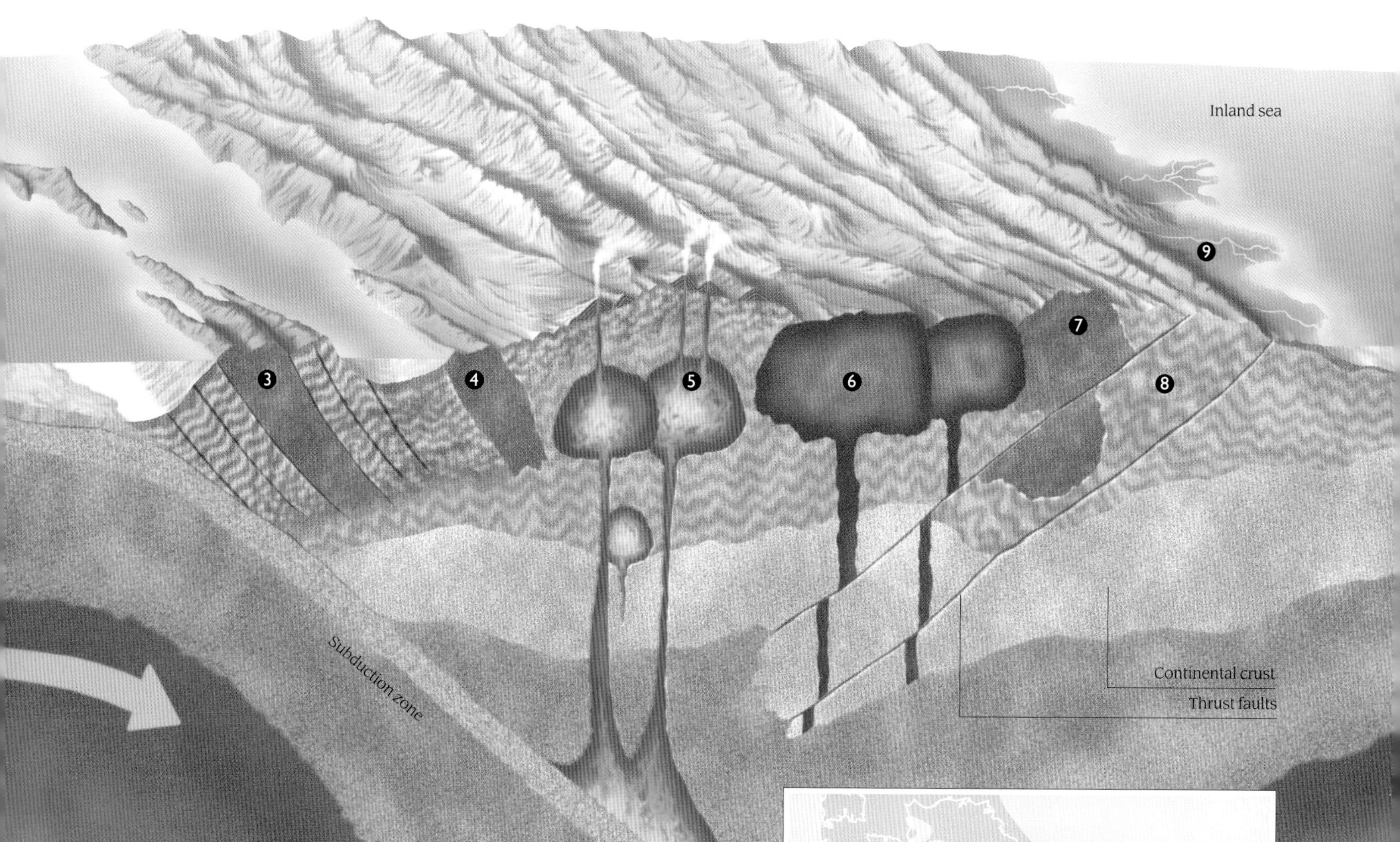

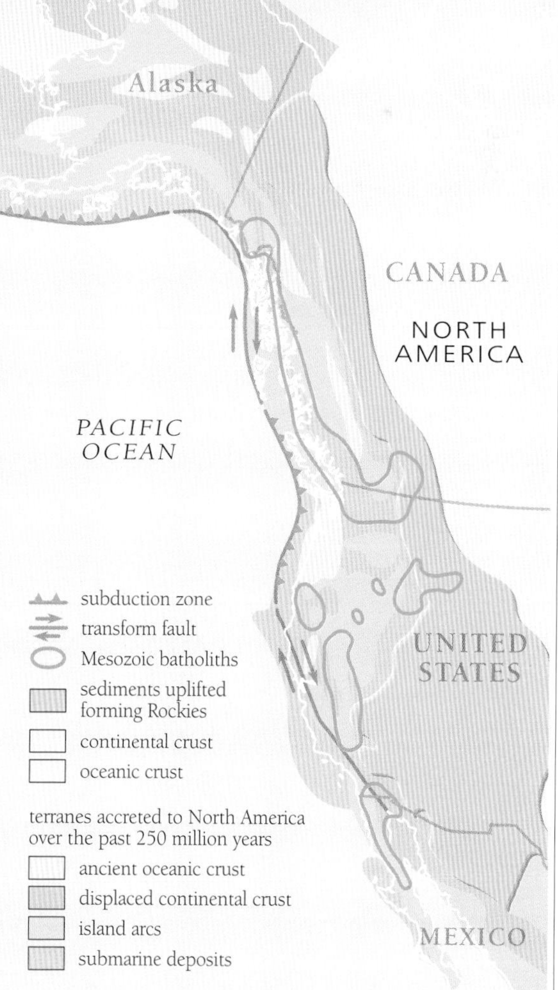

EXOTIC TERRANES

The Western Cordillera of North America, comprising the Rocky Mountains, the Sierra Nevada, and the various coast ranges from Washington through California, is made up of more than 200 exotic terranes, many of them too small to be shown on a smallscale map. They accumulated over a long period of geological time, with most of the buildup occurring between the Permian and the Cretaceous, although activity continued well into the Cenozoic. Each accretion added to the west coast of North America.

SIERRA NEVADA

The Sierra Nevada range (left) stretches some 350mi (600 km) through northern California. This exceptionally rugged territory was a formidable barrier to explorers, and its extreme complexity makes it difficult to study the relationships of different rock suites.

continent's edge. Made primarily of continental material with a preponderance of silica and alumina, they are less dense and more buoyant than the magnesium-rich substance of the ocean plate they ride upon. As a result they resist being dragged down by the descending plate and are scraped off against the edge of the continent.

The Western Cordillera of North America is a broad belt consisting of numerous mountain ranges extending from the Pacific coast to the Rockies. It is almost entirely composed of exotic terranes, each one separated from its neighbor by a fault. It is not yet possible to tell where they all came from, but their distinctive lithology shows that each had a different origin from its neighbors. Geologists have identified about 200 terranes from Mexico to Alaska, some small, others several hundred miles long. Most of this accretion took place between Permian and Cretaceous times. Paleomagnetic evidence suggests that some of these must have traveled from 600 to 1200mi (1000 to 2000km) before reaching the mainland. At least one terrane seems to have developed nearby as an offshore island arc in Paleozoic times and then been welded to the continent as the pattern of subduction changed. In more recent times the East Pacific oceanic ridge has been swallowed up beneath the North American continent. This adds to the complexity of the geology of the mountains of the Western Cordillera.

Today, old island chains and continental fragments lie scattered across the floor of the Pacific. It seems unlikely that the accretion of the American continents is finished.

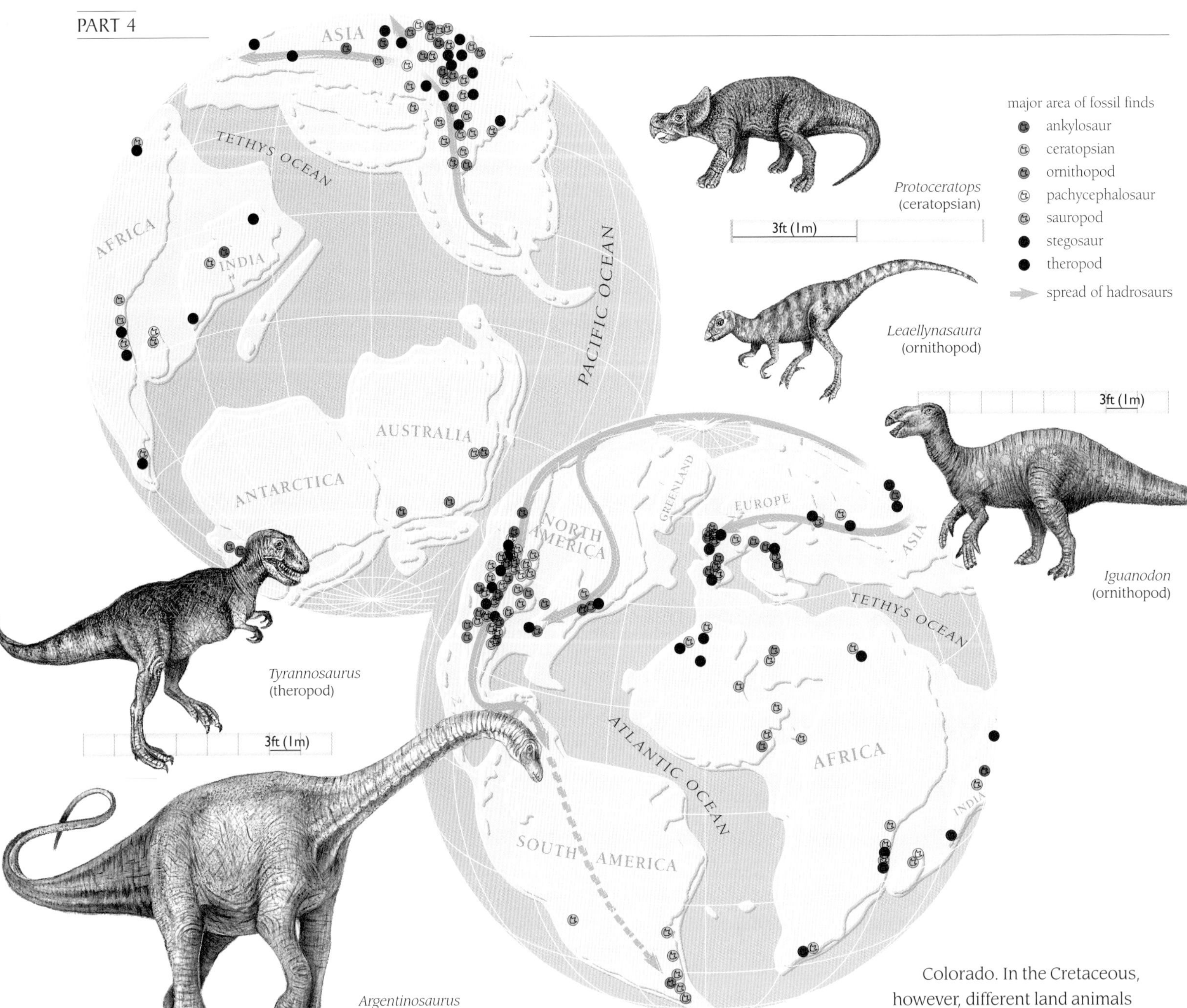

Protoceratops (ceratopsian)

Leaellynasaura (ornithopod)

Iguanodon (ornithopod)

Tyrannosaurus (theropod)

Argentinosaurus (sauropod)

AROUND THE WORLD

The Asian and North American faunas were similar, joined by the land bridge across what is now the Bering Strait. Here was the greatest variety of dinosaurs, including all the major groups. South America and Africa, in contrast, were the land of the giant plant-eating sauropods such as *Argentinosaurus*. The hadrosaurs (duck-billed dinosaurs) originated in Mongolia and colonized the northern hemisphere. Although an "advance guard" reached South America, the separation of landmasses stopped their spread into India, Africa, or Australia.

FROM the Cretaceous, for the first time in the history of life on Earth, zoogeography—the study of the distribution of animals—can be applied to animals that lived on land. Zoogeography investigates why a particular suite of animals lives on a particular continent and why it is distinct from a different suite of animals on another. It studies the barriers between the two populations and the various reasons why they do not mix. Before the Cretaceous, the distribution of animals was fairly even over the broad landmass that was Pangea. Fossils of very similar plant-eating pareiasaurs and mammal-like reptiles are found in Permian rocks from South Africa and Russia. In the Jurassic, the same dinosaurs lived both in what was to become Tanzania and in what became Colorado. In the Cretaceous, however, different land animals began to live on different continents. Just as today there are unique species on the isolated continent of Australia, in Cretaceous times new animals were evolving on the continents as they drifted apart. Within those continents were isolated landmasses separated by the shelf seas that spread everywhere at the end of the period, creating many more localized zoogeographic provinces.

The modern world is divided into five major zoogeographic realms. On the isolated landmasses of Cretaceous times, there would have been many more.

Just as interesting as the factors that separated the various animal populations are the factors that connected them. Nowadays northern Asia and North America are regarded as forming two distinct zoogeographic realms (areas where the animal life is distinctive). The modern Bering Strait forms a barrier between these two locations. In the Cretaceous period there was no Bering Strait, and northern Asia and the Western Cordillera of North America were a continuous landmass. It is hardly surprising, then, that strikingly similar dinosaur fossils have been found on the two continents.

It is also possible to trace the evolution of particular groups from one area to another. The ceratopsians—the horned dinosaurs—evolved in northern Asia. As the group expanded, it spread east, and at the height of their success they populated the forests of Canada and the western United States. Hadrosaurs (duck-billed dinosaurs) seem to have originated from iguanodontid stock somewhere in eastern Europe and then spread to Asia and to north America, where they became the most important plant-eating dinosaurs of the time. As the Cretaceous progressed, the hadrosaurs replaced the long-necked sauropods over most of the world—except in South America, which was an island continent at that time. It was its own zoogeographic realm, and long-necked sauropods remained the most important of the plant-eating dinosaurs until the end of the Cretaceous.

DESPITE the fact that climatic conditions throughout the world during the Cretaceous were fairly even, there were some differences in environmental conditions leading to differences in animal life. In the far north, in contemporary Alaska beyond the Arctic Circle, there was day-long darkness for part of the year. The climate was too warm for frost. The moist conditions were similar to today's Pacific northwest, and the vegetation consisted of small deciduous trees that shaded one another from the side. Dinosaurs migrated in and out of the area according to the season.

Small local differences in seasonal conditions contributed to the great diversity of Cretaceous life.

On the broad damp plains of Early Cretaceous northern Europe that existed before the whole area became flooded later in the period, there were vast swamps of horsetails and ferns, grazed by herds of *Iguanodon*. These conditions were not seasonal, and so there would have been no need to migrate.

The centers of the largest continents, such as northern Asia, had dry conditions. In the Gobi desert there are fossils of herds of the primitive ceratopsian *Protoceratops* killed suddenly by sandstorms. Some animals died in the act of fighting each other.

Australia at that time still lay within the Antarctic Circle. Here were climatic conditions similar to those in Alaska, with big animals migrating and little animals weathering the seasons. There were unique Australian Early Cretaceous creatures such as the hypsilophodont *Leaellynasaura* that had big eyes, probably as an adaptation to the long Antarctic night. Despite the presence of forests of podocarps, monkey-puzzles, and southern beech, a bit like modern South America, there was little or no coal formed. It was probably too dry for peat formation.

TOP CARNIVORE

Convergent evolution is an important concept in zoogeography—the same shapes of animals seem to develop independently in different areas that have similar conditions. *Tyrannosaurus* was one of the biggest carnivores that ever lived on land. The tyrannosaur family lived in North America and Asia. Meat-eaters just as big or even bigger, such as *Giganotosaurus*, evolved on the newly isolated continent of South America, but these evolved from the unrelated tetanurids. Conditions in both places favoured the evolution of giant meat-eaters.

Even within the same continent there were variations. In North America there were still areas of open dry land with clumps of conifer, unchanged since Jurassic times, allowing the survival of long-necked sauropods such as *Alamosaurus*. Other areas had more dense forests of newly evolved angiosperms, broadleaved trees like today's. These were the home of the duckbills and armored ankylosaurs.

Half of the species of dinosaurs currently known come from the late Cretaceous. One likely reason is that each isolated area could support many different dinosaur types. As the waters arose in the latter part of the period, small areas were cut off. Central Europe became a scatter of small islands. From these islands we find the fossils of dwarf dinosaurs. *Telmatosaurus* was a duckbill but it was only a third of the length of its mainland cousins. *Struthiosaurus* was an armored ankylosaur but was only the size of a small sheep. This pattern can be observed in the mammals of the Cenozoic, and in modern Shetland ponies, which come from a group of isolated islands off Scotland with limited resources.

Up to this time the Mesozoic flora had been dominated by primitive conifers like podocarps and araucarians, and also ginkgoes, tree ferns, cycads, and cycad-like plants. These still existed, but now there were also more modern-looking conifers such as pines and firs, and the flowering plants including oaks, ash poplars, sycamores, maples, willows, and birch. There were bushes of magnolia, cinnamon, viburnum, holly, and laurel. In the undergrowth the herbs were saxifrage, lilies, primulas, and heather. Climbing plants including grape vines and passionflower plants wound though all these forming a very distinct and modern-looking flora. Brightly-colored flowers now existed among the background green.

Flowering plants transformed the landscape of the Cretaceous, making it familiar to modern eyes. Oak, maple, and willow mingled with ginkgoes; magnolias, lilies, and viburnum all bloomed.

The presence of the dinosaurs may have been one ecological factor that contributed to the evolution of flowering plants. The main herbivores of the earlier Mesozoic were the sauropods, huge long-necked plant-eaters that scraped the needles from conifer trees and processed them in their gargantuan digestive systems.

Cretaceous ornithopods such as *Iguanodon* and the hadrosaurs, with their downward-flexed necks and their broad beaks, were more adapted to feeding from the ground. They were very efficient at clearing all the undergrowth in their feeding area. Under these conditions evolution favored the plants that could reproduce quickly, with seeds that were full of nourishment to give them a head start, and also the ability to propagate by underground stems even though the part of the plant above ground had been damaged or eaten.

Once flowering plants had evolved, their evolution and that of the dinosaurs probably went hand-in-hand. The large number of different flowering plants would have led to the development of a large number of different dinosaur species that could feed on the new varieties. This would have been another reason for the fact that half of all dinosaurs known came from the last 20 million years of the Cretaceous. There was another evolutionary implication: the highly nutritious nature of the new seeds meant that the dinosaurs would not have needed such a complex digestive system to break them down, and so the later plant-eaters were generally smaller than the great sauropods that went before.

However, in some areas the archaic cycads and conifers remained dominant. The narrow beaks of the ceratopsians were ideal for nipping out the leaves of cycads, and it was here that the sauropods clung to their existence until the end of the age of dinosaurs.

DINOSAURS OF THE DESERT

The skeletons of two late Cretaceous dinosaurs lie locked together in death. In the last moments of their lives they were engaged in a fight. The *Velociraptor* seized the head shield of the *Protoceratops* and slashed into its belly with the killing claws on its hind feet while the *Protoceratops* sank its beak into the abdomen of its attacker. Neither won the fight, since a sandstorm blew up so suddenly that it buried and suffocated them both.

The fossil record is nearly all based on the remains of sea-living animals. The reasons are obvious. Sediment is accumulating all the time at the bottom of the sea. Marine organisms fall into this sediment as they die. They become buried, and when the sediments turn to rock, then the remains become fossils. Land surfaces, on the other hand, are areas of erosion rather than of deposition. Dead animals on land are torn apart by scavengers and their

Many dinosaurs are known only from a single fragment. Complete skeletons are extremely rare.

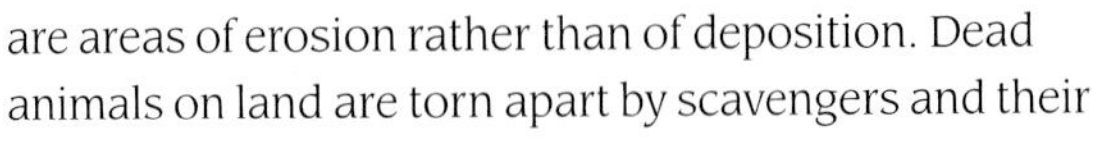

hard parts are broken down by the weather. Usually a land animal is preserved only if it falls into a river or a lake. This is the reason why dinosaur fossils are much rarer than the fossils of sea-living reptiles such as ichthyosaurs and plesiosaurs. Only about 3000 fossil finds of dinosaurs have ever been made.

However, there are a number of instances of good preservation on dry land. During the early history of dinosaur study, from the 1820s onwards, the scientists had only fragments from which to work. Because the pieces of fossilized bone and teeth appeared to be reptilian in origin, dinosaurs were at first classified as gigantic dragon-like lizards. This was the interpretation that prevailed until the discovery of the remains of more than 30 *Iguanodon* skeletons in a coal mine at Bernissart in Belgium in 1879. Miners tunneling through a seam of Carboniferous coal suddenly came across a concentrated mass of fossilized bones. It took two years to remove the blocks of stone containing these fossils, and another 30 years to make a thorough study of them.

This was the first time that scientists had complete skeletons to work from and on which to base their conclusions. The first reconstructions that were made from these remains were of animals that stood like kangaroos. A bipedal pose for an American duckbill, *Hadrosaurus*, had been proposed some 20 years before, but this had been on the basis of only a partial skeleton.

The complete *Iguanodon* skeletons were fully articulated, and showed animals with small heads, long tails, and hind limbs that were longer than the front. The king of Belgium at the time remarked that they looked rather like giraffes; indeed, in their upright pose, with their small browsing heads and long necks, they could easily have reached the tops of high trees. Modern studies, however, suggest that the *Iguanodon* spent most of their time on all fours, feeding from the ground.

Another example of fully articulated dinosaur remains comes from Mongolia. There, in late Cretaceous times, lived the small ceratopsian *Protoceratops*. These animals were so abundant that they must have dotted the landscape like sheep. Modern paleontologists often find their complete skeletons preserved in a lifelike position in desert sandstones. The inference usually drawn from such a discovery is that the animals were engulfed by sudden sandstorms and suffocated before they could find shelter. Such sandstorms are common in that area today. A testament to the suddenness of such an occurrence lies in the fossils of two ancient antagonists, a *Protoceratops* and a meat-eating *Velociraptor*, preserved coiled around each other in perfect condition, killed while they were ripping the guts and life out of each other.

CARBONIFEROUS ROCKS, CRETACEOUS DINOSAURS

In Early Cretaceous times a broad swampy plain spread across southern Britain and northern Europe. It was bounded on its northern edge by a ridge of Carboniferous limestones and coal measures. The limestones would have been subjected to the usual erosional process that produce sinkholes and caves. *Iguanodon* grazed the ferns and horsetails of the swamp, but when moving from one area to another they probably moved along the high ground in herds. There the unwary creatures may have lost their footing and fallen down sinkholes, to be found when the Bernissart coal miners tunneled through the Carboniferous rocks 135 million years later.

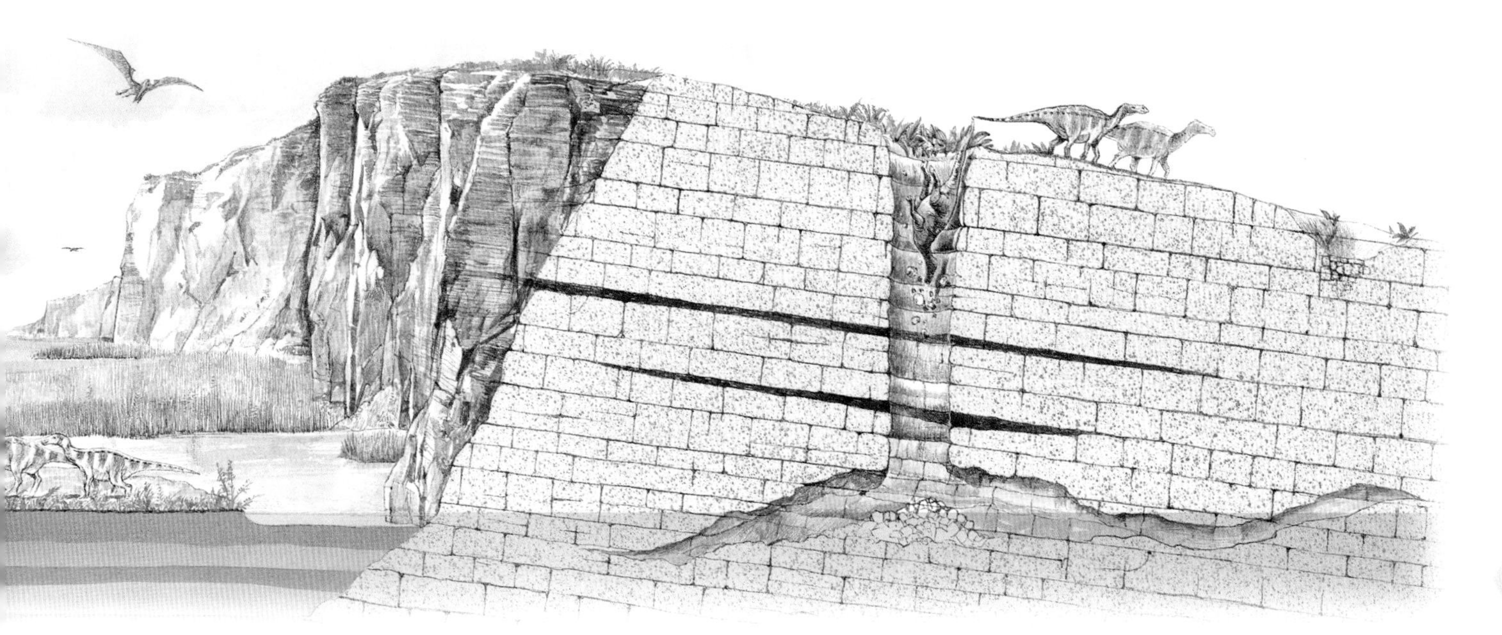

THESE were the closing days of the earlier part of the Mesozoic on the Wealden plain of northern Europe. Within a few tens of millions of years, all this area would be under shallow seas, and what remained as dry land would have completely different vegetation. Fields of horsetails, bordering ponds and channels of silvery water, stretched away to the limestone ridges misty on the horizon. Here and there clumps of conifers, cycadeoids and ferns stood out, rooted in slightly drier ground. Just above the water surface skimmed damselflies—flashes of bright color in the greenery. Among the stems and roots there seethed colonies of termites, the world's first social insects.

In Early Cretaceous times, northern Europe was an expanse of hot, steamy swamp, soon to be inundated and to re-emerge as a modern landscape full of flowering plants.

High in the sky wheeled the large pterosaur *Ornithodesmus*, keeping watch for any small prey, but the animals it saw were too big for it to tackle. Across the plain wandered a herd of *Iguanodon*, grazing the horsetails as they went. Between their feet scampered their smaller relatives *Hypsilophodon*, using the presence of the large dinosaurs as protection. Meat-eaters were around too. By a stream the long-snouted *Baryonyx* crouched, but it had no interest in *Iguanodon* meat—it waited for fish to swim by. Following the herd, though, was the fearsome theropod *Neovenator*. It posed no immediate threat...but it was not yet hungry.

5

1 *Iguanodon* (an ornithopod)
2 *Hypsilophodon* (an ornithopod)
3 *Neovenator* (a theropod)
4 *Ornithodesmus* (a pterosaur)
5 *Baryonyx* (a theropod)
6 *Araucaria* ("monkey puzzle")
7 Ginkgo
8 *Weichselia* (a fern)
9 *Equisetites* (a horsetail)
10 *Williamsonia* (a cycadeoid)

1

CRETACEOUS

4
6
7
8
3
9
2
10

THE MASS extinction at the end of the Cretaceous period was not the greatest in the history of the Earth, but it was certainly the most famous. It brought to an end the time of the dinosaurs, but it also extinguished the flying and the swimming reptiles, the belemnites and ammonites, the specialized bivalves of the oceans, and many groups of fish. The mass extinction is often referred to as the "KT event"—K from the German Kreide meaning Cretaceous, and T from Tertiary. It happened about 65 million years ago, bringing not only the Cretaceous but the whole Mesozoic to a close. It wiped out all land animals that weighed more than about 55lbs (25kg), and left the earth open for recolonization by completely different groups descended from the small survivors.

About 75 percent of species were wiped out at the end of the Cretaceous period.

What actually happened is still something of a mystery, but the theories can be divided into "catastrophist" and "gradualist."

Most significant of the catastrophist theories is that 65 million years ago the Earth was struck by a meteorite or a hail of comets. Shock waves would have wiped out anything in the immediate area, and towering sea waves would have swept over low-lying lands. Smoke and ash and dust would have spread throughout the atmosphere, cutting out the sunlight and inhibiting the growth of plants. Acid rain would have caused worldwide damage. In the disrupted ecosystems it would have been very difficult for many life forms to survive.

The meteorite impact theory was proposed in the 1970s with the discovery of a bed of iridium—an element rare on the surface of the Earth but common in meteorites—dating from the KT boundary, at different places on the Earth. Supporters of this hypothesis also cite the presence of grains of "shocked" quartz, which display the parallel welded fractures that only form under the heat and pressure of such an impact. Since then a deeply buried structure resembling a giant meteorite crater and dating from the KT boundary has been found off the coast of the Yucatán peninsula in Mexico. The crater measures about 110 miles (180km) in diameter, requiring an object with a diameter of 6mi (10km) to produce it; the sea waves generated would have swept inland for hundreds of miles in every direction around the Gulf of Mexico.

An alternative catastrophic theory is that the world suffered a great deal of volcanic activity at that time—exemplified in the Deccan Traps of India, which cover 190,000 mi² (500,000km²). The intensity of such an eruption would have had very much the same effect as the debris thrown up by a meteorite impact, and would even have produced beds of iridium concentrated from the material brought up from deep within the Earth. Significantly, the greatest mass extinction in the history of the Earth, that between the Permian and Triassic, was also accompanied by intense volcanic activity, in this instance producing the Siberian Traps.

An interesting angle on these two possibilities is that the Yucatán peninsula in Mexico and the Deccan Traps in India were exactly 180 degrees around the globe from each other at that time. It has therefore been suggested that the two events are linked. Perhaps a meteorite approaching the Earth was pulled apart by gravitational tidal forces; two large fragments struck the Earth 12 hours apart, with the one that fell in India setting off the volcanic activity. Alternatively, a huge strike in the Yucatán by a meteorite could have set up vibrations and resonances within the Earth that surfaced on the opposite side of the globe and caused the eruptions.

TRACK OF A COMET?

(Left) Two violent events at the end of the Cretaceous period were the meteorite impact in Mexico and the volcanic eruptions in India, one 180 degrees around the globe from the other. They may have been connected, in that the impact may have stimulated the eruptions, or else the Earth may have been struck by two meteorites 12 hours apart. Or the situation may have been just a coincidence. A possible meteorite crater of this age lies partly at the edge of the Indian subcontinent and partly at the edge of the Seychelles. These were a single landmass at the time.

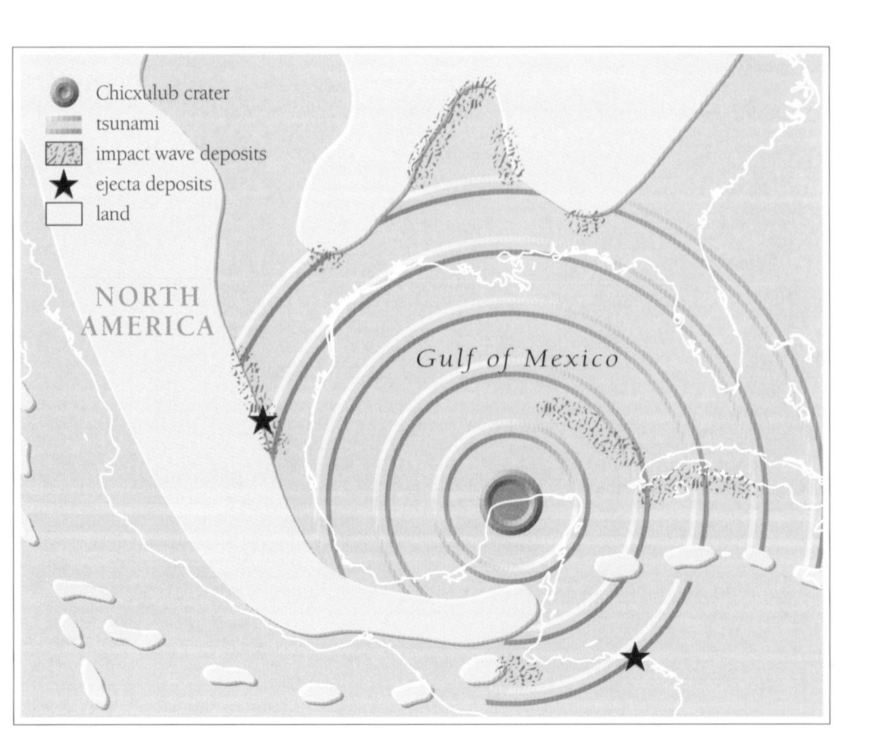

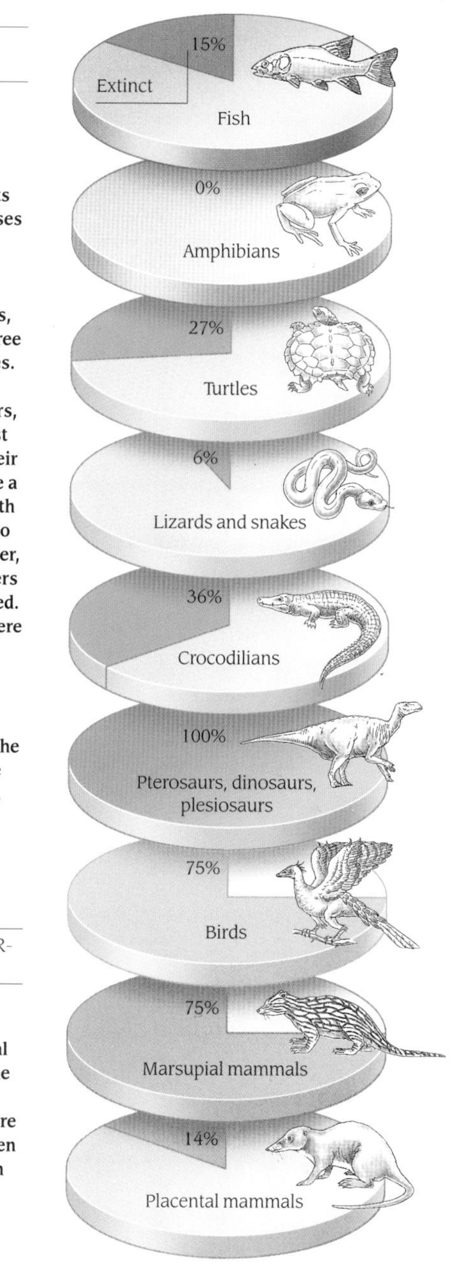

WINNERS AND LOSERS

Whatever it was that caused the mass extinctions at the end of the Cretaceous, it varied enormously in its effect on different classes of animals (right). Dinosaurs were completely wiped out, while their descendants, the birds, lost about three quarters of their species. In spite of their close relation to the dinosaurs, however, crocodiles lost only about a third of their number, and later made a good recovery along with the birds. Mammals also underwent a big turnover, with about three quarters of marsupials eliminated. Among the survivors were the vast majority of placental mammals as well as the fish. Only amphibians, which had been dominant during the Late Paleozoic but were eclipsed by the reptiles, seem to have remained entirely unaffected.

COMET SHOEMAKER-LEVY HITS JUPITER

In 1994 a comet passed Jupiter and was pulled apart by its gravitational forces. Fragments hit the planet in a series of impacts, as in this picture (left). This may have been what happened on Earth 65 million years ago, causing the mass extinctions.

EVIDENCE OF IMPACT

(Left) A massive, buried ringlike structure in the Gulf of Mexico, just offshore from the Yucatán Peninsula is consistent with a meteorite 6mi (10km) in diameter striking the Earth at the end of the Cretaceous. This crater is called the Chicxulub structure. Other pieces of evidence nearby include beds that may have been deposited by tsunamis (giant surge waves), and layers of glassy minerals that appear to have been melted and thrown over a wide area by a great explosion. Marine deposits from the region contain large quantities of terrestrial plant material, possibly swept out to sea by the back-wash of the tsunamis.

GRADUALIST theories of the KT extinctions are much less dramatic. At the very end of the Cretaceous period the sea level, which had been so high for so long, suddenly dropped. This change would have had a profound effect on the global climate. Winters became much cooler while summers became drier, putting intolerable pressure on the dinosaurs and other large animals that had existed in consistently favorable conditions for tens of millions of years. The lowering of sea levels would have opened up land bridges and connections between the continents—between their zoogeographic realms—and the resulting mix of populations could have spread diseases before individuals could develop resistance.

The climatic disruption caused by a meteorite impact could have resulted in a cold "impact winter" or in a "greenhouse effect." One may have followed the other. In either case the result would have been catastrophic.

Evidence cited in support of a more gradual extinction includes the timespan involved. The dinosaurs were apparently declining for about 20 million years before the very end of the Cretaceous, with only western North America supporting thriving dinosaur communities. In contrast, in the Pyrenees mountains between France and Spain, the dinosaurs seem to have disappeared about a million years before they did anywhere else. Many of the marine mollusks seem to have died out about 6 million years before the end of the period.

Catastrophic and gradualist theories are not mutually exclusive. It is entirely possible that the Earth was struck by a giant meteorite at the very end of the Cretaceous period, and that the severe environmental disruption it caused would have merely dealt the killing blow to a whole range of animals that were already declining.

The chief problem in coming closer to a satisfactory explanation is that the fossil record of land animals from the end of the Cretaceous period is very incomplete, and it is almost impossible to gain worthwhile statistical samples of the animal life of the time.

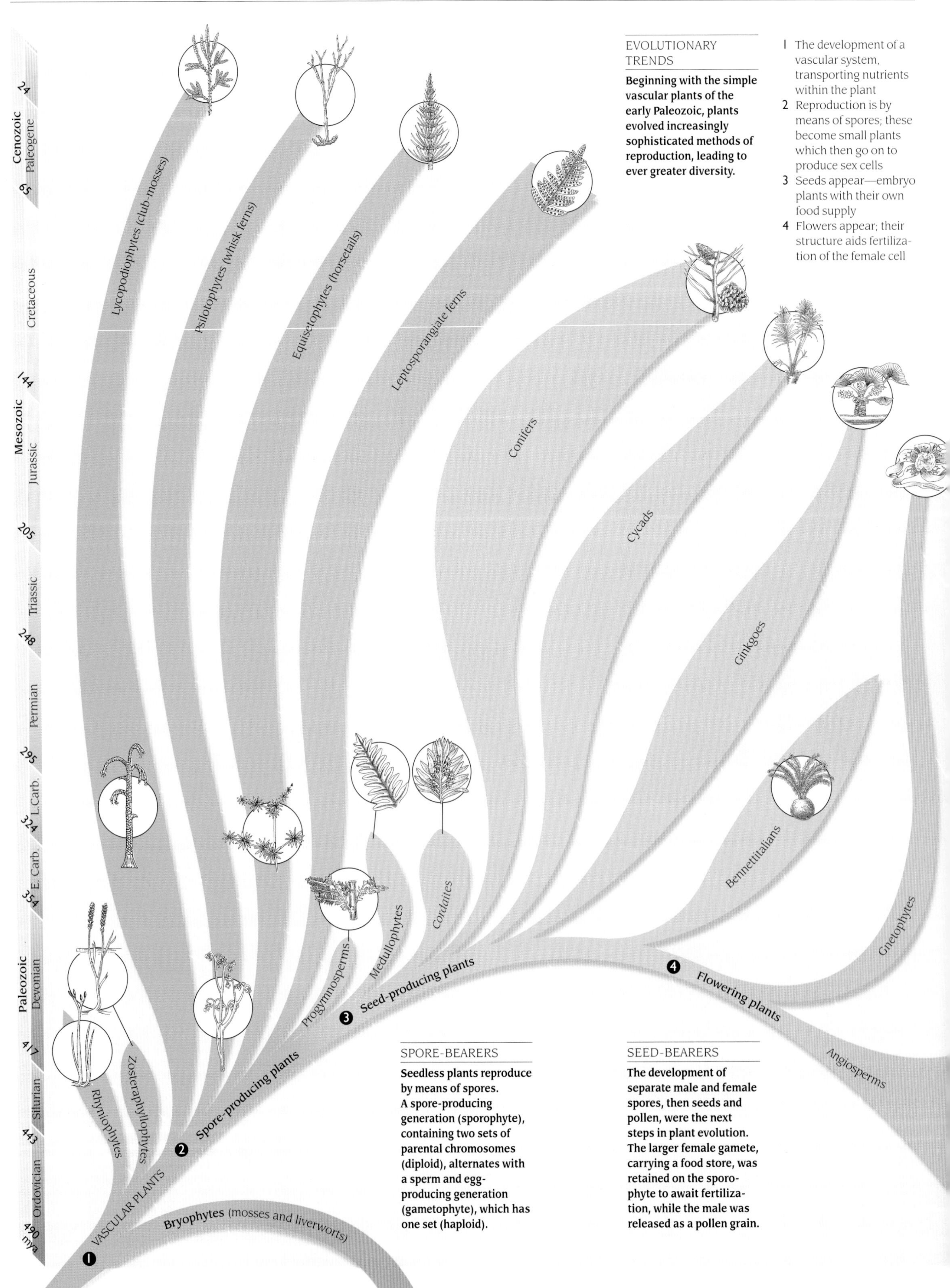

EVOLUTIONARY TRENDS

Beginning with the simple vascular plants of the early Paleozoic, plants evolved increasingly sophisticated methods of reproduction, leading to ever greater diversity.

1 The development of a vascular system, transporting nutrients within the plant
2 Reproduction is by means of spores; these become small plants which then go on to produce sex cells
3 Seeds appear—embryo plants with their own food supply
4 Flowers appear; their structure aids fertilization of the female cell

SPORE-BEARERS

Seedless plants reproduce by means of spores. A spore-producing generation (sporophyte), containing two sets of parental chromosomes (diploid), alternates with a sperm and egg-producing generation (gametophyte), which has one set (haploid).

SEED-BEARERS

The development of separate male and female spores, then seeds and pollen, were the next steps in plant evolution. The larger female gamete, carrying a food store, was retained on the sporophyte to await fertilization, while the male was released as a pollen grain.

THE EVOLUTION OF FLOWERING PLANTS

A CRETACEOUS FLOWERING PLANT

Magnolias are generally regarded as the most primitive living family of flowering plants. Their characteristic features include the very large number of parts such as petals and stamens, which are all separate (not fused to one another), and the spiral arrangement of the parts.

ANGIOSPERMS, or flowering plants, account for 80 percent of all living green plants. In the middle Cretaceous they emerged in a highly favorable climate and quickly took over the planet. It is not clear whether flowering plants evolved from gymnosperms or from seed-bearing ferns, but the earliest examples were probably woody shrubs or trees that existed 130 to 120 million years ago. Fossil flowers are scarce, and so the evolution of the flowers themselves is sketchy.

All of the parts of a flower—sepals, petals, stamens, and carpels—evolved from leaves that became specialized for reproduction. The earliest flowers developed from leaf-bearing shoots. Fertile "female" (ovule-bearing) leaves were found at the very top of the plant. These evolved to wrap around the ovule, forming the carpel, which protects the ovule and the seed from insects, dryness, and infection. The leaves in the next layer down were fertile but male, and these transformed themselves into stamens. Lower still, the leaves lost chlorophyll and became colored petals, some of which began to secrete nectar (a sugar). The only leaves remaining in their original form are the sepals. Grasses are flowering plants, too, though their flowers are less obvious.

Angiosperm "double fertilization" creates both a new plant and a self-contained food supply for it within the seed. Flowering plants thus invest less energy in reproduction until fertilization has occurred, and can reproduce faster. This may explain their rapid rise. Another possibility is that tall herbivorous dinosaurs influenced plant evolution by feeding on tall gymnosperms, which did not evolve defenses against lower-level grazers. When these dinosaurs were replaced by those who fed lower to the ground the gymnosperms declined. A new wave of insects, including bees and butterflies, followed the spread of flowering plants.

1.8
Neogene
24
Cenozoic
Paleogene
65
Mesozoic
Cretaceous
144
Jurassic
ANGIOSPERMS
Magnoliid complex
Non-monocot paleoherb complex
Monocots
Alismatales
Lilianae
Commelinanae
Tricolpates
Basal tricolpates
Caryophyllanae
Rosids
Asterids

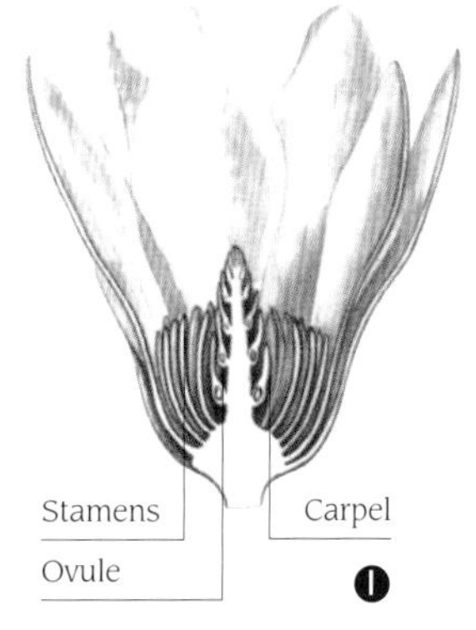

2

Stamen
Carpel
Ovules
3

SECRETS OF SUCCESS

A cross-section of a primitive flower such as a magnolia (*Magnolia heptapeta*, 1) shows the large number of specialized reproductive parts of a flowering plant. Each carpel holds an ovule, which becomes a seed (2) after fertilization. A more advanced flower such as the aster (3) has several ovules held within a fused carpel.

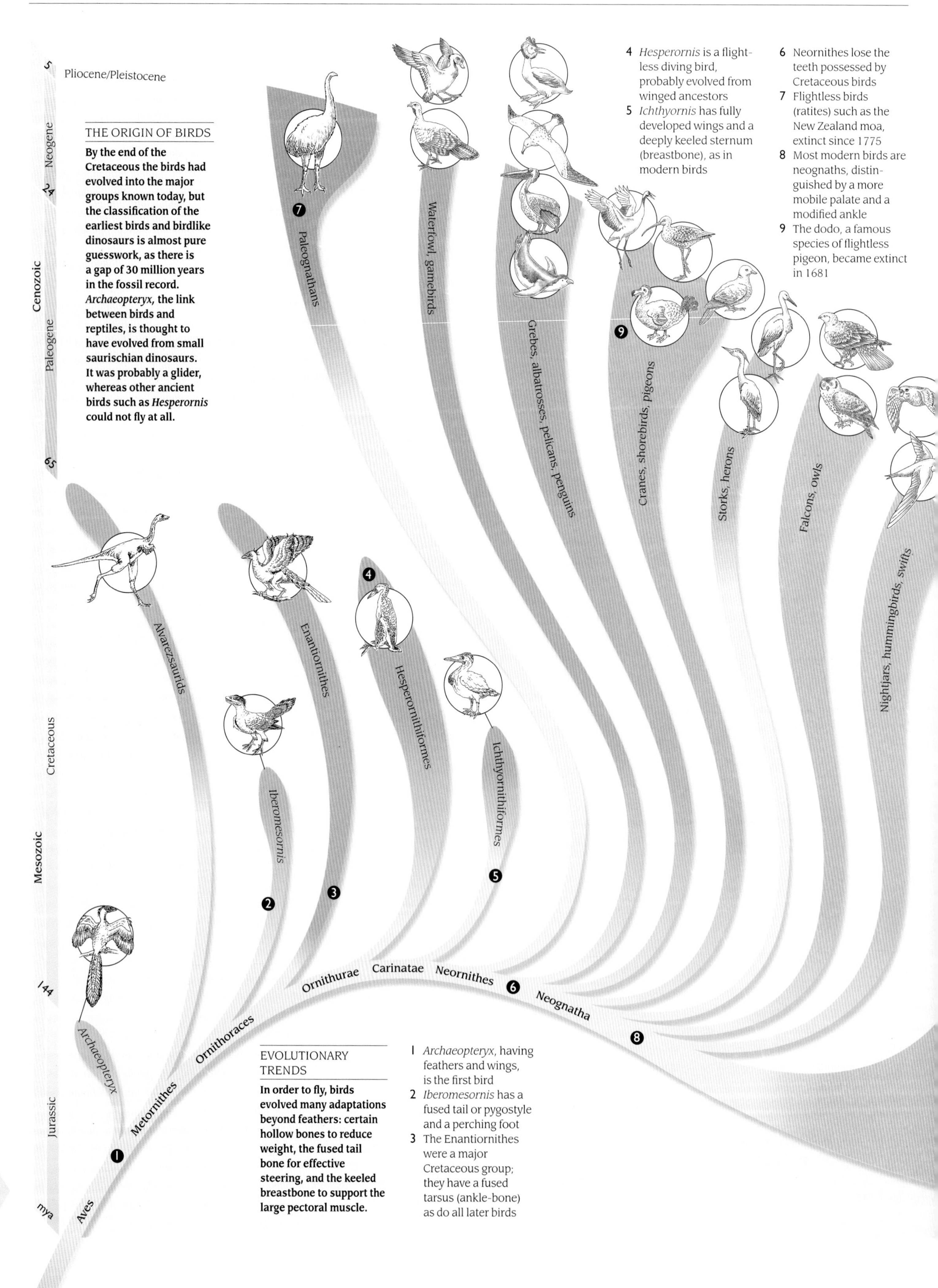

THE ORIGIN OF BIRDS

By the end of the Cretaceous the birds had evolved into the major groups known today, but the classification of the earliest birds and birdlike dinosaurs is almost pure guesswork, as there is a gap of 30 million years in the fossil record. *Archaeopteryx*, the link between birds and reptiles, is thought to have evolved from small saurischian dinosaurs. It was probably a glider, whereas other ancient birds such as *Hesperornis* could not fly at all.

EVOLUTIONARY TRENDS

In order to fly, birds evolved many adaptations beyond feathers: certain hollow bones to reduce weight, the fused tail bone for effective steering, and the keeled breastbone to support the large pectoral muscle.

1. *Archaeopteryx*, having feathers and wings, is the first bird
2. *Iberomesornis* has a fused tail or pygostyle and a perching foot
3. The Enantiornithes were a major Cretaceous group; they have a fused tarsus (ankle-bone) as do all later birds
4. *Hesperornis* is a flightless diving bird, probably evolved from winged ancestors
5. *Ichthyornis* has fully developed wings and a deeply keeled sternum (breastbone), as in modern birds
6. Neornithes lose the teeth possessed by Cretaceous birds
7. Flightless birds (ratites) such as the New Zealand moa, extinct since 1775
8. Most modern birds are neognaths, distinguished by a more mobile palate and a modified ankle
9. The dodo, a famous species of flightless pigeon, became extinct in 1681

THE EVOLUTION OF BIRDS

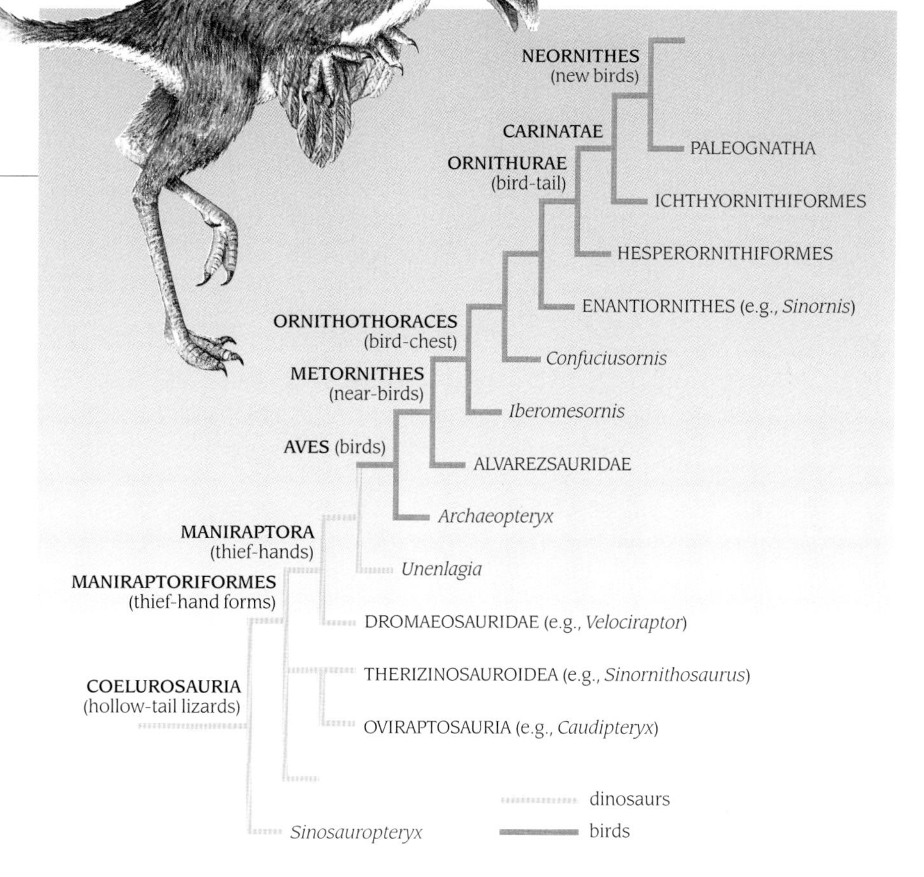

BIRDS evolved from dinosaurs: there is no doubt about that. In fact, the birds are so close to the dinosaurs that many scientists do not accept that birds should or can be classified separately. The dinosaur group that seems to have been closest to the birds was the group that included fast hunters like *Velociraptor* and *Troodon.* Their arms articulated like birds' wings; they had wishbones, long thin shoulder blades, hips like the hips of birds; and they ran on their hind legs in the same way that birds do.

Birds evolved in Jurassic times. *Archaeopteryx* was the earliest animal known that can be regarded as a bird. It had feathers arranged over its body and its wings; and, though not as agile as a modern bird, could probably fly well. It also had a dinosaur's jaws with teeth, claws on its hands, and a long reptilian tail.

Throughout the succeeding Cretaceous period all kinds of dinosaur-bird hybrid animals existed, but their exact relationships to one another are not really known. In the 1980s and 1990s fossils of all kinds of these animals began to be discovered in South America, Spain, and particularly in China. They all seem to show different degrees and different qualities of "birdness." Wings developed, but they were not all suitable for use in flying. Some wings had quite long feathers, but not long enough to support the animal in flight. If they were not used for flight, the bigger feathers may have been used for display, and evolved from small downy feathers that would have evolved as insulation for the body—necessary in a small, active, warm-blooded animal. Some had long lizard-like tails, whereas some had shorter tails with feather bunches, and yet others had tails like those of modern birds, just a stump called a pygostyle with a fan of feathers. Some of the flying wings were quite simple, but others showed the presence of the alula, a kind of feathered thumb that allowed for finer control during takeoff and landing. Some creatures had dinosaur-like toothy jaws while others had beaks.

The problem with the classification of all these is that all these features crop up in different combinations and at different times in different animals. Some of these bird-like dinosaurs were very large, one being over 6.5ft (2m) long and covered in feathers. In this instance the feathers were obviously not for flight but must have been for insulation. It seems likely that some dinosaurs did not die out—instead, they grew feathers and flew away!

PIECES IN A PUZZLE

The bird family tree is controversial but scientists can compare one type with another and judge how closely related they are, based on shared characteristics. On such a diagram—a cladogram—it is possible to plot the evolutionary steps towards modern birds. *Caudipteryx* (top), a recent discovery, is one of the intermediate stages; it was covered with primitive feathers but was unable to fly.

gfishers, woodpeckers, toucans, songbirds

STEPS TO FLIGHT

The evolving wing shows a variation in structures, not all of which would be suitable for flight. Feathers may have evolved first for insulation. Then longer feathers could have developed on the arms that allowed the animal to run faster and with more control over its balance. The first flying feathers may have allowed the animal to glide, and finally carefully controlled flapping flight would have been possible.

1 Forelimb of *Sinosauropteryx*, a theropod dinosaur
2 Flexible wrist of *Velociraptor*, a theropod dinosaur
3 *Unenlagia* could flap its rudimentary wings for balance while running
4 The longer flight feathers of *Archaeopteryx* made basic flight possible
5 *Eoalulavis* possessed an alula, a tuft of feathers attached to the thumb, for controlling slow flight

Alula

CRETACEOUS

The rise of the mammals

PART 5

THE TERTIARY

65 – 1.8 MILLION YEARS AGO

PART 5

THE TERTIARY

WITH THE BEGINNING of the Cenozoic era 65 million years ago, the Earth entered the comparatively modern age. The transition was marked by the disappearance not only of the dinosaurs but also other species, such as ammonoids, the great marine reptiles, and the calcareous nannoplankton that had formed the extensive chalk deposits of previous ages. Unlike older strata, Cenozoic sediments are mostly soft (with the exception of some carbonates and siliciclastics from the earliest part of the Neogene); this makes them comparatively easy to recognize, and they are easily accessible, with plenty of fossils to distinguish them from the underlying Mesozoic. The traditional system divided the Cenozoic into two periods, the Tertiary and the Quaternary, the latter being the most recent. Modern geologists have found it more useful, however, to work with individual epochs within these two periods, which cover very short intervals of time.

THE NAME Tertiary is derived from eighteenth-century studies by European miners and budding geologists, who identified a tripartite division of the rocks in mainland Europe and the British isles. These scientific pioneers considered that the igneous and metamorphic rocks, which formed the basement of mountain ranges such as the Alps, had been the first to form on Earth during the chaotic days of creation. These "primary" rocks were overlain by "secondary" sedimentary rocks, which were predominantly fossil-bearing, and were thought to have been deposited during the Biblical Flood. The third type of rock, found in the low hills surrounding many European mountains, consisted of loosely-packed, stratified limestones, clays, and sands full of fossils that looked very much like living species. For this reason, these "tertiary" sediments were thought to have been deposited after the Flood. However, it was while examining the fossils of mammals in Tertiary rocks that the great French comparative anatomist and geologist Baron Georges Cuvier hit on the idea of extinctions. Cuvier realized that these mammals were significantly different to anything contemporary, and must have vanished from the world long ago.

The Tertiary, which encompasses the Paleogene and Neogene periods, began 65 million years ago at the Mesozoic–Cenozoic boundary and ended 1.8 million years ago as the world grew much cooler and drier, ending the warm, tropical conditions that had dominated the Mesozoic and the early Tertiary. The ice ages of the Quaternary (specifically, its earlier period, the Pleistocene) were the product of a mass of cold water, the psychrosphere, forming deep in the world's oceans during the Tertiary as Australia finally separated from Antarctica. Other major events during the Tertiary included the formation of the Panamanian Isthmus, which redirected warm water from the Atlantic towards northern Europe. Both the Alps and the Himalayas also rose during this time as a result of continental collisions.

The Age of Mammals witnessed rapid and enormous diversification of a group of animals that had been unable to expand in the shadow of the dinosaurs.

ENVIRONMENTAL changes throughout the Tertiary led to a global radiation of mammals. Whales appeared in the newly cooled oceans, whose stock of small invertebrates organisms provided a rich food source for any animals that could feed on them in large enough quantities. On land, grasslands succeeded the tropical forests of the Mesozoic, providing an ecological niche into which large herbivorous mammals quickly expanded. Grass-eaters replaced many types of leaf-eaters, totally altering the composition of mammalian herbivore communities. Carnivores promptly appeared to feed upon the grass-eaters.

The Cenozoic Era has become known informally as "The Age of Mammals" due to the explosive diversification of this group as the dominant large-bodied animals. Mammals had first appeared during the late Triassic period of the Mesozoic, but there were few opportunities for them to expand into a world dominated by dinosaurs, which both competed for food (both flesh and plant) and preyed on smaller animals such as the early mammals. With the Cenozoic, all that changed. The small, mostly nocturnal, non-specialist mammals rapidly evolved into groups as diverse as whales, bats, and horses—nearly 100 families by the Eocene epoch. Elephants, cows, and cats and dogs made their appearance slightly later. Some of these animals were completely new; others had descended from lines that had existed in the Mesozoic. Not all of them survived into the succeeding Quaternary period; those that disappeared included long-legged ducks, giant carnivorous flightless birds, and *Indrichotherium* (formerly *Baluchitherium*), a member of the rhinoceros family and the largest mammal ever to have lived on land.

Among these expanding and diversifying mammals were the primates, which are characterized by their exceptional brainpower and social organization. The order began modestly, about 50 million years ago, with a tiny tree-dwelling animal that resembled a modern mouse lemur or tarsier; larger specimens weighed only 2.2lbs (1kg). Their descendants colonized almost every part of the world during the later Paleogene and early Neogene. From these evolved the entire line of anthropoids, the group that eventually gave rise to humans—a much later arrival.

About 35 million years ago, in the early Oligocene, the straight-tailed, narrow-snouted "Old World" monkeys of Africa and Eurasia appeared. The New World monkeys of South America, distinguished by their gripping tails and broad snouts, evolved later in the Oligocene from a group that had migrated out of Africa. Old World monkeys gave rise to the apes, many of which became ground-dwelling. One ape family, the Ramapithecidae, arose in Africa about 17 million years ago and ranged in weight from 44lbs (20kg) to 600lbs (275kg). Among this great diversity of apes, yet another new form appeared about 4 million years ago: the genus *Australopithecus*, an ape that walked upright. It was the oldest confirmed member of the human family.

Ian Jenkins

THE PALEOGENE

65 – 24 MILLION YEARS AGO

*T*HE PALEOGENE *is often overlooked in popular textbooks because it comes straight after the Cretaceous period and the demise of the dinosaurs. But much of what we see in our modern world had its origins in this important period of earth history. The most significant changes include the shift from the tropical forests of the Eocene to the cooler, open savannah of the Oligocene, and the accompanying reorganizations in the types of mammals that were abundant then. These global changes are linked to the formation of a cold ocean current that circled Antarctica. As South America and Australia rifted from Antarctica, the warm current that had flowed along the continental margins was deflected by the arrival of a cold one, leading eventually to the formation of ice sheets. Today's Antarctic ice caps and the modern diversity of living mammals are the two great legacies of this portion of Earth history.*

THE GREAT geologist Charles Lyell's work formed an early basis for the current form of the "Tertiary" that comprises all the Cenozoic (the last 65 million years) except for the Ice Ages that cover the past two million years, and for which the term Quaternary is used. Some experts subdivide the Cenozoic into two equal subdivisions: the Paleogene (65 to 24 million years ago) and the Neogene (24 million years ago to the present). It is somewhat confusing that the most fundamental changes in Cenozoic history happened between the Middle Eocene and earliest Oligocene, and not at the formal boundary of any of the Cenozoic epochs. The establishment of the six-epoch Cenozoic period (including the three-epoch Paleogene), and continued refinement of the stratigraphy of these rocks, allowed evolutionary changes that occurred throughout this part of Earth's history to be put into a clearer context.

The dramatic disappearance of the dinosaurs at the end of the Cretaceous left many ecological niches vacant.

The extinction of the dinosaurs at the boundary between the Cretaceous period and the Paleocene epoch changed the composition of the terrestrial vertebrate faunas completely. Ecological niches that had previously been filled by the dinosaurs were now available to be occupied by other animals. Roles such as those of large herbivores and large carnivores were taken up by the mammals. However, in the following epochs of the Paleogene, mammals defined many new ecological niches that had never before existed. Specializations such as those of anteater, grass grazer, and gnawer were originated by terrestrial mammals. Many of these bizarre-looking animals represent a response to what may be informally termed an evolutionary vacuum, in which ecological niches are made vacant by extinctions. However, during the early stages of this opportunistic refilling, the process of evolution produced a dazzling array of strange body plans and adaptations.

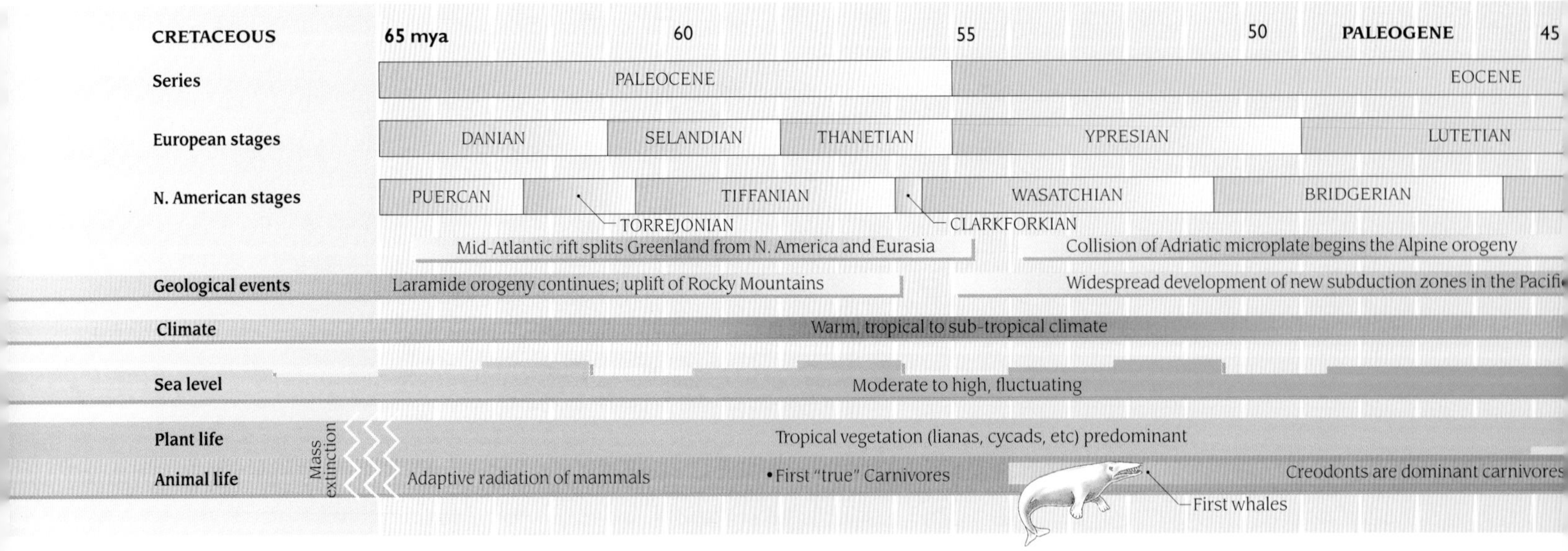

KEYWORDS

ALPINE OROGENY
ARCHAEOCETE
CARNIVORE
CONDYLARTH
CREODONT
HOT SPOT
LARAMIDE OROGENY
MAMMAL
MESONYCHID
RING OF FIRE
SEAFLOOR SPREADING
TETHYS SEAWAY
UNGULATE

CRETACEOUS | **65 mya** | 60 | 55 | 50 | **PALEOGENE** | 45

Series: PALEOCENE; EOCENE

European stages: DANIAN; SELANDIAN; THANETIAN; YPRESIAN; LUTETIAN

N. American stages: PUERCAN; TORREJONIAN; TIFFANIAN; CLARKFORKIAN; WASATCHIAN; BRIDGERIAN

Geological events: Mid-Atlantic rift splits Greenland from N. America and Eurasia; Laramide orogeny continues; uplift of Rocky Mountains; Collision of Adriatic microplate begins the Alpine orogeny; Widespread development of new subduction zones in the Pacific

Climate: Warm, tropical to sub-tropical climate

Sea level: Moderate to high, fluctuating

Plant life: Tropical vegetation (lianas, cycads, etc) predominant

Animal life: Mass extinction; Adaptive radiation of mammals; • First "true" Carnivores; First whales; Creodonts are dominant carnivores

This riot of mammalian evolution also extended into the marine realm, for it was during the Paleogene that the first whales appeared. Whales evolved from a group of Paleocene (early Paleogene) hoofed carnivores known as mesonychid condylarths. The earliest remains of whales come from Early–Middle Eocene rocks in Pakistan and Egypt. The seas of the Paleogene also witnessed other major events in the history of vertebrates, for it was then that modern sharks flourished. The cartilaginous skeletons of sharks soon disintegrate, but their hard teeth do not. Collecting fossil shark's teeth from Paleogene rocks has always been a popular pastime for amateur and professional alike.

In the middle of the seventeenth century the Danish physician Nicolas Steno, while dissecting the head of a large shark, noted similarities between its teeth and the curious "Tongue Stones" that had for centuries been dug out of soft rocks in cliffs on the island of Malta. In 1667 Steno published a book (*The Head of the Shark Dissected*) in which he argued that tongue stones were really the teeth of long-dead sharks that had become entombed during the biblical flood. His argument recognized that features of living creatures can be discovered in non-living objects now called fossils, and so Steno the physician became the world's first vertebrate paleontologist. However, the importance and historical legacies of the Paleogene rocks that gave Steno his material are recognized with increasing resolution today.

Major climatic changes occurred throughout the Paleogene Period. It was the start of a time of long-term cooling, although with numerous fluctuations. After an initial cool period during the Paleocene epoch, there was a return to tropical global conditions at the transition from the Paleocene to the Eocene. The collision of the Indian continental plate into the Asian continental plate in Late Paleocene/Early Eocene times (51 to 56 million years ago) is almost certainly linked to this—one of the warmest periods of the last 550 million years. This was followed by a significant alteration from the "greenhouse" global climate of the late Eocene to the icehouse conditions at the start of the Oligocene.

As THE Indian plate drifted north throughout much of the preceding Cretaceous period, volcanoes erupted in its path, waning as India collided with the vast Asian plate. This may have led to a passive global warming in the mid-Paleocene as carbon from organic matter in deep-sea sediments was exhumed along the uplifted margins of continents. The carbon came from the accumulated remains of huge amounts of dead organisms; plankton, fish, and microbes. This event eliminated the "carbon sink" effects of a part of the global sea volume and enhanced the greenhouse effect as more carbon, in the form of carbon dioxide gas, found its way into the atmosphere. According to current geological theory, global warming during the Paleocene–Eocene transition was a consequence of a combination of factors, rather than a single event. These factors also include the profuse volcanism in the North Atlantic as Greenland was detached from North America and Europe; ocean warming in high latitudes, weakening the atmospheric circulation; and the increase in productivity of organic matter in the oceans.

Volcanism and the release into the atmosphere of carbon from uplifted sediments contributed to global warming.

Throughout the Eocene and Paleocene, climates across the globe were generally warm and equable, so that many places—among them northwestern North America, southern Germany, and the London

THE AGE OF MAMMALS

The Paleogene is limited to just the last one and a half percent of the earth's existence and lasted for about 41 million years. During that time the mammalian fauna of the modern age began to take shape; while many of its component animals were the precursors of modern animals that are familiar today, many were strange and extraordinary. During this time, the mammals came to fill all the ecological niches freed by the extinction of the dinosaurs. Climate and vegetation changes also gave opportunities for diversification to new groups of creatures.

PALEOGENE

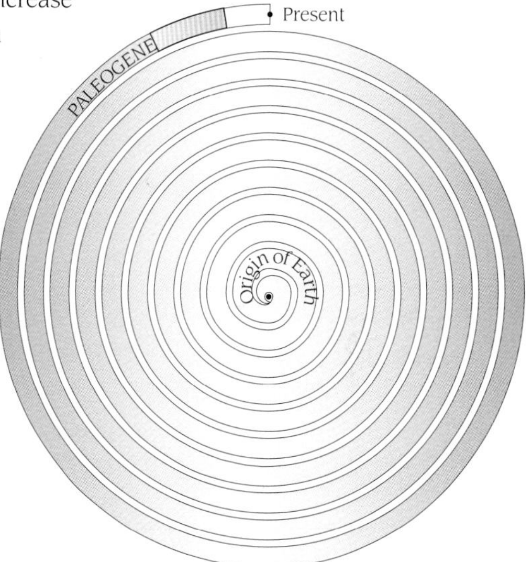

See Also

THE ARCHEAN: *Tectonic Plates; Seafloor Spreading and Ocean Ridges*
THE CRETACEOUS: *The Western Cordillera of North America*
THE NEOGENE: *Separation of Australia from Antarctica*
THE PLEISTOCENE: *The Ice Age*

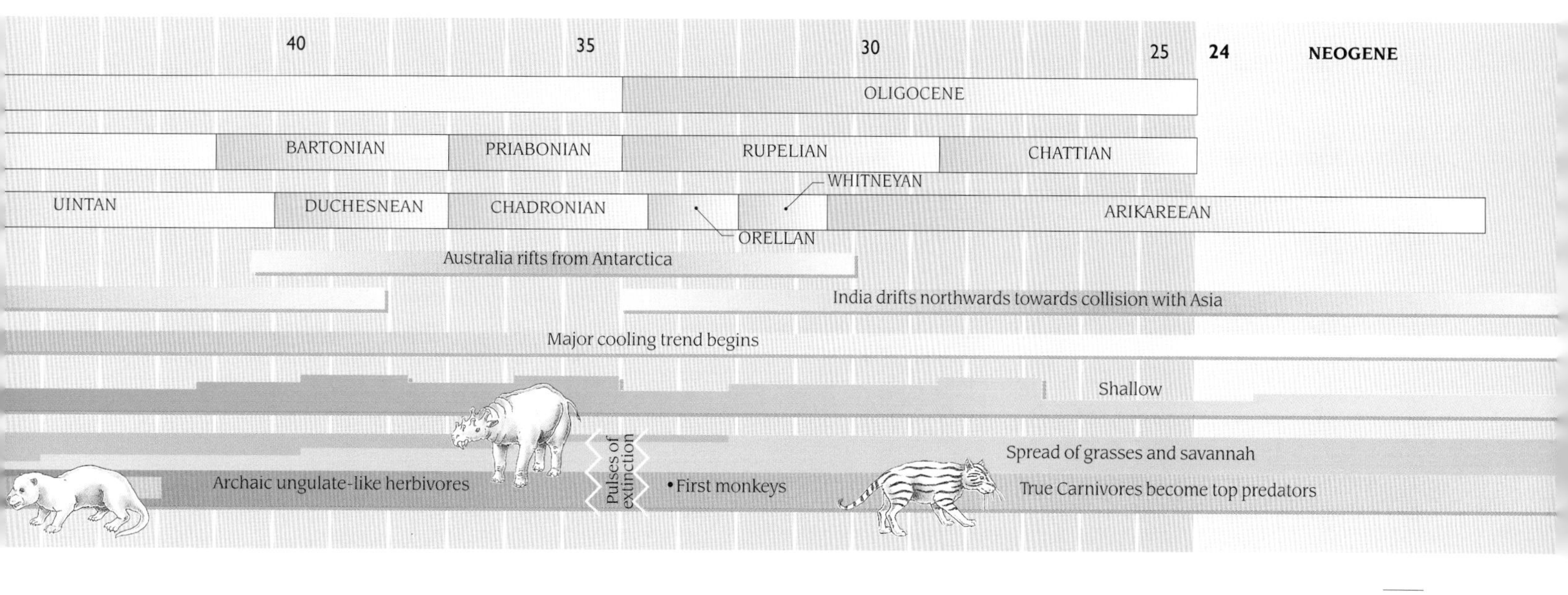

GIANT'S CAUSEWAY

(Left) The hexagonal basalt columns of the Giant's Causeway in Northern Ireland are evidence of the vast quantities of lava that poured from the fissures of the opening North Atlantic.

Basin, England—resembled the rainforests of modern Central America and Southeast Asia. Characteristic plant fossils include magnolias, citrus, laurels, avocados, sassafras, camphor, cashews, pistachios, mangoes, and tropical vines. Temperatures remained stable at about 68–77°F (20–25°C), varying by only 7–15°F (5–10°C) per year. Deltas, forests with tall canopies, and trees decorated by vines and lianas all existed in these habitats. Broad floodplains bordering meandering rivers were in evidence and would have been very much like areas such as the Amazon delta and basin of today. Fossils of the still living Nypa palms and cycads (sego palms) demonstrate the lush conditions.

Grasses had been present by the Late Paleozoic, but it was only in the late Oligocene that they began to flourish beyond their original niche in swamps and woodlands. This depended on the ability to survive intensive grazing by animals, which in turn required a quick and dependable method of reproduction. When grasses evolved that could be pollinated by wind instead of insects, they spread rapidly to form the great prairies of the world.

DURING the Paleogene the ancient Tethys Sea began to close as the African plate pushed up against Europe. This east-to-west body of water extended from southern Europe to the area of China, and was the precursor of the (much smaller) modern Mediterranean Sea. The Tethys Sea was a major oceanic realm and its effects on global climates must have been significant. The plate tectonic forces driving the closure of the Tethys eventually gave rise (in the subsequent Neogene period) to the great chain of Eurasian mountain ranges to which the Swiss Alps belong. Also part of the chain are the Pyrenees of Spain and France; the Pennines of Italy; the Carpathians of southeastern Europe; the Caucasus of Eurasia; the Himalayas of south Asia; and the Atlas of north Africa.

Africa pushed up against the edge of Europe, closing the Tethys Sea, building mountains from Spain to Tibet, and forming the Mediterranean.

Rifting, in contrast, was not a very widespread event in the northern hemisphere during the early Paleogene. About 65 to 60 million years ago, the individual continents of Laurentia and Baltica, which had formerly made up the giant northern supercontinent Pangea, were spreading apart at only some specific areas within 30° latitude of the North Pole. This site, between what is now Greenland and Europe, was the opening of the great Mid-Atlantic Ridge, a geological feature of the Atlantic Ocean that is still prominent today. The Atlantic was narrower than it is in modern times, but seafloor spreading as the mid-ocean rifts expanded caused the ocean to widen.

In the north Atlantic, as a result of the rifting, huge volcanoes were formed far out into the ocean, including one that became the modern island of Iceland. The regions that are now Northern Ireland and Scotland experienced intense volcanic activity, with lava flows 1mi (1.5km) thick forming extensive basalt plateaus. One of the flows that occurred in this region formed the Giant's Causeway in Northern Ireland. The strange, regular polygonal jointing that gives the characteristic shapes to the columns in this famous rock formation was due to the cooling of the lava flows at a very uniform temperature. Evidence for the same activity in eastern Greenland is found in volcanoclastic deposits (rocks composed of the cemented, fragmented remains of volcanic rocks) sandwiched in between near-shore marine sediments that are widespread in northwestern Europe.

During the Cretaceous, the Arctic Ocean basin had been largely isolated from the Atlantic because North America, Greenland, and Europe still formed one huge landmass. This isolation ended as continental rifting in the Atlantic continued to spread north. During the Paleogene the Mid-Atlantic Ridge gradually split along two forks that separated Greenland from North America on the west and from Europe to the east. At this point Greenland became a large island completely detached from the remainder of Laurentia.

The eastern border of what is now North America was feeling the powerful tensional effects of its separation from Europe and Africa as the Atlantic widened from the

Africa and Middle East
Antarctica
Australia and New Guinea
Central Asia
Europe
India
North America
South America
Southeast Asia
other land

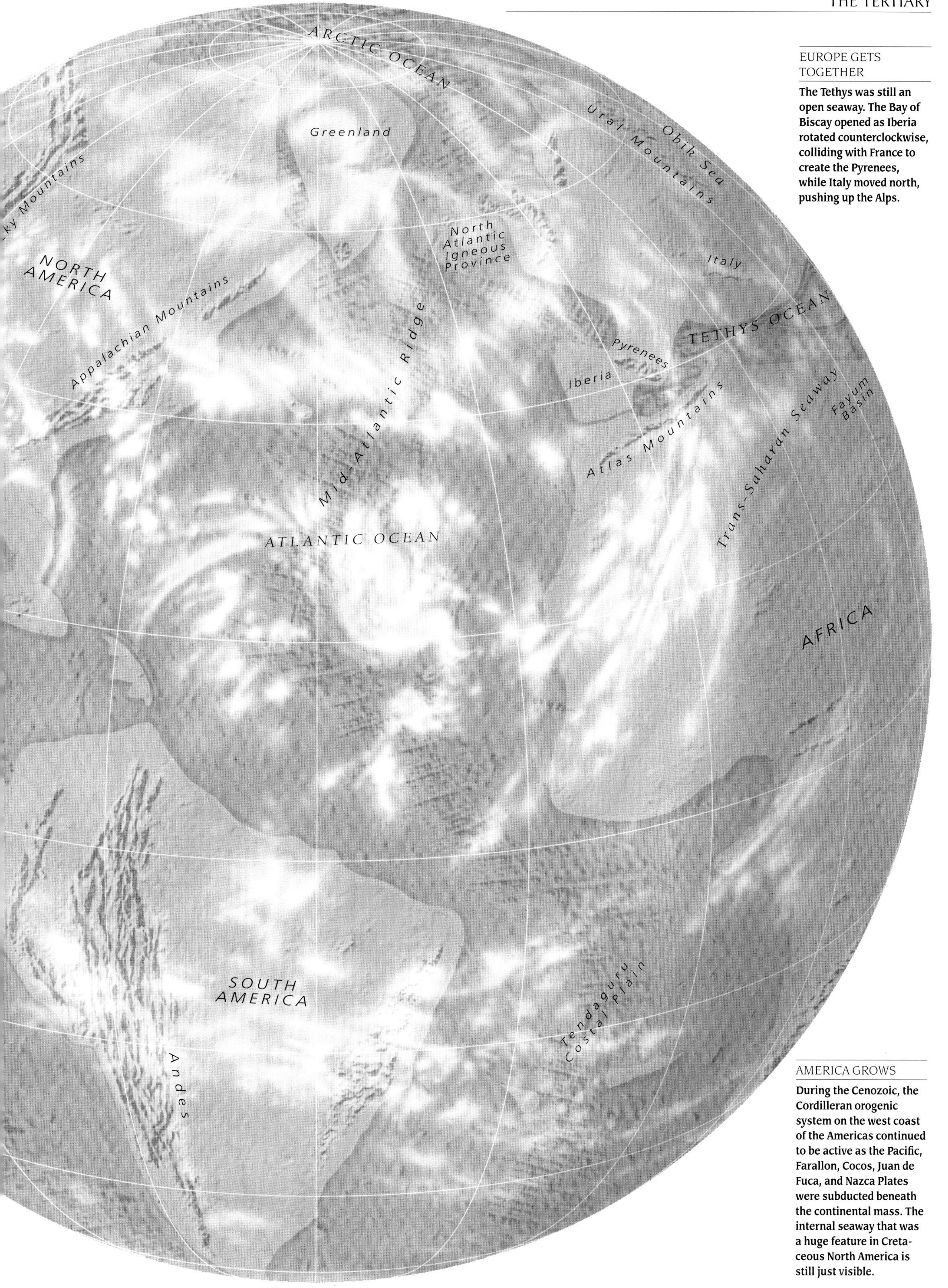

EUROPE GETS TOGETHER

The Tethys was still an open seaway. The Bay of Biscay opened as Iberia rotated counterclockwise, colliding with France to create the Pyrenees, while Italy moved north, pushing up the Alps.

AMERICA GROWS

During the Cenozoic, the Cordilleran orogenic system on the west coast of the Americas continued to be active as the Pacific, Farallon, Cocos, Juan de Fuca, and Nazca Plates were subducted beneath the continental mass. The internal seaway that was a huge feature in Cretaceous North America is still just visible.

mid-ocean ridge. In the west, tectonic compression as the North American plate overrode the Pacific plate gave rise to the Laramide orogeny, which continued well into the Cenozoic era. This event produced many of the geological features seen in the Rocky Mountains in the Cordillera region of North America. This subduction zone extended all the way along the length of South America, where the Andes continued to form.

THE LATE Eocene saw the final breakup of the continent of Gondwana with the separation of Australasia from Antarctica, and their subsequent isolation, as both moved in opposite directions. Geographical transformations elsewhere on this side of the world were brought about by the development of new subduction zones in southeast Asia, Japan, and the South Pacific, extending all the way to the Pacific coast of North America. Island arcs formed in the area from volcanic activity that gave the area its nickname, the Pacific "Ring of Fire," an epithet still valid today. The South China and Philippine seas appeared in the region at this time. In the late Paleocene, about 57 million years ago, there was still a significant seaway between the drifting Indian island and the southern margin of Asia, though mountains were already rising between the two. This seaway was part of the ancient Tethys sea. Farther east, the Turgai Strait linked the Tethys to the Arctic along the Ural Mountains.

As Australia rifted away, Antarctica became isolated over the South Pole and surrounded by a cold circumpolar current.

As the Australian continental plate drifted north, it lost contact with the landmass of Antarctica, allowing the formation of a circumpolar Antarctic current. In essence this event was the beginning of our modern climatic environment. Even before its separation from Australia, Antarctica had been centered over the South Pole but had remained warm because its shores had been lapped by warm waters from lower, subtropical latitudes. Plate tectonic movements caused an alteration of wind and oceanic currents, which stopped warm equatorial waters from getting to Antarctica because the cold ocean waters around the great southern continent began to flow in a circular direction around it, blocking the warmer waters from the north. This new distribution of cold waters was fully established in surface waters by the Oligocene.

In response to this cooling, the first sea ice began to form near the end of the Eocene and waters of near-freezing temperatures sank to the depths (cold water is denser than warmer water and so sinks). The freezing waters that sank to the deep sea around Antarctica then spread Northwards forming the psychrosphere—the deepest zone of the ocean, characterized by near-freezing water. The psychrosphere has had a major impact on deep-sea life and even now is a major physiological barrier to most oceanic life. At this time a major extinction is seen in the fossil record of deep-dwelling foraminiferans, microscopic calcite-shelled protistan

VOLCANIC ARCS

(Left) An eruption near Papua New Guinea. Australia's move north has had a major impact on the development of Indonesia as subduction continues to produce a range of curving seamounts.

animals that are abundant at various depths in the oceans. Other deep-dwelling marine organisms such as mollusks show extinctions as well.

These events were spread over a long period of time that extended from the mid-Eocene through to the mid-Oligocene. A sharp decline in animal life coincided with a major sea regression as the volume of ice over Antarctica increased rapidly.

ANOTHER change in marine ecology as a result of the formation of the psychrosphere was that of the re-expansion of scleractinian corals (which include all modern corals, as opposed to the rugose and tabulate coral groups that went extinct at the end of the Permian period). These stony or "true" corals survived the cooling episode to become the significant reef-builders of modern times. Larger marine animals were also affected, though not necessarily for the worse. The first whales appeared during the early Middle Eocene in the subtropical waters of the eastern Tethys. They were aquatic, carnivorous mammals and the precursors of the modern toothed whales. Towards the end of the Oligocene, as the world became colder, the baleen whales

The cooling that took place during the transition from the Eocene to Oligocene directed much of subsequent evolution.

Africa and Middle East
Antarctica
Australia and New Guinea
Central Asia
Europe
India
North America
South America
Southeast Asia
other land

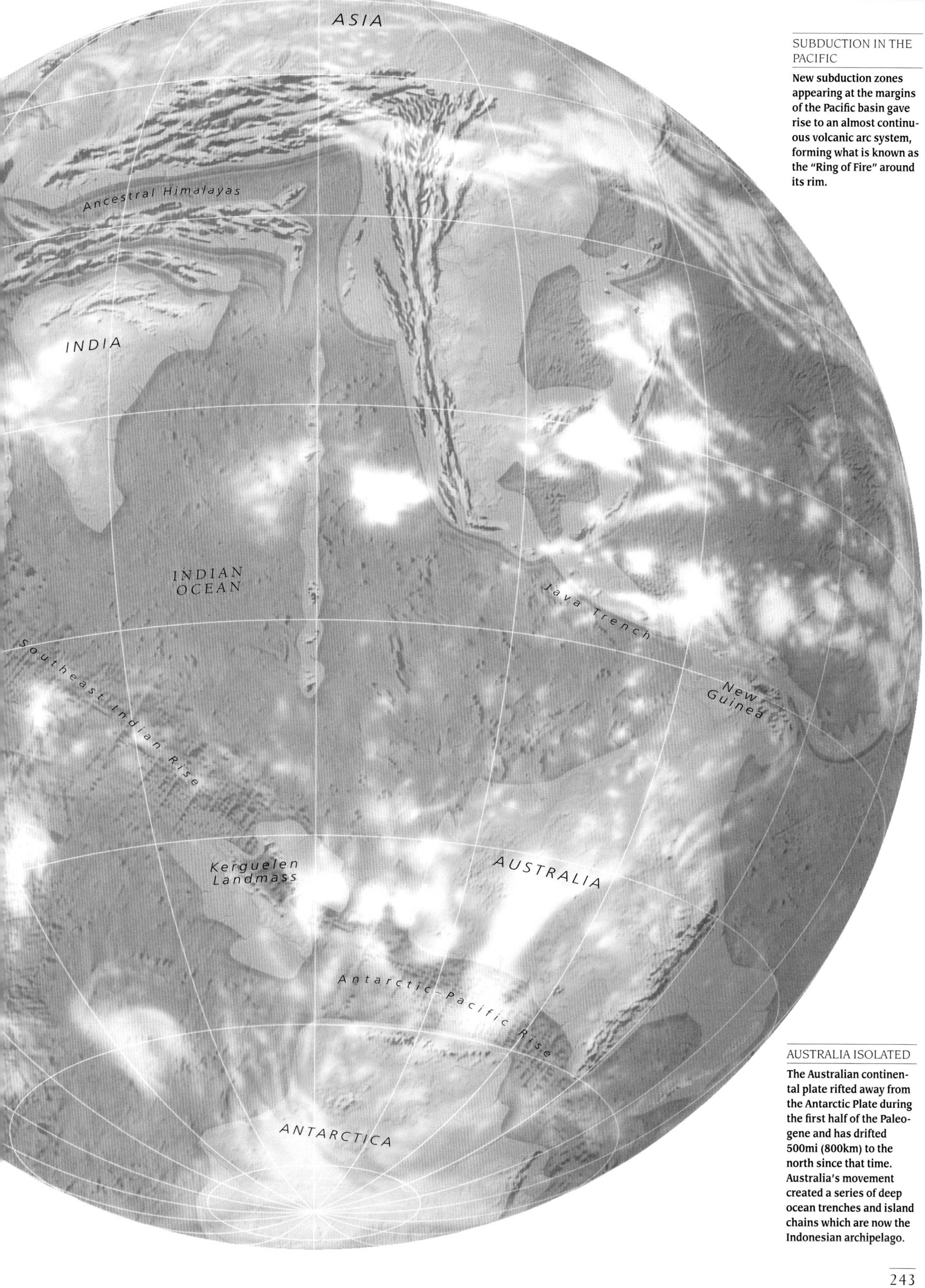

SUBDUCTION IN THE PACIFIC

New subduction zones appearing at the margins of the Pacific basin gave rise to an almost continuous volcanic arc system, forming what is known as the "Ring of Fire" around its rim.

ASIA

Ancestral Himalayas

INDIA

INDIAN OCEAN

Java Trench

New Guinea

Southeast Indian Rise

Kerguelen Landmass

AUSTRALIA

Antarctic-Pacific Rise

ANTARCTICA

AUSTRALIA ISOLATED

The Australian continental plate rifted away from the Antarctic Plate during the first half of the Paleogene and has drifted 500mi (800km) to the north since that time. Australia's movement created a series of deep ocean trenches and island chains which are now the Indonesian archipelago.

evolved, with their system of horny plates used to filter the planktonic organisms that would have thrived in the colder waters. Like all mammals, cetaceans are warm-blooded; to maintain their temperature they lay down fat, in the form of blubber, below the skin. Larger species—such as the baleen whales—have an advantage with their more favorable surface-to-volume ratio.

SOME 2mi (3km) below sea level, the Mid-Atlantic Ridge marks the boundary between the American and African plates. It extends nearly 10,000mi (16,000 km) in length from the Arctic Circle to the southern tip of Africa, and 1,000km (1600km) in width from its position at an equal distance from the continents east and west of it. At the crest of the ridge, a narrower zone about 50 to 75mi (80 to 120km) is the site of the underwater volcanic activity that causes the seafloor to "spread." Magma welling up from the crust is pushed away from the sites of eruption, building up new seafloor. In this way the basin of the Atlantic is estimated to be widening by about 0.5 to 4in (1 to 10cm) every year.

The Mid-Atlantic Ridge is a ten-thousand-mile-long underwater mountain range, created as volcanic material rises up from the mantle.

GRABEN STRUCTURE

Faults bounding the sides of a down-dropped graben extend below to the underlying magma chamber and form connections to the surface through which lava is extruded. These lava flows successively form new floors to the graben as the lateral extension continues, forming new oceanic crust.

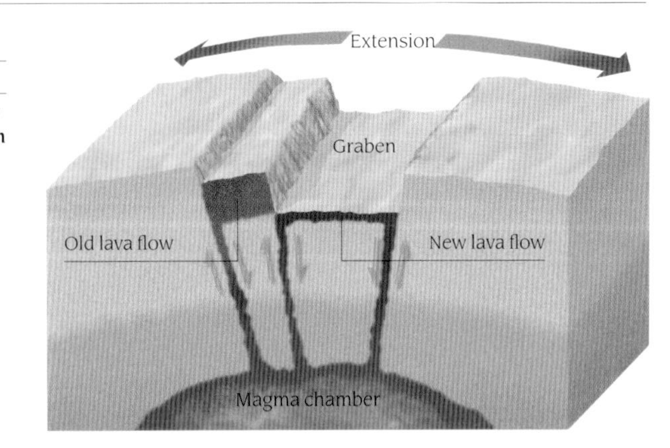

Iceland is a volcanic island of the Mid-Atlantic Ridge that has built up until it has risen above sea level, a boon for geologists. The "island," formed entirely from oceanic crust, is a contiguous part of the undersea mountain chain, rising more than 1.5mi (2.5km) higher than the rest of the ridge—enough to raise it above sea level and dry out the basaltic rock. Other volcanic islands along the ridge that have also risen above sea level are Ascension Island, St. Helena, and Tristan da Cunha, though none of these sits exactly on the Mid-Atlantic Ridge as Iceland does. All are associated with "hot spots" where molten material rises from the mantle, as much as 375mi (600km) below the surface. The hot spot directly under Iceland is responsible for the great buildup of lava that

MID-OCEAN RIDGE

The mid-ocean ridge of the North Atlantic is a vast feature that runs along the seabed for several thousand miles; there is one place, however, where it is exposed on the earth's surface—on Iceland. The steep-sided valley of the Thingvellir graben (below), where the basalt formation of the Mid-Atlantic Ridge surfaces, is a striking demonstration of tectonic plate divergence.

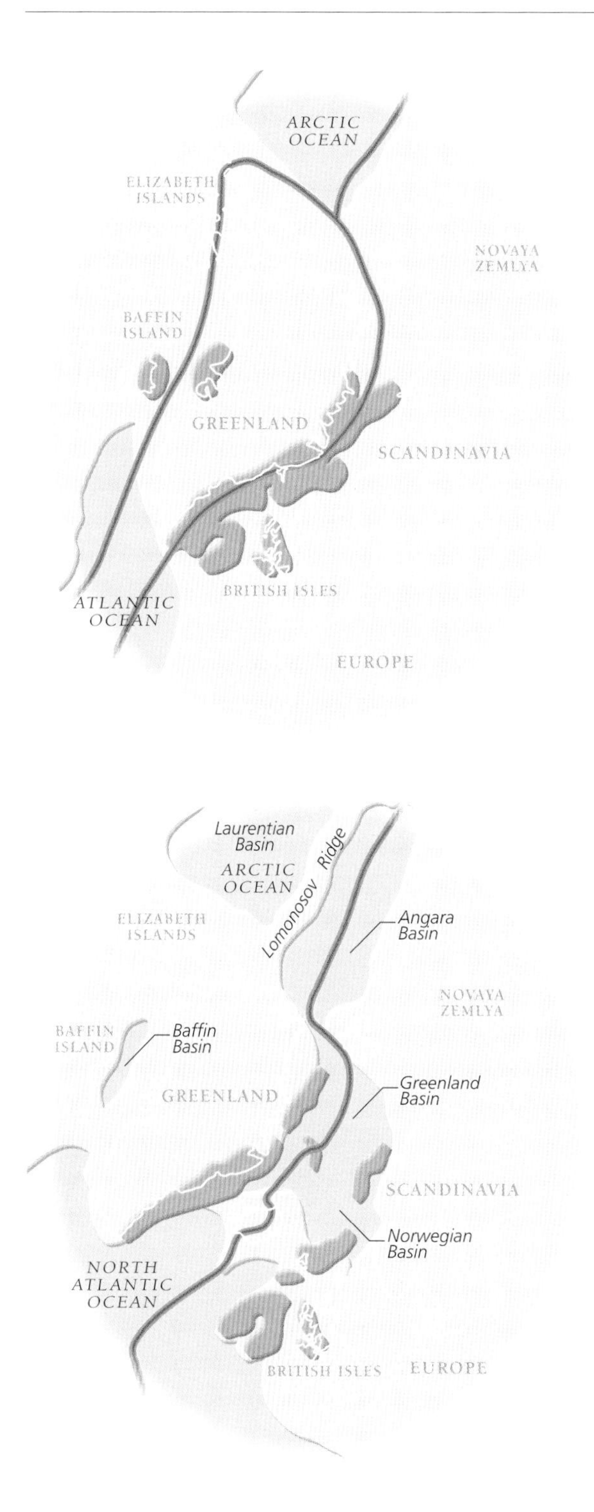

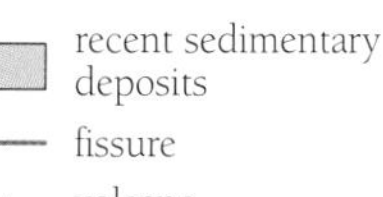

zone of rifting
volcanic deposits
continent
deep oceanic basin

Iceland detail
Tertiary basalts
Pliocene–Pleistocene basalts
Upper Pleistocene basalts
recent sedimentary deposits
fissure
volcano

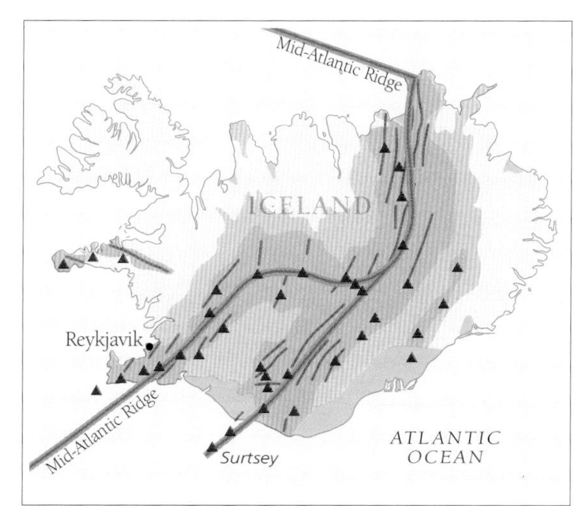

NEW OCEAN ARMS

The huge, rugged mountain range of the Mid-Atlantic Ridge that lies in the middle of the Atlantic Ocean is the site of the formation of new oceanic lithosphere. New crust spreads laterally, at a rate of about an inch (2.5cm) a year, carrying the continents with it, as magma—in the form of extruded lavas—moves outwards from the fissure. Between 63 and 52 million years ago there were enormous outpourings of volcanic material as a mantle hot-spot stretched the earth's crust. (These masses of basalt lava are preserved today in Northern Ireland and the Inner Hebrides of Scotland.) Ruptures opened up the narrow seaways between Europe, Greenland, and North America (top map), eventually isolating Greenland as a large island and connecting the Arctic Ocean with the Atlantic. In mid-Paleogene times, the western rift became inactive, but the eastern arm continues to widen today (lower map).

ZONES OF SEAFLOOR SPREADING

The mantle plume that initiated the rifting of Greenland from Europe still exists under Iceland. The progression of the North Atlantic constructive plate margin can be traced across the island by the zones of volcanic activity and the rifting structures that are found there. The whole island is volcanic in origin; the island of Surtsey surfaced in 1963.

constitutes the island. It was the same hot spot that produced huge lava flows covering hundreds of miles in eastern Greenland about 60 million years ago when Greenland began to separate from Europe.

The American geologist Harry Hess, author of the theory of seafloor spreading, noted that mid-ocean ridges are characterized by deep furrows, which in Iceland are visible as features of the landscape. Valleys, called grabens, are created where there are two fault zones on either side of a central block that slips down between them; this forms the furrow. Grabens form where crust has split and new lava has risen up. Submersible craft exploring the Mid-Atlantic Ridge have discovered the floor of the narrow rift zone to be covered with pillow lavas, the globular shapes that fresh lava often takes when rapidly cooled under water. Moreover, the valley floor is covered with fissures up to 30ft (10m) wide, parallel to the ridge axis, as if the bed of the sea is being pulled apart by enormous stresses.

ALSO in the early Paleogene, but much further to the southeast, another phenomenon related to the rifting in the northeastern region of the Atlantic ocean was that the seafloor of the ancient North Sea basin began to sag, accumulating thousands of feet of sediments in the center of this geological basin. The North Sea, touching the shores of nearby continents (what is now northern Europe), deposited extensive fine-grained marine sediments in southeastern Britain and the European mainland between northern France and Denmark. These deposits show that the conditions during the time of their deposition were those of subtropical environments. Later in the Paleogene, when sea levels rose, the North Sea flooded northern Europe as far east as the Ural Mountains.

Ebb and flow of the ancient North Sea left extensive marine deposits in Paris and London. Typically of the time, these were tropical.

Two deposits from the area are particularly important in the history of European geology. Sir Richard Owen in Great Britain studied fossils from the clay of the London Basin in detail. This legacy from the late Early Eocene yielded an array of exquisitely preserved fossils such as crocodiles, turtles, sharks, fishes, mammals, and even tiny birds—350 species in all. The animal and plant remains show it to have been a tidally-influenced subtropical shoreline that was intermittently forested. It was similar to a modern-day mangrove swamp, and some of its species have relatives in modern Malaysia.

The fossils of the Paris Basin limestones had been studied by the great comparative anatomist Baron Georges Cuvier, head of the National Museum of Natural History in Paris. They show a community of fossil animals similar to those of the London Clay. Cuvier's discovery and description of the ancient tapir-like fossil mammal *Palaeotherium* ("ancient beast") was the first of its kind and of great importance to the history of paleontology.

part of the Asian plate, the resulting undersea orogeny, in association with the fore arc (a submarine feature for long distances), produced a curving line of seamounts that now show as the Javanese Islands of the Indonesian arc. These include Java, Sumatra, Sulawesi, and New Guinea; they also include the volcano Krakatau, which last exploded in 1883. The Emperor–Hawaiian seamount chain, which had begun to form about 70 million years ago, began to curve as the spreading direction of the East Pacific Rise became more westerly at 43 million years ago, and new subduction zones and arc systems developed south of Japan and in the South Pacific.

The Mentawai Islands off the coast of Sumatra show a good exposure of Cenozoic strata. The rocks in these islands can be seen to have been strongly folded in the Early Cenozoic and then again during the Miocene. Terraces of uplifted coral reefs show that these islands have recently been rising; that they are still doing so is indicated by the geology of the Nicobar and Andaman Islands that extend towards Burma, where these islands become a continuous land feature and pass into the high mountain range of the Arkan Yoma west of the Irrawaddy delta. Far to the east of Java, out in the Pacific Ocean, the ridge of the Indonesian island arc rises as the island of Timor. This island is particularly interesting for the great height to which its successive coral reefs have been upheaved, some of which now stand well over 3900ft (1200m) above sea level. These reefs record the

OCEAN crust is created along mid-ocean ridges at rift zones and "destroyed" in deep-sea trenches where plates descend into the asthenosphere, to be re-melted and re-cycled through the mantle. Such areas are called subduction zones, and the majority of them are found around the margins of the Pacific Ocean, where they formed as a result of the oceanic lithosphere subducting beneath the continental plates of Asia and the western edge of the Americas. (The Philippine plate, one of the smaller plates in the world, is entirely surrounded by subduction zones, crushing it between the Pacific and the eastern margin of Asia.) The entire Pacific basin is characterized by intense volcanic and earthquake activity, earning it the nickname "the Ring of Fire." The same belt continues along the underside of Southeast Asia, at the northern edge of the Australian plate, and extends up into the Asian mainland, following the line of the Himalayas and across to the southern Mediterranean.

Most of the world's major subduction zones are found in the Pacific basin, forming a great "Ring of Fire" around its edges.

All of this activity began about 60 million years ago during the Paleogene as Australia began to pull away from Antarctica, taking with it New Zealand and New Guinea. India, a small continental fragment, was already heading north on its rendezvous with Tibet. As the embayed central portion of the northern rim of the Australian plate impacted with the projecting southwestern

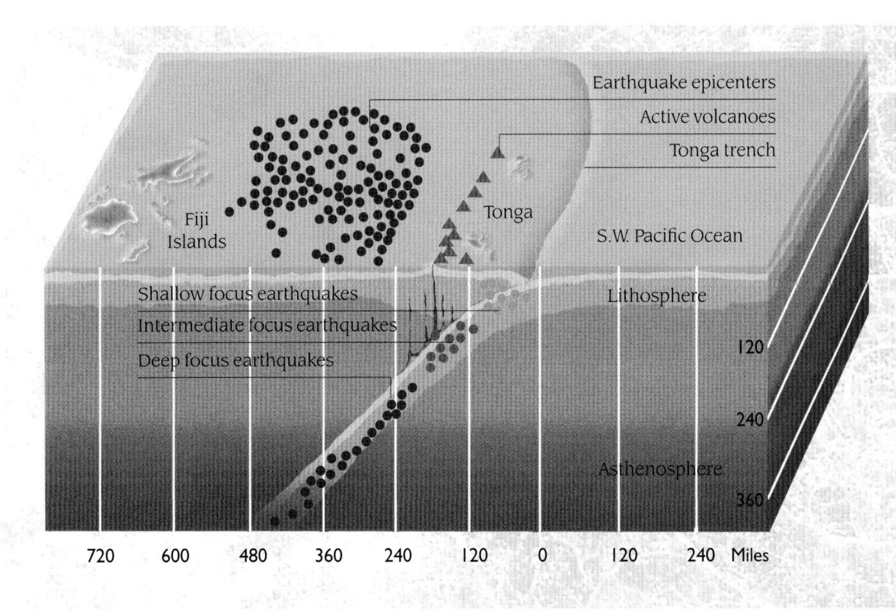

BENIOFF EARTHQUAKE ZONE

When the oceanic lithosphere sinks into the earth's interior at subduction zones, strains are set up which generate earthquakes. In 1954 the seismologist Hugo Benioff discovered that the differing depths of earthquakes (shallow, intermediate, and deep) were associated with the angle or "dip" of subduction zones, lying on a sloping plane corresponding to the plane of the descending plate. This characteristic earthquake pattern is very well illustrated by the Tonga region of the South Pacific. Away from the Tonga Trench, in a roughly northwest direction, the earthquake foci become increasingly deeper as the lithosphere descends. Earthquake zones associated with ocean trenches are known as Benioff zones and as a general rule they show that the angle of a subduction zone increases with increasing depth, while the magnitude or strength of earthquakes decreases with increasing depth. Consequently, it is shallow earthquakes that are the most powerful and dangerous; in ocean-continent destructive margins these tend therefore to be located nearer the shores of continents.

HOT-SPOT VOLCANO

Kilauea volcano (left) is one of the five shield volcanoes that form Hawaii, part of an island chain lying in the middle of the Pacific Ocean, far from any plate boundaries. It is a hot-spot volcano, producing the runny basaltic lava found at constructive margins.

changes in height of the islands during their development and also sea level changes.

The unity of this very long chain of islands, mountains, and submarine ridges is also demonstrated by its coincidence with a belt of negative gravity readings from the submarine trench that are particularly strong underwater. This anomaly was discovered in the 1920s by the pioneering Dutch geologist F.A. Vening Meinesz. By submerging his gravity-reading equipment, he overcame the distorting "pendulum" effects of the waves and eventually charted a well-defined negative gravity belt along almost 2500mi (4000km) of the Indonesian Island arc. He inferred this to represent a down-buckling and consequent thickening of the oceanic crust to form a mountain root.

Modern geophysical analysis, however, shows that the oceanic crust beneath these gravity anomalies is of the same thickness as transitional regions between the continental and oceanic margins.

Back-arc spreading—the extension and spreading of the ocean floor situated behind an island arc—may occur at convergent plate margins, and there is evidence to suggest that this is the result of very complex convection eddies taking place in the asthenosphere located above a subducting plate, or by the pulling away of the adjacent plate. Regional tension and normal faulting result from the effects of the rising mantle material, and also from stretching of the crust as it is "rolled over" by the friction and grip of the crust of the subducting plate. This back-arc spreading takes place behind the zone of subduction and, as a result of this tension, sags. It is now situated at a lower level than the surrounding continents and forms the incipient stages in a sea. Back-arc basins may attain huge dimensions. The floors of back-arc basins are generally young and sediments are thin; exposed rocks include fresh basalts and they cease to become active regions of spreading after a period of time—perhaps about 10 or 20 million years. Heat flow is often high but there is no well-defined ridge, as in the case of the Atlantic Ocean and its prominent mid-ocean ridge. Back-arc spreading, and the resulting basins, are considered to have been common throughout much of the Earth's history. The Sea of Japan is a back-arc marginal basin formed by spreading between Japan and the Eurasian margin and is now inactive; it appears to have begun forming during the early part of the Cenozoic.

The formation of the Arabian and Indian Oceans are very closely tied together. The Indian Ocean formed when the Indian plate eventually contacted the southern edge of the Eurasian plate; and the geographical region now occupied by the Arabian Sea—between the western edge of India and the eastern edge of the Arabian peninsula—came into existence. Ophiolites in the Gulf of Aden are dated to the Early Cretaceous and show that this is the time when the formation of the gulf began. This sequence was complete by about 20 million years ago.

PACIFIC RING OF FIRE

Where the edges of oceanic plates subduct beneath the margins of continental plates, deep trenches form as the oceanic lithosphere is carried down into the mantle. At depth it melts and water locked up in it also melts the overlying mantle. The molten rock collects, and being less dense, rises through the continental crust, erupting explosively at the surface. The huge plates that comprise the Pacific Ocean basin are bounded by subduction zones—and hence volcanoes and earthquake activity. Because of this, the resulting lines of volcanic activity are in the form of a gigantic arc that extends around most of the Pacific basin—the "Ring of Fire."

IN EUROPE, from the middle of Paleogene times, there were major changes to the physical geography. This was the origin and formation of the Alps mountain range. Geologically speaking, the Alps are rather young mountains. They were formed as a result of the northward movement of northern fragments of the African plate, causing impact with the European plate. The ancient Alps were part of a long string of mountain chains that ran from west to east along the southern border of Eurasia, from the Pyrenees on the border of France and Spain, the Carpathians in eastern Europe, and the Himalayas in Asia. The Atlas Mountains are their counterpart in northern Africa.

Europe's physical geography altered dramatically as Africa pushed up against the European plate, raising the Alps.

The highest of these ranges are the Himalayas, but all of these mountains formed as a consequence of the northward movements of Gondwana. They are positioned along what was until Paleogene times a part of the southern margin of the Eurasian landmass, bounded by the ancient Tethys Seaway. This history has been confirmed by the discovery within the Alps of ophiolites that represent Mesozoic and Early Cenozoic oceanic crust. These Alpine ophiolites occur mainly in the Pennine Alps between Italy and Switzerland—for example, at the base of the famed Matterhorn.

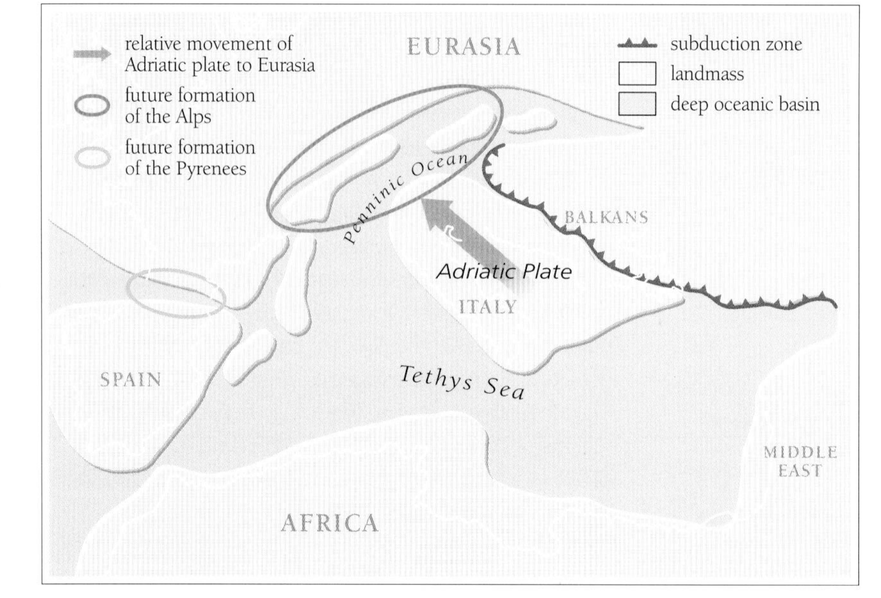

Interestingly, in most of the zone of orogenic activity at the junction of Africa and Europe, the African continent has not been sutured to the Eurasian continent. As a result, there is considerable seismic activity in this region as the mountain-building continues. This activity is apparent as earthquakes, especially in southern Italy and Turkey. Zones of seismic activity plotted on a map follow the line of mountain-building across the Mediterranean and the Middle East. Faults in these regions closely follow the earthquake zones.

THE ACTIVE tectonic zone running through the Mediterranean region is a reflection of the structure of the earth's crust in this area. Here, many small plates of lithosphere (the outer, rigid shell of the earth, including the crust) have been caught between the African and Eurasian shields, and these microplates have been pushed against one another and sandwiched between the African and Eurasian plates to produce a mosaic of tectonic activity. The Alps had their geological origins among the Iberian, Corsican, Sardinian, and especially the Adriatic microplates, which attached themselves to southern Europe. The Adriatic plate had originally been attached to the Eurasian continent in the region of the

Microplates carrying the Iberian and Adriatic peninsulas, as well as Corsica and Sardinia, contributed to the Alps' rise.

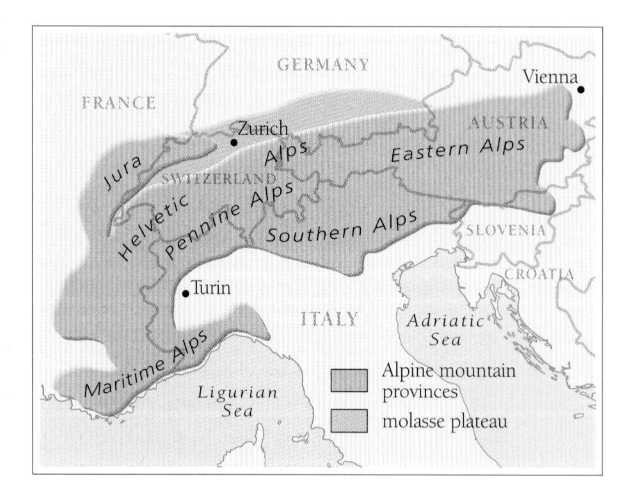

ORIGIN OF THE ALPS

As the Adriatic plate moved north it sutured Italy to Eurasia, closing the Penninic Ocean and uplifting the Alps. A similar collision by the Iberian peninsula created the Pyrenees. The Alps form a typical mountain chain with each zone representing a gradient of deformation. The igneous core lies in the Southern Alps, the Pennine zone is the metamorphic belt, while further north is the fold-and-thrust belt.

NEW PEAKS
The relative geological youthfulness of the European Alps (above) is reflected in their great height: this is because they have had less time to erode away than older mountain chains.

modern Balkans, and it carried the entire peninsula that eventually became known as Italy. About 45 million years ago, as the northern edge of the Adriatic plate was crumpled against the southern margin of Europe, the edge of the plate rode over the margin of the Eurasian craton. As the crust was thickened and buckled by this action it also received an accumulation of igneous rocks above the subducted plate. This complex crustal structure is the basis for the formation of the Alps.

During the first stages of the Alpine mountain formation, the emerging folds of crust were separated by patches of sea. (The Penninic Ocean had appeared in the region north of the Tethys, caused by rifting during the Triassic period.) Highly distinctive sediments accumulated in these folds; these were dark siliceous shales (indicating deep waters and the deposition of sand) and sandstones in random arrangements that were deposited between elongated deepwater submarine banks. These characteristic uplifted sedimentary rocks in the Alps are known to geologists as flysch. By Oligocene times, compression from the south had caused enormous recumbent (turned sideways) folds to rise as mountain arcs out of the old seaway and to slide over the land to the north. North of these complex folds lay a geographical depression that received these land-derived clastics (aggregated rocks), called molasse. A great plateau of this material lies on the northern edge of the Alpine system.

The Alps are formed into several mountain ridges that run more or less parallel to each other in an east-west direction. From the north these ridges are known as the Jura Mountains, the Helvetic Alps, the Pennine Alps, and the Southern Alps. The Helvetic Alps of Switzerland contain a particularly large volume of flysch rocks. The oldest flysch deposits now found in the Alps were deposited within the Penninic Ocean, which, during the Mesozoic, temporarily existed between the southern margin of the Eurasian plate and the north edge of the advancing Adriatic microplate. The collision between the two plates destroyed this ocean during the Eocene epoch, but it left deep marine waters standing to the south of the Alps, where younger flysch deposits accumulated. The Alpine orogeny continued to the Late Miocene, between 10 and 5 million years ago.

MOUNTAIN-BUILDING on the western margin of North America, which had begun at the end of the Cretaceous, continued into the Eocene. The ranges created by the Laramide orogeny extended from Mexico up to Canada. Tectonic activity also spread eastward, as far as the Black Hills of South Dakota. Between the local mountain ranges, such as the Wasatch and the Front Range of the rockies, lay broad basins. These received not only the erosional detritus from the nearby mountains, but also the discharge from their streams. Sediments and water accumulated in the subsiding areas throughout the Paleocene–Eocene, forming lakes and swamps. These basins were the Bighorn, Uinta, Washakie, and Green River basins. The Uinta Basin is the deepest of the Eocene lakes in this region. The Green River formation is notable for its rocks, which consist of almost 2000ft (600m) of freshwater limestones and very

In western North America, mountains had been raised from Mexico to Canada. Great lake basins lay between the ranges.

PALEOGENE

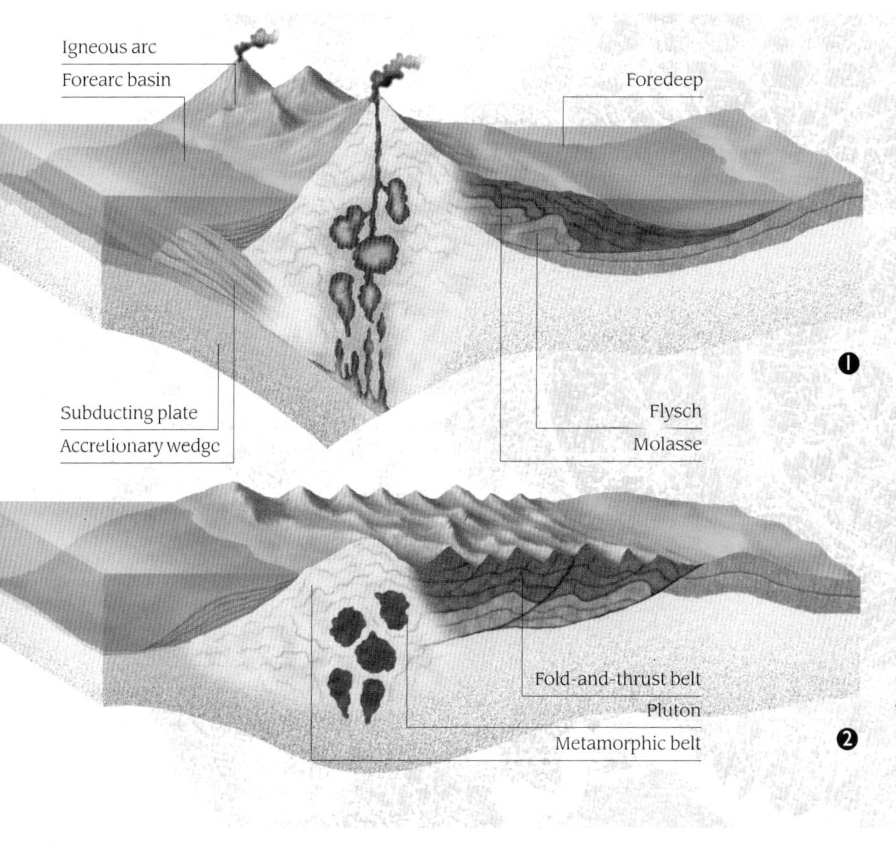

THE ANATOMY OF MOUNTAIN CHAINS

Mountains form at the collision zones of tectonic plates, either between two continental plates (as in the Alps and Himalayas) or between a continental and an oceanic plate (as in the Andes). As a subducting oceanic plate slides beneath a continental craton, immense shearing stresses are set up and slices of the surface are scraped off to accumulate in the angle between the two. This debris forms an accretionary wedge in front of the newly-forming mountain range (1). Magma from melting subducted material rises through the continental crust, forming a chain of volcanoes—the igneous arc—which is the core of the incipient mountain range; some magma crystallizes into large bodies of igneous rocks called plutons. Sediments accumulate in the sea on top of the accretionary wedge, which is known as the forearc basin. These sediments are mostly derived from the land. Beyond the mountains the crust is downwarped to form a foredeep. At first the foredeep fills with marine flysch deposits but as sediment transport from the mountains increases the sea retreats and flysch deposition is replaced by non-marine sediments called molasse.

The subducting plate is pushed into the region of metamorphism where extreme heat and pressure cause chemical and structural changes to the rocks. These metamorphosed rocks lie at the very deepest roots of mountains and, when eventually exposed in a mature mountain range, represent the most ancient rocks that are visible to geologists (2). This region forms a well defined area and is termed the metamorphic belt. Inland, the crumpled-up rocks are folded and ride over each other in great sheets, in an area known as the fold-and-thrust belt.

COLORFUL CLIFFS

Rocks of the Wasatch Formation in Wyoming were deposited by Eocene rivers. They preserve many mammal fossils.

fine laminated shales. The laminations in these basins are varves—thin sedimentary layers or pairs of layers that represent the deposition of a single year. From these it has been calculated that the Green River sediments were deposited over more than 6.5 million years.

THE WHITE River Formation is particularly well known for its fossils of aquatic marginal subtropical fauna. Flash floods in the Eocene deposited floodplain sediments that contain numerous fossil mammals, drowned and then preserved in sediment as it dried out. Large mammals that lived on the plains were also preserved, presumably due to being washed in during flash floods and also from dying near to the water's edge. Much of what is now known about the evolution of mammals in the Cenozoic of North America came from these rocks.

Some 18 million years after the extinction of the dinosaurs, mammals were thriving in and around the lakes of the Wyoming region.

Many other types of fossils were preserved in the lakes. Plant fossils confirm that the area was a subtropical environment, as do the presence of numerous fossils of frogs, pond turtles, lizards, boa constrictors, and crocodiles. These animals lived along the shores of the lakes and show a thriving and diverse lakeside vertebrate community, although the wonderfully preserved remains of freshwater fish are of particular interest.

The sediments in the Paleogene lakes in and around Wyoming were first studied during the nineteenth-century expeditions of O.C. Marsh of Yale University. In 1866 the field collector (and subsequently renowned academic) John Bell Hatcher discovered 23,000lbs (10,400kg) of immense animal bones. These bones were eventually recognized and named as those of enormous elephantine mammals called titanotheres ("titan beasts"). Graveyards have been found in ancient river channels, showing characteristic evidence of flash floods.

Titanotheres—properly known as brontotheres ("thunder beasts")—were browsers that ranged in size from that of a large pig to a modern elephant. They have, however, no real modern-day analogues, for they were equipped with ossicones (as in living giraffes) that were located on the top of the snout, above and behind the eyes, and on the cheeks. Some forms such as *Bathyopsis* sported huge, downward-curving canine tusks similar to some pigs today. The combination of skull horns, huge size, and lethal tusks probably made the adults of these animals almost invulnerable to the predators of the Eocene, such as the creodonts, which were much larger than the modern grizzly bear.

Titanotheres, creodonts, early "true" carnivores, early members of the horse family, ancient rhinos, and even mysterious groups such as the taeniodonts and tillodonts were in evidence, as were early primates, insectivores, and the archaic carnivores known as condylarths.

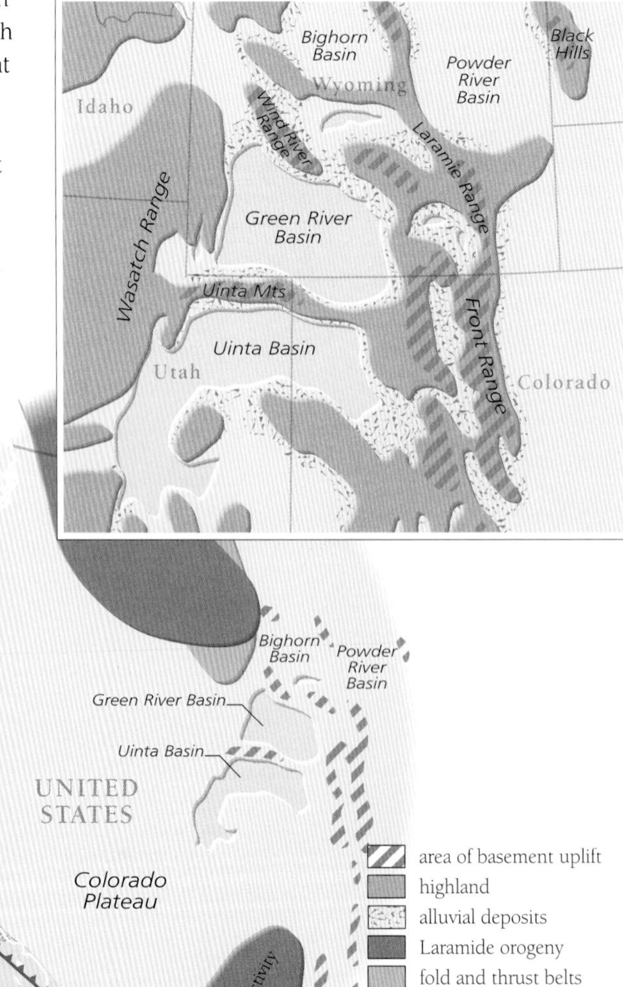

THE WYOMING LAKES

At the end of the early Eocene the angle of the west coast subduction zone changed from steep to shallow dipping, producing a band of volcanism and fold-and-thrust belts 900mi (1500km) inland. Farther east, sections of Precambrian crystalline basement rock were uplifted in north-south lying ranges. As water streamed off these elevated areas, basins in front of them formed lakes, which soon filled with outwashed sediments from the rapidly eroding mountains.

EOCENE HERBIVORES

The lakes of Wyoming and Utah are famed for their fauna of extinct ungulates, beautifully preserved in the fine sediments. During the Eocene the first hoofed mammals evolved to fill a wide variety of niches. The largest mammals at that time were the dinocerates ("terrifying horns"), which included the rhino-like *Uintatherium*. *Eohippus*, the "dawn horse," also appeared at that time.

MAMMALS had lived through the Mesozoic era as tiny burrowers and tree-dwellers. They existed in the shadow of the dinosaurs for 140 million years, until the end-Cretaceous extinction provided an enormous evolutionary opportunity. The evolutionary driving force that was behind a riot of adaptive evolution of the early Paleogene produced many land mammals that are no longer around and that look immensely different from those of today; most have no modern anatomical or ecological analogues whatsoever, which poses a real obstacle to investigating their individual lifestyles.

Mammals of the Paleogene were strange-looking to modern eyes and have no analogous forms among modern species of mammals.

Mammals during the Paleogene were bizarre by modern standards. Among the characteristic groups of primitive mammals from this period were the condylarths, which included abundant forms such as the huge tusked pantodont *Coryphodon*; small rabbit-deer-like hyopsodonts; and phenacodonts. They were stockily-built animals without claws, but a blunt hoof on each toe. Condylarths evolved some predatory types such as the superficially wolf-like *Mesonyx* and the immense "hyena-bear" *Pachyaena*. With hoofed toes and teeth that could rip, these impressive predators may be thought of as "wolf-sheep." They belonged to an assemblage of animals known as mesonychids, characterized by their five-toed feet—with hooves, rather than claws—and specialized front teeth.

But they did not have the niches of Paleogene predators all to themselves. Creodonts, the ancient cousins of true carnivores, were abundant at this time. They were more specialized than the giant mesonychids, having teeth that could cut, rather than rip flesh. Creodonts were among the more bizarre combinations of different animals, some resembling gigantic hybrids of bear, hyena, and lion. Creodonts such as *Patriofelis* and *Oxyaena* were low-slung, stocky animals with heavy tails, powerful jaws, and mouths full of slicing teeth. These huge early carnivores fed on the various ungulate-like (hoofed) browsers that were abundant in the subtropical Paleogene forests.

These strange herbivores included perhaps the oddest Paleogene mammals of all: the extraordinary taeniodonts, which were essentially part bear and part rodent. Taeniodonts such as *Psittacotherium* and *Stylinodon* possessed body proportions rather like that of an unusually thick-set bear, and had huge digging claws on the front feet, but the head was exceedingly deep and had an immense pair of open-rooted, extra-robust gnawing teeth, in which the bands of enamel gave rise to their name: taeniodont means "banded tooth."

SKULL BUMPS

The extraordinary 24in (60 cm) skull of the huge brontothere *Uintatherium* shows its array of horns, perhaps used in mating rituals, and its huge canine tusks, which were possibly used for fighting.

1 *Dolichorhinus* (a brontothere)
2 *Hyracotherium* (a very early horse)
3 *Hyrachyus* (an early rhinocerotid)
4 *Meniscotherium* (a condylarth)
5 *Amynodontopsis* (a small "running" rhinoceros)
6 *Uintatherium* (a brontothere)
7 *Phenacodus* (a condylarth)

1

2

3

4

5

6

7

PALEOGENE

DIETARY ADAPTATION

Bears and pandas are descendants of carnivores that also were ancestors to the dog, raccoon, and weasel families. While the cat branch of the carnivore family tree became specialized meat-eaters, the dog branch moved towards opportunistic omnivory and, in some animals such as the Kinkajou, near-vegetarianism. Bears also rely heavily on fruit and vegetables but, in general, their omnivorous diet has led to great success; bears are found on most continents, though there are just eight species. (Specializations tend to be fragile since they have evolved for one particular use; generalists can make a living off most food sources). The opportunism of some bears today (right) brings them into conflict with humans.

IN EVOLUTIONARY studies it is important to attempt to discover something about the way of life of extinct organisms. In the case of taeniodonts, any attempt to establish a modern analogue is flawed: there is no bear-like gnawer and there is no gnawing bear. They appear to have originated from insectivorous ancestors close to *Cimolestes*, an ancestral carnivore from the late Cretaceous. Taeniodonts occupied their bizarre niche of giant rodents along with a group called tillodonts. Tillodonts such as *Trogosus* were less powerfully built than taeniodonts and did not have such deep skulls, but their huge front teeth and the anatomy of their skulls gave them the appearance of gigantic rabbits. The subtle anatomical differences between tillodonts and taeniodonts probably meant that they fed on somewhat different food items; consequently there was probably less overlap in their diet and feeding habits than might seem the case. Taeniodonts had truly immense incisors, powerful jaws, and clawed feet that could have been used to dig up the roots and tubers that formed their diet. Fossils of both groups have been found in the same sediments; differences in size and dentition suggests that they lived side-by-side without competing for the same food. This is known as trophic partitioning.

Giant rabbit-like creatures such as Trogosus *shared the niche of giant gnawing rodents with the taeniodonts.*

The rodentlike adaptation, characterized by large gnawing front teeth, figured largely in another group of highly successful mammals, the multituberculates. Multituberculates ranged in size from a mouse to a lynx and had a wide range of ecological niches that spanned everything from burrower to arboreal gnawer. The skulls of the smaller ones looked a little like that of squirrels, as did their skeletons, but one group, the taeniolabids ("banded lips"), had chunky beaver-like heads and tough incisors. The name multituberculates arises from their

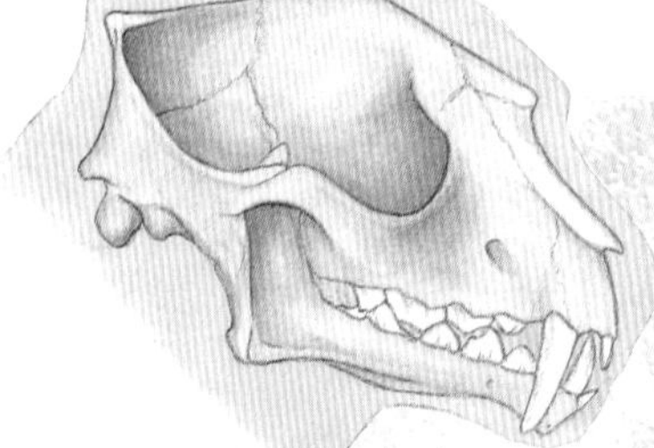

Dinictis (carnivore)

Entelodon (omnivore)

Merycoidodon (herbivore)

TEETH AND THE RISE OF MAMMALS

To paleobiologists, the one great feature that distinguishes mammals from all other vertebrates is the complex structure of their teeth and jaws, and much of the evolutionary success of mammals is due to the remarkably adaptive nature of their feeding apparatus. Mammals have complex teeth that occlude (interlock) in an extremely precise fashion, to give wear facets that are so predictable as to be practically diagnostic for some species.

The first incipient signs of the mammalian dentition was evident among the synapsid ("mammal-like") reptiles of the Triassic period. In these very advanced—for their time—mammalian precursors, wear facets had begun to appear as their teeth developed cusps and bumps on them. The teeth of synapsids had changed from a simple homodont dentition (where all the teeth are the same shape) to a heterodont dentition (where teeth are differentiated into, for example, incisors and canines) way back in Permo-Carboniferous times with animals such as *Dimetrodon*. Later, in the Triassic, the cynodont precursors of mammals added close-packed tooth occlusion to heterodonty and also developed novel arrangements of their jaw-closing musculature. Nevertheless, these animals still had not attained the mammalian condition of precise occlusion. This finally happened with the earliest true mammals in the Triassic and Jurassic, such as the shrew-sized *Morganucodon*. Jurassic multituberculate mammals developed their dental apparatus in the rodentiform (rodent-like) anatomical direction; but mammalian dental precision finally reached its modern level in the Jurassic and Cretaceous mammals that were themselves the precursors of Cenozoic large-bodied mammals.

In the 65 million years of the Cenozoic era, following the demise of the dinosaurs, a riot of mammalian dental evolution took place. After a slow start, tooth shapes became highly complex. Most early herbivores were browsing animals that fed on the leaves of trees and shrubs. Their molar teeth developed strong rounded cusps giving a lumpy crushing area on the biting surface. They are referred to as bunodont ("bun") teeth in recognition of their shape. The bunodont dentition has proved to be extremely successful ever since; it is found in a number of living mammals such as certain pigs.

Another type of molar tooth among Paleogene ungulates had sharp crescent-moon shaped cusps on the contact surfaces. This design is termed selenodont ("moon-toothed"). It was much less common among Paleogene mammal faunas than the bunodonts; however, selenodont dentition is superbly designed to grind up tough grass. When grazing animals evolved in the Oligocene in response to the expanding grasslands, the selenodont dental adaptation became very widespread.

It was during the Oligocene that another fundamental dental design was seen for the first time: the appearance of true rodents. A "rodentiform" dentition had been in evidence from the Triassic period onward, since the tritylodont cynodonts and multituberculates. However, neither of these animals had either the complex grinding molars of true rodents, nor the self-sharpening edges of the elongate incisor teeth that are so characteristic of rodents today. The adaptation of self-sharpening incisors is in part responsible for the incredible success of the rodents since the Oligocene epoch.

Dental innovation among the carnivores is represented by their characteristic molars that have become specialized for the efficient slicing and cutting up of flesh. These are the blade-like carnassial teeth; the most rearward upper premolar shearing against the most forward lower molar in a scissor action. This development accounts for the success of true carnivores as top predators. The canines in true carnivores always remained large and dangerous, although there were huge differences between the various groups.

TREETOP BROWSER

Indricotherium **provides an example of how ecological niches (defined by the animals that occupy them) are filled after the disappearance of the occupier. Several million years before the appearance of this animal, the niche of "long-necked superheavyweight browser" had been filled by the sauropod dinosaurs. After their extinction, a few mammals came close to achieving this adaptation, but only** ***Indricotherium*** **succeeded completely. It stood 26ft (8m) high but belonged to the rhinoceros family, rather than elephants or giraffes.**

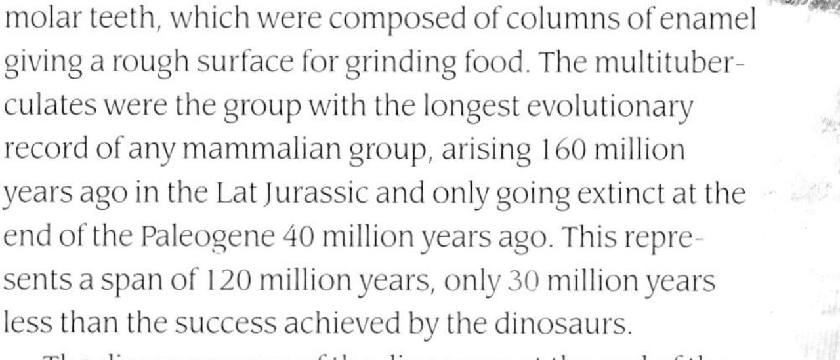

molar teeth, which were composed of columns of enamel giving a rough surface for grinding food. The multituberculates were the group with the longest evolutionary record of any mammalian group, arising 160 million years ago in the Lat Jurassic and only going extinct at the end of the Paleogene 40 million years ago. This represents a span of 120 million years, only 30 million years less than the success achieved by the dinosaurs.

The disappearance of the dinosaurs at the end of the Cretaceous had opened up many habitats and ecological niches, at the end of one of the most single-animal dominated ecosystems ever. There followed a rapid burst of mammalian adaptive evolution—much of it in the structures of the jaws and teeth—as these previously insignificant creatures evolved to take over the vast range of now unoccupied niches. Many of the anatomical 'results' may be informally viewed as 'evolutionary experiments' and this accounts for the sometimes bizarre forms that appeared. Many of these adaptations bear no resemblances to modern mammals and are quite mysterious and extraordinary.

HORNED BEAUTY

(Below) An example of the unusual "results" of evolution at this time is ***Arsinoitherium*****, the African counterpart of the dinocerates. This rhino-sized beast—named after the ancient Egyptian queen Arsinöe—looks superficially like a rhinoceros, but its teeth, skull, and limb bones show it to be unique. Indeed, it has its own family, the embrithopods.** ***Arsinoitherium*** **was a herbivore but not an ungulate; the latter, such as horses, walk on the adapted nails of their feet (hooves), and do not necessarily feed on plants.**

THE APPEARANCE of whales (mammalian order Cetacea) from terrestrial ancestors is one of the great evolutionary stories, supported in recent years by fabulous new fossil finds. Looking at a streamlined dolphin or an immense filter-feeding blue whale weighing 100 tons, it is difficult to imagine how this spectacular group of living mammals evolved from terrestrial ancestors, yet the fossil evidence for this remarkable transition is (perhaps surprisingly) very strong.

Fossil evidence indicates that whales evolved from land ancestors.

The disappearance of giant Mesozoic marine reptiles near the end of the Cretaceous period had created a series of evolutionary vacuua or ecological voids, which were filled by bony fish that subsequently attained a somewhat larger size. These, however, bear no relation to the whales. Unlike the situation on land, where mammals had very rapidly occupied the niches opened by the disappearance of the dinosaurs, the time lag between the disappearance of marine reptiles and the origin of whales is in the order of at least 10 to 15 million years, suggesting that a significant and gradual process of evolutionary transformation had to be undertaken before the new niche could be colonized.

Whales did not simply fill the vacant ecological niches of extinct Mesozoic marine reptiles.

Vestigial hind-limb

TOOTHED WHALES

Toothed whales of today's oceans resemble the ancient archaeocete whales of the Paleogene more than do modern baleen whales. The latter have become extremely specialized to feed on plankton, while toothed whales probably feed in a similar way to the predatory archaeocetes.

A plausible scenario of whales' origination suggests that terrestrial mesonychids such as the huge *Pachyaena*, and even proto-whales (archaeocetes) such as *Ambulocetus*, may have lived as seashore scavengers, much like the modern brown hyena (also called the Strand Wolf) of South Africa, which survives by scavenging animal remains and small marine animals from the Atlantic beaches (strands) of southern Africa. Over a period of time during the Paleogene, it may have been that competition from purely terrestrial mesonychids such as *Pachyaena* pushed their own descendants—the ancestors of archaeocetes—to live progressively more and more in the nearshore marine realm, from which the menace of huge marine predators had recently disappeared. The evolution of limbs that were adequate for swimming would have directed the evolution of archaeocete ancestors as they became more adept at living off marine animals. Finally the link with the terrestrial realm was broken and ancient cetaceans became fully marine.

It is useful to consider that living bears, especially polar bears, are proficient swimmers, and that seals, sea lions, and walruses arose from bearlike ancestors in the Oligocene–Miocene. In the light of such functional analogues, the terrestrial model of whale origins going back to the mesonychids seems entirely plausible.

THE EARLY evolution of whales is illustrated by partial skulls and skeletons of about five different types that range in age from early to mid-Eocene times. All of these animals were found in nearshore marine sediments in modern Pakistan, where, 50 million years ago, the northern shores of the ancient Tethys Seaway lay. These early proto-whales of Pakistan's Eocene are referred to as archaeocetes and were very different from modern whales in their anatomy and lifestyle.

In the northern reaches of the ancient Tethys Seaway (now in Pakistan), ancient whales flourished 50 million years ago.

Cetaceans certainly arose from mesonychid condylarths, the abundant archaic mammals of the Paleocene and Eocene. The broad-bladed, cusped zeuglodont teeth give the clue to this, as do the extraordinary anatomical changes from mesonychid to whale: body, limbs, ears, and progressive telescoping of the skull.

Ambulocetus ("walking whale") from the Middle Eocene—the earliest marine archaeocete known to date—had long femurs (thigh bones) and long paddle-like feet. It is interpreted as an extremely large otter-like or seal-like, foot-propelled, swimming animal that was about 11ft (3.5m) in length with a powerful, 2ft (0.6m) long skull. The limbs were adapted for swimming, though *Ambulocetus* was clearly mobile on land. It probably adopted a crouched posture and hauled itself around like an unusually agile seal. *Indocetus,* from about the same time, was similar, but its fused pelvic-backbone connection and tail structure show it to be more of a tail-propelled swimmer. All subsequent cetaceans were tail-swimmers, not paddlers.

A more recent find of a similar animal, *Rodhocetus,* casts new light on whale origins. The skull is long, powerful, and armed with the immense zeuglodont teeth of other archaeocetes, but its pelvis and backbone are much better preserved. It had very high spines, or processes, on its backbones and a pelvis that articulated directly with the backbone. It would appear that *Rodhocetus* supported its weight on land, and yet, at the same time, its neck was short, the thigh bone was small, and the pelvic backbones were free to move. These details show that this animal was intermediate between foot and tail-propelled swimmers. It could both swim and move well on land.

WHALE ORIGINS

During the Paleogene, the Tethys Sea ran in an east-west direction across what is now Eurasia. Along the shores of this ancient subtropical ocean lived the ancestors of whales. These were mesonychid condylarths and they probably resembled a mixture of bear and hyena. (Bears feed happily in water and hyenas scavenge along the coast of Southern Africa). The sediments deposited along the shores of the Tethys document the successive stages in whale origins as fossils show adaptations to land, semi-aquatic, and fully aquatic ways of life.

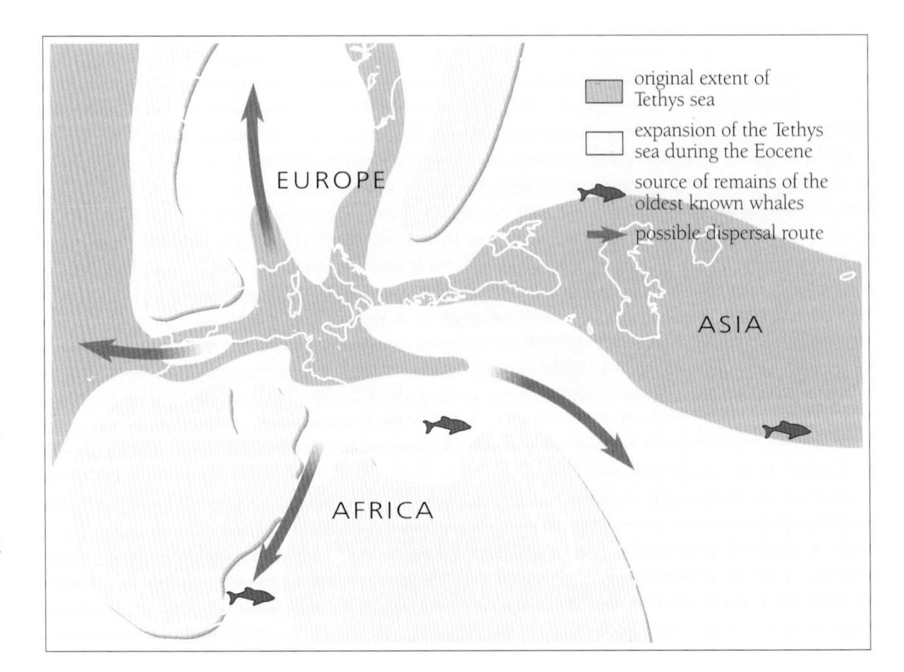

EOCENE GIANT

The archaeocete whale *Basilosaurus*, looking like a classic sea-serpent, was fully adapted to a marine existence.

Later Eocene whales had similar skulls—and, on the basis of their skull anatomy, similar predatory tendencies—to *Rodhocetus* and *Pakicetus*, but evolved long serpentine bodies up to 40ft (12m) in length. These were the basilosaurids, and they were the first really large whales to appear in the seas. The hindlimbs are still present, with all the elements still in place, though much reduced in size and of no use for swimming.

The first fully marine archaeocetes, such as *Protocetus*, had bodies that were probably less than 10ft (3m) long. Its features are echoed in the skulls of its evolutionary cousin *Pakicetus,* which exhibits a skull that is about 1ft (0.3m) long, with jaws that are also long, but not slim as those in dolphins. The jaws were very stout and the teeth that filled them were extremely impressive: huge triangular teeth reminiscent of the teeth of the modern great white shark, but less sharp and more robust, with a series of tough cusps cascading down the front and back edges. This tooth form is very characteristic of archaeocetes and is referred to as zeuglodont for its broad serrated shape. Both *Pakicetus* and *Protocetus* possessed this dentition.

Protocetus had a sinuous body with strongly reduced fore and hind limbs, but the hind limbs were still significant elements and the forelimbs were almost certainly sufficiently strong to allow some limited locomotion on land—in great contrast to modern whales. Recent finds in Pakistan have revealed quite extraordinary animals that clearly represent intermediates in the functional stages of cetacean evolution.

Basilosaurus

living whale suborders
Eocene whale families
ancestral carnivores

MYSTICETI (baleen whales)
ODONTOCETI (toothed whales)
DORUDONTIDS
ARCHAEOCETI
BASILOSAURIDS
PROTOCETIDS
CONDYLARTHS
MESONYCHIDS
ARCTOCYONIDS

WHALE RELATIONSHIPS

The cladogram above, a tree of interrelationships, is based on analyzing anatomical characteristics shared by related animals. It shows the mesonychid origins of the ancient archaeocete whales and the divergence of the groups of modern whales.

Pachyaena

Ambulocetus

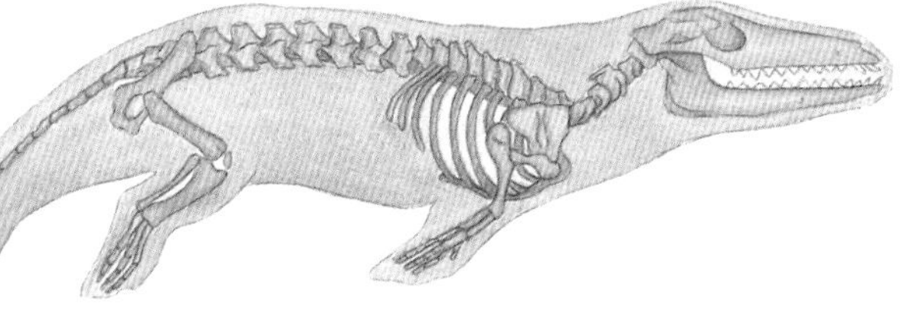

Protocetus

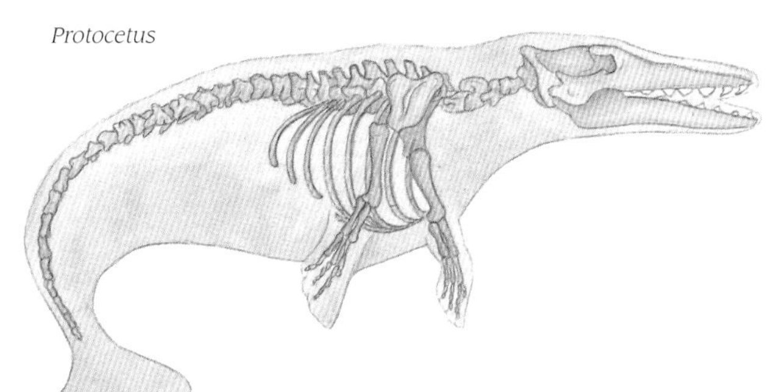

WHALE EVOLUTION

Computer analysis of fossil whale remains from the Tethys Sea, and their many shared anatomical features, show that whales originated from highly predatory mesonychid condylarths. The bear-sized *Pachyena* probably resembled these ancestors. Intermediate forms such as *Ambulocetus* show the functional stages in the transition from land-mammal to marine proto-whale such as *Protocetus*.

Zeuglodont tooth

AMONG the many sites in the world where fossils are found in abundance, a small number stand out. These precious sites include Lake Messel near Frankfurt, Germany, which is the source of a collection of Early Eocene fossils, about 50 million years old. In many ways the preservation of the animals from Lake Messel is perhaps the most perfect and extraordinary in the world of paleontology. Here are fossils found and preserved as nowhere else: hair, feathers, membranes, stomach contents, and even internal organs have remained intact in some cases.

A small, deep lake in what is now southern Germany provided perfect conditions for preserving a great variety of Eocene species.

The surrounding geology of the Messel lake site shows that it was a small, relatively deep lake in a fault valley with anoxic (non-oxygenated) bottom waters. Paleoenvironmental indicators, such as the types of fossils there, show that it stood in the middle of a lush tropical forest. Examples of these indicators include the plant fossils such as laurel, oak, beech, citrus fruits, vines, palms, rare conifers, and water lilies.The geological strata are mostly bituminous (organically derived tar) clays, which must have accumulated in the warm subtropical water.

The lake was unstratified, but its sediments reveal that conditions at the bottom were extremely unfavorable to life. Consequently, if the carcass of a dead animal drifted down through the warm waters to rest on the lake bed, there were no scavengers around to consume it. Poor oxygen content and extremely quiet conditions allowed the clays to embalm the body in a fine shroud of sediments. So many animals were preserved in these sediments that the carbon from their bodies formed the oil deposits that made the site valuable for industry in the nineteenth century and beyond.

MAMMALS OF MESSEL

Thirty-five different species of mammal have been found at Messel, including bats, insectivores, carnivores, ungulates, anteaters, primates, marsupials, pangolins, and rodents. There were oddities like *Leptictidium* that resemble no living animal, and creatures such as *Eurotamandua*, which is nearly identical to modern anteaters.

1 *Archaeonycteris* (the first known bat)
2 *Messelobunodon* (a primitive artiodactyl)
3 *Propalaeotherium* (a primitive horse)
4 *Leptictidium* (an insectivore)
5 *Paroodectes* (a miacid)
6 *Eurotamandua* (an anteater)
7 *Pholidocercus* (a hedgehog)
8 *Chelotriton* (a terrestrial salamander)

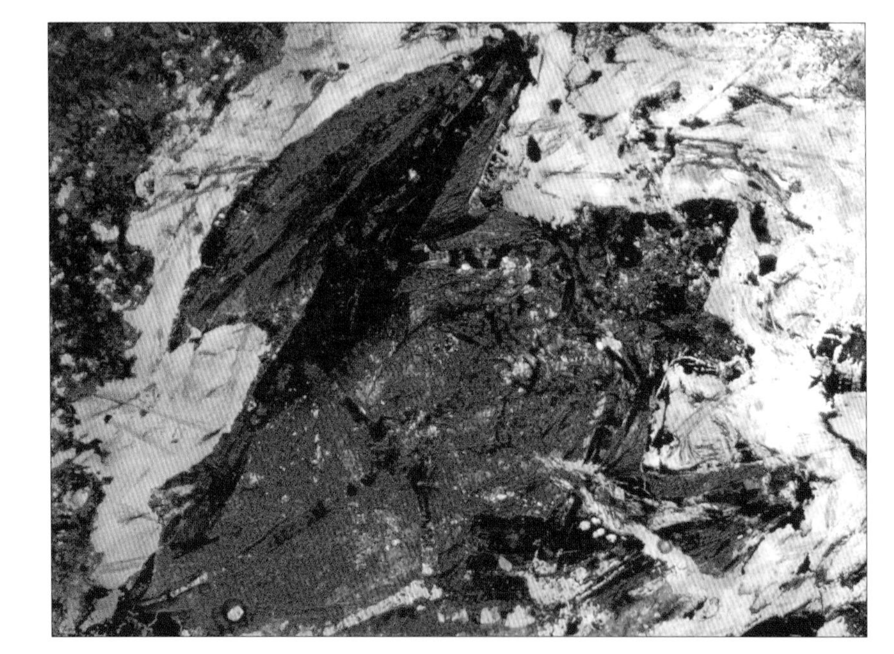

The Messel deposits preserve insects and other invertebrates, vertebrates, and plants that represent input from nearby terrestrial settings. Among the non-aquatic fossils, air-transported ones such as leaves, pollens, bats, birds, and insects predominate. Most of the small amphibians that are found here are concentrated near the mouth of a shifting drainage that fed the lake. These are critical considerations, for the remains of animals that lived in forests are exceedingly rare in the fossil record (such as the earliest cats, which probably lived in forests, then as now); this is because of forest conditions: an animal that dies and drops to the forest floor is rapidly eaten by the numerous scavengers that are present in the forest. The roots and shoots of trees and shrubs also disturb the forest floor, making it a mechanically active and destructive environment in which bones are quickly destroyed. In addition, forest floors are highly acidic and

EXCEPTIONAL PRESERVATION

The amazing preservation of the fossils from the site of Lake Messel allows unique insights into the anatomy, relationships, and ways of life of these long-dead species. Many of these are small forms that lived in or around trees—an environment that usually does not preserve such tiny remains; here though, the animals have been deposited in the still, anoxic waters of a deep forest lake and their delicate structures are found in perfect order, as in the bat *Archaeonycteris* (far left), and in many species of land birds.

dissolve bones readily. One aspect of major significance about the fossil site at Lake Messel is that it preserves a range of hitherto unknown forest animals from the European Eocene of 50 million years ago.

Two particular groups of animals at the Messel site have attracted the most attention from paleobiologists: terrestrial mammals and bats. A number of the mammals found at Messel are unique to this locality and provide critical information into mammalian history. The oldest anteater, *Eurotamandua*, comes from this site and is very like living anteaters except that it still has cheekbones, which have been lost in modern anteaters. *Eurotamandua*'s presence in Europe is a mystery, because all living anteaters are found only in South America. A similar evolutionary conundrum is *Eomanis*, the fossil pangolin from Messel, which belongs to a mammalian group found only in Africa and southeast Asia. It may be that both anteaters and pangolins originated in Europe despite now being absent from that continent.

Other terrestrial mammals from Messel include more unusual forms whose affinities with modern living mammals is often obscure. The early horse-like animal *Propalaeotherium* is only the size of a spaniel and shows the primitive arrangement of its feet with the presence of four little hooves in each foot. Its tooth crowns were low, showing that it fed on soft leaves and tropical fruits in the forests; it was a browsing animal and its affinities are not difficult to establish. The same cannot be said for the small bipedal insectivore *Leptictidium*, a mammal that stood only eight inches high. It had a short, strong trunk region, very long tail, and hind legs like a bandicoot. Its front limbs were too short to be used for walking, but instead of moving like a modern rabbit or hare, it seems to have run fast with alternating steps. This curious little creature has no analogous form among living animals. The Messel site preserves even bats, which are a great rarity in the fossil record, and even shows that one particular type fed exclusively on butterflies.

THIS might have been the scene in central Asia around the time of the Paleocene–Eocene boundary 55 million years ago. As today, much of the region is situated far from the ocean and so the ameliorating effects that the sea exerts on terrestrial climates was reduced. This resulted in seasonal extremes, with summers being ferociously hot and winters excessively cold. Despite this, lush forests were in abundance because of tropical global conditions. Open plains were in evidence but not yet to the extent that they were to reach from the Oligocene onwards. The forests provided huge amounts of plant food for browsing herbivorous mammals; these became commonplace and could attain immense sizes. In response to this great bounty, the scavengers and active hunters of the time assumed much larger body dimensions than modern predators.

Andrewsarchus ***might have been the largest land predator to have ever lived.***

At the edge of a subtropical forest, *Andrewsarchus*, a gigantic hooved mesonychid condylarth, scavenged a dead *Embolotherium* with its huge spike-like canines and tearing triangular premolars. Both animals were the size of a hippopotamus, with the embolothere weighing in at an estimated 3 tons. Almost equally massive predators closed in for a look, such as the *Sarkastodon*. About 10ft (3m) long, it greatly outweighed any living bear, and its slicing, creodont molars and robust, bone-cracking premolars suggest that it was a more carnivorous animal than our largely omnivorous modern bears. Another mesonychid, the lion-sized *Harpagolestes*, also approached the kill. The few areas of open plain in central Asia provided a hunting ground for packs of the hyenodont creodont *Paracynohyaenodon*. The marten-like miacid *Vulpavus*, one of a long line of "true" carnivores that would eventually replace the Paleogene giants, kept clear of the competition between these huge carnivores until they had eaten their fill, while tiny primates, a recently evolved order, remained safely in the trees.

1 *Harpagolestes* (a mesonychid)
2 *Paracynohyaenodon* (a creodont)
3 an early primate
4 *Sarkastodon* (a creodont)
5 *Andrewsarchus* (a mesonychid)
6 *Embolotherium* (a titanothere)
7 *Vulpavus* (a miacid)

1

2

3
4
5
6
7

THE EVOLUTION OF MAMMALS

MOTHER'S MILK

A major and unique feature that helps to define a mammal is the presence of mammary glands from which the live-born young suck milk.

Artiodactyls (even-toed ungu
Pantodc
Tillodonts
Taeniodonts
Condylarths
Ungu
Perissodactyls (odd-toed ungulates)
Mesonychids
Litopterns
Notoungulates
Astrapotheres
Hyraxes
Proboscideans (elephants)
Embrithopods
Sirenians (sea cows and manatees)
Desmostylians
EUTHERIA (placental
METATHERIA (marsupials)
Symmetrodonts
Morganucod
Monotremes

1.8 mya | 24 | 65 | 144
Neogene | Paleogene | Cretaceous | Jura
Cenozoic | Mesozoic

THE CHARACTERISTICS used to define a mammal (warm-blooded; giving birth to live young, which they suckle on milk [mammary] glands) are no use for interpreting the fossil record, since such features are not preserved. However, there are many clues that may be read from tough fossilized bones, almost as easily as from a recent dry skull.

Cladistic analysis—which looks at the recentness of common ancestry of animal groups and their relationships, by comparing anatomical data—demonstrates that the probable origins of true mammals lay among the mammal-like reptiles, the shrew-sized trithelodont cynodonts of the Middle Triassic. Certainly these little animals show some remarkable resemblances to mammals, especially in the structure of their skulls. Cladistic analysis of the tiny mammals of the Mesozoic era, however, is extremely contentious; for example, the living monotremes—egg-laying mammals of Australia—have much more primitive anatomy than many extinct Mesozoic mammal groups, such as the long-lasting multituberculates. One fact beyond doubt is that the living marsupials and placental mammals are the most advanced of all—that is, they have the most derived characteristics.

When cladistic analysis is applied to fossil mammals of the Cenozoic, certain relationships emerge. For instance, the anteaters, sloths, and armadillos are close cousins of the pangolins, while rabbits are found along with rodents and elephant shrews. Tree shrews, bats, and sugar gliders are actually rather closely related to primates, with which they all form a large group called the archontans. The ungulates (hoofed mammals) combine with the cetaceans, sirenians, hyraxes, elephants, and the aardvark to form the group Ungulata.

1 225 million years ago in North America, *Adelobasileus*, a shrew-sized animal, is the oldest mammal
2 The lineage splits into a series of still primitive forms such as multituberculates
3 The monotremes originate during the Late Jurassic
4 The marsupial–placental split occurs in the Early Cretaceous of North America with *Alphadon*, a marsupial.
5 Huge radiation of mammals as dinosaurs become extinct

Analysis of the fossil mammals of the Cenozoic also produces interesting patterns. For instance, the taeniodonts, the giant gnawing "bears" of the Paleogene, seem to be descended from a group that is close to the origins of the creodonts—the huge Paleogene cousins of the Carnivores. This ancestral group is probably the arctocyonids, small mammals that are among the very earliest of the condylarths. In another interpretation, the huge dinocerates—such as *Uintatherium*— come out alongside the the cetaceans and the perissodactyls. The clustering of whales with two groups that are obviously ungulate-like suggests that the cetaceans had origins among hoofed groups as well. This is supported by the close placement of the whales to the mesonychid condylarths, the primitive hoofed predators of the Paleocene.

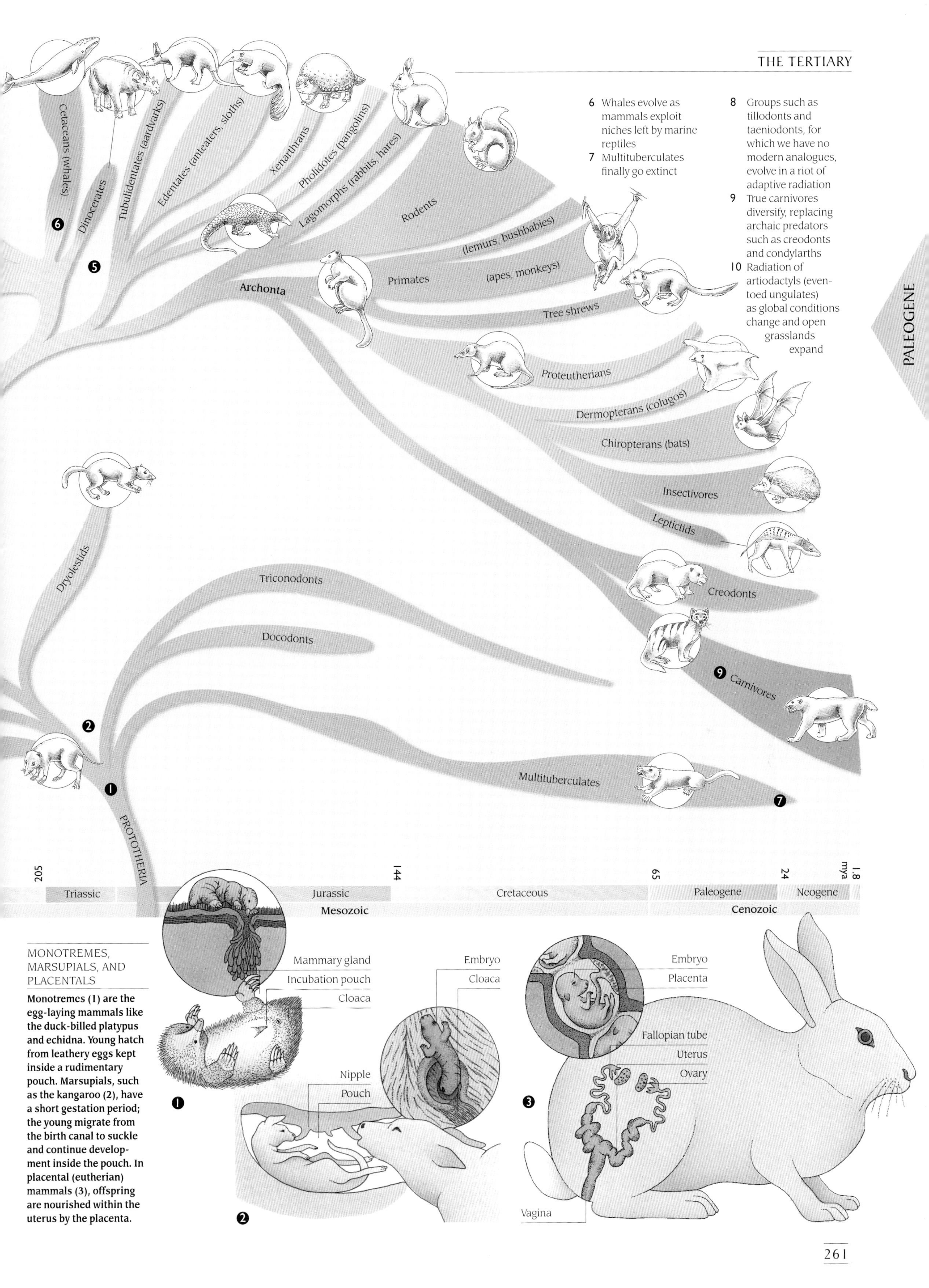

6 Whales evolve as mammals exploit niches left by marine reptiles

7 Multituberculates finally go extinct

8 Groups such as tillodonts and taeniodonts, for which we have no modern analogues, evolve in a riot of adaptive radiation

9 True carnivores diversify, replacing archaic predators such as creodonts and condylarths

10 Radiation of artiodactyls (even-toed ungulates) as global conditions change and open grasslands expand

MONOTREMES, MARSUPIALS, AND PLACENTALS

Monotremes (1) are the egg-laying mammals like the duck-billed platypus and echidna. Young hatch from leathery eggs kept inside a rudimentary pouch. Marsupials, such as the kangaroo (2), have a short gestation period; the young migrate from the birth canal to suckle and continue development inside the pouch. In placental (eutherian) mammals (3), offspring are nourished within the uterus by the placenta.

THE EVOLUTION OF CARNIVORES

ECOLOGICAL niches defined by the roles of top predator were opened up by the demise of the dinosaurs in the late Cretaceous period. Because the initial evolutionary radiation of early Cenozoic mammals did not include any large carnivores, a number of completely different animal groups were in a position to take over.

In South America the giant "terror birds" such as *Phorusrhacos* occupied this niche throughout the Eocene and most of the Oligocene, only meeting competition later with the advent of large mammalian carnivores—not placental mammals but marsupials. The jaguar-sized *Arminiheringia* posed a serious threat at least until the very latest Oligocene. By the start of the Miocene, the phorusrhacids were close to extinction and the marsupicarnivores soon came up against placental carnivores. Modern carnivorous marsupials are mainly in Australia and include the Tasmanian devil (*Sarcophilus*), a dasyurid, but no other much larger than a cat. The last medium-size marsupial predator on the planet, the Tasmanian wolf (*Cynocephalus*), became extinct in 1933.

Elsewhere, several groups of sizeable vertebrates competed for predatory niches. In Europe, "true" carnivores (*Carnivora vera*) vied with huge archaic creodonts and with an unusual group of fully terrestrial crocodilians. The latter had a relatively short evolutionary history, perhaps caught between competition from mammalian carnivores on land and the superbly adapted semiaquatic crocodilians in the rivers and estuaries.

In Asia, marsupials remained small, as did the earliest true carnivores. The main predators belonged to the creodonts and the mesonychid condlyarths. Many were exceptionally large and anatomically unusual: *Andrewsarchus* was the largest-ever mammalian land predator/scavenger, and *Sarkastodon* was a hulking, solitary, bear-like hunter. Another large mesonychid predator was *Pachyaena*. The wolf-sized creodont *Paracynohyaenodon* had the proportions of a low-slung dog and may have been a pack hunter, unlike *Sarkastodon*. All found easy prey among the browsing mammals and herbivores of Asia's lush Paleogene forests. However, many Asian predators seemed to have been better adapted for scavenging rather than full-fledged meat eating.

As the world cooled, subtropical forests disappeared and the biomass of herbivorous mammals was reduced. For large predators, the new, smaller, quicker herbivores yielded too little meat for the energy expended by chasing prey on short legs supporting heavy bodies. Longer legs are more efficient, and this may have been a factor in the subsequent success of true carnivores. They also had the advantage of being able to use their teeth to eat roots, nuts, and fruits; creodonts, whose teeth were only shears, could not. Paradoxically, it seems that true carnivores succeeded by not eating meat all the time.

CARNIVORE GROUPS

During the Paleogene, "true" carnivores were small miacids—marten-like animals that were very different from the hulking predators that existed then. These latter forms included mesonychids, creodonts, and, in South America, the borhyenid marsupials. The fundamental carnivore split into the vulpavines ("dog branch") and viverravines ("cat branch") occurred among the miacids some 60 million years ago. Living vulpavines (Canoidea) include the dog, bear, and weasel families, some members of which are omnivorous, while the more purely carnivorous viverravines (Feloidea) divide into four fundamental groups, including the superficially dog-like hyenids. Extinct feloids include the saber-toothed cats, while a number of canoid lineages have become extinct. Among these are the bone-crushing dogs or borophagines that filled the hyenid niche in North America, the immense amphicyonids ("bear-dogs"), and the nimravids or "paleo-sabers." Nimravids are essentially saber-toothed cats, yet they are anatomically more closely related to dogs than to cats.

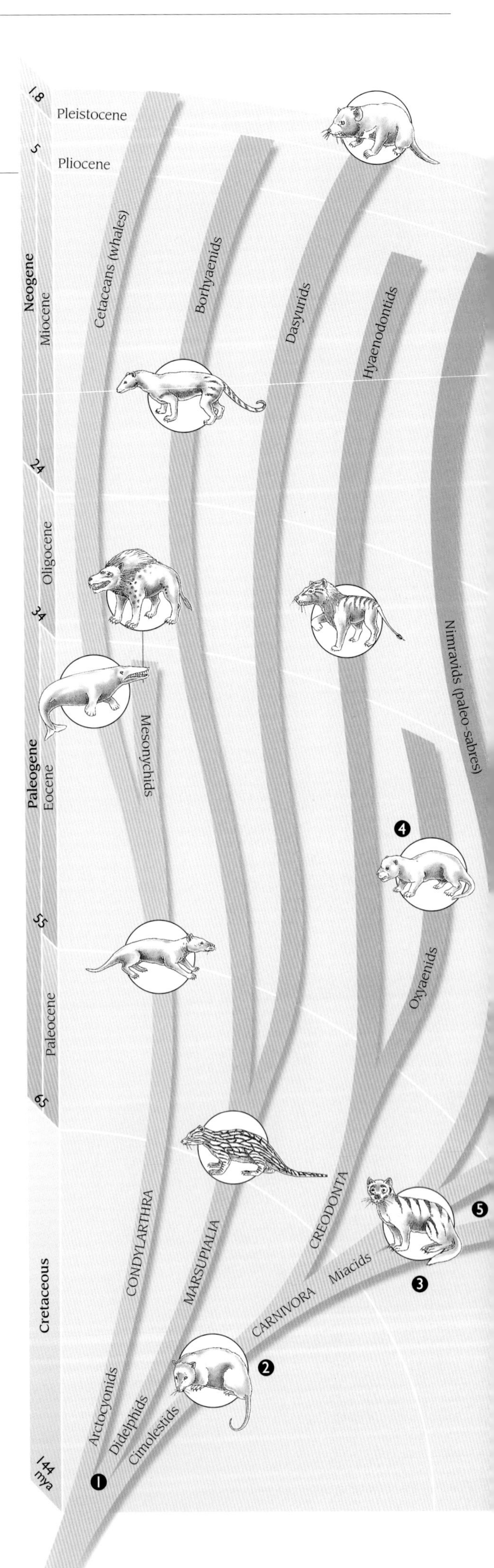

1 Various small meat-eating mammals, including didelphid marsupials and condylarths, appear
2 *Cimolestes* shows the beginnings of specialized molar teeth: the carnassial shear
3 Miacids are the first "true" carnivores
4 Creodonts, an archaic carnivore group, are the top predators
5 A split into "dogs" and "cats" takes place via the miacids, *Vulpavus* and *Viverravus*
6 The Canoidea and Feloidea split into a number of lineages
7 Saber-toothed cats represent the zenith of Cenozoic carnivorousness

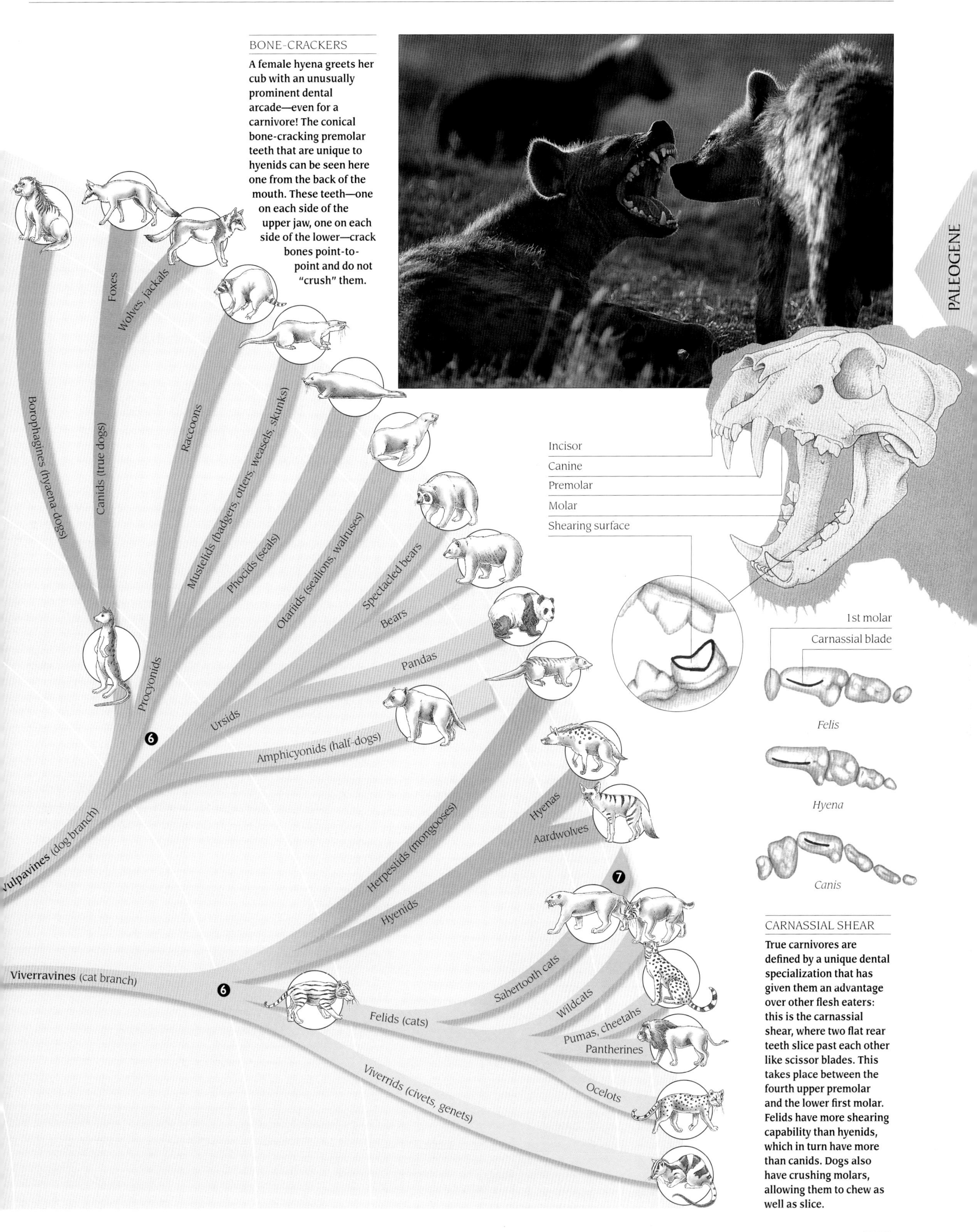

BONE-CRACKERS

A female hyena greets her cub with an unusually prominent dental arcade—even for a carnivore! The conical bone-cracking premolar teeth that are unique to hyenids can be seen here one from the back of the mouth. These teeth—one on each side of the upper jaw, one on each side of the lower—crack bones point-to-point and do not "crush" them.

CARNASSIAL SHEAR

True carnivores are defined by a unique dental specialization that has given them an advantage over other flesh eaters: this is the carnassial shear, where two flat rear teeth slice past each other like scissor blades. This takes place between the fourth upper premolar and the lower first molar. Felids have more shearing capability than hyenids, which in turn have more than canids. Dogs also have crushing molars, allowing them to chew as well as slice.

THE NEOGENE

24 – 1.8

MILLION YEARS AGO

THE NEOGENE period, which comprises the Miocene and Pliocene epochs, extends from 24 million years to 1.8 million years ago. During this time, the world experienced three events of great significance that are unique to this period: the evolution of modern mammals, including the appearance of the first humans, and the onset of a cycle of ice ages which dominated the Quaternary period that followed. The spread of grasslands, which had begun in the Paleogene, continued, leading to the diversification of grassland fauna.

During the Neogene, the modern groups of mammals that are familiar in today's ecosystems radiated and evolved from more archaic Paleogene ancestors. Most of the anatomically unusual Paleogene mammals went extinct, including all the large endemic land mammals of South America, and were replaced by more derived types. The modernization of the Cenozoic mammal fauna took place over a mere 24 million years—a very brief time in geological terms.

THE BOUNDARY between the Paleogene and the Neogene is not marked by any mass extinction or other historical event, but the latter period is distinguished by the occurrence of several significant biotic changes: the spread of herbaceous plants and grasses; the development of major new communities, typified by important changes in the teeth of herbivores; and the appearance of completely new families of mammals. The spread of grassland habitats was a result of climate change—the global cooling that was gradually taking place as ice sheets spread, first of all in Antarctica, and subsequently in the Arctic. Sea levels rose and fell and the temperature and humidity fluctuated, becoming cooler and drier overall. Local tectonic events caused parts of east Africa, western North America, and South America to receive less rainfall than they had previously enjoyed, with an impact on animal populations. New land-bridges formed that allowed the exchange of animals and plants.

During the Neogene the climate became increasingly cool and dry. Tropical forests diminished and grasses took over.

KEYWORDS

AUSTRALOPITHECINE
BALEEN
CETACEAN
GLYPTODONT
HOMINID
MARSUPIAL
MESSINIAN CRISIS
MYSTICETE
NOTOUNGULATE
PINNIPED
PRIMATE
UNGULATE

The most significant ecological pattern of the Neogene, prior to the cooling that foreshadowed the Pleistocene ice ages, was the redevelopment and diversification of ecosystems that were dominated by large herbivores. Although there was an increase in the extent of grasslands from the mid-Oligocene onwards, forests have remained one of the most characteristic types of vegetation on earth, for there has not been an expansion of open terrains at the expense of closed vegetation. Nevertheless, although forests have remained significant, one of the most evident outcomes worldwide during the Neogene was the diminishing of forests and the spread of more open woodlands, grasslands, and deserts.

PALEOGENE	24 mya · 20 · NEOGENE · 15
Series	MIOCENE
European stages	AQUITANIAN · BURDIGALIAN · LANGHIAN · SERRAVALLIAN
N. American stages	ARIKAREEAN · HEMINGFORDIAN · BARSTOVIAN
Geological events	Australian plate moves north; Indian craton collides with Eurasia; uplift of Himalayas Alpine orogeny continues; major deformation of Alps and Carpathians
Climate	Temperate boreal savannah, low rainfall
Sea level	Moderate
Plants	Spread of grasses and herbaceous plants adapted to dry habitats
Animals	Diversification of whales · Diversification of rodents, songbirds, snakes

That conditions were different from today was mainly a result of glacial cycles and not through major changes in continental distribution. Furthermore, complex regional variations and temporal (time) fluctuations are all notable in the climatic record of the Neogene period. These variations would have produced effects on local biotas that may have created rapid adaptive evolution in the animals of the area. Such changes, however, are extremely difficult to disentangle from the normal level of evolution that occurs over longer periods of time. The development of open vegetation certainly drove much of the evolution of Neogene mammals, which developed features that are strongly associated with an adaptive response to such ecosystems. These include increased development of high-crowned teeth that would withstand the wearing effect of the new high-silica content grasses; the ability to run fast and other locomotor characteristics related to movements in open habitats; larger body sizes in herbivores (as a result of the need to process high volumes of low-quality fodder); diversification of small herbivores that used burrows; and the development of a wide range of carnivores to feed on different herbivore classes.

The Neogene, despite its short 24-million-year history, showed rapid and profound evolution of life in response to widely fluctuating climates. In addition to the large herbivores, there was a tremendous radiation of small animals: songbirds, frogs, rats, and mice. They fed on the seeds of the grasses, or the insects that pollinated them, and, in the case of the rodents, burrowed in the dry terrain. In pursuit of the rodents came snakes.

In many ways the floras and faunas of the Neogene were similar to those of today, especially those of the later part of the period. Nevertheless, some of the mammals that inhabited the Miocene and Pliocene worlds are remarkable for their great distinctness from our recent mammals. According to a general rule of zoology, which was noted in the nineteenth century by Charles Darwin, invertebrates evolve more slowly than vertebrates. Compared with the modest changes in the fossil record of invertebrate life, the modernization of vertebrates during the brief span of the Neogene was extreme.

THE DEVELOPMENT of ice sheets over northern hemisphere continents began about 2.4 million years ago. It is interesting to note the enormous capacity of the polar ice sheets to overcome all other climatic effects. Variations in the amplitude (magnitude of variation) of the glacial cycles that characterize the Neogene, in the development of polar ice sheets and in other factors, resulted in a climatic history so complex that it is only just being unraveled now, using isotope geochemistry and novel paleontological approaches. However, there was certainly an overall global cooling as the world continued in its "icehouse" state, which also interacted with local factors such as mountain building and rifting events. The development and establishment of very steep thermal gradients from the poles to the tropics has been one of the major climatic shifts since the Eocene.

The complex climatic history of the Neogene is only just being unfolded.

Also associated with the global cooling was a significant global drying. This was in part due to reduced evaporation from the colder oceans and in part to the expansion of the polar ice caps and the formation of deep cold ocean waters (the psychrosphere); it is a form of physiological drought, in which water is present but unavailable to animals because it is in the wrong physical state: ice.

While studies of Neogene climate usually focus on the high latitudes of the northern hemisphere, patterns of climatic change during this time—and their effects on

See Also

THE PALEOGENE: *Psychrosphere, Grasslands, Whales, Mammals*
THE PLEISTOCENE: *Glaciation, Gulf Stream, Humans*
THE HOLOCENE: *Caribbean, Andes, East African Rift valley*

THE NEOGENE

The Neogene occupies less than one percent of the latest period of Earth's history. The most significant of the changes in ecosystems during this time had to do with explosive rates of evolution and expansion among mammals, which had a profound effect on future developments.

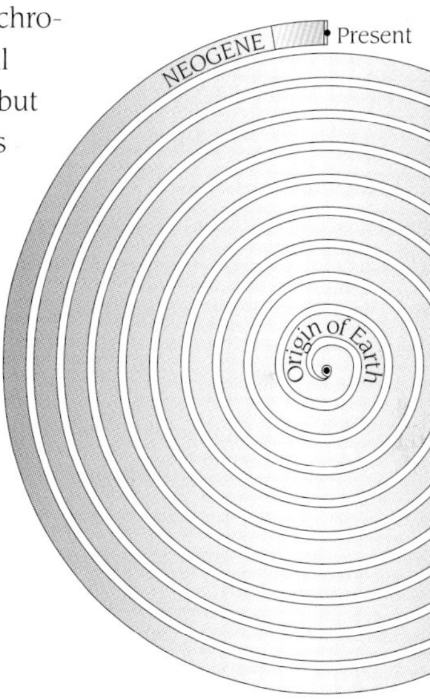

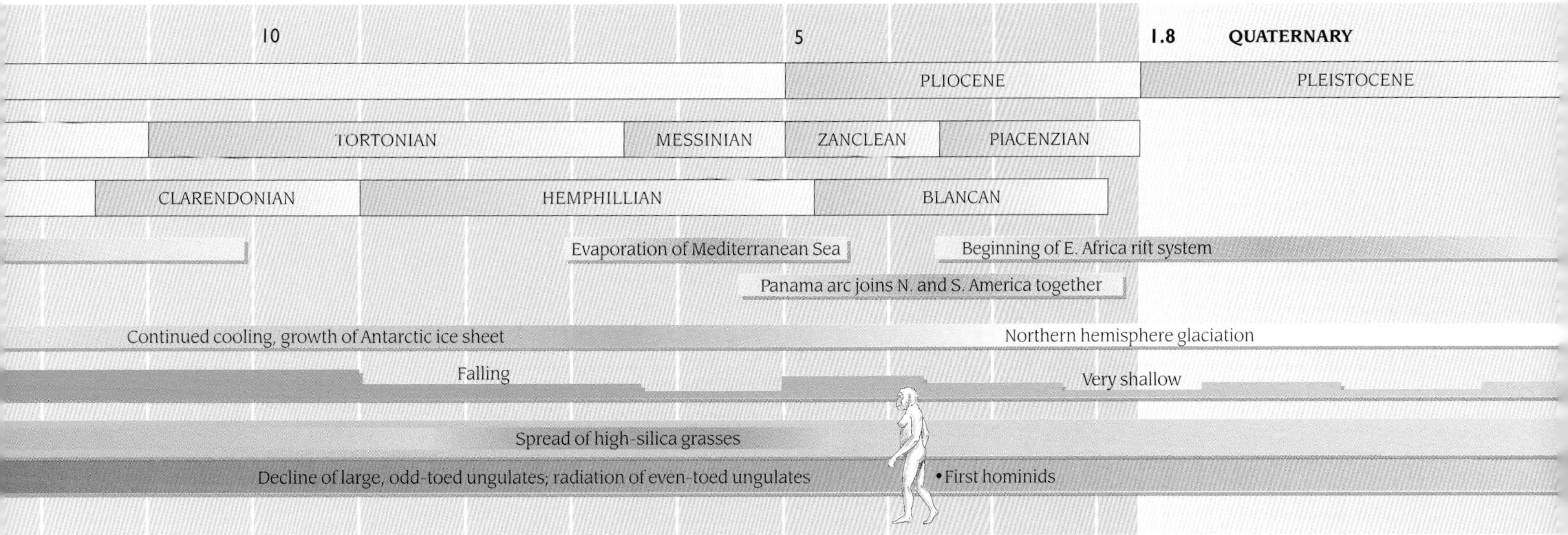

NEOGENE

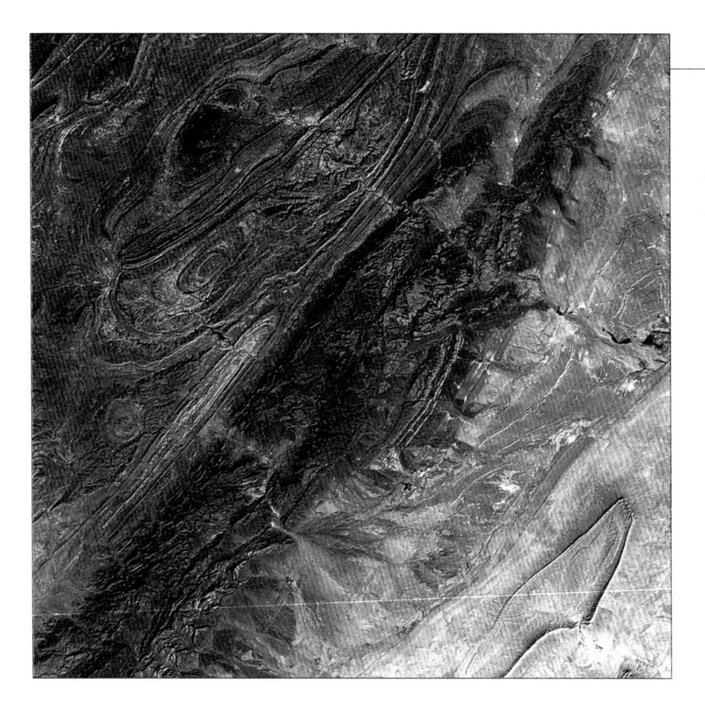

ATLAS MOUNTAINS

(Left) A satellite view of the Atlas range in North Africa shows the ridges and strongly folded rocks adjacent to this structure. These rocks were produced by the same plate movements that formed the Alps.

biotas—appear to have been equally complex in tropical latitudes. During phases of glacial aridity, the rainforests of central Africa and the Amazon basin shrank to a few isolated refuges, whereas those in Sumatra remained intact. Many low latitude regions experienced aridity when cold (glacial) periods occurred in higher latitudes, and moist conditions (as shown by high lake levels) appear to have coincided with interglacials. Disruption of the Southeast Asian rainforests also occurred repeatedly; this was due to high sea levels during the interglacial periods. The overall conclusion that can be drawn from the many independent studies of Neogene environments is that climates in the tropical regions of the globe were unstable, with a corresponding effect on the distribution and turnover of their biotas.

A GEOGRAPHICAL map of the Neogene world from about 6 million years ago—in the Late Miocene epoch—would look quite similar to that of today, with just a few major geographical differences. In the western hemisphere, the Rocky Mountains were rising in the western interior of the continent, on a wide expanse of uplift that was a leftover of the earlier Laramide orogeny in the Paleogene. Similarly, the modern Appalachians formed over the worn stumps of the ancient eastern ranges, relics of the Late Paleozoic collision of Laurentia with Baltica and then Gondwana. Continuing tectonic activity produced extensive flooding by the sea in southern California, while volcanoes formed the Cascade Range, and igneous rocks were extruded in a huge belt from Oregon all through Utah and into Arizona. The great Columbia River basalts are part of this formation. Mountain-building continued along the western edge of the Americas as the various Pacific plates pushed against the continents, raising the Andes range.

North America's Rocky Mountains were formed during this period, along with the ongoing rise of the steep Andes of South America.

The Panamanian land-bridge that now links South America with North America via the countries of Central America was slowly forming, and the Caribbean was still a natural extension of the Pacific rather than an embayment of the Atlantic as it is now. Some volcanic islands had risen in the Caribbean region; others remained underwater, where they accumulated great layers of limestone that later formed the distinctive karst landscapes of the Greater Antilles islands. This land-bridge is significant for another reason: it blocked the westward movements of warm waters from the southern Atlantic and redirected them to the northeast, forming the Gulf Stream.

DURING the Jurassic period, prior to the collision between the European and African plates, the Tethys—an Alpine Mesozoic ocean—had formed as an extremely long rift basin, because of the opening of the Atlantic Ocean and the eastward movement of Africa in relation to Europe. Closure of the Tethys began late in the Cretaceous period as the African plate continued to move. At the time of the Paleogene–Neogene boundary, the Tethys closed still further. What remained formed the modern Mediterranean Sea, with the Paratethys in existence to the east. Gradually, the volume of the Mediterranean Sea became much smaller as a result of the expansion of Antarctic ice sheets in the southern hemisphere, which lowered sea levels by as much as 165ft (50m) and cut off the Mediterranean from the Atlantic Ocean, causing the Mediterranean to dry up altogether at the end of the Neogene.

Mountain-building was far from finished in Europe, but changes in the seas were even more dramatic as the ancient Tethys closed.

The African plate hit Europe early in the Cenozoic era and is still continuing to shear in an eastward direction. The Mediterranean and North Africa were directly affected as a result of the fall and rise of sea levels and the movement of the African plate. The Atlas, Alps, Pyrenees, and Carpathians are all mountains that were thrown up as a result of this collision. Several microplates caught up in the zone between the African and European

- Africa and Middle East
- Antarctica
- Australia and New Guinea
- Central Asia
- Europe
- India
- North America
- South America
- Southeast Asia
- other land

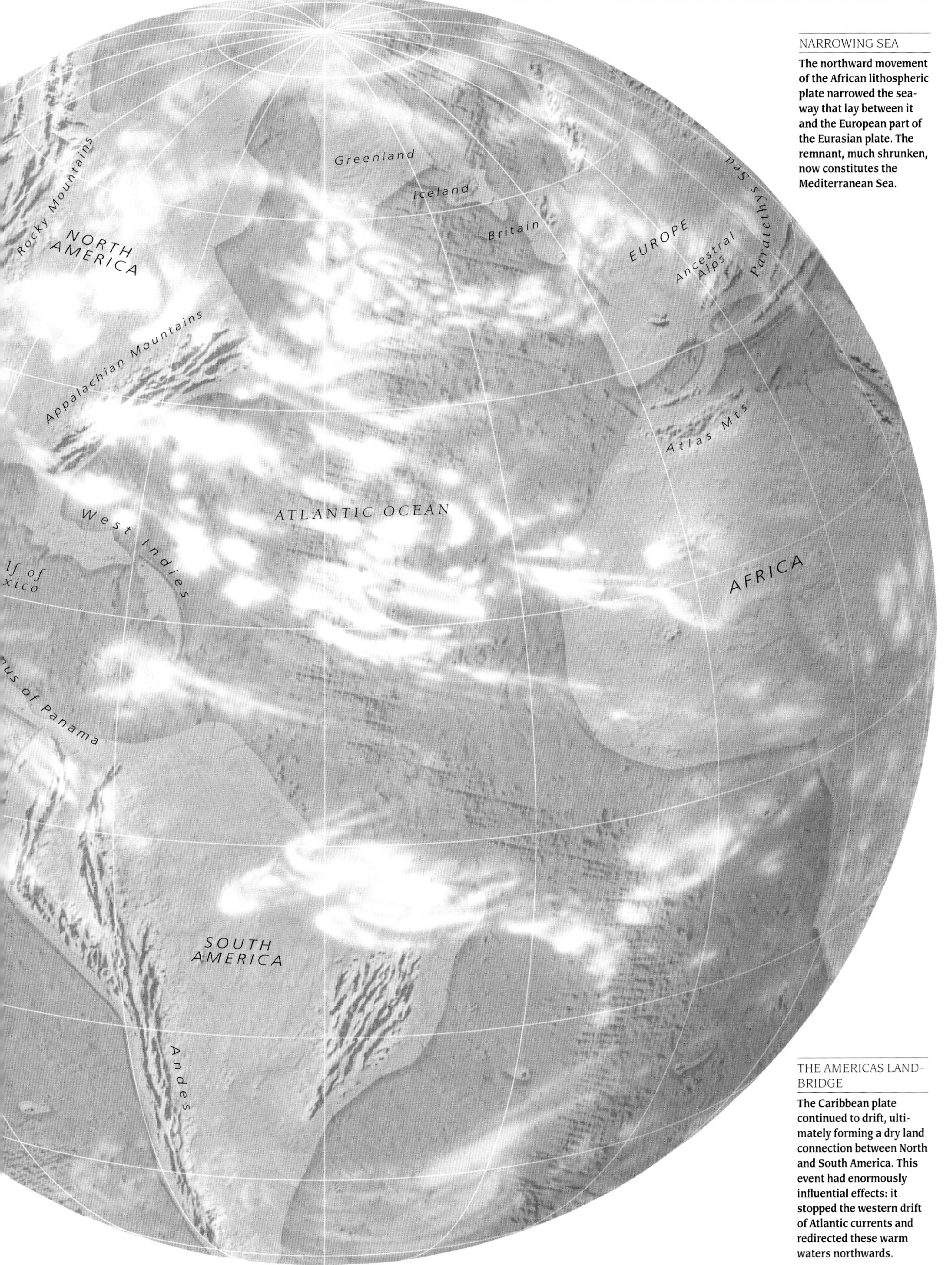

NARROWING SEA

The northward movement of the African lithospheric plate narrowed the seaway that lay between it and the European part of the Eurasian plate. The remnant, much shrunken, now constitutes the Mediterranean Sea.

THE AMERICAS LAND-BRIDGE

The Caribbean plate continued to drift, ultimately forming a dry land connection between North and South America. This event had enormously influential effects: it stopped the western drift of Atlantic currents and redirected these warm waters northwards.

plate became the Italian and Iberian peninsulas, and the islands of Corsica, Sicily, and Sardinia. Approximately 155mi (250km) thickness of European lithosphere was involved in building the Western Alps; deriving from what was mainly the upper elements of the Earth's crust. The shape and distribution of these mountains mark the deformations wrought by the impact. The west-to-east curve of the mountains can be seen clearly in aerial or satellite photographs. The process continues today, as these mountains are even now being bent into a vast S shape across North Africa, Europe, and Eurasia.

This lithospheric movement (of Africa) drove the Iberian plate and southern parts of Europe to the east. Spain and Portugal—the Iberian microplate—were wrenched out of what is now the Bay of Biscay, swung around, and thrown against western Europe. Crustal deformation related to the convergence between these plates stopped during the Miocene epoch.

The movement of the African plate caused large scale geological stretching in an east-west direction, forming areas of tensional strain throughout much of the European plate, with enormous shallow cracks forming more or less at right angles to the direction of tension. Because of the direction of shear, these cracks are oriented in a north-south direction visible in satellite photographs. They include the Rhône Valley and the Rhine Gorge.

The first signs of compressional tectonics in the region of impact between the Iberian plate and the European plate actually occurred about 75 million years ago in the Late Cretaceous. The transtensional basins—such as the Bay of Biscay—caused by these motions in the Cenozoic era were transformed into foreland basins as uplift occurred in the region of the Pyrenees. Most of these basins are now landlocked in northern Spain and southern France. This did not happen to the Bay of Biscay, which remains an offshore ocean basin on the western European continental shelf. The impact caused the crumpling up of the region between the Iberian plate and western Europe to form the Pyrenees Mountains.

TO THE east, the Indian continental plate was in the process of colliding with Eurasia, resulting in the uplift of the Himalayas. This process was not yet complete and the mountains were, as yet, only a low-lying range. However, their presence had a profound effect on the Asian continent as their great rivers, such as the Ganges, began to form in the Miocene. Farther north, the ancient inland Obik Sea and Turgai Strait, which had at times provided marine access between the Tethys and the northern ocean, were both now closed, but shallow seas covered large expanses of south and east Asia. The Arabian peninsula was all but an island, and Madagascar floated farther off the east coast of Africa, where rifts were just beginning to appear in the land as tectonic forces caused the continent to arch up by nearly 10,000ft (3000m). Australia was slowly moving north to its present position and was tectonically quiet, unlike the rest of the region, where volcanic islands were regularly emerging. Antarctica was surprisingly temperate at the beginning of the Miocene, with the great ice sheets only beginning to form between 15 and 10 million years ago.

The globe acquired two dramatic new features: the Himalayan mountains and the Late Miocene ice sheets that still cover Antarctica.

As the South Pole froze, the rest of the southern hemisphere was affected. There is considerable geological evidence that climates became cooler here between ten and five million years ago. Siliceous rather than calcareous sediments were abruptly deposited over larger areas of the southern oceans than before. The success of diatoms—microscopic algae with a silica "shell"—which produced these characteristic sediments indicate that upwelling of deep, cold ocean waters had increased, apparently because of steepening temperature gradients. This cooling was not a world-wide event at this time, but it had major consequences: as water became locked up in the great Antarctic ice caps, sea levels dropped to such a degree that Atlantic waters could not replenish the Mediterranean, and so it gradually dried out.

As the Pliocene epoch got underway, sea levels rose again, so that the seas stood well above the present level between about 4.5 and 3.5 million years ago. This sea level rise (transgression) left large volumes of marine deposits inland of the coastlines of areas such as California and eastern North America. Countries along the margins of the Mediterranean show some similar deposits from this period, as do the European countries that border the North Sea; Great Britain and Denmark are particularly good areas for the study of Pliocene marine sediments.

During this time of high sea stands, the environment of northern Europe was more equable than it is today. Evidence for this comes from the fossil faunas found here, especially the pollen record. Pollen analysis indicates that southeastern England was subtropical, or nearly so, which is warmer than the present climate. This is despite the fact that the world was—and still is—in "icehouse" mode. This pleasant climate came to an end with the start of the modern cycle of ice ages, a little more than three million years ago.

RIFTING IN EAST AFRICA

The massive East African rift system runs from north to south. It was in this region that fossils of hominoids—proto-humans—were preserved and discovered.

Arabian Peninsula

AFRICA

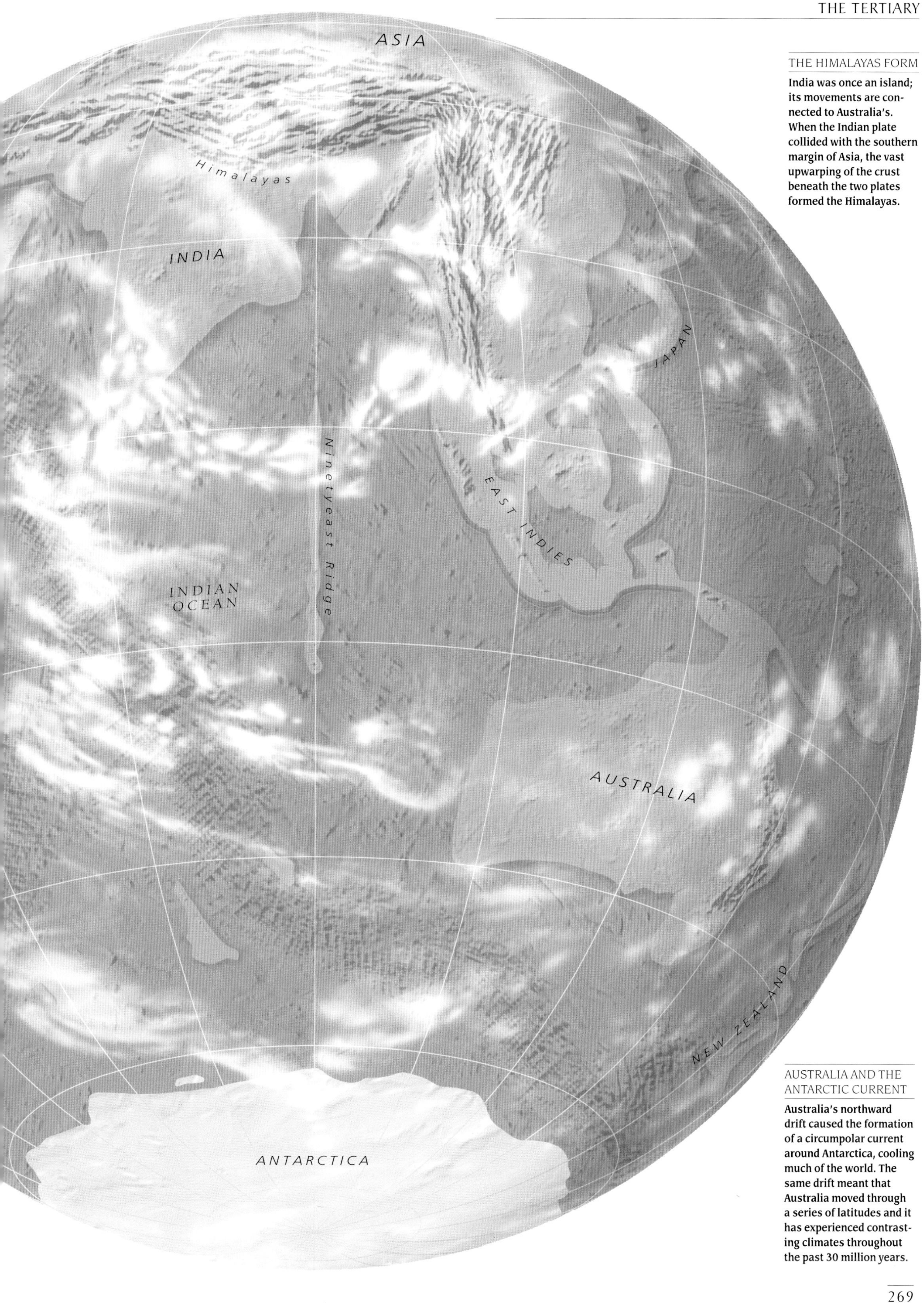

ASIA
Himalayas
INDIA
JAPAN
Ninetyeast Ridge
EAST INDIES
INDIAN OCEAN
AUSTRALIA
NEW ZEALAND
ANTARCTICA

THE HIMALAYAS FORM

India was once an island; its movements are connected to Australia's. When the Indian plate collided with the southern margin of Asia, the vast upwarping of the crust beneath the two plates formed the Himalayas.

AUSTRALIA AND THE ANTARCTIC CURRENT

Australia's northward drift caused the formation of a circumpolar current around Antarctica, cooling much of the world. The same drift meant that Australia moved through a series of latitudes and it has experienced contrasting climates throughout the past 30 million years.

THE FORMATION of the Himalayas began about 80 million years ago, late in the Cretaceous, when the Australian plate, carrying India, began to move north. It did not fully collide with the Eurasian plate until about 20 million years ago. The collision did not stop even there: the plate that carried India continued to push into Asia, eventually penetrating inland by 1500mi (2500km). The violence of this impact is probably unique in the long history of continental collision. It thrust up the high plateau of Tibet and pushed China and Mongolia to the east, creating a series of secondary mountains along the way.

The Himalayas began to form about 80 million years ago. It took 60 million years for the 1500-mile-long wedge of continent to push as far into the heart of Asia as we see it today.

The Himalayan front rises suddenly from the flat Ganges Plain. An effect called isostasy means that the Earth's crust under the Himalayas is very thick indeed. The crust is part of the lithosphere—the stiff outer layer of the planet—that floats on the dense partly molten rocks of the asthenosphere. As mountains are raised, their buoyancy on the surface of the asthenosphere is balanced by the formation of equally deep roots extending far down into the Earth. The young age of the Himalayas explains their great height: they have not had time to erode. Mount Everest, the world's tallest mountain, is about 27,700ft (8848m) high, and is still growing as India continues to crunch slowly into Eurasia. Even the Tibetan Plateau is 16,000ft (5000m) above sea level. At these high altitudes can be found segments of ancient seafloor called ophiolites. These are found in the Himalayas along with the remains of island-arc volcanics. Such characteristic geological features provide clues to the ancestry of this immense range of mountains: it appears that the ophiolites and island arcs were attached to the Indian craton shortly after the start of the Cenozoic era 65 million years ago, when the Indian craton was still far from Eurasia; thus the small craton must have collided with an island arc before it impacted with Eurasia.

The first point of impact was the northeast corner of India, which struck southeast Asia, slowing its progress, before the rest of it rotated to strike the southern margin of Eurasia. The Indian craton did not make contact with Eurasia near its present position until about 15 million years ago, when the oldest molasse deposits are dated. Much of the Himalayan orogeny has only taken place in the last 15 million years. These molasse deposits, because of the huge size of the Himalayan chain, cover immense areas of land. Much of the vast Ganges and Indus river deltas are located on molasse deposits, and it was the meltwater from the snow-capped peaks of the

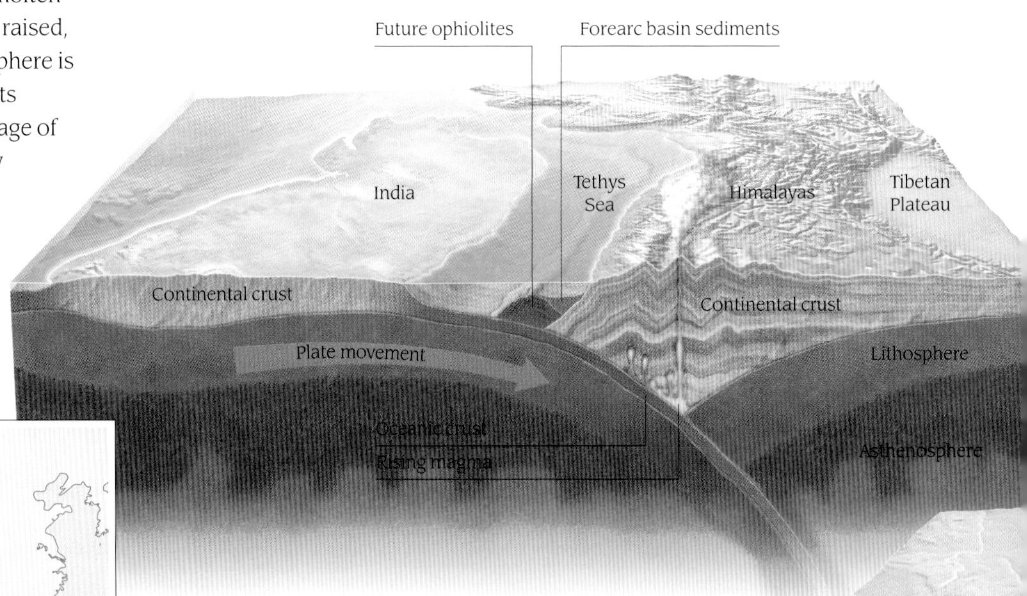

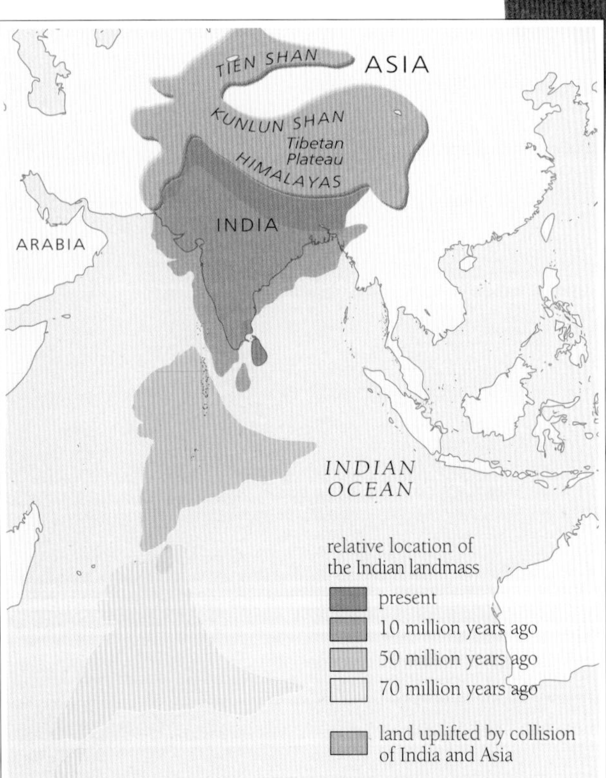

PASSAGE OF INDIA

The Indian craton, along with Australia and New Guinea, rode on the Australian plate. Originally a fragment of the supercontinent of Gondwana, it began its journey in the late Mesozoic, finally suturing with Eurasia about 15 million years ago. Fossil evidence shows that Cenozoic mammals arrived in India about 45 million years ago, probably as the northeast corner made contact with Indochina.

HIMALAYAS REPLACE THE TETHYS SEA

When the northern edge of the drifting Indian island collided with the southern edge of Asia (above), the vast west-east Tethys seaway that lay between them was completely obliterated in a mountain-building event. The intensely twisted, bunched-up remnants of the sea bed of this long-vanished ocean are now found distributed high among the Himalayas and their foothills.

THE TIBETAN PLATEAU

Geological theory suggests that the Himalayas and the Tibetan Plateau (above) owe their elevation to the isostatic uplift that comes from crustal thickening as the Indian plate is underthrust beneath the Eurasian plate. However, the lithospheric mantle below Tibet is relatively thin, possibly because part of it has fallen away into the asthenosphere.

new ranges—in proportion to the continent-wide expanse of the mountains—that formed these great rivers during the Neogene.

The timing of the arrival of the Indian craton at the southern margin of Asia may be gauged by the fossils from these areas. When the "Age of Mammals" began 65 million years ago, India was still an isolated continent in the middle of the southern ocean. New groups of mammals that evolved in Asia could not have reached India until the two continents neared each other or actually made geographical contact. The fossil record of the Indian peninsula demonstrates that the first Cenozoic Gondwanan mammals to arrive on it appeared in Middle Eocene times about 45 million years ago. At this time there was no mountain building taking place, and shallow seas covered much of the Indian craton. Nonmarine and shallow marine sediments (molasse) that were laid down on top of these limestones are of Late Miocene age, indicating that they must have formed after the thrusting and folding stage took place.

THE TIBETAN Plateau is cut by numerous fault lines, including the great Altyn Tagh fault, which trends west to east across all of China. It appears that the Indian craton is still squeezing the larger Asian landmass to the east, towards the Pacific Ocean. As this takes place, blocks of crust slide past each other horizontally along high-angle faults known as "strike-slip" faults. The Baikal and Shanxi rift systems of northern Asia may represent the extension of some of these huge slices of crust as others drag past them. Beneath the rifts the lithosphere is thinner than normal because the bottom part has sunk into the asthenosphere. The Baikal rift system seems to have appeared very soon after the initial point at which the corner of the Indian craton impinged onto Indochina. This is where northeast Asia may eventually separate from the heart of the continent.

Rifts, faults, and volcanoes are evidence of great stresses as the Earth's crust stretches and the Tibetan Plateau splits apart.

Typically of such a tectonically active zone, earthquakes are common: a major one near Tangshan killed a quarter of a million people as recently as 1976. China's historical records indicate an even worse quake in the mid-sixteenth century in which 800,000 perished.

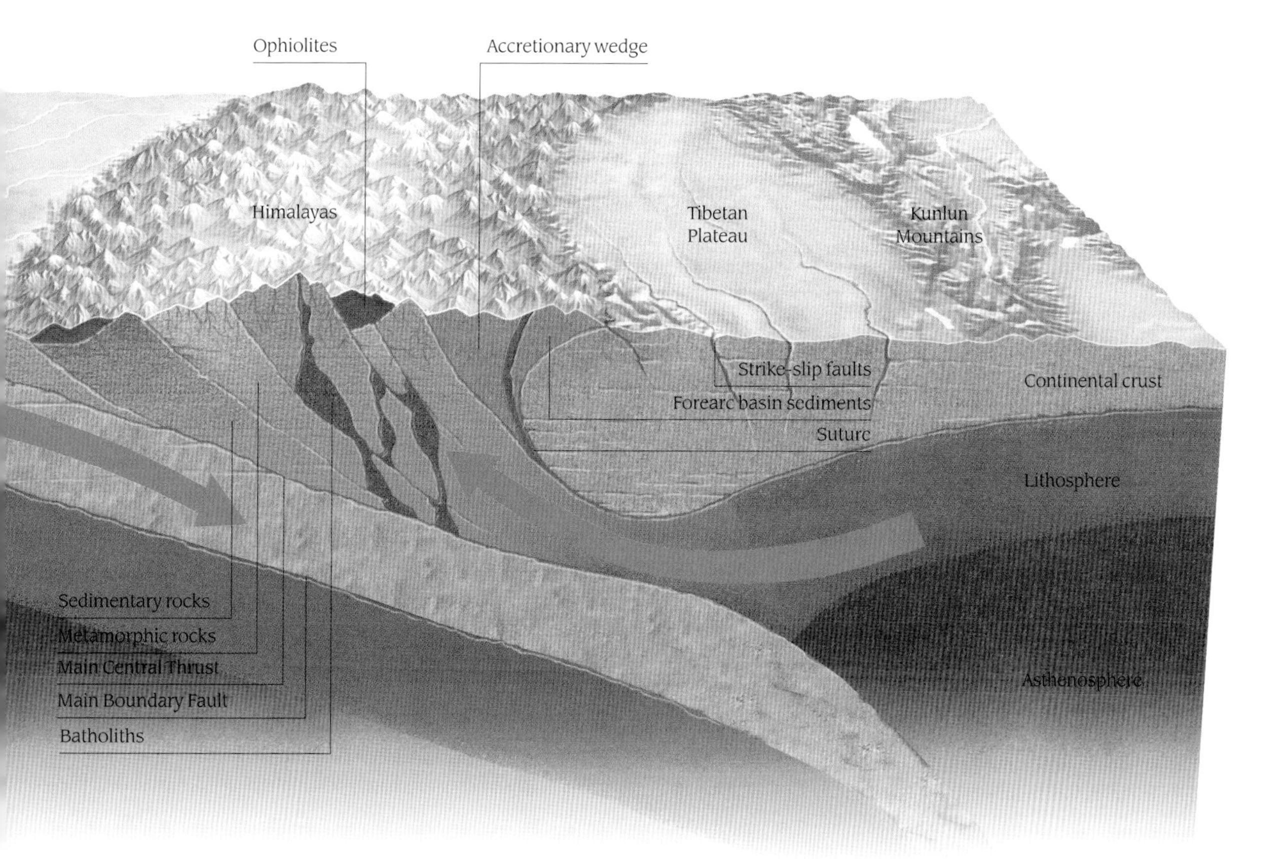

FORMATION OF THE HIMALAYAS

As India reached the Eurasian craton, subduction of the plate ceased as the Indian craton began to be wedged beneath the margin of Eurasia. The terrane south of the suture is a segment of Indian crust that was apparently attached to the Eurasian craton before the remainder of the Indian craton fractured along the Main Central Thrust and began to slide northwards. A band of ophiolites (uplifted sea crust) lies along the line of the suture. Strike-slip faults to the north of the Himalayas formed as a result of pull-apart forces resulting from friction between the subducting Indian plate and the underside of the Eurasian plate. Today movement has shifted to a new fault, the Main Boundary Fault.

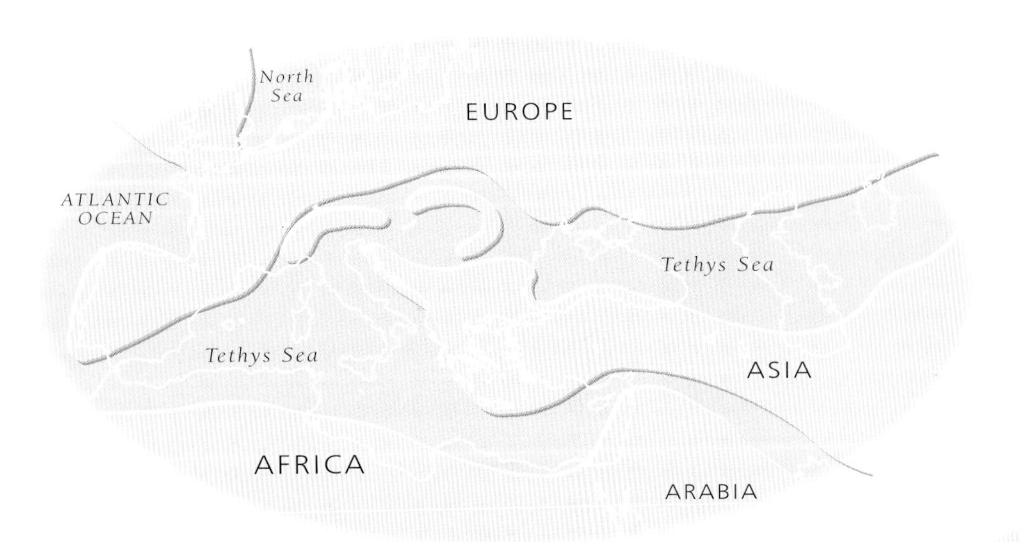

AN OPEN SEAWAY

The present Mediterranean Sea occupies an area that was rifted apart by the breakup of Pangea in the Mesozoic. Early in the Miocene, 20 million years ago (left), the Tethys was an open seaway, with a narrow inlet to the Atlantic Ocean and extended two long arms eastwards into Asia where it connected with the waters of the Indo-Pacific.

THE TETHYS CLOSES

As Africa and Eurasia collided the Tethys was cut off from the open ocean. By mid-Miocene times mountain building isolated the eastern arm, creating the Paratethys. This received water from Eurasian rivers, becoming a brackish inland sea. Near the end of the Miocene the Atlantic link closed and the Mediterranean dried up (below).

THE NORTHWARD movement of the giant African lithospheric plate closed the remains of the ancient east-west Tethys Seaway. The only remnants of this ocean that are still in evidence are the Mediterranean, Black, Caspian, and Aral seas, all of which stretch in a line from Europe and across Eurasia. Between 18 and 14 million years ago, in the Miocene epoch, the creation of a land corridor between Africa and Eurasia allowed the migration of terrestrial animals—primarily mammals—across the landmass in the region of southern Spain and northern Africa. This event marked the end of the eastward connection between the Tethys Sea and the waters of the Indo-Pacific region.

The Mediterranean Sea formed during the Neogene, dried to a near-desert as global sea levels fell, and then filled up again.

By Early Miocene times, the Tethys was divided into a northern and a southern arm. During the period of continental collision, in which the Dinaride and Hellenide mountain ranges of Yugoslavia and Greece, as well as the Taurus mountains of southern Turkey, rose up as barriers, the two arms of the Tethys became separate seas. The Mediterranean was born out of an isolated body of water known as the Paratethys; this seaway received fresh water from river systems in Eurasia and survived as a brackish inland sea until the end of the Miocene.

At the end of the Miocene epoch about 6 million years ago, there was a major event in the Mediterranean that has come to be known as the Messinian crisis. The geological mechanisms of this event were conducted on a grand scale. Several million years before this time, the polar ice caps of the Antarctic continent had begun to expand rapidly. A reasonably accurate explanation of the role of ice in the Messinian crisis may be illustrated by modern examples. The expansion of sea ice today is often cited as reducing the water available in liquid form in the world's oceans, but in fact the opposite happens. Ice is less dense than liquid water and therefore takes up more volume, not less. Since this is the case, the melting—not formation—of sea ice would cause sea level to fall.

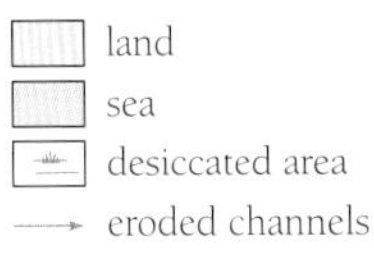

However, the role of continental (land) ice is a different matter. For ice sheets to cover large areas of a land mass, the temperature must drop considerably to below a critical level. Once this is achieved, water volume is indeed lost from the seas as it is transferred—in the form of ice—to the land. As a result, huge volumes of ice accumulate on land and the sea level falls. The Antarctic continental landmass is an immense body of land and its ice volume is correspondingly large. Geological and paleontological evidence in the form of siliceous deposits of radiolarians and changes in foram types show the cooling of the Antarctic continent and its surrounding waters early enough in the Miocene to make the fall of global sea levels plausible in time for the Messinian crisis.

SALT CRUST

The Mediterranean may once have resembled the Devil's Golf Course in Death Valley, California (below). This crust was left behind when a lake evaporated as the climate warmed after the last ice age. Halite (rock salt) was found near the center of the Mediterranean, confirming that it had once been an evaporite basin.

The expansion of the Antarctic polar ice sheets reduced sea levels greatly—by as much as 165ft (50m)—a detail that is also shown by geological and fossil evidence. As sea levels regressed (fell) this meant that contact between the Atlantic Ocean and the Mediterranean, at the site of what is now the Strait of Gibraltar, was severed. The connection between these bodies of water must have taken place in a region where the Africa–Europe land-bridge stood higher than the seabed. When the sea level was lower than the land barrier, the Mediterranean lost its main source of water. Rivers such as the Rhône and the Nile still emptied into the Mediterranean, but from this point on its rate of evaporation was higher than its water replenishment rate. The shallow basin structure of the seabed further exacerbated the evaporation rate. By about 5 million years ago the Mediterranean had dried out entirely: this was the Messinian crisis.

The first geological clue to a dry spell in the lifespan of the Mediterranean was the 1961 discovery of pillar-like domes of salt under the seafloor, which were found using seismic profiling. These were extremely unusual structures and looked very much like the salt domes of Jurassic age that had been found in the Gulf of Mexico. Their presence in the Mediterranean presented a conundrum, for if these pillar-like structures were indeed salt domes, then they could only have formed by the near-total or total evaporation of the sea. Evidently, the beaches of today where so many Europeans frolic were once a salt desert.

In 1970, the presence of evaporites under the seabed of the Mediterranean was confirmed during drilling. These evaporites were anhydrite (the rock compound of calcium sulfate) and were of Late Miocene age. Furthermore, the presence of gravels that were characteristic of shallow marine or non-marine conditions were discovered in the drill cores. Finally, the discovery of halite (rock salt) in the center of the Mediterranean provided the final line of evidence that indicated the Mediterranean to have dried out in the Neogene. Halite is highly soluble and does not precipitate until after all other salts in a solution have done so. It is invariably found in the center of an evaporite basin.

Fossils also provided some supporting data. Freshwater ostracodes (sandgrain-sized arthropods) were discovered in the sediments of the rock cores taken from the drilling operations. These animals, which occur in the latest Miocene record of the seafloor, are observed today in many habitats—including desert pools. In addition, deep valleys filled with Pliocene sediments were found beneath the recent river beds of European rivers such as the Rhône and the Po. The combination of ostracodes and deep river valleys suggest that some rivers flowed rapidly downward to reach the floor of the dry Mediterranean basin to form brackish water pools where ostracodes thrived seasonally.

The Messinian crisis was quick: if the Mediterranean were cut off from the Atlantic today, the rate of evaporation would cause it to dry out in a thousand years. Equally, when the Atlantic flooded back in over the straits of Gibraltar, it was no trickle; it must have rushed in as a waterfall more impressive than Niagara Falls.

ARAL SEA

The modern Aral Sea lies deep in the center of Asia, far from the coastline, yet it represents a small remainder of the huge east-west Tethys Seaway that closed during the Cenozoic. The Aral Sea itself is rapidly drying out in modern times.

MEDITERRANEAN LAKES

Gradually the Paratethys drained into the desiccated Mediterranean basin, leaving behind a series of small lakes, three of which remain today: the Black, Caspian, and Aral Seas (below). Evidence of the drying of the Mediterranean is offered by seismic profiling of the seabed which reveals the presence of salt domes, while the presence of deep canyons is a clue to ancient rivers.

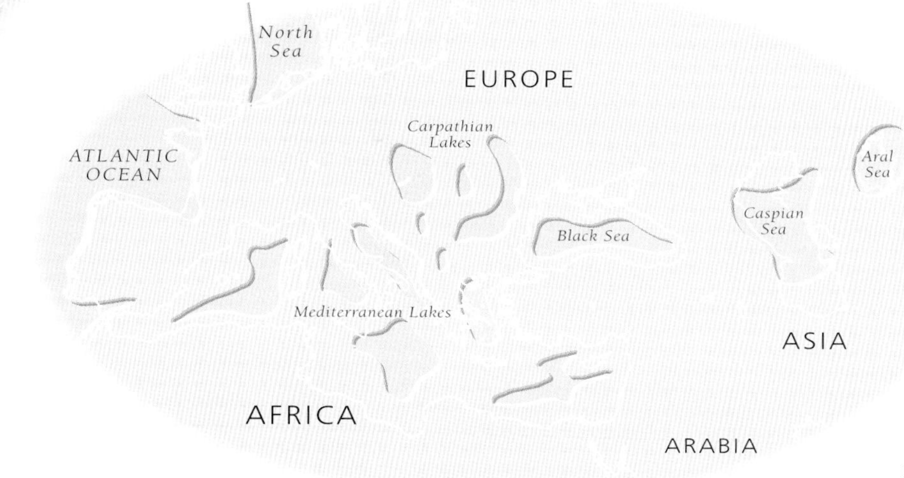

ASIA

NEOGENE

WALRUS

(Right) Except for their large tusks, walruses are similar in appearance to sea lions, with whom they share their bear-like ancestors.

BALEEN WHALES

(Above left) Originating in the Miocene, these whales are the most highly specialized that have ever lived. This is due to the filter-feeding adaptations of their huge jaws and comblike tooth plates. The delicate but stiff fibers (baleen) can strain tons of marine invertebrates in just a few hours of feeding time.

WHALES had begun to evolve in the Tethys more than 50 million years ago. The closing of the Tethys in the Neogene forced them into new habitats, and the change from warm to cold climate may have driven the great radiation of whales at this time—also their separation into tropical marine and freshwater (river) dolphins. It is worth recalling that the most productive areas of ocean are the cold ones, near the poles; here phytoplankton flourish, and so do the fish that eat them. Both toothed whales and baleen whales may have evolved their dentition in order to maximize the abundant food supply they discovered in cold waters. Lacking hair, whales developed a thick layer of blubber (fat) to insulate against them against cold waters and maintain their internal temperature at a mammal-like 97° to 99°F (36 to 37°C) in the core of their muscles. The size of the larger whales is also an advantage in keeping warm, following the general observation that body size corresponds to coldness of climate: this improves the surface-to-volume ratio. It may also explain why dolphins do not live in cool waters.

Whales had begun evolving more than 50 million years ago from mesonychids and archaeocetes, but they diversified greatly in the Neogene.

Miocene whales and dolphins occupy a critical point in the evolutionary lineage of marine mammals. The rocks of the Calvert Formation in the American state of Maryland preserve a diverse assemblage of marine mammals from the Miocene that included toothed whales, dolphins, baleen whales, and even early members of the seal and sea lion group. *Kentriodon* from the Calvert Formation was a small dolphin precursor at less than 6.5ft (2m) long. Unlike modern dolphins, in which the base of the skull is asymmetrical to aid in echolocation, the skull of *Kentriodon* is perfectly symmetrical. It may have been that this small dolphin did not possess the acute echolocation capacity of modern dolphins.

There was considerable diversity among Miocene dolphins, as the presence of the extremely long-snouted *Eurhinodelphis* confirms: its upper jaw was more than twice as long as that of *Kentriodon*. The group of primitive whales to which *Kentriodon* belongs is the eponymous Kentriodontidae. From this predominantly Miocene group, various lineages sprang that ultimately gave rise to modern dolphins, killer whales, porpoises, belugas, and narwhals. All of these are odontocetes—or toothed whales—can be grouped into five families, each representing a major radiation of a subtly different type. Of 76 cetacean species, 66 belong to this suborder. Another odontocete, the carnivorous sperm whale, had an impressive precursor in the Miocene named *Orycterocetus*. This animal was not of the same immense size as the modern sperm whale but showed the characteristically narrow lower jaw and tough conical teeth.

The most gigantic of all living vertebrates are the mysticetes—baleen whales— which include animals such as the right whale and the immense blue whale, which grows to 80 to 90ft (24 to 27m) long, and weighs up to 150 tons. The first evolutionary flowering of baleen

Potamotherium

Allodesmus

Enaliarctos

NEOGENE PINNIPEDS

***Potamotherium* is an Oligocene otter that is known from many well-preserved skeletons from Europe. It seems to be close to the seals, and some paleobiologists consider it simply an aquatic mustelid. *Allodesmus* was a 7ft (2m) early member of the sea lions with a lower jaw and forearms that were unusually robust and powerful. Its clumsy locomotion on land may have been more front-flipper dominated than in living sea lions. *Enaliarctos* was also an ancestral sea lion that was probably not yet fully aquatic.**

whales in the Micocene is represented by animals such as *Pelocetus.* Baleen whales strain plankton using the hair-like mats of their baleen plates, which are in fact highly modified teeth, despite the commercial name "whalebone." The thickness and number of plates varies in species. Baleen does not fossilize, however; in living mysticetes such as the gray whale this material requires considerable body reserves in order to stay functional. Consequently, the skull bones of living mysticetes show a characteristic pattern of grooves where huge blood vessels were located that provided sufficient nutrients for the baleen. The same pattern of blood vessels is found in the skull of *Pelocetus*, confirming it as a mysticete.

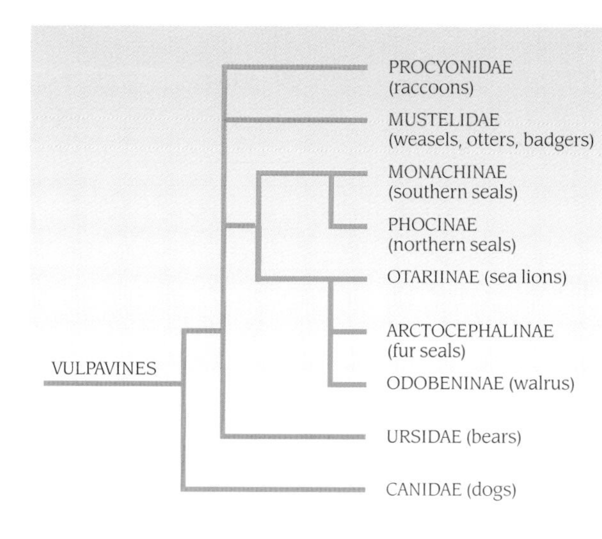

PINNIPED ANCESTORS

The ancestry of seals, sea lions and walruses has been intensely debated. Paleontologists originally viewed pinnipeds as diphyletic: two subsets derived from separate ancestors. Zoologists, however, normally view them as monophyletic: a natural group stemming from a single shared ancestor. This latter view is supported by molecular data. Anatomical evidence used by paleontologists suggested that sea lions originated from a bear-dog ancestor via the enaliarctines, and that seals came from an otterlike member of the mustelids. Recent fossil interpretations support a monophyletic origin for pinnipeds, arising from the vulpavine (dog branch) carnivores via the Ursidae (bears and relatives).

related Carnivores

suborder PHOCOIDEA

suborder OTARIOIDEA

SEALS, sea lions, and walruses, a group traditionally known to zoologists as the pinnipeds ("wing-footed" mammals), are another wonderful product of Neogene marine diversification. Their origin is currently a matter of considerable debate: fossils show origins from bear-dog-like mammals and their intermediate forms. Paleobiologists have identified fossils that suggest that seals and sea lions are from different lineages that converged on a very similar body pattern as they adapted to life in similar environments.

Seals and sea lions appear to have evolved separately but developed the same adaptations.

The oldest known pinniped is from Early Miocene rocks in North America. *Enaliarctos* shows anatomical affinities with the ancestors of bears, such as well-defined slicing carnassial teeth—very unlike the smaller cone-shaped teeth of modern fish-eating sea lions. It had fully developed flippers, although these were not as flattened as in modern pinnipeds. The early sea lion *Thalassoleon* comes from Late Miocene rocks in North America and is more like a modern sea lion in many anatomical details, such as the undifferentiated, single-cusped conical teeth characteristic of living pinnipeds.

The walruses also have their earliest representative in the Miocene, an animal known as *Imagotaria*. It had yet to evolve the massive tusks and foreshortened skulls of living walruses, but there are many indications of the relationship, such as the very simplified cheek teeth. Sea lions and walruses appear to have had their origins in the Pacific basin, whereas the seals seem to have originated in the Atlantic–Mediterranean region. The enaliarctines were restricted to the Miocene, and perhaps were unsuccessful in the long term due to competition from the more specialized seals and sea lions.

Along with the enaliarctines were the marine desmostylians, which were also restricted to the Miocene. Desmostylians were originally known from strange columnar teeth linked like a series of chains. Skeletal remains eventually revealed the nature of these animals, which belong to their own zoological group and have left no descendants. *Palaeoparadoxia* is a well-chosen name for an animal with a shovel-headed skull with long teeth projecting outwards from the tip of the lower jaw and a body that looks like a mixture of seal and elephant with huge splayed feet and crouched limbs. The sleek seals, sea lions, and walruses had an obvious advantage.

AUSTRALIAN HERBIVORES

A view of Australia during the mid-Miocene depicts the heavily-bodied herbivorous marsupials *Neohelos* and *Palorchestes*. Both were diprotodonts with well-developed molars and expanded front incisors. *Neohelos* was a cow-sized browser, analagous to the large African herbivores; abundant fossils suggest that it lived in herds. *Palorchestes* had a higher skull but a very low snout, which suggests that it had a tapir-like proboscis.

1 *Wakeleo*
2 *Neohelos*
3 *Palorchestes*
4 Bettong (*bettongia moyesi*)

AUSTRALIAN fauna during the Neogene period was mostly composed of marsupials. Australian marsupials are more varied and diverse than their South American counterparts simply because they have always been the major mammalian group on this continent. Their evolutionary and faunal history is closely interlinked with that of plate tectonics, migrations, and geographic isolation. The separation of Australia from Antarctica was initiated at the very base of the Cenozoic, about 60 million years ago during the Paleogene epoch. As Australia moved northwards, so did New Zealand and New Guinea, both of which are part of the same lithospheric plate. During the Eocene, Australia broke away from Antarctica and effectively isolated the marsupials that were on this landmass. Their isolation has been the critical factor in the direction of their evolution, and—with no outside influences from migrating invaders—they have evolved very differently from the rest of the world. In this way, island Australia was like Noah's legendary ark.

Characteristic Australian faunas, dominated by marsupial mammals and terrestrial lizards, were the result of geographic isolation.

WALLACE'S LINE

The northward drift of the Australian plate brought together two originally separate biotas, the Australian and the Asian. The contact zone between them, in the Malay Archipelago, was first recognized by the naturalist Alfred Russel Wallace, independent co-founder of natural selection with Charles Darwin. It was in 1859—the same year that Darwin published *The Origin of Species*—that Wallace published a paper on the biogeographic line of demarcation that was subsequently known by his name. Wallace's studies originally placed this line to the north of the island of Sulawesi, but later in his career he noted that the biotas of Sulawesi were dominated by Asian elements, and shifted this line to a position south of Sulawesi accordingly: in fact, only one marsupial mammal from Australia has reached Sulawesi. As for the diverse Asian fauna to the north, only a few species of shrews, two monkeys, a deer, a pig, and a porcupine have reached Sulawesi. The sea constitutes a major barrier to the passage of animals. Plants fare better, and as a result the Australian and Asian floras intermingle over a much broader zone.

The theory of plate tectonics also supplies an explanation for the narrow zone of overlap in the region of Sulawesi. The collision of New Guinea with the Asian plate took place only about 15 million years ago, and it was afterwards that the Indonesian Islands emerged to act as a series of stepping stones to allow the limited intermingling of species. Characteristic Asian mammals include the small-toothed palm civet and the big, furry civet known as the binturong. Australian mammals are mainly marsupials.

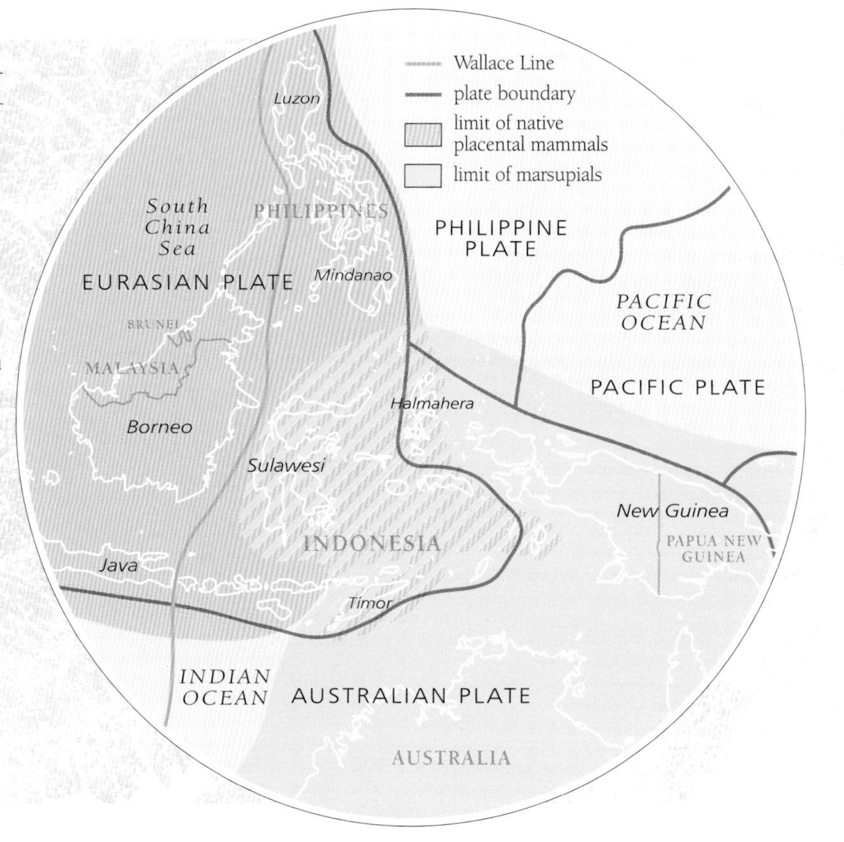

During the past 2 million years, the interior of Australia dried out as the northward drift of the plate brought it into hotter equatorial regions. Widespread dune construction coincided with very dry periods that began about 300,000 years ago, and sediment cores from Lake George show these gradually changing climates in some detail: it turns out that there was glaciation in the highlands of Tasmania far to the south. The Tasmanian region has not been explored in any real depth, and animals that lived in harshly glaciated areas tend to be poorly preserved; consequently there are not many fossils available from Tasmania.

The same cannot be said for Australia, which, despite its drier climate, provided enough characteristic Neogene grasslands to allow the evolution of new herbivorous animals. The faunas of Australia evolved some highly characteristic forms that were functionally and ecologically analogous—and therefore evolutionarily convergent with—placental mammals elsewhere. They were similar to those of the large mammal communities of modern Africa, apart from being composed almost entirely of marsupials. Perhaps the only major difference was that there were no truly gigantic marsupials the size of elephants. Most other ecological niches were present, such as the role of large low-browsing herbivores. In Africa today this is filled by buffalo, rhinos, and hippos; in the Neogene of Australia, these roles were taken up by cow-sized marsupials such as *Neohelos, Palorchestes, Diprotodon, Euryzygoma,* and *Zygomataurus*.

The first fossil finds of some of these animals came from the Wellington cave in the 1830s. The great British comparative anatomist Sir Richard Owen looked at these scrappy fossil remains and called them *Diprotodon* ("two first teeth"). At about the same period the zoologist Gerard Krefft examined them, but was forced to send them back to Owen in England as a result of various reasons, one being an embarrassing lack of funds.

It was not until 1892 that the unearthing of huge numbers of *Diprotodon* fossils shed new light on its relationships. Because of the nature of the *Diprotodon* skull, and his interpretation of the origin of a large fleshy proboscis (nose), Owen originally thought that *Diprotodon* was a member of the living pachyderms (elephants). Because marsupials are adapted to drought, whereas elephants require plenty of water, Owen thought that ancient Australia must have been much wetter and more humid than it actually was. Once *Diprotodon* was firmly identified as a gigantic herbivorous marsupial—wombat-like, but as big as a rhino, making it the largest marsupial ever known to have lived—the true climate of Neogene Australia was also understood.

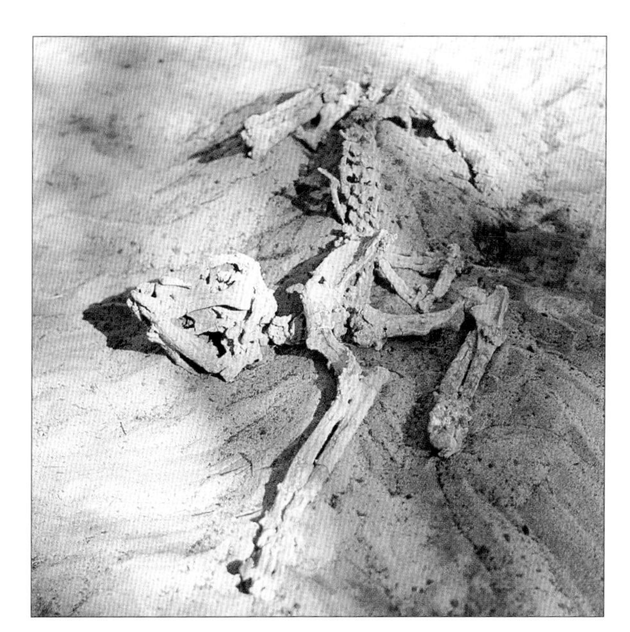

A SPECIALIST KILLER

The beautifully preserved remains of the marsupial predator *Thylacoleo* show the strong boxlike skull and extremely powerful bones of the forearms. *Thylacoleo* was jaguar-sized but more powerfully built. The large curved claw on its thumb seems to have been designed to grab its diprotodont prey. Unlike true carnivores, the killing teeth are a pair of daggerlike incisors rather than the canines. There is no crushing element in *Thylacoleo*'s dentition, and microscope analysis of its tooth wear also confirms that it was a specialist killer.

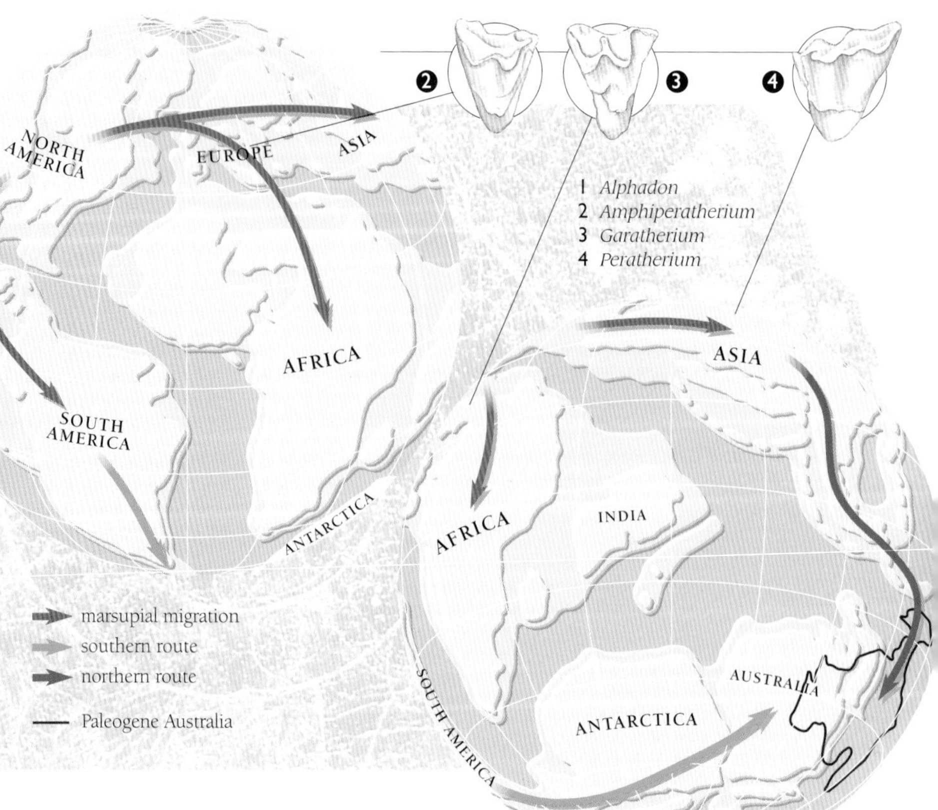

ROUTES TO AUSTRALIA

Marsupial mammals had their evolutionary origins in Early Cretaceous North America and subsequently spread across the globe, as evidenced by the discovery of opossum-like marsupial molars in many locations (1–4). Today, there are two major assemblages, one in South America and the other in Australasia; for many years this split distribution led to much debate, with a northerly dispersal route being preferred. However recognition of continental drift has led to the acceptance of a southern route. Between the Early Cretaceous and the Early Paleogene, there were land connections (via island chains) between North and South America, South America, and Antarctica, and North America and Europe. Eocene rocks in Antarctica have yielded a fossil marsupial that shows affinities with the American marsupials. Although a marsupial fossil has been recovered from Oligocene rocks in Asia, its affinities with European marsupials and its late age—when marsupials were already established in Australia—still suggests that the marsupials dispersed along a southern route during the Cretaceous–Cenozoic transition.

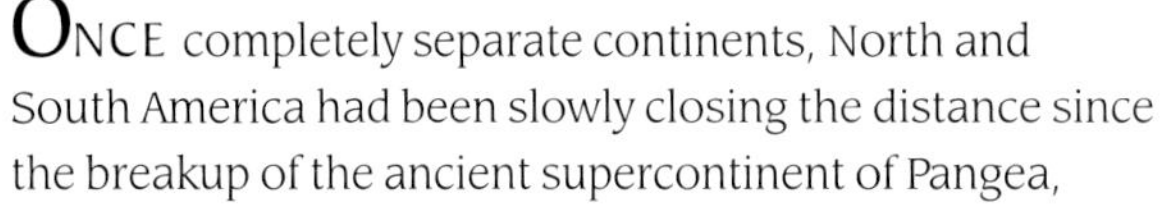

ONCE completely separate continents, North and South America had been slowly closing the distance since the breakup of the ancient supercontinent of Pangea, when Africa—not South America—had been attached to the edge of North America. By the time of the early Neogene, the gap across the water was narrow enough so that a few species of mammals were able to swim across or float on logs. Then, a little more than three million years ago, a piece of land corresponding to modern Nicaragua broke off from southern Mexico and drifted to the east side of the Yucatán peninsula. When the fragment of Panama slotted into place between Nicaragua and the edge of South America, the land-bridge was complete, and the two continents were joined again. South American faunas at this time resembled those of Australia in the abundance of marsupials, which had originated from the Early Cretaceous *Kokopellia* in North America. In Cretaceous to Early Cenozoic times they had migrated en masse to South America across the islands and archipelagos between North and South America.

When Panama formed a land-bridge between North and South America, a two-way animal exchange occurred.

As the land connection between these two continents was broken during the subsequent plate movements, marsupials in South America diversified greatly in their isolation, although they did not attain the same level of diversity as the marsupials in Australia, because there was a considerable community of placental mammals in South America as well. The latter also diversified along different lineages than the North American placentals, and by the Early Cenozoic the fauna of South America was radically different from that of North America. This situation was about to change again.

With the formation of the Panamanian land-bridge about three million years ago, an exchange of fauna between the two continents was again possible. This transfer of animals has been called the Great American Interchange. Monkeys, armadillos, anteaters, sloths, porcupines, and opossums, which had previously been unknown in North America, arrived from the south. Among them were giant species that were many times the size of their modern descendants. Fossils of the giant ground sloths, such as *Megatherium,* are numerous, especially in western North America, where they were highly successful.

In South America the immigration was even more extensive, including various forms of horses, pigs, deer, bears, tapirs, rhinos, camels, squirrels, rats, skunks, cats, and dogs. Between one and two million years ago, the faunal composition of South America was complex and extremely rich, with a wide variety of different body forms occupying many different ecological niches.

A NORTHERN WAVE

Immigrant mammals from North America depressed the diversity of the South American mammal groups a little, but overall the newcomers simply added to the diversity of South American mammalian faunas by a process of insinuation, that of occupying non-competing niches. The diversity of most groups recovered their former levels, but notoungulates and litopterns became extinct. Both North and South America show similar levels of extinction of invading genera.

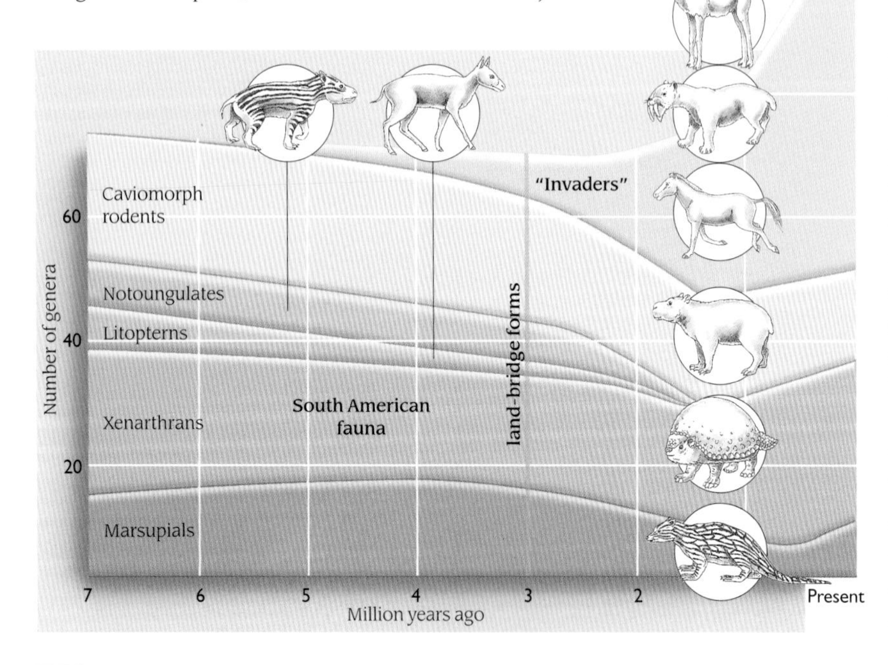

MANY groups of animals that had been unique to South America disappeared completely during the Late Pliocene and the Pleistocene. These included ungulates (hoofed) mammals such as the superficially camel-like litopterns, the bizarre notoungulates that looked like rhino-sized rodents, xenarthrans such as glyptodonts and ground sloths, and a number of carnivorous marsupial types such as the borhyenids. This extinction event coincided very closely with the formation of the Panamanian land-bridge, but the reasons for it are complicated and not fully understood.

During the period that followed the Great American Interchange many South American mammals disappeared forever.

The standard explanation for these large-scale mammalian extinctions was that intensive competition from the "superior" northern migrants devastated the "inferior" southern mammals that were, in some way, less adaptable. However in recent years considerable work has been done on this area of paleontology and this view has largely been overturned. Indeed, what seemed to have happened was much more complex. At one biological level, this classic story seems to be confirmed, as 50 percent of the present day mammals are derived from immigrant North American families, whereas only 21 percent come from South American families.

The total number of mammalian genera in South America actually increased after the land-bridge connection, and this increase consisted of North American immigrants that infiltrated the communities of the time without causing extinctions; that is, they found or defined themselves new ecological niches that were not occupied by South American mammals. This is an important point since—as was stated—the South American mammals had very different anatomy from their North American counterparts, which would have involved different lifestyles. The North American invaders depressed the diversity of South American forms a little, but mainly they added to the overall diversity of the southern fauna. Invaders from the north included jaguars, saber-toothed cats, elephants, squirrels, deer, wolves, rabbits, bears, and horses. Few of these would share ecological niches with the huge ground sloths like *Megatherium* that used curved claws to process vegetation from the trees, or the glyptodonts that grazed the plains.

Litopterns and notoungulates were already in decline before the invaders arrived, and the astrapotheres and pyrotheres had already more or less disappeared, but the surviving lineages died out much later, around 11,000 years ago, along with their so-called competitors, the invading mastodont elephants and horses. Furthermore, the weird glyptodonts—massively armored animals the size of a small automobile—gigantic ground sloths and toxodonts (rhino-like grazers), caviomorph rodents, armadillos, porcupines, opossums, anteaters, and "normal" sloths also migrated north in a reverse event. There was no gradual replacement of South American forms by North American ones—the numbers of South American genera went down from 26 to 21 during the event, but then rose back up to 26 afterwards. The Pleistocene extinctions cannot simply be explained as a result of invasion from the north.

The climate was also changing. The rising Andes created a rain shadow and, as aridity increased, the tropical rainforests of Brazil and the woodlands of Patagonia were being replaced first by savannah, and later by steppe; by the mid-Miocene faunal diversity was already declining as the climate deteriorated and habitats vanished.

BIG BROWSER

Among the distinctive South American order known as the edentates are the extinct herbivorous xenarthrans—the glyptodonts and the tree-browsing ground sloths, or megalonychids, such as *Hapalops* (below). Some of the ground sloths were enormous; the largest, *Megatherium*, measured over 20ft (6m) from head to tail. They used their powerful arms and huge curving claws to pull down branches. They were successful in colonizing North America, finally dying out about 10,000 years ago.

A TIME OF SAVANNAHS

In order to utilize the newly-opened savannah plains, many animals evolved longer legs and a more energy-efficient locomotion; there was even one long-legged rhinoceros, *Hyracodon*. The faster herbivores naturally directed the subsequent evolution of many species of fleet-footed carnivore, such as the leopard (right). In-depth studies of the ecosystems in the African savannah are crucial because they give vital insights into the similar ecosystems that came into prominence from the Oligocene onwards. In fact, the number of genera of savannah mammals known from their fossil remains in the later Miocene of North America is comparable in diversity to modern Africa.

MIOCENE climate changes had a profound effect on large ecosystems; sea level dropped because of an increase in the ice volume at the Antarctic polar ice cap, following the separation of Australia. In addition, reduced evaporation from the cooler seas meant less rainfall. This led to a global expansion of savannahs and semi-deserts at the expense of the forests. The Oligocene had markedly colder winters than the preceding Eocene, a trend that continued into the Miocene and still prevails today. Pollen records from Europe suggest correspondingly dramatic shifts in vegetation and changes in mammals. The changes wrought by the circumpolar Antarctic current affected the terrestrial realm as much as it did the marine world; fossil evidence indicates a huge turnover of mammal types. The change from a "greenhouse" global climate to an "icehouse" was a pivotal moment in mammalian faunal history, as many of the archaic herbivores of the Paleogene were replaced by more modern ungulate groups.

Cool, dry conditions caused forests to shrink and opened up new habitats.

Pigs, horses, rhinos, deer, cattle, and other hoofed mammals are all classed as ungulates. They have a long and distinguished fossil history that demonstrates the great adaptability of the group as a whole, with many different lineages responding adaptively to changing environments, ecosystems, and climates throughout the entire Cenozoic era. Their abundance and success has driven the evolution of modern terrestrial carnivores, such as the dog and cat families, and also been a major factor in recent human evolution.

Early Paleogene terrestrial ecosystems were marked by a very low diversity of extremely large-bodied herbivores and a predominance of closed (forest) vegetation. This was in great contrast to Late Cretaceous terrestrial ecosystems with open areas—but no grass—and an abundance of gigantic dinosaurian herbivores. In this light, the dominant ecological theme of the Neogene has been the redevelopment and diversification of ecosystems that include very large herbivores and more open, lower-biomass vegetation. However, Neogene terrestrial ecosystems differ greatly from those of the Late Cretaceous. The latter were devoid of grasses, which had yet to evolve, and open habitat plant biomass was provided by conifers, tree-like flowering plants such as the magnolias, and an abundance of cycads. The open habitats of the Neogene period some 41 million years later were colonized by grasses, a plant that can withstand low rainfall and prefers an open situation in which its seeds can disperse widely. The ecological niches occupied by dinosaurs of between 2 and 60 tons dinosaurs were now taken by mammals weighing from 220lbs (100kg) to15 tons. They became extremely diverse as grasslands expanded and provided new feeding opportunities. Several groups, such as antelopes, cattle, and horses, evolved species adapted for long distance running with innovations in dentition—extra long molars—to resist the wear that the diet of harsh grasses imposed. Rodents, passerine birds (songbirds), frogs, and snakes also benefited from the changed conditions.

The evolution of vegetation during the Neogene does not show a simple trend toward the expansion of grasslands at the expense of closed vegetation; the evolutionary changes have been more varied in their patterns. Nevertheless, the most evident outcome worldwide during the Neogene has been the diminishing of forests and the concomitant expansion of woodlands, grasslands, and

THE HORSE'S MOUTH

The teeth of this fossil *Hipparion*—a stocky, short-legged Pliocene equid—display the rounded cusps and complex enamel patterns characteristic of the group. (The front of the jaw is to the left and the tongue side is to the top). The general tooth pattern in horses is termed selenodont— where the cusps are sharp crescent-shaped ridges across the crown. Modern horses have hypsodont—tall-crowned —teeth to withstand continuous wear. The dentition of horses gradually changed as they changed from browsing to grazing.

deserts. Although the mammalian grazers of Neogene grasslands appear much less impressive than the huge sauropod dinosaurs of the Late Cretaceous some 50 million years earlier, it should be remembered that mammals have a very much higher metabolic rate than dinosaurs and therefore require up to eight times as much food in a given period. Furthermore, the extraordinarily specialized dentition of herbivorous mammals is probably the most efficient plant-processing equipment that has ever existed in the entire fossil record of terrestrial vertebrates. In view of this, the amount of plant material eaten by the mammalian herbivores of the Neogene period was almost certainly comparable to the amounts consumed by members of the Late Cretaceous herbivorous dinosaur communities.

PRIMATES, the group that includes lemurs, marmosets and their relatives, monkeys, apes, and humans, in general prefer forest habitats to savannahs. Indeed, most are tree-dwelling. Dating from the Eocene, the oldest group includes the "Old World" apes and monkeys of Africa and Asia —the catarrhines ("narrow snouts"). Shortly afterwards another distinct group, the platyrrhines ("broad snouts") reached South America. Catarrhines are made up of two groups, the cercopithecoids (monkeys) and the hominoids (apes), which include orangutans, gorillas, chimpanzees, and the two hominid genera—*Homo* (humans) and *Australopithecus*. Primates are unusual mammals with distinctive traits such as binocular vision, mobile fingers and toes, and large brains relative to body size; the bipedal hominids—and particularly humans—are unusual primates. Although humans share 99 percent of their genetic material with the African apes, there are notable physical differences.

Humans belong to the mammal order known as primates—an essentially arboreal group. At some point during the Miocene their ancestors left the trees for the plains.

The evolutionary history of humans is, along with dinosaur studies, perhaps the most popularized area of vertebrate paleobiology and one that is more frequently

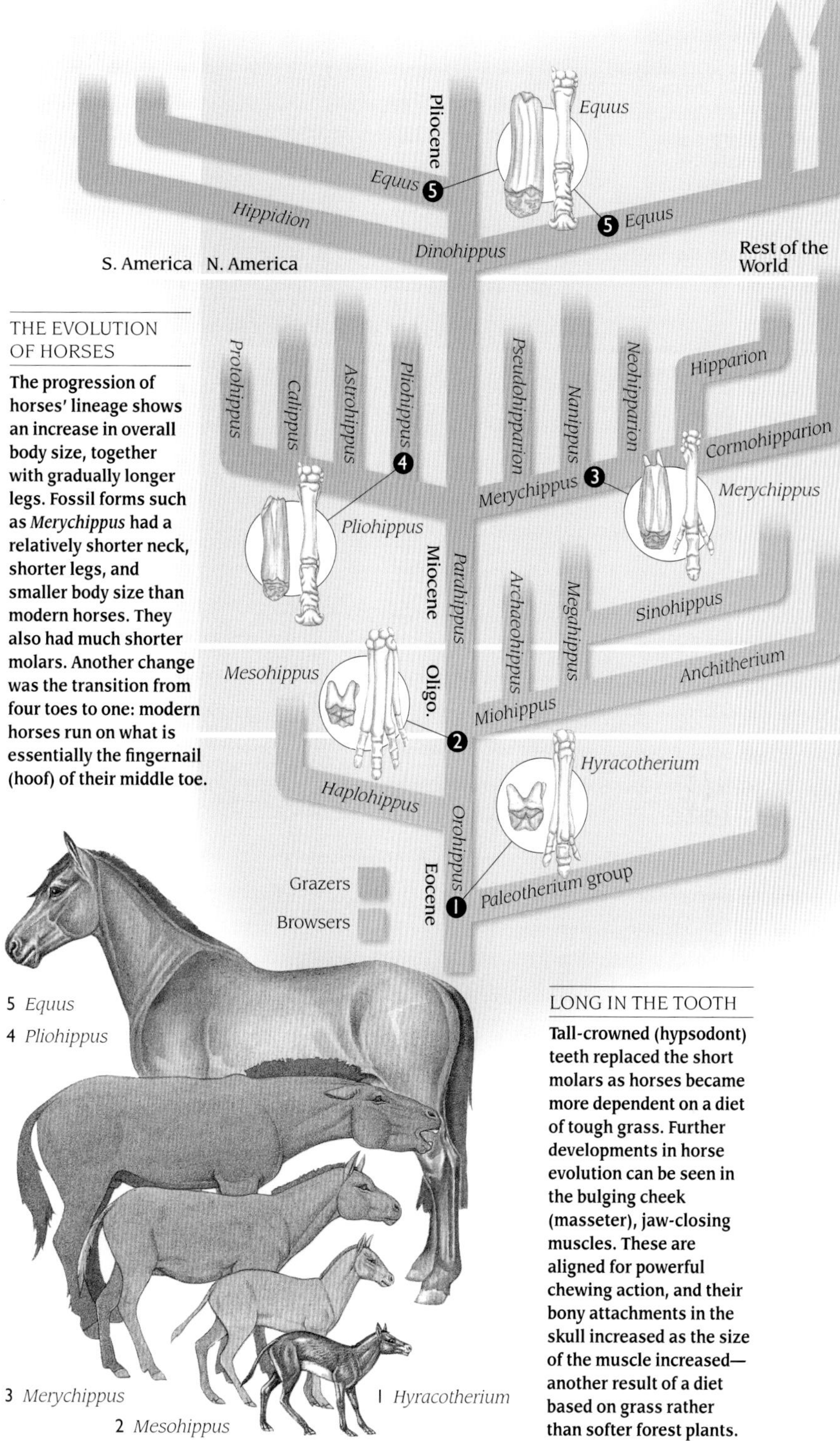

THE EVOLUTION OF HORSES

The progression of horses' lineage shows an increase in overall body size, together with gradually longer legs. Fossil forms such as *Merychippus* had a relatively shorter neck, shorter legs, and smaller body size than modern horses. They also had much shorter molars. Another change was the transition from four toes to one: modern horses run on what is essentially the fingernail (hoof) of their middle toe.

LONG IN THE TOOTH

Tall-crowned (hypsodont) teeth replaced the short molars as horses became more dependent on a diet of tough grass. Further developments in horse evolution can be seen in the bulging cheek (masseter), jaw-closing muscles. These are aligned for powerful chewing action, and their bony attachments in the skull increased as the size of the muscle increased—another result of a diet based on grass rather than softer forest plants.

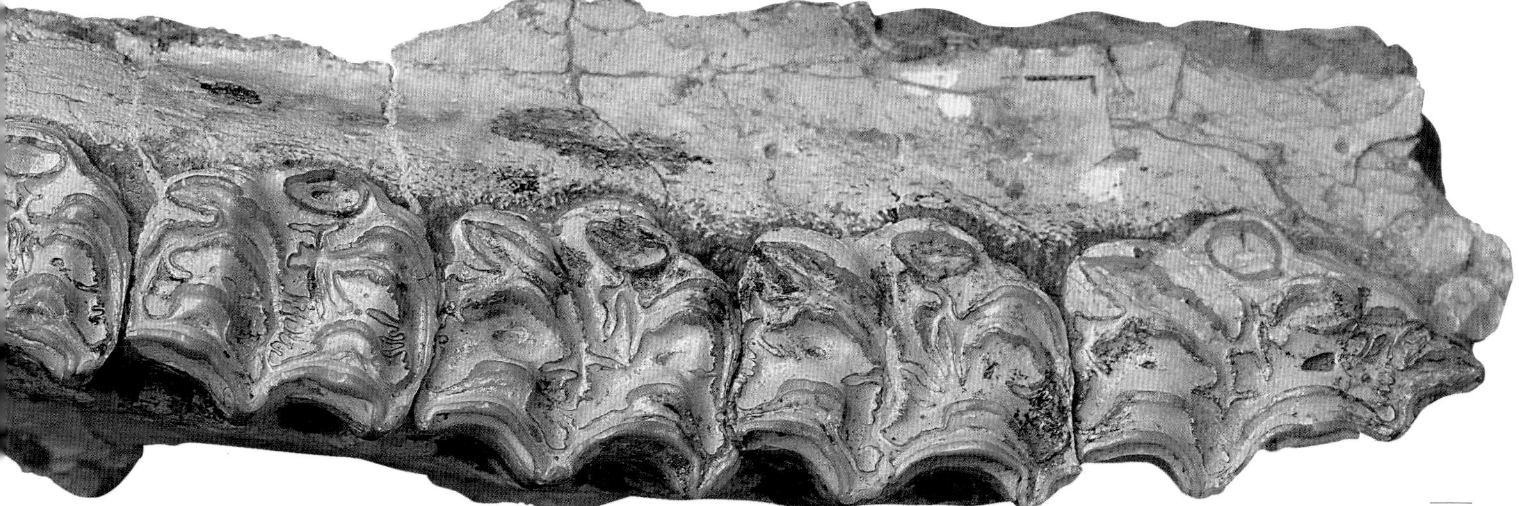

HOME OF HUMANS

East Africa's Rift Valley is the location of large collections of inland waters which have provided animals in this region with liquid and food resources for millions of years. Such areas attract most animals that live in the near vicinity, including humans, who have existed by the lakes for more than a million years. Geological conditions along the margins of the waters have allowed the preservation of hominid fossils used to document this still-contentious history. The most famous discoveries were made at Hadar, Omo, Koobi Fora, Olduvai, and Laetoli.

OLDUVAI GORGE

Olduvai Gorge (below), as much as 325ft (100m) deep and about 30mi (50km) long, is a canyon cutting into a shallow basin within the Serengeti Plain of north Tanzania. It was here in 1959 that Mary Leakey discovered *Australopithecus* (now *Paranthropus*) *boisei*, a fossil hominid dated at 1.8 million years old.

in the media than any other. Yet, in spite of the amount of interest and research, the fact remains that hominid remains are not common—hominids were a relatively small component of paleocommunities—so the fossil record of human origins is very patchy. Nevertheless, over the past century, a picture of hominid evolution has begun to take shape as a result of pioneering anatomical studies, archaeological approaches, cladistic analysis, and innovative investigations into the function and biomechanics of various parts of the hominid skeleton.

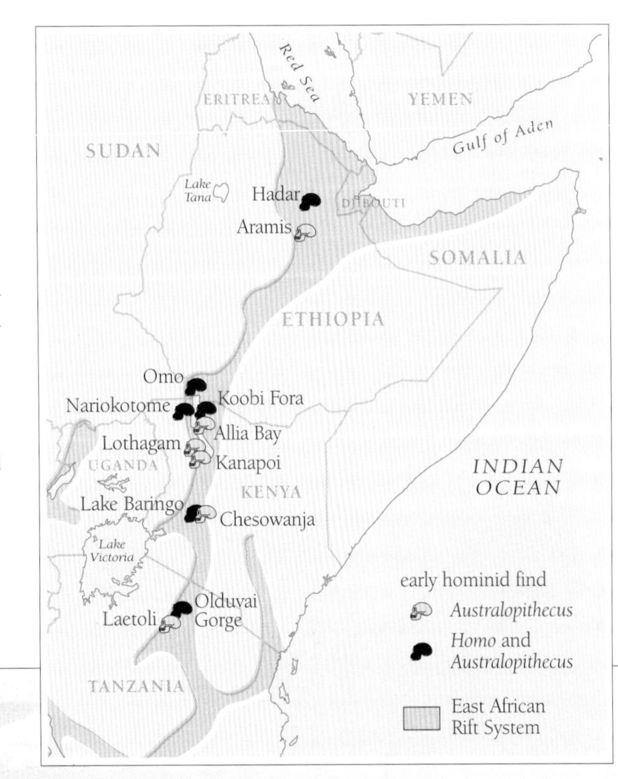

THE FIRST fossil specimen of a human was found in Gibraltar in 1848, but it was at the time misidentified. The first reasonably complete skeleton was found in 1856 in the Neander Valley in Germany, at just about the time Charles Darwin was completing *The Origin of Species*. It was a slouched and somewhat deformed specimen that was subsequently named Neandertal man. Until more primitive remains were found in Africa, the mistaken interpretation of this skeleton gave rise to the persistent but inaccurate idea of ancient, brutish "cave man" as occupying the intermediate position between apes and modern humans.

The discovery of adult australopithecine fossils in East Africa confirmed their position as an early human ancestor.

In 1924, the South African anatomist Raymond Dart discovered and announced a skull dating from the very Latest Pliocene, about 2 million years old, which he named *Australopithecus africanus* ("African Southern Ape"). Known as the "Taung child," it was a child's skull with rather ape-like features and a small braincase. Dart's claim that the skull was an intermediate stage between apes and humans sparked fierce debate and won general acceptance only years later. Some argued that it was a juvenile ape, but large numbers of australopithecine bones, including adults, have been discovered since then, with most finds coming from the Rift Valley of East Africa.

HEAVY-DUTY SKULL

Within the australopithecines there are two apparent evolutionary trends. Some individuals are termed "gracile", with light bones and modest teeth and jaws, while others are "robust". The massive cranium of this robust individual is characterized by exceptional jaw power. It is a strongly-buttressed face with flat cheeks, strong cheekbones, and a bony ridge (sagittal crest) on top of the skull. The gracile line gave rise eventually to *Homo*, but the robust forms disappeared without descendants.

FIRST TOOLS

These tools (below right) from Olduvai Gorge are grouped under the name Oldowan Industrial Complex. Dating from around 2.4 million years ago, they represent the dawn of human technology. They are simply flaked cores, made from pebbles or chunks of rock, which were used for chopping and scraping.

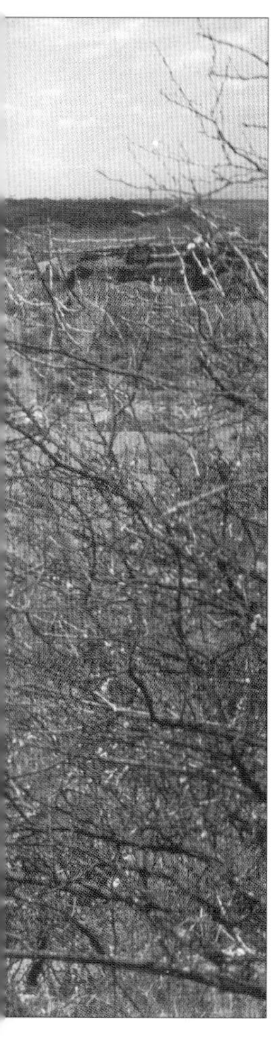

The line leading to modern humans includes as many as twelve species of *Australopithecus*. Until 1990, the australopithecines had been assigned to one genus, *Australopithecus*, but since then, new finds suggest that there are in fact three genera: *Ardipithecus,* the oldest known human at 4.4 million years old; *Australopithecus* (the gracile australopithecines); and *Paranthropus* (the robust ones). Regarding *Homo,* current views indicate perhaps six species (or seven, if the Neandertals are assigned their own branch of the evolutionary tree), and the search for the oldest human being is in many ways like the holy grail of fossil studies. Nevertheless, *Ardipithecus* is currently the oldest known human fossil; it has relatively large canine teeth, narrow molars, and thin enamel, indicating a diet of leaves and fruit; these teeth are more hominine than any of the living great apes. As for the australopithecines, they may be viewed as the sister group to *Homo*, a situation confirmed by cladistic analysis. Another ancient hominine termed *Australopithecus anamensis* appears to be intermediate between *Ardipithecus* and later species and is a specimen in which the lower leg bones suggest that it was bipedal.

HOWEVER, the most substantial remains of fossil hominids were the many skeletons belonging to a species named *Australopithecus afarensis* ("Southern ape from the Afar basin"), one of which was discovered by Donald Johanson at Hadar in Ethiopia in 1974. It was given the nickname "Lucy" and is possibly the most famous of our human ancestors. The skeleton of Lucy was 40 percent complete—an extraordinary amount of preservation for such fossils. She lived about 3.18 million years ago and was about twenty when she died. She walked on two legs, but with slightly bent limbs. Analysis of fossil pollen and animal bones indicate that her environment ranged from open grassland to woodland. Individuals of *A. afarensis* were about 3.2 to 4ft (1 to 1.2m) tall—just under four feet in height at the maximum, with a brain only 25in^3 (415cm^3) in volume and ape like features; but *A. afarensis* is fully human in the way the dental arcade is fully rounded and does not have straight sides as in apes. Along with *A. afarensis*, another australopithicene termed *Australopithecus anamensis* has very recently shed light on the origins of one of the most characteristic of all human features: bipedalism. Very recent findings show that the wrist bones of *A. afarensis* and *A. anamensis* retain anatomical specializations for knuckle walking, where the wrist is bent back and weight taken on the backs of the second of the three rows of finger bones. This is critical evidence for interpreting the evolution of human gait. It suggests that the early humans went through a knuckle walking stage in which the "ancestor" was already partially terrestrial. It also shows that knuckle walking was common to both humans, gorillas and chimps rather than being a special adaptation in gorillas and chimps only.

The skeleton of "Lucy", a member of the species A. afarensis, *throws light on one of the earliest stages of human evolution.*

Australopithecines such as Lucy demonstrate the possible evolutionary lineages taken in the course of the rise of modern humans. But there were other australopithecines present in Plio–Pleistocene South Africa that were very different and that must have constituted a distinct element in the terrestrial "ape" ecology of the time. *Paranthropus boisei* had a huge, flattened face with immense cheekbones, stood up to 5.5ft (1.6m) tall and probably weighed about 110lbs (50kg). *Australopithecus robustus* was similarly proportioned but did not have the "helmet" face of *P. boisei*. Neither of these animals is on the line to humans, because their jaws and teeth, which give evidence of a tough diet of grasses, show specializations such as the massive premolars and molars that preclude the possibility of their being human ancestors.

THE COOLING of the Earth's climate that led to the expansion of open grasslands was the root cause of the development of a savannah-adapted land fauna, particularly artiodactyls (even-toed ungulates). In North America, a component of this fauna, which was well established by Pliocene times, was a variety of deer-like herbivores. Some of these were "true" deer or cervids, some, like *Synthetoceros* were protoceratids, a sister group to camels, and some—like the tiny *Merycodus,* only slightly bigger than the living water chevrotain—were antilocaprids, related to the bovids (cattle, sheep, and goats). The antlers of these animals were very different in shape from those of modern deer and antelope, but probably functioned in a similar way. Although similar to modern ecosystems, Neogene grassland habitats were still recognizably different and supported some bizarre creatures, such as the twin-horned *Epigaulus*, a burrowing rodent which lived in much the same way as modern marmots; its small horns are not found on any living rodent and their function remains obscure. Carnivores, too, spread out on to the open plains. In North America there were no bone-cracking hyenas, as in modern Africa, and this ecological niche was occupied by the massive-jawed "hyena-dogs" or borophagines such as *Osteoborus,* which fed on carrion.

The savannah habitat of Pliocene North America contained a an abundance and variety of ungulates.

1 *Cranioceras* (a cervid)
2 *Neohipparion* (an equid)
3 *Synthetoceras* (a proceratid)
4 *Megatylopus* (a camelid)
5 *Osteoborus* (a borophagine)
6 *Merycodus* (an antilocaprid)
7 *Epigaulus* (a rodent)
8 *Pseudaelurus* (a felid)

8
4
5
3
6
7

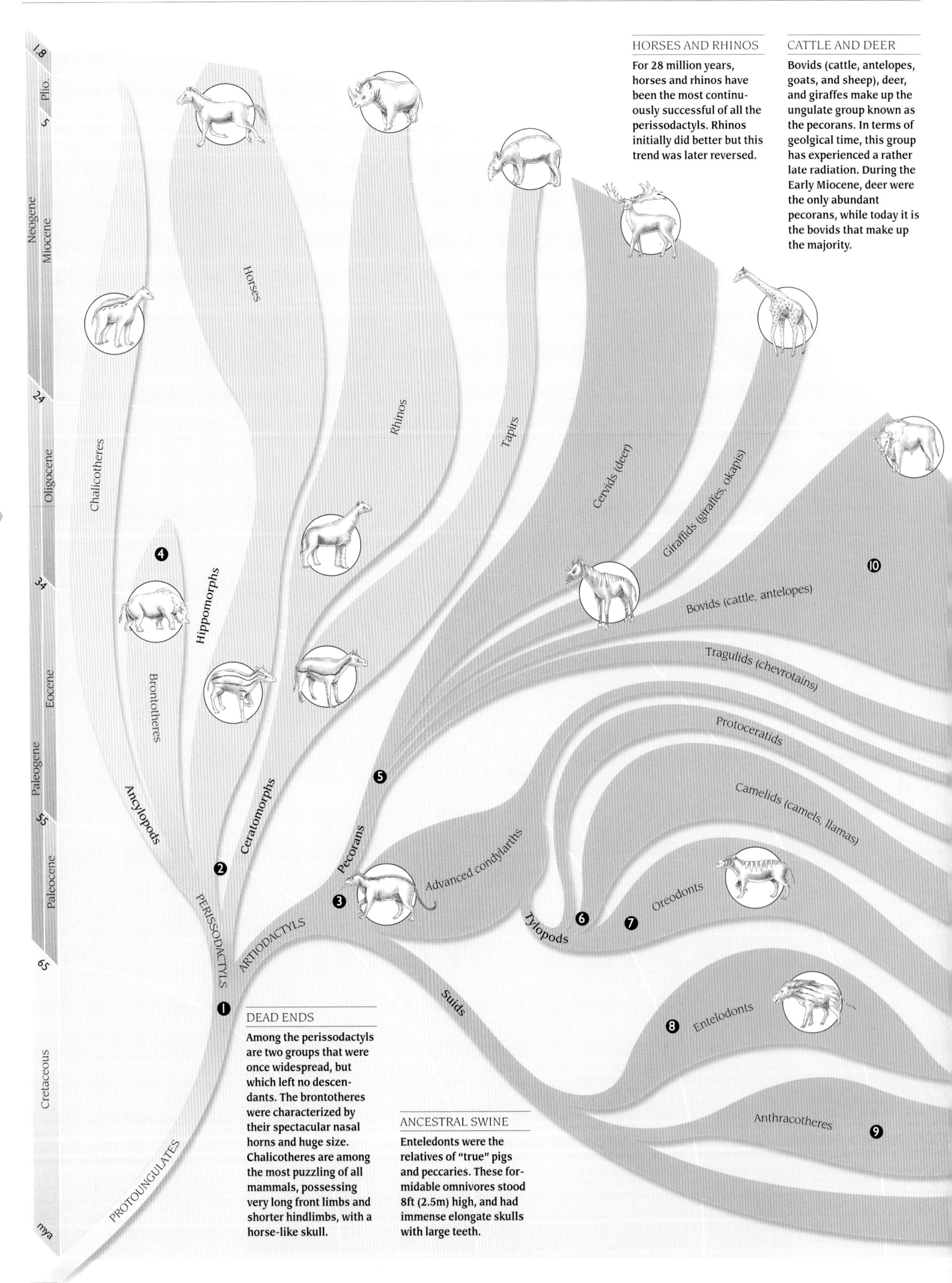

HORSES AND RHINOS

For 28 million years, horses and rhinos have been the most continuously successful of all the perissodactyls. Rhinos initially did better but this trend was later reversed.

CATTLE AND DEER

Bovids (cattle, antelopes, goats, and sheep), deer, and giraffes make up the ungulate group known as the pecorans. In terms of geolgical time, this group has experienced a rather late radiation. During the Early Miocene, deer were the only abundant pecorans, while today it is the bovids that make up the majority.

DEAD ENDS

Among the perissodactyls are two groups that were once widespread, but which left no descendants. The brontotheres were characterized by their spectacular nasal horns and huge size. Chalicotheres are among the most puzzling of all mammals, possessing very long front limbs and shorter hindlimbs, with a horse-like skull.

ANCESTRAL SWINE

Enteledonts were the relatives of "true" pigs and peccaries. These formidable omnivores stood 8ft (2.5m) high, and had immense elongate skulls with large teeth.

THE EVOLUTION OF UNGULATES

WATER BABY

People often suppose tapirs to be relatives of the pigs, but they are far more closely related to horses. Tapirs live in the forests of South America and Asia and spend much time in the water.

UNGULATES are hoofed herbivores, including cattle, pigs, tapirs, camels, rhinos, and horses—all of which have been economically important to humans for centuries. Their diversity, and the numbers of fossils and living species, testify to huge success throughout the Cenozoic. On the family tree of mammals, ungulates belong—with cetaceans (whales), aardvarks, hyraxes, sirenians (sea-cows and manatees), and elephants—to the group known as the Ungulata. There are two orders of ungulates; the perissodactyls and the artiodactyls, meaning odd-toed and even-toed respectively.

Like many other mammals, ungulates originated among one of the many groups of condylarths in the Late Cretaceous. The earliest known condylarth genus, *Protungulatum* ("first hoof-bearer"), shows a dietary shift from the pattern of other Late Cretaceous placental mammals. The cusps of its teeth are blunt and bulbous (bunodont), which improves the capacity for grinding and crushing. Early ungulates show this bunodont dentition, but later forms developed selenedont teeth with crescent-shaped cusps. Horses, in particular, evolved complex, high-crowned (hypsodont) teeth, to cope with coarse vegetation.

The artiodactyls dealt with their diet in a different way: rumination. There are several families of "true"ruminants, the cervids (deer), musk deer, tragulids, giraffids, antilocaprids (pronghorns), and bovids. Pigs, peccaries, and hippos are non-ruminants, while camels are "pseudo-ruminants." Ruminants all possess a rumen or fore-stomach from which they can regurgitate food that has been partially broken down by digestion, for another chewing inside the mouth; this is known as "chewing the cud." Then the food is re-swallowed to pass through further chambers of the stomach for maximum extraction of nutrients.

The oldest artiodactyl is the Early Eocene animal *Diacodexis*, a rabbit-sized animal reminiscent of living small deer such as the muntjac, whose remains are found in North America, Europe, and Asia. Its limbs were gracile (long and thin) and were clearly specialized for running. Its body plan is echoed by all later artiodactyls.

Neogene ungulate faunas developed from Paleogene communities in a number of ways. Entelodonts, oreodonts, anthracotheres, and tapirs were all reduced in diversity and abundance, and the huge catapult-horned brontotheres had vanished. Ecological niches formerly occupied by these Paleogene animals were adopted by rhinos, deer, camels, and anthracotheres. Artiodactyls became more numerous, widespread, and diverse than perissodactyls. Today there are 79 genera of artiodactyls and only six of perissodactyls. Modern groups such as deer, cattle, and antelopes further radiated in the Late Neogene but, mostly, the main evolutionary origin of the basal forms of these groups was in the Paleogene. Origins in the Neogene were restricted to the genus level.

The standard view among scientists was formerly that the Oligocene plains and forests of North America and Asia were dominated by early horses and rhinos, while from mid-Miocene times onwards the camels, pigs, and cattle rose to prominence, but this scenario has been challenged in the last two decades. It would seem that the patterns of evolutionary radiation and extinction in both groups run more or less in parallel, rather than in opposition, and that each of these groups was evolving independently in response to environmental stimuli.

EVOLUTIONARY TRENDS

The number of grazers that feed on grasses and have high-crowned teeth is much greater than in the past. There has also been a decline in the numbers of really gigantic herbivores; huge animals such as brontotheres, dinocerates, entelodonts, and giant indricothere rhinos have vanished.

1 Ancestral ungulates divide into two groups
2 First adaptive radiation of perissodactyls
3 First adaptive radiation of artiodactyls
4 Gigantic brontotheres are important herbivores
5 Radiation of pecorans
6 Tylopods diversify
7 Oreodonts are the most abundant North American grazers
8 Entelodonts (stilt-legged pigs) abundant in North America
9 Amphibious anthracotheres very common worldwide
10 Huge adaptive radiation of bovids

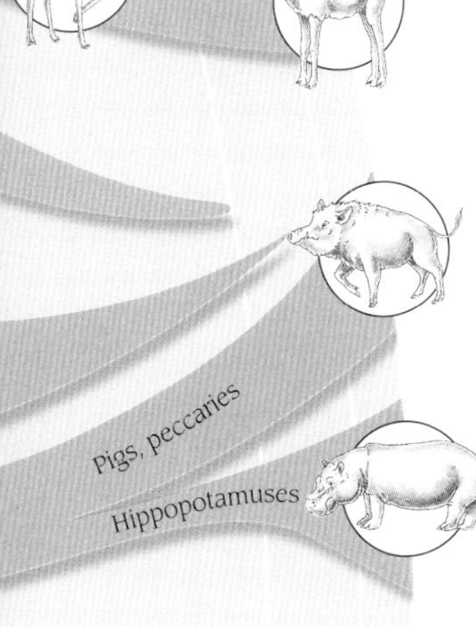

CAMELS AND THEIR RELATIVES

Camels are one of the groups that make up the tylopods ("padded foot"). Camels evolved and diversified exclusively in North America, until the end of the Miocene, and were much more varied and abundant than today. One of the later forms, *Alticamelus*, had an enormously long neck and occupied a giraffe-style ecological niche. Swift-footed camel types remained in South America and their descendants are still there—in the form of llamas, alpacas, guanacos, and vicuñas.

ODD AND EVEN

In ungulates the general mammalian foot has been modified, with the alteration of the ankle joints and the loss of one or more digits (brown). Perissodactyls have one or three toes and the animal's weight is sustained mainly through the middle toe. In artiodactyls (two or four toes) the weight is transmitted down the two middle toes. Rhino and hippo feet are splayed, for weight bearing, while in running animals the metapodial bones (blue) are elongated and fused. All ungulates walk on the tip of their toenails (hooves).

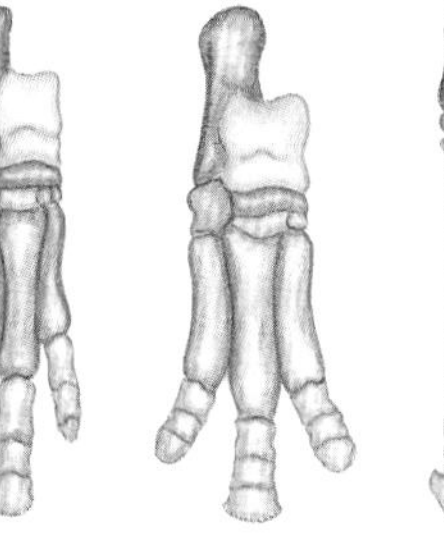

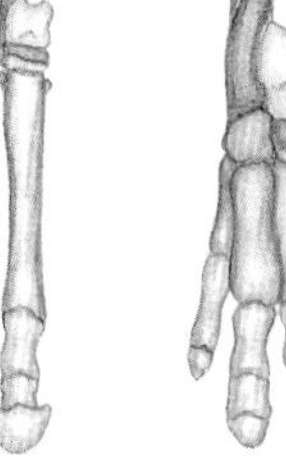

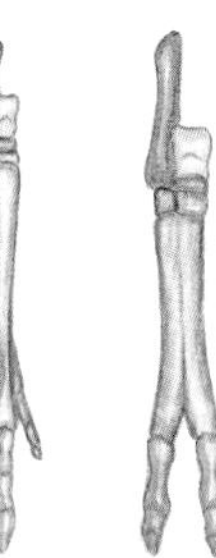

THE EVOLUTION OF PRIMATES

MAMMALS are characterized by their large brains and intelligence. Among them, the brain expansion of one group in particular has allowed them to occupy entirely unique ecological niches. These are the primates—including lemurs, bush-babies, monkeys, apes, and humans—which first appeared about 50 million years ago. Their special adaptations relate to agility in trees and the complex social interactions of forest life; exceptional brain-power, keen eyesight with stereoscopic vision, and extended parental care. Some are terrestrial. Most primates live in forested areas, where few fossils are preserved; as a result the fossil record of primates is not as good as that of large-bodied savannah-dwelling mammals such as ungulates, for example. Despite this, fossils do allow a glimpse of their evolutionary history.

The earliest confirmed primate is *Altiatlasius*, which is based on just ten cheek teeth from the Late Paleocene of Morocco. This tiny animal was an omomyid, one of an extinct group, mainly from the Eocene of North America and Europe, whose members looked like modern mouse lemurs or tarsiers. Most omomyids were larger than *Altiatlasius*, weighing up to 2.2lbs (1kg). The most abundant early primates were the adapiformes such as *Smilodectes*, which resembled modern lemurs and were found almost worldwide from the Eocene to the Miocene. Adapiformes, omomyids, lemurs, lorises, and tarsiers are grouped together under the heading "prosimians"—a descriptive rather than a phylogenetic term, as the group is paraphyletic, with two clearly separate lineages.

There is controversy as to whether the adapiformes or the omomyids are the basal primate group (the most anatomically primitive) in terms of primate phylogeny and which group are most closely related to the anthropoid ("human-like") suborder. Omomyids were more tarsier-like than the adapiformes; this similarity to tarsiers supports the hypothesis that tarsiers plus omomyids (tarsiiforms) are the nearest relatives of the higher primates.

EVOLUTIONARY TRENDS

The radiation of primates began in earnest at the start of the Miocene, when diversification became very rapid. Based on the anatomy of fossil primates, it is clear that ground-dwelling was a late adaptation. The origin of primates, via animals such as *Altiatlasius*, came from a small arboreal form. Terrestrial monkeys only appeared at the very end of the Miocene epoch, less than seven million years ago. Semi-bipedal terrestrial apes appeared a little after this. The development of bipedalism can be linked to the expansion of open habitats.

1. *Altiatlasius* is the oldest protosimiiform
2. The adapiformes become the most abundant primates
3. Omomyids, the most primitive group of primates, diversify
4. Anthropoids appear
5. Anthropoids split into the platyrrhines and catarrhines
6. Platyrrhines split into cebids and atelids
7. Many cercopithecoid lineages evolve
8. Initial divergence of hominoids—terrestrially adapted "apes"
9. Lemurs undergo an extensive radiation and produce a giant form—*Megaladapis*
10. Appearance of advanced hominids

PROSIMIANS

This ring-tailed lemur shows the generalized body plan of lemurs; it is an unspecialized anatomy that seems to be very similar to that of many early primates such as the adapid *Smilodectes*. Lemurs, which are all confined to Madagascar, and other prosimians, share many primate features with the anthropoids, but they also have more primitive mammalian features, such as greater reliance on their sense of smell.

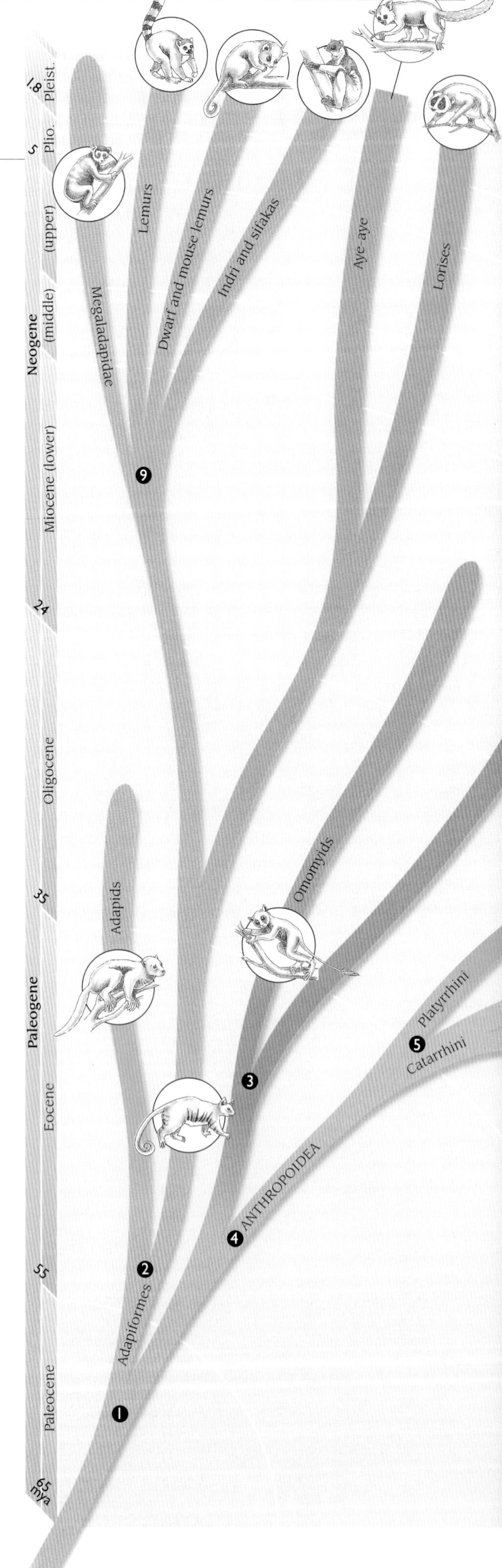

ARBOREAL APE

This orangutan and her offspring have the elongate arms and short legs characteristic of tree-dwelling primates. The orangutan is the only "great ape" found in Asia.

The anthropoids occupy the final branches of the primate cladogram and comprise the monkeys and apes, including humans. Anthropoids are made up of two groups which evolved separately in the New World (mainly South America) and the Old World respectively: these are the platyrrhines (literally "broad-nosed") and the catarrhines. The New World platyrrhines, such as the marmosets and capuchins, have widely spaced nostrils that face forward, and many have a prehensile (gripping) tail; all are arboreal. The catarrhines have two main branches; the Old World monkeys (Cercopithecidae) and the apes (Hominoidea). They are characterized by narrow snouts and nonprehensile tails; many venture onto the ground.

Currently, the most ancient known anthropoid—and the smallest primate ever, weighing a bare 0.3oz (10g)—is *Eosimias sinensis,* a 40-million-year-old fossil from the Middle Eocene of China. It is a critical link between the anthropoids and prosimians and shows that the divergence between these two stocks occurred earlier than the Oligocene, as previously thought.

Tarsiers
Marmosets
Tamarins
Capuchins
Spider and woolly monkeys
Howler monkeys
Sakis
Titis and Night monkey
Proboscis monkeys, langurs
Colobus monkeys
Guenons
Macaques
Mandrills
Cebidae
Atelidae
6
Cercopithecoidea
7
Hominoidea
8
Proconsul
Gibbons
Kenyapithecus
Sivapithecus
Orangutans
Dryopithecus
Gorillas
Chimpanzees
Hominidae
10
Australopithecus
Homo

MONKEYS

Monkeys and apes together are known as simians; they have bigger brains relative to body size than prosimians. Monkeys tend to be smaller than apes, with long—often prehensile (gripping)—tails. Apes are usually tailless. Monkeys also retain long, potentially dangerous canine teeth, especially the baboon, which is a rather unusual terrestrial monkey.

APES

Apes (hominoids) tend to be more terrestrial in their way of life than true monkeys, and most spend a considerable amount of their time in family groups foraging for food on the ground. A number of fossil apes exist, including *Proconsul*, a generalized early ape; *Sivapithecus*, a close relative of the modern orangutans (the only fully arboreal ape); and *Dryopithecus*, a possible relative of the modern African apes.

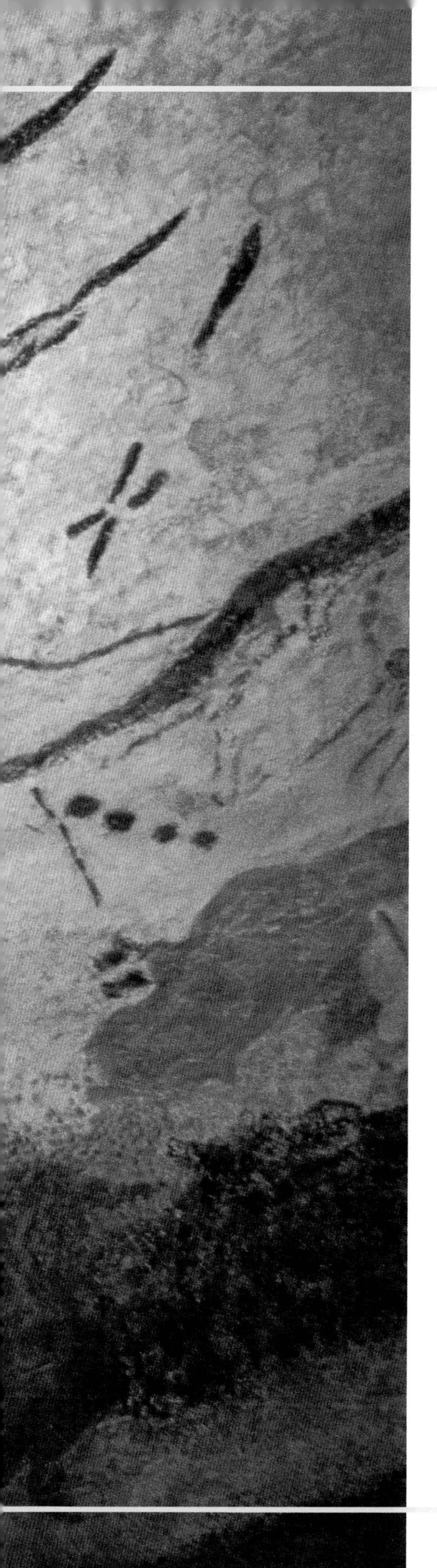

Towards modern times

PART 6

THE QUATERNARY

1.8 MILLION YEARS AGO TO THE PRESENT

THE QUATERNARY comprises the last 1.8 million years of Earth history. It covers the transition from paleontological science (the study of fossils) to archaeology (the much more recent time period involving the study of the remains of people and their civilisations). The adaptive radiation and colonization of the planet by the human race that took place during this period is, undoubtedly, of the greatest significance to us, its members. The impact on the Earth of human expansion is probably, at least in the short term—geologically speaking—important too. But the dominant theme of the Quaternary has been climate change.

Although the temperature of the Earth has fluctuated throughout its history, the range has been relatively small. Currently, we are experiencing one of a number of warm intervals punctuating a period of exceptional cold that has lasted for 2 million years—an ice age. Ice ages have occurred throughout geological time; there were at least four during the Precambrian, one in the Ordovician and one in the Late Carboniferous and Permian. On the whole, however, the great ice ages of the past were a result of continental drift, as ancient continents moved across the poles and froze, eventually being released from the grip of ice as the geography changed again.

The major events of the Quaternary period were widespread glaciation, particularly in the northern hemisphere, and the spread of human beings across the globe.

The present appearance of the globe, with its familiar distribution of continents and its polar ice caps is, in fact, unusual in terms of paleogeography. In the Tertiary, the separation of Antarctica caused it to become cut off from warm air and ocean currents by a circular current of cold water. The Arctic Ocean, surrounded by landmasses, also became isolated from warm currents as the Isthmus of Panama joined the continents of North and South America and produced the Gulf Stream, which carried moist air and higher precipitation to northern latitudes. The result in both hemispheres was a rapid growth in ice sheets, with those in the north advancing and retreating periodically in a way that is unique in Earth's history. Previous glacial episodes in ancient Earth history were non-repeated, isolated, but often long-lived events. Geological evidence shows that the glacial cycles that characterize the Quaternary have never occurred before this time.

MODERN EDUCATION takes the events of glaciation as evident, but many scholars of the last century were dubious and believed that the Earth's landscape had been shaped by great catastrophes such as the biblical Flood. Similarly, until it was realized that animal extinctions had occurred not just once but repeatedly, fossil remains were attributed to animals that had been drowned. Invoking divine action to explain natural phenomena was a normal idea in nineteenth century society, given the the philosophical attitudes and limited scientific techniques of the time. A glacial theory was only accepted in the 1830s, when the Swiss-American naturalist Louis Agassiz provided robust interpretations of geological features such as the embankments of debris termed moraines, *roches moutonnées* ("sheep rocks") with their nubby texture, and hanging valleys whose open ends drop into larger valleys—all testament, not to a deluge, but to the power of glaciers to permanently sculpt the landscape.

Quaternary glaciations also affected the oceans. As the ice sheets waxed and waned, there was a resulting dramatic rise and fall in sea level—as much as 300ft (100m). The cyclic deepening of global oceans was reflected in the formation of deep-water black shales and siliceous sediments resulting from an increase in cold-water radiolarians—silica-shelled microscopic planktonic organisms. Successive changes in sea level are visible in stepped coral reefs on the shores of the Caribbean islands. Reefs and beaches formed at different times were exposed when sea levels fell, or covered when sea levels rose, with corals that grew in shallow water now found deep below the surface.

IN GEOLOGICAL TERMS, the Quaternary world was much as it is today, with all its geographical features in nearly identical positions. But when the continental glaciers were at their furthest extent, with ice extending in Europe as far as northern Italy, and in North America as far as New York, the fall in sea levels caused land-bridges to appear. The Bering land-bridge that connected Siberia and Alaska, allowing the passage of animals between Asia and North America, is no longer present, but fossils from Quaternary sediments record two-way migrations between Asia and North America. Similarly, the European ice sheets allowed the migration of animals to the British Isles, where the fauna during the Quaternary was similar to that of mainland Europe.

During the periodic retreat of the ice sheets, or interglacials, these connections were lost and the biotas developed differently. However, the temperate conditions of the interglacials (as in the present day) temporarily allowed warm-adapted animals to move north. The presence of fossil lions, hyenas, and hippopotamuses in Britain shows how warm these interglacials were. The geology of the British Isles records repeated transitions from Arctic steppe-tundra through boreal birch and conifer forest to temperate broad-leafed forest and back again. It is likely that hundreds of feet of ice will cover all this again in less than 10,000 years.

Land-bridges also made possible the rapid spread of modern humans from their origin in East Africa. By the beginning of the Holocene, the time of permanent settlement and the development of agriculture, all other hominid species were extinct and the world population was 4–5 million; today it is about 10 billion. People have become a powerful force on the Earth. We have gained a great deal of knowledge about its past and its workings but, so far, we have not used that knowledge particularly wisely. Our actions in just the last 150 years have brought the extinction rates of animals to an all-time high and are now having an impact on the global climate. We should consider ourselves as stewards of the planet if we are not to prove the undoing of huge numbers of its creatures—ourselves included.

Ian Jenkins

THE PLEISTOCENE

1.8 – 0.01 MILLION YEARS AGO

THE PLEISTOCENE epoch, which immediately preceded the current Holocene, lasted from approximately 1.8 million years ago until about 10,000 years ago. Among the scientific disciplines it spans the transition from paleontology (the study of very ancient life forms as fossils) to archaeology (the study of the remains of humans and civilization) as primitive humans evolved into more advanced beings. These humans developed complex social structures that gave rise to civilized societies, which have left tangible archaeological remains.

To paleontologists and geologists, the most characteristic feature of the Pleistocene was the repeated cycle of glaciation and deglaciation as huge ice sheets advanced, covering a third of the northern continents, and then retreated. Other epochs had had episodes of glaciation, but what was unique in the Pleistocene was the rapid reversal between glacial and interglacial conditions. As a direct result, there were many repeated extinctions of plants and animals, both terrestrial and marine. These glaciations also caused major redistributions of life.

ON TODAY'S Earth, approximately 14 percent of the total land area is covered by ice sheets or underlain by ice-cemented rock, and about 4 percent of the entire ocean surface is shrouded by a thin layer of ice that fluctuates seasonally. During the past 2 million years, as much as 25 percent of the land and up to 6 percent of the oceanic surface area have been covered or underlain by ice. This global ice mass is referred to as the cryosphere and is composed of sea ice, permafrost, and glaciers. The Pleistocene glacial events represent an entirely different set of climatically controlled processes from those that had existed for much of Earth's history. Because the Pleistocene glacial cycles have occurred only in the past 2 million years or so, their effects have been superimposed on much older rocks. Glacial geological features are found in all areas that have been subjected to the effects of glaciers, regardless of the nature of the underlying strata. As the ice sheets formed and then melted, alternately locking up and releasing immense volumes of glacial water, changes in terrestrial water supply caused the formation of many distinctive glacial landscapes. There were repeated episodes of various land and lake redistributions, one legacy of which we see today—for example, in the Great Lakes of North America. Many of these landscape features are geologically young.

Ice covered 25 percent of the land and 6 percent of the oceans—three times more ice than today.

KEYWORDS

BERING LAND BRIDGE

GULF STREAM

HOMO SAPIENS

INTERGLACIAL

MITOCHONDRIAL EVE

MORAINE

OUT OF AFRICA HYPOTHESIS

PRECESSION

TUNDRA

NEOGENE	1.8 mya – 1.7 – 1.6 – 1.5 – 1.4 – 1.3 – 1.2 — PLEISTOCENE
Series	EARLY/LOWER
European stages	CALABRIAN
North American (mammalian) stages	IRVINGTONIAN
Glacial periods (Europe)	DONAU (1.7–1.4); Donau/Günz interglacial (1.4–1.2); GUNZ (from 1.2)
Glacial periods (N. America)	
Geological events	Ice sheets cover about 30% of continents in the northern hemisphere; Continuing subduction of Pacific plates beneath N. American plate
Sea level	Very shallow, fluctuating
Archaeological periods	LOWER PALEOLITHIC (Oldowan); First *Homo ergaster*; Oldest use of fire •
Animal life	• First *Smilodon* saber-tooth cats; • First mammoth

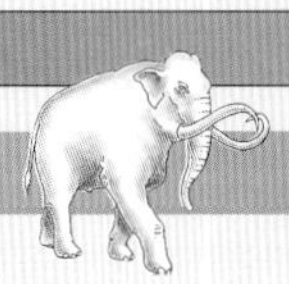

Sir Charles Lyell, the great British geologist, defined and named the Pleistocene in his historical four-volume book *Principles of Geology*, which was first published in 1833. The Pleistocene was formerly named the Newer Pliocene, a name that stood until 1837. Lyell based his definition of the Pliocene epoch on its marine fossil faunas. Pliocene strata contained fewer species that are alive today than the more recent Pleistocene epoch. The Swiss–American paleontologist Louis Agassiz, on the other hand, was the first person to realize that the very characteristic landforms of Europe were the result of glaciers in the region, and he did not define any particular epochs as Lyell did. Agassiz is remembered as a paleontologist and geomorphologist, Lyell as a general geologist and theorist.

Many nineteenth-century naturalists were skeptical about glaciation, and usually called on biblical accounts of floods and other divinely-directed phenomena to explain unusual geological features such as "erratic" boulders that lay far from their origins. It was only when a substantial amount of incontrovertible evidence was compiled by the geologists of the time that natural explanations for glacial geology superseded religious interpretations. It was in the 1830s that Agassiz, in the face of much opposition, insisted that glaciers had been responsible for forming much of the landscape of the modern world, from the steep valleys of his native Switzerland to the Great Lakes of North America.

The Plio–Pleistocene glaciation of the northern hemisphere also provided an explanation for the disjunct distribution of some species, such as the magnolia tree, which had puzzled naturalists. Moving south through Europe ahead of the ice sheets, this plant had its progress blocked by the glaciers that had developed in the Alps and Pyrenees and disappeared, leaving populations in the Americas and Asia. With the realization that glaciations could redistribute populations, the significance of these geographical curiosities was finally appreciated.

DRAMATIC environmental changes on land were associated with the latest episode of the Pleistocene ice ages, which involved the growth and decay of very large ice sheets in the mid-latitudes of the northern hemisphere. These episodes included ice sheets covering North America and the Canadian Rocky Mountains, northwest Europe and the British Isles, the Arctic Islands off the north shores of Canada, and much of the northern regions of Siberia. During the last glacial period, about 18,000 years ago, the huge ice sheets in northwest Europe and in North America eventually reached their maximum extents. Following an initial period of very slow decay, they began to decay very rapidly at about 14,000 years ago. This stage stopped sometime between 11,000 and 10,000 years ago, but then continued. The ice had more or less disappeared in Europe by about 8,500 years ago and in North America by about 6,500 years ago.

Dates of the glacial cycles are approximate, because movement of the ice varied across the globe.

Along with the growth and extension of large ice sheets in the northern hemisphere, many of the earth's principal environmental zones, or biomes, were displaced toward the equator. Immediately south of the ice sheets in the northern

See Also

THE PALEOGENE: *Psychrosphere, Grasslands, Carnivores*
THE NEOGENE: *Himalayas, Panamanian land-bridge, Hominids*
THE HOLOCENE: *Caribbean, Andes, East African rift valley, Glacial relicts*

A SMALL SLICE OF TIME

The Pleistocene epoch represents the last 0.039 percent of the Earth's history. If the whole of Earth's existence was compressed into one year, then the Pleistocene epoch occupies only the final twelve minutes of December 31! Yet, during this geologically brief timespan, huge changes have been wrought on the Earth, especially in its biosphere and climate. As far as it has been possible for geologists to tell, the cyclic glaciations that have taken place throughout the Pleistocene epoch are unique to that period, never having occurred in such a repeated fashion before.

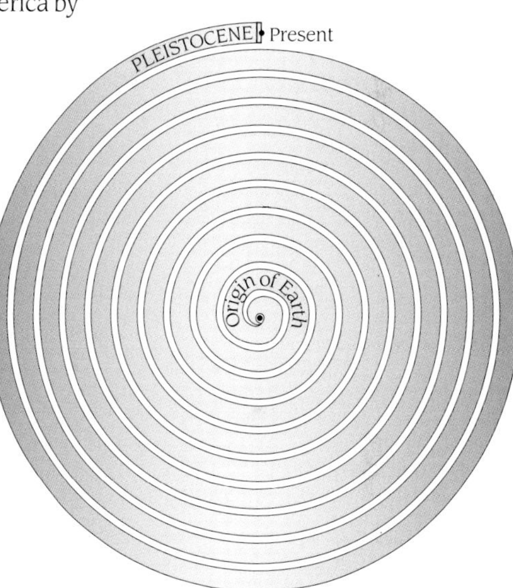

PLEISTOCENE

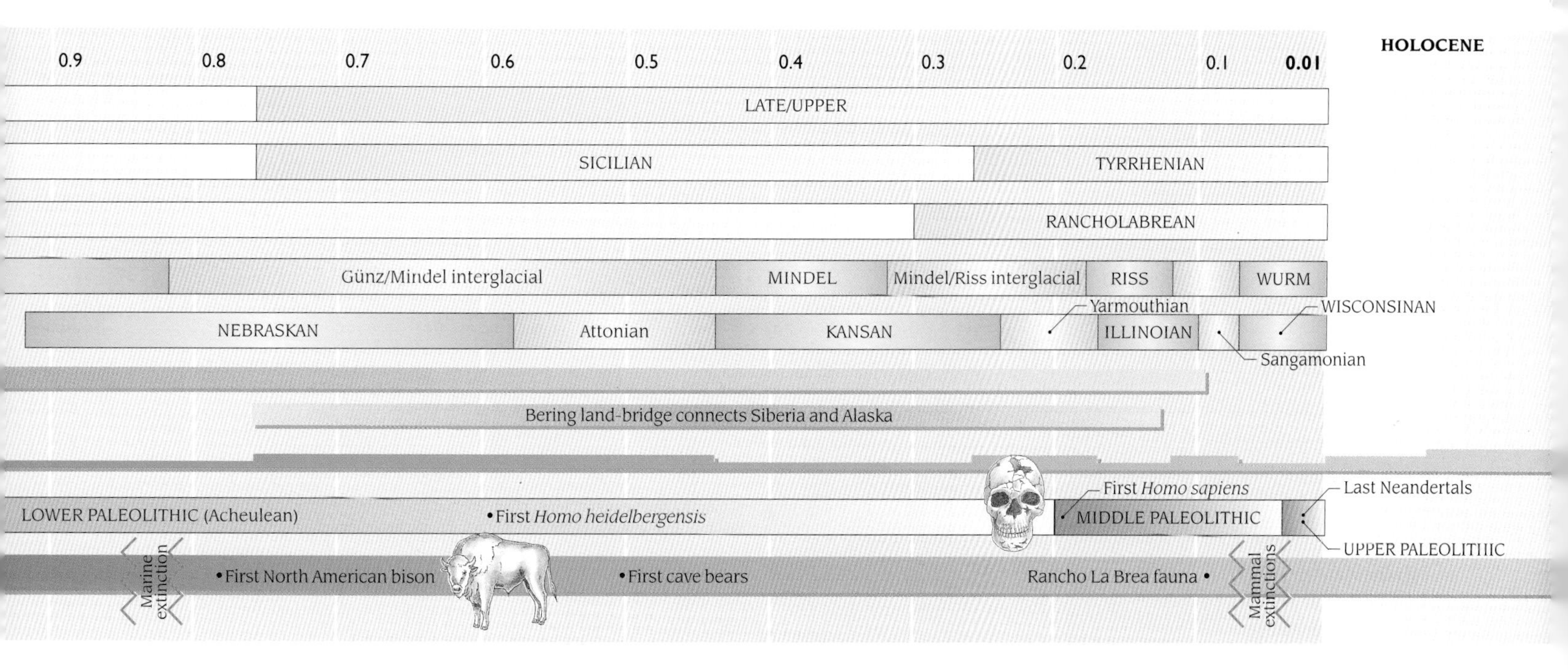

GLACIATION AND ISOSTATIC ADJUSTMENT

When ice sheets expand during glacial events, they cause both the lowering of sea levels through a decrease in water volume as ice mass increases on the continents, and isostatic depression of the land beneath them. Isostasy is the mechanism by which the Earth's crust stays in gravitational equilibrium as it floats on the denser, more plastic mantle beneath. So, in the same way that icebergs are supported by seawater—the ice below the surface weighing less than the water it displaces—the large volume of less dense rock in continental crust gives it buoyancy, with the thicker mountain regions being balanced by deep roots. As mountains erode, the root rises in compensation. Similarly, when ice sheets retreat there is gradual uplift of the depressed crust as the load is removed. This can be considerable: as much as 1100ft (330m) over the last 10,000 years in the Hudson Bay area of Canada, for example. However, such uplift is a delayed response, and a large amount will remain to be completed after the total withdrawal of the ice sheet. As the ice melts during warm interglacial periods, sea levels rise and flood into the isostatically depressed areas; the Baltic Sea is one result of this. The relationship between the rising and falling of sea levels and continents is complex, but it results in a characteristic geological feature known as raised shorelines. A "staircase" of these raised beaches can be found in some places high above current sea levels. Raised shorelines are particularly evident in northern Canada, Scandinavia, and Spitsbergen, Norway, all of which were heavily loaded by Pleistocene ice sheets and are still rebounding even now.

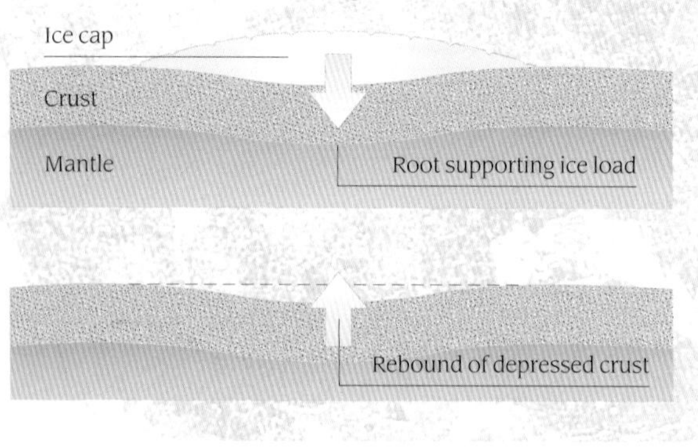

PLEISTOCENE

hemisphere lay the tundra. This sparsely vegetated zone, where winter temperatures may be as low as –134°F (–57°C), extended as far south as northern France in Europe and at least 100mi (150km) south of the ice sheet margin in central North America. Tundra is found today, particularly in the Siberian steppes and the Canadian Arctic; vegetation there rarely grows more than a few inches high and is low in nutrients. Several inches below the soil surface is permafrost, a layer of ice which remains permanently frozen, even during the summer. Ice-age tundra was probably an even harsher place, and the permafrost extended many tens or hundreds of feet below ground level. Here, unable to penetrate, surface water cut deep channels as it ran off.

As the immense ice sheets expanded southwards from the Arctic regions, the latitudinal variations in climate were compressed towards the equator as ice covered the north. Climates thus varied more over shorter distance in a north–south direction than before. Biological zones were located closer together and there was less distance between distinct faunas. Climate zones shifted south, bringing rain to what had been (and have become once again) arid regions, such as eastern Africa.

It was here that the first humans appeared approximately 300,000 years ago, an event that may prove to be one of the most consequential of all in the history of the planet. At the time, however, humans were merely a timid new entry onto the evolutionary scene, which was dominated by large, ferocious, swift-moving carnivores. "By all the biological laws, these ill-armed, gangling beasts [humans] should have perished," in the words of the biologist Loren Eiseley; but they flourished instead.

A MAP of the Pleistocene world is essentially the same as that of today, although some differences are evident. For example, the Bering land-bridge that stood between Siberia and Alaska was in existence and—importantly—allowed passage between Asia and North America. Many of the fossils from Pleistocene deposits record a transfer of animal types between the two continents. This land-bridge existed on and off from about 75,000 to 11,000 years ago and would have been covered with tundra vegetation. Enormous herds of grazing animals such as musk ox and reindeer moved across the bridge into Alaska and Canada. Although the continental ice sheets were widespread, at times there was an ice-free corridor to the west, down which the first humans to reach America traveled in pursuit of the great herds.

Ice sheets covered most of the northern hemisphere, but beneath them the distribution of the land masses was much the same as it is today.

The advancing ice sheets diverted the north Atlantic Gulf Stream to the south, forcing it towards Spain, and the Mediterranean "sluice" between the straits of Gibraltar, making northern Europe still cooler. In places farther afield, the drop in sea levels as a result of water being locked into the ice sheets, had geographical effects as some islands were reconnected to their nearest continents. The area of the Caribbean islands, for example, was increased as the drop in sea levels exposed more of their previously submerged offshore land areas. At the height of the last glaciation the sea levels around Barbados were as much as 400ft (120m) lower than today.

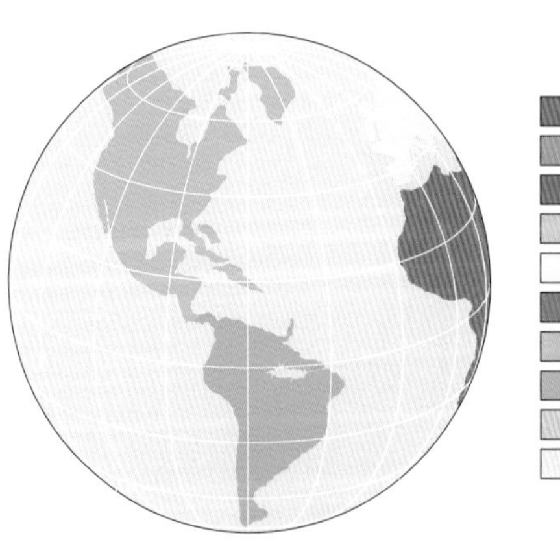

WEST COAST TECTONICS

In western North America, regional uplift produced the Rocky Mountains. Along the coast, subduction and volcanism continued along the San Andreas Fault.

Lake Bonneville
Gulf of California
PACIFIC OCEAN
East Pacific Ri

TWO TYPES OF ICE

The northern hemisphere was covered in both pack (sea) ice and continental ice. Sea ice forms in a relatively short period, but continental ice sheets grow over hundreds of years during episodes of intense cold.

DIVERTED WATERS

By the time the Panamanian land-bridge had been in existence for three million years, its redirection of warm Atlantic currents carried moist air to northern latitudes, increasing precipitation and freshwater runoff into the ocean. This resulted in the rapid growth of continental ice sheets.

CONTINENTAL GLACIER

The only present-day continental glaciers are over Antarctica and Greenland. Glacial ice (left) flows from the ice sheet into the sea at Jakobshavn in west Greenland.

In the east of the globe, mountain glaciers covered the newly uplifted Himalayas and extensive meltwater lakes lay in the vast areas of Eurasian tundra. Although the most dramatic glaciation was in the north, the Antarctic ice sheet expanded to about double its present area, with a great increase in the amount of sea ice. The circumpolar current that had formed in the late Paleogene dragged the pack ice eastwards. Small ice sheets grew in the Andes and the mountains of Australia and New Zealand. At this time of low sea levels Australia was joined to New Guinea; the East Indies were also continuous land, with extensive river systems. Their modern island shapes appeared as mountain ranges.

APART from the periodic waxing and waning of the huge ice sheets, which had occurred in other periods (notably the Ordovician and the Permo–Carboniferous), most physical parameters of the Earth's atmospheric, hydrospheric, and biotic systems had long been established in patterns that continue today. With the Pleistocene, a number of sudden temperature falls occurred. It was clear, by the end of the nineteenth century, that the concept of an "ice age," which had replaced that of "the deluge," was not a simple event, but that there had been more than one glacial episode. Eventually a chronology of four major glaciations and three interglacials was widely accepted. Evidence for these episodes comes from geology, geochemistry (oxygen isotopes in Greenland ice cores), and fossils, and even tree-ring dating techniques. Since the 1950s, radiometric dating has enabled fairly precise dates to be obtained.

The modern ice age is a complex event, with pulses of glacial expansion (maxima) separated by partial retreats (minima).

The last glacial maximum (LGM)—episode of widespread ice cover—took place in the Pleistocene epoch about 20,000 years ago, but in subsequent millennia the earth warmed and the ice sheets of the northern hemisphere retreated northward into the ancient Arctic circle. Our planet is currently in the middle of an interglacial period. Such periods last only a short time but have been shown to have extremely fast rates of biotic and physical change—in fact, faster than ever before.

Climatic instability typifies the closing stages of a glacial period and the opening of an interglacial, and shows how fast climates can change in certain circumstances. This instability was first noted in cyclic lake sediments in Denmark. The sediments dating from the transition between the glacial and current interglacial exhibited some rather unusual features: tundra was replaced as vegetation stabilized the shores of the ancient lakes and productivity increased in the water column. But this process, reflecting the increasing warmth of the climate, was evidently interrupted. The clays show a return to cold climates as heavy clays became evident again in the sediment cores; these features were found in cores of the same age throughout the entire European continent.

The cold episode was severe enough to cause the regrowth of the northern glaciers, which left their mark on Scandinavia and Scotland. This cold spell was called the Younger Dryas event because the fossil leaves of *Dryas octopetala*—mountain aven, an Arctic–alpine plant—were found in abundance in the clay layers. Radiocarbon dates show that the Younger Dryas took place between about 11,000 and 10,000 years ago. At a number of sites in continental Europe, a shorter period of cold again took place about 12,000 to 11,800 years ago; it is called the Older Dryas.

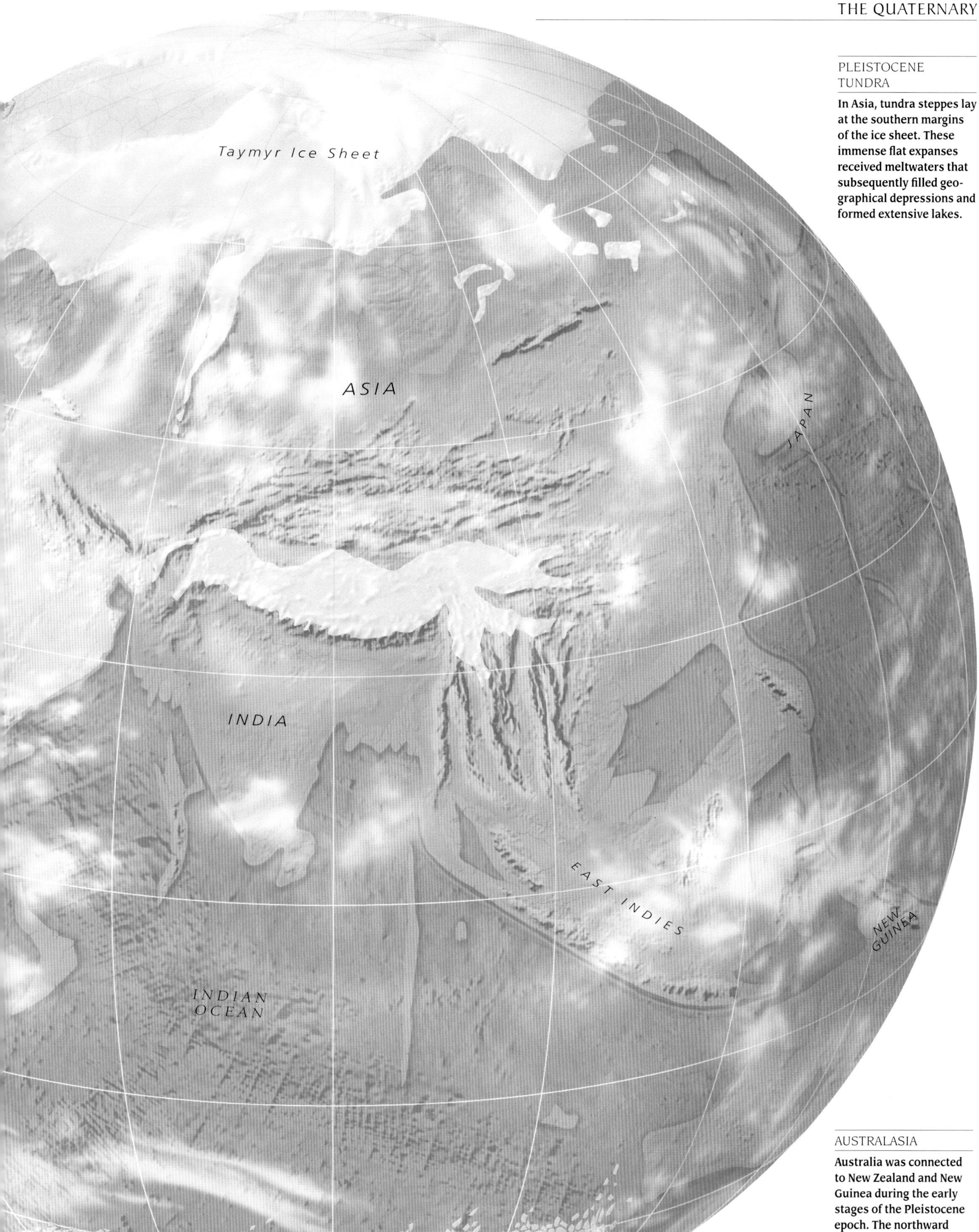

PLEISTOCENE TUNDRA

In Asia, tundra steppes lay at the southern margins of the ice sheet. These immense flat expanses received meltwaters that subsequently filled geographical depressions and formed extensive lakes.

AUSTRALASIA

Australia was connected to New Zealand and New Guinea during the early stages of the Pleistocene epoch. The northward drift of the Australian plate caused the uplift of seamounts that eventually formed the Indonesian island chain. At this time, though, a large amount of land was above sea level.

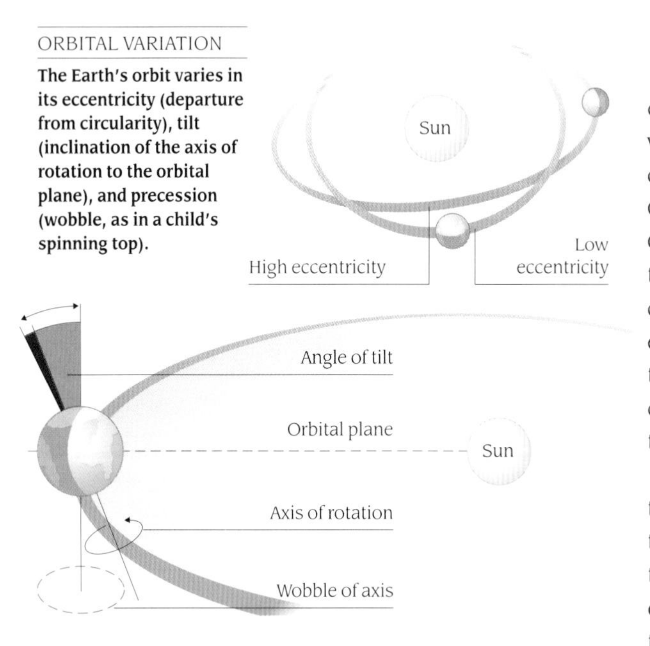

ORBITAL VARIATION

The Earth's orbit varies in its eccentricity (departure from circularity), tilt (inclination of the axis of rotation to the orbital plane), and precession (wobble, as in a child's spinning top).

THE EARTH'S atmosphere, oceans, biosphere, and ice sheets are part of one huge global system in which these components are very strongly coupled. A change in any one part will lead to changes in the others. Each of the different components of the global climatic system operates on an entirely different timescale. The atmosphere adjusts to changes very quickly, more so than the other components—this occurs over a period of just weeks, whereas the oceans and the biosphere adjust more slowly, over periods of hundreds to thousands of years, and ice sheets are slowest of all to respond. Changes to ice sheets tend to occur over tens of thousands to hundreds of thousands of years.

Ice ages are slow-moving phenomena, taking many thousands of years to respond to global changes.

The major expansion of global ice cover that occurred during the late Ordovican and the Permo–Carboniferous was connected to the changing position of the supercontinent of Gondwana in relation to the South Pole. Glacial conditions migrated from North Africa in the late Ordovician, to South Africa during the Carboniferous, and to Australia during the Permian. Because climatic changes brought about by plate movements occur extremely gradually, on a scale of millions of years, plate tectonics theory cannot be used to explain the repeated cycles of glaciation that are are uniquely characteristic of the Pleistocene epoch.

This phenomenon puzzled early geologists. In 1876 the British geologist James Croll suggested that long-term changes in the amount of solar radiation reaching the Earth were controlled by rhythmic variations in its orbit around the Sun, and that these were responsible for the periodic climate changes that European geologists had only begun to discover. However, it was not until 1941, when the Yugoslavian astronomer Milutin Milankovitch calculated the magnitude and frequency of the changes in solar radiation received by the Earth as a result of orbital changes, that there was a mechanism to account for Croll's proposal.

Milankovitch identified three orbital processes that would control these changes: the tilt of the Earth's axis, the eccentricity of the Earth's orbit, and the precession of the equinoxes. The Earth's axis is not at right angles to the plane of the Earth's orbit about the Sun, but inclined at an angle of about 23.5°. This angle varies from between 24.5° and 21.5° in cycles of about 40,000 years; when it is greatest, there is the largest difference in seasonal heating at any latitude.

The Earth's orbit is not circular, and sometimes it is more strongly elliptical than other times. The length of this eccentricity cycle is about 100,000 years. The Earth

ICE FLOE

Pack ice is usually only a few meters thick. Ocean movements can cause it to break up, with ice at the edge of the pack floating free as floes (above, off the coast of Spitsbergen). Ice floes drift under the influence of winds, unlike icebergs, which move with ocean currents.

MEASURING THE ICE AGES

Foraminiferans (forams) are microscopic single-celled animals that form a globular, calcareous skeleton around their cell wall. They form part of the ocean plankton, and, upon death, sink to the sea floor to accumulate along with other, benthic (bottom dwelling), types. Over millions of years they are preserved in sea bed limestones that often seem to contain nothing but forams. Oxygen is present in sea water as the stable isotopes oxygen-16 and oxygen-18, and these are incorporated into the skeletons of forams in the same proportion in which they occur in the surrounding sea water. The relative proportions of these two oxygen isotopes in the fossil forams found in deep-sea cores fluctuates widely, with the ratio shifting towards heavier values during glacial maxima.

Initially, scientists thought that these fluctuations reflected variations in water temperature; however, it was found that similar fluctuations were exhibited not only by planktonic forams but also the deep sea benthic forams. This is the region in the oceans of the freezing-water layer that has existed since the formation of the psychrosphere 35 million years ago. It is now known that temperature has only a minor effect on the oxygen-isotope ratio in the foram skeleton, and that this fluctuation rather reflects the oxygen isotope ratio of the oceanic waters in which the foram lived; which varies greatly with the size of continental ice sheets.

During glacial episodes, more of the lighter oxygen-16 falls as snow. This accumulates in the glaciers, leaving a higher proportion of the heavier isotope in the ocean. Recognition of the fact that this high proportion of oxygen-18 in foram skeletons corresponds with an increased volume of glacial ice was a critical moment in establishing the use of forams (such as the planktonic *Globorotalia* species, right) as paleoclimatic indicators. Correlation with periods of magnetic reversal, as shown by geomagnetic analysis of the rock cores, calibrates the chronology of climatic cycles that foram isotope analysis displays. In this way, the waxing and waning of the Pleistocene ice sheets can be precisely dated.

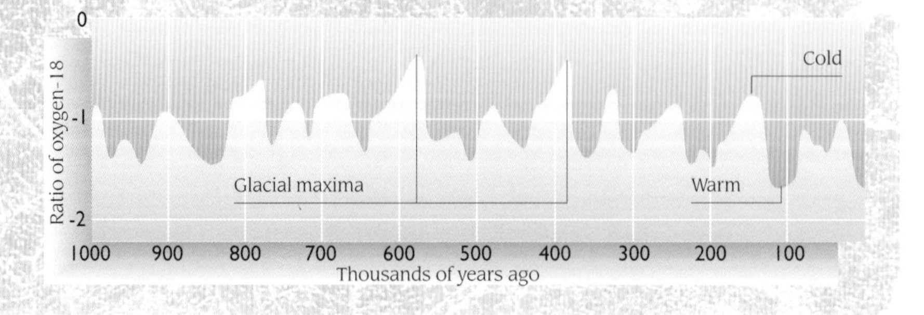

also wobbles slightly on its axis—a phenomenon known as precession— due to the gravitational effects of the Moon and Sun on the Earth's equatorial bulge, which changes the timing of the solstices. These are relative to the position that the Earth occupies in its elliptical path around the Sun. About 11,000 years ago, the Earth's nearest point to the Sun occurred when the northern hemisphere was tilted towards the Sun (northern hemisphere summer), rather than during the northern hemisphere's winter as is the case today. The length of the precession cycle is about 23,000 years.

Analysis of the very long-term variations in the Earth's climate show that the pattern of change between 800,000 years ago and today is composed of three dominant rhythms: one with a period of 100,000 years, one of 40,000 years, and one of 20,000 years. From this it can be concluded that variations in the Earth's orbit are the main determinants of long-term patterns of climate.

However, Milankovitch cycles cannot be reconciled with some geological data. The temperature change that takes place between the coldest parts of glacials and the warmest parts of interglacials is about 4–5°C, but the difference in the intensity of solar radiation reaching the Earth due to orbital variations is not enough to change the global average temperature by more than 0.4–0.5°. There are also changes in the climatic rhythms through time: before 800,000 years ago, the 40,000 year rhythm was the governing rhythm and after this time the 100,000 rhythm was dominant. Finally, there is no fundamental change in the "orbital pacemaker" which might explain the progressive intensification of the ice ages over the past 2 to 3 million years. As a result, there is not yet a completely satisfactory explanation for why the Pleistocene glaciation cycle began.

Feedback mechanisms in climatic systems have major effects. Growth of ice sheets increases the albedo or reflectivity of the Earth's surface, with a net cooling effect larger than the consequence of Milankovitch cooling alone. Cooled waters absorb more carbon dioxide than warmer waters and decrease the greenhouse effect, further contributing to global cooling. Other effects include the partially isolated Arctic Ocean in the northern hemisphere, which began to spill its cold waters into the north Atlantic as Greenland rifted off North America. In the southern hemisphere, the circumpolar Antarctic current had been supplying cold global waters in the form of the psychrosphere for more than 40 million years. All of these effects have combined to produce the extraordinary glacial cycles of the past 2 million years, although the exact mechanism by which this occurs has yet to be fully explained.

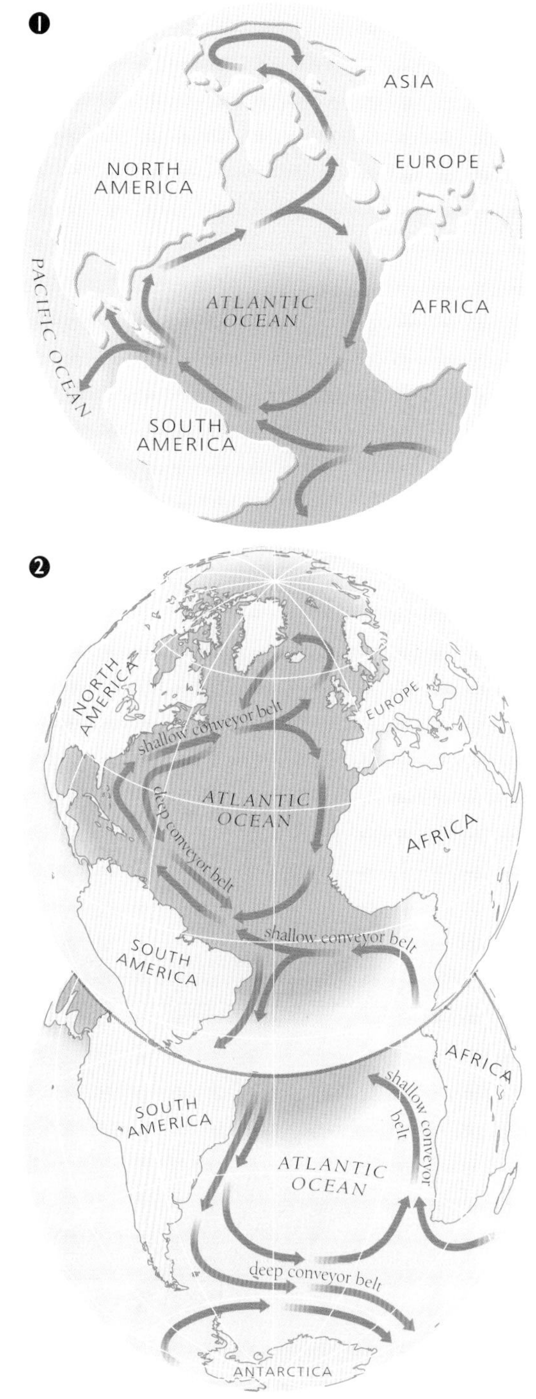

ATLANTIC CIRCULATION

One possible cause of Quaternary glaciation was the changing circulation pattern of the Atlantic Ocean. During the early Pliocene (1), before the Isthmus of Panama formed, warm Atlantic waters flowing north mixed freely with Pacific waters. This would have made the Atlantic waters less saline, and therefore more buoyant, than today. They may then have flowed into the Arctic Ocean, warming the polar regions, before cooling made the current dense enough to sink. The creation of the isthmus cut off access to the Pacific, and the drying effect of the trade winds blowing off the Sahara increased the salinity of the Atlantic. Today (2) these denser waters sink north of Iceland, forming a loop known as the oceanic conveyor belt (lower maps), isolating and cooling the whole Arctic region as it brings cold water down from the North Pole to the South in deep-water currents.

shallow ocean current
deep ocean current
high-salinity water

Firn (compacted snow)

Frost-shattered peaks

Scree

Lateral moraine

Crevasses

Medial moraine

Glacier snout

Gravel outwash

Braided streams of meltwater

U-SHAPED VALLEY

(Right) The passage of huge continental glaciers leaves wide, shallowly scoured valleys. Corrie lakes (also known as cwms or cirques) lie in the valley of Cwm Idwal in Snowdonia, North Wales.

THE UNIQUE cyclic glaciations during the Pleistocene epoch effectively complicated the geological evolution of the areas or regions that they touched. They not only left characteristic geomorphological features of their own making, but obscured much of the underlying geology as well. Glaciers in mountain valleys do not leave a record of their presence in the rock strata because these highland regions are areas of erosion rather than of deposition. In contrast, continental glaciers do leave their mark. An example of this is seen in modern Greenland and Antarctica, which are almost entirely covered by vast ice sheets.

Evidence for the former presence of glaciers is found in the characteristic landscape features that remain.

The Greenland and Antarctic glaciers are thickest at their centers. Gravity causes them to flow outwards from these high points as the typically riverlike glaciers that are more familiar in less remote places, such as the Alps. These glaciers, which flow in geographical channels, are called valley glaciers when they are very long relative to their width and may prove to be confluent with other valley glaciers. Cirque glaciers (or "corries" in Gaelic and "cwms" in Welsh) are relatively small bodies of ice that occupy hollows high on the sides of mountains. They occur in places where the temperature is only just low enough to maintain a permanent snow field (firn), and where there is insufficient snow and ice to support a valley glacier.

The geological evidence of past glaciations is represented by a suite of features that result from the peculiarities of glaciers. Glaciers move rocks, erode rocks, and move sediments in their direction of flow. Rocks entrained at the base of a glacier erode the substrate still further to give highly characteristic scratches on rock surfaces that are called glacial striations. When glaciers flow over rock "knobs" they tend to pluck blocks from the lee side of the knob due to pressure release at this location of the rock. These plucked blocks are entrained within the bottom region of the glacier where they then act as grinding tools by cutting striations into the bedrock. The combination of lee-side plucking and up-glacier side-smoothing of bedrock knobs produces the widespread rockforms of smoothed and striated hummocks known as *roches moutonnées*.

Uplifted areas provide high-velocity streams that cut deeply into rock. In complex glacier systems, where relatively minor tributary glaciers feed into more rapidly flowing trunk glaciers, the trunk (main) glacier tends to erode down into the bedrock much more rapidly than the tributary (side-branch) glacier. As a result of this, the trunk valley is deeper than the tributary valley, which now forms a hanging valley in which its open end is high up

GLACIAL EROSION

(Above) Ice is usually considered a solid, but behaves as a slow-moving liquid, influenced by gravity and slope. In mountain glaciers there is gradual downward flow of ice. The erosive force of moving glaciers is evident from the landscape that they leave behind: U-shaped troughs, hanging valleys from tributary glaciers left suspended, sharp-edged arêtes, isolated horns—produced as cirques grow at the glacier head—and debris such as till and moraine.

GLACIAL ERRATICS

(Left) Glaciers entrain, carry, and deposit rocks of different geological provinces from distant regions; often from as far as 300 miles away (500km). These "erratics" appear as randomly scattered, slightly striated boulders that are usually of recognizably different lithologies to those of the surrounding valley strata. Although the source of most erratics is unknown, the origin of some can be determined. By studying them, it is possible for geologists to trace the path of an ice flow.

Roche moutonnée
Cirque (corrie) with moraine-dammed lake
Hanging valley
Pyramidal peak or horn
Arête
U-shaped valley
Ridge of lateral moraine
Kettle hole
Esker
Drumlins
Moraine-dammed lake
Outwash plain

on the side wall of the trunk valley. V-shaped valleys are eroded by valley glaciers into U-shaped valleys because the glacier erosion occurs across the whole cross-section of the valley instead of being concentrated by rivers only on the center line of the valley.

MORAINE is transported debris that had originally been entrained by a glacier and has now been dumped following its retreat. Moraines mark the present or former positions of ice margins. They come in various types, such as end, or terminal moraines, where material carried by the ice is deposited at the terminus of the glacier, while recessional moraines are formed as the ice retreats, each representing a temporary halt. Lateral moraines occur as ice erodes the sides of a valley; when it melts, the accumulation of rubble is left next to the valley walls and medial moraines are created where the lateral moraines of two valley glaciers join. If the ice advances across a moraine the sediments can become contorted and folded; this feature is called a push moraine.

Cape Cod on the northeast coast of North America is an example of an unusually large medial moraine.

Moraines are scattered over large areas of both North America and Eurasia. Some of these moraines are of spectacular sizes; one extends out into the north Atlantic as Cape Cod. Terminal moraines often flank depressions in the country rocks of a landscape, and were made as the glaciers retreated. Some of these huge depressions became the Great Lakes of North America. The Hudson Bay of Canada is not bounded by a terminal moraine; it is an arm of the ocean that spread into a region where the crust had been depressed by the thickest part of the continental glaciers in North America, and has not yet rebounded back to its higher original level.

During the erosion of a subglacial rock bed, the process of crushing and abrasion during glacial transport produces debris with a wide range of sediment grain sizes, from silt to large boulders; this may be deposited subglacially as a sediment termed till. Unlike moving water and wind, ice cannot sort the sediment it carries, so till is characterized by being composed of unsorted mixtures of many particle sizes; in addition many pieces are striated and polished by the action of the glacier. Till that has become lithified due to burial, compaction, and chemical alteration is termed tillite. Both tills and tillites are extremely characteristic of the past influences of glaciers.

Glaciers tend to block normal river drainage paths; during the recession of the Pleistocene ice sheets, ice-dammed lakes came into a temporary existence. However, permanent lakes also receive meltwaters and glacially transported sediments. Glaciers are often

PLEISTOCENE

influenced by seasonal climatic changes; as winter approaches, meltwaters cease as they are locked up as ice and the input of water into lakes diminishes, but upon warming in the spring, meltwaters flow freely again and fill up the lakes. This annual cycle leads to the accumulation of annual layers, called varves, in sedimentary rocks. Furthermore, icebergs containing debris "calve" (divide) into many glacial lakes, and any debris entrained within them is released as they melt. These released rocks, called dropstones, drop into the finer lake-bottom sediments. They are a sure indicator of previously glaciated terrains and may be easily identified in the field.

POSTGLACIAL LAKES

Ice sheets over large areas of the Canadian Shield disturbed drainage patterns. Today, postglacial lakes (right) fill hollows that were eroded by ice or blocked by moraine left by retreating ice sheets. (Below) Other lakes, such as Lake Bonneville, developed in basins south of the glaciers; these pluvial lakes resulted from the increased runoff and rain.

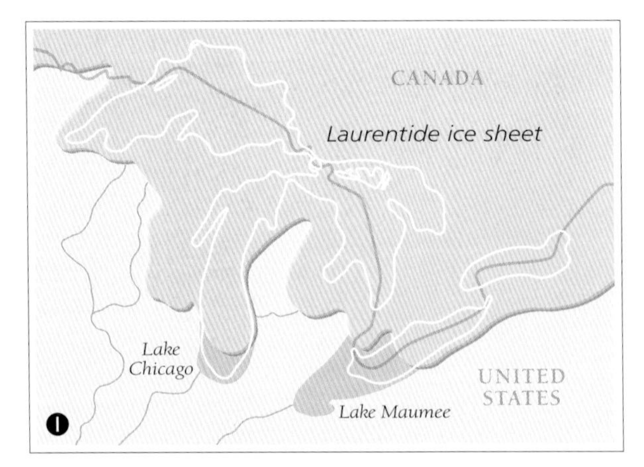

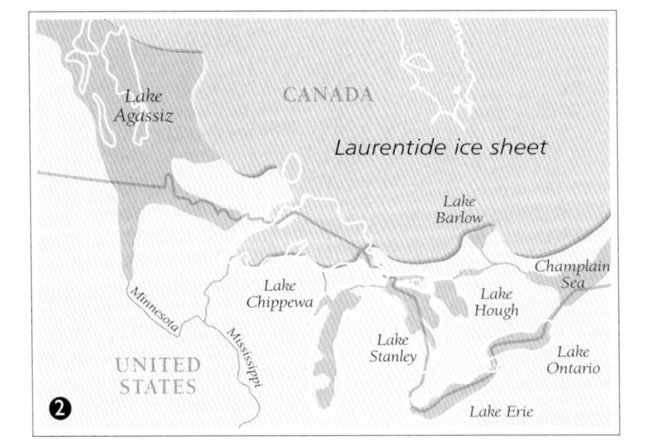

EVOLUTION OF THE GREAT LAKES

(Right) The Great Lakes are the remnants of huge ice lobes that extended southwards from the margin of the Pleistocene ice sheets. The ice sheets formed depressions that were subsequently filled with meltwaters during their retreat about 14,000 years ago (1). By 10,000 years ago (2), all the modern Great Lakes had begun to fill except Lake Superior, whose basin still lay beneath the ice. The Great Lakes are just a few of an enormous number of lakes formed by continental glaciation in the shield region of North America. More lakes were created in this area by glacial processes than by all other geological processes combined. However, the size of the Great Lakes greatly surpasses that of all the others.

SCIENTIFIC investigations of Pleistocene and Holocene geology in the region of the Great Lakes, and examinations of their possible origins, began in the mid-nineteenth century. The Swiss–American paleontologist Louis Agassiz of Harvard University went to Lake Superior in 1850 and described some aspects of its geology in great detail. It was Agassiz who was the instigator of the ideas regarding the mechanisms of glaciation and its effects on the land: these effects had left a characteristic geomorphology, and Agassiz was the first to recognize the classic signs. His researches were followed in later years by Lawson (1893) and F.B. Taylor (1895 and 1897), both of whom studied the glacial geology of the Great Lakes area. Agassiz gave his name to the largest of the ancient proglacial lakes (lakes formed directly adjacent to the snout of glaciers), Lake Agassiz, which covered a vast area of south-central Canada, with an arm extending south into North Dakota.

The Great Lakes of North America (and similar lakes elsewhere in the world) were formed as a result of Pleistocene glaciation.

The age of the Great Lakes was unknown in the last century, and despite immense advances in geological science, is still largely unknown. This is because they were successively excavated by the ice sheets, and each excavation obliterated any previous evidence that might be used to date the system. However, evidence from rocks surrounding the Great Lakes shows that these inland bodies of water were not in existence prior to the Pleistocene epoch. This means that they are not more than

about 1.7 to 2 million years old. The present floors of the Great Lakes were previously lowlands. The glaciers that were extending southward from their northern site of origin flowed slowly into these lowlands, scouring them deeply and isostatically depressing the crust (by up to 1100ft/330m in the Hudson Bay area). As the glaciers withdrew, over a period of 10,000 years or so, during the interglacial periods, freezing meltwaters collected in these depressions and began to form the present Great Lakes system.

During the glaciation and meltwater events, huge blocks of ice enclosed in glacial clastics collected to the south of the lake basins and gradually began to melt. As a result of this mosaic pattern of melting ice, a characteristic hummocky topography was formed in the regions of what is now mid-northern Michigan. This unusual landscape is composed of irregular hills and small lakes known as kettles. At a later stage in the Pleistocene epoch, sands that had been deposited by the melting waters along the southern shores of lake Michigan were later carried eastwards by North American weather systems. These sands were deposited as dunes around the southern bend of the lake, where some can still be seen today. Because the resulting Great Lakes are so big, their shorelines show tilt when measured, and even rates of isostatic uplift can be determined. The lakes also controlled ice movements by providing more channels of lower relief than the surrounding countryside into which the glaciers preferentially flowed due to the effects of gravity. The geological history of the Great Lakes after the Wisconsinian glacial stage is a complex one of changing meltwater outlets and the presence or absence of ice barriers. The Wisconsinian stage glaciations of the Pleistocene occurred in North America only, and this was about 800,000 years ago, at a time when Europe was somewhat less glaciated. The development of changing topographies in and around the Great Lakes system caused the formation of different lakes at different times throughout the Pleistocene epoch. Examples of these now-vanished ancient lakes are Lake Milwaukee, Lake Leverett, Lake Maumee, and Lake Saginaw.

The geomorphology of the terrain that surrounds each lake is different and gives each one its own particular shape. Such individual peculiarities of the lake geomorphology also extended to glacial lakes in Europe; here much smaller basins were formed which now include the Saalin basin in the Netherlands and the Elsterian basin in Germany. However, the sheer size of the American continental plate meant that there was a much greater potential to form extensive glacial lake systems than in the more circumscribed area of Europe. This area also features a structure that has become famous: Niagara Falls. The dramatic waterfall at Niagara, which lies between Lake Erie and Lake Ontario, came about when the retreating ice of the last glacial uncovered an escarpment that was formed by southerly tilted, mechanically resistant strata (dolomite, a hard limestone). Water from the Niagara River flowed over the edge of this escarpment— which is supported underneath by the solid Lockport Limestone series—and undermined the weaker shales below, causing a southward, upstream erosional retreat of the falls. The Pleistocene glaciations that covered North America have given this area a number of magnificent landscape features, such as the Great Lakes and Niagara Falls, yet these natural wonders are still quite young in geological terms.

THE LAKE MISSOULA FLOOD

(Right) The channelled scablands of Washington State, USA, are a complex of interlaced deep channels called coulees. They were formed when a large ice lobe advanced southward and blocked the Clark Fork River—a major tributary of the huge Columbia River. (1) Ice dams trapped waters, which banked up to form Lake Missoula in Western Montana. As the glacier retreated, the ice dam broke, releasing a vast and unparalleled flood; the water swept across the Columbia plateau, stripping away soil and cutting deep channels (2).

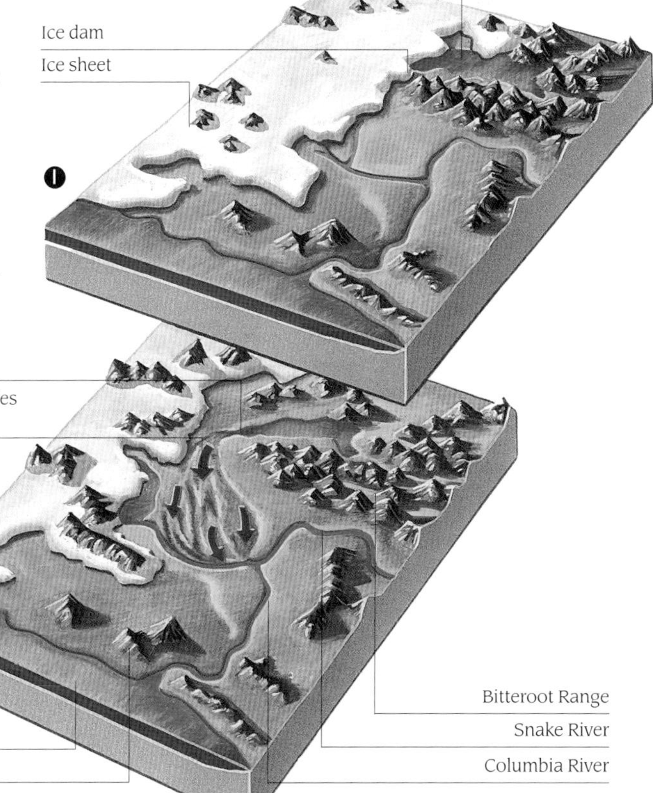

PLEISTOCENE

THE GLACIAL cycles of the Pleistocene epoch brought about great differences in the geographical distributions of mammals during the cold and warm phases as the ice sheets waxed and waned. This is particularly well documented for small mammals such as rodents and mustelid carnivores (weasels, ferrets, and stoats). In the latter part of the Pleistocene, many mammals that had formerly occurred together became separated due to the increased patchiness and disjunction of their habitats. These tundra faunas are classed as "disharmonious" faunas because they contain animals that no longer co-exist. Tundra steppes probably extended to a much greater area than today and supported a fauna of large, medium, and small animals, all of which were adapted to the harsh, cold environments, or could migrate away from the winter tundra with ease. The expansion of the Pleistocene ice sheets had a marked effect on animal life in the northern hemisphere, and one of these effects was the formation of an extensive tundra biota, the legacy of which still remains with us today.

The tundra was inhabited by a wide range of cold-adapted animals.

The tundra of today's planet is found enclosing the Arctic Circle, north of the tree-line, and in smaller areas in the southern hemisphere on the sub-Antarctic islands. Alpine tundra also occurs above the tree-line on high mountains, including those in the tropics. For tundra vegetation there is an extremely short growing season and only cold-tolerant plants can survive; typical tundra plants are mosses, lichens, sedges, and dwarf trees. The main large animals that live there are reindeer (caribou) and musk ox. Small herbivores include snowshoe hares, voles, and lemmings. In addition, many birds migrate to the tundra from the south in the summer in order to feed on the huge warm-season insect population.

Feeding on these are the carnivores, which in modern tundra ecosystems are represented by Arctic fox, wolves, falcons, hawks, and owls. Fossils of animals that derive from what appears to be Pleistocene tundra steppe environments include a number of animals that still inhabit this ecosystem, such as reindeer, elk, hares, wolves, voles, and ermine. The latter two mammals lived under the thin seasonal coverings of snow rather than attempt to burrow into the iron-hard soil to any great extent.

The tundra of Pleistocene times was more widely populated than today's, and by animals that are no longer seen anywhere on Earth. Some of the most significant of these were not native. In response to interglacial–glacial oscillations, mammals either migrated or expanded their ecological tolerances to climatic and/or vegetational changes. Associations of some mammals with certain types of environments and plants suggest that many have altered their preferences or tolerances. For example, the hamster is now found on Asian steppe but has been recovered as fossils from sediments deposited in forested environments during interglacial periods. Modern wild horses are typically associated with open environments, but fossil wild horses are also found in forested habitats, as well as grasslands. Canids had their evolutionary origin in North America and crossed the Bering land-bridge into Asia where they continued their success as predators of open habitats. Those animals that immigrated into North America across the land-bridge were cold-adapted, such as mountain sheep, musk oxen, moose, lions—and humans. Oxen of different types were especially prevalent, but mammoth elephants were less common.

MAMMAL RANGES

(Right) The traditionally migratory herbivores such as elk, woolly mammoth, and woolly rhinoceros had much greater biogeographical distributions than most contemporary Pleistocene carnivores, possibly as a result of greater adaptability. Saber cats were dominant in North America, but Europe was inhabited by dirk-toothed and scimitar-toothed cats.

MAMMOTH BONES

(Above) The widespread incidence of mammoth remains, such as these in South Dakota, has provided good information about their appearance. The Siberian form was smaller than the American, with males being about 10ft (3m) high at the shoulder.

Woolly mammoths are usually imagined scratching a living in snowy wastes devoid of vegetation. This is misleading, for such an environment could probably not support herds of these animals, with their enormous food requirements. It seems likely that woolly mammoths inhabited the boreal (forested) regions outside the tundra, where food would have been available at least part of the year; and also the warmer and more equable margins south of the steppes, where a richer vegetation was more readily available. Very short mosses and lichens could not have been grazed by tall-standing mammoths, and their trunks could probably not have gained a grip on small low-lying vegetation. Oxen, on the other hand, with their downward-pointing necks, grazing dentition, flexible lips, and broad muzzles, were particularly well adapted to graze efficiently on the extremely short tundra vegetation. This factor partially accounts for their abundance in these harsh environments. The same functional consideration of ecology holds true for the reindeer and elk. These large-bodied animals provided considerable bounty for proficient carnivores such as wolves that were capable of bringing them down in the hunt.

WOOLLY COATS

(Below) Large-bodied animals lose heat more slowly than smaller creatures, so that large body size is generally advantageous in cold climates. Nonetheless, animals the size of mammoths required extra insulation against the extremes of the Pleistocene ice ages. Smaller animals may burrow or hibernate, but huge ones cannot do this. Continuous exposure to such cold conditions resulted in the super-thick, hairy coats of mammoth, bison, and rhinos.

1 *Coelodonta antiquitatis* (woolly rhinoceros)
2 *Megaloceros giganteus* (giant deer)
3 *Mammuthus primigenius* (woolly mammoth)

Several hundred miles south of the tundra margin, in the region which is now California, lay a radically different biome. Despite the northern glaciation, California was still relatively warm, and it was home to its own distinctive fauna. Some of these are familiar today, while others are mysterious and often spectacular extinct representatives of modern animal groups. There is a considerable amount of information about this rich ecosystem due to an important geological feature on the west coast of North America: the San Andreas Fault. This fault is one feature of a major system that extends in a broadly north–south

The San Andreas fault is the visible boundary between the North American and the Pacific plates.

3

PLEISTOCENE

trend along the western side of the whole of the Americas. This huge line, several thousand miles long, is the demarcation zone of collision between the constituent plates of the Pacific lithosphere and that of the American continent. From the north these Pacific plates are: the Kula plate, the Farallon plate, the Nazca plate, the Phoenix plate, and, to the extreme southern tip of South America, the Antarctic plate. As the spreading ridge of the East Pacific Rise pushes it east, the subduction of the Nazca plate beneath the South American plate is largely responsible for the formation of the Andes Mountains.

In the California area of the eastern Pacific, the North American lithospheric plate is heading south while the Farallon plate that forms the northeastern margin of the Pacific plate goes north. These plate margins exist as the San Andreas transform fault and scrape past each other, over a stretch of land several hundred miles long. This is known as a strike–slip fault, or one in which the two opposing blocks of lithosphere are passing sideways to each other, not over or under either block. Occasionally, over time, the contacts between the rock surfaces of these two lithospheric plates lock and grab, and as a result of the continued plate movements they flex and bend before snapping back into position. The release of this massive stress is felt as earthquakes. Measurements show that the plates are moving at a velocity of about 2in (5cm) per year; the past average was 0.4in (1cm) per year. According to these figures, Los Angeles should be parallel to San Francisco in 25 million years or so.

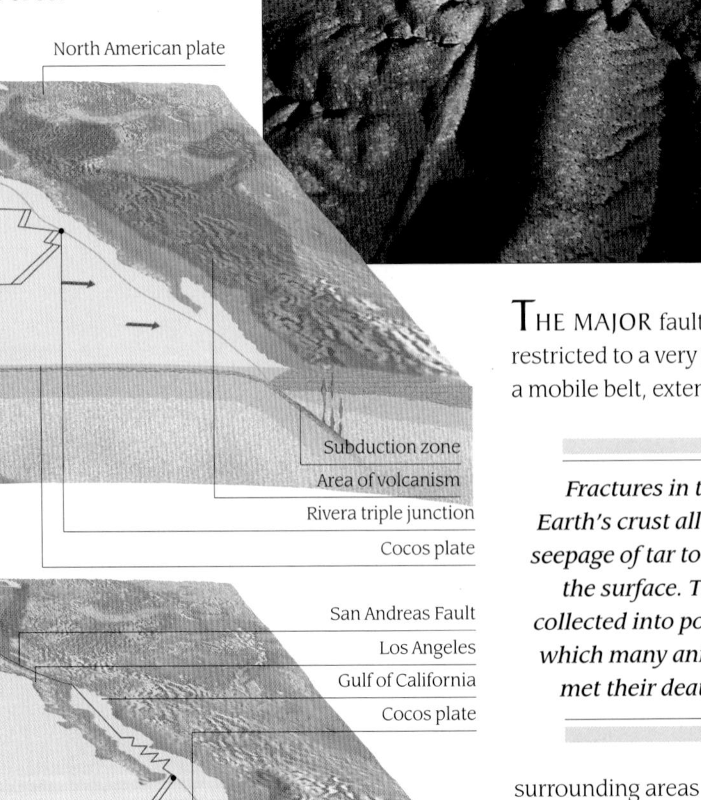

THE MAJOR fault zone of the San Andreas Fault is restricted to a very narrow belt, sometimes described as a mobile belt, extending south into the Gulf of California where there are a number of smaller, stepwise transform faults in the seabed. Strike–slip faults such as the San Andreas also form shallow basins as offset portions of the two lithospheric plates stretch apart, causing subsidence between them. These terrestrial basins receive sedimentary input from the surrounding areas and act as centers of deposition; consequently they have a high potential to preserve the remains of the animals that lived in those low-lying regions. The San Andreas Fault extends through the lithosphere. Many of its smaller microfaults, extending outwards into the surrounding rocks of both the partly undersea Farallon plate and the terrestrial North

Fractures in the Earth's crust allowed seepage of tar to reach the surface. This collected into pools in which many animals met their deaths.

SAN ANDREAS FAULT

(Left) In some places, the exposure of the San Andreas transform fault is particularly evident. The relative motions of the plates can be determined by careful mapping and measurement of the separate sides of the fault. It is at times when the two sides of the fault grip, lock, and part explosively that earthquakes occur.

EVOLUTION OF THE SAN ANDREAS FAULT

(Below left) About 25 million years ago the East Pacific Rise began to be subducted under the North American plate (1). As the rise was consumed, the Farallon plate was reduced to two remnants and the San Andreas Fault developed, lengthening as the Rivera and Mendocino triple points moved north and south. Between 4 and 3 million years ago (2), the fault migrated inland and a segment of Mexico—including Baja California—became fragmented and joined to the Pacific plate. The San Andreas and Mendocino faults are both transform faults. The Mendocino connects the Juan de Fuca spreading ridge to the subduction zone beneath the North American plate, and the San Andreas connects the Juan de Fuca Ridge and a second divergent boundary in the Gulf of California.

OIL DEPOSITS

(Below) At Rancho La Brea an earlier series of faulted Tertiary deposits is unconformably overlain by Pleistocene deposits. Impermeable layers in the Tertiary rocks have allowed oil to collect; it escapes to the surface via numerous fractures.

American plate, intersect underground oil reservoirs. These reservoirs are the result of the accumulation of decaying organic material. An abundant animal community has left its trace in the form of decaying matter, which typically collects in low-lying areas such as estuaries and deltas before becoming entombed under layers of deposition.

Under the ground at depth, chemical changes due to pressure, temperature, and entrapment change the decaying organic matter into oil. The semi-liquid organic matter is then released at the surface due to the rupture of the underground oil reserves. These are mostly located in permeable beds, some of which are at shallow depths in geologically recent Pliocene deposits. Oil oozes to the surface through these faults in the sediments, gradually becoming more viscous on contact with the air, as the result of oxidation and the loss of its volatile constituents, and becomes tar, or asphalt. It pools at the surface, where (because it is water-repellent) it is covered by rainwater. These water-covered tar deposits look superficially like waterholes, but any animal that comes to drink at them runs a serious risk of becoming trapped. Fur and feathers are easily clogged by the tar, and animals may stray too far out into the waterhole in order to escape predators or to take a drink.

Due to the continuous shifting of the ground in the region of the San Andreas Fault, a Pleistocene tar pit now lies exposed in the middle of urban Los Angeles. The tar pit at Rancho La Brea is perhaps best known for its spectacularly preserved mammals such as giant ground sloths, mammoths, bison, large-headed "dire" wolves, and truly gigantic condor-like vultures that are known by the appropriate term of teratorns or "terror birds." But the species most often associated with Rancho La Brea is the saber-toothed cat *Smilodon*. It is usually represented as lion-sized, with unusually long teeth, but recent analysis has shown some unexpected features of this highly successful predator. It was up to 50 percent heavier than a fully-grown lion, and the bones of its front limbs were twisted in response to a massively expanded front limb musculature. Shorter hind limbs made it incapable of running down its prey in the same manner as a modern lion. Instead, this animal relied on brute strength to bring down its prey—mammoths and ground sloths—which were far bigger than any modern lion could tackle. The great strength of the forequarters of *Smilodon* and its remarkable teeth functioned together: it grasped its huge slow-moving prey in a powerful immobilizing embrace to minimize any movement that could break the saber teeth. These were plunged into the prey and used in conjunction with the lower jaws to lever a huge chunk of flesh out of the prey in what has been called the saber-toothed shear-bite. The victim rapidly bled to death—almost immediately if the bite were on the neck.

SABER-TOOTH CAT

The tar pits of Rancho La Brea in the center of Los Angeles preserve a diverse community of plains-dwelling animals. Some of the most abundant fossils, and perhaps the most famous, are thosc of the spectacular saber-toothed cat *Smilodon*. This powerful predator used its immense canine teeth to penetrate the thick hides of its huge prey.

ISLAND LIFE: BIG AND LITTLE

The Panamanian Isthmus turned South America from an immense island into a conjoined continent with North America, bringing previously isolated animals into contact with invaders. The isolation of animals on islands has a major impact on their evolution; some biological features of this are characteristic and can be identified in the fossil record. The evolution of miniaturized animals is one such characteristic and is a major demonstration of how evolution can work. Geographical isolation allows a gene pool to become different from the "normal stock" of animals from which the island types descended while still in contact with the original continental population. However, the key to understanding island dwarfing is to consider the effects of a confined area in which many requirements for life are reduced to critical levels: a drastically reduced foraging and breeding space, diminished supply of nutrients and fresh water, and limited terrain in which to escape predators. Generally, island biological effects are most marked on terrestrial animals such as mammals and reptiles; birds are less prone to geographical isolation by a sea barrier.

In terms of evolution, isolation works wonders: it leads to dwarfing of big animals and gigantism in small ones. Herbivores may become small because of limited food resources, but small animals such as rodents become large only when stranded on islands that have an absence of predators. The Pleistocene epoch was the time of the giant lemur *Megaladapis* on the island of Madagascar, a bear-sized animal that fed on fruits and leaves and evolved as the result of a lack of really large predators. A living example of giantism is the Komodo Dragon, the largest species of lizard. An absence of large mammalian predators on the Komodo Islands of Indonesia has resulted in its becoming the top carnivore there.

The effect of island giantism is complemented by the effect of island dwarfism, such as in the dwarf elephant, whose fossils are found on the island of Malta. Elephants stranded during the Pleistocene, facing few predators, developed dwarf strains, some no bigger than a St. Bernard dog. Other examples are dwarf mammoths on Wrangel Island off the coast of Siberia, and the more widely known Shetland pony.

ANOTHER direct effect of the subduction zone between the Nazca and South American plates is the Isthmus of Panama. This land connection between North and South America—which makes up Central America —is essentially a mountain chain that is more or less confluent with the Andes mountains in South America: the mountain-building episode that uplifted the Andes also threw up the Panamanian isthmus. However, the connection between North and South America was not a complete one, but rather a series of long islands and promontories stretching between these two continents.

The formation of the Gulf Stream has had major implications for the climate, ecosystems, and habitats of western Europe.

The formation of the Panamanian land-bridge only became complete during the Pliocene, about 3 million years ago, with a major impact on terrestrial vertebrate biotas of both North and South America. There were equally momentous effects that had impact on distant regions of the planet. Up until this time, the huge oceanic currents of the Atlantic ocean had flowed from east to west (due to the rotation of the Earth and its atmospheric directions) and passed more or less unhindered from the Atlantic Ocean into the Pacific Ocean by way of the open seaway between the islands of the Central American mountain chain. But when the Panamanian Isthmus completely formed it blocked the passage of Atlantic waters into the Pacific and redirected these warm equatorial waters northeastwards along the edge of North America. This diversion is the basis of the warm ocean current known as the Gulf Stream that bathes western Europe and the British Isles.

The importance of the circulation of the Earth's ocean currents coupled with those of the atmosphere explains the remarkable rapidity with which some of the climatic changes of the Pleistocene epoch have come about, and the initiation of the Gulf Stream is just one example of this. The warmth of the north Atlantic Ocean arises from warm, saline waters traveling north at intermediate depths of about 800m, eventually surfacing near Iceland. They then cool and sink (cold water being denser and

WATER BOUNDARY

(Above) This image of the Atlantic Ocean from space shows the boundary between the warm, fast flowing water of the Gulf Stream (lower half) and the cooler, calmer coastal waters of the eastern USA. The Gulf Stream has been crucial in the development of northern hemisphere ecosystems. It has not only kept Europe much warmer than its high latitude would otherwise indicate, but also prevented a southward extension of the sea ice during the Pleistocene ice ages. Therefore, the North Atlantic ice sheet never spread into southern Europe.

THE IRISH DEER

(Right) The so-called "Irish Elk" was, in fact, a deer; but with antlers spanning as much as 10ft (3m), it was truly a giant. The skull is unusually wide in order to accommodate the stresses imposed by such huge antlers, as are the neck bones. The antlers alone must have required a considerable energy intake to maintain their health and growth.

ISLAND CLIMATES

Hippopotamus fossils from British interglacial Pleistocene deposits (c.120,000 years ago) testify to the heating effects of the Gulf Stream on this region, but by 20,000 years ago, glacial ice had spread over Scotland, northern England, and most of Wales and Ireland, with a corresponding change in the fauna. The fossil remains of cold-adapted mammals such as bison, reindeer, and elk have been found in some abundance—evidence of a much colder climate than today. Lower sea levels allowed passage of these mammals from mainland Europe, and occasional land-bridges allowed some British mammals to cross into Ireland, including, by the end of the period, the Irish giant deer *Megaloceros* (below).

heavier than warm water). Before the Isthmus of Panama was completed the water would have been less saline, having mixed with water from the Pacific, and would have reached the Arctic Ocean before sinking. When the isthmus closed, the drying effects of the trade winds led to greater evaporation and higher salinity, which caused the water to sink sooner, thus isolating and cooling the Arctic. This cooling may have initiated the ice age.

During the Pleistocene glaciations, the formation of immense ice sheets in the north had extremely significant effects on the Gulf Stream which at that time had been in existence for less than one-and-a-half million years; initially ice floes occupying much of the north Atlantic pushed the Gulf Stream towards western Europe and the Iberian peninsula. However, during the glacial maxima, very strong latitudinal temperature gradients set up by the formation of extensive continental and oceanic ice sheets strengthened the trade winds of the northern hemisphere and pushed the warm equatorial currents that formed the basis of the Gulf Stream back into the southern hemisphere; these currents were also blocked from reaching Europe by the southward extension of the fully-formed ice sheets. Fossil evidence gained from marine invertebrates such as forams and radiolarians, from Atlantic sediments, suggest that the "conveyor belt" of currents that initiated the formation of the Gulf Stream were essentially switched off during major glacial episodes, further reducing heat transfer to the north and contributing to the formation of ice sheets.

During those times when the Gulf Stream was at its most active, Europe and the British Isles were kept much warmer than their northerly position would otherwise allow—an effect that is still very much with us today. It also, at times, had a major impact on animal life in the British Isles, the climate of these small islands and their mammalian fauna changing rapidly and dramatically under the competing influences of the cold polar front and the warm Atlantic waters.

Faunal diversity was greater on the European continent than in Britain which, in turn, was greater than that of Ireland. British temperate/interglacial faunas—prior to the last interglacial—are very similar to continental faunas. Until the early mid-Pleistocene there was also contact between Ireland and mainland Britain via land-bridges, as indicated by the distribution of large carnivores such as hyenas and bears. The earliest indications of British isolation are in the last interglacial (120,000 years ago), in which animals found in Europe—pine voles, extinct rhino, horses, and humans— are absent in Britain. British Last Cold Stage (LCS) faunas were similar to those of Europe, suggesting unimpeded migration. But the Irish LCS faunas are impoverished (no woolly rhinos or humans) and imply no connection with mainland Britain. Continued isolation, and thus faunal impoverishment, is indicated by the lack in Ireland today of such animals as voles, frogs, and snakes, which are present on the mainland.

IRELAND
GREAT BRITAIN

northern limit of hippo (last interglacial)
southern limit of ice sheet (last glacial advance)
possible land bridges

Quaternary mammalian sites
limestone areas with cave deposits
river terrace deposits
bogs and fens
marine crags

PLEISTOCENE

BECAUSE of the physical nature of the environments in which early humans lived, and the low preservation potential of these regions for fossils, the evidence for human evolution is patchy. The early stages, especially, are poorly known, and theories for the origins and subsequent dispersal of anatomically-modern humans are among the most contentious of all scientific arguments of any discipline. However, to begin with, most paleoanthropologists agree that there are two separate lines of hominid evolution, the australopithecines and *Homo*, which took place in east and southern Africa between approximately 5 and 2 million years ago. All these hominid species followed the same basic design: they were essentially small bipedal apes with small brains, large cheek-teeth for processing tough plant foods, and slightly curved hand and foot bones—suggesting a partly arboreal lifestyle.

About 1.9 million years ago in Kenya, a new, taller, upright hominid appeared.

Then, about 1.9 million years ago, in the latest Pliocene, a new hominid species arose, the fossils of which showed significant anatomical advances over those of the fossil hominid *Homo habilis* (the oldest fossils of which are close to 2 million years old). The most perfectly preserved specimen of this new hominid species was collected in 1984 by Richard Leakey near Lake Turkana in Kenya. The specimen—known as "Turkana boy" and given the species name *Homo ergaster* ("workman"), though originally called *Homo erectus* because of its upright stance—was that of an adolescent male hominid. He stood about 5.3ft (1.6m) tall and had a brain volume of around 51in^3 (830cc). The skull was clearly more primitive than that of *H. sapiens* (modern humans) in its large eyebrow ridges, heavy lower jaw, wide nasal opening, rounded facial region, and lack of a bony chin. Nevertheless, the skeleton appears to be more or less modern, and it walked upright on its two legs. *H. ergaster* sites show that this hominid manufactured advanced tools and weapons, ate meat, and that it foraged and hunted in groups in a cooperative manner. Similar fossils have been found in North Africa, Asia, Indonesia, and Europe, mostly dating from about 1.25 million years ago, and it is these later specimens that are assigned to *H. erectus*. In 1995 a specimen of *H. erectus* was found in China. It has also been dated at 1.9 million years, suggesting that the species has a longer evolutionary history than previously thought and that *H. erectus* emigrated from Africa closer to 2 million years ago—twice as long as was once believed.

Homo ergaster/erectus improved on the technology of its predecessor, *H. habilis*. Most of the tools used by the latter are simple and rough, consisting of no more than rounded pebbles, typically with only a single cutting edge; they are Oldowan tools, named for the site at Olduvai Gorge in East Africa. A number of *Homo erectus* sites in Europe have produced characteristic sharp-edged tools—known as Acheulean, after Saint-Acheul in France—and associated human remains that date from about 780,000 to 530,000 years ago. The presence of the tools may indicate that during the Middle Pleistocene of Africa and Europe there was a unique radiation of humans that were more derived than *Homo erectus*, but ancestral to subspecies of *Homo sapiens* of Europe.

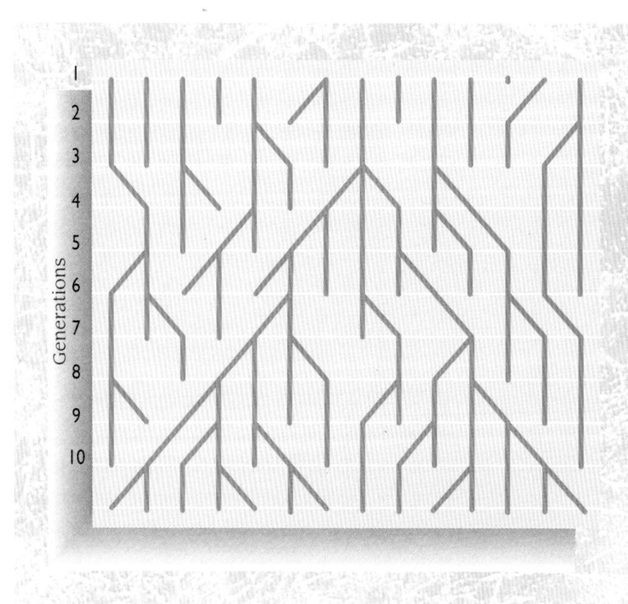

MITOCHONDRIAL EVE

The mitochondrion is one of the semi-autonomous organelles found within the eukaryotic cell and contains its own DNA. Mitochondrial DNA is passed on only through the maternal line (red). Lineages that just produce males (blue) disappear, so that, in time, it is possible for a population of diverse MtDNA types to be replaced by a descendant population of only one type, through a process of random variation.

OUT OF AFRICA

(Right) Modern humans, *Homo sapiens*, arose in Africa c.150,000 years ago and dispersed throughout the Old World and into Australia c.50,000–35,000 years ago. As they did so, they probably caused the disappearance archaic humans such as *H. erectus*. Some 18,000 years ago the last glacial was at its height, with maximum spread of ice sheets and low sea levels. These conditions were an added catalyst for the development of a diversity of specialized hunting techniques, tool and weapon development, and survival strategies.

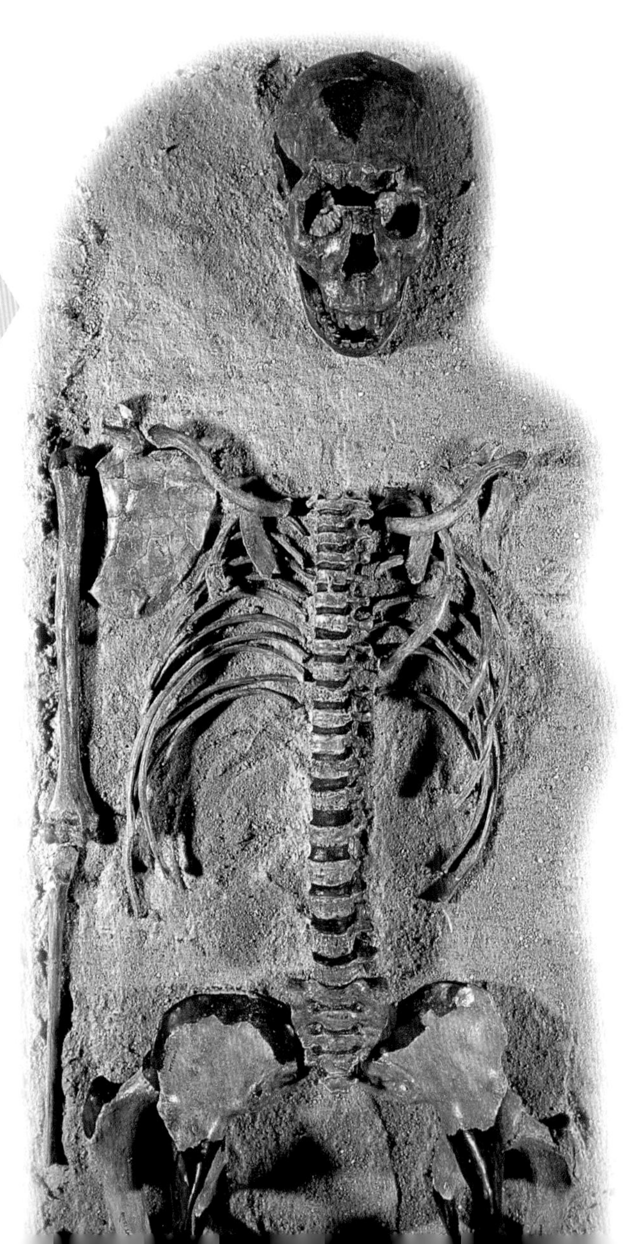

"UPRIGHT MAN"

(Left) *Homo erectus* ("upright man") was the first geographically widespread human being. Its skeleton shows anatomical advances such as a larger cranial volume, suggesting a larger brain. This skeleton, of its ancestor *H. ergaster*, which is one of the most complete early hominids ever found, stands about 5.3ft (1.6m) high and is that of a male, as shown by the pelvic structure. He appears to have been about 12 years old when he died.

IN 1987 and 1991, studies of the molecular DNA of *Homo sapiens* made some fundamental claims: that all modern humans are closely related, that is, they share a common ancestor; that the ancestor originated in Africa; that this origin occurred no more than 200,000 years ago; and that archaic species made no contribution to the modern gene pool. The studies relied on that portion of our genetic material known as mitochondrial DNA (mtDNA), which is found in the mitochondria of cells. MtDNA evolves more rapidly than nuclear DNA; it can be used as a molecular clock to show the differences between populations due to accumulated genetic mutations that have taken place since a past point of divergence. The longer two populations have been separate, the greater the amount of genetic difference. MtDNA is inherited only from one's mother, so the hypothetical ancestor became known as "Mitochondrial Eve" and the hypothesis the "Out of Africa" hypothesis.

Studies of mitchondrial DNA show that the first modern humans originated from a common African ancestor, then spread to all parts of the globe in a new wave of migration.

The scientists examined gene diversity among, initially, 147 people from different parts of the world and then a further 189 people, including 121 Africans from six sub-Saharan regions; they found that there was only about 0.4 percent variation among the individuals (indicating a recent origin), with a greater diversity among the Africans (reflecting their longer period of evolution). By comparison to the genetic variation among our closest relatives, the apes, this is very small, suggesting a very recent divergence of modern human populations. Both studies proposed that modern humans arose in Africa around 200,000 years ago, with a common female ancestor from whom we all derive our mitochondria; and that from this origin there were two main branches of the phylogenetic tree, one leading to six sub-Saharan mtDNA types and the other to everyone else.

Furthermore, if the descendants of that "mother" had bred with existing populations of archaic humans, ancient mtDNAs would have been incorporated into the gene pool. This has never been found in modern

BIFACE AXE

Acheulean tools (named for their first discovery at St Acheul in France) are characterized by having a long axis down which both sides have been worked to produce a crenulated cutting edge.

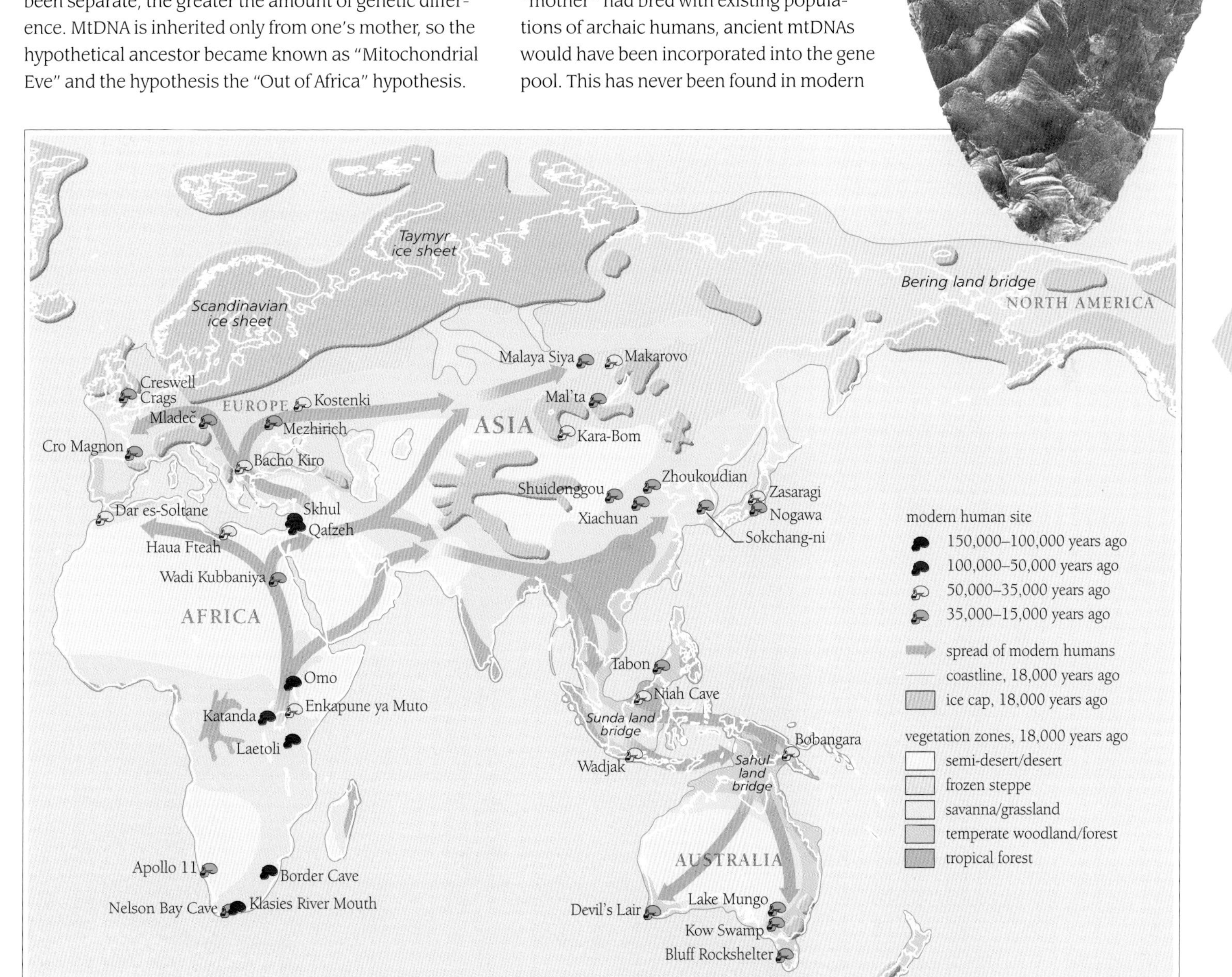

samples. The implication of this is that when modern humans spread out of Africa they replaced, rather than interbred with, the established *Homo erectus* populations that had preceded them.

As might be expected, the "out of Africa" model faced considerable opposition and criticism of some of its techniques. The major alternative—and opposite—viewpoint is the theory of multiregional evolution, by which *Homo sapiens* is supposed to have emerged throughout the Old World (Africa, Europe, and Asia) through the gradual evolutionary change of the established archaic populations, without either significant migration or replacement of existing populations taking place.

Variations on the multiregional model are more extreme. One suggests a scenario of geographically isolated groups dating back a million years, implying a deep genetic division between them, which is supposed to account for the range of human diversity known as "races." Although human populations have been isolated at times, this has never been for long enough to create genetic barriers to mating. Despite persistent attempts to categorize humans by race, this classification has no biological value: the fact is that more genetic variation exists within a population than between populations.

A multiregional mode of evolution, with a common origin in *Homo erectus* ancestors that left Africa at least a million years ago, would display extensive mtDNA variation in modern populations. This has not been shown to be the case. By contrast, and although it remains controversial, the single-origin model has been confirmed again and again by newer studies corroborating the genetic data. The fossil evidence is also strong.

AN EARLY OCCUPATION

Meadowcroft is a rock-shelter located about 30mi(48km) south of Pittsburgh, Pennsylvania. Its deep deposits contain 11 major layers, and radiocarbon dates indicate that the site was occupied intermittently from at least 14,000 (and perhaps even 17,000) years ago until about 250 years ago, giving it the longest history of occupation in the New World.

Skulls from Omo (Ethiopia), Laetoli (Tanzania), Border Cave, and Klasies River Mouth (both South Africa), dating to between 150,000 and 100,000 years ago, are all those of recognizably modern humans (*H. sapiens*), albeit with some archaic traits. Similar remains have been discovered in caves at Qafzeh and Skhul, in Israel; these specimens have high, short braincases with fairly vertical foreheads, only slight brow ridges, well-defined chins, and a cranial capacity of about 95in^3 (1550cc)—all modern features. They have been dated to 100,000–90,000 years ago and provide the earliest known evidence for the presence of modern humans outside Africa.

The oldest known *Homo* fossils in Europe were found at Atapuerca in northern Spain and date from 800,000 years ago. These fossils show marked differences with their African contemporaries and were given the name *H. heidelbergensis* by paleoanthropologists (after a

THE FIRST AMERICANS

(Below) The question of when the first humans crossed the Bering land-bridge and entered North America is hotly debated. The most widely accepted view is that a passage was found between the Cordilleran ice sheet in the east and the Laurentide in the west, about 13,000 years ago. However, some sites south of the ice, like Meadowcroft Rockshelter in Pennsylvania, have produced earlier settlement dates, and there are claims that some sites in Brazil and Chile are as much as 33,000 years old.

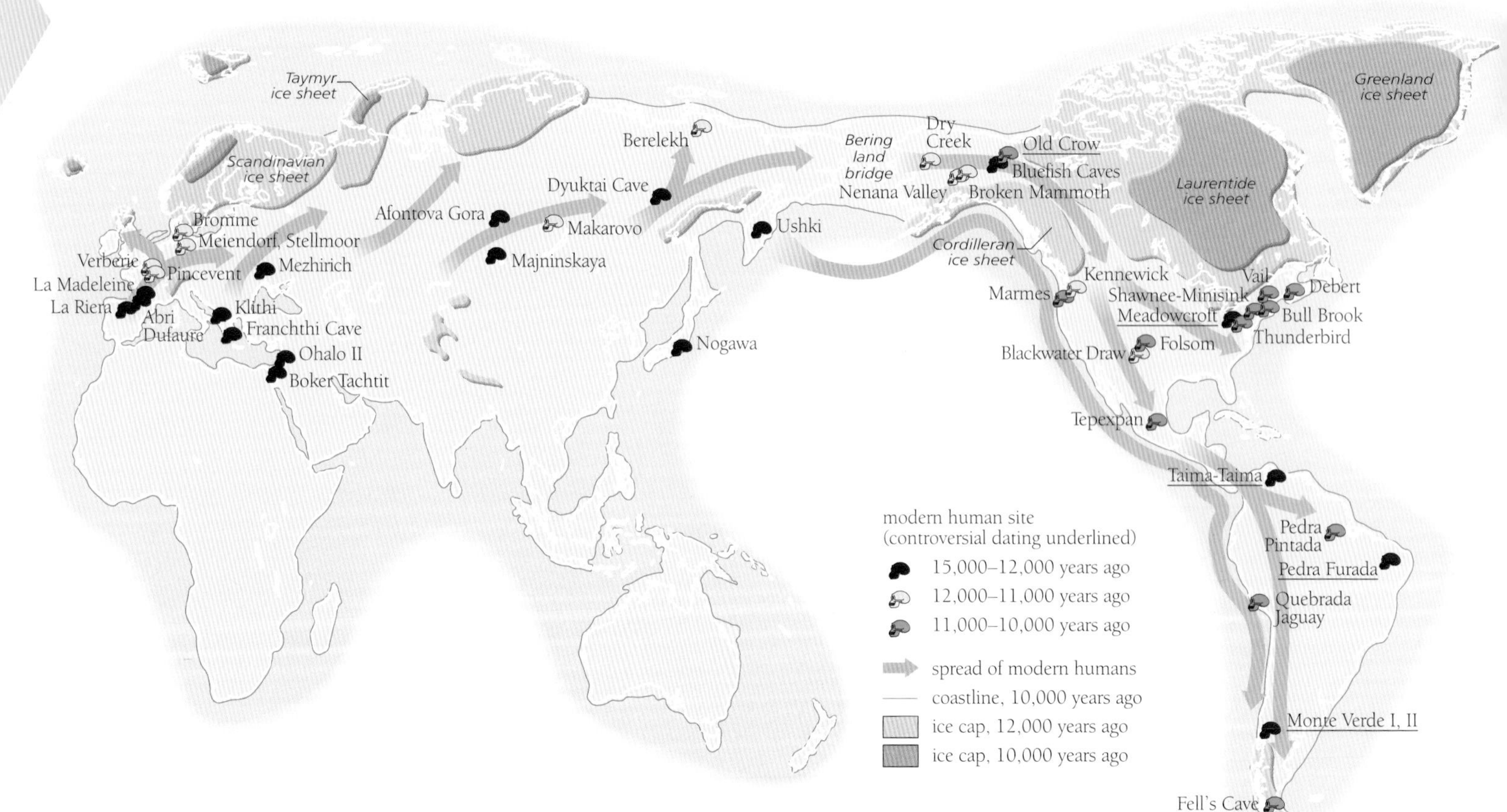

specimen discovered at Mauer, Germany). Over thousands of years, archaic humans in Europe developed characteristic features, termed Neandertal (after the Neander Valley, Germany, where they were first identified), many of which reflect adaptation to cold climates. Typically, they possess a powerful, stocky, short-limbed frame with a large projecting face and robust teeth. When *Homo sapiens* subsequently spread into Europe from the Afro-Asian region about 40,000 years ago, they displaced the local Neandertal populations, which had disappeared by 30,000 years ago. Modern humans reached southern Siberia by 40,000–35,000 years ago and Australia about 32,750 years ago. From Siberia they migrated across the Bering land-bridge into America.

DURING the period of the last major glacial, humans occupied much of the Earth. Unlike earlier forms of humankind they responded to the particular demands of their environment primarily through cultural and technological adaptation. One of these behavioral adaptations was complex cooperative hunting of very big game using weapons. The expansion of human populations and their effectiveness as hunters has been blamed, to a great extent, for the disappearance, between 12,000 and 10,000 years ago, of many species, particularly large herbivores, on many continents. This idea has become known as the "overkill hypothesis."

There is a coincidence between the spread of modern humans across the globe and the disappearance of large herbivores.

In North America 33 genera (about 73 percent) of large terrestrial mammals died out, including all the proboscideans (mammoths, elephants, and mastodonts), many horses, all camels, tapirs, and the huge, heavily-armored glyptodonts and the immense ground sloths. Many deer species became extinct, as did many of the predators that lived off this abundant source of meat. They included the North American lion (often referred to as the American giant jaguar), the saber-toothed cat *Smilodon,* and the scimitar-toothed cats such as *Homotherium*. In Australia five species of marsupials vanished, as did the giant monitor lizard; in South America 46 genera died out, including many of the forms characteristic of this continent such as litopterns, notoungulates, and edentates. Losses in Europe were less severe although woolly rhino, giant deer, and woolly mammoth became extinct; others—such as hippo and hyena—simply contracted their geographical ranges.

It is suggested that large animals are particularly at risk from intensive hunting because of their extremely long gestation period and slow growth, which means they attain sexual maturity only after many years. Losses would inevitably have greater effects than on small, fast-growing, highly fecund animals. Humans' excessive hunting of these animals eventually reached a point at which the population was not of viable breeding size, as removal of individuals exceeded the birth rate.

HALL OF THE BULLS

The modern humans that reached Europe are often referred to as Cro-Magnon people; they brought with them a characteristic array of advanced tools and covered the interiors of caves in France and northern Spain with equally characteristic paintings. The Hall of the Bulls in the cave of Lascaux, in south central France, is a significant source of archaeological, anthropological and semi-paleontological data. It provides a record of some of the animals which shared these environments with ancient humans about 16—17,000 years ago and corroborates fossil finding of aurochs (wild cattle), bison, horses, ibex and reindeer that come from this area and date from this time. A wounded bison suggests that these ancient humans hunted large-bodied animals as part of their food intake.

Supporters of the overkill hypothesis point out the good correlation between the spread of human populations and the extinctions, and that the only prey to suffer were the large-bodied animals that were attractive food sources. That the big game of Africa were not affected to the extent of similar populations elsewhere is explained by the fact that humans had evolved in Africa alongside their prey species, who had time to adapt to the threat they presented. In those parts of the world where humans were new, animals didn't recognize them as predators and were easy targets. In addition, humans are "switching predators": that is, as soon as one source of food disappears, they will turn to another.

There is also circumstantial evidence that climate changes, at least in Europe, had little effect on many animals and that earlier glacial retreats did not necessarily cause extinctions. Unfortunately, a lack of archaeological evidence of kill sites, and the fact that humans entered Australia (and possibly also America) long before the bulk of the extinctions there took place, reduces the strength of the overkill hypothesis as the main cause of extinctions of large Pleistocene herbivores. Recent re-analysis of the dating and geographical spread of the extinctions in North America suggests that what happened was in reverse from the overkill hypothesis. Although the fact that non-prey species also died out is an argument for the climate-change model, it also supports a combined theory or "keystone species" hypothesis. According to this, the disappearance of large herbivores, as a result of hunting, had an impact on habitats that was detrimental to smaller animals. It was probably a combination of environmental factors and hunting that led to the demise of many of the larger animals.

ONE OF the richest sources of Pleistocene mammal remains is the Rancho La Brea tar pits, on the outskirts of modern-day Los Angeles, where, mistaking them for water holes, unwary animals became mired in the sticky deposits. The quantity and diversity of animals found there is astonishing; the remains of over 4000 individuals of 40 different mammal species and over 100 species of birds have been recovered. The ancient landscape, about 2 million years ago, would have been similar in many respects to open grassland habitats of today. Grasses and broadleaf trees flourished, along with shrublike plants such as the still-living Californian sage. Plants and animals showed a diverse warm-weather habitat. The largest animals there were the huge mammoths, some of which displayed immense upwardly-curving tusks. These were an indirect indicator of the high nutrient load of this ecosystem, since tusks of such size require considerable upkeep in the form of food intake. Mammoths are not, however, the most numerous large herbivores at the tar pits of La Brea; these were the high-withered bison. Bison fossils are found in such abundance that their dentition has a produced some startling information regarding the age structure of their community. Dental anatomy shows that individuals are always of a distinct age: 1 or 2 or 3 years old, but not 1.5 or 2.5 years old. This, and the patterns of wear on their teeth, show that the bison were coming to La Brea only at specific times of year, and were absent during other periods. In other words, the bison were migrating just as they did until recently, before they were almost wiped out in the last 150 years or so.

Animals struggling in the treacherous tar deposits at Rancho La Brea would have attracted the attention of predators and scavengers.

1 *Bison antiquus*
2 *Canis dirus* (dire wolf)
3 *Teratornis* (vulture)
4 *Smilodon* (saber-toothed cat)
5 Imperial mammoth
6 Coyote
7 Heron

1

2

One of the unusual things about the fossil assemblage at Rancho La Brea is the disproportionate number of carnivores, which make up nearly 90 percent of individuals. Most abundantly represented is the dire wolf, closely followed by the saber-toothed cat *Smilodon*, of which thousands of partial specimens and three fully preserved specimens are known. Preservation of animals at different growth stages demonstrate how the little canine "milk teeth" of the cubs were replaced by the immense 9in (23cm) curving "sabers" of the adults. Along with *Smilodon* was the giant short-faced running bear, with meat-shearing teeth and powerful jaw muscles; it stood 6ft (1.8m) tall at the shoulder and probably weighed close to a ton. The kills of these powerful predators would have been observed by the vulture *Teratornis* ("frightful bird"), which had a 13ft (4m) wingspan. Even larger vultures are known from elsewhere in California.

❸

❹

❺

❻

7

THE EVOLUTION OF HUMANS

HOMO *sapiens*—anatomically modern humans— are defined by a thick cranial vault; a foramen magnum (the hole where the spinal column joins) situated directly beneath the skull; a reduced "snout"; a large brain; small cheek teeth, a well-developed chin; and no eyebrow ridges. *Homo sapiens* may claim a lineage of hominini ancestors, a small, specialized group within the primate subfamily Homininae, which separated around 5 million years ago from the gorillas and chimps that form the rest of the family Hominidae.

Cladistics show australopithecines to be the sister group to *Homo*. They had been assigned to one genus, *Australopithecus*, but in 1990 new finds showed that there are three genera: *Ardipithecus ramidus*, the oldest known hominid at 4.4 million years old; *Australopithecus;* and *Paranthropus*. *Ardipithecus*, from Ethiopia, has relatively large canine teeth and narrow molars with thin enamel, indicating a diet of leaves and fruit. These teeth are more hominine than any of the living apes. The foramen magnum is placed forward, showing that *Ardipithecus* must have walked on two feet.

Australopithecus anamensis was found by Maeve Leakey in sediments of 4.1–3.9 mya near Lake Turkana in Kenya. Anatomically closest to *Ardipithecus*, it is considered the most primitive australopithecine. *A. afarensis* remains have come from the Hadar region of Ethiopia, the most celebrated of which was the 40-percent-complete skeleton of a young female, nicknamed "Lucy," dating to about 3.18 million years ago. Later australopithecines include *A. africanus* and *P. robustus* from southern Africa; *P. boisei* and *P. aethiopicus* from eastern Africa. The *Paranthropus* species are, in many respects, the most interesting of all the australopithecines. These "robust" forms have broad shield-like faces, huge molar teeth, and a heavy ridge along the cranial midline that anchored powerful jaw muscles adapted to a tough diet.

There is great controversy regarding the recognition of several species of *Homo*. Modern anatomical and biomechanical studies show that the spectrum of variations among modern humans is greater than that of these fossil species. Many scientists regard the seven species of *Homo* as merely geographical variants of the normal anatomical variations that are usually encountered in any

HUMAN ANCESTORS

The human family tree appears rather sparse in that it depicts a taxonomically restricted view of a single group of fossils. Furthermore, humans are not very diverse and their fossils are extremely rare. Because the evidence is so meagre, any tree is, at best, a summary and many of the connections are open to debate. The main division is between the small-brained australopithecines and the large-brained, fully bipedal *Homo*.

1 Divergence from chimpanzees
2 Divergence and radiation of australopithecines
3 Appearance of robust australopithecines
4 Appearance of early *Homo* species
5 Acquisition of obligate bipedalism
6 Divergence of *Homo sapiens* subspecies
7 Coexistence of Neandertals and modern humans
8 Extinction or subsumation of Neandertals into the modern human gene pool

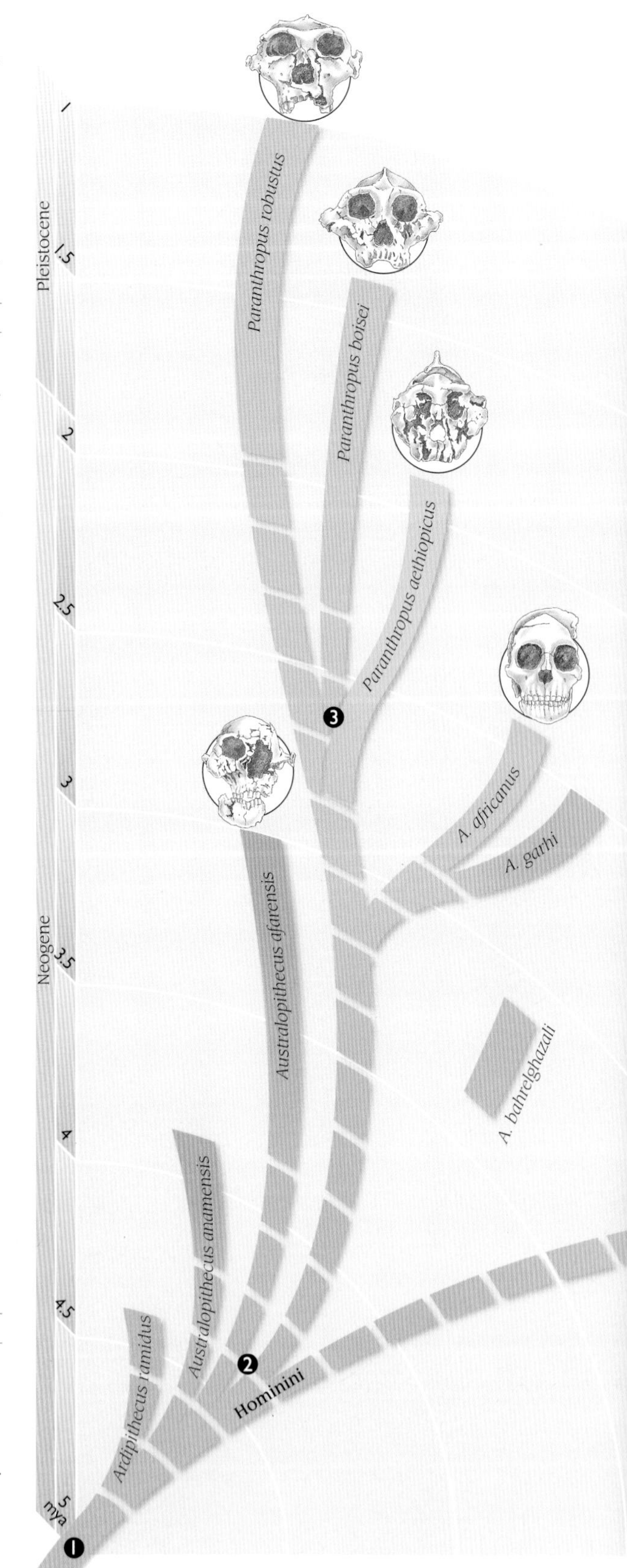

BEST FOOT FORWARD

Along with intelligence and manual dexterity, the most characteristic feature of humans is the bipedal form of locomotion, an important adaptation to a savannah environment (left). In order to pull the upper torso into an upright posture and position it directly over the legs, the buttock muscles must be large enough to lever the weight of the body into such a position. Accordingly, human buttocks are much bigger than those of the chimpanzee and gorilla. These muscular arrangements are reflected in the comparative anatomy of hominid pelvises; the human pelvis is more bowl-shaped than that of other hominids.

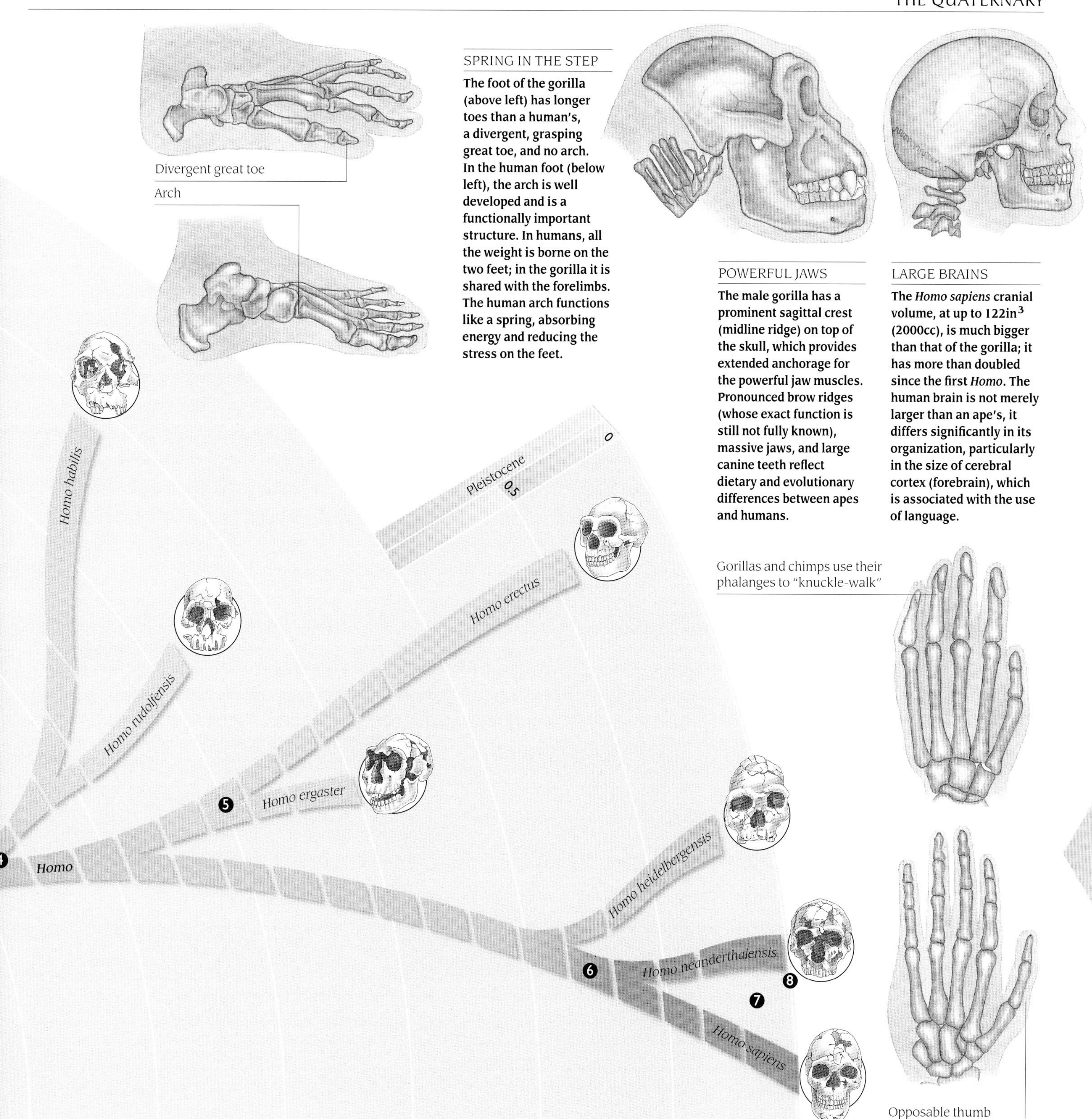

SPRING IN THE STEP

The foot of the gorilla (above left) has longer toes than a human's, a divergent, grasping great toe, and no arch. In the human foot (below left), the arch is well developed and is a functionally important structure. In humans, all the weight is borne on the two feet; in the gorilla it is shared with the forelimbs. The human arch functions like a spring, absorbing energy and reducing the stress on the feet.

POWERFUL JAWS

The male gorilla has a prominent sagittal crest (midline ridge) on top of the skull, which provides extended anchorage for the powerful jaw muscles. Pronounced brow ridges (whose exact function is still not fully known), massive jaws, and large canine teeth reflect dietary and evolutionary differences between apes and humans.

LARGE BRAINS

The *Homo sapiens* cranial volume, at up to 122in^3 (2000cc), is much bigger than that of the gorilla; it has more than doubled since the first *Homo*. The human brain is not merely larger than an ape's, it differs significantly in its organization, particularly in the size of cerebral cortex (forebrain), which is associated with the use of language.

population. Cladistic analysis gives the following order of derivation (advances in anatomy) for the generally recognized species of *Homo: H. habilis, H. rudolfensis, H. erectus*, *H.ergaster, H. heidelbergensis, H. neanderthalensis,* and *H. sapiens.*

H. habilis and *H. rudolfensis* came from Lake Turkana and Lake Rudolf in Kenya about 2.4 to 1.5 million years ago, and many anthropologists consider them to be the same. *H. habilis* was certainly the first of our line, while *H. erectus* was the first geographically widespread human species. It had a similar form in Africa, *H. ergaster*, which—again—some anthropologists consider to be the same animal. Recent discoveries in Spain put *H. heidelbergensis* (780,000 to 500,000 years ago) intermediate in form between *H. erectus* and *H. sapiens*. They may indicate a unique radiation of humans that were more derived than *H.erectus* but possibly ancestral to the Neandertal subspecies of *H. sapiens*. Neandertals have a more powerful but narrower jaw, thick brow ridges, wide nasal openings, a projection at the back of the skull and powerful cheek teeth. This suggests that the Neandertals were a separate species from modern *Homo,* since the level of development seen in these features is far outside the range seen in modern human populations.

PRECISION GRIP

The bones of fossil hominids progressively show an increased ability to oppose the thumb with the hand (above). The development of this "precision grip" has given humans the greatest manual dexterity among the primates. The opposability of the thumb necessitates strong muscles and these give us our characteristic "ball" of the thumb.

THE HOLOCENE

10,000 YEARS AGO TO THE PRESENT

LIKE the rapid climate fluctuations of the Pleistocene, modern conditions are not representative of typical patterns. The Earth is currently in the middle of an interglacial stage, which means that another ice age is likely to commence in about 5,000 to 10,000 years—unless our actions prove to have upset the equilibrium of the globe too extensively for this to occur. It is this unprecedented rapid change, driven by the agricultural and industrial activities of humans, that has made the Holocene unique in the Earth's history.

One of the problems in predicting future climate changes is the timescale involved. Accurate study of the climate began only in the mid-seventeenth century, and direct studies of the atmosphere only over the past few decades. Enormous swings in climate, the geosphere, and the hydrosphere can take place in shorter periods than this, as the geological record shows. Without data collected over a long time it is impossible to say anything about long-term developments. It is clear, however, that life on Earth is under threat from humans.

THE END of the Pleistocene epoch—and consequently the beginning of the Holocene epoch—is often taken as the melting of the great northern hemisphere ice sheets to approximately their present extent; it also included a component of the rising sea levels that took place at about the same time. This would mean that the Pleistocene ended about 8,000 years ago, but there is some argument about the date of this boundary. It is considered that the Pleistocene–Holocene boundary should be located at the mid-point in the warming of the oceans, in which case the ice age would have ended between 11,000 and 12,000 years ago. Despite these arguments, carbon-14 dating of old terminal moraines has shown that cold spells have recurred periodically in the Holocene epoch. In other words, the Holocene is an interglacial period. Climatic instability is characteristic of the ending of glacial periods and the formation of warmer interglacial times and represents the delicate balance between atmospheric, hydrospheric, and biotic global systems and their vulnerability to the effects of disturbances in our solar system. For example, it is now well known that there is a good correlation between periods of minimum sunspot activity and episodes of colder conditions on Earth. A well-documented period of cooler, drier conditions in the northern hemisphere occurred between 1540 and 1890 AD, when temperatures were often 2° to 4° cooler than today—enough to cause crop failures and freeze parts of major rivers in northern Europe. This difference may be slightly exaggerated by the current trend to global warming.

Comparatively warm, but with recurring periods of cold, the Holocene may be best regarded as an interglacial period.

KEYWORDS

BIODIVERSITY
BIOGEOGRAPHIC REALM
BIOME
GENE POOL
GLOBAL WARMING
GREENHOUSE EFFECT
K/T BOUNDARY EVENT
MASS EXTINCTION
OZONE LAYER
RELICT POPULATION

As the massive Pleistocene ice sheets in the northern hemisphere melted and retreated, sea levels rose spectacularly, which has had the added effect of causing truly

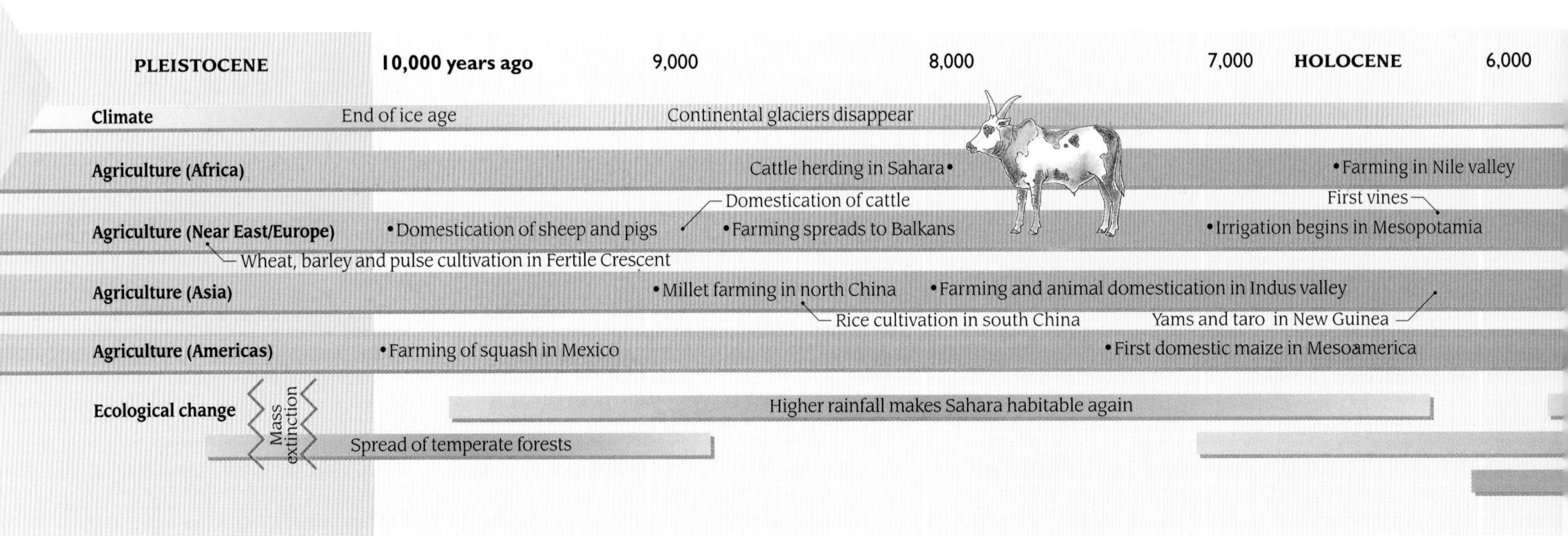

immense changes to topography and landscape. One place where this has been particularly evident since the last ice age is the Amazon region. Areas off the northeast coast of the continent of South America, which had previously been aerially exposed terrestrial habitats, were inundated by the rising Atlantic and submerged to a depth of about 390ft (120m), as they are today.

The British Isles have been more or less the same as they are now for most of the Holocene epoch: the melting of the Arctic polar ice sheets raised sea levels so that a deep body of water, the English Channel, once again stood between the British Isles and mainland Europe. The resulting isolation allowed animals in Britain to begin evolving differently from those in Europe, but because of the very short span of geological time—only 10,000 years—dramatic change is not yet apparent.

TODAY'S climate is still typical of the end of a glacial episode and the beginning of an interglacial. Modern conditions may be compared with those in the recent geological past from data gleaned from sedimentary, geochemical, isotope, and fossil analysis. The current interglacial has had major effects on the world's biotas. However, in the past 10,000 years, human influences have added to those of climatic effects and produced some changes of their own. For example, human clearing of forests in the first half of the Holocene may have resulted in the formation of many blanket mires in western Europe. Around the Mediterranean there is evidence of just how much oak woodland has been cleared by humans over the last few thousand years. This tends to mask the possible effects of climate on this region, and it is difficult to say which effects have natural causes and which are human-made.

Some changes of climate may be attributed to the current interglacial period, but the unprecedented effects of human activity must also be taken into account.

Data from such areas show that there has been an overall cooling of the world during the past 5000 years, but that this trend has not been a simple gradual pattern; it has taken place in a series of steps. One of the most pronounced of these sudden temperature falls in Europe occurred about 500 BC, causing a sharp increase in the rate of formation of peat bogs over the entire area. This event has left a permanent mark upon the stratigraphic profile of many bogs: a dark oxidized peat is suddenly replaced by the almost undecomposed vegetable matter typical of a fast-growing bog. Such stratigraphic indicators are useful for understanding the process of recent climatic and biological interactions that have taken place over the past few thousand years. Unfortunately, they tend to indicate only very localized changes in environmental conditions, and may simply be associated with local drainage patterns. More useful are oxygen isotopes taken from ice cores in Greenland, which accurately document climatic conditions over a wider area.

Some biological consequences of cooling climate over the past 5000 years may be seen in the changing distributions of plants and animals. In Europe, for example, plant species such as the hazel extended much farther north than they do today. Animals were similarly affected: the pond terrapin apparently had a very wide distribution in northwestern Europe, which is at odds with its current restricted distribution in the extreme south. In areas where a significant human population has existed for such a long time, it is difficult to be sure that such diminishing ranges of flora and fauna are really the result of climate or simply due to human interventions. In the case of Europe it is worthwhile to consider the effects wrought by ancient civilizations,

DATING THE RECENT PAST

Tree rings are a powerful tool for analyzing recent prehistory. The technique of dendrochronology relies on the comparison and correlation of tree rings in living and dead trees to standardize a time sequence for the past few thousand years. The relative thickness of individual rings faithfully records growing seasons and their conditions: whether they were unduly harsh and whether intervening winters were particularly cold. Compared with rock strata, tree rings are only useful for a very short time range, but they provide an accurate chronology for the most recent Holocene.

See Also

THE ARCHEAN: *Plate Tectonics*
THE JURASSIC: *Rift Valleys*
THE NEOGENE: *Grasslands, Herbivores, Land Bridges*
THE PLEISTOCENE: *Ice Age, Humans*

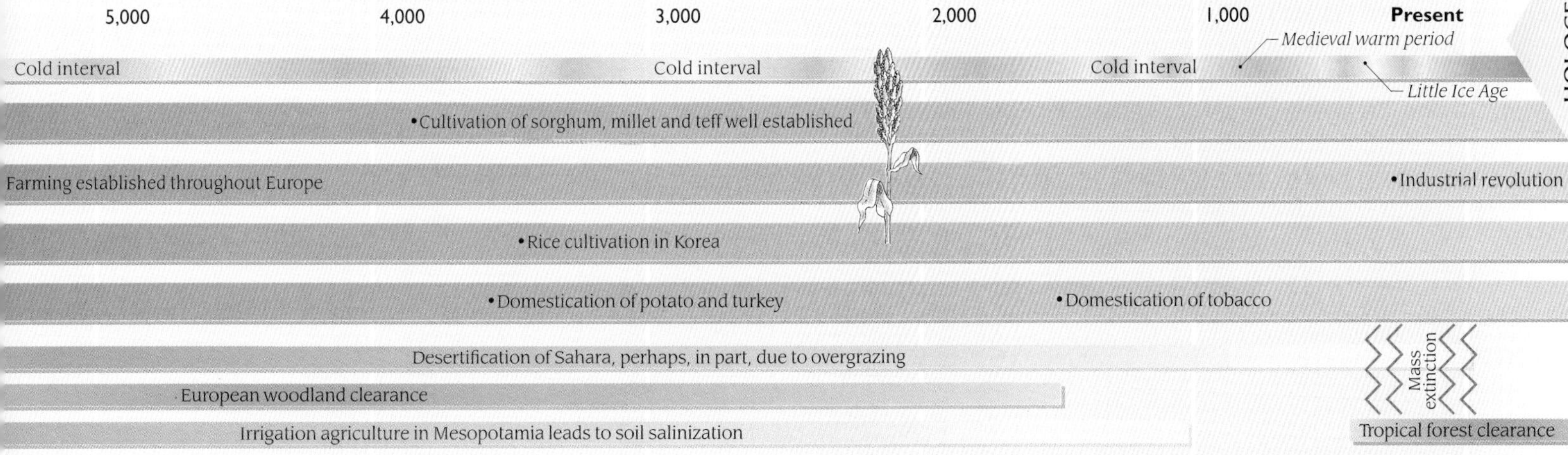

THREAT FROM SPACE

Impact craters from meteorite strikes have have been found at more than 100 locations around the world. The largest of these, 6mi (10km) in diameter, is believed to have struck the Yucatán peninsula of Mexico 65 million years ago, at the end of the Cretaceous period, leaving a crater 112mi (180km) across and causing mass extinctions around the globe. The crater shown here, in the desert of Arizona, is much smaller, and the effects of this impact would have been much more localized. The devastation caused by a future strike would depend on the size of the object and where it landed. At the very least, its destructive power would be comparable to that of a volcano erupting, which disrupts climate patterns over whole continents for months afterwards. Astronomers now track the path of meteorites passing near the Earth to try to predict if and when one might strike again. It is not yet clear what could be done to avert an impact if it were predicted.

50 MILLION YEARS FROM NOW

In North America, continued largescale regional activity associated with the San Andreas Fault will cause southern California to shear off and drift north, eventually colliding with or near Alaska.

PACIFIC OCEAN

tions, thousands of years after the first populations of modern humans and Neandertals.

Large mammalian carnivores show little change in their ranges purely as a result of climate fluctuations, but they have been affected by human activity in almost every part of the world during the past 10,000 years. Big cats became extinct in Europe about 35,000 years ago; the largest living European wild cat is the Iberian lynx, which stands a mere 16in (40cm) tall. Bears and wolves are almost totally absent, except a few individuals in the extreme east of the region. All around the world, large animals such as elephants, lions, and polar bears have remained only in sparsely populated areas. The dominant large animals of recent times are humans. However, humans are mammals, and the fossil record shows that mammals rarely survive more than four million years.

EXTRAPOLATING from what is known of the past movements of the continental plates is useful in predicting the future of the world. Based on this, a geologist might expect to see the following changes (though there is always a possibility that events may take an unforeseen turn—if there were a large meteor strike, for instance). About 30 million years from now, the movement of continental crustal slivers at the San Andreas fault in California will bring Los Angeles level with San Francisco, and eventually most of the state will drift north to collide with Alaska, producing California-like peninsulas as it bumps along. This will raise new mountain ranges all along the western edge of North America and as far as the Rocky Mountains, while the impact with Alaska will produce a Himalaya-sized range there. Meanwhile, older mountains will erode to much lower levels; these include the Appalachians and the Rockies. In another ten million years, the region just to the east of the Rockies would become wetter.

A million years is a mere blink in the span of the Earth's history. What might the world be like 50 million years from now?

Widening of the Atlantic along the Mid-Atlantic Rift will push North America farther away from Africa and western Europe. Iceland will become one in a straight line of volcanic island chains over the "hot spot" on the rift. Also, the Central American land bridge may fragment and disperse, re-establishing connection between the Atlantic and Pacific oceans. This event will alter more than geography: it will also have a powerful effect on climate in the western hemisphere, as the Gulf Stream circulation of warm water between eastern North America and western Europe is disrupted. Colder water from the Pacific may enter the system. As sea level rises in the region, the probable result of global warming and melting polar ice, the basins of the Amazon and Paraná rivers will become flooded, and tropical rainforest will regrow in the area.

On the other hand, the continued widening of the Atlantic Ocean will increase circulation between the warm salty waters of the Atlantic and the cool, less salty waters of the Arctic. At present the Arctic is a partially enclosed ocean with its salinity kept low by the constant influx of fresh water from surrounding rivers. As a result there is a constant floating ice cap over the North Pole. With the widening of the Atlantic, warmer saltier water will circulate, discouraging the formation of permanent floating ice. With no permanent ice cap, climates in the far north will be much milder than today.

Africa and Middle East
Antarctica
Australia and New Guinea
Central Asia
Europe
India
North America
South America
Southeast Asia
other land

DE-ICING OF THE ARCTIC

Removal of ice on the Canadian shield will uplift Hudson Bay, draining it and increasing Canada's land area. With the ice gone in a warmer climate, forests will appear in the Arctic Circle.

IN THE SOUTHERN HEMISPHERE

As the Caribbean plate continues to move, the Panamanian land bridge will disappear, separating North and South America, and rejoining the Caribbean to the Pacific. This event will also disrupt ocean circulation.

Western Europe will be pushed east as the Atlantic widens, infilling the North Sea with sediment from the delta of the Rhine River. The Mediterranean should disappear entirely as Africa pushes north, and be covered by a towering, intensely folded mountain range like the Himalayas. Similarly, eastern Africa splits off and drifts north, crashing into Iran and raising yet another tall mountain chain. The part of Africa between the rift valleys will become a partially submerged plateau, with isolated parts appearing as "islands" like the present-day Seychelles. The presently youthful Himalayas will still be tall, as India continues to be subducted beneath the edge of the Asian continent. Higher sea levels will flood the Ganges and Indus plains, making India appear more of a peninsula. Australia has moved so far north that it has collided with southeast Asia, raising steep mountains between them, and completing the great landmass of Africa-Eurasia, a new supercontinent. However, supercontinents do not remain so, and this one is already fragmenting into rift valleys along Lake Baikal and Lake Balkhas in central Asia.

Climates and biomes will have undergone many changes during this time. Antarctica, for instance, will have turned green as its ice melted, and forest will once again cover a great belt across northern Europe. Many species will have died, however, as continents moved through different zones, and in and out of ice ages.

TEMPERATURE increases associated with the current interglacial period have been supplemented by a rise that began in the early part of the nineteenth century as the Industrial Revolution gained momentum and atmospheric pollutants—chiefly in the form of burned coal products—began to fill the air with an extra load of carbon dioxide gas. In addition, there seems to be a gradual increase in global temperatures that began sometime in the mid-nineteenth century, although the upward trend was interrupted between 1940 and 1970 and shows as a plateau on a temperature graph. In the past 30 years, however, temperatures have risen sharply.

"Global warming" is a normal aspect of an interglacial episode, but it has clearly been accelerated by the burning of fossil fuels.

There is much concern about the "greenhouse effect," by which the increased quantity of carbon dioxide gas in the atmosphere acts as a blanket in the atmosphere, keeping heat close to Earth and raising the global temperature. Some thinkers argue that the greenhouse effect may be a natural phenomenon in the development of our planet and nothing to do with the appearance of industrialized societies, while others are adamant that the only possible explanation is human activity and the increased level of carbon dioxide gas in the atmosphere.

Higher temperatures and levels of atmospheric carbon dioxide would probably be very beneficial for plants, but not for humans. It is generally assumed that global warming would melt the polar ice caps, causing sea levels to rise and flood coastal regions around the world, with a catastrophic loss of living space and farmland. Ice is less dense than water, and so the melting of sea ice—which is already part of the sea's volume—would not raise sea levels; it would lower them very slightly. If continental ice at the poles melted, it would be a totally different matter. Continental ice is not a part of the sea volume. Melting of the Antarctic ice cap would add enormously to the sea volumes and sea levels might increase by approximately 200ft (60m). On the other hand, global warming might actually expand the Antarctic ice cap: with higher temperatures, evaporation of seawater increases, building up more water vapor in the atmosphere and increasing snowfall. It is difficult to predict which outcome is more likely.

If sea levels do rise with global warming, they will bring major regional climate changes. Areas with high rainfall, such as northwestern North America, will become wetter, again due to increased evaporation of warmer water. Tropical regions will experience more severe monsoons, because these storms are drawn inland by warm landmasses, which will have become warmer. Finally, higher temperatures will exacerbate the tendency to dryness in the interiors of large continents, compounding the expansion of desert as trees are cut down and erosion wears out the land.

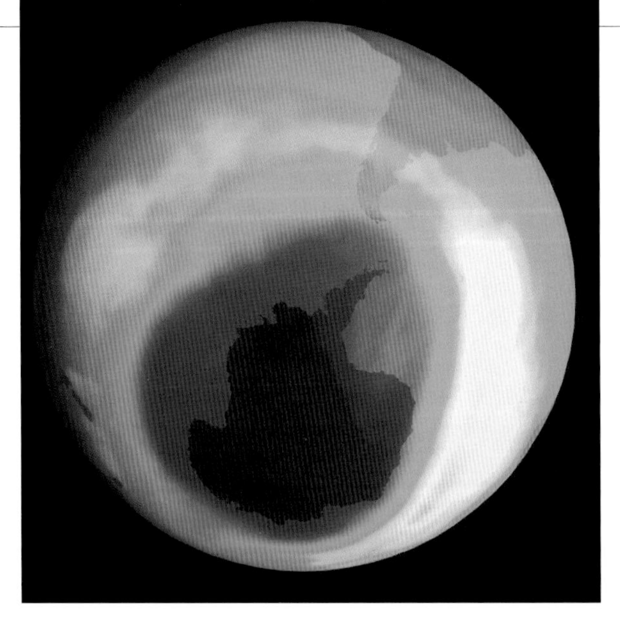

THE OZONE HOLE

A hole in the ozone layer, which blocks harmful ultraviolet radiation from the Sun, is visible over Antarctica each spring and seems to be growing; by 2000 it was the size of the United States.

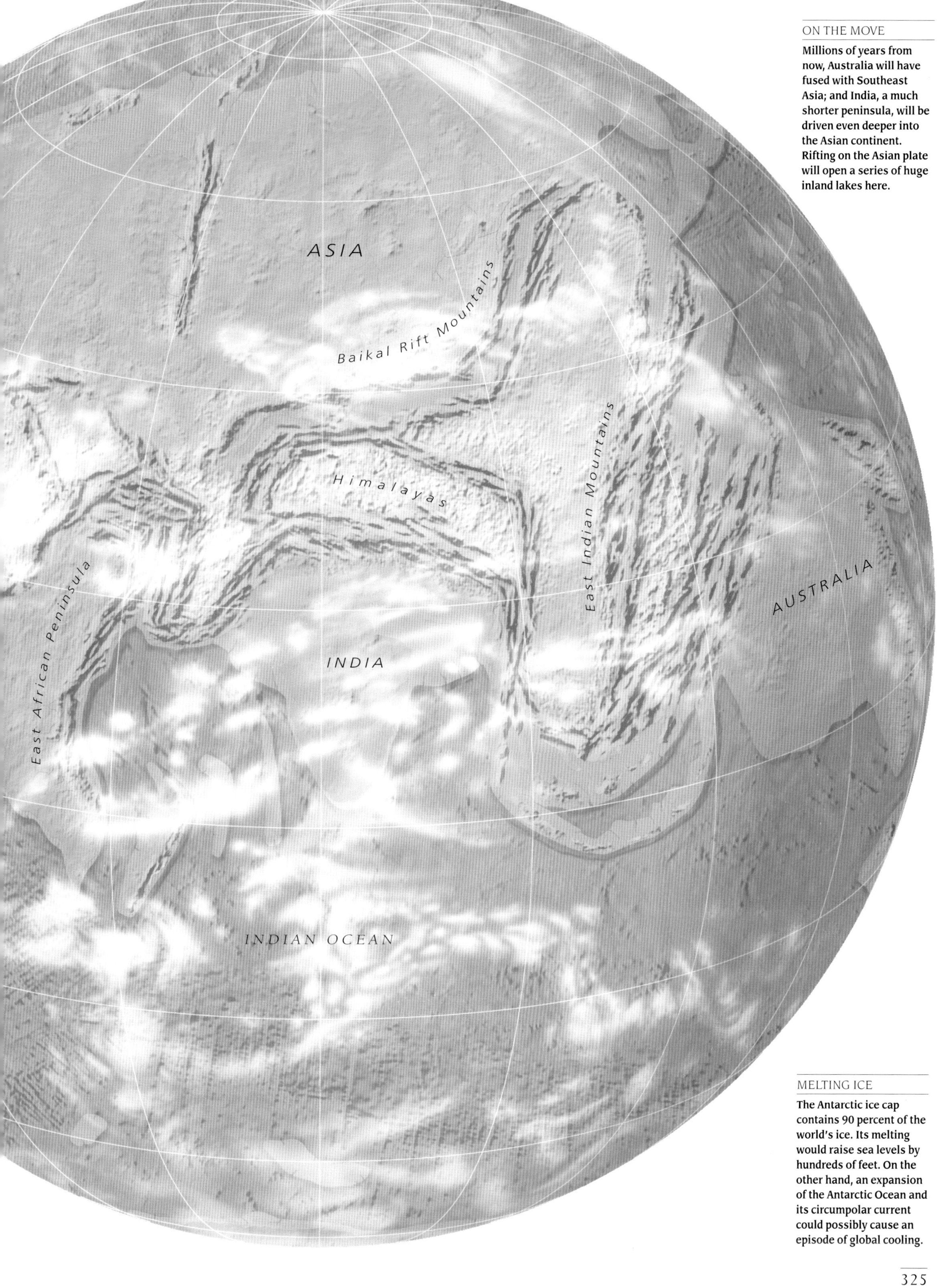

ON THE MOVE

Millions of years from now, Australia will have fused with Southeast Asia; and India, a much shorter peninsula, will be driven even deeper into the Asian continent. Rifting on the Asian plate will open a series of huge inland lakes here.

ASIA

Baikal Rift Mountains

East Indian Mountains

Himalayas

East African Peninsula

AUSTRALIA

INDIA

INDIAN OCEAN

MELTING ICE

The Antarctic ice cap contains 90 percent of the world's ice. Its melting would raise sea levels by hundreds of feet. On the other hand, an expansion of the Antarctic Ocean and its circumpolar current could possibly cause an episode of global cooling.

MOUNT KILIMANJARO

The volcanic mountain of Kilimanjaro is Africa's highest peak, formed during the Pleistocene by rifting in the Great Rift Valley of eastern Africa. This makes it a relatively young mountain in geological terms—and, typical of areas where rifting is occurring, Kilimanjaro was active until very recently.

If the greenhouse effect does not interfere with the onset of the next glacial interval, will there be another ice age? If so, then it is the northern hemisphere that would be most affected as ice sheets again descend on the major industrialized and food-producing regions of some of the wealthiest countries on the planet. Technology may become sufficiently developed to shield these nations from climate conditions; climate-proof research stations on Antarctica show that this is possible. Nevertheless, agriculture and life in western Europe, Scandinavia, Russia, and North America would suffer.

MODERN Africa provides a dramatic picture of tectonically-driven change in progress. The eastern part first attracted the attention of geologists in the 1920s because of its unique geomorphology, tectonic significance, and associated volcanic activity. Inland southern Africa is a classic flat plateau of wide topography and very high elevation. A large part of southern Africa is more than 6500ft (2000m) above sea level—although the Kalahari basin is lower than sea level—and this huge high expanse is characterized by an even more celebrated suite of impressive geological features that have drawn the attentions of geologists from most countries: the African Rift Valley system.

Signs of a future split in Africa are already apparent in the great Rift Valley of the east, a product of faulting.

The term "rift valley" was introduced in 1921 by J.W. Gregory, who recognized the rift valley of East Africa as a product of faulting. He defined the term as a long strip of land that had been let down between normal faults, or a parallel series of step faults. Parallel fault zones are about 30–50mi (50–80km) apart, and the down-dropped central portion is displaced down by about 10,000ft (3000m).

The Great African Rift Valley extends 1850mi (3000km) south from the end of the Red Sea through Ethiopia, Kenya, and Tanzania in a long fracture in the lithosphere of the African continental plate. A western arc goes from the southern border of Sudan and is occupied by Lakes Albert, Edward, Kivu, Tanganyika, Rukwa, and Malawi, which have filled some of the deepest rifts. One of the most famous landmarks of the region is Mount Kilimanjaro, rising about 19,000ft (5900m) above sea level.

Modern rift valleys are closely associated with uplifted crust that may be major lithospheric domes or long undulating upwarps. The width of rift valley faults (30–50mi/50–80km) approximates the thickness of the rifted crust. Rifts are usually bounded by major normal faults that may occur in "steps" and that produce frequent shallow earthquakes. Hot springs, associated with the high heat flow produced by the molten material rising from the mantle, may also be present. As mantle material rises the lithosphere bulges. Crust at the surface stretches and cracks open, forming the rift valleys.

Rifting in East Africa is part of a regional process that dates back to the Tertiary period, when Arabia began to separate from Africa. The East African rift is part of a

❷

❸

Extensive tool finds by riverside in Stone Age site

triple junction involving the Red Sea and the Gulf of Aden. It is very likely that the East African rift will become the zone at which African splits entirely, tens of millions of years hence. The Indian Ocean will flood in between the halves, forming a narrow extension of the ocean, as the Red Sea has flooded between Arabia and Africa.

VOLCANIC rocks in rift valley systems are usually extruded as outwardly-directed plateau flows. An example is the Trap Basalts that form much of the Ethiopian highlands. The total volume of rocks that are considered to be directly associated with the Eastern Rift in Kenya has been estimated to lie in the region of approximately 143,000 mi^3 (600,000 km^3), and there is an even greater volume in the Ethiopian Plateau. The oldest volcanic rocks in the Great Rift Valley system are from Ethiopia and have been dated to about 30 million years of age. These volcanic rocks have proved to be of immense importance, since they can be assigned highly accurate dates by isotope analysis and radiometric geophysical techniques. Their dating has provided extremely well-defined stratigraphical chronologies for the fossils that have been found in associated sedimentary rocks of the region. The geological conditions in the rift valley—eruption, uplift, and erosion—have alternately preserved and exposed layers of life dating back several million years. In parts of the rift system in southern Ethiopia, northern Kenya, and northern Tanzania, hominid and other faunal remains have been discovered; and examples of the former have shed new light and provided fundamental data on the early origins of anatomically modern humans.

Volcanic rocks, which are abundant in the Great Rift Valley, have been important in the search for human origins in the region.

From Plio–Pleistocene times onward, the lakes that formed in the downfaulted regions of the rift valleys have attracted human habitation. Then as now, the water supply would have supported vegetation and drawn all wildlife in the vicinity to drink at the plentiful watering-holes. Such regions are always a gathering point for large animals. In addition, these lakes and smaller bodies of water were essentially geographical regions of sediment deposition. Sediments at the bottom of the lakes, left undisturbed by humans throughout thousands of years, preserved the remains of animals very well. The more quickly the bodies were buried, the better the preservation. Layers of ash from erupting volcanoes in the valley added their protection. Modern Lake Turkana, which is more than 185mi (300km) long, and comparable in width to the English Channel, occupies an even larger basin in the generally arid eastern part of the Great Rift Valley. It formed in the vicinity of an ancient lake where the four-million-year old *Australopithecus anamensis,* the oldest australopithecine hominid, might have gone for water.

1
Lava flows
Traces of fire by river
Lava cobbles washed into rivers are later used to make tools

New layer of volcanic ash
Old lava flow
Acheulean hand axes exposed by river erosion
Vegetation colonizes layer of volcanic ash

EARLY AND RECENT

(Below) An intriguing assortment of fossils comes from the Chemoigut Formation in Kenya, almost equally spaced between Olduvai Gorge and Lake Turkana. (1) Rocks of the formation were deposited 1.5 million years ago. Fossils of crocodiles and antelope are common in lakeshore sediments, and early Oldowan stone tools are present. (2) Strata of volcanic ash 250,000 years old covered the area in deep deposits, cut into by channels of rainwater. Hand axes found along these channels are of the advanced Acheulean type of the Late Paleolithic ("Stone Age"). (3) About 5000 years ago, the modern landscape of the rift valleys began to take shape. Bones and artifacts suggest that a "factory" industry producing stone tools was operating by then.

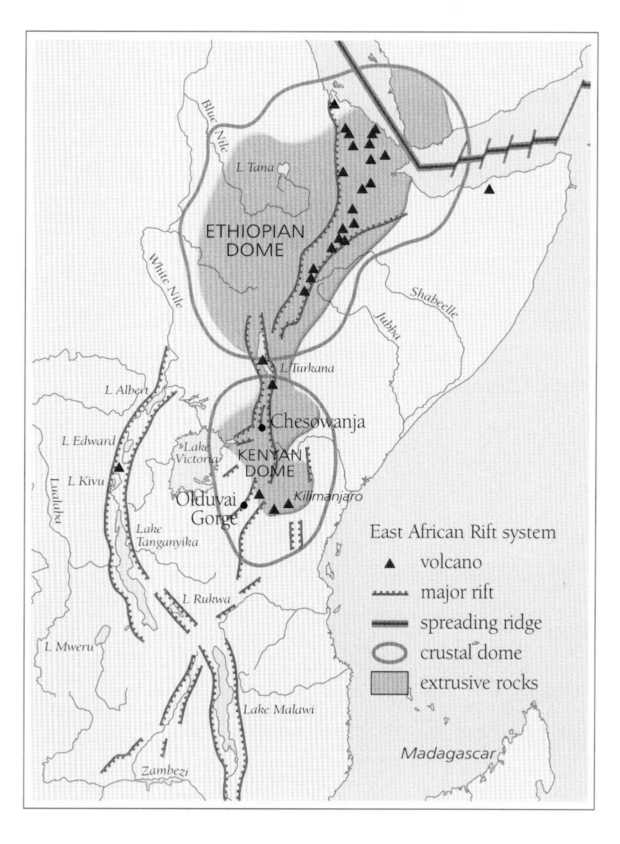

THE RIFT VALLEY

(Right) Rifting of the Earth's crust in eastern Africa was the direct result of the formation of two huge domes, the Ethiopian and Kenyan, during the Neogene. As the overlying crust expanded above the domes, tension caused the rifts to form. They have become progressively deeper since Miocene times, and have accumulated sediments and water. It was beside these lakes that the first hominids lived, and their fossils were preserved in the sediments.

CONTINUING tectonic activity in South America has produced a very different landscape. The Andes range stretches along the entire western margin of South America, from the Caribbean sea in the north to the Scotia sea in the far south—a distance of some 6200mi (10,000km), making the Andes the longest continuous mountain chain in the world. This great range is about 250mi (400km) wide and has a maximum elevation of about 23,000ft (7000m) above sea level. Cerro Aconcagua on the border between Chile and Argentina, at 22,834ft (6960m), is the highest mountain on the continent and in the entire western hemisphere.

Mountain-building is still in progress along the entire length of the west coast of South America. The Andes are one of the youngest ranges in the world.

Above all other features, the Andes are characterized by a huge number of very high and frequently very active volcanoes. Due to the bulge of the earth at the equator, the top of the immense volcano Cotopaxi in Ecuador, at a modest elevation of 19,347ft (5897m), is farther from the center of the earth than the top of Mount Everest, whose peak is 29,028ft (8848m) above sea level.

Most mountains are the product of collision between continents, but the Andes arose—and are still rising—as a result of subduction between the ocean plate of the southern Pacific and the continental plate of South America. A major tectonic system extends in a north-south trend along the western side of the whole of North, Central, and South America, beginning with the Juan de Fuca plate in the north, which is linked to mountain-building and earthquakes along the west coast of North America. This long coastal span is the demarcation zone of collision, where the ocean plates are being pushed beneath the edge of the continents by the spreading of the Pacific ocean ridges. The subducting Nazca ocean plate beneath South America is mainly responsible for the formation of the Andes Mountains; the Cocos and Antarctic plates to the north and south are also involved.

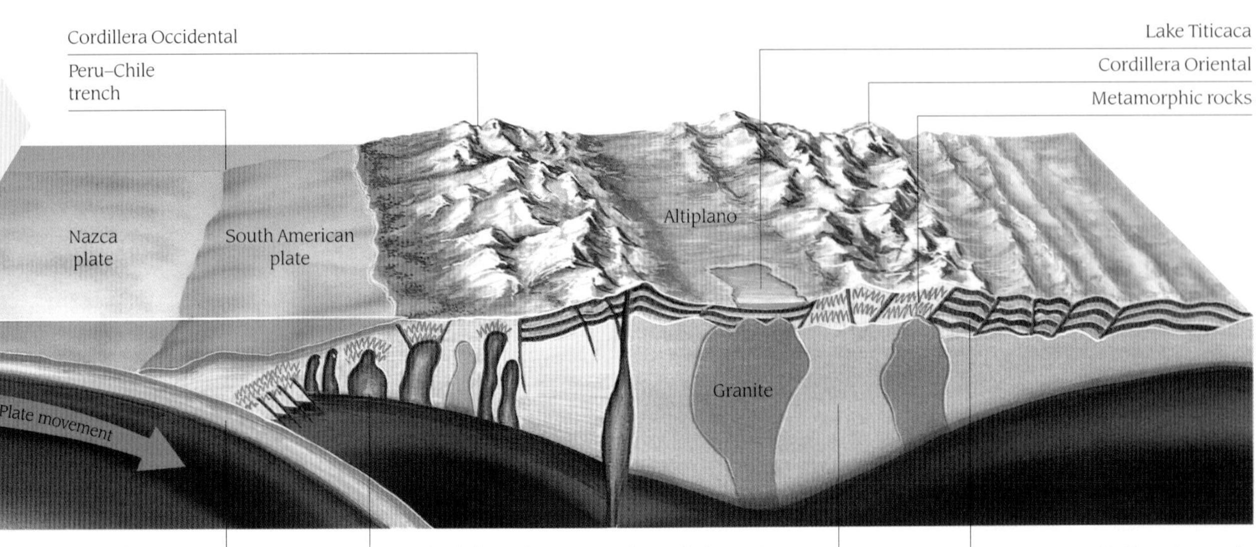

ANDEAN CORDILLERA

Subduction of the Nazca plate under the western edge of the South American plate is forming the Andes (left). Ocean crust and seawater are pushed to great depths, where they melt into a lighter magma than the surrounding material. The subducted material rises, pushing through the edge of the continental crust, and erupting as volcanoes. The range is unusually wide, because the plate subducts at a very low angle, causing the center of igneous activity to move inland by 125mi (200km).

YOUNG MOUNTAINS

The Andes (above) are still forming, and they have had little time to erode. Both factors account for the exceptionally jagged shape of their numerous peaks.

As the southern plates curve downwards into the upper reaches of the asthenosphere beneath the western edge of the South American plate, the constituent minerals of these oceanic crustal portions are melted progressively at greater depth. These minerals are often less dense—and therefore lighter—than the minerals that make up the main portion of the mantle. As they melt, they rise up from deep in the mantle, collecting and pooling under the edge of the South American plate. This process extends far inland into the continent: about 500 miles (800km), which accounts for the unusual width of the Andes chain. Eventually these hot, liquid mantle materials, derived from the long-subducted Pacific oceanic crust, burst through the upper continental crust of the South American plate as volcanoes. These are still active throughout the entire cordillera system, from Mount St. Helens in the northwest to the Caribbean island of Montserrat, which spewed ash for a whole year in the late 1990s, through the length of the Andes to the volcanoes at the tip of Chile.

The Andes gave their name to a type of lava, andesite, which is the most commonly erupted type of lava in the region, where it is a major constituent of the Earth's upper crust. Andesite is a fine-grained igneous rock formed under certain conditions of magma crystallizing in the upper crust, and its presence is associated with highly explosive volcanic activity. Less viscous, slower-moving basalt lavas of the Hawaiian volcanoes tend to ooze rather than run.

Another geological peculiarity of the Andes is the presence of gigantic granite batholiths ("deep rocks")—immense globules of ancient magma that are located deep in the uppermost layer of the Earth's crust. This is the location of the fractional melting that produces andesite lavas. The Patagonian Batholith at the southern tip of South America is about 620 by 620mi (1000 by 1000km).On the other hand, the ancient seafloor fragments known as ophiolites, which are typical of mountains produced by continental collisions, are rare in the Andes, because most of the seafloor has been pushed under the edge of the continent rather than raised up. Dating of the Andean batholiths show that the events of subduction in this area began about 130 million years ago, sometime in the Early Cretaceous period. This was when the South Atlantic ocean ridge began to form, and the Pacific correspondingly began to shrink in size and the Pacific ocean floor started to converge with South America.

THE CARIBBEAN region formed as a segment of Pacific oceanic crust overrode a mid-portion of the Atlantic crust along a zone that now extends northward from South America east of the Lesser Antilles (from the Virgin Islands down to the coast of Venezeula). The small plate so formed is called the Caribbean plate. The Greater Antilles—Cuba, Jamaica, Puerto Rico, and Hispaniola—are a northeastern extension of the Andean mountain chain, which extends out the north coast of the continent and into the Caribbean sea; the Lesser Antilles are a volcanic chain that formed later, but both are intricately linked to the Andean system.

One of the last features of the modern globe to take shape, the Caribbean region is likely to be a short-lived phenomenon.

During glacial and interglacial episodes throughout the Pleistocene epoch, sea levels fell by as much as 420ft (130m) as the ice sheets removed water from the oceans. But sea level increased as the melting of the polar ice caps inundated much of this high terrain, leaving the islands, and parts of the old Cordilleran mountains, exposed. The Greater Antilles display distinctive karst landscapes formed by the erosion of huge limestone beds typical of the region. Weathering of rock occurs particularly rapidly in the tropics, where the high humidity and rainfall help any overlying vegetation to decay faster. Sinkholes are another feature of karst landscapes, often measuring 60 to 180ft (100 to 300m) in diameter. Even larger depressions are known as cockpits, and these may be up to 125ft (200m) deep, as in the Cockpit Country of Jamaica. The karst landscapes there and in Cuba have been flooded by rising sea levels to produce a maze of underground caverns connected to the sea, decorated by

PANAMA
VENEZUELA
Nevado del Ruiz
Nevado de Huila
COLOMBIA
Nevado de Cumbal
Cotopaxi
ECUADOR
Sangay
BRAZIL
PERU
L. Titicaca
BOLIVIA
El Misti
Guallatiri
PARAGUAY
Antofall'a
Copiapó
CHILE
ARGENTINA
Aconcagua
Tupungatu
Tinguiririca
Azul
Villarrica
Osorno
Minchinmavida
Cerro Hudson
Cerro Lautaro
Monte Burney

major volcanoes
subduction zone
Andean cordillera

metallic ores
c copper
g gold
i iron
s silver
t tin
o other

nonmetallic minerals

MINERAL RICHES

The Andes are rich in minerals (right), both metallic and non-metallic, derived from geochemical changes that take place deep in the heart of the mountain belt. Uplift exposed the minerals to native Andeans, whose mined wealth was the target of European greed in the sixteenth century.

THE CARIBBEAN

(Below) The Caribbean formed over 125 million years at a subduction zone along the Pacific coast between North and South America. (1) About 100 mya, the subduction zone along the Farallon plate changed direction, and an island arc, the proto-Antilles (including Cuba, Puerto Rico, and Hispaniola), shifted east, beginning to consume the Atlantic plate. The northern arc collided with the Yucatán to form Nicaragua; the rest of Central America was a volcanic arc formed when the Antilles struck Cuba, creating a new subduction zone in the Farallon plate and isolating a piece that became the Caribbean plate. The Greater Antilles gradually separated. (2) By 10 mya, the Farallon plate had left two remnants, the Juan de Fuca and Cocos plates. About 3.5 mya, the landbridge separated the Caribbean and Cocos plates. The Lesser Antilles arose where the Atlantic plate is being subducted under the Caribbean.

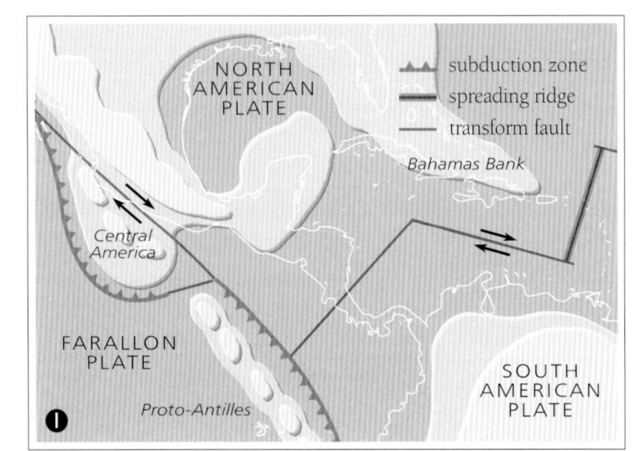

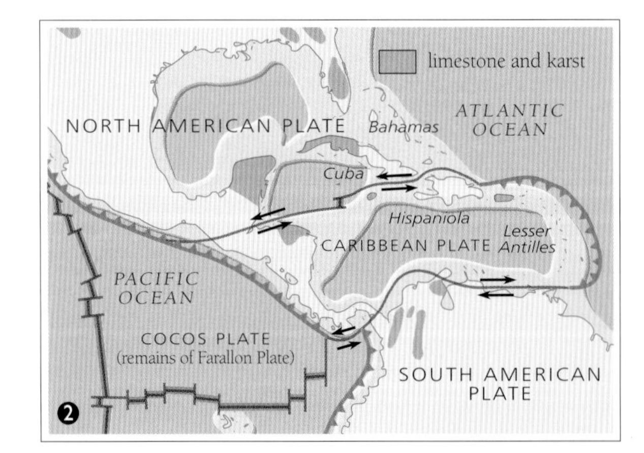

stalactites and stalagmites. A highly specialized light-hating biota has developed in these sea caves.

The development of slow-growing reefs in the Caribbean region shows how sea levels have changed in that area since the Pleistocene ice ages. Coral stops growing when exposed to air, and resumes when covered again with water, with younger segments growing on top of the earlier. As a result, a series of coral units are superimposed upon each other rather like sedimentary strata, following the pattern of rising and falling sea levels, and can provide clues to the past offshore conditions of the Caribbean islands. These "stranded" reefs are especially good on the island of Barbados.

During glacial episodes of the Pleistocene epoch, temperatures in the Caribbean region were near the lower limit for the growth of coral reefs, inhibiting their spread. The effect of this temperature drop can be measured by comparing the numbers of coral species in the modern Caribbean with those of the western Pacific and the Indian Ocean: whereas there are about 60 surviving Caribbean species, there are nearly 600 in the Pacific and Indian oceans of today. Such changes reflect the increased melting of ice sheets about 18,000 to 12,000 years ago. This melting decelerated from about 12,000 to 10,000 years ago, and speeded up again afterwards. Rates of sea level rise were as high as 8.2ft (2.5m) per hundred years about 9500 years ago.

KARST ISLAND

(Above) Cuba is an island jammed onto a large limestone platform, the Bahamas Bank, about 55 million years ago. Its karst landscape resulted from the passage of rainwater through cracks in the limestone, where either the rock itself or the rocks that fill the fissures are solvent in rainwater. Intricately angled fissures, long-term erosion, and solution have an effect not seen in other geological formations, producing relatively small but very sharp peaks. Karst derives its name from an outcrop in the former Yugoslavia.

SOUTH of the Caribbean is the huge delta of the Amazon River, which occupies the Atlantic side of the Andes chain. Here salt waters from the marine realm mix with fresh waters running off the higher regions inland to form enormous brackish-water environments. The rise and fall of sea levels during the past two million years has had profound effects upon life and geography in this region. Andean tectonic margins have also had a considerable effect on this region and its natural history. Active margins are the site of mountain building, but passive margins are low, hence the land slopes toward passive margins. In this case, the north part of South America is sloping down from the Andes mountain range to the Atlantic ocean on the eastern side of the continent. Much of the drainage in the Amazon basin is due simply to gravity, which has acted in concert with tectonic events in South America to produce the great expanse of Amazonian tropical rainforest that has been a feature throughout most of the Holocene epoch.

South America's Amazon Delta and its great rainforest are a side effect of regional tectonics and the fluctuating sea levels of the Pleistocene.

At first glance, a continental drainage map of the last 10,000 years might look somewhat haphazard, but in fact there are some consistent features: the largest rivers (those with the highest drainage) are always situated on sloping terrain where there is high rainfall. The Amazon discharges 3,675,691cu ft (104,083cu m) per second, approximately three times more than Africa's Congo River. The Amazon is a good model for an idealized continental river system, with a large collecting area in a nearby mountain belt, a trunk stream (river) flowing across a stable land platform or shield, and a passive margin to collect its output. The preglacial drainage of North America was also similar to this system, and what we see today is a feature that has only been fully developed in the Holocene: the river's main trunk arises in the Andes and flows across the downward-sloping passive margin of northern South America toward the Atlantic Ocean. More than a thousand smaller tributary rivers flow into the main trunk, many of which are major rivers in their own right. During the rainy season, water from high up in the Andes causes the main trunk to flood, covering up to 12mi (20km) of forest on each of its sides to a depth of up to 32ft (10m).

The glacial Amazon Delta was previously very much bigger because of the lowered sea levels that had caused an increase in the areal extent of the continent of South America. With most recent melting of the Arctic ice sheets about 10,000 years ago, this lower-lying area was submerged to form the current topography. Future rifting of the South American shield will cause further changes. The trunk of the main drainage system will be rerouted, draining new areas, and the new ecosystem will be most favorable to trees with water-resistant adaptations, such as mangroves.

Ecological changes have stimulated a number of innovations in the animals that inhabit the Amazon region. Perhaps the most unusual of these is found in the blind Amazonian river dolphin. Unlike marine dolphins, which have smoothly sloping foreheads, the forehead of the Amazonian dolphin slopes almost vertically down to its beak, and its beak is about twice as long. Its eyes are so reduced as to be almost non-existent: vision is useless in the muddy water of Amazonian estuaries, cluttered by millions of close-packed underwater tree roots. Echolocation is especially sharp in this curious dolphin, which can detect an object the size of a matchstick head in the huge volume of an Olympic swimming pool.

THE AMAZON BASIN

(Right) Amazon topography has always been connected with rising and falling sea levels. At times of glacial maxima, low sea levels exposed enormous areas of lowland tropical rainforest around the Amazon basin, whereas when the ice retreated, the high sea stands flooded these areas and diminished the areal extent of tropical lowland forests greatly, reducing them to isolated "islands." One product of submerging has been the estuarine Amazonian rainforest, in which most of the plants are adapted to living partially submerged in brackish water.

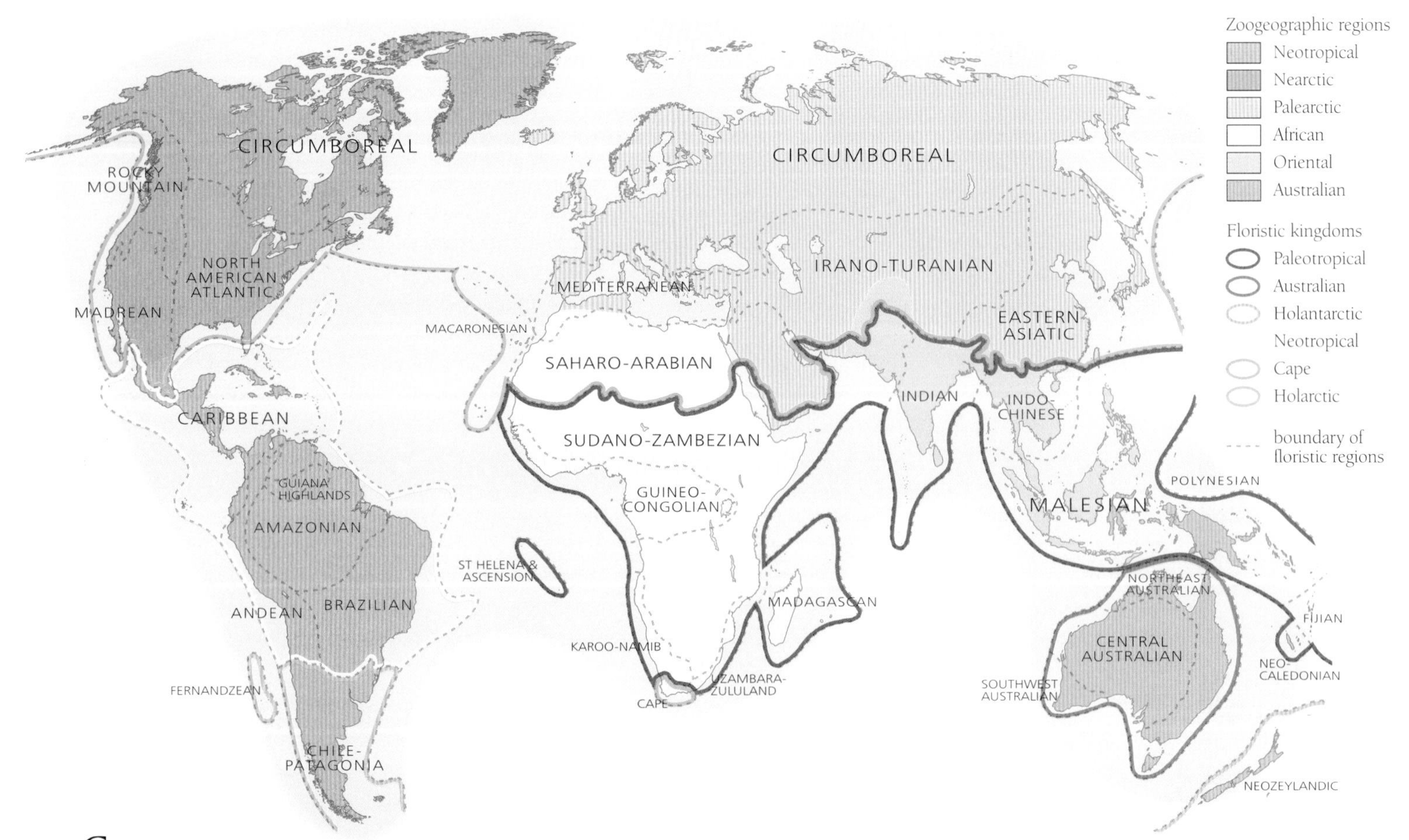

GLACIAL cycles were responsible for the distribution and population densities of plants and animals as they stood at the beginning of the Holocene. Biogeographic realms are the divisions of these biotas, shaped by the locations of the continents—mostly stable throughout the Cenozoic—and the recent climates on each continent, which have changed dramatically as ice advanced and retreated around the world. Biogeographic realms are based on the most prevalent groups of plants and animals in each. Among fauna, these tend to be the larger vertebrate animals such as birds and mammals, while in the floral realm they are usually angiosperms (flowering plants).

Climate and the physical barriers of mountains and oceans have shaped the presence of plants and animals in the world.

There are six recognized modern biogeographic realms, based on the division into Old World and New World refined by the nineteenth-century naturalist Alfred Russel Wallace, which in turn corresponds to the separation of the continents in the Cenozoic. South America and Africa were isolated earlier than the other continents and have very distinctive mammalian faunas; that of India and southeast Asia, while similar to Africa's, has diverged enough to merit its own realm, the Oriental; and the Australian, the most recent, is unique in being dominated by marsupial mammals. The Nearctic and Palearctic realms (North America and Eurasia), where the Pleistocene ice sheets were most extensive, both lost many species and have remained impoverished.

A species may be found in only one place (endemic), in all places (pandemic), or in a variety of places (cosmopolitan). Rodents are pandemic; they are present in all biogeographic realms. Marsupials are endemic, found in the Neotropics and Australia. Perissodactyls and artiodactyls (hoofed mammals), elephants, carnivores, and lagomorphs (rabbits) are cosmopolitan, living everywhere except Australia; as are primates, which originally occurred in all but the Nearctic and Australian realms.

Animals and plants that survive in icy conditions tend to be drought-resistant, because frozen water is not available. Such plants grow too slowly and sparsely to support large herbivores. Animals in this ecosystem must

BIOGEOGRAPHY

(Above) The biogeographic realms of the Earth are characterized by distinctive plants and animals. Temperature is the most important factor controlling what lives in each realm: more species, with greater diversity, live in tropical climates than in cold ones. Barriers imposed by land (such as mountains) and water also shape biogeography. Both climate and barriers may change over time.

STARLINGS

The European starling (left) is remarkably adaptive to new environments, which has not been good news for the native birds of North America, which it displaced. The starling was introduced into New York state in 1905. In just 50 years (below), it could be found in almost every part of North America, ousting the American bluebird as it spread. Huge flocks of these noisy birds are now a familiar sight from northern Canada down to the Gulf of Mexico and across to the Pacific coast.

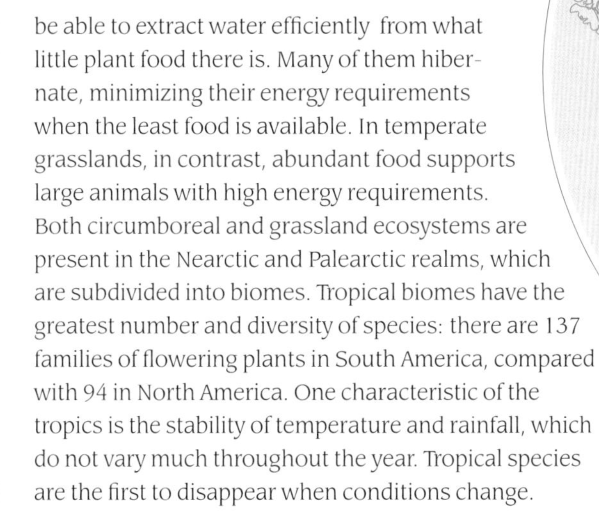

AUSTRALIAN REALM

(Below) The Australian biogeographic realm is particularly well defined. Its fauna is dominated by large lizards and marsupial mammals. Placental mammals have, for the most part, been introduced by Europeans in the last 200 years. The Australian flora is also distinctive for its high proportion of sclerophyllous plants, which have thick waxy cuticles to resist prolonged drought.

be able to extract water efficiently from what little plant food there is. Many of them hibernate, minimizing their energy requirements when the least food is available. In temperate grasslands, in contrast, abundant food supports large animals with high energy requirements. Both circumboreal and grassland ecosystems are present in the Nearctic and Palearctic realms, which are subdivided into biomes. Tropical biomes have the greatest number and diversity of species: there are 137 families of flowering plants in South America, compared with 94 in North America. One characteristic of the tropics is the stability of temperature and rainfall, which do not vary much throughout the year. Tropical species are the first to disappear when conditions change.

THE CHANGES wrought by plate movements and glacial cycles are now being compounded by human interference, as we (often unwittingly) distribute plants and animals into entirely different regions. Australia's isolation 50 million years ago contributed to the evolution of its unique marsupial mammals such as the kangaroo. Placental mammals have only been introduced by humans in relatively recent times.

Biogeographic realms of the future are being shaped today by human activity, which is having a profound effect on biomes.

Because of its long-standing isolation there are still very few big mammalian carnivores in Australia. In the past, considerable numbers of large reptilian predators in the dry climate tended to exclude sizeable mammalian predators. This was simply due to low primary productivity, which restricted the numbers of large herbivores and carnivores (which feed on herbivores). The introduction of domestic dogs about eight thousand years ago displaced two mammalian carnivores, the Tasmanian devil and the Tasmanian wolf (thylacine), from the mainland to the Tasmanian island, and contributed to the extinction of the Tasmanian wolf.

Introduced species may overrun an ecosystem as well as displacing its native inhabitants. The colonization of Australia and New Zealand by Europeans provides more examples. Australia had no hoofed grazing animals such as sheep and cattle, because its native grasses were more fragile than those in the North American and European biomes. The import of these animals quickly wrecked Australia's grasslands. Replanting the region with tougher imported grasses did no good, because they were poorly adapted to the hot, dry climate with its summer bush fires. Rabbits, another introduced species, experienced a population boom and depleted the ground cover still further—to the extent that a viral disease, myxomatosis, was deliberately introduced in the 1950s to reduce their numbers. Yet more destruction of the original Holocene ecosystem was caused by the introduction of decorative plants such as the prickly pear cactus, which invaded vast stretches of open plains and water holes in the tropical north, taking over 65 million acres (25 million hectares) before an effective biological control for it was found.

STRAWBERRY TREE

(Left) The strawberry tree shows the swollen red fruit from which it derives its name. This tree has a broad distribution across southern Europe and a single relict population in Ireland. It was probably the result of distribution during the last Ice Age of the Pleistocene.

MODERN biogeographic realms are complicated by relict populations: lingering groups of animals or plants that were once widespread in an area but whose numbers have now greatly diminished. In Europe, these relicts may be directly attributed to the effects of the Pleistocene glaciations. Many species that were widely distributed in the past now exist in only a few "islands" of favorable climate or physical conditions. They are not necessarily species with long evolutionary histories, because some climatic changes took place relatively recently. Plants and animals that lived on or near to ice sheets would have had to be adapted to the freezing conditions, which extended almost as far south as the Mediterranean. Since the beginning of the Holocene interglacial, cold-adapted animals have been restricted to the very coldest areas of Europe. These are invariably at higher altitudes among mountain ranges, and consequently the same animals that were once found close to the subtropical Mediterranean have moved north, up into Scandinavia, Iceland, and Scotland. Certain species have become extinct in northern areas and are represented in European biotas as glacial relict populations of extremely restricted distributions. A good example is the tiny primitive insect known as the springtail. *Tetracanthella arctica* is a dark blue springtail, about 0.05 inches (1.5 mm) long, that lives in the surface layers of the soil and in clumps of moss and lichens, where it feeds on dead plant tissues and fungi. It is particularly common in coastal Greenland, all of Spitsbergen and Iceland, and a few areas of northern Canada. But outside these truly arctic regions it is known to occur in only two regions: the Pyrenees Mountains between Spain and France, and the Tatra Mountains on the borders of Poland and Czechoslovakia, with very isolated finds in the Carpathian Mountains to the east in Romania. In these mountain ranges, *Tetracanthella arctica* is found

Mountaintops and other isolated habitats, such as caves, are ideal refuges for species that once had a much wider distribution.

Norwegian mugwort
springtail
strawberry tree
magnolia
gorilla
dung beetle
tulip tree

GORILLAS

Gorillas (left), strictly speaking, are not relict populations but disjuncts: descended from a common ancestor, they split into highland and lowland groups on either side of the range of the ancestor, which then became extinct. Today the highlanders and lowlanders are kept separate by the geographical barrier of the Zaire (Congo) River.

RELICT SPECIES

(Above) Norwegian mugwort, magnolias, and dung beetles as well as springtails, tulip, and strawberry trees all appear in more than one population, but in far-flung locations. All of these are old species that originated when the continents were still close enough together to permit distribution across land areas that have now drifted far apart.

SPRINGTAILS

(Right) Springtails are primitive arthropods with a long evolutionary history. Modern populations are mostly found in Greenland, Iceland, and Spitsbergen, with two relict populations in continental Europe.

in arctic and subarctic conditions at altitudes of around 6600ft (2000m). It is clear that it would have been impossible for this tiny insect to colonize the Pyrenees and Tatras from its main habitation so much further to the north, from which it is separated by open water and large expanses of warm low-lying land—conditions that would be fatal to this animal. Springtails are quickly killed by low humidity as well as higher ambient temperatures; and it is extremely unlikely that springtails were transported from such inhospitable regions to such high isolated areas by humans or their domesticated livestock.

The most likely explanation for the presence of springtails in the Pyrenees and the Tatras is that they represent remnants of a much wider distribution in Europe during the Ice Ages. This animal expanded as Pleistocene ice sheets covered the continent, and survived only in a few pockets of suitable environmental conditions after the ice retreated north. It is surprising that little *Tetracanthella arctica* has not been discovered at high altitudes in the Alps, much closer to home, where suitable conditions prevail. It may be that it once occupied a glacial relict position there but has died out in very recent times.

A plant example of a glacial relict is the Norwegian mugwort, *Artemisia norvegica*, a small Alpine plant that is now restricted to two areas of Scotland and very small regions of Norway and the Ural Mountains in Russia. During the last glacial period and immediately afterwards, the Norwegian mugwort was very widespread, but it was heavily curtailed as the post-glacial European forests expanded. It is almost certain that there are several hundred glacial relicts of this sort from both the plant world and animal world existing in Eurasia, including many forms that—in contrast to the springtails—could have traveled more easily.

One example of organisms that fit this criterion is the mountain or varying hare *Lepus timidus*. The varying hare derives its name from its appearance, which varies with the seasons: its fur is white in winter but tinged with blue for the rest of the year. It has a circumboreal distribution (Scandinavia, northern Japan, Siberia, and Northern Canada) and is closely related to the common Brown hare. The varying hare is found as relict populations in Ireland, the Southern Pennines in England—climates that are not particularly cold—and as a small population in the Alps. A factor in its distribution is that it is apparently poor at competing directly with the brown hare but is much better adapted to the cold. Yet another curious relict is the dung beetle *Aphodius holderi*. This big beetle is found in the high Tibetan plateau and as recent fossils dating from the middle of the last glaciation from a gravel pit in southern England, not far from the river Thames.

TULIP TREES

(Right) There are only two species of tulip tree known today, found in two widely separated locations: eastern North America and southeast Asia. This scattering suggests that they once had global distribution and the modern populations are relicts.

SUCCESS in evolutionary terms is defined differently for different species: mammals that survive for a million years may be said to be successful, whereas the average for marine species is 10 million years. In general, animals with highly specialized adaptations tend to go extinct most easily, whereas those that occupy a more general niche are often the ones that survive mass extinctions. Rodents, for instance, can feed on absolutely anything and live anywhere, as can canids, which are the most general of carnivores. Saber teeth, however, while unsurpassed for preying on large, slow, thick-skinned herbivores, are evolutionarily fragile: when large herbivores die out due to climatic and floral changes, saber-toothed carnivores cannot adapt quickly to different prey, and so they are also lost. Nevertheless, the repeated recurrence of saber teeth among Cenozoic carnivores shows that it is an excellent design for this specialized lifestyle.

Whether or not a species is successful in the long term depends to a large extent on environmental factors, including the presence of other animals.

Other species are important to the cycle of population growth of a given animal (or plant). There is a close relationship between predator and prey in any population. Rodent populations grew explosively from the Oligocene onward as global cooling caused savannahs and prairies to spread. The rodents flourished in these habitats but were of no use to big cats and other large carnivores, which could not get at them in their burrows. Into this niche came a new line of small, long-bodied, highly specialized carnivores. Nearly all of them belonged to the mustelid family of true mammalian carnivores. They included weasels, stoats, otters, badgers, skunks, and wolverines, but it was only the stoats and weasels that developed the underground burrowing mode of hunting. They have been highly successful for millions of years.

One mustelid, the black-footed ferret of North America, was once abundant due to the huge populations of the fat burrowing rodents known as prairie dogs which were present on the North American plains of the last 200 years. Throughout the eighteenth and nineteenth centuries, prairie dogs were hunted until their population crashed, and the black-footed ferrets promptly went into a sharp decline. Demand for their fur reduced their numbers further until they were thought to be totally extinct in the wild. Then, in 1986, ten individuals were found in Wyoming, and captive breeding was begun.

Another mustelid, the stone marten of Europe, has survived overhunting and colonized urban areas, where it is known to eat rubber parts from car engines.

BLACKFOOT FERRET

(Left) This mustelid's dependence on prairie dog prey nearly caused its extinction when prairie dog populations on the North American plains were nearly wiped out.

CAPE VULTURES

(Below) Vultures feed on dead carcasses but need large carnivores to open them. Vulture populations in South Africa nearly crashed when farmers all but eliminated hyenas, which crack bones into splinters small enough for vultures to use as food for their chicks. The chicks lacked calcium and consequently failed to develop.

EVOLUTIONARY fragility seems to be as much a function of behavioral adaptation as of anatomical suitability, but this is very difficult to infer from fossils. The stone marten and the black-footed ferret, though successful, illustrate vulnerability at the hands of humans. Another familiar example of this is the cheetah, *Acinonyx jubata*. Its oldest fossils are about 3.5–3.0 million years old and come from southeastern Africa. The modern species is restricted to Africa, though fossils are found in Asia and the Middle East as well. Living cheetahs feed exclusively on Thomson's gazelle, which has profound implications for their evolutionary prospects.

In the well-known case of the cheetah, being the fastest predator of all time is no guarantee of being able to eat and survive.

In Europe the cheetah was present until as recently as 500,000 years ago. Fossils show it to have been much larger than the modern African version—an adaptation to the colder climate—and possibly even faster, if it was not slowed down by its greater weight. It is curious that the skeleton of the living Himalayan snow leopard, with its small head and slim limbs, resembles that of the modern cheetah more closely than any other cat, suggesting that extinct cheetah-like forms with similar proportions could have lived in the cold mountain ecosystems of Plio–Pleistocene Europe. Some South African cheetahs hunt in irregular terrain, despite their reputation for preferring open plains.

A large relative of cheetahs, *Miracinonyx inexpectatus,* lived in North America from about 3.2 million years ago until 20,000 years ago. Like the European cheetah, it was big; early *Miracinonyx* are very close anatomically to modern pumas and may even be ancestral to them. However, it was less specialized than the living cheetah, which raises the question of what it might have eaten. Its prey were probably larger versions of today's pronghorn antelope and similar bulky grazers.

The living cats most anatomically similar to the cheetah are the puma and the snow leopard—both fairly chunky animals. The basic structure of their skeletons is similar, but the differences arise from the cheetah's adaptation for super-high-speed pursuit. It departed from the cat pattern to become a more doglike sprinter with only partly retractable claws (for better grip), stiff legs, and a small skull. It has retained the highly flexible catlike spine to make it probably the swiftest land predator ever.

What it does not have is strength, as might be expected from its slim lithe build. Cheetahs are adapted to chase only the very fastest of prey—animals that no other carnivores can catch. But they are easily driven away from their kills by more powerful animals such as even a lone hyena. Their bursts of speed are exhausting, and if several successive kills are wrested from them, they become too fatigued to sustain further chases.

At this point an individual cheetah faces potential disaster, especially a female with cubs, which may soon starve if their mother cannot provide food. When prey becomes scarce, as when local ecosystems change, the whole species is affected, because cheetahs are not adapted to kill larger, more powerful animals. Over an extended period of time—perhaps times of low prey density—populations crash and may go extinct. So cheetahs are enormously vulnerable to ecological disturbances, like the various saber-toothed cats before them.

The presence of humans does not help the situation of the cheetah either, with farming, hunting, and loss of habitat areas, all of which limit the cheetah's options for survival.

TOO FAST TO LIVE?

The cheetah is so specialized for high-speed pursuit of only the swiftest prey that it has not developed muscle power, and consequently is vulnerable to the theft of its hard-won food by bigger, stronger competitors. Habitat loss and the diminished number of prey—increasingly as the result of human activity in its homelands—put the cheetah under evolutionary pressure to adapt or face extinction. Its time will probably run out.

LYNX AND HARE

In the Arctic regions of Canada, the snowshoe hare is the preferred prey of the Canadian lynx, whose diet consists of 80 to 90 percent snowshoe hare. Populations of these two mammals mirror each other closely, and when harsh winters cause a reduction in the numbers of hares, the lynx population also drops sharply; conversely, an abundance of hares following a mild year results in an upturn in the number of lynxes. This parallel fluctuation in the numbers of prey and predator can be seen on the graph (left).

INDIVIDUAL groups of organisms have always been at risk in the natural world because of the changes that can occur in the Earth's geosphere, hydrosphere, and atmosphere. Changes in any of these systems can utterly disrupt established organisms, and mass extinction events show this to have happened at various times throughout the Earth's long history. Despite at least five well-documented mass extinctions, and a considerable number of smaller extinction events that have taken place during the past 500 million years, life has recovered eventually.

Extinctions are part of nature, but they are increasingly becoming human-made.

EXTINCT FAUNA OF MAURITIUS

Europeans who arrived on Mauritius in the sixteenth century encountered the dodo, a large flightless bird of the pigeon family, unique to Mauritius. It had no natural predators on the island and so had discarded its ability to fly; in fact, it lacked even the instinct to run away. Dodos were thus easy prey for humans, who hunted them for meat, and for the dogs who accompanied Dutch settlers in the mid-seventeenth century. Cats and rats helped to dispose of the eggs and chicks while men and dogs hunted the adult birds. By 1680 the dodo was extinct. The same fate befell the blue pigeon (pigeon hollandaise), which could fly, but whose eggs were hunted from its treetop nests by the monkeys introduced by colonists. Another loss was the massive domed giant tortoise, weighing up to 100lbs (45kg). Hunted by humans, it also became the prey of imported pigs, which killed young tortoises and dug up eggs from the sand in which they were incubating.

Continued disruption of the global environment by humans may be one step beyond the abilities of other organisms to bounce back. It is extremely unlikely that the consequences of human activities could wipe out microbes, fungi, algae, and other organisms that have been found in environments as extreme and far apart as several miles below the ocean bed and the edges of the stratosphere. But for larger, more conspicuous organisms such as trees, mammals, birds, and reptiles, the outlook is very different. These organisms require large habitable areas of complex micro-environments that are often very delicately balanced. One species of algae has spores that simply recolonize other rock surfaces away from areas of human-generated disturbance. Larger animals cannot do this. Once their habitats have gone for good, so are they. Examples include many types of bird, mammal, reptile, and tree that live in the forested areas of South America, which have been deforested by unrestricted logging.

Humans have had a major part to play in the evolution of animal life for the past 500,000 years or more. Our impact has by no means been confined to the time span since the Industrial Revolution: the dodo, a large flightless dove, became extinct about 100 years before industrialization of the northern hemisphere began, and its home was the Indian Ocean island of Mauritius, far from the scene of the first factories. Like other small islands scattered around the globe, Mauritius had probably been visited by sailors from various regions, but it was uninhabited by humans when Europeans arrived in the early sixteenth century. Forests of ebony covered the mountains, and flocks of giant turtles numbering in the thousands lay on the beaches. The island was dominated by birds, many of which had no predators, and so had no instinct for self-defense. The dodo was just one Mauritian species that vanished.

❶ ❷ ❸ ❹

THE DISAPPEARANCE of species through habitat loss or reduction is an indirect effect of human interference. Human effects on organisms may also be direct, as in the continued hunting of whales for blubber and associated products, which has reduced many populations to critically low levels. When so few individuals are left, there is insufficient variation in their gene pool (the total genetic material of the population), and fewer opportunities for adaptation to changing circumstances. As the population dwindles, inbreeding reduces the number of healthy offspring, and damaging mutations begin to appear in subsequent generations.

Hunting not only reduces the number of individual animals, it also diminishes the gene pool, limiting the variety and adaptability of a species.

THREAT TO SEALS

(Above) Seal bones in the Arctic. Excessive hunting of seals for their fur has caused a collapse in some populations, while pollution of their habitat is endangering others.

BISON RANGES

(Below) Before the arrival of Europeans, bison grazed from one end of North America to the other; by 1875 they were concentrated in two groups in some of the more remote terrain. Most surviving bison today are in herds of fewer than 500 animals.

This is the problem facing endangered species with small populations, and the human experts who try to restore the numbers of individuals to viable levels.

As their habitats are interfered with, similar animals may experience contrasting fortunes: for instance, the black-footed ferret of the North American plains did not have access to the new urban ecosystems that have become available to the European stone marten. If the roles of these mustelid carnivores were reversed, the outcomes would probably have been very similar, although the longer legs and tree-dwelling habit of the marten would have given it different advantages.

Other species seem all but impervious to the effects of humans—even to humans' concerted efforts to eliminate them. Rats and cockroaches fall into this category. Unlike the highly specialized cheetah and black-footed ferret, most "pests" are able to retain full populations in spite of environmental disturbances because there is nothing specialist about their lifestyles. To judge from their fossil record, which shows that they have survived at least two mass extinctions, frogs, small lizards, and chelonians (turtles and tortoises) are also highly stable animals that sail through major changes in ecosystems. All of these non-endangered animals are relatively small.

Large animals—like the dodo—make conspicuous targets. Four species of North American bison once lived and grazed from Oregon as far east as Pennsylvania in herds of a million animals or more, grazing on areas more than 1000 square miles (2600 square kilometers). As the plains were settled by Europeans and their descendants, the great open ranges disappeared, and bison were wantonly shot; "Buffalo Bill" Cody personally dispatched 4862 animals in a single year. With the full encouragement of the United States government, which wished to subdue the Native Americans who depended on bison for meat and hides, at least 75 million animals were killed between 1850 and 1880. The Oregon and

1 Mauritius blue pigeon (pigeon hollandaise)
2 Broad-billed parrot
3 Domed Mauritian giant tortoise
4 Dodo
5 Mauritian red rail

5

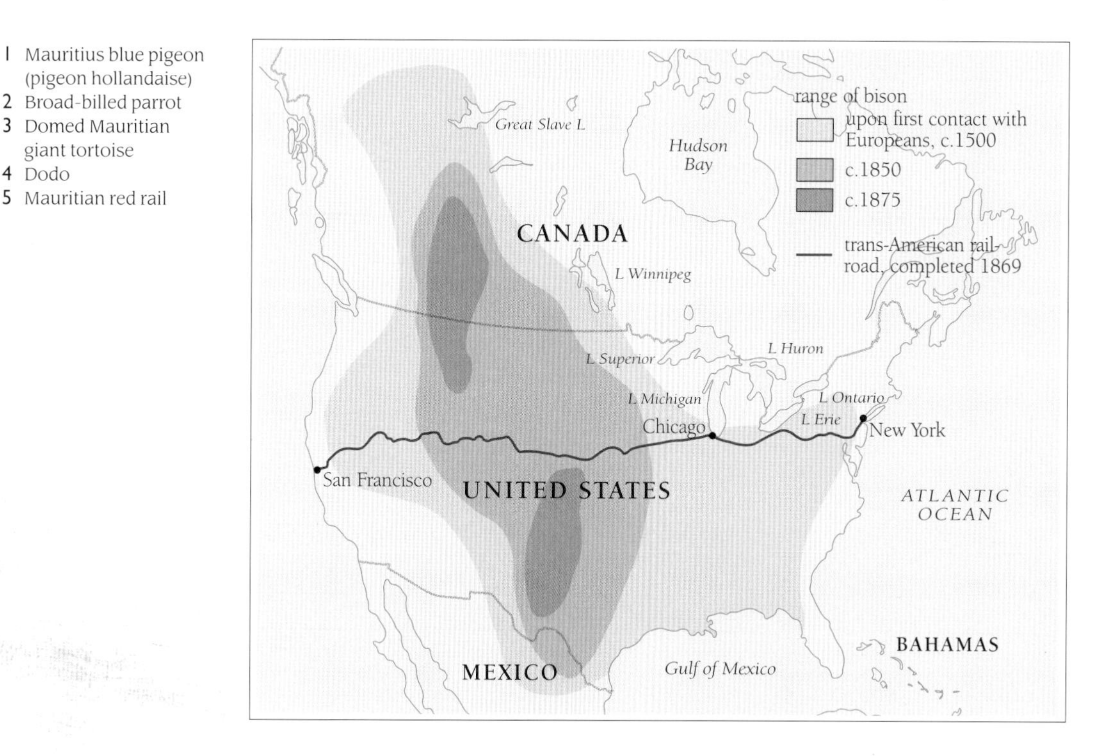

Pennsylvania species went extinct, but the two remaining American bison were successfully protected and conserved from the beginning of the twentieth century. About 30,000 Great Plains bison (*Bison bison bison*), the smallest species, survive today, while the Wood bison (*Bison bison athabasca*) of the north remains endangered.

Large animals are conspicuous and often beautiful, which helps to bring them sympathetic attention from the public. Less spectacular species are in a more perilous position. They include many species of insects and plants which are becoming scarce as humans devastate their habitats. Because a worm or moss is less visible than a magnificent tiger or redwood tree, the disappearance of the former tends to be discovered too late.

ONE group of animals may yet prove to have a greater effect on global climates than the ice ages in which they evolved: humans. By taking materials from nature for food, shelter, clothing, tools, and other goods, people have been altering their environment for more than a million years. At first, the impact made by small bands of hunter–gatherers was no more than that of most other large animals. It was only in the most recent 10,000 years that technological advances by humans began to have profound effects on the landscape and environment.

Human impact on the environment began with settled farming: the axe, the plow, and livestock.

The first of these innovations was settled farming, which began about 8000 years ago in the Middle East, southeast Asia, China, and Central and South America. Large areas of wild plants were lost as land was plowed up for crops or used to graze domesticated livestock. Entire forests began to be cut down for fuel and timber. Terraces were built on hillsides, and irrigation schemes diverted the courses of rivers in order to water crops. The food surpluses engendered by these technological advances allowed human populations to grow dramatically, increasing the impact on the environment, and eventually sending people in search of ever more land to support them. With European colonization after 1500 AD, new ground came under cultivation throughout the Americas, Australia, and New Zealand. Between 1700 and 1850, farmland around the world doubled from 655 million acres (265 million hectares) to 1.3 billion acres (537 million hectares).

The consequences of these activities are only now becoming fully understood. The area of the Earth that is suitable for growing food is limited: half of the planet's surface is covered by ice, snow, desert, and mountains. Most of the world's population lives on and farms a mere 21 percent of the land, which has come under increasing pressure, particularly in the last 200 years. As much as a third of this land is at risk of becoming non-productive due to overplanting and overgrazing.

Forests have decreased by one-third since the advent of agriculture, and are disappearing faster than any other biome, at a rate of up to 50 million acres (20 million hectares) a year as trees are cleared to support industry as well as farming. Much of this is tropical rainforest,

HARVEST DEITY?

This stone carving (left) from Hungary shows a male, possibly a deity, with a sickle. It dates from the third or fourth millennium BC, when cereal farming was well established in Europe.

THE SPREAD OF AGRICULTURE

(Below left) Farming appears to have arisen independently at different stages in several parts of the world. Climate played a part, with increased rainfall at the end of the last ice age encouraging the growth of wild cereals in the Middle East, where archaeological evidence of farming is especially rich. Mexico, northern Asia, and the Balkans were also sites of very early developments.

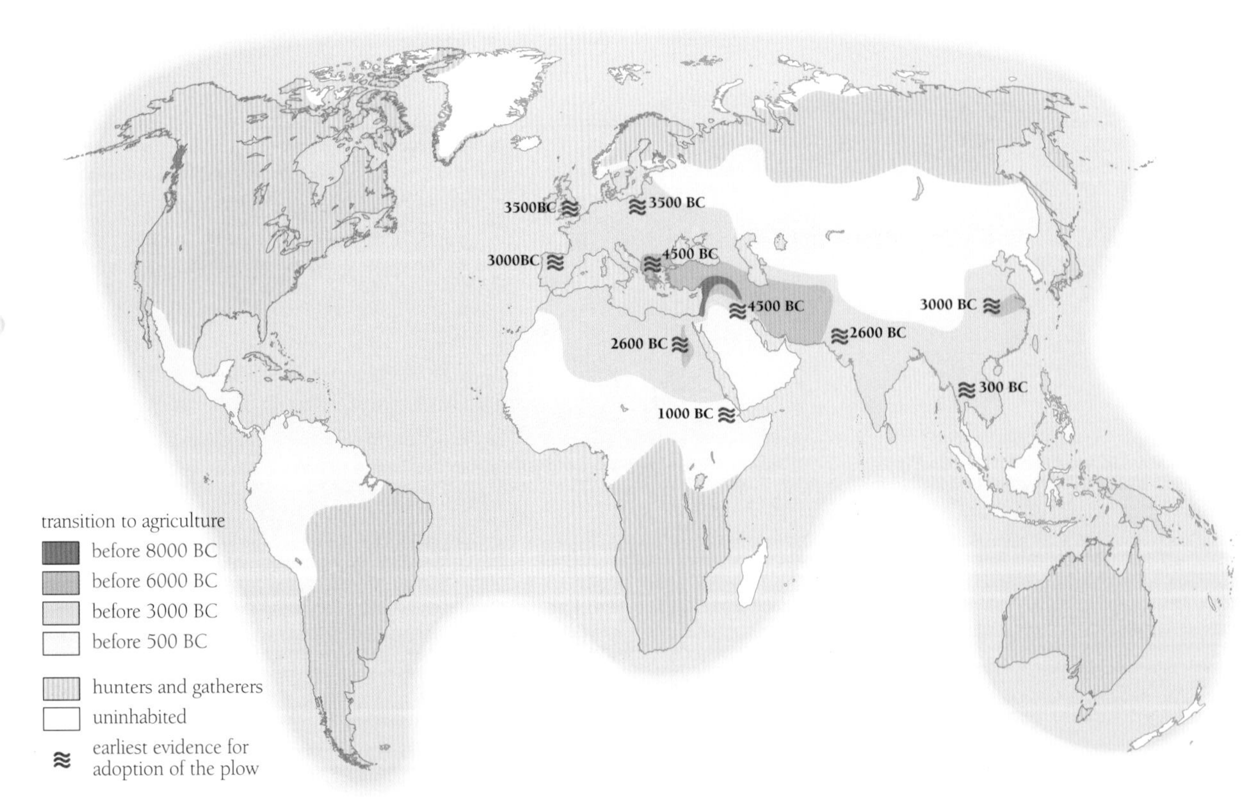

DEFORESTATION AND DESERTIFICATION

(Right) Human activity has caused the loss of as much as one-third of the world's forests in the past 10,000 years. Clearing of forests causes the loss of mature trees and associated plants. Topsoil loses its binding and the land is rapidly denuded, a process almost impossible to reverse. Runoff and sediment clogs rivers, and land is turned into desert. Huge areas of the world are now under this threat.

remaining tropical rainforest
area of tropical rainforest deforestation
deforested area at risk of desertification
area at risk of desertification
true desert

CUT DOWN

Clearing of forest (below) contributes to global warming, as plants trap carbon dioxide gas that is otherwise released into the atmosphere. Deforestation is thus a crucial environmental issue.

which supports at least 50 percent of all species of plants and animals now living. Forests have taken hundreds or thousands of years to develop to their present complex maturity; even if an area is replanted when logging is over, the new forest will take many generations to come to maturity, and species may be lost in the meantime. Removal of trees also kills the understory plants, and as roots that previously bound the top surface of the soils are removed, water runs off more easily. Intensive grazing by animals has a similar effect, badly damaging land by loosening the soil. Eroded topsoil is degraded, leaving a barren area with almost zero productivity.

Stripped of the vegetation that holds them together, soils are washed away by rain, and deep gulleys form, reducing the area that can be cultivated. The eroded soil silts up rivers, waterways, and dams, interfering with the water supply; or it blows away, sometimes forming massive dust and sand storms, as in Saharan Africa and on the central plains of North America in the 1930s. Attempts to replace topsoil are not effective and often make the problem worse; the added topsoil undergoes further denuding and erosion, while plants find it difficult to root themselves. Rapidly decaying plant matter is washed away and clogs nearby streams. In arid and semi-arid areas, eroded land quickly turns into desert. In the 1990s the United Nations estimated that desertification affected nearly one-third of the world's land surfaces and could threaten the livelihoods of 850 million people. Access to water supplies from major bodies of water is becoming a sensitive political issue from the arid Middle East and Africa to parts of Asia and North America.

However much humans and other modern species are affected by these events, they may be regarded as secondary: even if most species disappeared, new ones would evolve to take their places. The other great issue of ecology is how much the planet itself is being damaged.

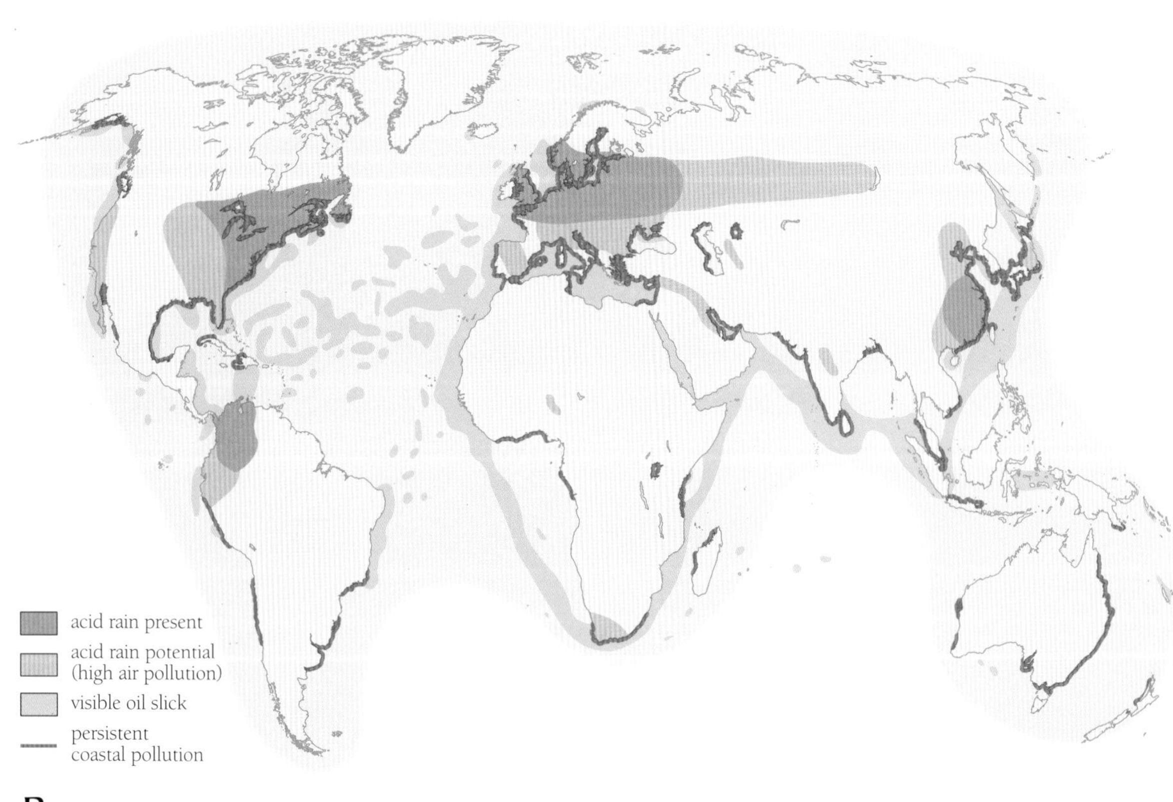

THE GREENHOUSE EFFECT

(Right) Energy from the Sun is absorbed by the Earth and radiated back into space. However, most of the heat is prevented from escaping by "greenhouse gases" such as carbon dioxide, carbon monoxide, and methane in the lower atmosphere. In natural conditions (1), most carbon dioxide—produced through decay, respiration, and chemical weathering—is absorbed by trees in forests and by phytoplankton in the seas. (2) As humans create more and more carbon dioxide through the burning of fossil fuels, it accumulates in the atmosphere. There it absorbs the outgoing solar energy, trapping heat so that global temperatures begin to rise.

BEGINNING with largescale industrialization in the nineteenth century, human activities moved beyond overuse of land to the pollution of entire ecosystems. Not only the land but the whole atmosphere, hydrosphere, and geosphere are at risk. Thousands of tons of hazardous and toxic chemicals are poured into the atmosphere every day, mostly from factories and vehicle emissions. These wastes combine to produce acid rain and smog, and contribute to the depletion of the ozone layer and to global warming.

Pollution now affects all of the biosphere: air, water, and land. No part of the Earth is still beyond reach of contamination.

Acid rain forms from sulfur dioxide, nitrogen oxides, and hydrocarbons. It increase the slight natural acidity of rain, increasing in turn the acidity of the soil or water on which it falls. Toxic elements such as aluminum and cadmium are leached from the soil and absorbed by plants, which lose leaves and then branches; whole forests have died in Eastern Europe and northeastern

DIFFERENT FORMS OF POLLUTION

(Above) There are few places left on Earth that are not affected by the major forms of pollution. Oil slicks or industrial and sewage pollution follow the coast of every continent, while inland, acid rain is the plague of industrial areas, and is spreading as air pollution becomes more prevalent. Land areas not affected by industrialization are often at risk of desertification.

GREAT BARRIER REEF

Australia's Great Barrier Reef (right) is one of the most fragile ecosystems on Earth. Its 400 species of coral support another 5500 species, all of which are vulnerable to even slight changes in the water temperature.

OIL SLICKS AND SEABIRDS

Maritime accidents have a disastrous effect on marine life (left). Seabirds and shore birds are particularly vulnerable to oil slicks, which are a major cause of mortality. Petroleum destroys the natural coating on their feathers, which lose their waterproofing; oil-slicked birds are unable to float. Ingestion of oil is usually fatal to those birds that do not drown.

North America. Those plants that survive are left in a weakened condition, unable to tolerate drought, frost, or disease. Acid rain also kills fish and invertebrates, leaving lakes and rivers "dead." Fresh water is a particularly precious resource; only 0.25 percent of all the water on the planet is of use to humans. The rest is too salty.

The ozone layer in the upper atmosphere, 9 to 12mi (15 to 20 km) above the Earth, absorbs harmful ultraviolet radiation from the Sun. Chemicals such as chlorofluorocarbons (CFCs) react with the ozone, causing it to break down. CFCs were extensively used in aerosols and refrigerants until the 1980s, and they will remain in the atmosphere for many decades. Their presence has thinned the ozone layer, as seen in a "hole" over Antarctica in the spring. In 1987 the hole was the size of the United States, and it appears to be increasing, letting in ever more ultraviolet radiation.

Chlorofluorocarbons are among the "greenhouse gases," which also include carbon dioxide and methane. They trap heat from the Earth that would otherwise be radiated back out into space, thus raising the temperature of the planet. Without these gases, it is thought that the average temperature around the world would currently be 0.4°F (−18°C). Increased levels of greenhouse gases in the atmosphere over the past hundred years seem instead to have raised the temperature by an average of 1°C (0.5°C). This sounds minor, but most scientists agree that if carbon dioxide levels continue to rise at the present rate, the Earth could warm up by 5.4° to 10°F (3° to 5°C). The last comparable temperature change was during the last ice age—and it took place 10 to 100 times more gradually. This may prevent the onset of the next ice age, or modify it, but consequences are likely to be severe, bringing worse flooding and storms to some areas and drought to others.

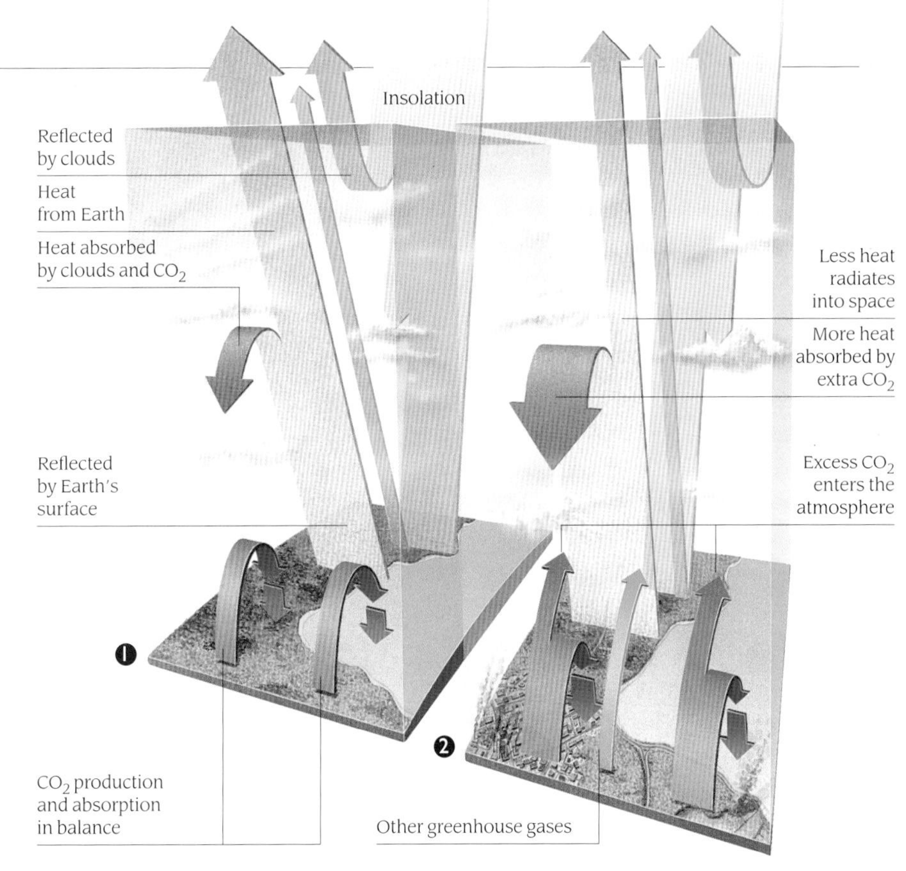

The oceans have an immense capacity to soak up and recycle many substances, including carbon dioxide. However, the oceans are being abused on no less a scale than the land and air, receiving forms of waste that are considered too toxic for land disposal: nuclear waste has been dumped at sea since the harnessing of atomic energy in the 1940s. The problem of safe disposal remains one of the great drawbacks of converting to nuclear energy to reduce the burning of fossil fuels on which the industrialized nations are so dependent.

Given the fluctuations that have taken place in the hydrosphere and geosphere during the past 3 billion years, can humans really put the entire planet at risk? If the hydrosphere, atmosphere, and geosphere are considered separately from the biosphere, then the answer is probably "No." Even if all life became extinct, the physical cycles of the planet would continue much as they have for the last four billion years or so, at least until the Sun burns out, another five billion years from now. But it would be supremely ironic if the species that has evolved to this level, priding itself on its "superiority" to other animals by virtue of its great brainpower, turned out to be the one that wipes out the others on which it depends.

MODERN-DAY EXTINCTIONS

PHOTO OPPORTUNITY

The plight of the endangered blue whale, and other large mammals, has captured the public imagination. Yet even more at risk of extinction are many smaller, less celebrated creatures, such as rats and bats, fish, and numerous invertebrates.

EXTINCTIONS are part of the normal meta-cycle of life. The fossil record is full of species that appeared and then vanished—about 95 percent of all that have ever existed. Factors such as body size and level of specialization of diet are important in directing which groups become extinct under "normal" circumstances. This probably accounts for a steady rate of extinction of one species per year over the past 3.5 billion years. In addition there have been mass extinctions due to catastrophic events such as meteorite impacts, or to periodic global climate changes. The latter have always been particularly devastating to the richly diverse plants and animals of tropical regions.

Modern extinctions are almost entirely the direct or indirect consequences of human action. Some species are hunted; others die out as their habitats are polluted, or cleared for farming or development. Yet others cling to existence in dwindling populations too small to be viable for breeding, especially for large animals that produce only one offspring at a time, at infrequent intervals.

According to the most conservative estimate of the distinguished American biologist Edward O. Wilson, in the 1990s species were becoming extinct at a rate of three per hour, or 27,000 per year. Within 30 years this figure may rise to *several hundred species per day*, with consequences that are truly terrible to contemplate. Every plant that goes extinct, for instance, may take with it as many as 30 insects and animals that depend on it for food.

No one is sure how many species currently occupy the planet, but the figure is believed to be between five and ten million. In the past 400 years, 611 species of animals alone are known to have disappeared, representing 1.8 percent of all mammals and one percent of birds at the absolute minimum, but possibly more. Birds and mollusks account for most animal extinctions, with mammals in third place. Of mammals, rodents and bats are disappearing fastest and have the highest number of endangered species. Primates also figure prominently among animals at risk, although there have not yet been any proven cases of extinction among them.

The Earth itself is resilient; life in one form or another will probably go on. After the latest mass extinctions, new species of rapid-spreading opportunists will arise to fill the empty niches. Weeds are a familiar example. So are starlings, a successful but unpopular bird; also cyanobacteria, which flourish in polluted water, lowering its oxygen content, and smelling foul. If humans survive the devastation we have wreaked, we may not enjoy sharing the planet with such species while waiting for wild flowers, dolphins, and pandas to re-evolve. But the human population itself will most likely be severely depleted as well, as our own habitats turn to desert or become flooded by the sea as polar ice melts, and food becomes scarce in many parts of the world.

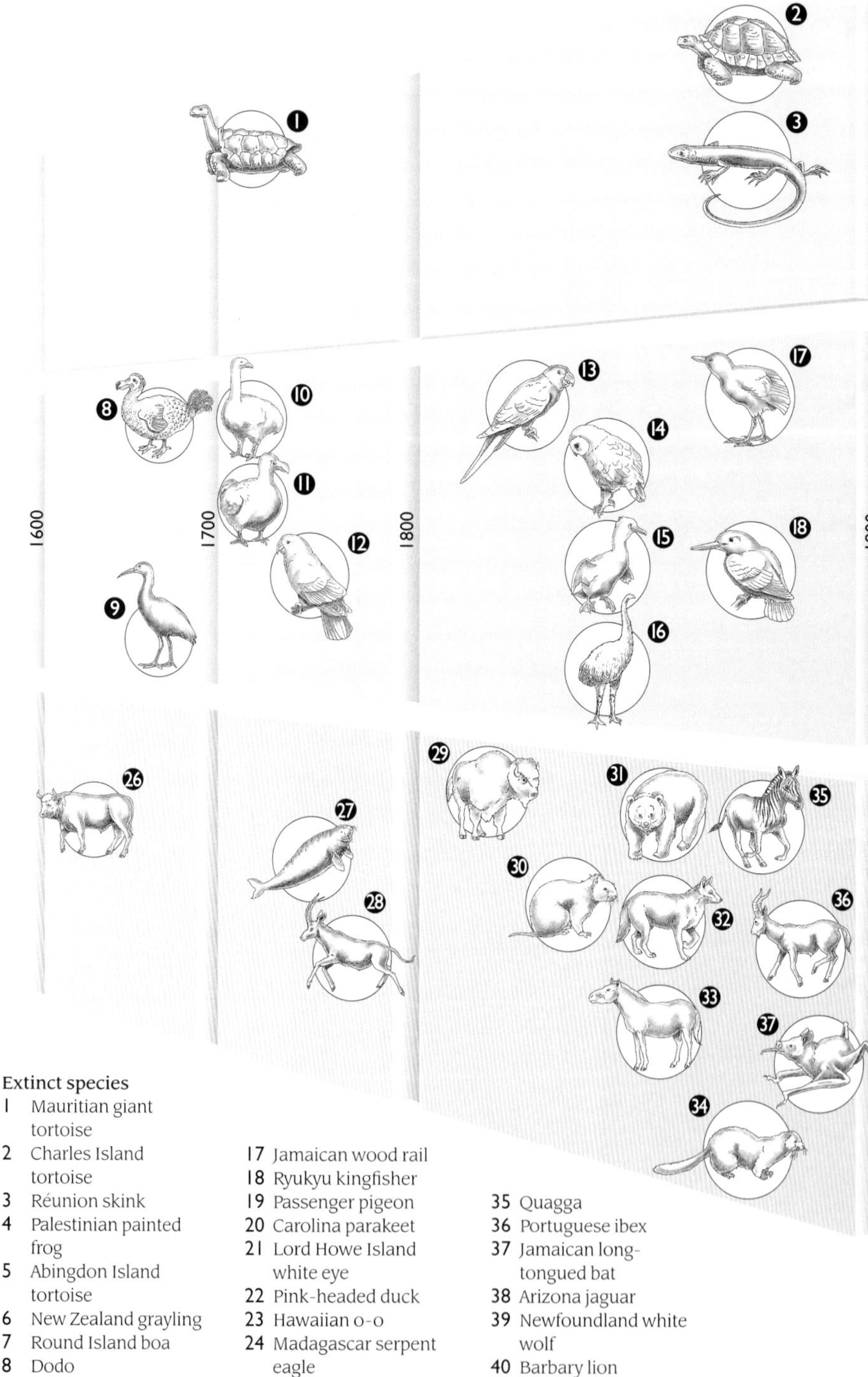

Extinct species

1 Mauritian giant tortoise
2 Charles Island tortoise
3 Réunion skink
4 Palestinian painted frog
5 Abingdon Island tortoise
6 New Zealand grayling
7 Round Island boa
8 Dodo
9 Mauritian red rail
10 Elephant bird
11 Réunion solitaire
12 Guadeloupe amazon
13 Green and yellow macaw
14 Rodriguez little owl
15 Steller's spectacled cormorant
16 Giant moa
17 Jamaican wood rail
18 Ryukyu kingfisher
19 Passenger pigeon
20 Carolina parakeet
21 Lord Howe Island white eye
22 Pink-headed duck
23 Hawaiian o-o
24 Madagascar serpent eagle
25 Kauai nukupuu
26 Aurochs
27 Steller's sea cow
28 Blue buck
29 Eastern bison
30 Hispaniolan huita
31 Atlas bear
32 Falklands wolf
33 Tarpan
34 Sea mink
35 Quagga
36 Portuguese ibex
37 Jamaican long-tongued bat
38 Arizona jaguar
39 Newfoundland white wolf
40 Barbary lion
41 Greater rabbit-bandicoot
42 Syrian onager
43 Schomburgk's deer
44 Thylacine
45 Bali tiger
46 Toolache wallaby
47 Caribbean monk seal
48 Mexican silver grizzly bear

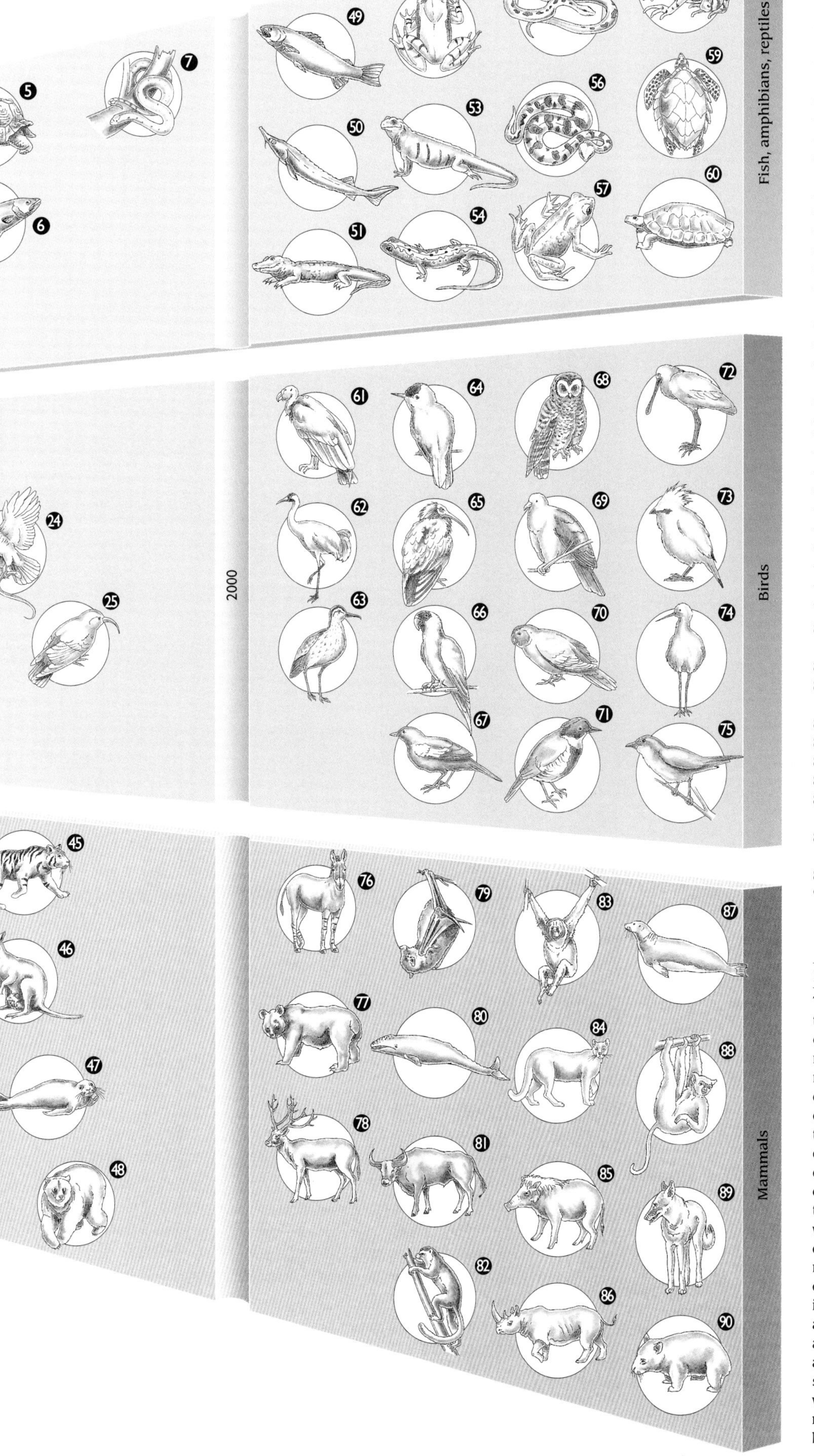

Critically-endangered species

49 Totoaba
50 Common sturgeon
51 Chinese alligator
52 Tinkling frog
53 Grand Cayman blue rock iguana
54 Santa Cruz long-toed salamander
55 Antiguan racer
56 Milos viper
57 Golden toad
58 Mallorcan midwife toad
59 Hawksbill turtle
60 Western swamp turtle
61 Californian condor
62 Whooping crane
63 Eskimo curlew
64 Juan Fernandez firecrown
65 Northern bald ibis
66 Spix's macaw
67 Seychelles magpie-robin
68 Spotted owl
69 Pink pigeon
70 Night parrot
71 Gurneys pitta
72 Black spoonbill
73 Bali starling
74 Black stilt
75 Zapata wren
76 African wild ass
77 Baluchistan bear
78 Père David's deer
79 Rodriguez flying fox
80 Gray whale (Asian stock)
81 Kouprey
82 Broad-nosed gentle lemur
83 Sumatran orang-utan
84 Florida panther
85 Visayan warty pig
86 Black rhino
87 Mediterranean monk seal
88 Golden crowned sifaka
89 Ethiopian wolf
90 Hairy nosed wombat

RECORDING LOSS

The fossil record shows that the average lifespan of a species is 5 to10 million years. With 5 to 10 million living species, the extinction rate should be one a year. While it is hard to know exactly the current rate of loss it is clear that it is well in excess of "normal". Over the last 400 years there were 611 documented extinctions, but this record excluded many creatures including most invertebrates—which account for 95 percent of all animals. Today, there are over 5000 threatened species listed, but only a very small proportion of recognized species have been evaluated.

GLOSSARY

A

ABYSSAL PLAIN The floor of the ocean where it forms a broad plain between 2 and 4mi (3 and 6km) below sea level.

ACADIAN OROGENY The mountain-building event that formed the northern Appalachians in the Devonian period. In Europe it is known as the CALEDONIAN OROGENY.

ACANTHODIAN A jawed fish of the Silurian–Carboniferous with a short, blunt head and prominent spines in front of each fin.

ACCRETION In PLATE TECTONICS, the building up of a CONTINENT when volcanic ISLAND ARCS become fused to the edge of a landmass.

ACCRETIONARY BELT (wedge) A part of a CONTINENT formed by ACCRETION of ISLAND ARCS, TERRANES, and fragments of emergent OCEANIC CRUST.

ACHEULEAN An early Pleistocene culture of stone tool-making consisting of roughly-worked stone blades. It has been attributed to early *Homo erectus* or *Homo habilis*.

ACID RAIN Rain that has become acidic due to the presence of dissolved substances such as sulfur dioxide. This may be caused by volcanic eruptions or, in modern times, industrial pollution.

ACRITARCH A planktonic microalga, usually with an ornamented envelope, that existed from the Proterozoic through the early Cenozoic. Most acritarchs were probably related to DINOFLAGELLATES.

ACTINOPTERYGIAN A member of the subclass of fish that have fins with radiating supports. Most living fish are actinopterygians.

ADAPIFORME A member of a group of primitive lemurs that lived in early Tertiary times.

ADAPTATION In EVOLUTION, the changing of an organism's structure or behavior to allow it to live a particular lifestyle in a particular environment, such as ducks' webbed feet.

AGNATHAN *See* JAWLESS FISH.

AGRICULTURAL REVOLUTION The change in farming practices in Europe in the late eighteenth and early nineteenth centuries, when scientific practices began to be applied to the cultivation of large areas and food production increased dramatically.

ALBEDO The amount of light reflected from a body, particularly from different areas of Earth or from the moon or a planet.

ALGA (plural algae) The most primitive form of plant, consisting of a single cell or a mass of cells but without a plumbing system. Seaweeds are an example of algae.

ALLEGHENIAN OROGENY A continuation of the ACADIAN OROGENY that occurred in the Late Paleozoic as three continents encroached on LAURENTIA, forming the ancient Appalachians. The European extension of this event is known as the Hercynian orogeny.

ALLOCHTHONOUS TERRANE *See* EXOTIC TERRANE.

ALLUVIAL FAN A fan-shaped sheet of eroded material washed down from an upland area and deposited over a flat plain.

ALPINE OROGENY The Tertiary collision of Europe with Africa, closing the Tethys Ocean between them and raising the Alps.

ALTAY SAYAN FOLD BELT The mountain system in Central Asia which was pushed up during the accretion of southern Siberia and Mongolia to northern Siberia.

ALULA A group of feathers on a bird's wing, in the position of the thumb, that helps maneuverability in flight.

AMINO ACID An organic compound based on NH_2 and COOR that forms the basis of PROTEIN and hence the basis of all living things. There are about twenty different types of amino acid.

AMMONITE One of a group of AMMONOID CEPHALOPODS common in the Mesozoic, having mostly coiled shells and extremely complex suture lines. Their range and rapid evolution make them ideal INDEX FOSSILS.

AMMONOID The extinct group of CEPHALOPODS to which AMMONITES belonged, along with goniatites and ceratites.

AMPHIBIAN The most primitive form of TETRAPOD, which passes a larval stage in the water and a mature stage usually on land. Frogs and newts are examples.

ANAPSID A member of one of the major subclasses of REPTILES, defined by the absence of holes in the skull behind the eye socket. Anapsids are regarded as the most primitive of the reptile groups; turtles are examples.

ANASPID A member of an order of JAWLESS FISH of the Silurian and Devonian which lacked the heavy head armor of many early fish.

ANDESITE A fine-grained IGNEOUS rock, gray in color, consisting predominantly of oligoclase or FELDSPAR. It is particularly abundant in the Andes, for which it was named.

ANGARALAND The continent formed by the collision of the individual islands of Kazakhstania and Siberia during the Permian. It in turn became attached to LAURASIA as the Uralian Seaway closed.

ANGIOSPERM The technical term for a flowering plant—one that bears its seeds in a box such as a pod or a fruit.

ANGULAR UNCONFORMITY A type of stratigraphic UNCONFORMITY in which overlying horizontal rock strata are separated from the older, tilted, eroded beds below.

ANKYLOSAUR One of a group of four-footed ORNITHISCHIAN dinosaurs characterized by an armored covering on the back, and either a bony club on the tail or an array of protective spikes at each side.

ANNELID WORM An INVERTEBRATE with an elongated body which is a muscle sack subdivided into segments with a recurrent set of organs. Annelids existed from the Cambrian.

ANOMALOCARIDID A Cambrian marine INVERTEBRATE predator. It had a huge head on which there was a pair of compound eyes on its upper surface and a rounded mouth with two spiny appendages below.

ANOXIA The state in which water contains less than 0.003fl.oz. of oxygen per 0.22 gallon (0.1ml oxygen per liter), the threshold below which animal life diminishes significantly.

ANTHROPOID One of a group of highly derived PRIMATES that appeared in the Paleogene period. It includes apes, monkeys, and all HOMINOIDS.

ANTLER OROGENY The mountain-building event that produced a contemporary range from Nevada to Alberta, North America, in Late Devonian and Early Carboniferous times.

APLACOPHORAN A worm-like marine mollusk lacking a shell and foot. Only modern species are known.

APPALACHIAN OROGENY The prolonged Late Paleozoic mountain-building event produced by the long collision between Laurentia (North America), Baltica (northern Europe), and Gondwana. It encompassed the TACONIAN, ACADIAN, and ALLEGHENIAN orogenies, which built the Appalachians.

ARCHAEOCETE A member of a family of early whales from the Early Tertiary. They had teeth of different shapes and sizes, and most had long serpentine bodies.

ARCHAEOCYATH A cuplike sessile marine animal of the Cambrian. It was a possible relative of DEMOSPONGES.

ARCHAEOPTERYX The earliest known bird, found in Upper Jurassic rocks of Germany, which retained many of the anatomical features of its ancestors, the DINOSAURS.

ARCHEAN The first eon of geological time, comprising about 45 percent of Earth's history, from 4.55 billion to 2.5 billion years ago.

ARCHOSAUR A member of a group of DIAPSID REPTILES encompassing the crocodiles, the DINOSAURS, the pterosaurs, and the birds.

ARTHROPOD An INVERTEBRATE animal with an articulated external skeleton covering a segmented body which bears serial pairs of articulated appendages. Arthropods first appeared in the Cambrian and have survived to the present; they include spiders, mites, CRUSTACEANS, centipedes, and insects.

ARTIODACTYL An even-toed, cloven-hoofed, grazing or browsing UNGULATE, such as pigs, deer, and cattle. *See also* PERISSODACTYL.

ASTEROID A minor planet in orbit around the Sun. Most lie between the orbits of Mars and Jupiter and vary in diameter from about 10mi (16km) to over 500mi (800km).

ASTHENOSPHERE The mobile portion of the Earth's inner layers, about 30 to155mi (50 to 250km) below the surface, on which the TECTONIC PLATES move.

ATMOSPHERE The envelope of gases surrounding the Earth, keeping it warm enough for life to survive, and filtering out harmful ultraviolet rays from the Sun. The most prevalent gas is nitrogen (78 percent), and the most important for life are oxygen (21 percent) and carbon dioxide (0.03 percent).

AUSTRALOPITHECINE A member of a group of Plio–Pleistocene HOMINIDS with anatomy intermediate between apes and humans.

AUTOTROPH Any organism that is a producer of food: that is, plants.

AVALONIA A continent that coalesced in the early Paleozoic and was amalgamated into LAURENTIA and BALTICA in the Late Paleozoic. Its components included modern eastern Newfoundland, the Avalon Peninsula and Nova Scotia (in North America), southern Ireland, England, Wales, and some fragments of continental Europe: parts of northern France, Belgium, and northern Germany.

B

BACK ARC BASIN The sea behind the arc of islands formed by volcanic action as one of the Earth's plates overrides another.

BACK REEF The landward side of a reef, including the area behind the reef crest and shelf lagoon.

BACTERIOPLANKTON PLANKTONIC bacteria.

BACTERIUM One of a group of unicellular prokaryotic microorganisms related to the fungi. Bacteria appeared between 3.5 and 3.8 billion years ago and are one of the most successful life forms on Earth.

BALEEN A horned comb-like structure in the mouths of the toothless whales, used for filtering small animals from the seawater.

BALTICA A continent of the Paleozoic and Mesozoic which included eastern and northern Europe surrounding the Baltic Sea.

BANDED IRONSTONE A Precambrian rock of alternating iron-rich and iron-poor layers.

BARNACLE A sessile filter-feeding marine CRUSTACEAN with a multi-plated calcareous shell. Barnacles originated in the Silurian and became widespread in the Triassic.

BASALT A dark IGNEOUS rock in OCEANIC CRUST, consisting mostly of PLAGIOCLASE, pyroxene, and glassy substances.

BASE PAIR A pair of nucleotide bases, joined by hydrogen bonds, that holds together the double strands of DNA and some RNA. Their units are pyrimidines (thymine, cytosine, or uracil) and purines (adenine or guanine), which are components of nucleic acids.

BASIN A low-lying geographic area that tends to collect sediments from higher land around it, thus building up rock sequences.

BATHOLITH A large, typically irregularly shaped body of intrusive IGNEOUS rocks, often granitic, with an exposed surface area of more than 40mi^2 (100km^2).

BED A layer of SEDIMENTARY ROCK that is distinct from the layers above and below.

BEDDING PLANE The surface that separates one bed of SEDIMENTARY ROCK from the next.

BELEMNITE One of a group of Mesozoic squid-like CEPHALOPODS.

BENIOFF ZONE The steeply inclined zone of seismic activity that extends down from an OCEAN TRENCH toward the ASTHENOSPHERE. Such zones mark the path of a TECTONIC PLATE being subducted at a destructive plate margin. The foci of earthquakes become deeper toward the non-subducting plate, reaching more than 370mi (600km) deep.

BENTHIC A bottom-dwelling aquatic organism.

BERING LAND-BRIDGE The span of land that is exposed periodically across the Bering Strait, connecting North America and Asia.

BIG BANG The theory that proposes the origin of the entire Universe (including space, matter, energy, time, and the laws of physics) about 15 billion years ago in the explosion of an extremely hot, dense body. The debris of the explosion moved radially away from the point of origin, cooled, and eventually formed galaxies and stars.

BIODIVERSITY A measure of the variety of the Earth's species, of the genetic differences within species, and of the ecosystems that support those species.

BIOGENESIS The principle that living things can only evolve out of living things like themselves rather than being created spontaneously or transformed from other things.

BIOGEOGRAPHIC PROVINCE An area with a distinct suite of animals and plants produced by geographical barriers that prevent the inhabitants from mixing with neighboring floras and faunas. Provinces may be recognized by the distribution of unusual life forms such as marsupials, which exist exclusively in Australia in modern times.

BIOMASS The total mass of living organisms in a defined area.

BIOME A broad community of plants and animals shaped by common patterns of vegetation and climate. Grassland, desert, tundra, and rainforest are examples.

BIOSPHERE The part of the Earth that supports life, beginning with the lower ATMOSPHERE and extending through the surface (land and water) to the upper fraction of CRUST.

BIOSTRATIGRAPHY The division of rocks into zones based on their fossil contents.

BIPEDALISM The ability to walk on two legs.

BIVALVE An aquatic MOLLUSK, covered with a calcareous shell of two valves and lacking a distinct head, dating from the Cambrian.

BLACK SMOKER A jet of hot mineralized water rising from vents in the ocean floor where RIFTS form. The color is from dissolved sulfides of iron, zinc, manganese, and copper.

BOROPHAGINE Hyena–dog: a member of a group that evolved from the Tertiary VULPAVINE carnivores and split from the canids (true dogs) in the late Paleogene.

BRACHIOPOD A solitary bivalved sessile marine animal that catches food particles with tentacles on a loop-shaped organ (lophophore). Brachiopods evolved in the early Cambrian before the true MOLLUSKS.

BRECCIA A coarse-grained SEDIMENTARY ROCK made up of angular fragments.

BRONTOTHERE A member of a group of rhinoceros-like ungulates, some of them very large, that existed in early Tertiary times.

BRYOPHYTE A simple land plant with leaves and a stem but no VASCULAR system, such as mosses and liverworts.

BRYOZOAN A colonial sessile marine INVERTEBRATE that traps food particles with tentacles occurring on a lophophore.

BURGESS SHALE The most famous of Cambrian LAGERSTATTEN, located in the Middle Cambrian rocks of western Canada.

C

CALCITE A mineral consisting of calcium CARBONATE.

CALCRETE A layer of LIMESTONE formed on or below the surface of soil caused by the evaporation of CALCITE-rich ground water. It is sometimes called caliche or kunkar.

CALEDONIAN OROGENY The mountain-building event that formed the Scottish Highlands and the Norwegian mountains during the Devonian period, when Baltica and Laurentia collided.

CAMBRIAN EXPLOSION The phenomenal radiation of marine animals during the Cambrian. The Cambrian–Precambrian boundary marked the appearance of nearly all animal phyla, and also a unique array of extinct creatures that cannot be assigned to any recognized phylum.

CARBON CYCLE The sequence of chemical reactions by which carbon circulates through the ecosystem. Carbon is a major component of LIMESTONE, often deposited as the shells of living things. Carbon from carbon dioxide is taken up by plants during PHOTOSYNTHESIS, producing carbohydrates and releasing oxygen into the atmosphere. The carbohydrates are used directly by plants—or by animals that eat plants—in respiration, releasing carbon dioxide back into the atmosphere.

CARBONATE A salt of carbonic acid. Carbonates are common in minerals, and are the principal constituents of SEDIMENTARY ROCKS such as LIMESTONE. The most widespread carbonate minerals are CALCITE, aragonite, and dolomite.

CARBONATE COMPENSATION DEPTH The ocean depth at which the rate of CARBONATE precipitation equals that of dissolution.

CARNASSIAL SHEAR The scissor action of specialized, blade-like molar or premolar teeth in CARNIVORES such as cats and dogs, an evolutionary development that increased their efficiency in cutting flesh.

CARNIVORE Generally, an animal that eats flesh; technically, the term applies only to

MAMMALS of the order Carnivora, encompassing cats, dogs, weasels, bears, and seals.

CARPEL The female reproductive organ of flowering plants from which the fruits and seeds grow.

CATARRHINE A member of the Old World group of broad-nosed monkeys from which humans descended. *See also* PLATYRRHINE (New World monkeys).

CENOZOIC The most recent era of geological time, which began 65 million years ago. It encompasses the Tertiary and Quaternary periods and includes the present day.

CEPHALOCHORDATE A lancelet: a small, scaleless, fishlike, primitive CHORDATE with a NOTOCHORD and a nerve cord but no brain. Cephalochordates arose in the Cambrian.

CEPHALON The head portion of a TRILOBITE.

CEPHALOPOD An advanced marine MOLLUSK, with a large brain and eyes, that evolved in the Cambrian. The foot is developed into a jet-propulsion system and tentacles.

CERATOPSIAN One of group of four-footed ORNITHISCHIAN dinosaurs characterized by the presence of an armored shield and an arrangement of horns on the head.

CERCOPITHECOID A member of a family of primitive CATARRHINE monkeys dating from late Tertiary times.

CETACEAN A member of the mammalian order Cetacea, encompassing the whales, dolphins, and porpoises.

CHAETETID A DEMOSPONGE, possessing a rigid calcareous skeleton, that first appeared in the Ordovician.

CHALK A pure form of LIMESTONE, formed from the deposits of microscopic shelly animals.

CHANCELLORIID A sessile COELOSCLERITOPHORAN of the Early Paleozoic whose skeleton consisted of rosette-shaped spiny hollow sclerites surrounding a sack-like body.

CHELICERATE An ARTHROPOD such as a water scorpion, spider, mite, or horseshoe crab. Chelicerates appeared in the Ordovician and have bodies subdivided into a head end with six pairs of appendages, the first pair of which are grasping jawlike chelicerae, and a tail portion.

CHEMOSYNTHESIS A process of organic matter production with the use of chemical redox reactions as an energy source. Bacteria are the principal chemotrophs.

CHERT A rock formed of non-crystalline silica.

CHITIN An ORGANIC substance that forms insects' shells and human fingernails.

CHITINOZOAN A problematic Ordovician–Carboniferous PLANKTONIC microorganism with a chitinous casing which is a chain of retortlike individuals; possibly a marine animal egg.

CHITON *See* POLYPLACOPHORAN.

CHLOROFLUOROCARBON A group of synthetic, nontoxic, inert gases containing chlorine, fluorine, carbon, and sometimes hydrogen, used in refrigerants, which accumulate in the upper atmosphere and break down ozone.

CHLOROPHYLL The green pigment in plants. Its function is to produce carbohydrates for use as food from carbon dioxide and water.

CHLOROPLAST The structure in a plant cell that contains CHLOROPHYLL.

CHONDRICHTHYAN A jawed fish, first known from the Silurian, whose skeleton consists entirely of cartilage; a shark is an example.

CHORDATE A DEUTEROSTOME with an anterior nerve cord, a cartilaginous rod (notochord) which is replaced by the vertebral column in higher chordates, and gill slits in the throat. Chordates have existed since the Cambrian.

CHROMOSOME A thread-shaped string of DNA that exists in the cells of living things and contains the GENES.

CIRCUMPOLAR CURRENT An ocean current that flows around one of the Earth's poles. In the present there is an important circumpolar current around Antarctica.

CIRQUE (also corrie, cwm) An armchair-like hollow in a hillside that was once the source of a GLACIER and has been widened and deepened by the weight of the ice.

CLADISTICS A method of classification that assigns organisms to taxonomic groups by assessing the extent to which they share characteristics. *See* TAXONOMY.

CLADOGRAM A chart that shows the evolutionary relationships of organisms or groups of organisms by comparing the numbers of features they have in common.

CLASTIC A rock made up of fragments of other rocks or their minerals (such as quartz).

CLASTIC WEDGE A large deposit of clastic sediments produced by uplift nearby. The thicker end of the wedge, and its sediments, occurs closer to the source, and the thinner end further away.

CLAY A very fine-grained SEDIMENTARY ROCK, usually with plastic properties.

CLUBMOSS A primitive VASCULAR PLANT related to the ferns. Nowadays they are small and insignificant, but in late Paleozoic times they grew as trees 300ft (100m) tall.

CNIDARIAN A primitive jellylike aquatic INVERTEBRATE with stinging cells (nematocysts) on their tentacles. Cnidarians have existed since the latest Proterozoic and include hydras, corals, sea anemones, and jellyfish.

COAL An organic SEDIMENTARY METAMORPHIC rock consisting largely (more than 50 percent) of the carbon remains of plant material. This must accumulate under water or be rapidly buried to prevent oxidation and decay. The depth at which the coal is buried in rock strata, and the resulting pressure, produces either soft, low-grade coal (peat, LIGNITE) or hard, high-grade (anthracite).

COCOS PLATE A small tectonic plate in the eastern Pacific Ocean, bounded by the Galapagos Ridge, the East Pacific Rise, and the landmass of Central America.

COELENTERATE *See* CNIDARIAN.

COELOSCLERITOPHORAN A Cambrian marine animal whose body was covered with hollow scalelike or spinelike SCLERITES. It is a possible ancestor of ANNELIDS, MOLLUSKS, and BRACHIOPODS.

COLD SEEP An area of seafloor on which cool mineral waters seep out through pores and crevices in the rocks.

COMET A planetary body consisting mostly of water, ice, and rock fragments. When its orbit brings it close to the Sun, the water from the ice boils away and forms a tail.

COMMUNITY An entity of interrelated organisms inhabiting a limited area.

CONDYLARTH One of an order of herbivorous PLACENTAL mammals that constituted the majority of MAMMALS in the early Tertiary.

CONGLOMERATE A SEDIMENTARY ROCK that consists of rounded lumps of pre-existing rock cemented together. It is essentially a fossilized shingle beach.

CONIFER A GYMNOSPERM tree, such as fir and pine, that reproduces by means of a cone.

CONODONT A primitive Paleozoic–Mesozoic eel-like swimming marine vertebrate with multiple phosphatized conical teeth.

CONSUMER An animal that eats PRODUCERS or other consumers.

CONTINENT A body of relatively buoyant terrestrial CRUST. The Earth's continents lie on average 2.8mi (4.6km) above the ocean floor and range in thickness between 12 and 37mi (20 and 60km). The oldest continental rocks found so far are approximately 3.8 billion years old. At the heart of each continent are one or more masses of ancient rock called CRATONS or shields, surrounded by successively younger MOBILE BELTS of FOLD mountains. The edges of continents may be flooded to form CONTINENTAL SHELVES.

CONTINENTAL DRIFT A theory, generally attributed to the German meteorologist Alfred Wegener, that postulates the early existence of a single supercontinent that eventually broke up, beginning to drift apart about 200 million years ago—and its components, the continents, are still drifting. Modern research has established that this is the result of SEAFLOOR SPREADING, driven by CONVECTION within the Earth's MANTLE.

CONTINENTAL SHELF A gently shelving, submerged part of a continent's margin that extends from the coastline to the top of the continental slope. Most sedimentation occurs on this part of the ocean floor.

CONULARIID A marine sessile animal of the latest Proterozoic that had a four-sided, elongate pyramidal mineralized skeleton and a four-lobed folded lid.

CONVECTION The movement of a fluid by heat. Hot fluid is less dense than cold, and so it rises; cool fluid descends. Convection currents drive the wind systems of the world as well as PLATE TECTONICS.

CONVERGENT EVOLUTION The phenomenon

by which animals with no recent common ancestry evolve similar shapes or habits through adaptation, enabling them to live the same lifestyle in a similar environment. Ichthyosaurs (reptiles), sharks (fish), and dolphins (mammals) have developed the same shape through convergent evolution, although they are unrelated.

CONVERGENT PLATE MARGIN A region of the LITHOSPHERE in which LITHOSPHERIC PLATES are pushed together and crustal surface area is lost. It may be caused by SUBDUCTION, in which lithosphere is consumed into the MANTLE, or by crustal shortening or thickening, in which slices of lithosphere are stacked upon each other as thrust slices.

COPE'S LAW The principle of EVOLUTION articulated by the American paleontologist Edward Drinker Cope (1840–97), which states that over the course of time all animals tend to evolve larger body sizes.

CORAL Any of a group of marine INVERTEBRATES of the class Anthozoa in the phylum CNIDARIA, and a few species of the class Hydrozoa (hydroids). Corals secrete a skeleton of calcium CARBONATE extracted from water. They occur in warm seas at moderate depths with adequate light. Corals live in a symbiotic (mutually beneficial) relationship with algae, in which the algae obtain carbon dioxide from the coral and the coral receives nutrients from the algae. Most corals form large colonies, although there are some species that live singly, as did early corals, which appeared in the Cambrian. Their accumulated skeletons make up REEFS and atolls.

CORDILLERA A series of parallel FOLD mountain ranges.

CORE The innermost portion of the Earth's surface below the MANTLE, more than 1800mi (2900km) deep, which is believed to be composed mostly of iron, with a solid center surrounded by a molten layer.

CRANIATE Another name for a VERTEBRATE. The term includes animals such as hagfishes which do not have vertebrae but do have the group's specialized skull features.

CRATON A mass of ancient METAMORPHIC ROCK at the center of a continent which is so distorted and compacted that it cannot be deformed further. It is the stable heart of the continent.

CREATIONISM A theory that claims that the world was created by a supreme being not more than 6000 years ago, as stated in the Bible, and that species have individual origins and are unchanging. Developed in response to Darwin's theory of EVOLUTION, it is not considered factual by most scientists.

CREODONT A member of the Creodonta, an order of large carnivorous MAMMALS of the early Tertiary and a sister group to the still-living Carnivora. There were two main lineages, the OXYENIDS and the HYENODONTS.

CRINOID (sea lily) A member of the ECHINODERM group, related to starfish, and usually anchored to the sea bed by a stalk.

CROSS-BEDDING (current bedding) Inclined planes in SEDIMENTARY ROCKS caused by strong currents of water or wind during deposition. For example, in a typical delta, where a flowing river drops its sediment load on reaching the deeper water of the sea, there are more or less horizontal or very gently shelving topset beds; inclined foreset beds (the delta front); and gently sloping bottomset beds which meet the flat seafloor in front of the delta. Current flows in the direction of the downward-sloping strata. A similar pattern (DUNE BEDDING) develops where wind forms sand dunes in a desert.

CRUST The outermost portion of the Earth's lithosphere, above the MANTLE. There are two forms: dense OCEANIC or MAFIC crust, and lighter continental or FELSIC crust.

CRUSTACEAN An aquatic gill-breathing member of the ARTHROPOD class Crustacea. Crustaceans evolved in the Cambrian and include crabs, lobsters, shrimps, woodlice, and barnacles. The segmented body usually has a distinct head, thorax, and abdomen, and is protected by an external skeleton made of protein and chitin hardened with calcium carbonate.

CTENOPHORE A radially symmetrical marine invertebrate possessing paddlelike comb plates (giving the common name comb jelly), which propel the animal forward. Ctenophores appeared in the Cambrian.

CUTICLE The hard noncellular protective surface layer of many INVERTEBRATES such as insects. it acts as an EXOSKELETON to which muscles are attached, reduces water loss, and provides defensive functions.

CUVETTE An area of inland drainage between mountains in which beds of SEDIMENTARY ROCK accumulate.

CYANOBACTERIA (blue-green algae) Primitive single-celled organisms structurally similar to bacteria. They are sometimes joined in colonies or filaments. Cyanobacteria are among the oldest known living things, more than 3.5 billion years old, and belong to the kingdom Monera. Their development of PHOTOSYNTHESIS changed the Earth by contributing more oxygen to the atmosphere, allowing advanced life forms to develop. Cyanobacteria are widely distributed in aquatic habitats, on the damp surfaces of rocks and trees, and in the soil. The name derives from the coloration, which is produced by the pigments CHLOROPHYLL and phycocyanin.

CYCADS A diverse group of seed-bearing plants common during the Mesozoic, resembling palms in appearance.

CYCLOTHEM A sequence of beds of SEDIMENTARY ROCK that show that they were deposited in cycles: for instance, LIMESTONE formed in the sea, followed by SANDSTONE deposited as a river encroaches, followed by COAL as plants grew on the river bank, followed by limestone as the sea encroached again.

CYNODONT One of a suborder of flesh-eating MAMMAL-like REPTILES with doglike teeth.

D

DARWINISM A popular name for the theory of EVOLUTION proposed by the British natural scientist Charles Darwin (1809–82). His main argument, now known as the theory of NATURAL SELECTION, concerned the variation that exists between members of a sexually reproducing population. According to Darwin, those members with variations better fitted to the environment would be more likely to survive and breed (survival of the fittest), subsequently passing on these characteristics to their offspring. Over time, the genetic makeup of the population would change; and, given enough time, a new SPECIES would form. Existing species would thus have originated by evolution from older species. *See also* CREATIONISM.

DEGASSING The process by which gas escapes from a body or a substance. Early in the Archean the Earth was heated to liquefaction and lost a large volume of gases into space.

DEMOSPONGE A class of sponges in existence since the Cambrian. The skeleton is built from spongin (a flexible protein similar to keratin), siliceous spicules, or rigid calcium carbonate, or a combination of these.

DESERTIFICATION The creation of deserts by climate change or by artificial processes including overgrazing; the destruction of forest belts; exhaustion of the soil by intensive cultivation with or without the use of fertilizers; and salinization of soils due to mismanaged irrigation.

DETRITUS FEEDER A consumer that ingests sediment to feed on organic matter; predominantly bacteria.

DEUTEROSTOME ("last mouth") An animal whose embryonic mouth becomes the anus as another mouth develops in the adult. ECHINODERMS, HEMICHORDATES, and CHORDATES are all deuterostomes.

DIAGENESIS The formation of SEDIMENTARY ROCKS from buried sediment at low temperatures. Two processes are involved: first, the particles of the sediment are compacted, and then they are cemented together by minerals.

DIAPIR A dome-shaped rock structure formed by a layer of rock being squeezed up through the overlying beds. This only occurs when that layer consists of a rock that is plastic under pressure, such as salt.

DIAPSID A member of one of the major subclasses of REPTILES. The diapsids are defined by the presence of two holes in the skull behind the eye socket. The lizards and snakes and the ARCHOSAURS are diapsids.

DICYNODONT A member of a suborder of MAMMAL-like REPTILES with a pair of prominent canine teeth.

DIFFERENTIATION (biology) The process by which cells in developing tissues and organs become increasingly different and specialized, giving rise to more complex structures that have particular functions.

DIFFERENTIATION (geology) The formation of distinct types of IGNEOUS ROCK from the same magma. Minerals crystallize at different temperatures and pressures and some accumulate before others, resulting in rocks of different compositions. The primary differentiation of all the planets, from masses of homogenous molten rock to layered planets with cores, occurred in a similar way.

DIKE A small sheetlike intrusion of IGNEOUS ROCK that cuts through pre-existing strata, formed by molten material pushing its way through a crack and solidifying there.

DINOCERATE A member of an order of rhinoceros-like MAMMALS dating from early Tertiary times. The most spectacular of these had three pairs of horns and a pair of tusks.

DINOFLAGELLATE A marine or freshwater unicellular EUKARYOTE with membrane-bounded nuclei and two flagella of different length, common as planktonic or symbiotic algae; Dinoflagellates arose in the Silurian.

DINOSAUR One of a group of large REPTILES that existed during the Mesozoic era. They are distinguished by the arrangement of hip bones, which were either birdlike (ornithischian) or reptilian (saurischian), and an upright stance.

DIVERGENT EVOLUTION The EVOLUTION of closely related species in different directions, often as a result of diverging lifestyles, eventually leading to the appearance of two very distinct evolutionary lines.

DIVERGENT PLATE MARGIN A LITHOSPHERIC PLATE boundary at which two plates are moving apart and there is upwelling of MANTLE-derived material to create new CRUST. It is associated with MID-OCEAN RIDGES, axial rifts, and active submarine volcanism.

DNA Deoxyribonucleic acid. The main chemical constituent of CHROMOSOMES.

DOLLO'S LAW The evolutionary principle proposed by the Belgian paleontologist Louis Dollo (1857–1931), which states that once a structure is lost or changed it will not reappear in new generations.

DOLOMITE A SEDIMENTARY ROCK consisting of a mineral of the same name, which is a form of magnesium carbonate.

DREIKANTER A rock that has been eroded into a three-sided shape by the wind.

DRUMLIN An elongated hillock composed of sediment deposited by a GLACIER. Its longer axis is parallel to the ice's movement.

DUNE A mound of sand, usually found on a beach or in a desert, built up and driven along by the wind.

DUNE BEDDING The CROSS-BEDDING resulting from the depositional action of wind during dune construction in deserts.

E

ECCENTRICITY (orbital) The degree to which the orbit of an object such as a planet or moon departs from that of a circle.

ECHINODERM One of a group of spiny-shelled DEUTEROSTOME marine INVERTEBRATES, characterized by fivefold radial symmetry, a calcareous internal skeleton or skeletal plates, and hydraulically-powered "tube-feet" equipped with suckers. Echinoderms incorporate the starfish, the brittle stars, the sea urchins, and the crinoids.

ECOLOGY The study of animal and plant communities and their relationships to their surroundings.

ECOSYSTEM An integrated ecological unit consisting of the living organisms and the physical environment (biotic and abiotic factors) in an area. Ecosystems may be largescale or smallscale: the Earth is an ecosystem, and so is a pond. The passage of energy and nutrients through an ecosystem is its FOOD CHAIN.

EDENTATE A member of an order of MAMMALS that encompasses the anteaters, sloths, and armadillos. Edentates lack teeth.

EDIACARA FAUNA An assemblage of late Precambrian fossils known from the Ediacara region of Australia, and consisting of wormlike and seapenlike organisms.

ELEMENT (chemical) A substance that cannot be divided into simpler components and whose atomic structure is constant.

ENDOPTERYGOTE A member of the subclass of insects that have a larval form very different from the adult form. A butterfly with its caterpillar larva and its winged adult is an example. The opposite is an exopterygote.

EOCRINOID An Early Paleozoic stemmed sessile ECHINODERM, a possible ancestor of crinoids.

EPEIRIC SEA An extensive shallow inland sea.

EPICONTINENTAL SEA *See* EPEIRIC SEA.

EPIPELAGIC An organism inhabiting the upper (PHOTIC) zone of the WATER COLUMN.

ERRATIC BOULDER A boulder that has been transported and deposited by a GLACIER and so is different from the rocks around it.

ESKER A winding ridge of MORAINE left behind by streams flowing beneath a sheet of ice.

EUKARYOTE An organism with a complex cell structure of a clearly defined nucleus containing DNA and separated from the rest of the structure by a membrane, and with specialized organelles such as mitochondria. Eukaryotes include all organisms except bacteria and cyanobacteria, which are PROKARYOTES.

EURYPTERID An aquatic Ordovician–Devonian chelicerate resembling a modern water scorpion.

EVAPORITE A SEDIMENTARY ROCK or bed formed from minerals precipitated from water as a lake, or an inlet of the sea, dries out.

EVOLUTION The process of biological change by which organisms come to differ from their ancestors. The idea of gradual evolution (as opposed to CREATIONISM) gained acceptance in the nineteenth century but remained controversial into the twenty-first because it contradicted many traditional religious beliefs. The British naturalist Charles Darwin (1809–1882) assigned the major role in evolutionary change to NATURAL SELECTION (that is, environmental pressures acting through competition for resources). The current theory of evolution (neo-Darwinism) combines Darwin's theory with the genetic theories of Gregor Mendel and Hugo de Vries' theory of MUTATION. Evolutionary change may have long periods of relative stability interspersed with periods of rapid change (PUNCTUATED EQUILIBRIUM).

EXOPTERYGOTE A member of the subclass of insects that grow by a series of molts, so that the juvenile form is similar to the adult. A newly-hatched grasshopper, for example, is like a miniature version of the adult. The opposite is an ENDTOPTERYGOTE.

EXOSKELETON The hard shell of an insect or similar animal.

EXOTIC TERRANE A comparatively small piece of "foreign" LITHOSPHERE fused to the edge of a continent.

EXTINCTION The complete disappearance of a species or other group of organisms, which occurs when its reproductive rate falls below its mortality rate. Most extinctions in the past occurred because species could not adapt quickly enough to natural changes in the environment; today they are primarily due to human activity.

EXTRUSIVE ROCK An IGNEOUS ROCK, the product of an eruption, that appears on the surface of the Earth as opposed to forming within it.

F

FACIES An assemblage of different SEDIMENTARY ROCKS that are spatially connected and related to the same geological event, so that they represent local conditions.

FAMMATINIAN OROGENY A phase of mountain building in South America during the middle Ordovician, following the accretion of the Andean Precordillera to Gondwana.

FARALLON PLATE A TECTONIC PLATE in the eastern Pacific that was subducted beneath the North American plate during Tertiary times.

FAULT A fracture in a body of rock along which one mass of rock moves against another. Normal faults are caused by extension of the CRUST and may occur in pairs to form GRABEN. Faults also occur along thrusts where there is crustal shortening (reverse faults) and where there is lateral movement of adjacent blocks with little or no vertical slipping (STRIKE-SLIP or TRANSFORM FAULTS).

FELDSPAR Any of a group of alumino-silicate rock-forming minerals.

FELSIC CRUST The light-colored, low-density IGNEOUS ROCKS, with a high FELDSPAR and silica content, that form continents. GRANITE is a felsic rock abundant in CRUST.

FILTER-FEEDER A consumer that strains organic particles or dissolved organic matter from the surrounding water.

FIRN Fallen snow that has partially frozen on top of a GLACIER but has not formed ice.

FLAGELLATE A collective name for unicellular organisms that move by means of flagella.

FLOWERING PLANT *See* ANGIOSPERM.

FLYSCH Thick deposits of SANDSTONE and SHALE eroded from newly-arisen mountain ranges. The term is sometimes restricted to deposits to the north and south of the Alps.

FOLD Originally horizontal strata of SEDIMENTARY ROCK that have been warped and bent by mountain-building. Folds may form mountain massifs by compressional deformation, as in the case of the Alps and the Appalachians.

FOLD BELT A long, narrow zone of CRUST in which there has been intense deformation and development of FOLDS. Such belts usually develop along continental margins associated with CONVERGENT PLATE MARGINS. They have also been recognized on Venus.

FOLD-AND-THRUST BELT The inland zone of a mountain chain, characterized by FOLDS and THRUST FAULTS.

FOOD CHAIN The levels of nutrition in an ecosystem, beginning at the bottom with primary PRODUCERS, which are principally plants, to a series of CONSUMERS—herbivores, carnivores, and decomposers.

FOOD WEB A more complex food chain, with several species at each level, so that there is more than one PRODUCER and more than one CONSUMER of each type.

FORAMINIFERAN An order of single-celled PROTOZOANS, mostly marine, with tests (shells) usually of CALCIUM CARBONATE and perforated by pores and reinforced with minerals. They evolved in the Cambrian.

FOREARC BASIN An elongate BASIN of deposition that lies between an ISLAND ARC and the wedge of ACCRETION behind it.

FOREDEEP The deep area on the oceanic side of an ISLAND ARC.

FORMATION The basic stratigraphic unit: a body of rock with distinctive lithographic features that can be mapped.

FOSSIL The remains of a once-living creature found preserved in rocks. Fossils may be part of the creature, or the shape of the creature turned to stone, or even just traces such as footprints or worm burrows.

FRAGMENTATION A process of splitting of a continent into smaller pieces during RIFTING and following spreading.

FUNGUS A EUKARYOTE that forms spores as slender septate tubes (hyphae) and lacks any motile stage during its life cycle. Fungi have existed since the Proterozoic.

FUSULINIDS A group of calcareous coiled FORAMINIFERANS, abundant during the Carboniferous and Permian periods, many of which were spindle-shaped.

G

GABBRO A coarse-grained IGNEOUS ROCK similar to BASALT in composition but formed below the Earth's surface. It contains FELDSPAR.

GAMETE The reproductive cell of an organism that fuses with one from another organism during sexual preproduction.

GASTROPOD A snail: a univalved MOLLUSK possessing a well-developed head and foot, and asymmetrical development of its shell and inner organs. Gastropods appeared in the Cambrian.

GENE The basic unit of inheritance that controls a characteristic of an organism. It may be considered as a length of DNA organized in a very specific manner. Genes mutate and recombine to produce the variation on which NATURAL SELECTION acts.

GENE POOL The mix of genetic material in breeding populations of organisms.

GENUS The sixth division under the Linnaean system of classification for living things, consisting of a number of similar or closely related species. Similar genera are grouped into families.

GEOSPHERE The solid portion of the Earth, as distinct from the ATMOSPHERE or BIOSPHERE.

GEYSER An eruption of hot water from the ground. Groundwater boils below the surface by volcanic heat, and the expanding superheated steam pushes water out of a vent. This releases the pressure and the rest of the water column boils explosively, blasting the water high into the air.

GLACIAL MAXIMUM A period during an ICE AGE in which the glaciation is most extensive.

GLACIER A thick mass of ice and compacted snow that moves slowly downhill and persists year-round for up to 100 years.

GLACIOEUSTASY Fluctuations in global sea levels caused by changes in the quantity of sea water as ice caps grow or melt.

GLOBAL WARMING A change in the Earth's climate involving an increase of overall temperatures, sometimes as a result of natural processes, but mostly attributed in the present to the GREENHOUSE EFFECT. Such a change is not regular and may show itself as unpredictable weather conditions caused by the increased temperature gradient between permanent ice caps and the warmer conditions of the rest of the world. The United Nations Environment Program estimates that by the year 2025 global warming will cause average world temperatures to rise by 2.7°F (1.5°C), with a consequent rise of 8in (20cm) in sea level caused by the melting of polar ice.

GLOSSOPTERIS A type of seed-bearing fern that characterized the BIOGEOGRAPHIC PROVINCE of GONDWANA at the end of the Paleozoic.

GNEISS A coarse-grained METAMORPHIC ROCK that exhibits alternating bands of dark and light materials. Such rocks have formed at depth within MOBILE BELTS inside the Earth. Granitic gneisses are frequently associated with granites in continental CRATONS.

GONDWANA The supercontinent incorporating today's southern continents—Africa, South America, Australia, and Antarctica—and also India, Madagascar, and New Zealand during the Paleozoic and Mesozoic.

GONIATITE A member of a group of CEPHALOPODS similar to the modern NAUTILUS, with a characteristic jagged pattern of suture lines.

GRABEN A geological structure in which a section of rock has descended between parallel FAULTS. On the surface this may show up as a RIFT VALLEY.

GRANITE A hard, coarse-grained IGNEOUS ROCK that consists mainly of QUARTZ and FELDSPAR, often with mica or other colored minerals. Most regions of granite have resulted from the crystallization of molten MAGMA, though some granites may have formed through the metamorphism of other, existing rocks. The extrusive equivalent is rhyolite.

GRAPTOLITE A Cambrian–Carboniferous filter-feeding colonial HEMICHORDATE. Graptolites lived near the surface of the oceans, like modern PLANKTON, and are known from fossils of their simple outer skeletons. Most died out at the end of the Silurian period.

GRASSES A group of Cenozoic ANGIOSPERM plants characterized by long straplike leaves and underground stems.

GRAZER In a TROPHIC WEB, a herbivorous CONSUMER that crops along the substrate.

GREENHOUSE EFFECT The gradual increase of the temperature of air in the lower ATMOSPHERE, which is believed to be due to the buildup of gases such as carbon dioxide, OZONE, methane, nitrous oxide, and CHLOROFLUOROCARBONS. These trap solar radiation absorbed and re-emitted from the Earth's surface, preventing it from escaping into space and thus overheating the Earth.

GREENSTONE A SEDIMENTARY ROCK formed on the Earth's surface in Precambrian times when the atmosphere was deficient in oxygen. The green color is derived from minerals that form under low-oxygen conditions.

GREYWACKE A poorly sorted, dark, very hard, coarse-grained SEDIMENTARY ROCK with angular particles.

GULF STREAM The ocean current that travels from the Gulf of Mexico northeast across the North Atlantic—where it becomes the North Atlantic Drift—bringing warm conditions to the west coast of Europe.

GYMNOSPERM An informal designation for a seed-bearing plant that does not protect its seeds in an enclosed fruit. Conifers and ginkgoes are examples.

H

HADROSAUR A member of a group of ORNITHOPOD dinosaurs of the Cretaceous, characterized by a broad ducklike bill.

HAECKEL'S LAW An evolutionary principle, now modified, that states that the young of a species resemble the adults of their ancestors (ontogeny recapitulates PHYLOGENY).

HALKIERIID An Early Paleozoic COELOSCLERITOPHORAN with two shells, a sluglike foot, and an upper surface covered with spiny and scalelike SCLERITES.

HANGING VALLEY A side valley that enters a glaciated U-SHAPED VALLEY some way up the valley side.

HEMICHORDATE A primitive wormlike marine DEUTEROSTOME with a NOTOCHORD. The body is subdivided into a head shield, collar, and trunk perforated by gill slits. Hemichordates appeared in the Cambrian.

HERBIVORE Any animal that eats plants. Unlike CARNIVORE, the term is not restricted to any particular group of animals.

HERCYNIAN OROGENY A mountain-building event during the Late Paleozoic that emplaced many of the granitic masses now found in western Europe. North America's equivalent was the ALLEGHENIAN OROGENY.

HETEROTROPH An organism that is a CONSUMER —one that takes in ORGANIC compounds as food because it cannot synthesize them itself from simple inorganic substances. Heterotrophs include bacteria, fungi, protozoa, and all animals. Plants, which are AUTOTROPHS (producers), are the main exception.

HEXACORAL One of a group of tentacle-feeding CORALS, characterized by their hexagonal symmetry, that replaced the Paleozoic RUGOSAN corals during the Mesozoic. Hexacorals still live today.

HIRNANTIAN ICE AGE The first Phanerozoic glaciation, which occurred at the very end of the Ordovician (Hirnantian stage).

HOMINID A late HOMINOID, characterized by its upright gait (bipedalism), and a member of the line from which modern humans evolved about 4 or 5 million years ago.

HOMINOID A PRIMATE such as one of the lesser apes (gibbons and siamangs), great apes (orangutans, gorillas, and chimpanzees), and humans.

HOMO The genus to which humans belong. Two species are recognized besides the modern *H. sapiens*: *H. habilis*, the first toolmaker, and *H. erectus*, which was the first to spread around the world from Africa.

HOMOLOGOUS STRUCTURE A structure that is similar in different species, suggesting a common ancestor, but serves a different function in each, such as arms and wings.

HORST BLOCK A geological structure, the reverse of a GRABEN, in which a section of rock is raised between two FAULTS. The surface feature of this may be a flat-topped hill.

HOT SPOT A place at which a MANTLE PLUME causes hot MAGMA to rise toward the base of the Earth's CRUST, producing high heat flow and volcanism at the surface. Iceland and the Hawaiian islands lie above hot spots.

HYDROSPHERE The part of the Earth's structure made up of water. This includes the oceans, the ice caps, and the gases in the ATMOSPHERE.

HYDROTHERM A deep, hot (900° to 7200°F/500° to 4000°C) marine spring of mineral waters, often the site of a BLACK SMOKER.

HYOLITH A marine bivalved Paleozoic INVERTEBRATE, a possible relative of BRACHIOPODS.

HYPSILOPHODONT One of a group of small ORNITHOPOD dinosaurs built for fast running.

I

IAPETUS OCEAN The ocean that existed between LAURENTIA, AVALONIA, and BALTICA before they came together to form Euramerica; it is sometimes known as the Proto-Atlantic Ocean because it lay between what is now North America and Europe.

ICE AGE An extended period of cold climate during which there is an increase in the surface area of GLACIERS and ice caps. There have been several ice ages in the Earth's history, with the most recent occurring in the Pleistocene epoch from 1.8 million years ago to about 12,000 years ago.

ICE SHEET A continental GLACIER that spreads out from an area of extreme cold rather than downhill from a mountain. Antarctica and Greenland are covered by ice sheets.

ICHNOLOGY The study of FOSSIL footprints and burrows.

ICHTHYOSAUR One of a group of marine REPTILES of the Mesozoic that resembled dolphins in appearance.

IGNEOUS CORE The solidified molten material in the middle of a mountain range, formed at very high temperatures.

IGNEOUS ROCK Any rock formed by the solidification of molten MAGMA. There are two main types: intrusive igneous rocks that formed underground, and extrusive igneous rocks that formed from lava that erupted at the Earth's surface. The former, such as GRANITE, are coarse-grained, while the latter, such as BASALT, are fine-grained.

IGUANODONTID One of a group of plant-eating ORNITHOPOD dinosaurs.

INDEX FOSSIL A FOSSIL whose presence indicates a rock's age. Useful index fossil species are short-lived and widespread; GRAPTOLITES and AMMONITES are examples. Index fossils are also known as zone fossils.

INDUSTRIAL REVOLUTION The rise of the use of machinery in industry, which began in the late eighteenth century in Britain.

INSECT One of a group of air-breathing ARTHROPODS that appeared in the Devonian. The body is subdivided into a head, a chest with three pairs of legs, an abdomen, and one or two pairs of wings.

INTERGLACIAL A period of milder climates occurring within an ICE AGE.

INVERTEBRATE An animal with no backbone—95 percent of all animal species.

ISLAND ARC A chain of volcanic islands that develops at the edge of an ocean trench, where volcanoes are generated by the melting of a subducting plate (see SUBDUCTION).

ISOSTASY The balance between parts of the Earth's surface caused by differences in density, based on the principle that the rocks of the CRUST "float" on those of the underlying MANTLE. Ocean crust is made from dense BASALT, whereas upper continental crust is mostly low-density FELSIC rocks with deep "roots" to compensate for its lightness.

ISOTOPE One of a number of forms of a chemical ELEMENT that has the same number of protons in its nucleus but a different number of neutrons and thus different physical properties.

J

JASPER A METAMORPHIC ROCK formed chiefly on siliceous deep-marine sediments.

JAWLESS FISH A fishlike CRANIATE lacking jaws, trunk bones, and, in most cases, paired fins. Jawless fish appeared in the Ordovician.

JUAN DE FUCA RIDGE An ocean ridge off the west coast of Canada, an isolated part of the East Pacific Rise that is now being subducted beneath the North American Plate.

K

KARST A LIMESTONE landscape, characterized by extreme dryness and EROSION of the exposed rocks into blocks (CLINTS) separated by deep gullies (GRYKES), caused by chemical breakdown of CALCITE in the limestone.

KERGUELEAN LANDMASS A submerged continental plateau in the southern Indian Ocean.

KETTLE HOLE A depression formed in an area of MORAINE—rocky debris left by a retreating GLACIER. A block of glacier ice left stranded eventually melts, leaving this structure.

KOMATIITE An extrusive IGNEOUS rock composed of peridotite that was widespread in Archean times and was the precursor of BASALT rocks in the Earth's crust.

KT BOUNDARY EVENT The boundary between the Cretaceous and Tertiary periods about 65 million years ago, marked by a MASS EXTINCTION in which the DINOSAURS perished, among others. One possible explanation is a METEORITE strike on the Earth, which left a 110mi (180km) crater in the Gulf of Mexico.

L

LAGERSTATTEN Localities containing FOSSILS preserved much better than is usual.

LAMARCKISM A theory of EVOLUTION proposed by the French biologist Jean-Baptiste de Lamarck (1744–1829), by which traits acquired by an individual during its lifetime could be passed down to offspring. It was discredited by the work of Charles Darwin.

LARAMIDE OROGENY A mountain-building event that took place in western North

America in late Cretaceous times and contributed to the Rocky Mountains.

LAURASIA The supercontinent that represented the northern part of PANGEA, consisting of the landmasses that now constitute North America, Europe, and northern Asia.

LAURENTIA The supercontinent that broke up to form North America and parts of Europe.

LAVA Molten rock material that rises from the Earth's interior, as in a volcanic eruption. BASALT is a typical lava.

LEPTOSPORANGIATE FERNS "True" ferns. The general term "ferns" includes several groups of similar appearance which are not MONOPHYLETIC in origin.

LICHEN A symbiotic organism consisting of a FUNGUS and CYANOBACTERIUM or green alga; only modern species are known.

LIGNITE A soft brown form of COAL.

LIMESTONE A CARBONATE SEDIMENTARY rock consisting primarily of CALCITE, which may be derived from solution in seawater or by the accumulation of animal shells.

LITHIFICATION The hardening of sediment until it forms rock.

LITHISTID DEMOSPONGE A marine DEMOSPONGE with a rigid skeleton. Lithistids have existed since the Cambrian.

LITHOSPHERE The solid outer layer of the Earth, about 60mi (100km) deep, consisting of the CRUST and the uppermost part of the MANTLE. It is segmented into plates floating on the more fluid ASTHENOSPHERE.

LITOPTERN One of a group of extinct South American UNGULATES, some of them horselike, from Tertiary times.

LOBE-FINNED FISH A bony fish belonging to the subclass sauropterygii. Lobefins are distinguished from ray-finned fish by the fleshy lobes that support the fins. They are considered ancestors of the AMPHIBIANS and hence of all land VERTEBRATES.

LOWLAND A part of land where accumulation processes prevail over destruction.

"LUCY" The first known FOSSIL skeleton of the early human ancestor *Australopithecus afarensis*: an adult female AUSTRALOPITHECINE discovered in Ethiopa in 1974.

LYCOPOD/LYCOPSID See CLUBMOSS.

M

MAFIC CRUST The relatively heavy rock material that underlies the oceans. The principal constituents are magnesium and FELDSPAR.

MAGMA Molten rocky material that forms in the Earth's lower crust or MANTLE. When it solidifies, it is known as IGNEOUS ROCK; when it erupts at the surface, it is LAVA.

MAMMAL Any member of the vertebrate class Mammalia, which contains about 4000 species. Their most distinctive characteristic is mammary (milk) glands in the female. There are three orders: PLACENTAL, MARSUPIAL, and MONOTREMES. Placentals are the most common, monotremes the least.

MANICOUAGAN EVENT A meteorite impact in Quebec, Canada, at the end of the Triassic.

MANTLE The section of the Earth's structure that lies between the thin outer crust and the core. At almost 1800mi (2900km) thick, the mantle comprises the greatest part of the Earth's volume. Like that of other terrestrial planets, it is composed of dense silicates of iron and magnesium, whereas the mantles of the gaseous outer planets are thought to be mostly hydrogen.

MANTLE PLUME An upwelling lobe or jet of hot, partly molten material rising within the Earth's mantle. Plumes are believed to give rise to volcanic islands away from the edges of continental plates, such as Hawaii.

MARGINAL SEA A semi-closed sea attached to a continent and formed during RIFTING and early spreading.

MARSUPIAL A member of an order of MAMMALS that nurture their immature young in a pouch. Kangaroos and wombats are among the few modern marsupials, but in Tertiary times the group was widespread.

MASS EXTINCTION The EXTINCTION of a significant proportion of the Earth's organisms over a wide area in a short span of time.

MEGALONYCHID A member of a group of extinct giant ground sloths dating from the late Tertiary and Quaternary.

MESONYCHID A member of a group of primitive omnivorous UNGULATE MAMMALS dating from the early Tertiary. They included the wolf-sized *Mesonyx* and the giant *Andrewsarchus*.

MESOPELAGIC The middle zone of the WATER COLUMN and the organisms that inhabit it.

MESOZOIC The era of geological time from 248 to 65 million years ago, encompassing the Triassic, Jurassic, and Cretaceous periods.

MESSINIAN CRISIS The biological disruption caused by a lowering of sea levels at the end of the Miocene. The Antarctic ice sheets expanded and the Mediterranean dried up.

METAMORPHIC BELT An elongated region of METAMORPHIC ROCKS exposed at the core of an ancient FOLD mountain chain.

METAMORPHIC ROCK A rock, usually SEDIMENTARY in origin, that is subjected to such heat or pressure that it recrystallizes into new minerals without leaving the solid state. If it melted at any time, the result is an IGNEOUS ROCK.

METAZOAN Any multicellular animal whose cells are organized in tissues—that is, all animals except PROTOZOANS. Metazoans appeared in the latest Proterozoic.

METEORITE A piece of interstellar rocky debris, particularly one that has fallen to Earth.

MIACID A member of a group of primitive carnivorous MAMMALS from the early Tertiary, which diversified into VULPAVINES and VIVERRAVINES—the dog and cat branches.

MICROPLATE A small LITHOSPHERIC plate, usually composed mainly of FELSIC rocks.

MID-OCEAN RIDGE The raised feature on the ocean floor where new OCEANIC CRUST emerges, spreading laterally on each side. It forms a long ridge with volcanoes, HYDROTHERMS, and RIFT VALLEYS along the crest. Such ridges are frequently offset by TRANSFORM FAULTS which assist the MANTLE's attempts to "bend" the brittle margins of LITHOSPHERIC PLATES. Ridges may develop above CONVECTION cells in the mantle.

MILANKOVITCH CYCLE The cycle of changes in the Earth's movements (ECCENTRICITY of the orbit, PRECESSION of the rotational axis, and obliquity), invoked first as an explanation for ICE AGES in the late eighteenth century, and revived by Milutin Milankovitch.

MINERAL A naturally formed inorganic chemical substance with a particular chemical composition. A rock consists of crystals of several different types of mineral.

MINERALIZED SKELETON A skeleton made of minerals: mostly CARBONATES, phosphates, and silicon oxides.

MITOCHONDRIA A microscopic structure in the living cells of PROKARYOTES that provides energy for the cell's working. It may be a descendant of small bacteria that were trapped within larger ones.

MITOCHONDRIAL DNA (mtDNA) DNA found in mitochondria. It evolves more rapidly than nuclear (ordinary) DNA, so it can be used to trace the divergence of populations, and is passed down only through the maternal line (see MITOCHONDRIAL EVE). The small variation in mtDNA found in humans supports the out of Africa hypothesis of human origins.

MITOCHONDRIAL EVE The nickname for the hypothetical female ancestor who was the source of all MITOCHONDRIAL DNA in humans today.

MOBILE BELT A region of intense geological activity located along a plate margin. Mobile belts are characterized by volcanism, seismic activity, and mountain-building.

MOHOROVICIC DISCONTINUITY (Moho) The boundary between the Earth's crust and its mantle at which the speed of seismic waves increases sharply. The depth of the Moho varies, from about 6mi (10km) below the ocean floor to 20mi (35km) below continents and 40mi (70km) below mountains.

MOLASSE An assemblage of usually coarse-grained, non-marine SEDIMENTARY ROCKS formed by rapidly eroding new mountains.

MOLLUSK Any of the INVERTEBRATE phylum Mollusca, which appeared in the Cambrian.

MONOPLACOPHORAN An early single-valved marine MOLLUSK possessing a caplike, bilaterally symmetrical calcareous shell.

MONOPHYLETIC A group that contains all the descendants of a single common ancestor.

MONOTREME A member of an order of MAMMALS that lay eggs. The echidna and platypus are the only living monotremes.

MORAINE Rocky debris picked up by a GLACIER, carried along and deposited elsewhere.

MUDSTONE A fine-grained SEDIMENTARY ROCK formed by the consolidation of mud. It resembles SHALE but lacks distinct fine bedding.

MULTITUBERCULATE A member of an order of primitive rodentlike MAMMALS from the Mesozoic and early Tertiary. They may have been the first herbivorous mammals.

MUTATION A change in the genetic makeup of an organism produced by an alternation in its DNA. Mutations, the raw material of EVOLUTION, result from mistakes during replication (copying) of DNA. Only beneficial mistakes are therefore favored by NATURAL SELECTION.

MYSTICETE A BALEEN whale.

N

NANNOPLANKTON Microscopic plankton.

NATURAL SELECTION The primary mechanism of EVOLUTION, first stated by Charles Darwin, by which gene frequencies in a population change through certain individuals producing more descendants than others. Because most environments are slowly but continuously changing, natural selection enhances the reproductive success of individuals that possess favorable characteristics. The process is slow, relying on random variation in the genes of an organism due to MUTATION and on genetic recombination during sexual reproduction.

NAUTILOID A member of a subclass of CEPHALOPODS with coiled shells, related to the AMMONITES and GONIATITES. Abundant in the Early Paleozoic, they are all but extinct.

NEANDERTAL A member of the subspecies *Homo sapiens neanderthalensis*, closely related to modern humans (*H. sapiens sapiens*), preceding them for much of the Pleistocene. Named after Germany's Neander valley where they were discovered.

NEKTONIC An organism that actively swims in the WATER COLUMN, as opposed to floating.

NEOLITHIC The "New Stone Age," a culture characterized by the use of advanced stone tools and the development of agriculture around the end of the Pleistocene ice age.

NEVADAN OROGENY The mountain-building episode along the west coast of North America during the Jurassic and Early Cretaceous periods, which contributed to the Western Cordillera.

NEW RED SANDSTONE The sequence of terrigenous sedimentary rocks deposited in the supercontinent of Laurasia during the Permian and Triassic periods.

NEW WORLD MONKEYS *See* PLATYRRHINES.

NICHE The ecological role of a species; the total combination of environmental factors to which a species is adapted. In biological terms, the space in a particular environment that is occupied by a particular creature on account of its lifestyle.

NONCONFORMITY A type of stratigraphic UNCONFORMITY that separates a bed of rocks from crystalline rocks below.

NON-SEQUENCE A gap in the stratigraphic succession which arises because deposits from a particular time were never laid down, or because these deposits have subsequently eroded completely. The existence of a gap is proven by other paleontological evidence.

NOTHOSAUR A member of a group of swimming reptiles from the Triassic, and a forerunner of the PLESIOSAURS.

NOTOCHORD A flexible support along the length of the body of certain wormlike animals. The notochord is ancestral to the vertebral column in VERTEBRATES.

NOTOUNGULATE A primitive South American hoofed MAMMAL. *See also* UNGULATE.

NUEE ARDENTE A fast-moving "glowing avalanche" consisting of hot volcanic ash, fine dust, molten lava fragments, and hot gases associated with a volcanic eruption.

O

OBIK SEA A shelf sea that existed in the early Tertiary across part of Russia to the east of the Ural mountains.

OCEAN RIDGE *See* MID-OCEAN RIDGE.

OCEAN TRENCH The deepest part of an ocean, an elongated depression pulled down as one plate slides beneath another during the process of PLATE TECTONICS. ISLAND ARCS usually form at the rim of an ocean trench.

OCEANIC CRUST The relatively heavy basaltic rock, an average of 5mi (8km) thick, that underlies the oceans. Its main constituents are magnesium and feldspar, and the bottom layer gives way to GABBRO and perioditic rocks at the MOHOROVIC DISCONTINUITY.

ODONTOCETE A toothed whale.

OIL SHALE A fine-grained SEDIMENTARY ROCK, formed from the lithification of mud, rich in organic matter, easily split into fine layers or flakes, and combustible.

OLD RED SANDSTONE The sequence of terrigenous SEDIMENTARY ROCKS deposited during the Devonian period from the newly arisen Acadian–Caledonian mountain range, forming thick repeating beds of SANDSTONE.

OLDOWAN CULTURE A stone-tool-making culture that existed in Africa's OLDUVAI GORGE (Tanzania) in the early Pleistocene.

OLDUVAI GORGE A site in Tanzania, in the East African Rift Valley, where there have been many important finds of fossil HOMINIDS since the 1970s, including "LUCY."

OMOMYID A member of a family of primitive lemurs from early Tertiary times.

ONYCHOPHORAN Common name velvet worm: a terrestrial segmented invertebrate with numerous rigid telescopic legs. Velvet worms appeared in the Carboniferous.

OOLITE LIMESTONE formed by tiny particles of CALCITE precipitated out of seawater.

OPHIOLITE An assemblage of rocks—mostly BASALT, GABBRO, and JASPER—representing the remains of OCEANIC CRUST pushed onto land during continental collision and OROGENY.

ORDER A grade of zoological classification. A class may contain several orders, and an order several families. The primates are an order within the class Mammalia and contain several families such as the Omomyidae and the Hominidae.

ORDOVICIAN RADIATION A rapid increase of animal diversity, biomass, and size, as well as the appearance of new groups of CORALS, BRYOZOANS, BRACHIOPODS, TRILOBITES, OSTRACODES, and other INVERTEBRATES and CHORDATES, in the Early and Middle Ordovician.

ORGANIC Anything that contains carbon, except CARBONATES and the oxides of carbon; thus, organic substances include all living things and their products.

ORIGINAL HORIZONTALITY, PRINCIPLE OF The principle of geology that states that all strata are laid down horizontally.

ORIGINAL LATERAL CONTINUITY, PRINCIPLE OF The principle of geology that states that similar rock strata separated by any erosional feature, such as a valley, were originally laid down together.

ORNITHISCHIAN The herbivorous "bird-hips," one of two major groups of DINOSAURS—though not the line from which *ARCHAEOPTERYX* and the birds evolved.

ORNITHOPODA A line of Jurassic–Cretaceous two-footed plant-eating dinosaurs descended from the ORNITHISCHIANS. It included *Camptosaurus, Hadrosaurus*, and *Iguanodon*.

OROGENY An episode of mountain-building.

OSTEICHTHYAN A jawed fish whose skeletal cartilage is ossified partly or completely. Osteichthyans appeared in the Devonian.

OSTEOSTRACAN A member of a group of early jawless fish, with well-defined skeletons and paired fins, dating from the early Paleozoic.

OSTRACODE A microscopic aquatic CRUSTACEAN that appeared in the Ordovician.

OUT OF AFRICA HYPOTHESIS The widely accepted theory that human beings evolved in Africa and then spread throughout the world, as opposed to their evolving from an already widespread ancestral stock (the little- accepted "multiregional hypothesis").

OVULE A reproductive structure in seed plants that develops into a seed once fertilized.

OZONE LAYER A layer of the atmosphere, particularly rich in ozone gas, that filters out ultraviolet rays from the sun, preventing GLOBAL WARMING and the GREENHOUSE EFFECT. It is vulnerable to atmospheric pollution.

P

PACK ICE Blocks of floating ice compacted together to form a solid surface on the sea.

PALEOASIAN OCEAN The ocean that separated Siberia and eastern GONDWANA in the latest Proterozoic and early Paleozoic.

PALEOLITHIC The "Early Stone Age," an Early Pleistocene culture represented by the earliest and most primitive stone tools. *See* OLDOWAN CULTURE.

PALEOMAGNETISM The study of the condition of the Earth's magnetic field and its properties in the geologic past. This magnetic field leaves an impression in the rock formed at the time, and this gives clues as to the position of poles and continents in history.

PALEOTHETIS OCEAN A body of water, forerunner of the THETIS OCEAN, that existed as a vast embayment into PANGEA almost separating LAURASIA from GONDWANA in the middle and late Paleozoic.

PALEOZOIC The era of geological time from 545 to 248 million years ago, encompassing the Cambrian, Ordovician, and Silurian periods (the early Paleozoic) and the Devonian, Carboniferous, and Permian periods (the late Paleozoic).

PALEOZOIC FAUNA The animals produced by the ORDOVICIAN RADIATION and subsequent diversification. Most of them (TRILOBITES, GRAPTOLITES, and others) vanished by the end of the Paleozoic, but a few (CEPHALOPODS, ECHINODERMS) persisted into modern times.

PANGEA The late Paleozoic–early Mesozoic supercontinent comprised of every major continental landmass.

PANTHALASSA OCEAN The single ocean that covered the northern hemisphere in the Paleozoic and early Mesozoic. It was a predecessor of the Pacific Ocean.

PARAPHYLETIC A group evolved from more than one ancestor (the opposite of MONOPHYLETIC). The dinosaurs may be paraphyletic because the two major groups (saurischians and ornithischians) may have evolved independently.

PARATETHYS The shallow sea in the area of the Black and Caspian seas in late Tertiary times.

PAREIASAUR A member of a group of herbivorous REPTILES from the Permian period. They were big heavy animals and may have been closely related to the ancestors of the turtles.

PECORAN A member of the group of even-toed UNGULATES encompassing deer and giraffes.

PELAGIC Describing an organism that inhabits the open sea, including forms that are free-swimming (NEKTONIC) and those that float passively (PLANKTONIC).

PELLET CONVEYOR A natural water-purification system that evolved in the Cambrian, when microscopic ZOOPLANKTON began to remove the organic waste of other animals from the top of the water to the seabed, where DETRITUS FEEDERS could make use of it.

PELYCOSAUR A member of the most primitive group of mammal-like REPTILES, many of which had sails on their backs.

PENTASTOME A parasitic CRUSTACEAN that appeared in the Cambrian. It is also known as the tongue worm for its flat soft body covered by a soft ringed cuticle.

PERISSODACTYL An odd-toed UNGULATE.

PERMAFROST The permanently frozen soil and subsoil in the Arctic and sub-Arctic regions of the Earth.

PETROLEUM Crude oil, which forms from decayed ORGANIC matter found concentrated in rock structures called traps, and is extracted as the raw material for industry.

PHANEROZOIC The geological eon from the beginning of the Paleozoic onwards, when the first identifiable fossils formed.

PHOSPHORITE A SEDIMENTARY ROCK consisting of phosphate minerals as guano, shells, or bacterial deposit.

PHOTIC ZONE The depth to which light penetrates a body of water sufficiently to permit PHOTOSYNTHESIS—up to about 300ft (100m).

PHOTOSYNTHESIS The process by which a plant extracts the energy of sunlight and uses it to manufacture food from water and carbon dioxide in the atmosphere.

PHYLOGENY The sequence of changes that occurs in a given species or other taxonomic group during the course of EVOLUTION.

PHYLUM A category of organisms that consists of one or more similar or closely related classes. Related phyla are grouped together in a kingdom. Chordata and Mollusca are two examples of phyla.

PHYTOPLANKTON An algal PLANKTON. Phytoplankton consist mostly of ALGAE and carry out almost all photosynthesis in the oceans. They are the basis of the FOOD CHAIN.

PINNIPED A member of the group of carnivorous MAMMALS that encompasses the seals and walruses.

PLACENTAL A member of the order of MAMMALS that nurture their young in the uterus before birth. This encompasses all modern mammals except the MARSUPIALS and the egg-laying MONOTREMES.

PLACODERM A Silurian–Devonian jawed cartilaginous fish with a head covered by rigid bony shields.

PLACODONT One of a group of marine REPTILES, mostly slow-moving shellfish-eaters and some with turtle-like shells, from Triassic times.

PLANKTON Small, often microscopic free-floating organisms that inhabit the top layer of the water column and are an important food source for larger animals. They include PHYTOPLANKTON and ZOOPLANKTON.

PLATE TECTONICS The theory that invokes the movement and interaction of LITHOSPHERIC PLATES as an explanation for CONTINENTAL DRIFT, SEAFLOOR SPREADING, volcanism, earthquakes, and mountain-building.

PLATYRRHINE A member of the NEW WORLD MONKEY group, characterized by a narrow nose and usually prehensile tail, which were not involved in the evolution of humans (*see also* CATARRHINES).

PLAYA An enclosed flat BASIN in a desert, usually occupied in part by an ephemeral (seasonal) lake or lakes. When these dry out, they form EVAPORITE deposits.

PLESIOSAUR A carnivorous swimming REPTILE of the Mesozoic. Plesiosaurs had turtle-like bodies and paddles and long necks.

PLIOSAUR One of a group of large-headed, short-necked Mesozoic swimming REPTILES.

PLUME *See* MANTLE PLUME.

PLUTON A body of intrusive IGNEOUS ROCK that has formed beneath the Earth's surface.

PLUVIAL LAKE A lake formed by rain.

POLAR WANDERING The slight changes in position of the Earth's magnetic poles due to CONTINENTAL DRIFT and PLATE TECTONICS.

POLARITY REVERSAL The reversal of the Earth's magnetic field at intervals of 10,000 to 25,000 years, leaving magnetic stripes in seafloor rocks. *See* PALEOMAGNETISM.

POLYCHAETE A mostly marine ANNELID worm that appeared in the Cambrian. Each segment has a pair of fleshy flaps bearing a bundle of bristles (setae).

POLYPLACOPHORAN A multivalved marine MOLLUSK possessing a bilaterally symmetrical, calcareous multiple shell. Polyplacophorans evolved during the Cambrian.

PRECESSION In astronomy, the apparent slow motion of the celestial poles, largely due to the wobbling of the Earth's rotational axis induced by the gravitational pull of the Sun and Moon. The axis gradually changes direction over a cycle of about 26,000 years; this is why the equinoxes occur earlier each succeeding year. *See* MILANKOVITCH CYCLE.

PRECORDILLERA A South American terrane that split from the Appalachian margin of Laurentia (future North America) during the Cambrian. The Andes later formed there.

PREDATOR An animal that kills and eats others.

PRIMATE A member of the highly derived order of MAMMALS that contains the lemurs, the monkeys, the apes, and humans.

PROBLEMATIC FOSSIL A fossil of an organism that has no apparent affinity with any modern phylum. Problematic fossils are especially abundant in Cambrian strata.

PRODUCER An organism that produces organic matter by modifying light (PHOTOSYNTHESIS) or chemicals (chemosynthesis).

PROKARYOTE A very simple cell in which the genetic material is not confined to a nucleus but spread through the cell structure. Only primitive bacteria and cyanobacteria are prokaryotes; all other organisms are EUKARYOTES.

PROSIMIAN One of the basal members of the PRIMATE order, including tarsiers, lemurs, and tree shrews.

PROTEIN A complex organic compound made up of AMINO ACIDS and forming the bulk of a living thing.

PROTOCTIST A member of the kingdom Protoctista (sometimes called Protista), a category for all organisms that are neither bacteria, animals, plants (nor fungi, according to some scientists): that is, they are algae, protozoa, and slime molds (and some fungi).

PROTOSTOME "First mouth": an animal in which the embryonic mouth develops into the adult mouth. It includes most bilaterally symmetrical invertebrates.

PROTOZOAN A unicellular organism, one of the earliest EUKARYOTES.

PSILOPHYTE An old name for ancient VASCULAR PLANTS. It now refers to a number of primitive TRACHEOPHYTES of different origins, namely rhyniophytes, zosterophyllophytes, and trimerophytes.

PSYCHROSPHERE The near-freezing waters of the deepest part of the ocean, formed by CONVECTION, which causes the cold water at the poles to sink.

PTEROBRANCH A mostly colonial dendritic sessile HEMICHORDATE from the Cambrian.

PUNCTUATED EQUILIBRIUM A pattern of EVOLUTION in which periods of comparative stability are interspersed with bursts of increased VARIATION and the formation of new species. The duration of the respective periods varies greatly under different environmental circumstances.

PYCNOGONID (sea spider) An articulated marine INVERTEBRATE, with a thin body and jointed legs, that appeared in the Devonian.

PYGIDIUM The tail portion of a TRILOBITE.

PYRITE A gold-yellow mineral of iron sulfide, and an important source of sulfur and iron.

PYROCLASTIC A SEDIMENTARY ROCK consisting of fragments of volcanic material.

Q

QUARTZ One of the most widespread SILICATE minerals in the Earth's CONTINENTAL CRUST and the chief component of CLASTIC SEDIMENTARY rocks. It is mostly silicon dioxide (SiO_2).

QUATERNARY The era of geological time that encompasses the Pleistocene and the Holocene epochs, and so covers the last ICE AGE and the whole of human history.

R

RADIATION (ADAPTIVE) The process by which a lineage evolves different forms, allowing members to adapt to different lifestyles in different environments.

RADIOACTIVE DECAY The process in which a radioactive element sheds neutrons and changes its atomic number, thereby becoming a completely different substance.

RADIOCARBON DATING A type of RADIOMETRIC DATING using carbon-14, which has a very short half-life, and can be used to date younger rocks (up to about 70,000 years).

RADIOCYATH An Early Cambrian RECEPTACULITID with multi-rayed heads.

RADIOMETRIC DATING The technique used to estimate the age of a rock or mineral by calculating how much of its radioactive matter has decayed since it was formed.

RAISED BEACH A coastal landform consisting of a flat shelf at some height above sea level, indicating the position of sea level at some time in the past. Raised beaches are a product of glaciation and the weight of ice.

RARE EARTH ELEMENTS Chemically active metals such as yttrium, lanthanum, and lanthanides that are rare in the Earth's crust.

RECEPTACULITID A problematic Paleozoic marine sessile calcareous organism with egg-shaped skeletons built of elements arranged in whorls around the central axis.

REDBED A bed of terrigenous SEDIMENTARY ROCK that has oxidized through exposure to air, turning its iron components rusty red. Red beds are often associated with SANDSTONE.

REEF A CARBONATE deposit, formed by accumulated skeletons of CORAL, that forms an important marine habitat. Fringe reefs build up on the shores of continents or islands, the living animals mainly occupying the outer edges; barrier reefs are separated from the shore by saltwater lagoons as much as 18mi (30km) wide; atolls surround lagoons and form where an extinct volcano has subsided.

RELICT POPULATION A group of animals or plants that survive in a limited area after having been much more widespread.

REPTILE Any member of the VERTEBRATE class Reptilia, which includes snakes, turtles, alligators, and crocodiles. Reptiles evolved from AMPHIBIANS in the Carboniferous period. Some ancient forms, such as the PLESIOSAURS and ICHTHYOSAURS, lived in the sea; modern reptiles live on land. They are cold-blooded and reproduce by means of a hard-shelled egg, the device that allowed them to colonize land.

RHABDOSOME The protective casing for a complete colony of GRAPTOLITE zooids.

RHEIC OCEAN The ocean that separated AVALONIA and GONDWANA in the early Paleozoic.

RIFT An elongated depression formed by the downward movement of an area of land between parallel systems of faults. Rifts occur in regions of crustal extension, where lithospheric plates are diverging and a continent is breaking apart. Rifts tend to form valleys.

RING OF FIRE The seismically active borders of the Pacific Ocean, shown by the frequency of earthquakes and the abundance of volcanoes, caused by the Benioff zones produced by the subduction of the Pacific plate, the Cocos plate, and the Nazca plate.

RNA Ribonucleic acid, a nucleic acid present in all cells. Several different type of RNA play a part in the mechanisms by which DNA directs the synthesis of PROTEINS in a cell.

ROCHE MOUTONNEE An exposed rock that has been polished on one side and pulled apart on the other by the passage of a GLACIER over it.

RODINIA The Precambrian supercontinent that consisted of parts of all the modern continents except Africa.

RUGOSAN An extinct solitary or branching horn-shaped modular CORAL from the Ordovician and Permian.

RUMINANT An animal that chews the cud, such as a cow.

S

SALT A compound consisting of a metal and a base, as is formed when an acid has its hydrogen replaced by a metal. Common salt, NaCl, is the sodium salt of hydrochloric acid.

SALT DOME A DIAPIR formed from salt. As a bed of rock salt is compressed it deforms plastically and rises through the overlying beds, twisting them upwards.

SAN ANDREAS FAULT A TRANSFORM FAULT along the coast of California that gives rise to the many earthquakes in the area. Its continued movement will probably cause that part of California to shear off and drift away over the next few million years.

SARCOPTERYGIAN *See* LOBE-FINNED FISH.

SAURISCHIAN A "lizard-hipped" dinosaur, from which (in spite of the name) all birds descended. Saurischians included both meat-eaters and long-necked plant-eaters.

SAUROPOD A member of a group of plant-eating SAURISCHIAN dinosaurs characterized by their long necks.

SAVANNAH A landscape of tropical grasslands with scattered trees, typical of the area between the Earth's equatorial rainforests and the tropical desert belt.

SCAVENGER An animal that feeds on the flesh of dead animals killed by other animals.

SCHIST A METAMORPHIC ROCK (such as mica) that tends to be split into layers by increased temperature and pressure, causing the FELSIC and MAFIC constituents to separate. This gives metamorphic rock a banded appearance with alternate layers of felsic and mafic crystals.

SCLERACTINIAN A member of the order (Scleractinia) to which most corals have belonged since the Paleozoic. It includes modern corals.

SCLERITE A scalelike or spinelike hollow element of a skeletal cover (SCLERITOME).

SCLERITOME A skeletal cover consisting of isolated SCLERITES.

SEAFLOOR SPREADING The process by which the ocean floor grows as new crust emerges and moves laterally away from MID-OCEAN RIDGES. Observation of this in the 1960s, combined with the theory of CONTINENTAL DRIFT, generated the idea of PLATE TECTONICS.

SEAMOUNT An isolated submarine uplift more than 3250ft (1000m) high.

SEDIMENTARY ROCK A rock formed by the accumulation and solidification of layers of fragments to form a solid mass. There are three types: CLASTIC, such as SANDSTONE, in which the fragments are derived from the breakdown of pre-existing rocks; BIOGENIC, such as COAL, in which the fragments are derived from once-living things; and chemical, such as rock salt, which is formed from crystals precipitated from solution in water.

SEED FERNS An extinct group of varied Carboniferous plants that reproduced by means of seeds rather than spores. However, they were not technically "ferns" at all, just fernlike in appearance.

SEISMIC WAVES Vibrations sent out by an earthquake. They take a number of forms, including P-waves, which are waves of compression; S-waves, which cause the shaking action; and L-waves that travel on the surface and cause the damage.

SEISMOLOGY The study of earthquakes and of the passage of vibrations through the Earth.

SEPTUM A wall of bone or shell that divides a hollow in a skeleton into separate chambers.

SESSILE Describing an immobile organism that lives on the sea floor.

SEVIER OROGENY An episode of IGNEOUS and FOLD-AND-THRUST activity north of California during the Cretaceous period.

SERIES A stratigraphic unit consisting of rocks deposited or emplaced during an epoch.

SHALE A fine-grained SEDIMENTARY ROCK, formed from the solidification of mud, easily split into fine layers or flakes.

SHELL BED A CARBONATE or phosphate bed consisting of fossil shells.

SHELF A continental margin that is covered with water, forming shallow seas.

SHELF SEA A sea that covers a continental shelf and is much shallower than true ocean. The North Sea is an example.

SHIELD Another term for a continental CRATON.

SHOCKED QUARTZ A mineral such as QUARTZ or FELDSPAR with closely-spaced microscopic layers within it, caused by the intensely high pressures of an impact shock, as when a meteorite strikes the Earth.

SIDEROPHILE ELEMENT A chemical element that has an affinity for the metal phase; for example, iron or nickel. During the Earth's formation the siderophile elements sank towards the core.

SILICATE A MINERAL that is a compound of metallic elements with silicon and oxygen.

SILICICLASTIC A SILICATE and MINERAL sediment derived mostly from weathering land.

SILL An emplacement of IGNEOUS ROCK, intruded between beds of SEDIMENTARY ROCK.

SIPHUNCLE A canal that runs through all the chambers of a NAUTILOID or AMMONITE shell, adjusting the air pressure to affect buoyancy.

SOLAR NEBULA The cloud of dust and gas from which the solar system eventually condensed after the BIG BANG.

SONOMA OROGENY A mountain-building event at the Permian–Triassic boundary as an eastward-moving ISLAND ARC collided with the Pacific margin of North America.

SPECIATION The process by which new SPECIES appear and change over time.

SPECIES The basic level of taxonomic classification; a group of organisms that can interbreed and produce fertile offspring. Related species are grouped together in a genus.

SPHENOPSID (Equisiophyta) A group of spore-bearing plants such as the giant horsetail *Calamites*, common in the Late Paleozoic.

SPICULE A tiny needle-like calcareous or siliceous structure, forming part of the skeleton of an INVERTEBRATE animal.

SPONGE A primitive sessile aquatic multicellular animal with an aquiferous system and a body enclosed by a covering tissue. Sponges appeared in the latest Proterozoic.

SPORANGIUM The structure on a plant that holds the SPORES.

SPORE A reproductive plant body that consists mostly of a cell with half the viable number of CHROMOSOMES. It must unite with another spore before growing into a plant.

STABLE ZONE A region of the Earth's crust that is not subject to OROGENY or other deformational processes. Stable zones are typically found within continental interiors, away from margins and MOBILE BELTS.

STAGE A stratigraphic unit smaller than a series or epoch.

STRATA Layers or beds of SEDIMENTARY ROCK.

STRATIGRAPHY The study of the relationships, classification, age, and correlation of rock STRATA that lie at or near to the surface of the planet. The succession of these rocks enables scientists to build up a geological history of the Earth.

STRIKE-SLIP FAULT A FAULT in which one body of rock moves sideways rather than vertically in relation to the next.

STROMATOLITE A laminated structure formed in quiet water when a layer of filamentous ALGAE traps sedimentary particles, chiefly CARBONATE. Another layer of algae grows on this sedimentary surface, trapping another layer, so building up a dome shape or a column. Fossil stromatolites are known from Precambrian times, when there were no other life forms to disturb their growth.

STROMATOPOROID One of a group of extinct calcified marine SPONGES that built REEFS.

SUBAERIAL Taking place on land. REDBEDS are subaerial SEDIMENTARY ROCKS.

SUBDUCTION The movement of one LITHOSPHERIC plate as it slides beneath another into the MANTLE and is consumed. The process is an integral part of PLATE TECTONICS.

SUBDUCTION ZONE The inclined region where subduction of the LITHOSPHERE occurs.

SUPERCONTINENT A continent made up from the amalgamation of more than one continental mass.

SUPERNOVA The explosion that results when the structure of a star collapses.

SYMBIONT An organism that coexists with and depends on another organism.

SYNAPSID A member of one of the major subclasses of REPTILES, including mammal-like reptiles and MAMMALS. They had a characteristic skull with an extra opening on each side. *See* DIAPSID.

SYSTEM (geological) A stratigraphic unit consisting of all the rocks deposited or emplaced during a geological period.

T

TABULATE A form of CORAL that existed from the early to the late Paleozoic.

TACONIC OROGENY An early phase of the APPALACHIAN OROGENY, which took place during the Ordovician during the accretion of ISLAND ARCS to Laurentia.

TAENIODONT A member of a group of primitive MAMMALS from the earliest Tertiary.

TARDIGRADE A microscopic INVERTEBRATE that evolved in the Cambrian. It possessed four segments covered with a firm cuticle and bearing a pair of telescopic legs each.

TARDIPOLYPOD A Cambrian marine wormlike INVERTEBRATE with a segmented body and multiple telescopic legs.

TAXONOMY The classification of organisms into groups (taxa). The basic unit is the SPECIES, progressing up through genus, family, order, class, phylum, and kingdom.

TECTONIC PLATE A section of LITHOSPHERE that moves as a unit during PLATE TECTONICS. Usually a plate grows along one edge, at a MID-OCEAN RIDGE, and is destroyed along another, at an OCEAN TRENCH.

TECTONOEUSTASY Global sea level fluctuation caused by absolute changes in the quantity of seawater as MID-OCEAN RIDGES grow.

TELEOST A group of fish with a bony skeleton, small rounded scales, and a symmetrical tail. Most modern fish are teleosts.

TERRANE A relatively small block of the Earth's crust that is distinct from those around it.

TERRIGENOUS Relating to rocks or sediment that have been formed of material eroded from landmasses.

TERTIARY The era of geological time between the Mesozoic and the Quaternary. It encompasses the last 65 million years of Earth's history except for the last two million or so.

TETHYS SEAWAY (or Ocean) An oceanic region that existed as a vast embayment into PANGEA, almost separating LAURASIA from GONDWANA. It was lost as Africa and India closed with Europe and Asia, leaving behind the Mediterranean, Black, Caspian, and Aral Seas.

TETRACORAL *See* RUGOSAN.

TETRAPOD Any VERTEBRATE that is not a fish. Although the name means "four-footed," the classification also covers whales, birds, and snakes, whose ancestors were real tetrapods with Devonian origins.

THALLOPHYTE A primitive plant, such as a seaweed, in which the body (thallus) is not divided into roots or stems or leaves, or any of the other features associated with more advanced plants.

THECODONT A crocodile-like Triassic REPTILE ancestral to the dinosaurs.

THERAPSID A member of a group of advanced mammal-like REPTILES.

THEROPOD A member a group of meat-eating SAURISCHIAN dinosaurs.

THRUST FAULT A low-angle fault produced by compression. In mountain-building, thrust sheets (large slices of rock) can slide horizontally over underlying rocks for great distances.

TILL An unsorted mixture of clay and cobbles deposited by GLACIERS in MORAINE.

TILLITE A rock formed by the lithification of TILL.

TILLODONT A member of a group of primitive plant-eating MAMMALS from the early Tertiary, probably closely related to the TAENIODONTS.

TITANOTHERE *See* BRONTOTHERE.

TOMMOTIID A problematic marine INVERTEBRATE of the Cambrian that was covered with phosphatized ribbed SCLERITES.

TORNQUIST SEA A sea that occupied the western part of BALTICA in the early Paleozoic.

TRACE FOSSIL A fossilized trail, track, borehole, burrow, or footprint.

TRACHEOPHYTE A multicellular land plant with distinct tissues and organs and developed vascular system.

TRADE WINDS The prevailing winds that blow towards the equator, caused by hot tropical air rising and drawing in the cooler air from north and south. They blow from the southeast and the northeast, deflected in these directions by the Coriolis effect of the Earth's rotation.

TRANSFORM FAULT A geological fault produced at a MID-OCEAN RIDGE and formed as adjacent TECTONIC PLATES slide past one another.

TRANSGRESSION The gradual encroachment of sea over a land area.

TRAP A stairlike structure formed from successive basaltic LAVA flows over a wide area, as seen in the Deccan and Siberian traps.

TRILOBITE A Paleozoic marine ARTHROPOD that scavenged on the bottom of the shallow seas. Trilobites had plated, segmented bodies with many Y-shaped legs—they resembled modern wood lice. They died out at the end of the Permian but are abundant as fossils in Paleozoic rocks.

TRIPLE JUNCTION A point at which three LITHOSPHERIC plates meet. Ocean ridges often meet at triple junctions, which accounts for the often jagged nature of continental margins.

TROPHIC WEB A system of continuous chains of SPECIES where each link is a species consumable by subsequent species; this web transforms the energy in an ECOSYSTEM.

TUNDRA A landscape of stunted seasonal vegetation, snow-covered in winter and flooded in summer, caused by PERMAFROST and typical of far northern continental areas.

TURBIDITY CURRENT Moving water that contains suspended sediment, making it denser than surrounding water, and causing it to flow at a lower depth along the seafloor.

TYLOPOD A member of a group of even-toed UNGULATES that includes the camels.

U

ULTRAMAFIC CRUST Heavy crust, like MAFIC crust but containing even less silica.

UNCONFORMITY A break in the sequence of deposition of SEDIMENTARY ROCKS, formed when a sequence of rocks is raised above sea level and eroded, and then submerged so that deposition resumes.

UNGULATE Any four-legged hoofed mammal.

UNICELLULAR An organism whose entire body consists of a single cell.

UNIFORMITARIANISM The principle that the natural laws and processes that form present-day rocks and landscapes have remained the same over time, so that ancient geological formations and their processes may be interpreted by observing analogous formations and processes in the world today. This is expressed as "The present is the key to the past." However, the rate at which processes operate may have been different in the distant past, and their relative importance may have changed also.

UNIRAMIAN An ARTHROPOD whose appendages are not branched.

UPWELLING (marine) The movement of deep water, usually off the coast of a continent, that brings nutrients closer to the surface where PLANKTON and other organisms feed.

URALIAN OCEAN The ocean separating Siberia and BALTICA in the early Paleozoic.

UROCHORDATE A sea squirt—a mostly sessile, sacklike marine CHORDATE lacking both chord and notochord when adult. Sea squirts appeared first in the Carboniferous.

U-SHAPED VALLEY A valley that has been ground downwards and sideways by the weight of a GLACIER so that it has a flat bottom and vertical sides.

V

VARIATION A difference between individuals of the same SPECIES, found in any sexually reproducing population, due to genetic or environmental factors or their combination.

VARVE A thin layer of sediment deposited in a glacial lake. GLACIERS melt at different rates between seasons, and the meltwater carries away a different load of TILL. Varves build up as annual cycles of coarse and fine material, which can be used by geologists to investigate a glaciated area.

VASCULAR PLANT A plant with a plumbing system that can carry food and water around its body. All plants more advanced than mosses are vascular plants.

VERTEBRATE Any animal that has a backbone. There are approximately 41,000 vertebrate species, including mammals, birds, reptiles, amphibians, and fish.

VESTIMENTIFERAN A wormlike marine INVERTEBRATE that appeared in the Silurian.

VIVERRAVINE The cat branch of the CARNIVORES: a group of primitive meat-eating MAMMALS that evolved from MIACIDS in the early Tertiary, and the line from which hyenas, mongooses, civets, and all felids (cats) evolved.

VOLATILE ELEMENT Any of the elements that show an affinity for the ATMOSPHERE, including hydrogen, nitrogen, carbon, oxygen, and the inert gases (helium, argon, neon, krypton, and xenon).

VOLCANIC ARC *See* ISLAND ARC.

VULPAVINE The dog branch of the CARNIVORES: a group of primitive meat-eating MAMMALS that evolved from MIACIDS in the early Tertiary, and diversified into bears, foxes, wolves, mustelids, seals and sea lions, pandas, and all true dogs (canids).

W

WALLACE'S LINE The boundary between the BIOGEOGRAPHIC PROVINCES of Australia and southeast Asia, which runs through the strait between Bali and Lombok.

WATER COLUMN A vertical section through the sea or a lake, highlighting the differences in properties of the water at different levels.

WEATHERING The chemical or physical processes by which exposed rock is broken down by rain, frost, wind, and other elements of the weather. It is the beginning of EROSION.

WILLISTON'S LAW The evolutionary principle which states that serially arranged structures in animals, such as teeth and legs, will become fewer and take on new functions as new species evolve. For example, MAMMALS have fewer ribs than do their fish ancestors.

WIWAXIID An extinct COELOSCLERITOPHORAN.

XYZ

XENARTHRAN A member of an order of MAMMALS that includes the armadillos, the anteaters, and the sloths.

ZEUGLODONT Having teeth shaped like arches, as in early whales.

ZONE The shortest time unit used in geology.

ZONE FOSSIL *See* INDEX FOSSIL.

ZOOGEOGRAPHY The study of the distribution of animal life, the animal assemblages of particular areas, and the barriers between distinct BIOGEOGRAPHIC REALMS.

ZOOID An individual, clonally produced unit of a modular animal.

ZOOPLANKTON The animal component of PLANKTON: chiefly PROTOZOANS, small CRUSTACEANS, and the larval stages of MOLLUSKS and other INVERTEBRATES.

FURTHER READING

PARTS 1 & 2

Cairns-Smith, A.G. *Seven Clues to the Origin of Life.* Cambridge, England: Cambridge University Press, 1985.

Cone, J. *Fire Under the Sea.* New York: William Morrow & Co, 1991.

Conway Morris, S. *The Crucible of Creation: The Burgess Shale and the Rise of Animals.* Oxford; New York; Melbourne: Oxford University Press. 1998.

Darwin, C. *On the Origin of Species by Natural Selection.* London: John Murray, 1859.

Decker, R. and Decker, B. *Mountains of Fire.* Cambridge, England: Cambridge University Press, 1991.

Dixon, B. *Power Unseen: How Microbes Rule the World.* New York: WH Freeman and Company, 1994.

Fortey, R. *The Hidden Landscape: A Journey into the Geological Past.* London: Pimlico, 1993.

Glaessner, M. F. *The Dawn of Animal Life.* Cambridge: Cambridge University Press, 1984.

Gould, S. J. *Wonderful Life: The Burgess Shale and the Nature of History.* New York: Norton, 1989.

Gross, M. Grant. *Oceanography: A View of the Earth.* Englewood Cliffs, NJ: Prentice-Hall, 1982.

Hsu, K.J. *Physical Principles of Sedimentology: A Readable Textbook for Beginners and Experts.* New York: Springer Verlag, 1989.

McMenamin, M. A. S. and D. L. S. McMenamin. *The Emergence of Animals. The Cambrian Breakthrough.* New York: Columbia University Press, 1990.

Margulis, L. and Schwartz, K. 1998. *Five Kingdoms: An Illustrated Guide to the Phyla of Life on Earth.* (3rd ed.) New York: WH Freeman and Company.

Norman, D. *Prehistoric Life.* London: Boxtree, 1994.

Sagan, D. and Margulis, L. *Garden of Microbial Delights: A Practical Guide to the Subdivisible World.* Dubuque, IA: Kendall-Hunt, 1993.

Schopf, J.W. *Major Events in the History of Life.* Boston: Jones and Bartlett, 1992.

Stewart, W. N. and G. W. Rothwell. *Palaeobotany and the Evolution of Plants* (2nd edition). Cambridge: Cambridge University Press, 1993.

Rodgers, J.J.W. *A History of the Earth.* Cambridge, England: Cambridge University Press, 1993.

Whittington, H. B. *The Burgess Shale.* New Haven: Yale University Press, 1985.

Wood, R. *Reef Evolution.* New York: Oxford University Press, 1999.

PARTS 3 & 4

Alvarez, W. *T. Rex and the Crater of Doom.* Princeton, NJ: Princeton University Press, 1997.

Bakker, R.T. *The Dinosaur Heresies.* New York: William Morrow & Co, 1986.

Brusca, R.C. and Brusca, G.J. *Invertebrates.* Sunderland, Mass.: Sinauer Associates, 1990.

Currie, P.J. and Padian, K. *Encyclopedia of Dinosaurs.* San Diego: Academic Press, 1996.

Dingus, L. and Rowe, T. *The Mistaken Extinction: Dinosaur Evolution and the Origin of Birds.* New York: W.H. Freeman and Company, 1997.

Erwin, D.H . *The Great Paleozoic Crisis: Life and Death in the Permian.* New York: Columbia University Press, 1993.

Feduccia, A. *The Origin and Evolution of Birds.* New Haven: Yale University Press, 1996.

Fraser, N.C. and Sues, H-D. *In the Shadow of the Dinosaurs: Early Mesozoic Tetrapods.* Cambridge, England: Cambridge University Press, 1994.

Kenrick, P. and Crane, P. *The Origin and Early Diversification of Land Plants.* Washington, DC: Smithsonian Institution Press, 1997.

Lambert, D. *Dinosaur Data Book.* New York: Facts on File, 1988.

Lessem, D. *Dinosaur Worlds.* Hondsale, Pennsylvania: Boyd's Mill Press, 1996.

Long, J.A. *The Rise of Fishes.* Baltimore, MD and London: The Johns Hopkins University Press, 1995.

Savage, R.J.G. and Long, M.R. *Mammalian Evolution: An Illustrated Guide.* London: British Museum of Natural History, 1987.

Thomas, B.A. and Spicer, R.A. *The Evolution and Paleobiology of Land Plants.* London: Croon Helm, 1987.

PARTS 5 & 6

Alexander, David. *Natural Disasters.* London: University College Press, 1993.

Andel, T. van. *New Views of an Old Planet.* Cambridge, England: Cambridge University Press, 1994.

Goudie, A. *Environmental Change.* London: Clarendon Press, 1992.

Hsu, K.J. *The Mediterranean Was a Desert.* Princeton, NJ: Princeton UP, 1983.

Johanson, D.C. and Edey, M.A. *Lucy: The Beginnings of Humankind.* New York: Simon and Schuster, 1981.

Lamb, H.H. *Cimate, History and the Modern World.* London: Routledge, 1995.

Lewin, R. *The Origin of Modern Humans.* New York: Scientific American Library, 1993.

McFadden, B.J. *Fossil Horses.* Cambridge, England: Cambridge Univesity Press, 1992.

Pielou, E.C. *After The Ice Age: The Return of Life to Glaciated North America.* Chicago: University of Chicago Press, 1991.

Prothero, D.R. *The Eocene-Oligocene Transition: Paradise Lost.* New York: Columbia University Press, 1994.

Stanley, S.M. *Children of the Ice Age: How a Global Catastrophe Allowed Humans to Evolve.* New York: W.H. Freeman and Company, 1998.

Tattersall, Ian. 1993. *The Human Odyssey: Four Million Years of Human Evolution.*

Tudge, C. *The Variety of Life: A survey and a celebration of all the creatures that have ever Lived.* Oxford, England: Oxford University Press, 2000.

Young, J.Z. *The Life of Vertebrates* (2nd ed.) Oxford, England: Oxford University Press, 1962.

ACKNOWLEDGMENTS

AL Ardea London
BCC Bruce Coleman Collection
C Corbis
NHM Natural History Museum, London
NHPA Natural History Photographic Agency
OSF www.osf.uk.com
PEP Planet Earth Pictures
SPL Science Photo Library

2-3 © Kevin Schafer/C; **3** NHM; **10–11** & **12–13** Royal Observatory, Edinburgh/AATB/SPL; **16** NASA/SPL; **18t** Bernhard Edmaier/SPL; **20–21** © NASA/Roger Ressmeyer/C; **23** Dr. Ken Macdonald/SPL; **26–27** © Buddy Mays/C; **28** SPL; **30** Sinclair Stammers/SPL; **32** E.A. Janes/NHPA; **33** M.I. Walker/NHPA; **35** © W. Perry Conway/C; **36** CNRI/SPL; **37** Volker Steger/SPL; **38** © Stuart Westmorland/C; **39** © Manuel Bellver/C; **40** Bruce Coleman Inc.; **42** Manfred Kage/SPL; **45** © C; **47t** A.N.T./NHPA; **47b** RADARSAT International Inc.; **48** © Kevin Schafer/C; **50–51** © Ralph White/C; **51** SPL; **54** Sinclair Stammers/SPL; **55** Martin Bond/SPL; **59** © James L. Amos/C; **60** Image Quest 3-D/NHPA; **61** © Kevin Schafer/C; **65** © Stuart Westmorland/C; **66–67** & **68–69** Paul Kay/OSF; **71** Andrey Zhuravlev; **74** P.D. Kruse; **77** Digital image © 1996 C: Original image courtesy of NASA/C; **78** Andrey Zhuravlev; **80t** S. Conway Morris, University of Cambridge; **81** © Raymond Gehman/C; **85** Andrew Syred/SPL; **87** © Raymond Gehman/C; **90t** © David Muench/C; **90b** Breck P. Kent/OSF; **92–93** Rick Price/Survival Anglia/OSF; **94** & **96** Sinclair Stammers/SPL; **99t** P.D. Kruse; **99c** Andrey Zhuravlev; **102** © James L. Amos/C; **106** Laurie Campbell/NHPA; **108** © Ralph White/C; **110** Jens Rydell/BCC; **112** Sinclair Stammers/SPL; **115** Breck P. Kent/Animals Animals/OSF; **116** Sinclair Stammers/SPL; **120** Norbert Wu/NHPA; **122–123** & **124–125** Alfred Pasieka/SPL; **130** Jane Gifford/NHPA; **132** Jon Wilson/SPL; **132–133** © Jonathan Blair/C; **135t** NHM; **135b** & **139** © James L. Amos/C; **144** Oxford University Museum of Natural History; **144–145** © Patrick Ward/C; **147** Trustees of The National Museums of Scotland; **149** Richard Packwood/OSF; **155** © David Muench/C; **157** Tony Craddock/SPL; **162** George Bernard/SPL; **168** Tony Waltham/Geophotos; **169b** NHM; **170** Brenda Kirkland George, University of Texas at Austin; **171** © Buddy Mays/C; **172** Hjalmar R. Bardarson/OSF; **177** © Jonathan Blair/C; **178–179** & **180–181** François Gohier/AL; **187t** © Scott T. Smith/C; **187b** NHM; **188** © David Muench/C; **189t** Jane Burton/BCC; **189b** © Kevin Schafer/C; **193** C. Munoz-Yague/Eurelios/SPL; **199t** © C; **200** Jane Burton/BCC; **201** NHM; **202** François Gohier/AL; **204** © James L. Amos/C; **210** Ken Lucas/PEP; **212** © Michael S. Yamashita/C; **216** Ron Lilley/BCC; **218–219** U.S. Geological Survey/SPL; **220** Martin Bond/SPL; **223 & 224** François Gohier/AL; **224–225** Louie Psihoyos/Colorific; **228–229** © C; **231** SPL; **234–235** & **236–237** Jeff Foott/BCC; **240** John Mason/AL; **242** Digital image © 1996 C: Original image courtesy of NASA/C; **244** Patrick Fagot/NHPA; **246** © Douglas Peebles/C; **248–249** Dr. Eckart Pott/BCC; **250** Tony Waltham/Geophotos; **251** NHM; **252t** S. Roberts/AL; **253t** John Sibbick; **253b** NHM; **254** Bruce Coleman Inc.; **256** © Jonathan Blair/C; **260** AL; **263** Anup Shah/PEP; **266** CNES, 1986 Distribution Spot Image/SPL; **270–271** © Michael S. Yamashita/C; **272–273** © Liz Hymans/C; **273** Digital image © 1996 C: Original image courtesy of NASA/C; **274** François Gohier/AL; **275** B & C Alexander/PEP; **277b** BCC; **280** Ferrero-Labat/AL; **280–281** NHM; **282–283** © Sally A. Morgan; Ecoscene/C; **283t** & **283b** NHM; **287** G.I. Bernard/NHPA; **288** Nigel J. Dennis/NHPA; **289** Andy Rouse/NHPA; **290–291** & **292–293** F. Jalain/Robert Harding Picture Library; **298** Peter Steyn/AL; **300–301** Simon Fraser/SPL; **302** Wardene Weisser/AL; **303** David Woodfall/NHPA; **304–305** M. Moisnard/Explorer; **306** François Gohier/AL; **308** Kevin Schafer/NHPA; **309 NHM**; **310l** NASA/SPL; **310r** inset Jane Gifford/NHPA; **311** Chris Collins, Sedgwick Museum, University of Cambridge; **312l** Volker Steger/-Nordstar-4 Million Years of Man/SPL; **313** NHM; 314 J.M. Adovasio/Mercyhurst Archaeological Institute; **315** © Gianni Dagli Orti/C; **318** © Peter Johnson/C; **321** Sheila Terry/SPL; **322** inset © Charles & Josette Lenars/C; **324** NASA/SPL; **326** Matthew Wright/Been There Done That Photo Library; **328–329** © Galen Rowell/C; **330–331** Tom Bean; **331** Luiz Claudio Marigo/BCC; **332–333** A.N.T./NHPA; **333** Felix Labhardt/BCC; **334t** © Mike Zens/C; **334b Adrian** Warren/AL; **335t** © Robert Pickett/C; **335b** © Eric Crichton/C; **336c** Steven C. Kaufman/BCC; **336b** © Clem Haagner; Gallo Images/C; **336–337** Gunter Ziesler/BCC; **339** Jeff Foott/BCC; **340 Erich** Lessing/Archiv für Kunst und Geschichte; **340–341** © Yann Arthus-Bertrand/C; **342** David Woodfall/NHPA; **342–343** D. Parer & E. Parer-Cook/AL; **344** Mark Conlin/PEP.

GENERAL ACKNOWLEDGMENTS

We would like to thank Dr. Robin Allaby of the University of Manchester Institute of Science and Technology (UMIST) and Dr. Angela Milner of the Natural History Museum, London for their specialist help, and John Clark, Neil Curtis, and Sarah Hudson for editorial assistance.

INDEX

Page numbers in *italic type* refer to timechart entries and picture captions; these pages will usually also contain text references to the subjects illustrated. Page numbers in **bold type** refer to main entries, while index entries in SMALL CAPS can also be found in the Glossary.

N

O

P

Q

R